Construction Contracting

Construction Contracting

Seventh Edition

A Practical Guide to Company Management

Richard H. Clough
Glenn A. Sears
S. Keoki Sears

WILEY

JOHN WILEY & SONS, INC.

Library of Congress Cataloging-in-Publication Data:

Construction contracting: a practical guide to company management / Richard H. Clough,
    Glenn A. Sears, S. Keoki Sears.—7th ed.
        p.   cm.
    Includes bibliographical references and index.
    ISBN 0-471-44988-1 (cloth)
    1. Construction industry—Management.   2. Construction industry—Subcontracting.
    I. Sears, Glenn A.   II. Sears, S. Keoki.
    TH438.C618 2005
    692'.8—dc22   2004018703

Printed in the United States of America
10   9   8   7   6   5   4   3   2   1

In memory of Richard Hudson Clough, 1922–2004

A gentleman, scholar, educator, author, colleague, and respected friend. He will be missed.

Contents

Preface

Nearly 50 years ago, an emergency moved Richard Clough from academia to the running of a large construction company. In his struggle to make the transition, he searched for information that would help him in making informed decisions regarding bonding, insurance, company organization, labor law, labor relations, and a dozen other aspects of the industry. Nothing was available. Later, when he rejoined the University of New Mexico faculty, he wrote the first edition of *Construction Contracting* in an effort to fill that void. Professor Clough was also instrumental in developing one of the first civil engineering construction programs in the nation.

Seven editions and more than 40 years later, this book still provides contractors, design professionals, owners, and students with up-to-date essentials for construction contracting. Over the years, the book has been used at nearly 200 universities and is found on the desks of successful contractors across the country. This seventh edition includes the latest information regarding labor laws, taxes, contracts, insurance, and forms of business ownership.

Construction education has become a major curriculum in universities, colleges, junior colleges, and vocational-technical schools in the United States. The Associated Schools of Construction has nearly 100 member schools. The men and women graduating from these programs work throughout the world for contractors, architects, engineers, developers, and governments. Many of these constructors have used previous editions of this book in school and now keep those copies at their sides in industry.

There are those who say our book is not exciting reading. However, the same people tell us that ours is the best book of its type on the market. Try as we may, we can't avoid the facts. Presenting everything one needs to know in order to run a construction company in one volume means lots of facts—over 650 pages of them. We have included an extensive index so that this will be a convenient reference book. We have also included end-of-chapter questions to help readers reflect on the facts, as well as Internet addresses to assist in further study of a given topic.

For generations men, machines, materials, and money have been the four Ms of construction. The efficient use of these four resources is the essence of construction management. However, in recent years there has been an important change in these basic resources. Women now constitute an important and growing part of the construction industry. They occupy responsible positions in the field trades and at all levels of management. At times, the authors use the word *he* or *him* as a singular pronoun for *foreman, project manager, superintendent,* or *design professional*. Such use of the masculine gender is done solely for the sake of readability and has no inference of gender. The authors of this book recognize the important contribution that women have made and are making to the construction industry.

We are pleased and proud to have Keoki Sears join us as a coauthor on this edition. Keoki is a graduate of Virginia Tech and Stanford University and has extensive construction experience. He has built numerous heavy construction projects in Hawaii and across the mainland and has managed telecommunications construction projects in Malaysia, Spain, and Germany.

We could not have written a book like this without the help of a great many people. Among those who made exceptional contributions are Kenneth Simonson, economist for the Associated General Contractors of America, Sam Conlee of the Kinney Agency, Daniel Correnti of the Government Services Administration, Craig Springe of La Plata County Building Department, Steve Rockenstein of CH2M Hill, Robert Clark of Stanford University, and many, many others. Thanks to all of you for your advice and help. Special thanks to Mary Sears for editing the manuscript.

Glenn A. Sears
GSears@Stanfordalumni.org
S. Keoki Sears
KSears@Stanfordalumni.org

Durango, Colorado
March, 2005

List of Figures

Chapter 1

The Construction Industry

1.1 THE CONSTRUCTION PROJECT

Humans are compulsive builders. We have demonstrated throughout the ages a remarkable and continually improving talent for construction. As knowledge and experience have been gained, humanity's ability to build structures of increasing size and complexity has increased enormously. In this modern world, everyday life is maintained and enhanced by an impressive array of construction, awesome in its diversity of form and function. Buildings, tunnels, highways, pipelines, dams, docks, canals, bridges, airfields, and a myriad of other structures are created to provide us with the shelter, goods, and services we require. As long as there are people on earth, structures will be built to serve them.

The projects within the construction industry are expensive, complex, and time-consuming undertakings. A structure must be designed in accordance with applicable codes and standards, culminating in working drawings and specifications that describe the work in sufficient detail for its accomplishment in the field. The building of a structure of even modest proportions involves many skills and materials and literally hundreds of different operations. The assembly process must follow a more or less natural order of events that, in total combination, constitute a complicated pattern of individual time requirements and sequential relationships among the various segments of the structure.

To some degree, each construction project is unique, and no two are ever quite alike. In its specifics, each structure is tailored to suit its environment, arranged to perform its own particular function, and designed to reflect personal tastes and preferences. The vagaries of the construction site and the infinite possibilities for creative and utilitarian variation of even the most standardized building product combine to make each construction project a new and different experience. The contractor sets up a "factory" on the site and, to a large extent, custom-builds each job.

The construction process is subject to the influence of highly variable and often unpredictable factors. The construction team—which includes various combinations of contractors, owners, architects, engineers, workers, sureties, lending agencies, governmental bodies, insurance companies, material dealers, and others—changes from one job to the next. All of the complexities inherent in different construction sites, such as subsoil conditions, surface topography, weather, transportation, material supply, utilities and services, local subcontractors, and labor conditions, are an innate part of the construction project.

As a consequence of the circumstances just discussed, construction projects are typified by their complexity and diversity and by the nonstandardized nature of their production. Despite the use of prefabricated units in certain applications, it seems unlikely that field construction can ever adapt completely to the standardized methods and product uniformity of assembly-line production.

1.2 ECONOMIC IMPORTANCE

For many years, construction has been the largest single production industry in the American economy. It is not surprising, therefore, that the construction industry has great influence on the state of this nation's economic health. In fact, construction is commonly regarded as the country's bellwether industry. Times of prosperity are accompanied by a high national level of construction expenditure. During periods of recession, construction is depressed and the building of publicly financed projects is often one of the first governmental actions taken to stimulate the general economy. A high level of construction activity and periods of national prosperity are simultaneous phenomena; each is a natural result of the other. The effect of the construction industry on the general economy is well illustrated by some figures recently reported by the Associated General Contractors of America. For every billion dollars of new construction, about 47,000 jobs are created in construction, supplies, and service industries. Of the total of 47,000 jobs created, almost 13,000 are in the construction industry itself. Some facts and figures pertaining to construction in the United States are useful in gaining insight into the tremendous dimensions of this vital industry. The total annual volume of new construction in this country at the present time is approximately $900 billion. The annual expenditure for construction normally accounts for about 8 percent of the dollar value of our gross national product. Approximately 80 percent of construction is privately financed, and 20 percent is paid for by various public agencies. Each dollar spent on new construction in the United States generates approximately $3.60 in economic activity in other services and industries. Each additional million dollars spent on new construction creates almost 50 jobs. The U.S. Department of Labor presently indicates that construction contractors directly employ more than 6.7 million workers during a typical year. Construction industry employment now represents approximately 6 percent of the private non-farm employment in the United States. If the production, transportation, and distribution of construction materials and equipment are taken into account, construction creates, directly or indirectly, about 12 percent of the total gainful employment in the United States.

1.3 THE OWNER

The owner, public or private, is the instigating party for whose purposes the construction project is designed and built. Public owners range from agencies of the federal government, through state, county, and municipal entities, to a multiplicity of local boards, commissions, and authorities. Public projects are paid for by appropriations, bonds, tax levies, or other forms of financing and are built to meet some defined public need. Public owners must proceed in accordance with applicable statutes and administrative directives pertaining to the advertising for bids, bidding procedures, construction contracts, contract administration, and other matters relating to the design and construction process.

Private owners may be individuals, partnerships, corporations, or various combinations thereof. Most private owners have structures built for their own use: business, habitation, pleasure, or otherwise. However, some private owners do not intend to become the end users. The completed structure is to be sold, leased, or rented to others.

In the main, most owners relegate by contract the design of their projects to professional architect-engineer firms and the field construction to construction contractors. However, there are some owners who, for various reasons, elect to play an active role in the design and construction phases of their projects. For example, some owners develop their own designs, or at least substantial portions of them. With respect to construction, some owners choose to act as their own construction managers or perhaps even to perform their own construction. Many industrial and public owners have established their own construction organizations that work actively and closely with the design and construction of their projects.

1.4   THE ARCHITECT-ENGINEER

The architect-engineer, also known as the design professional, is the party, organization, or firm that designs the project. Because such projects involve architectural or engineering design, or often a combination of both, the term *architect-engineer* is used in this book to refer to the design professional, regardless of the applicable specialty or the relationship between the designer and the owner.

The architect-engineer can occupy a variety of positions with respect to the owner for whom the design is done. Many public agencies and large corporate owners maintain their own in-house design capability. In such instances, the architect-engineer is a functional part of the owner's organization. The traditional and most common arrangement is one in which the architect-engineer is a private and independent design firm that produces the project design under contract with the owner. In other instances, the owner contracts with a single party for both design and construction services, which is referred to as design-build. In such a case, the architect-engineer is a branch of, or is affiliated in some way with, the construction contractor.

There are also arrangements in which large industrial firms have chosen to reduce their in-house design staffs and have established permanent relationships with outside architect-engineers. Such "corporate partnerships" call upon the architect-engineer to provide a broad range of design, engineering, and related services on an open-ended basis. Such arrangements are said to work to an owner's advantage by fostering a team approach and reducing litigation between the parties.

1.5   THE PRIME CONTRACTOR

The prime contractor, also known as the general contractor, is the business firm that is in contract with the owner for the construction of the project, either in its entirety or for some specialized portion thereof. The prime contractor is the party that brings together all of the diverse elements and inputs of the construction process into a single, coordinated effort.

The essential function of the prime contractor is close management control of construction. Ordinarily, this contractor is in complete and sole charge of the field operations, including the procurement and provision of necessary construction materials and equipment. The chief contribution of the prime contractor to the construction process is the ability to marshal and allocate the resources of labor, equipment, and materials to the project in order to achieve completion at maximum efficiency of time and cost. A construction project presents the contractor with many difficult management problems. The skill with which these problems are met determines, in large measure, how favorably the contractor's efforts serve its own interests as well as those of the project owner.

1.6   THE SUBCONTRACTOR

A subcontractor is a construction firm that contracts with a prime contractor to perform some aspect of the prime contractor's work. The extent to which a prime contractor subcontracts portions of its work depends on the nature of its business organization and the type of construction involved. There are instances in which the job is entirely subcontracted, with the general contractor providing only supervision, job coordination, and perhaps general site services. At the other end of the spectrum are those projects in which the general contractor does no subcontracting, choosing to do the entire job with its own forces. In the usual case, however, the prime contractor performs the basic operations and subcontracts the rest to various specialty contractors. Subcontracting is used much more extensively in housing and building construction than in engineering and industrial projects.

When the prime contractor engages a specialty firm to execute a particular portion of the overall construction program, the two parties enter into an agreement called a subcontract. No contractual relationship is established thereby between the owner and the subcontractor. The prime contractor, by the terms of its contract with the owner, assumes complete responsibility for the direction and control of the entire construction program. An important part of this responsibility is coordinating and supervising the work of the subcontractors. When the general contractor subcontracts a portion of the work, this contractor remains completely responsible to the owner for the total project and is liable to the owner for any negligent performance of the subcontractors. However, the courts have ruled that in the absence of provisions in the general contract holding the prime contractor responsible for the negligence of its subcontractors, the contractor cannot be held liable for damages caused by the negligence or collateral torts of its subcontractors.

Everyday economic facts have confirmed the subcontract system to be efficient and economical in the use of available resources. The operations of the average general contractor are not always sufficiently extensive to afford full-time employment of skilled craftsmen in each of the several trade classifications needed in the field. Hence, these contractors are able to keep only a limited nucleus of full-time employees and hire from the local pool of skilled labor as additional needs arise. By subcontracting, the prime contractor can obtain workers with the requisite skills when they are needed, without having to maintain an unwieldy and inefficient full-time labor force. By the same token, subcontractors are able to provide substantially full-time employment for their workers, thereby affording an opportunity for the acquisition and retention of the most highly skilled and productive construction workers. Another common problem solved by subcontracting involves projects requiring construction equipment the prime contractor does not have. In many cases it is economically preferable to subcontract a certain part of the project rather than attempt to acquire the necessary equipment. Qualified subcontractors are usually able to perform their work specialty more quickly and at a lesser cost than can the general contractor. In addition, many construction specialties have particular licensing, bonding, and insurance requirements that are not met by the prime contractor.

Public construction contracts frequently limit the proportion of the total construction that the prime contractor is allowed to subcontract. For example, several federal agencies have set such limitations. In addition, some states have established statutory restrictions on the subcontracting of public works in those states. Such limitations on construction subcontracting are intended to circumvent potential problems associated with extensive subcontracting, which can seriously complicate the overall scheduling of job operations, lead to a serious division of project authority, fragmentize responsibility, make the coordination of construction activities difficult, weaken communication between management and the field, foster disputes, and be generally detrimental to job efficiency. Obviously, the extent to which these difficulties may actually occur is greatly dependent on the experience, organization, and management skill of the prime contractor involved.

1.7 CONSTRUCTION CATEGORIES

The field of construction is as diversified as the uses and forms of the many types of structures it produces. However, construction is commonly divided into four main categories, although there is some overlap among these divisions and certain projects do not fit neatly into any one of them. In general, contracting firms specialize to some extent, limiting their efforts to a relatively narrow range of construction types. Specialization is usual and necessary because of the radically different equipment requirements, construction methods, trade and supervisory skills, contract provisions, and financial arrangements involved with the different construction categories. The four main divisions—residential, building, engineering, and industrial construction—are described in the following paragraphs.

Residential Construction

Residential, or housing, construction includes the building of single-family homes; condominiums; multiunit town houses; low-rise, garden-type apartments; and high-rise apartments. Design of this construction type is done by owners, architects, or the builders themselves. Construction is performed by the owner, builder-vendors, or independent contractors under contract with the owner. This category of construction is dominated by small building firms and normally accounts for about 40–45 percent of new construction during a typical year. Historically, residential construction has been characterized by instability of market demand and is strongly influenced by governmental regulation and national monetary policy. Housing is an area of construction typified by periodic high rates of contractor business failures. A large proportion of housing construction is financed through private financial institutions. Mortgage funds are provided to these financial institutions by private and government lenders, including Fannie Mae (Federal National Mortgage Association), Freddie Mac (Federal Home Loan Mortgage Loan Corporation), and Ginnie Mae (Government National Mortgage Association). Mortgages are often insured by the FHA (Federal Housing Authority) and the VA (Veteran's Administration).

Building Construction

Building construction includes buildings in the commonly understood sense, other than housing, that are erected for institutional, educational, light industrial, commercial, social, religious, governmental, and recreational purposes. In normal business years, private capital finances most building construction, which normally accounts for 25–30 percent of the annual total of new construction. Design of this construction category is predominantly done by architects, with engineering design services being obtained as required. Construction of this kind is generally accomplished by prime contractors or construction managers who subcontract substantial portions of the work to specialty firms.

Engineering Construction

Engineering construction is a very broad category and covers structures that are planned and designed by engineers. This category includes those structures whose design is concerned more with functional considerations than aesthetics and which involve field materials such as earth, rock, steel, asphalt, concrete, timbers, and piping. Most engineering construction projects are publicly financed and account for approximately 20–25 percent of the new construction market. Engineering construction utilizes major items of construction equipment, and projects of this type are characterized by large spreads of power shovels, tractor-scrapers, pile drivers, draglines, large cranes, heavy-duty haulers, paving plants, rock crushers, and associated equipment types. Engineering construction is commonly divided into three subgroups: highway and airfield, heavy, and utility construction.

Highway and airfield construction covers clearing, excavation, fill, aggregate production, subbase and base, paving, drainage structures, bridges, traffic signs, lighting systems, and other such items commonly associated with this type of work. Heavy construction is usually construed to include sewage- and water-treatment plants, dams, levees, pipe and pole lines, ports and harbor structures, tunnels, large bridges, reclamation and irrigation work, flood control structures, and railroads. Utility construction mostly involves work performed for municipalities, such as the construction of sanitary and storm drains, curbs and gutters, street paving, water lines, electrical and telephone distribution facilities, drainage structures, and pumping stations.

Industrial Construction

Industrial construction includes the erection of projects associated with the manufacture or production of commercial products or services. Such structures require a highly technical approach and are frequently built by large, specialized contracting firms that do both the design and field construction.

Although this category accounts for only 5–10 percent of the annual volume of new construction, it includes some of the largest projects built. The design responsibility for this type of construction is predominantly that of the engineer. Petroleum refineries, steel mills, chemical plants, smelters, electric power–generating stations, heavy manufacturing facilities, and ore-handling installations are examples of industrial construction. In the United States a very large proportion of this category is privately financed.

1.8 PROJECT FINANCING

By Owner

The owner makes the necessary financial arrangements for the construction of most construction projects. This normally requires obtaining the funding from some external source. In the case of public owners, the necessary capital may be obtained via tax revenues, appropriations, or bonds. A large corporate firm may obtain the funds by the issuance of its own securities, such as bonds. For the average private owner, funding is normally sought from one of several possible loan sources—banks, savings and loan associations, insurance companies, real estate trusts, or government agencies.

Where construction funding is obtained by commercial loans, the owner must typically arrange two kinds of financing: (1) short-term, to pay the construction costs, and (2) a long-term mortgage. The short-term financing involves a construction loan and provides funds for land purchase and project construction. A construction loan usually extends only over the construction period and is granted by a lending institution with the expectation that it will be repaid at the completion of construction by some other loan such as the mortgage financing. The mortgage loan usually applies for an appreciable period such as 10 to 30 years. The first objective of the project owner is to obtain the long-term or permanent financing from a lender. Once this commitment is arranged, the construction loan is normally easy to obtain.

When the mortgage lender has approved the long-term loan, a preliminary commitment is issued. Most lenders will not give the final approval until they have reviewed and approved the project design documents. On final approval, the mortgage lender and the borrower enter into a contract, by the terms of which the project owner promises to construct the project according to the approved design and the lender agrees to provide the funds for the stated period of time.

When the long-term financing has been arranged, the construction loan is obtained. Commercial banks and savings and loan associations are common sources for such financing. To obtain this loan, the owner may be required to name its intended design firm and the contractor that will do the work. Lending institutions typically limit their construction loans to no more than 75–80 percent of the estimated cost of construction. When the construction loan has been approved, the lender sets up a "draw" schedule, which specifies the rate at which the lender will make payments to the contractor during the construction period. Typically, the short-term construction loan is paid off by the mortgage lender when the construction is completed. This cancels the construction loan, and the owner is left with the long-term obligation to the mortgage holder.

By Builder-Vendor

A builder-vendor is a business entity that designs, builds, and finances the construction of structures for sale to the general public. The most common example of such structures is tract housing, for which the builder-vendor acquires land and builds housing units. This is a form of speculative construction, whereby the builder-vendors act as their own prime contractors, build dwelling units on their own accounts, and employ sales forces to market their products. Hence, the ultimate owner incurs no financial obligation until the structure is finished and a decision to buy is made.

In much construction of this type, the builder-vendor constructs for an unknown owner. Most builder-vendors function more as construction brokers than contractors per se, choosing to subcontract all or most of the actual construction work. The usual construction contract between owner and prime contractor is not present in such cases because the builder-vendor occupies both roles. The source of business for the builder-vendor is entirely self-generated, as opposed to that of the professional contractor that obtains its work in the open construction marketplace.

By Developer

A developer acquires financing for an owner's project in two different ways. A comparatively recent development in the construction of large buildings for business corporations and public agencies is the concept of design-finance. In this case, the owner teams up with a developer firm that provides the owner with a project design and a source of financing for the construction process. This procedure minimizes or eliminates altogether the initial capital investment of the owner. Developers are invited to submit proposals to the owner for the design and funding of a defined new structure. A contract is then negotiated with the developer of the owner's choice. After the detailed design is completed, a construction contractor is selected and the structure is erected.

The second procedure used by developers is currently being applied to the design and construction of a wide range of commercial structures. Here, the developer not only arranges project design and financing for the owner but is also responsible for the construction process. Upon completion of the project under either of the two procedures just discussed, the developer sells or leases the completed structure to the owner.

1.9   THE CONTRACT SYSTEM

The owner of a proposed construction project has many options available regarding how the work is to be accomplished. It is true that public owners must conform to a variety of statutory and administrative requirements in this regard, but the construction process is a flexible one, offering the owner many procedural choices.

In the usual mode of accomplishing construction work, the prime contractor enters into a contract with the owner. The contract describes in detail the nature of the construction to be accomplished and the services that are to be performed. The contractor is obligated to perform the work in full accordance with the contract documents, and the owner is required to pay the contractor as agreed. Experience shows that owners can often reduce their construction costs by giving more thought to the type of contract that best suits their requirements and objectives. The chances for a successful construction process are enhanced by thoughtful and thorough study by the owner before the process is initiated. The consideration of risk during field construction is a critical issue. Typical contract choices force either the contractor or the owner to bear most of the risk. Each of these contract types has its advantages, but there are variations that apportion construction risks to the party that can best manage and control them.

The means by which the prime contractor is selected, the form of contract used, and the scope of duties thereby assumed by the contractor can be highly variable, depending on the preferences and requirements of the owner. The prime contractor may be selected on the basis of competitive bidding, the owner may negotiate a contract with a selected contractor, or at times a combination of the two approaches may be used. The entire project may be included within a single general contract, or separate prime contracts for specific portions of the job may be used. The contract may include project design as well as construction, or the contractor's responsibility may be primarily managerial.

The following sections discuss the various ways in which owners can select prime contractors and the different types of contracts commonly used between the two parties.

1.10 COMPETITIVE BIDDING

The owner can select a prime contractor to construct a project on the basis of competitive bidding, negotiation, or some combination of the two. A substantial proportion of the construction work in the United States is done by contractors who obtain their work in bidding competition with other contractors. In such cases, the work is normally awarded to the lowest responsible bidder. The competitive bidding of many public construction projects is a formal procedure for public agencies.

The contracting firm that obtains its work by competitive bidding occupies an essential position in the construction industry. Equipment and know-how are its stock in trade, and it acts as a broker for the necessary construction materials and labor. It is often said that competitive bidding for construction work is the biggest gamble in the business world. When bidding a project, a contractor must estimate how much the structure will cost at a time when it exists only on paper. To this cost is added what seems to be a reasonable profit, and the contractor guarantees to do the project for the quoted price. If its bid is selected by the owner, the contractor must complete the project for the contracted amount. If the actual costs of construction should exceed this figure, the resulting loss is borne by the contractor.

In the construction industry, competitive bidding is traditional and is widely used. Competitive bidding is used to encourage efficiency and innovation by the participating contractors, thereby providing the owner with a constructed project of specified quality at the lowest possible price. This mode of contractor selection has served its purpose well but, like other contracting procedures, does have its weaknesses. In particular, the bidding process places the prime contractor and the owner in adversarial positions, which can lead to undesirable side effects. For example, the owner sees the specifications as the minimum quality requirement and the contractor sees them as the maximum requirement. Evidence can be offered to show that competitive bidding can, and sometimes does, lead to the selection of incompetent contractors, excessive claims by the contractor against the owner, disputes and litigation between the two parties, bid shopping, and other problems.

Two types of competitive bidding are used in the United States, open and closed. The predominant form is open bidding, in which all contractors use the same proposal form provided with the bidding documents, with the bids being opened and read publicly. In each case the proposal amount is the contractor's final offer, and there is no subsequent bargaining or negotiation. This procedure is sometimes referred to as the "hard-bid" approach. In closed bidding, sometimes used by private owners, there is no prescribed proposal form, and there is no public opening. The competing contractors are required to submit their qualifications along with their bids and are encouraged to tender suggestions as to how the cost of the work might be reduced. The owner selects one of the proposals submitted, or perhaps interviews those contractors whose proposals appear most advantageous and negotiates a contract with one of them.

It is interesting to note that the American sense that the low bidder is normally the successful bidder is not shared in many other parts of the world. Some very good arguments can be advanced to show that the lowest bid is not necessarily the best value for the owner, and there are some interesting variations used in other countries. Comparing the competing proposals and using the average of the bids received is common. For example, in one European country the work is awarded to the bidder whose bid is nearest to the average of all bids received. The bid that is greater than but nearest to, the average of all bids received, but is still below the owner's estimate, wins the contract in an Asian nation. In another European country, the successful bidder is the one nearest the average after the highest and lowest bids have been rejected.

Competitive bidding can also be used where the successful contractor is determined on a basis other than the estimated cost of the construction. When the contract involves the payment of a fee to the contractor, the amount of the fee is sometimes used as a basis of competition among contractors. For instance, construction management services are sometimes obtained by an owner using the fees proposed by the different bidders as one basis for the contract award.

1.11   THE NEGOTIATED CONTRACT

As pointed out previously, competitive bidding can have drawbacks associated with it and is not necessarily the optimal method of selecting a contractor to perform all classes of construction work. In many instances it can be advantageous for an owner to negotiate a contract for a project with a preselected contractor or small group of contractors. It is common practice for a private owner to forgo the competitive bidding process entirely and handpick a contractor on the basis of reputation and overall qualifications to do the job. The forms of negotiated contracts are almost limitless because such agreements can include any provisions mutually agreeable to both parties and that are best suited to the particular work involved. Most negotiated contracts are of the cost-plus-fee type, a subject more fully discussed in Chapter 6. Negotiated contracts have traditionally been limited to privately financed work because competitive bidding was, until recently, a legal requirement for most public projects.

Extensive use of negotiated contracts has been traditional in some areas of the construction industry, such as in residential and industrial construction. However, recent years have seen the increasing application of negotiated contracts across the board in the private sector. This can only be interpreted as a sign that owners are increasingly finding that such arrangements are in their best interest. A large proportion of the annual work volume of many contractors is now made up of negotiated contracts.

1.12   COMPETITIVE PROPOSALS

A method of acquisition referred to as *Competitive Proposals*, which began with passage of the Competition in Contracting Act of 1984, is now used by the federal government. This Act was passed to eliminate the noncompetitive, sole-source contracts that were often used by the U.S. Department of Defense and to provide a uniform set of procurement regulations for use by all executive branch agencies. This legislation placed the use of Competitive Proposals on the same level as sealed bidding, allowing federal contracting officers to use either method. Under the Competitive Proposal process, contracting officers prepare a request for proposals in which they state the relative importance of price and technical merit and list the technical evaluation criteria and their relative weights. After receiving initial proposals, the contracting agency advises each contractor within the competitive range on how it can improve its proposal from the standpoint of both technical approach and cost and then invites each offeror to submit a proposal revision. Selection is based on price and technical factors, with the technical issues evaluated by a panel of government technical experts. Federal construction-awarding agencies currently use Competitive Proposals on a majority of projects. National contractor organizations originally objected strenuously to the use of this procedure, seeing it as constituting a threat to the long-established process of public works bidding and contract awards to low bidders. More recently, their views are mixed, with some members appreciating their advantages under this system.

1.13   LINEAR CONSTRUCTION

Traditional in the construction industry and still widely used, *linear construction* is the procedure of design, bidding or negotiation, and construction, following one another in consecutive order. Using this process, also referred to as design-then-construct, each phase is finished before the next one begins. Thus, it can be seen that the linear procedure is not always an efficient one with regard to the total time required for the design and construction of a project. Although the consecutive completion of one step before initiating the next is acceptable to some owners in today's market, it is not acceptable to others. A number of financial considerations dictate the earliest possible completion date for many construction projects. For instance, the owner of a resort hotel may believe that an early opening with the resulting early cash flow from room rentals offsets additional construction costs. It is possible to reduce the total design-bidding-construction time required for some projects by releasing successive phases of the structure for bidding and construction as the design proceeds.

1.14   PHASED CONSTRUCTION

Phased construction, also known as fast-tracking, refers to the overlapping accomplishment of project design and construction. As the design of successive phases of the work is finalized, these work packages are put under contract, either with a single general contractor or with a series of different contractors. In the first case, each successive work package is done by the general contracting firm or awarded by it to a subcontractor. In the second instance, each phase is put out for bids and awarded to a series of different prime contractors. In either instance, early phases of the project are under construction while later stages are still on the drawing boards. This procedure of telescoping design and construction can appreciably reduce the total time required to complete a project.

Based on the premise of reducing construction duration, owners are increasingly choosing fast-tracking to help counteract high interest rates and to afford earlier beneficial usage of the structure. However, fast-tracking has received harsh criticism because it emphasizes time rather than quality and can sometimes take longer than the usual sequential process when applied to complex projects. It has been pointed out that the final construction cost is unknown at the start of fast-tracked construction, and if bids for subsequent phases of the work come in over budget, redesign options to reduce cost are very limited. Despite these problems, however, fast-tracking has proved to be a successful and desirable owner option when properly applied and managed.

1.15   THE TEAM APPROACH

An appreciable share of the private construction market is now using the "team approach." When this procedure is followed, the private owner selects the architect and building contractor as soon as the project has been conceived. The three parties now constitute a team that serves to achieve budgeting, cost control, time scheduling, and project design in a cooperative manner.

Using the team approach, the owner assembles the key players, architect and contractor, to study the proposed project. The team determines the project scope and budget, and the designer develops preliminary drawings, on the basis of which the contractor makes conceptual cost estimates. As the process continues, the designer prepares the final drawings and specifications and the owner makes the necessary financial arrangements. After financial commitments and required permits are obtained, actual construction begins. The designer and contractor work closely together, modifying the design and drawings as may be required. The process offers the owner the advantages of time savings, cost control, and improved quality.

Where the team approach is used, the owner is mainly responsible for setting the goals and parameters of the project as well as providing the funding. The architect is responsible for developing the functional, aesthetic, and technical features of the building and for preparing drawings and specifications. The contractor contributes its expertise in building materials, construction methods, and project scheduling. Such a procedure gives the owner some control over the process and helps the owner to make intelligent decisions about design, materials, and other project considerations.

1.16   CONSTRUCTION CONTRACT TYPES

A myriad of contract forms and types are available to owners for accomplishing their construction needs. All call for defined services to be provided under contract to the owner. The scope and nature of such services can be made to include almost anything the owner wishes. The selection of the proper contract form appropriate to the situation is an important decision for the owner and is deserving of careful consideration and consultation. In this regard, public owners must work within the constraints placed on them by applicable law.

The construction contract can be written to include construction, design-construct, construction management, or design-manage services. These concepts are discussed in the following sections.

1.17   CONSTRUCTION SERVICES ONLY

A large percentage of construction contracts provide that the general contractor will have responsibility to the owner only for the accomplishment of the field construction. Under such an arrangement, the contractor is completely removed from the design process and has no design input. Its obligation to the owner is limited to constructing the project in full accordance with the contract terms.

Where the contractor provides construction services only, the owner may have an in-house design capability. The more common arrangement, however, is for a private architect-engineer firm to perform the design in contract with the owner. Under this latter arrangement, the design professional acts essentially as an independent contractor during the design phase and as an agent of the owner during construction operations. The architect-engineer acts as a professional intermediary between the owner and contractor and often represents the owner in matters of construction contract administration. Under such contractual arrangements, the owner, architect-engineer, and contractor play narrowly defined roles, each performing a particular function semi-independently of the others.

1.18   THE SINGLE PRIME CONTRACT

Under the single-contract system, the owner awards construction of the entire project to a single prime contractor. The distinctive function of the prime contractor is to coordinate and direct the activities of the various parties and agencies involved with the construction and to assume full, centralized responsibility to the owner for the delivery of the finished project within the specified time. Customarily, the prime contractor constructs certain portions of the job with its own forces and subcontracts the rest to various specialty contractors. The general contractor accepts complete accountability, not only to build the project according to the contract documents but also to ensure that all costs associated with the construction are paid.

   The general contractor is the only party having a contractual relationship with the owner and is fully responsible for the performance of the subcontractors and other third parties to the construction contract. It is worthy of note that when a prime contractor sublets a portion of the work to an independent specialty contractor, the general contractor has a nondelegable duty to the owner for the proper performance of the entire work and remains liable under the contract with the owner for any negligent or faulty performance, including that of the subcontractors. If the work is not done in accordance with the contract, the general contractor is in breach of contract and is liable for damages to the owner, regardless of whether the faulty work was performed by the contractor's own forces or by those of a subcontractor.

1.19   SEPARATE PRIME CONTRACTS

When separate prime contracts are used, the project is not constructed under the terms of a contract between the owner and a single prime contractor who exercises centralized control over the entire construction process. Rather, the project is constructed by several different prime contractors, each being in contract with the owner and each being responsible for a specified portion of the total project. For example, the owner of a building to be constructed may decide to award separate contracts for general construction; plumbing; heating, ventilating, and air-conditioning; and electrical work. Each of these four contractors functions as a prime contractor. Each is in contract with the owner, and each performs independently of the others. Subcontracting by any or all of these four prime contractors is still possible, of course.

   Separate contracts are now widely used to satisfy a number of owner needs and requirements. Natural division points of the work or subdividing the work into stages or specialty areas can give the owner considerable flexibility in letting the work out to contract. A common example can be found in bridge construction, where separate prime contracts are commonly awarded for the foundations and piers, approach structures, bridge super-structure, and painting. Phased construction often makes use of multiple prime contracts. The use of separate contracts by breaking up the work into smaller packages can also result in contracts of shorter duration, the ability to obtain highly skilled specialty contractors, and better prices through increased competition. One of the principal advantages for the owner stemming from the use of separate contracts is the saving of the markup on work that would otherwise be subcontracted. When the entire work is awarded as a single contract, the general contractor includes in its price a markup on all subcontracted work as a fee for its management and coordination of the subcontractors. Separate contracts can reduce this cost or eliminate it altogether. Nevertheless, critics of the multiple-contract approach maintain that, for a number of reasons, the system raises the owner's total construction costs.

The use of separate contracts can be very troublesome unless carefully prepared contract documents are used and the work is rigorously planned, scheduled, coordinated, and controlled. If strong, centralized management control is not exerted over the separate prime contractors, the several possible advantages of separate contracts will likely turn out to be illusory. The owner, architect-engineer, a prime contractor, or a construction manager may assume this responsibility, or each contractor may, by contract, be given the duty of coordinating its work with those of the other primes. Many courts follow the rule that under separate contracts, the owner is responsible for the overall coordination of the project construction unless the responsibility to coordinate is expressly delegated to another. One approach has been to have all the prime contracts but one include an assignment clause that establishes the remaining prime contractor as the lead contractor. That party then has project coordination and interface responsibility. The construction manager normally shoulders the coordination responsibility where this form of construction is being used (see Section 1.22). In any event, if the party assuming responsibility for overall project direction does not possess the experience and skill required to perform the demanding and specialized management functions needed, the use of separate contracts may become a troublesome procedure indeed. Of particular importance is job delay caused by the inadequate performance of one of the prime contractors. In a separate-contract system, allocation of liability for project delay can become extremely complex, and responsibility may devolve to that party having the duty of overall management control and coordination.

A few states have enacted statutes requiring that designated separate contracts be awarded on state-financed public works. Many other public jurisdictions make the use of separate contracts optional under administrative authority. The application of separate contracts has proven to be very troublesome at times for some public agencies, causing contract-letting difficulties, litigation, and extra construction expense. Experience seems to indicate that public agencies might advantageously be let free to select a contract system appropriate to project circumstances rather than be restricted by law to a specific procedure.

1.20   DESIGN-BUILD

When the owner contracts with a single firm for both design and construction, this is referred to as a design-build or a design-construct project. The term *turnkey* is sometimes used to refer to a special case of design-build in which the contractor performs a complete construction service for the owner. This would be the case when the contractor obtains project financing, procures the land, designs and constructs the project, and hands it over to the owner ready for occupancy. The design-build concept has been enjoying increasing acceptance in the United States in recent years. It is estimated that 40 percent of the non-residential construction in Europe and 60 percent of the nonresidential projects in Japan are design-build. The increasing use of this approach by owners is largely due to economies of cost and time that can be realized by melding the two functions of design and construction. Injecting contractor expertise and advice into the design process offers the possibility of achieving both time and cost savings for the owner. Because phased construction can be utilized under a design-build contract, the owner will likely have beneficial use of its structure well before it would under the more traditional linear construction arrangement.

Design-build utilizes different types of contracts, depending on the construction involved. Housing and standard or prefabricated buildings usually use a negotiated lump-sum contract. Design-construct is extensively used with industrial construction, usually utilizing negotiated contracts with some form of cost-plus-fee arrangement and a guaranteed maximum cost. There are times when two separate contracts are drawn between the owner and the contractor-designer, with the design contract based on a cost-plus arrangement and the construction contract written on a lump-sum basis. The basic feature of the design-build method is that it entails single responsibility for the entire project. Actually, it is an expansion of the single-contract system to include design as well as construction.

A key aspect of design-construct is the team concept. The owner and the designer-builder work closely together in the planning, design, cost control, scheduling, site investigation, and possibly even land acquisition and project financing. In addition, design-build has the potential of significantly reducing project cost and delivery time while avoiding design-construction conflicts by having only one contractor accountable for the entire project. Moreover, by having design excellence as a major factor in the contractor selection process, the owner will receive a high-quality finished project.

A design-build firm is typically selected by one of three methods. In the first case, a single design-build firm is selected on the basis of its past experience, management, and quality and a contract is negotiated. The Design-Build Institute of America (DBIA) refers to this method as direct selection. In the second case, an owner prepares a short list of qualified design-build firms that make proposals, which are judged on the basis of past performance, managerial skill, references, preliminary design solutions, price, and personnel committed to the project. This is a form of competitive negotiation and is used by private and public owners. The third method is selection through a design competition in which a short list of experienced design-build firms are invited to make priced design proposals. These proposals are then scored against an evaluation scheme, and a firm is selected.

Although fixed-price contracts are appropriate for some projects, others may use fixed-price-incentive contracts. These contracts provide for shared savings between the owner and contractor to encourage cost savings from the contractor's originally proposed guaranteed maximum price.

1.21   PRE-ENGINEERED METAL BUILDINGS

Recent years have seen a tremendous increase in the application of a specialized form of design-construct. Referred to as "pre-engineered" or "systems" building, this construction mode is now applied to a wide range of building types, ranging from industrial applications to office buildings, retail centers, institutional buildings, and government-owned facilities.

Intended to provide an architecturally appealing building to meet an owner's specific requirements, this is a design-build arrangement in which the contractor is a dealer in metal structural systems and is directly affiliated with the manufacturer. The contractor, architect, and building manufacturer work closely together to create a finished project that best serves the owner's needs. Owners have found that pre-engineered metal buildings offer a competitive price along with much design flexibility. Such metal buildings now account for more than half of the low-rise commercial and industrial construction market in this country. The Metal Building Manufacturers Association (MBMA) is now a national trade association of such producers.

Initially, pre-engineered buildings consisted of standard modules, but in today's market essentially every metal building system is custom-engineered to meet the owner's specific requirements. The producer now has the ability to custom-fabricate systems to fit specific project criteria. Manufacturers now fabricate complete metal buildings and multistory structural systems, including the framing, metal roof, walls, finishes, partitions, and necessary subsystems. These preengineered packages are presized, precut, and pre-engineered to assemble into a fully integrated building unit. These parts are shipped to the job site, where the building is assembled by a local contractor. Such contractors are metal building "builders" or "dealers" who are formally affiliated with the manufacturer and are usually full-service contractors assigned to serve a given geographical area. Many metal building contractors also market their products and services as subcontractors to other general contractors. Metal buildings can be manufactured using many different exterior wall materials. A single manufacturer usually supplies a complete package of integrated building components, which arrive at the job site ready for assembly and erection. Systems building has proven to have a number of advantages: speed of construction, design flexibility, high quality, economy, minimal maintenance, and relative ease of future expansion. Collectively, these features have led to a widespread acceptance of metal building systems by construction buyers and specifiers.

1.22   CONSTRUCTION MANAGEMENT (CM)

The term *construction management* (CM) refers to the providing of professional management services to the owner of a construction project with the objective of achieving high quality at minimum cost and time. Such services may encompass only a defined portion of the construction process, such as field construction, or may include total project responsibility. The essential objective of this approach is to treat project planning, design, and construction as integrated tasks within a construction system. The construction participants, by working together from a project's inception to its completion, aim to serve the owner's best interests. By striking a balance between construction cost, project quality, and completion schedule, the management team strives to produce for the owner a project of maximum value within the most economical time frame.

Construction management services are performed by design firms, contractors, and professional construction managers. Such services range from merely coordinating contractors during the construction phase to broad-scale responsibilities over project planning and design, design document review, construction scheduling, value engineering, cost monitoring, and other management services. The demand for construction management services has increased greatly in recent years as owners have identified new needs, financial approaches, and construction technologies.

Program management is the application of the CM process to one or more projects in order to provide standardized technical and management expertise on each project.

In a technical sense, an owner who engages a firm to function as an arm of its own staff and act as an agent for a fee without being at risk is said to be obtaining "agency construction management." Such firms are financially liable only for their fees, do not contract directly with the prime contractor or subcontractors, and assume none of the responsibility of the general contractor. Owners who engage a manager to replace the general contractor and accept some risk are obtaining "at-risk construction management." Selection of a construction manager by the owner is sometimes accomplished by competitive bidding, using both fee and qualifications as the basis for awarding a contract. In the usual instance, however, the construction management contract is considered to be a professional services contract and is negotiated. Such contracts usually provide for a fixed fee plus reimbursement of management costs.

1.23   CURRENT STATUS OF CONSTRUCTION MANAGEMENT

At the present time construction management (CM) is extensively utilized in this country by both private and public owners. During its early evolutionary period, there was some confusion as to just what the duties and responsibilities of CM were, what type of business firm was best qualified to perform such services, and how traditional construction relationships and responsibilities were changed when the system was used. Serious problems arose concerning the lack of a practical code of ethics, standards of practice, uncertainty about the liabilities of the participants, and the absence of standard contract documents to cover the wide range of CM services being provided by engineers, architects, contractors, and other parties. However, time and experience have helped to stabilize this form of construction accomplishment, and the Construction Management Association of America (CMAA) has developed an accepted professional identity for construction managers and uniform standards for CM practice. In 1986, CMAA published its standards of practice for construction managers. These standards serve as a guide to CM services and propose measurements of performance. They describe the component elements of CM practice and define the scope of CM services. CMAA has also developed a series of standard CM contract documents. Still to be solved is the problem of persuading individual states to adopt uniform procedures in licensing construction management firms.

1.24   DESIGN-MANAGE

Design-manage is a term sometimes used to refer to a more comprehensive single-source construction service for the owner than the CM process discussed in the previous section. In the usual CM process, the owner separately contracts for design services and for the professional management of the construction process. Design-manage is an arrangement whereby the owner enters into a single contract for both design and CM services. In this arrangement, a single entity both designs the project and acts as a construction manager during the construction phase. A design-manage arrangement is often the result of a joint venture (see Section 2.17) between a design firm and a construction management enterprise.

With design-manage, the construction is performed by a number of independent contractors in contract with either the owner or the design-manage firm, but with the design manager planning, administering, and controlling the construction process. As is the case with design-construct and construction management, design-manage can, and usually does, utilize phased construction.

1.25   CONSTRUCTION BY FORCE ACCOUNT

Using the method of force account, also known as accomplishing work by day labor, is another instance in which the owner acts as the prime contractor. In this case, the project is normally constructed for the owner's own use. With this method, owner-builders may choose to do the work with their own forces, providing the necessary field supervision, materials, construction equipment, and labor. However, owners may choose to have the work performed entirely by subcontract, subletting individual segments of the work to specialty contractors. In such a case, owners assume the responsibility of coordinating and directing the work of the subcontractors.

Many studies have been conducted, mostly by public agencies, to compare the costs of construction accomplished by competitive bidding and by force account. These studies have clearly demonstrated that for all but very small projects, the force-account procedure is generally more expensive than the competitive-bid method. In addition, the quality of the work is usually better when done by a professional contractor.

There are several good reasons why contract construction is usually cheaper and better than that done by force-account means. A qualified prime contractor has a wide background of experience and is intimately familiar with materials, construction equipment, and field methods. The contractor maintains a force of competent supervisors and workers and is adequately equipped to do the job. Construction is a specialized business, and a contractor must be adept at it in order to survive. All these factors lead to efficiencies of cost and time that owners find difficult to match unless their own organizations already include trained construction forces.

In the usual instance, public construction is contracted out on a competitive-bid basis as required by policy or law. However, there is a continuing issue of government at all levels—federal, state, and local—doing some of its own construction work and being in competition with the private sector. Even though day labor construction by a public agency is ordinarily limited to maintenance, repair, small jobs, and emergencies, there are still many cases of public agencies performing substantial construction with their own forces. The construction industry has attempted to minimize such government competition with the private sector by getting legislation passed to require contracting out by public owners.

It must be noted at this point that the term *force account* is sometimes used in a different context than that just described. While a contractor is proceeding with work under contract, it is not unusual for the owner to find it necessary to modify or add to the work. Additional payment to the contractor in such an instance is often established on the basis of a lump-sum or unit prices. There are times, however, when the contractor is authorized to proceed on some kind of a cost-plus arrangement. This is sometimes referred to as proceeding with the extra work on a force-account basis.

1.26   GENERAL CONDITIONS CONSTRUCTION

In the traditional construction process, general contractors customarily provide certain common job services, not only for their own forces but for their subcontractors as well.

These services, called general conditions construction or support construction, include many items normally required and described by the General Conditions section of the project specifications. They involve such services as temporary heat, access to the project, hoisting, weather protection, guardrails, stairways, fire protection, job fencing, drinking water, sanitary facilities, job security, job signage, trash disposal, and job parking. When separate contracts are used, the contractor designated as the general contractor usually provides these services for the entire project, including work performed by the other prime contractors.

There are instances in which general conditions construction is the only part of the construction process actually performed by the general contractor. This would be true, for example, when the contractor, builder-vendor, or owner-builder subcontracts the entire project. Under some construction management arrangements, the construction manager provides support construction services.

1.27   SMALL AND DISADVANTAGED BUSINESS ENTERPRISES

For a number of years federal, state, and local governments have applied various forms of construction procurement to assist economically and socially disadvantaged contractors. The Small Business Reorganization Act of 1997 mandated that the Small Business Administration assist small businesses through the Small Business Enterprise (SBE) program, Small Disadvantaged Business (SDB) program, and businesses located in Historically Underutilized Business Zones (HUBZones) in obtaining a larger share of public works construction contracts. Such actions are directed toward establishing various forms of bidding advantages for small and disadvantaged contractors. One procedure in this regard is the use of "set-asides," whereby certain public construction contracts are designated as being available only to such businesses. Various forms of goals have been established, assisted by imposed quotas and bid penalties applied to nonqualified contractors. Another scheme has been to require prime contractors bidding on designated public contracts to subcontract at least a designated percentage of their work to small businesses or disadvantaged contractors.

These efforts have resulted in a plethora of public programs with differing regulations and certification procedures. In the process, confusion has been created for participants in these programs and problems have plagued their administration. A subject of continuing controversy, these programs have been troublesome and difficult to manage. Enforcement of the various regulations has been inconsistent and has met with considerable opposition, court challenges, and criticism. Much evidence has been presented to show that meeting established goals for small and disadvantaged contractor participation has substantially increased the cost of many associated public projects.

Recent court decisions, however, have imposed some measure of control over such goal-oriented contracting programs. For example, in its 1989 *Croson* decision, the U.S. Supreme Court ruled that the City of Richmond's program mandating that 30 percent of the value of its construction contracts go to minority-controlled firms was illegal and constituted an unconstitutional form of reverse discrimination against white-owned businesses. The Court's view was that government set-aside programs based on race are unconstitutional unless they are tailored to remedy past discrimination. Even then, the plan has to be limited to the specific minority groups who are shown to be victims of prior discrimination. The

Croson decision tightened the rules that public bodies must follow in conducting programs with race- or gender-based goals by requiring that governments first prove discrimination against local firms. Although the *Croson* case probably will not completely eliminate minority set-asides, it has cleared the way for legal challenges and has forced public bodies to reexamine the bases for their programs.

Despite controversy and problems with such public programs, however, there is no doubt that they have made it easier for small and disadvantaged firms to establish and conduct active and viable construction businesses. In so doing, public agencies have clearly increased the business opportunities for such contractors. The National Association of Minority Contractors (NAMC) has rendered valuable assistance in helping minority contractors to make the necessary contacts and to meet established certification requirements. Experience with these construction programs has enabled public agencies to substantially improve the rules and administration of such procurement procedures and to establish better certification rules and criteria for the prequalification of small and disadvantaged businesses.

1.28 SEASONALITY IN CONSTRUCTION

The volume of construction under way in this country at any one point in time varies according to a number of factors, a major one being the season of the year. Although seasonal fluctuations vary with geographical location and type of construction activity, on a national basis summer is the peak season and winter is a slack period for almost all segments of the construction industry. The primary reason construction is such a seasonal business is that inclement weather and low temperatures affect certain key construction operations. An undesirable result of this uneven work volume is serious shortages of skilled workers during the warm-weather months and extensive unemployment during the cold-weather months. Work does not stop during the winter months, of course, but there are layoffs, reductions in crew sizes, fewer project starts, and weather delays. All told, these add up to significant reductions in the workforce.

Illustrative of this point is the fact that total construction employment in this country during a typical year will vary approximately 25 percent from the summer to the winter months. It must also be realized that, like total employment, average weekly hours worked follow a seasonal pattern in the construction industry. Overtime work is common during the summer months, whereas less than full workweeks are frequent during the winter. Seasonality in construction is a serious national problem, resulting in inefficiency, increased costs of production, and the wasteful misuse of valuable skilled labor. Peak-season bottlenecks and the resulting inflationary pressures have led to public concern. Several government groups have studied the problem and recommended actions to provide fuller utilization of construction labor. Recommendations have included the more effective use of science and technology, improved scheduling of public contracts, and the relationship of national manpower policy to the stabilization of construction employment. Private action has also been encouraged to better regularize the employment of construction workers throughout the year. Many foreign countries with severe winter climates have been forced to find ways to lengthen their construction seasons. This has been accomplished by a variety of public policies involving licensing, resource allocation, loans, and the payment of government subsidies of one type or another to encourage winter work.

1.29 LICENSING

Because construction can affect the public interest, there are special laws pertaining to construction that are designed to protect public health and safety. These measures include zoning regulations, building codes, environmental regulations, building permits, field inspections, safety and health regulations, and the licensing of contractors and of skilled workers in certain construction crafts. Note that the discussion of licensing in this section is concerned only with the licensing of a construction contractor per se by a contractors' license board. Such a license is separate and apart from a corporate license to do business as a corporation in a given state and from an occupational license from a county or city.

The licensing of contractors is not universally required, but many states and local governments do require that some or all contractors doing business within their jurisdictions be licensed as such. Some of the licensing statutes or ordinances are solely revenue-raising measures. Under these particular laws, the payment of a license fee confers the right to conduct a construction business but requires no further conditions. Generally, however, the statutes provide that the contractor must not only pay a fee but must also meet certain minimum qualifications. These statutes establish a board of registration or other regulatory body that administers the law, accepts applications for licensing, gives examinations where required, issues the licenses, collects the fees, and generally enforces the provisions of the licensing law.

Licensing requirements vary widely among those areas that have such laws. In most cases, however, when the law requires a license, it applies equally to general construction and specialty work, even when subcontracted. Most of these laws require licenses only for contractors whose annual volume of business exceeds certain designated amounts. Almost all require that a license be obtained in advance of any bidding within the state, with certain minor statutory exemptions. Some areas exclude from licensing those contractors that do work financed by federal funds or performed for the federal government on government-owned land.

Certain jurisdictions issue various classes of licenses, differentiated in accordance with maximum size of contract or annual volume of business. In addition, licenses may be issued that are valid for, and apply only to, specific construction types such as general engineering construction, general building construction, or any of several classifications of specialty work. In a number of cases, a contractor must pass an examination before the license is issued.

It is very important that a contractor be properly licensed where this is a legal requirement of the area in which the work is to be performed. Most licensing laws provide that acting or offering to act in the capacity of a contractor without a valid license is a misdemeanor. Also important is the fact that a contractor not licensed at the time it contracts to perform construction services may not sue to collect for work that has been performed or for breach of contract. Contracts executed by unlicensed contractors are normally void as a matter of law, and such contracts cannot be enforced in court or in arbitration. In addition, the contractor may have no right to file or claim a mechanic's lien against the owner's property (see Section 9.34).

With regard to licensing, where it is required, it is important to note that the construction company must be licensed in the form in which it contracts or offers to contract. For example, if a licensed partnership changes to a corporation, most licensing laws provide that the partnership license is not applicable and that the corporation must be licensed as such. Another instance is that the law in some areas makes it unlawful for a joint venture to act as a contractor without being licensed as such, even though each member of the joint venture is individually licensed. There are now some states that require any party providing construction management services in those states to possess a valid state contractor's license.

Many political jurisdictions require the licensing, registration, or certification of workers in certain crafts. Plumbers and pipe fitters, electricians, welders, riggers, elevator installers, sprinkler fitters, and hoisting machine operators are examples. Required qualifications differ somewhat with the geographical area.

1.30 LICENSE BONDS

A few states and many local governments require all licensed contractors, including subcontractors who operate within their jurisdictions, to post permanent surety bonds with the appropriate government authority. These are in the nature of performance and payment bonds, a subject discussed more completely in Chapter 7, and they serve for the protection of the public. The public authority or other party may bring process against such a bond in the event of unpaid debt or malfeasance on the part of the contracting firm. The bond also serves to guarantee that the contractor make proper payment for required permits and inspections.

It is a somewhat more common practice just to require bonds of certain contractors whose work is intimately associated with public health and safety. The following is a representative list of trades requiring bonded specialty contractors:

Boiler installation

Curb, gutter, and sidewalk installation (on public property)

Demolition

Electrical elevator installation

Gas fitting

House moving

Plumbing

Sign erection

In many areas firms proposing to do such work must post bond with the designated public authority in the required amount before commencing operations.

1.31 BUILDING CODES

Universally applicable are building codes as adopted by the several states, counties, and municipalities. Passed into law to protect the public health and safety, building codes are rules that control the design, materials, and methods of construction, and compliance with such codes within their jurisdictions is mandatory. These codes cover the construction, alteration, repair, demolition, and removal of new and existing buildings, including service equipment. Most of these local codes are based, in whole or in part, on the International Building Code. The purpose of changes, where made, has generally been to accommodate local needs and conditions and to make the requirements more stringent than provided for by the national code itself. There are, of course, codes that are not based on national codes but have been specifically written and devised for a particular locality. The codes of some large cities are of this type. In addition, there are special-purpose codes, such as those now adopted by several states, that pertain to prefabricated housing. These state laws regulate operations of housing manufacturers within the state and set standards for such housing coming into the state from other manufacturers. Mobile home manufacturing is currently regulated by a preemptive federal code.

Various boards, commissions, building departments, and other public bodies administer and enforce the provisions of their codes.

Prior to 2003, there were three model building codes—the National Building Code, the Uniform Building Code, and the Standard Building Code—each finding favor in different parts of the country. The organizations behind these three codes have consolidated to form the International Code Council (ICC). For further information on this subject see www.intlcode.org.

In addition to the International Building Code, the ICC also publishes the following specialty codes:

International Plumbing Code

International Mechanical Code

International Fire Code

International Fuel Gas Code

International Zoning Code

International Residential Code

International Property Maintenance Code

International Energy Conservation Code

International Private Sewage Disposal Code

ICC Electrical Code Administrative Provisions

The National Electric Code is published by the National Fire Protection Association and for many years has been the only code used throughout the United States. Because there is only one code for the whole country, the ICC adopted it in the Administrative Provisions mentioned in the preceding list.

To ensure that applicable codes are adhered to within their corporate limits, as well as to serve other control purposes, political subdivisions normally require various kinds of permits, field inspections at certain stages of construction, and test reports. A general building permit is a universal requisite, requiring the filing of complete drawings and specifications prepared by a registered architect-engineer with a designated public office. These documents are reviewed for design conformance with the applicable codes by the responsible building authority. Permits for plumbing, electrical work, heating equipment, signs, air-conditioning, elevators and escalators, and refrigeration systems are also normally required. In addition, occupancy permits are sometimes required after the completion of a building, necessitating an inspection to ensure compliance with building code standards.

Conformity with the applicable building code is primarily the responsibility of the design professional. In the usual sense, the contractor undertakes to construct in accordance with contract documents and is not directly concerned with building codes or potential violations thereof. However, the contractor has a legal duty to notify the owner or design professional concerning any deviation from the requirements of the applicable building code that comes to the contractor's attention. The contracting firm is obviously responsible for its own construction methods.

1.32 CONTRACTOR ORGANIZATIONS

A large number of trade, technical, and professional organizations serve the diverse interests of the construction industry. These associations represent the design professionals, general and specialty contractors, home builders, manufacturers and distributors of building materials and construction equipment, insurance and surety companies, financial interests, and others.

There are many associations of contractors throughout the country. Among the national associations that represent general contractors, the senior organization is the Associated General Contractors of America (AGC), whose membership includes building, heavy, highway, and utility contractors. The Associated Builders and Contractors (ABC) is made up of member contractors who operate open shops, also referred to as merit shops. The National Association of Home Builders (NAHB) is a national group representing the housing industry. Many highway and heavy contractors belong to the American Road and Transportation Builders Association (ARTBA).

These associations perform a number of valuable services for their members, such as providing information about, and assistance with, labor negotiations, monitoring both local and federal legislation, sponsoring safety and apprenticeship programs, providing tax information, holding conferences, promoting public relations, and serving as clearinghouses for construction information. These organizations strive to maintain the business and ethical standards of contracting at a high level and to establish the integrity and responsibility of their members in the public mind.

Both local and national specialty contractor organizations also function to promote the mutual benefit of their members and to bring their combined resources to bear on common problems. These associations usually parallel the craft jurisdictions of the building trades and are represented by aggregations of specialty contractors such as electrical, mechanical, utility, masonry, or roofing contractors. There are usually several such groups functioning within a city or other local area. These regional groups are frequently affiliated with national organizations such as the National Electrical Contractors Association, the National Association of Plumbing-Heating-Cooling Contractors, the Mechanical Contractors Association of America, the National Utility Contractors Association, the Mason Contractors Association of America, and the National Roofing Contractors Association. The Associated Specialty Contractors is a coalition of nine national associations of construction specialty contractors. The American Subcontractors Association is a nationwide organization that represents subcontractors in all segments of the building construction industry and concentrates on business, contract, and payment issues affecting all subcontractors.

1.33 MANAGEMENT IN CONSTRUCTION

On the whole, construction contractors have been slow in applying proven management methods to the conduct of their businesses. Specialists have characterized management in the construction industry as being weak, inefficient, nebulous, backward, and slow to react to changing conditions. This does not mean that all construction companies are poorly managed. On the contrary, some of America's best-managed businesses are construction firms, and it may be noted with satisfaction that the list of profitable construction companies is a long one. Nevertheless, in the overall picture the construction industry is at or near the top in the annual rate of business failures and resulting liabilities.

There are several explanations given for why the construction industry has been slow in applying management procedures that have proven effective in other areas. Construction projects are unique in character and do not lend themselves to standardization. Construction operations involve many skills and are largely nonrepetitive. Projects are constructed under local conditions of weather, terrain, transportation, and labor that are more or less beyond the contractor's control. The construction business is a volatile one, with many seasonal and cyclical ups and downs. Construction firms, in the main, are small operations, with the management decisions being made by one or two persons.

The conclusion cannot be drawn, however, that management problems in the construction industry are uniquely different from those in other industries. The complexity of the product and the lack of production standardization do indeed lead to difficult management problems, but these are not necessarily more complicated than those in other business fields. Many industries are characterized by a large number of relatively small firms. Many areas of business experience wide seasonal or cyclical variations in the demand for their products and services. Construction is certainly not the only industry experiencing keen competition. Ineffective management in construction is not, therefore, the inevitable result of pressures and demands peculiar to the industry. As a matter of fact, the presence of such pressures and demands emphasizes the need for astute management in the construction industry.

It is debatable as to how many construction firms are well managed and whether many such enterprises enjoy any conscious management at all. The inescapable conclusion, however, is that skilled management and business survival go together. That this maxim has not been recognized by all construction firms is amply demonstrated by the high incidence of financial failure in the industry.

1.34 BUSINESS FAILURES IN CONSTRUCTION

Construction contracting is one of the riskiest of all business types, and every year many contracting concerns fail for a variety of reasons, causing millions of dollars in losses. The construction contracting business has the second highest failure rate of any business, exceeded only by restaurants. In a typical year, about 12 percent of all contracting firms will go out of business and be replaced by a similar number. Although companies of all sizes, whether old or new, fail each year, statistics show that the most likely firms to fail are those of small size and limited experience. For the past several years construction contractors have accounted for a disproportionate number of business failures in this country. For example, during a recent year in which construction accounted for 8 percent of the gross national product, contractors accounted for approximately 22 percent of all financial failures and 18 percent of the resulting liabilities. Studies disclose that the underlying management weaknesses, in descending order of importance, are as follows:

Incompetence

Lack of company expertise in sales, financing, and purchasing

Lack of managerial experience

Lack of experience in the firm's line of work

Fraud, neglect, and disaster

Consideration of these factors makes it clear that the financial success of a construction enterprise depends almost entirely on the quality of its management. Many outward manifestations of these root causes can be cited for business failure: seasonal business variations, weather hazards, job delays, inadequate sales, competitive weaknesses, heavy operating expenses, low profit margins, cash flow difficulties resulting from slow payments, and overextension. It is obvious, however, that these business inadequacies are simply indicative of poor business management.

QUESTIONS

1. The Engineering News Record publishes a list of the 400 largest contractors each year. Are these 400 companies responsible for a majority of the construction work?

2. Is most construction work financed by public or private funds?

3. Name the major players that make up the construction industry

4. Why do subcontractors occupy such an important place in the construction process?

5. Who are some of the major financial institutions that furnish mortgage money to the residential construction market?

6. Construction is often subdivided into residential, building, engineering and industrial construction. Which one is responsible for the largest share of the construction market?

7. What is the difference between construction financing and permanent financing?

8. What is the difference between a builder-vendor and a developer?

9. From an owner's prospective, contrast the differences between competitively bidding your project versus negotiation with a single contractor.

10. Why has the federal government moved away from competitive bidding and towards competitive proposals?

11. From an owner's prospective, contrast the differences between linear construction and phased construction.

12. Design-build has become increasingly popular. What are some of the advantages?

13. What are the differences between Construction Management at risk and Construction Management agency?

14. What situation might cause a large owner to choose not to do construction work with his own construction crews?

15. What is meant by general conditions work on a construction project?

16. What is the relationship between contractor licensing and access to courts in case of a dispute?

17. Name three national contractor organizations.

18. What are some of the reasons so many construction contractors fail each year?

Chapter 2

Business Ownership

2.1 ALTERNATIVE FORMS

The traditional forms of business ownership used by construction contractors are the individual proprietorship, the partnership, the corporation, and, recently, the limited liability company. Selection of the proper type depends on many considerations and is a matter deserving careful study. Each form of business organization has its own legal, tax, and financial implications that must be thoroughly investigated and understood. The selection of the ownership type most advantageous and appropriate for an owner's particular circumstances is a business decision that too often does not receive the serious study it deserves. Each type of organization has advantages and disadvantages and must be carefully evaluated in light of the individual context. The advice of a lawyer and a tax specialist is desirable when setting up a new business or when changing the form of a going concern.

2.2 THE INDIVIDUAL PROPRIETORSHIP

The simplest business entity is the individual proprietorship, also called sole ownership. It is the easiest and least expensive procedure for establishing and administering a business and enjoys a maximum degree of freedom from governmental regulation. No legal procedures are needed to go into business as a sole owner, with the exception of obtaining the required insurance, registration with appropriate tax authorities, and possible licensing as a contractor. The owner has the choice of operating under his or her own name or under a company name. In some jurisdictions a company name must be registered with a designated public authority.

The proprietor owns and operates the business, provides the capital, and furnishes all the necessary equipment and property. As the sole manager of the enterprise, the owner can make any and all decisions unilaterally and can act immediately. Title to property used in the business may be held in the name of the proprietor or in the name of the firm, if they differ. All business transactions and contracts are made in the owner's name. This mode of doing business offers many advantages, such as possible tax savings, simplicity of organization, and freedom of action. Other than the usual obligation to complete outstanding contracts and settle financial obligations, there are no legal formalities barring termination of the business.

As an individual proprietor, however, the owner is personally liable for all debts, obligations, and responsibilities of the business. This unlimited liability extends to personal assets even though they may not be directly involved in the business. Although management is immediate and direct, the owner alone must shoulder all of the burdens and responsibilities that accompany this function. A proprietorship has no continuity in the event of the death of the owner, unless there is a direction in the proprietor's will that the executor continue the business until it can be taken over by the person to whom the business is bequeathed. In the case of sickness or absence of the proprietor, the business may suffer severely unless there is a competent person available to take over. Entrepreneurs who form proprietorships can raise money for business purposes only through personal contribution, borrowing, or the sale of company assets. The owner must pay income taxes at normal individual rates on the full earnings of the business, whether or not such profits are actually withdrawn. Such earnings are added to income from other sources, and tax is computed on the whole.

2.3 THE GENERAL PARTNERSHIP

A general partnership is an unincorporated association of two or more persons who, as co-owners, carry on a business for mutual profit. The principal benefits to be gained by such a merger are the concentration of assets and personal credit, equipment, facilities, and individual talents into a common course of action. The pooling of financial resources results in the increased bonding capacity of the business (see Section 7.13), offering the possibility of a greater scope and volume of construction operations than would be possible for the partners individually. Each general partner customarily makes a contribution of capital and shares in the management of the business. Profits or losses are usually allocated in the same proportion as the distribution of ownership. If no such agreement is made in the articles of partnership, however, each partner receives an equal share of the profits or bears an equal share of the losses, regardless of the amount invested.

For most purposes a partnership is not recognized by the law as being an entity separate from the partners. For example, a partnership pays no income tax, although it must file an information return. However, a partnership can operate under a company name and for certain purposes a partnership is recognized as a separate legal entity. In many jurisdictions, for example, a partnership can own property, have employees, and sue or be sued. The Internal Revenue Service has ruled that bona fide members of a partnership are not employees of the partnership. Rather, such persons are deemed to be self-employed individuals whose remuneration is not subject to the Federal Unemployment Tax Act, the Federal Insurance Contributions Act, or to income tax withholding. Partners usually receive drawing accounts against anticipated earnings or annual salaries that are considered to be operating expenses of the partnership. In any event, partners must pay annual income taxes at the normal individual rates on their salaries and on their allocated shares of the partnership profits, whether or not these are actually withdrawn.

A partner's share of the profits in a partnership can be assigned. However, an individual partner cannot sell or mortgage partnership assets. Nor can a partner sell, assign, or mortgage an interest in a functioning partnership without the consent of the other partners. Partners act as fiduciaries and, commensurately, have an obligation to act in good faith toward one another.

2.4 ESTABLISHING A PARTNERSHIP

State laws regarding partnerships are not completely standardized, although most states have adopted the Uniform Partnership Act more or less verbatim. It is customary and advisable to draw up written articles of partnership, although a binding partnership can, in certain instances, be formed by oral agreement. It is always advisable to obtain the services of a lawyer to assist in the preparation of the articles of partnership, which is a contract between the partners. Ruling out statutory requirements, these articles may include almost anything the partners believe to be desirable. It is usual that articles of partnership, signed by the parties concerned, contain complete and explicit statements concerning the rights, responsibilities, and obligations of each partner. Although such articles must be individually tailored for each specific case, the following list indicates the kinds of provisions typically included in partnership agreements.

1. Names and addresses of the partners
2. Business name of the partnership
3. Nature of the business, with any limitations thereof
4. Location of the business
5. Date on which business operations will commence
6. Contribution of each partner in the form of capital, equipment, property, contracts, goodwill, services, and the like
7. Division of responsibilities and duties between partners
8. Any statement requiring partners' full-time attention to partnership affairs
9. Division of ownership as it affects allocation of profits and losses
10. Voting strength of each partner
11. Drawing accounts or salaries of the partners
12. Any restriction of management authority of individual partners
13. Specification of need for majority or unanimous decision on management questions
14. Payment of expenses incurred by a partner in carrying out partnership duties
15. Rental or other remuneration for use of a partner's own property or for personal services
16. Arbitration of disputes between partners
17. Record keeping and inventories
18. Right of each partner to full access to books and audits
19. Rights and responsibilities of the individual partners upon dissolution
20. Procedure in the event of an incapacitating disability of a partner
21. Extraordinary powers of surviving partners, such as option to purchase the interest of a deceased or withdrawing partner
22. Provision for final accounting of business in the event of death, retirement, or incapacity of any partner
23. Provision that an executor or spouse of a deceased partner continue as partner
24. Termination date of the agreement, if any

Unless there is specific agreement to the contrary, general partners are expected to devote full time to the affairs of the partnership. Partners cannot normally withdraw capital from the partnership unless provision for this is made in the partnership agreement. By mutual consent of all the partners, any provision of a partnership agreement can be modified or deleted or new provisions can be added at any time.

2.5 LIABILITY OF A GENERAL PARTNER

The liability of a general partner for the debts of the partnership is based on two legal principles. First, each general partner is an agent of the partnership and can bind the other partners in the normal course of business without express authority. Second, partners are individually liable to creditors for the debts of the partnership. The implications of these two points are discussed in the following paragraph.

An agency exists when one party, called the principal, authorizes another, called the agent, to act for the former in certain types of business or commercial transactions. An agency relationship exists by agreement, and an agent can be appointed by the principal to do any act the principal might lawfully do. The principal is liable for all contracts made by the agent while the agent is acting within the scope of his actual or apparent authority. The principal is also responsible for nonwillful torts (civil wrongs) the agent commits in the furtherance of the principal's business. A general agent is one empowered to transact all of the business of his principal. Each full member of a partnership is a general agent of the partnership and has complete authority to make binding commitments, enter into contracts, and otherwise act for the other partners within the scope of the business. This is subject, of course, to any limitations set forth in the partnership agreement.

In most states the responsibility of a partner for the contractual obligations of the partnership is joint with the other partners, meaning that all partners are necessary parties to any contractual action. The liability of a partner for torts committed by the partners in the ordinary conduct of business affairs is joint and several. In any such action, every member of the firm is individually liable and the injured party can proceed against all parties jointly or against any of the partners he chooses. Each partner assumes unlimited personal liability to third persons for the claims, contracts, and debts of the partnership. If any one partner is not able to pay his proper share of company liabilities, creditors can force the remaining partners to pay the share of the first. A creditor of a partner can attach the latter's interest in the profits of the business but cannot proceed against partnership property unless the partnership is being dissolved. Under the agency principle, notice to any one partner is considered to be notice to all.

It is apparent from the foregoing that very careful judgment should be exercised in the selection of business partners. Each general partner accepts unlimited financial responsibility for the acts of the other partners and underwrites the debts of the enterprise to the full extent of his personal fortune. If a partner withdraws from a going partnership, he remains personally liable for the partnership obligations outstanding as of the date of the withdrawal. To protect himself from future partnership debts, the retiring partner should give personal notice to firms doing business with the partnership and publish public notice of his departure. In most states the withdrawing partner can discharge himself of all liability by an agreement to that effect between himself, the firm's creditors, and the partners continuing the business.

This partnership liability issue has given rise to a relatively new form of partnership, the limited liability partnership (LLP). All other characteristics of the partnership remain the same, except that partners are shielded from liability beyond the assets of the partnership itself.

2.6   DISSOLUTION OF A PARTNERSHIP

One of the major weaknesses of the partnership form of business organization is its automatic dissolution upon the death of one of the partners. However, this weakness can be circumvented by making provision in the partnership agreement that the business will continue and that the surviving partners will purchase the decedent's interest. Dissolution may also be precipitated by bankruptcy, a duration provision in the original partnership agreement, the decision of a partner to withdraw, the insanity of a partner, a court of equity decree, or the mutual consent of all parties. Dissolution is not a termination of the partnership but simply a restriction of the authority of the partners to those activities necessary for the conclusion of the business. It has no effect on the debts and obligations of the enterprise. Voluntary dissolution is often accomplished by a written agreement between the partners. This agreement provides that the partners will undertake to complete all contracts in progress, settle company affairs, and pay all partnership obligations from the proceeds and assets of the business.

In the settlement of partnership debts, outside creditors enjoy a position of first priority. If partnership assets are insufficient to satisfy the outstanding obligations, the partners must personally make up the difference, except in the case of an LLP. After the creditors have been satisfied, partnership assets are used to repay any loans or advances that were made by partners above and beyond their capital contributions. Then follows the return to the partners of their capital investments. If additional assets remain, these are treated as profits and are distributed accordingly. When construction contracts involving considerable time for their completion are involved, the clearing of partnership accounts is often subject to substantial delay pending payment of all debts, receipt of all accounts receivable, and discharge of all contract obligations of the partnership.

2.7   THE LIMITED PARTNERSHIP

A general partner is a recognized member of the firm, is active in its management, and has unlimited liability to its creditors. The general partnership has been the basis of discussion in the preceding sections. In contrast, a limited partner is one who contributes cash or property to the business and commensurately shares in the profits or losses, but provides no services and has no voice or vote in matters of management. Unlike general partners, limited partners are liable for partnership debts only to the amount of their investments in the partnership. This is true, however, only as long as their names are not used as part of the firm name and they do not participate in management. Limited partners can be held fully liable if the true nature of their participation in the partnership has been withheld from creditors. The interest of a limited partner is assignable.

Under the Uniform Limited Partnership Act, adopted by most states, a limited partnership may be formed by two or more persons, at least one of whom must be a general partner. A contract is drawn up between the partners that clearly defines their individual duties and financial obligations. Also stipulated may be the length of time the agreement stays in effect and the date by which the contributions of each limited partner are to be returned. In addition, a limited partnership certificate must be signed and filed with a designated public office and published as required by state statute. The operation of a limited partnership, as well as its establishment, must be in strict accordance with the laws of the state in which it is formed. The partnership is not automatically dissolved by the death of a limited partner, as in the case of a general partner.

A limited partnership is used principally as a means of raising capital. It is a vehicle for getting money into a business while providing immunity from any personal liability for the investor. The limited partner is not a creditor of the business, but an owner of a share of the firm, and is entitled to a share of the profits while the business is operational and to a return of capital and undistributed profits on its dissolution. Obtaining new investment capital by bringing in additional general partners, each of whom shares in the management, can result in serious problems in regard to company control and decision making. The addition of a limited partner may also be preferable to borrowing the desired funds from a bank or other source. Yet one who lends money to the partnership can be in a better position than a limited partner. A lender to whom a firm is in debt can possibly exert some control over the firm, whereas any show of management interference by a limited partner can cause a loss of the limited partner's immunity from personal liability. Moreover, the lender enjoys a higher priority than a limited partner on dissolution of a partnership.

It should be noted that high corporate income tax rates have generated considerable interest in limited partnerships as an organizational alternative to the corporate form of business ownership. In such a case, an investor in a company is a limited partner rather than a stockholder. Income from a partnership is taxed only at the individual investor's rate rather than being taxed twice, at the corporate and individual levels. Limited partners, like stockholders, do not have personal liability for company debts and obligations. Publicly traded master limited partnerships (MLPs) some of which are listed on the major security exchanges have attracted much attention. This form of organization serves principally as a capital pooling technique whereby small companies or individual investors can pool their resources to initiate large projects. A sponsoring corporation normally serves as the general partner in an MLP.

2.8   SUBPARTNERSHIP

A general partner can enter into an agreement with an outsider, called a subpartner, whereby the latter will share in some designated way that general partner's profits or losses derived from the partnership's business activities. The subpartner is not a member of the firm. Consequently, he performs no active function, has no voice in the management of the partnership, and has no contractual relationship with the partners other than the one with whom he made the profit- or loss-sharing arrangement. Because the subpartner has contracted to participate in the individual partner's share only, he is not personally liable to creditors of the partnership.

2.9   THE CORPORATION

There are several classifications of corporations: public and private, profit and nonprofit, quasi-public, and foreign and domestic. The following discussion considers only the private corporation that a contractor would establish for ordinary business and profit motives.

A corporation is an entity created by state law, that is composed of one or more individuals united into one body under a special or corporate name. Corporations have certain privileges and duties, enjoy the right of perpetual succession, and are regarded by the law as being separate and distinct from their owners. A corporation is authorized to do business, own and convey real and personal property, enter into contracts, and incur debts in its own name and on its own responsibility. It sues and is sued in its corporate name. The principal advantages of this form of business organization are the limited liability of its owners (stockholders), the perpetual life of the company, the ease of raising capital, the easy provision for multiple ownership, and the economic benefit that owners pay taxes only on profits (dividends) actually received. In a corporation, unlike a partnership, the owners (shareholders) can also be employees.

A corporation is formed under state law by a prescribed number of shareholders filing a certificate of articles of incorporation with the appropriate state department. The corporation comes into being upon receipt of its corporate charter from the state of domicile. The powers of a corporation are limited to those enumerated in its charter and to those that are reasonably necessary to implement its declared purpose. The provisions of the corporate charter are usually made very broad so that the new corporation will be authorized to do many things that may become necessary or desirable in future years. These powers include the drawing up of bylaws pertaining to the day-to-day conduct of business. Acts of the corporation are governed and controlled by its articles of incorporation, its bylaws, and applicable state law.

Corporations are regulated by the various states, each having a corporation code that prescribes the formal process that must be followed by the organizers in obtaining a charter. These legal requirements differ somewhat from state to state, and it is always important to secure competent legal guidance when establishing a new corporation or modifying an existing one. In any case, fees must be paid and substantial quantities of information filed with the proper officials. Certain detailed requirements must be satisfied regarding the incorporators' residence and financial structure.

A corporation is dissolved by the surrender or expiration of its charter. Dissolution of a corporation merely means that the corporation ceases to exist. State laws govern the means by which the business affairs of a dissolved corporation must be settled.

2.10 THE FOREIGN CORPORATION

A business corporation is created under the laws of a specific state and within that state is known as a domestic corporation. In all other states it is considered to be a foreign corporation. The state of incorporation may be chosen not on the basis of location of work done by the company, but rather because of favorable corporation laws. The incorporated contractor that wishes to extend its operations beyond the borders of its state of incorporation must apply for and receive certification to do business as a corporation within each additional state. Usual requirements for the certification of a foreign corporation are the filing of a copy of the corporate charter, designation of a local state resident as an agent for the service of process, and the payment of a filing fee.

The acts of a foreign corporation within another state are considered to be unlawful until the corporation has become certified to do business as a corporation in that state. Hence, the incorporated contractor must be careful to comply with the corporation laws of those states in which it does business. Failure to be properly certified as a foreign corporation may render construction contracts unenforceable and eliminate the right of lien. In addition, an uncertified foreign corporation may not have access to the state courts. A foreign corporation is, of course, subject to the control of the state within which it does business. For example, if franchise or income taxes are levied on businesses of that state, the foreign corporation must pay such taxes in accordance with its total business in the state.

2.11   STOCKHOLDERS

The ownership of a corporation is exercised through shares of stock that entitle the owners thereof to a portion of the business profits and the net assets on liquidation. This does not mean the shareholders own the corporate assets, because the corporation itself holds legal title to its own property. However, each share of stock does represent a share in the ownership of the corporation. Under the laws of most states, a single individual can own all the stock of a corporation. The courts recognize a clear distinction between the individual and his corporation.

Ordinarily, shares of stock in a corporation are freely transferable, which is one of the advantages of incorporation. Stock certificates are negotiable instruments that may be transferred by endorsement and delivery in the same manner as promissory notes. However, it is not an uncommon desire that the ownership of stock of an incorporated business be limited to members of a family or a small circle of associates. In this way control of the company is maintained within a select group. Such a corporation whose stock is not available to the public is referred to as a closed or closely held corporation. When corporate stock is available to the general public, the company is called a publicly owned or public corporation. Any restriction on the ownership or transfer of stock can be accomplished by stamping the nature of the restriction directly on the face of the stock certificate. Another means is to provide in the bylaws that before stockholders can sell their stock to an outside party, they must first offer their stock for sale back to the corporation.

When a corporation is established, its articles of incorporation authorize the issuance of a certain dollar value of stock (the authorized capital stock). The actual capital stock is the amount actually paid to the corporation for its outstanding stock and does not necessarily equal the authorized capital stock. The stock may be common or preferred or both. Preferred stock is subject to less risk than common stock because it has a certain preference as to dividends and distribution of assets on liquidation of the corporation. However, preferred stock usually carries no voting privileges. Common stock generally entitles its owners to one vote per share and presents at least the possibility of greater ultimate profits to its owners.

Stockholders have certain rights and privileges that are based on state statute, the charter of the corporation, and the terms of ownership as stipulated on the stock certificate. These rights include the preemptive right to subscribe new stock issues, to carry out reasonable inspection of the corporation's books and records, to bring suit against the corporation, to share ratably in declared dividends, to receive a share of the net assets on dissolution, and to enact bylaws.

The profits of a corporation are subject to federal and state income taxes at the corporate rate. Such taxes are paid by the corporation before any profits can be distributed as dividends to the stockholders. The stockholders are then taxed individually on any such dividends distributed on a cash basis. Such double taxation is one of the biggest drawbacks associated with a corporate form of business. Because a corporation cannot deduct dividend payments to its stockholders for tax purposes, both the corporation and its shareholders pay taxes on the same income. Dividends can be paid only out of earned surplus and may be declared only if such action does not impair the position of corporation creditors. Such dividends, or surplus distributions, are declared as a fixed sum per share of outstanding stock. This payment is generally made quarterly but may be declared at such intervals as decided on by the board of directors.

An outstanding advantage of the corporate system is the privilege of an individual shareholder's being immune from personal liability for corporate debts. With very few exceptions, stockholders have no personal liability for obligations of the corporation, and their responsibility is limited to their investment in the corporation's stock. Important exceptions to this generality, however, are the personal guarantees of the shareholders of a closely held corporation often required by that corporation's bonding company (see Chapter 7). When this is the case, the shareholder's liability is not limited to the amount invested in the corporation. A stockholder is not an agent of the corporation and cannot act in any way for, or on behalf of, the organization.

2.12   CORPORATE DIRECTORS AND OFFICERS

The voting shareholders elect a board of directors that exercises general control over the business and determines the overall policies of the corporation. The directors must conduct the firm's affairs in accordance with its bylaws and are ultimately responsible to the stockholders. Directors have no authority to act individually for the corporation and are not agents of the corporation or of the stockholders. Their powers may be exercised only through the majority actions of the board. Directors occupy the position of fiduciaries and are required to serve the corporation's interests with prudence and reasonable care.

The articles of incorporation normally provide for annual meetings of the shareholders and for periodic meetings of the board of directors. Minutes of these meetings must be kept and filed. Failure to observe such formalities can lead to serious problems in the event of claims or lawsuits. Chronic disregard of corporate technicalities may result, for legal purposes, in the business being treated as a partnership or sole proprietorship rather than as a corporation.

The president, vice president, secretary, and treasurer are the corporate officers, normally appointed by the board of directors, who carry out the everyday administrative and management functions. The officers are empowered to act as agents for the corporation. The president is authorized to function for the firm in any proper way, including making contracts, whereas the lesser officers generally have more limited authority. Officers also serve in the capacity of fiduciaries and may be held personally liable for corporation losses caused by their neglect or misconduct.

2.13   THE S CORPORATION

If a corporation meets certain criteria specified by the Internal Revenue Service (IRS), none of which restricts the size of business volume, profits, or capital, it has the option of being taxed as a partnership rather than as a corporation. If such an election is made, the company is referred to as an "S corporation." To be eligible, the corporation must meet certain criteria including being a closely held corporation having no more than 35 stockholders who must be U.S. citizens, estates, or some trusts, and there being only one class of stock outstanding. Partnerships and other corporations cannot be shareholders. An eligible corporation must elect to be treated as an S corporation, and stockholders must consent by signing prescribed forms that are filed with the IRS. Once the taxation election is made, this status must continue until it is properly terminated. This may be done at any time by a majority vote of the stockholders. Once terminated, the corporation is not allowed to reelect S status for five years unless approved by the IRS.

Shareholders in an S corporation enjoy typical corporate privileges such as limited shareholder liability and the ability to easily transfer ownership. The use of an S corporation can be especially beneficial in the early years of a business when operating losses occur and substantial equipment must be acquired, when there is high company profitability with no additional capital needs, and for family income tax planning. The prime attraction is that profits or losses of the corporation are taxed directly on the shareholder's tax return instead of at the corporate level. Losses incurred by S corporations, with some limitations, can be offset by the taxpayer shareholder against taxable income from other sources. In contrast, a regular corporation pays taxes at the corporate level and the stockholder then must pay a second tax on any profits withdrawn as dividends. S corporations are not subject to the penalty tax applied by the Internal Revenue Service when earnings are retained in a corporation beyond its reasonable needs. Moreover, the IRS cannot assert that unreasonably high compensation is being paid to an employee-shareholder of an S corporation.

2.14 EMPLOYEE STOCK OWNERSHIP PLAN (ESOP)

An ESOP is a form of deferred compensation to employees that is required by law to be invested primarily in the stock of the employer corporation and is funded by tax-deductible contributions made by the employer. Company employees pay no tax on their stock ownership until they withdraw their benefits from the plan, usually upon retirement. This is a tax-advantageous way in which to transfer ownership in a closely held corporation.

A company ESOP can provide benefits to both the employer and its employees. An ESOP is similar to other stock bonuses or profit sharing plans, but differs in that it invests its assets primarily in company stock. Such a plan can be designed to accomplish different purposes, such as to promote company growth by providing a new source of equity capital, to improve benefits for long-term employees, to allow the retirement of company owners, to accomplish a change of company ownership, or to retire outstanding stock presently owned by stockholders. By selling company stock to an ESOP trust, the employer determines the quantity of stock to be divested and the rate at which it will be sold to the employees.

When an ESOP proposal is accepted by the firm's stockholders and board of directors, there are two ways in which the plan can be implemented. The choice of procedure depends on the purposes to be served by the plan. One form of ESOP is that in which the corporation makes a series of periodic tax-deductible contributions to the trust for the benefit of the participating employees. These company payments replace traditional profit-sharing distributions made to individual employees. If the plan borrows no money from outside sources, such company payments are limited to 25 percent of the total payroll. The second form of ESOP is one in which the trust obtains a third-party loan to buy the shares of stock offered. The loan is retired by a series of tax-deductible company contributions, as in the first procedure. If funding is borrowed, the company payments cannot exceed 25 percent of covered payroll plus interest paid on the loan.

Under either of the two procedures, the privately held company is obligated to repurchase the shares of employees who retire, are disabled, die, or terminate employment for other reasons. The shareholders of the sponsoring corporation can defer capital gains recognition when they sell their stock to an ESOP, provided that the ESOP owns 30 percent of the corporate stock and the seller reinvests the sale proceeds in securities of other domestic corporations.

An ESOP can be advantageous for some companies but can present serious problems for others. A privately held corporate firm must thoroughly understand all of the risks and drawbacks that may be involved. A wide range of issues must be carefully considered before making the decision to implement such a plan.

2.15   THE LIMITED LIABILITY COMPANY (LLC)

A relatively new and rapidly growing form of business entity, the limited liability company (LLC) can offer legal, tax, and economic advantages over the traditional forms of business ownership. An LLC is a statutory entity that combines the limited personal liability feature of a corporation and pass-through tax benefits. An LLC is formed by two or more members with an LLC agreement written in conformity with the applicable state statute. This agreement is much like a partnership agreement or a firm's articles of partnership. Many states have amended their requirements to allow a single-owner LLC.

A single-owner LLC is taxed at the federal level as a sole proprietorship, whereas an LLC owned by two or more entities is taxed as a partnership. As such, the income from an LLC is classified as earned income rather than dividends, a factor that may have negative implications for some owners.

Simplicity is one of the positive features of an LLC. It is typically easier to start an LLC than a corporation, and it has lower filing fees. An LLC has many fewer formal requirements than a corporation, with no requirement for regular shareholder and director meetings and the attendant meeting notices and minutes. An LLC differs from a corporation in that it has a finite life and can be dissolved with the death, bankruptcy, or retirement of a member. As a result, the LLC agreement must provide for these circumstances. The termination conditions of an LLC agreement are important because the IRS uses these as a means of differentiating between an LLC and a corporation for tax purposes.

Owners of an LLC can include corporations and non-U. S. citizens, features unavailable in S corporations. Some states have variations of the LLC in the form of limited liability partnerships (LLP) and limited liability limited partnerships (LLLP).

2.16   CONSTRUCTION CONTRACTING FIRMS

The construction industry is one of the easiest to enter. Any interested person can get started with little or no capital investment. There are many large firms in this industry, but they are far from dominant. Family-owned businesses that work close to home are the general rule. There are approximately 700,000 business firms in the United States that classify themselves as being construction contractors of one type or another. Of these, about 77 percent are individual proprietorships, 6 percent are partnerships, and 17 percent are corporations. From these figures it is easy to see that a large majority of construction firms are small businesses. This relates to the fact that the business scope of the usual individual proprietorship is entirely dependent on the owner's own resources and personal credit.

Figures made available by the U.S. Department of Commerce reveal that about 80 percent of construction establishments have annual receipts of $1 million or less. Almost one-half have no payroll as such, being operated by proprietors, partners, or business associates who do the work themselves. Approximately 10 percent of the contractors account for about 80 percent of the annual dollar volume of construction. There are approximately 13,000 construction firms in the United States that have a gross annual business of more than $10 million.

2.17   THE JOINT VENTURE

As a means of spreading risk and pooling resources for certain projects, it is common for two or more contracting firms to unite forces through the medium of a joint venture. The members of a joint venture can be sole proprietorships, partnerships, or corporations, but the joint venture itself is a separate business entity. Joint venture agreements can take many forms, but the usual joint venture may best be described as a special-purpose partnership. Although there is a difference in scope, similar legal principles are applied generally to both partnerships and joint ventures. The determination of the relationship depends on the particular facts of the case and the agreement on which the joint relationship is based. Properly used, joint ventures are well-proven devices for pooling resources and facilities and for spreading construction risks. A joint venture creates an entity that is bigger and stronger than any of its parts acting alone. This arrangement is often used to enable contractors to obtain a job that is too large for any one of the individual participants. In addition, a joint venture can sometimes be desirable when a contractor is contemplating work of a different type or in a new geographical area.

A joint venture relates to only a single project, even though its completion may take a period of years. Normally, the participating contractors submit to the project owner a joint venture proposal in which all the interested parties are named and in which all, jointly and severally, are directly bound to the owner for the performance of the contract. Each coventurer is thereby individually liable for the performance of the entire contract and for the payment of all construction costs.

A joint venture serves to combine the resources, assets, and skills of the participating companies. Such a combination offers the multiple advantages of pooling construction equipment, estimating and office facilities, personnel, and financial means. Each joint venturer participates in the conduct of the work and shares in any profits or losses. Many of America's largest structures have been built by joint ventures. Hoover Dam, constructed by an association of six companies, is an early example. Succeeding years have seen joint ventures develop into common practice in the construction industry.

The participants in a joint venture enter into a written agreement that defines the aims and objectives of the association and clearly delineates such matters as the advance of working capital, obligations and rights of each party, the percentage interest of each participant, which contractor will play the leadership role of project sponsor, limitations of liability, settlement of disputes among the coventurers, the division of profits and losses, the specific responsibilities and contributions of the individual parties, the details of contract administration and project management, the supervision of accounting and purchasing, the procedure in case a coventurer defaults on its commitments, and the termination of the agreement. For a joint venture to exist in a legal sense there must be (1) a contract, (2) a common purpose, (3) a community of interest, and (4) an equal right of control. Writing a joint venture agreement is an essential and technical task requiring the services of a lawyer and expert tax counsel. An important consideration is that a joint venture can select its own accounting procedure for tax purposes without regard to the methods used by the individual venture members. Prior to bidding, it may be necessary to obtain a contractor's license for the joint venture, prequalify the joint venture for bidding, and file a statement of intent to submit a joint venture proposal with the owner.

QUESTIONS

1. What are the basic advantages of (1) a sole proprietorship, (2) a partnership and (3) a corporation?

2. The form of business ownership impacts the taxes, liability and finances of a construction company. Describe how these vary between a partnership and a corporation.

3. What is the difference between a general partner and a limited partner?

4. What is meant when an incorporated contractor is classified as a foreign corporation?

5. Discuss the differences between and reasons for a closed corporation and a public corporation.

6. A contractor has been operating as a sole proprietorship for many years, has expanded in size and is considering changing to either an S corporation or an LLC. Discuss the pros and cons regarding the contractor's decision.

7. How do the responsibilities of the board of directors differ from the officers of a corporation?

8. An Employee Stock Ownership Plan (ESOP) has several advantages for both owners and employees. Discuss these advantages.

9. How does a Limited Liability Company (LLC) differ from a partnership?

10. Large construction projects are often done by joint ventures. What are the advantages and disadvantages of this type of organization?

11. Why is construction such an easy industry to enter?

Chapter 3

Company Organization

3.1 THE NEED FOR AN ORGANIZATION

The conduct of a construction business involves a number of separate and diverse activities. Cost estimating, bidding, contract negotiation, procurement, project planning and scheduling, construction methods, equipment management, insurance, surety bonds, material control and storage, accounting, payrolls, field cost control, labor relations, project safety, supervisor training, and many other functions are involved in the business of contract construction. It is the role of the contractor to conduct and coordinate these activities into a smoothly running and profitable plan of action. The basic ingredient in this entire process of business management is the contractor's company organization. Organization is not an end in itself, but rather a means to achieve company objectives.

In carrying out the necessary actions of its business, the construction contractor must continually cope with complex and shifting organizational problems. The contractor faces the need for devising a suitable field organization for each of its projects and a general company organization that acts in a support capacity for the total of company operations. In the construction industry, it is common practice for the contractor to locate all of its business functional groups together in a central office, together with the necessary project support services, and to locate on a given project site such management capability as the need dictates. This chapter discusses company organization as it applies to the head or regional office. Project organization is discussed in Chapter 10.

3.2 ORGANIZATION BASICS

With the inception of any private business, company ownership must first establish the fundamental purpose of the enterprise, define the overall scope of operations, set long-term objectives, and establish a general plan of action. Once these basics are formulated, a company organization can then be designed and created that will implement these plans and work toward the goals identified. An indispensable ingredient for a successful enterprise is an efficient, well-trained, and vigorous company organization.

The establishment of an effective company structure, one of the principal functions of a firm's management, is the subject of this chapter. Organization is the foundation for the operating management of a business and is the structure that makes a successful and profitable business possible. Organizing refers to the company structure, particularly to the establishment of the functions and duties of its several parts. It is the process of determining the responsibilities and scope of authority of each position in the management structure and defining how each company segment interrelates with the others. A primary objective of any organizational plan is to establish an effective operating routine for each element of the company with a minimum of direction from above.

3.3   GENERAL CONSIDERATIONS

Organizing a business may be described as determining the individual job positions that are required, defining the duties and responsibilities of each such position, and establishing the working relationships between these positions. If this is properly done, company members will be well aware of the parts they are expected to play in company affairs and what is expected of them. The main tasks of a contractor's organization are to obtain construction contracts, to plan, direct, and control the many elements associated with field operation so that the best combination of operational economy and time efficiency is achieved, and to collect and manage the resulting revenues. To perform these tasks, the contracting firm must create an organizational structure that is especially suited to meet the peculiar needs and demands of its specific mode of operation.

The organizational framework must be sufficiently stable to ensure action while being reasonably flexible and adaptive. Within the necessary metes and bounds of company discipline, a certain degree of personal freedom of action should be allowed. It is very important that those who shoulder the decision-making responsibilities of the company are relieved of excessive detail. To this end, people in responsible charge should not be expected to act on every matter arising within their general jurisdictions. Rather, matters that are repetitive or routine should be referred to subordinates.

Authority, *responsibility*, and *duty* are terms useful to the discussion of organization principles. Authority may be defined as the ability to act or make a decision without the necessity of obtaining approval from a superior. Authority may be delegated to others. Responsibility implies the accountability of a supervisor for an assigned function or duty. Although responsibility may be assigned to subordinates, the supervisor still remains accountable in full to his own superior. A duty is a specifically assigned task that cannot be relegated to another. When devising a company organization plan, the individual positions of authority and responsibility must be identified and the duties of each participant defined.

Management must protect against both under- and overorganization. A balance must be struck between overhead salaries and the monetary benefit actually realized from those salaries. It is a serious organizational failing to burden too few with too much. It is equally undesirable to have an organization top-heavy with half-productive administrators and supervisors.

3.4   PRINCIPLES OF ORGANIZATION

No one organizational pattern can possibly be appropriate for every construction firm. An operational plan for a highway contractor will not likely fit the needs of a design-construct industrial contractor or a contractor acting as a construction manager. Each company must devise an organizational plan that best suits its own peculiar operation. There are, however, certain well-recognized principles of organization that can be applied by any contracting firm that wishes to formulate an efficient organizational plan for its business. No business is so small that accepted methods of organization cannot be profitably applied. The mere act of making a formalized analysis of the necessary tasks, determining how they relate to the company as a whole, and specifying who is responsible for each task creates a clear under-standing of departmental interrelationships, working procedures, and the combined effect they have on the company's ability to conduct business effectively. An organizational plan removes confusion, indecision, buck-passing, duplication of efforts, and neglect of duties. The following simple steps are suggested for the development of an effective company organization. The accomplishment of these steps should include extensive discussion and consultation with everyone concerned.

1. List every duty for which someone must be made responsible.
2. Divide the listed duties into individual job positions.
3. Arrange these positions into an integrated functional structure showing lines of supervision.
4. Staff the organization.
5. Establish lines of communication.
6. Prepare a manual of policies and procedures.
7. Implement the plan and adjust as necessary.

3.5   LIST OF DUTIES

The conduct of a contracting business involves certain duties regardless of whether the business is large or small. These duties will be carried out by only a few persons in a small organization, whereas many people will be involved in a larger company. In making up a list of company duties, the question immediately arises as to how detailed such a list should be. The answer is that a level of breakdown appropriate for the size of the company and the number of employees involved should be used. It seems reasonable to suggest that the larger the company, the finer should be the subdivision of duties. The more detailed the thinking concerning all of the requirements for sound operation, the less likely it is that something essential will be overlooked. The following list indicating the various duties to be performed is used here for illustrative purposes and is not intended to be complete, nor is it meant to apply to a specific construction company.

Executive

Banking

Construction loans

Financial structure

Legal matters

Business organization

Company organization

Auditors and audits

Public relations

Industry associations

Labor policy

Contract negotiation and execution

Investment

Personnel relations and policies

Long-range planning

Setting company objectives and goals

Reviewing and approving business plans

Reviewing and approving financial and investor reports

Investor relations

Setting estimating and bidding policy

Salaries, bonuses, pensions, and profit sharing

Legislative matters

Capital improvements

Scope of operations

Physical facilities

Operating procedures and policies

Accounting and Payroll

General books of account

Subsidiary records

Cost records and reports

Financial reports

Tax returns and payments

Understanding and adjusting to new tax regulations and their implications

Payment of invoices

Billing

Collections

Assignments

Bank deposits

Personnel records

Payrolls and records

Wage and personnel reports to public agencies

Office services

Procurement

Requisitions

Purchase orders

Subcontracts

Change orders

Negotiating material and service contracts

Reviewing material and service contracts for budget compliance

Inventories

Ordering and control of stores

Expediting

Licenses

Insurance, project and company

Subcontractors' insurance

Owner's contract bonds

Bonds from subcontractors

Releases of lien

Guarantees and warranties

Routing and scheduling materials

Building permits

Checking and approval of invoices

Information on prices and sources of supply

Verification of quantity and quality of deliveries

Estimating

Decision to bid

Visiting the site

Obtaining bidding documents

Mailing out bid invitations

Pre-bid conferences

Quantity takeoff

Subcontract and material quotations

Soliciting and ensuring that minority participation goals are achieved

Incorporation of applicable environmental compliance regulations

Pricing

Checking estimate

Preparation of proposal

Bid bond

Delivering proposal

Bills of materials and subcontractors

Engineering

Project planning

Construction schedules

Project cost accounting

Project monitoring

Project cost breakdowns for pay purposes

Periodic project pay requests

Preparation of change orders

Material Safety Data Sheets (MSDS)

Shop drawings

Project cost reports

Field and office engineering

Safety policies and procedures

Accident reports to insurance companies

Relations with owners and architect-engineers

Labor relations

Quality assurance/quality control

Construction

Hiring labor crews

Supervision of construction

Coordination of subcontractors

Timekeeping

Work quantity measurement and tracking

Project cost data

Project accident reports

Project safety programs

Project progress reports

Construction methods

Storage of materials on project sites

Scheduling construction equipment

On-site environmental regulation compliance

Yard Facilities

Receipt, storage, and warehousing of project materials

Maintenance of Material Safety Data Sheets (MSDS)

Maintenance and repair of construction equipment

Storage of construction equipment

Maintenance and issue of stores

Issue, receipt, and repair of hand tools

Transportation

Equipment rental

Prefabrication and subassembly

Spare parts

3.6 DIVISION OF DUTIES

After the duties have been listed, the next step in the development of an organizational plan is to subdivide them into groups so that the duties in each group can become the assigned responsibilities of a single individual. To illustrate, suppose the business is a small partnership consisting of two partners and an employed bookkeeper. One partner is in charge of the office and the other supervises the field operations. The three people concerned must, in a collective way, carry out all the duties listed. Both partners acting together would normally carry out the executive duties. The bookkeeper would perform the accounting and payroll tasks. The office partner could perform all duties related to procurement and estimating and be responsible for the bookkeeper. The field partner's duties could include those associated with engineering, construction, and yard facilities. It is obvious that the duties can be distributed among the three participants in any way desired. The experience, education, expertise, and talents of individuals are normally the basis for the allocation of company responsibilities. It is important that every duty be assigned and, conversely, that every position created include responsibility for a specific list of duties. The list of duties for any given position is known as its "job description."

From the foregoing discussion, it is clear that a member of a small firm will normally be responsible for a much broader range and diversity of things to do than a member of a large company. It is an intrinsic characteristic of a large organization that most of its members have relatively narrow job responsibilities and tend to be considerably more specialized. For instance, a small firm may have one person who, single-handedly, does all the takeoff, pricing, bidding, and purchasing for the company. A large contractor may well employ several people to accomplish the same things, each being involved with only a limited aspect of the total process.

3.7 ORGANIZATIONAL STRUCTURE

The organizational procedure followed thus far ensures that, in total combination, the employment positions established will accomplish each and every duty that has been identified as necessary for the proper operation of business. It is now desirable to link these positions together into an integrated operational structure. It is common practice in the construction industry to establish company jurisdictions or departments. Each department is roughly equivalent in authority and, although interrelated, each operates semi-independently of the others. This is a functional form of organizational structure. The organization is formed by partitioning the work to be done into major functional areas and developing departments around each of these areas. The functional system of management has the advantage that an individual or group can specialize to some extent in some particular aspect of the business. Semiautonomous departments are created, each of which performs a specialized function. How far this *horizontal division* is carried depends on the size of the company involved and the wishes of its management. Each department is assigned a specific area of responsibility (estimating, for example) and is headed by a manager who has training, experience, and skill in that particular aspect of the business.

Each department thus created is then divided vertically. *Vertical division* refers to the establishment of lines of supervision, with each individual along a line being accountable to the person above and acting in a supervisory capacity to those below. The lower a position appears on the organizational ladder, the more limited is the responsibility and authority of the person concerned.

3.8 ORGANIZATION CHARTS

Organization charts present the company's organizational structure in pictorial form, showing every position of responsibility and all lines of supervision and authority. They provide an understanding of the company's structure at a glance. The organizational chart is a particularly efficient means to establish clearly in the minds of the employees involved their individual positions within the overall company, the identities of their supervisors, as well as those whom they supervise, and the nature of their duties. It constitutes an established, permanent reminder of job responsibilities. Such a chart also underscores for employees the fact that the business is well organized and that top management knows at all times who is responsible for what. A company organization chart is well worth the necessary thought and effort in that it requires thinking through all company interrelationships and the fixing of responsibilities. It is also very useful in fitting new personnel into the organization.

Figure 3.1 shows a typical organizational structure of a small contracting company being operated as an individual proprietorship. An organization chart for a small partnership is presented in Figure 3.2. Figure 3.3 is a typical organization chart for a moderately large corporate firm. The organizational structures of contractors can be infinitely varied, and the charts shown here are intended only to illustrate frequently used schemes. The organizational structure and the assignment of duties depicted in these charts are for illustrative purposes only, and the allocation of duties is not intended to be complete. Note that for each position shown, a person's name appears together with a list of major job responsibilities. In addition, each member of the team reports to only one superior.

3.9 STAFFING

The organizational structure devised must now be staffed—that is, a person must be assigned to each of the positions created. A company's success is made by its people, and it is impossible to overemphasize the importance of selecting the right person for each position. In the construction industry, supervisory personnel are typically selected on the basis of

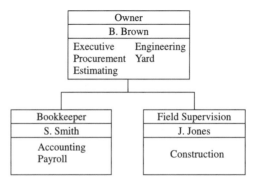

Figure 3.1 Organization chart for small individual proprietorship.

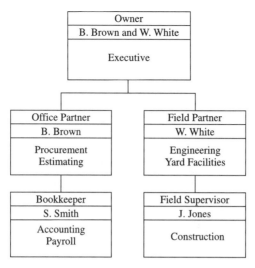

Figure 3.2   Organization chart for small partnership.

their field construction knowledge or technical ability rather than management training or experience. The reasons for this are obvious, but it should be pointed out that the selection of an individual for a supervisory position based only on construction competence is no guarantee that the most effective person will thereby be obtained.

Admittedly, detailed knowledge of the construction process is an important attribute for a construction manager, but other qualifications are also required. Positions on the lower rungs of the organizational ladder are concerned primarily with specific and technical details. Correspondingly, it is entirely appropriate that people be selected for such positions based largely on their job knowledge, because their management function is limited. However, in selecting individuals for progressively higher positions, increasing attention must be paid to managerial ability and general experience in the industry. Experience has shown that technical ability alone does not ensure managerial success.

3.10   RESPONSIBILITY AND AUTHORITY

A responsibility is a personal assignment for which one is held accountable. In a business venture such as a construction firm, each employee is charged with specific responsibilities that are attached to the company position occupied and for which the individual is personally answerable. Authority is the power to act and make the decisions that are necessary in the process of meeting one's assigned responsibilities. Authority commensurate with responsibility must apply at all operating levels. When responsibility for a specific aspect of company operations is assigned, adequate authority must be conveyed at the same time.

Delegation is the process of assigning specific responsibilities and authorities to individual employees commensurate with their training and experience. The obvious objective of delegation is to ensure the effective functioning of the organization as a whole. Basically, delegation is the management function that links the needs of the company with the activities of the individuals who perform the work. Depending on how well the patterns

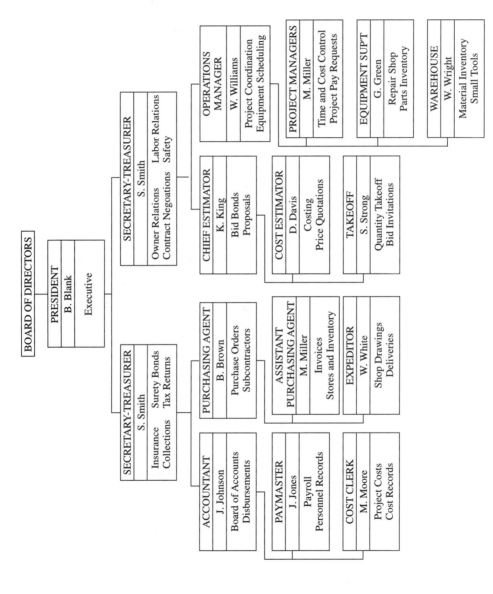

Figure 3.3 Organization chart for moderately large corporation.

of delegation mesh with the basic talents and preferences of the individual employees, it can either enhance the organization or detract from its effectiveness. Delegation defines the organization and has a tremendous impact on all company employees. Delegation creates the lines of company authority and establishes the emotional atmosphere of the firm as it applies to individual development and advancement. Delegation should emphasize not only productivity but also the quality of the work environment, sensitivity to the needs of employees for their personal satisfaction and fulfillment, and the enhancement of morale in the workplace.

In every instance, authority must be given to the individuals concerned to do their specific jobs in any way they choose as long as the results are satisfactory and the procedures conform to established company policies. Responsibility and authority for each company operation must be specifically delegated to an individual and the necessary resources provided. Top management must be able to hold one person fully responsible for each aspect of company operations.

At this stage of development, the organization chart has established a business team that, in a collective and coordinated manner, will do all that is required. There should be no misunderstanding among participants concerning job duties, responsibilities, authorities, or company position.

3.11 COMMUNICATIONS

The proper functioning of a business depends on the exchange of information of many kinds, both within the firm itself and with external agencies. A good system of company communications is essential for smooth and profitable operation. Top management must be kept apprised of job costs and job progress. Procurement personnel must receive purchasing information concerning the materials and subcontractors required for a new project. The project superintendents must be kept advised of contract changes such as drawing revisions and change orders. Procurement people must keep the yard and the project managers aware of the delivery status of materials and must be told of the nature and extent of delivery shortages or damage. Information about job accidents must be conveyed by project managers to company management. The payroll department must be informed concerning hirings and layoffs. Information on back charges against material suppliers and subcontractors must get to the accounting department. These are only a few examples of the necessary communications within a company. It is also important to maintain clear communications with external agencies such as owners, architect-engineers, banks, insurance companies, sureties, material dealers, subcontractors, and governmental agencies.

The establishment of set procedures can help in maintaining repetitive and routine communications. The next section discusses the manual of policies and procedures, one of the principal purposes of which is the description of routine communicative processes.

Periodic meetings of various groups within company management are a necessity. These meetings provide opportunities to exchange ideas, resolve misunderstandings, and decide on future courses of action. In addition, this mode of communication helps to establish team spirit. Brainstorming sessions of such groups have proven to be extremely effective in producing new ideas and management innovations.

To disseminate company information of general interest, bulletin boards and a company publication are very effective. Issued periodically, the publication can include matters such as the firm's safety record, new projects, personnel changes, company policies, and other pertinent information.

3.12   THE MANUAL OF POLICIES AND PROCEDURES

An organization chart is a very useful and effective management device, but it alone cannot completely describe the total workings, details, and interrelationships of the organization. There is an obvious need for further development and description of the company plan. Once the organizational structure has been established, written company policies and operating procedures can be prepared that augment the organization chart and can be used in conjunction with it. These policies and procedures can be set forth in a manual made available to all company personnel concerned. If a loose-leaf manual is used, it is easy to make revisions and add new sheets. In larger organizations, manuals are often issued in separate sections, each of which pertains to the operation of a single department. With the advent of inexpensive computer networks, companies may now make these resources available on-line for improved employee access and real-time document amendment.

Decision making is the essence of management. Some business situations requiring decisions are unique and must be handled separately and individually. Other situations recur with some regularity. Company rules and regulations, or policies, provide standard decisions for such repetitive cases. Policies serve as guides for action by all levels of company management and provide uniform and consistent guidance in the handling of problems that recur frequently. Obviously, policies are also designed to meet specific situations. Such rules and regulations enhance the effectiveness of the organization and are important to the conduct of everyday business affairs.

Operating procedures establish general rules governing communications, the flow of paperwork, and other routine company operations. Such rules remove the element of decision making from office routine and allow duties to be assigned to the lowest practical management level. The very act of reducing these procedures to writing helps to clarify ambiguities, removes areas of overlap, and reveals discrepancies and other shortcomings in the organizational plan. The written procedures ensure uniformity of action, are valuable in training new personnel, and reduce the need for close supervision. The first step in writing any such set of company procedures is generally to have the supervisor of each department consult with his personnel and then write down the procedures that apply to his own area of responsibility. These various procedures are then incorporated into a single set of rules through a series of interdepartmental meetings in which the proposed procedures are coordinated and adjustments are made.

The manual must be explicit concerning the keeping of records. Records of various kinds are indispensable to the conduct of a business. Government agencies require certain kinds of business records to be kept. In addition, company records pertaining to project costs, shop drawings, equipment maintenance, inventory, job progress, estimating, personnel, and other aspects of the business are of real importance. Although maintaining too few records can be costly, excessive record keeping can become equally expensive. It makes no sense to keep records, other than those required by law, that are not actually used and whose potential value does not at least equal their cost of preparation. Record forms should be carefully selected to yield a maximum of information with a minimum of effort and clerical time. The manual can include samples of all standard record forms with explanations and illustrations of their use. Printed or electronic forms for records and communications are great labor savers. They simplify the task of adhering to company policies and procedures and make it possible to process information automatically or to utilize clerical people with less supervision. Users are able to understand and digest information more quickly if it is repeatedly presented in the same format. If most staff members have access to a company's computer network, using electronic forms to input information into a master database may offer the greatest efficiency. Such systems can provide management with powerful and timely reporting while significantly reducing administrative costs.

Another matter to be covered by the operating manual concerns routine company reports—who prepares them, when they are prepared, and to whom they are to be routed. Effective decision making depends to a large degree on the timely and continuous flow of needed management information. Reports on project costs, accident experience, current financial status, cash projections, and similar matters present vital operating information in a condensed and summary form.

3.13 THE OPERATING CHART

A valuable adjunct to the organization chart and the company manual is an operating chart, an example of which is shown in Figure 3.4. This figure is not intended to be complete or to apply necessarily to the operations of any given construction firm, but it does illustrate the type of information that can be conveyed by such a chart. It serves to establish and define the working relationships between company personnel. It shows who participates, and to what degree, when a given activity is performed. Shown across the top are all positions that are involved in a given function and the extent and nature of their involvement. Listed on the left side are the functions for which an individual is responsible and the nature of this responsibility. This information serves as an amplified job description.

3.14 EMPLOYEE HANDBOOK

Either as a part of the manual of policies and procedures, or as a separate document, every company should prepare and distribute an employee handbook that describes the company's human resources policies. By putting such information in writing, company employees will be better informed concerning their expectations and what their employer expects from them. A human resources handbook can do much to minimize poor employee morale, turnover, absenteeism, and low productivity. When specific policies are established, those who enforce them are provided with invaluable guidance in making future employee-related decisions.

The handbook should describe every facet of the company's established human resources procedures. An important point here is the weakening of the traditional employment-at-will arrangement, which holds that an employer can terminate employment at any time and the employee may resign at any time. Many contractors still regard employment as an at-will relationship. However, recent years have seen the at-will attitude come under legal attack. Lawsuits filed by disgruntled employees over the circumstances of their discharge are receiving favorable consideration by the courts. To avoid wrongful discharge lawsuits, a disclaimer should be included in the handbook, on the employment application, and on all other employee-related forms to the effect that employment can be terminated

Company operating chart — **Figure 3.4**

Function \ Position	President	Secretary-Treasurer	Vice President	Accountant	Paymaster	Cost Clerk	Purchasing Agent	Ass't Purchasing Agent	Expediter	Chief Estimator	Cost Estimator	Takeoff Engineer	Operations Manager	Project Manager	Equipment Superintendent	Warehouse Person
Estimating																
Decision to Bid	A		R							O			I	I	I	
Pre-bid Conference			I							I	O		I	I	I	
Quantity Takeoff												O				
Pricing	R		A							A	O					
Markup	R		A							O						
Bid Security	I		I							O						
Proposal			A							O						
Purchasing																
Bill of Materials			R				A	I			O			R		
Subcontracts		A					A	O						R		
Purchase Orders		A					A	O						R		
Expediting								R	O					I		

O - Originator; does the work
R - Must review
A - Must approve
I - Must be informed or advised

Figure 3.4 Company operating chart.

without cause. Such disclaimers may also note that the handbook is compiled solely to provide information and is not the basis of a contractual relationship. The human resources handbook and all other employee-related forms should be reviewed by the company's legal counsel before being distributed.

In preparing a human resources handbook, attention should be given to the following major subject areas:

Employment Status. This section should describe the company's employment categories such as part-time, full-time, contract, hourly and salary employment, to whom and under what conditions access will be granted to employee files and personal information, compensation and salary administration, performance evaluation, promotion, seniority issues, and the procedure for changing employee information records.

Benefits. This section should describe the company's benefits programs such as paid time off, sick leave, health and welfare programs and insurance, education assistance, rehabilitation programs, and worker's compensation.

Timekeeping and Payroll. This section should include information on pay periods, timesheet procedures and regulations, overtime criteria, bonuses and pay raises, payroll deductions, policy on payroll advances, direct deposit, and wage garnishing.

Work Rules and Conduct. This section should broadly cover subjects like employee attendance, work schedules, safety policies, drug and alcohol testing programs, "open door" and sexual harassment policies, equal opportunity employment, employee dispute resolution initiatives, the use of company equipment and materials, and disciplining.

In an overall sense, the handbook should stress the company's desire to maintain an atmosphere of mutual concern and respect regarding working conditions and personnel relationships. National contractor associations such as the Associated General Contractors of Ameica (AGC) and the Associated Builders and Contractors (ABC), and human resources organizations such as the Society of Human Resource Management (SHRM) can provide their contractor members with valuable guidance in developing an employee handbook.

3.15   PLAN IMPLEMENTATION AND ADJUSTMENT

To plan and formulate a company organization requires considerable time and effort. All members of the team must be consulted, and many decisions have to be made concerning a wide range of details and relationships. When the organizational plan has reached an advanced stage, a meeting of company personnel should be called to discuss and debate the pertinent issues. Following consensus, the plan can be finalized and implemented. There is then a period of adjustment and readjustment as the plan is modified and made to work smoothly and more efficiently. In this regard, it must be recognized that no organizational plan can apply very long without being corrected and revised. As business conditions evolve and as company size, personnel, and attitudes change, so must company organization be altered to suit current circumstances.

3.16 MAKING THE ORGANIZATION WORK

There are further considerations on which action must be taken if the company organization is to function and produce to its full potential. A positive environment and a favorable work atmosphere are basic factors in a company's employee relations program. A number of relevant concepts are briefly discussed in the following paragraphs.

Motivation. People within the organization must be encouraged, inspired, and impelled to do what has to be done. The key factor in employee motivation is the human need for recognition and self-esteem. In this regard, money is not the only compelling element. Working with other people is the primary means of securing a feeling of recognition and importance; people generally want to feel that they are part of a team. A good manager will build a closely knit, effective operating group. To this end, employees should be given opportunity to display their talents and capabilities, and there must be a suitable reward system or other outward manifestation of achievement recognition.

Personnel Development. People must acquire or improve the skills, attitudes, and abilities necessary to do the work assigned to them. A plan for the personal improvement of company personnel, based on an appraisal of individual strengths and weaknesses, is an important management responsibility. There are many ways in which this can be accomplished, such as by encouraging continuing education through home study or attending formal courses and seminars. Paying all or a part of the attendant costs and giving time off from work can be powerful stimuli. Encouraging committee participation, providing job rotation, and coaching on the job can also be effective.

Training of Replacements. Good management encourages the development of young and upcoming talent. People who die, retire, or leave the company must be replaced. Even in cases of sickness or other temporary absence, someone must fill in. The person in a position of responsibility who, because of a dominant personality or fearing loss of job security, does not develop likely successors among subordinates, is not a good manager, and is not serving the best interests of the company.

Decision Making. A common source of management difficulty is the failure of some member of the organization to make business decisions promptly and positively. Decisions cannot be rushed and must receive due consideration, but undue delay can often make a difficult situation worse. Decisions rarely please everyone, but indecision, procrastination, and vacilation serve only to exacerbate the matter. At each level of the organization there should be one individual whose function it is to decide on operational courses of action. This person should have the benefit of full consultation and recommendations, but the responsibility for decisions rests here, and the individual must act accordingly.

QUESTIONS

1. In building a contractor's organization, what primary considerations or objectives must the contractor consider?

2. Authority, responsibility, and duty are terms frequently used in discussing organizational principals. What is the basic difference between employee responsibilities and employee duties?

3. In an organizational structure, what basic attributes diminish as one moves lower down the organizational ladder?

4. What management function links the needs of the company with the activities of the individuals who perform the work?

5. Name and discuss three important objectives of an organization chart.

6. The organization chart is an important company document. To ensure organizational effectiveness, what additional documents must be made available to all employees, and what basic issues should these documents address?

7. What important legal considerations must be contemplated when preparing an employee handbook?

8. What is meant by at-will employment? What effect might changes in this relationship have on project related hiring in the construction industry?

Chapter 4

Drawings and Specifications

4.1 THE ARCHITECT-ENGINEER

The design of construction projects is performed by architects and/or engineers, the division of responsibility between the two depending on the nature of the construction involved. As a general rule, both residential and building construction are primarily of an architectural nature, with structural, electrical, mechanical, and other engineers providing supportive services as may be required. Highway, heavy, utility, and industrial projects are predominantly engineering projects, with architectural services obtained if needed. There are many firms that designate themselves as architect-engineers and perform planning and design work of both classifications. In this text, the term *architect-engineer* is used to designate the organization, person, firm, or team that performs the project design, whether it is architectural, engineering, or a combination of both.

The architect-engineer can occupy a variety of positions relative to the owner and contractor. Typical among these are the following:

In-House Capability. Owners sometimes have their own in-house design capability. Some large industrial firms and many public agencies that have continuing construction needs maintain their own design departments. In such cases, the architect-engineer is an integral part of the owner's organization. The construction is usually accomplished by a construction firm or firms under contract with the owner.

Owner-Client. In the traditional pattern, the architect-engineer is a private professional design firm that performs its function under contract with the owner. Normally the construction is performed by a contractor or contractors also under contract with the owner. There is no formal relationship between the contractor and the architect-engineer.

Design-Construct. In this instance, the owner contracts with a single party for both design and construction. Although the architect-engineer and the contractor may be related in a variety of ways, three are mentioned here as being commonly used: The contractor may have its own design capability with architects and engineers on its payroll; the architect-engineer can be a corporate affiliate or subsidiary of the contractor; or the contractor and architect-engineer, both independent firms, can form a joint venture for a given project or contract.

Construction Management. Contractually, this is much the same arrangement as with owner-client. However, the contracts between owner and architect-engineer and between owner and construction manager (CM) provide for extensive cooperation between the architect-engineer and the construction manager from the inception of a project.

Design-Manage. With this arrangement, the owner signs a single contract for both design and construction management services. A design-manage firm may perform both the design and management functions, the architect-engineer can be a corporate affiliate or subsidiary of the CM, or there can be a joint venture between the architect-engineer and the CM.

Several associations of construction designers perform a very valuable service in providing ethical, professional, and business guidance to the individual architect-engineer. Professional organizations such as the American Institute of Architects, the National Society of Professional Engineers, and the American Consulting Engineers Council act to promote and protect the image and reputation of the design profession and to assist their members in achieving higher professional, ethical, business, and economic standards.

4.2 SELECTION OF THE ARCHITECT-ENGINEER

When the owner has an in-house design capability, there is no need to choose an architect-engineer to design the project. Without that capability, some selection process must be followed. Three options for the owner in selecting a design firm are to (1) call for submissions from interested designers, with selection based on professional qualifications alone, (2) call for submissions that include designation of fees, which become a part of the evaluation criteria, and (3) call for bids for professional services and select on the basis of price alone.

Traditionally, professional societies of architects and engineers have opposed competitive bidding and price competition as a means of selecting firms to perform professional design services and have included in their codes of ethics prohibitions against competitive bidding by their members. These organizations advocate that design professionals be selected by the owner on the basis of their qualifications and competence, with the fee to be determined through negotiation. Objection to competitive bidding for planning and design services is voiced on the grounds that it is not in the client's or the public's best interests.

However, in 1978 the U.S. Supreme Court ruled that the National Society of Professional Engineers' ethical ban on competitive bidding for engineering services was not permissible under the Sherman Antitrust Act. As a result of this decision, the professional societies may not bar competitive bidding procedures if those in practice wish to engage in such methods. Consequently, it is not now considered unethical to submit or invite such bids. Professional societies, however, may encourage owners to continue using the traditional selection and fee-setting procedures. Professional designers are not required to submit bids, and owners may, if they wish, shun competitive procedures in their procurement of professional design services. For instance, most agencies of the federal government are barred from using competitive bidding in the procurement of design services by the Brooks Act (1972). This Act mandates the negotiation of design contracts at fair and reasonable prices on the basis of demonstrated competence and qualifications for the type of professional service required. Many states have passed mini-Brooks Acts that set forth professional selection and negotiation procedures for state-financed projects.

Nevertheless, recent years have witnessed a pronounced shift away from the traditional selection of design professionals exclusively on the basis of qualifications. Following the 1978 Supreme Court decision on competitive bidding by design professionals, there has been a substantial movement by public owners, other than the federal and some state governments, toward the use of price as one criterion in the selection of architect-engineer firms. The procurement codes of many state and local governments now make the design fee very much a part of the selection process. Certain public agencies presently issue fee schedules that apply to their selection of design professionals. In the private sector the fee amount now usually plays an important role in the selection procedure.

With regard to the selection of architect-engineer firms to design projects for certain public agencies, there are now requirements that designated construction design contracts be set aside for small businesses. Such set-aside projects are identified as those requiring up to a specified maximum estimated professional fee or specified maximum estimated project cost. Small design businesses are identified by the U.S. Small Business Administration as those with annual gross billings of less than a specified maximum average value.

4.3   SERVICES PROVIDED BY THE ARCHITECT-ENGINEER

The scope of services required of the architect-engineer is subject to considerable variation, depending on the needs and wishes of the owner. Basic to such services, however, are ascertaining the needs and desires of the owner, developing the design, preparing the several documents required for contractor bidding or negotiation as well as for contract purposes, aiding in the selection of a contractor, and making an estimate of construction costs. In this regard, an early action in the initiation of a proposed construction project is the making of an estimate of its probable final cost. Such conceptual and feasibility cost estimates are standard procedure for most construction projects and provide the owner with invaluable information concerning the necessary financing for a project. A conceptual cost estimate is the usual starting point for a construction project and is often based on the standardized system developed by the Construction Specifications Institute, as shown in Appendix B of this book. In the case of competitive bidding, the documents prepared by the architect-engineer are referred to as the "bidding documents" during the bidding phase. After the contract is signed, these documents are called the "contract documents."

The scope of services provided to the owner by the architect-engineer during field construction depends on the needs and preferences of the owner. The architect-engineer's responsibility to the owner may cease when the contract documents are finalized and delivered. However, the owner may require full construction-phase services, including project inspection, the checking of shop drawings, the approval of periodic payments to the contractor, the issuance of a certificate of completion, and the processing of change orders. Although the architect-engineer is not a party to the usual construction contract between the owner and the contractor, this contract often conveys certain powers to the architect-engineer, such as the authority to decide contract interpretation questions, judge performance, condemn defective work, and stop field operations under certain circumstances.

A standard form of design contract published by the American Institute of Architects (AIA) and widely used for building construction is reproduced in Appendix A. This and other standard forms and documents reproduced in this text are for illustrative purposes only. Because AIA documents are revised from time to time, users should obtain from the AIA the current editions of the documents reproduced herein.

4.4   CONTRACTOR INPUT INTO DESIGN

Although the function of the contracting firm is seldom project design as such, its experience and expertise can make it a valuable member of a planning and design team. In the traditional and still predominant linear construction process, input from the contractor into the design process does not usually occur. There are occasional instances in which the owner or architect-engineer obtains the consultation services of a contractor during the planning and design phases, but this is much more the exception than the rule. However, extensive contractor input into design is a normal part of design-construct, construction management, and design-manage, where the team concept prevails.

Where the input of experienced construction people into the planning and design of the project is a part of the procedure, the contractor typically enters the picture at an early date. This party does not do the design but provides continuing advice concerning general site planning, local work practices, labor availability and costs, material availability, delivery times, and alternative work methods and procedures. The contractor prepares cost estimates and construction and procurement schedules and participates in the value-engineering program. The contractor's knowledge of prices and availability of materials and services is a valuable source of information to those making design decisions. As one who is familiar with performance, maintenance, and installation costs, the contractor can help assess life-cycle costs and benefits. Under the right circumstances, contractor input into the design process can result in substantial benefits to the owner.

4.5   FEE FOR DESIGN SERVICES

When an architect-engineer, acting as a private practitioner, performs a design service for a client, the fee may be determined in one of several ways, of which the most commonly used are the following:

Percentage of construction cost

Multiple of salary cost

Multiple of salary cost plus nonsalary expense

Fixed lump-sum fee

Total expense plus professional fee

Hourly or per diem charge

Fees for the design professional on publicly financed projects are often subject to statutory or administrative limitations. There are instances in which the design contract between the owner and the architect-engineer provides that payment of the fee is contingent in some way on the project cost being within the budget established by the owner. For example, such contracts may provide that the architect-engineer will not be entitled to any increase in fee when project redesign is required to keep construction costs within an agreed-upon amount. It is common for design contracts to provide that the architect-engineer's fee be paid in installments as designated phases of the designer's services are completed.

4.6   RESPONSIBILITY TO THE OWNER

Although the owner and the architect-engineer are usually joined together by contract, their exact relationship depends somewhat on the duties being performed by the architect-engineer. In the preparation of construction documents, the architect-engineer firm functions primarily as an independent contractor, but its role is more that of an agent of the owner during the construction phase. The architect-engineer has a fiduciary obligation to its client, requiring fairness, trust, and loyalty. The design professional must avoid any conflict of interest that could work to the disadvantage of the owner.

The common-law standard of care applicable to architect-engineers is the same as that for other professionals, such as doctors and lawyers. The services of experts are sought by the public because of their special skills. As such, professionals have a duty to exercise ordinary skill and competence in carrying out their functions, and a failure to discharge this duty will subject them to liability for negligence.

As a professional, the architect-engineer is required to exercise care and diligence in carrying out its responsibilities. Learning, skill, and experience are expected to the degree customarily regarded as being necessary and sufficient for the usual practice of that profession. By the contract of employment, the architect-engineer implies that it possesses ordinary skill and ability and that it will carry out the design with promptness and a reasonable exactness of performance. The designer is responsible for the adequacy of the materials, components, and systems that are selected and specified. The architect-engineer also bears the responsibility for preparing design documents that are in conformance with applicable building codes, setback requirements, zoning regulations, and environmental regulations. It is not expected that the architect-engineer will produce a perfect design or even satisfactory end results. However, the law expects the design professional to carry out the tasks it has undertaken within reasonable standards of care, skill, and performance. If the owner suffers loss or injury because such standards are not met, the architect-engineer is liable.

Architect-engineer liability to the owner may arise through breach of contract or tort responsibility. More specifically, the designer can be held liable to its client for violation of a specific contract provision, because of an express or implied warranty in the design contract, or because of negligence in the performance of the architect-engineer's duties under the contract. If the design professional has given an express or implied warranty of the sufficiency of the design or of the reasonable suitability of the structure for the purpose intended, then the architect-engineer is strictly liable for damages caused by a breach of the warranty and it is not necessary for the owner to prove any specific negligence. Although there is a split in authority, most jurisdictions have generally held that the design professionals have not given an implied warranty that their plans and specifications are adequate for a specific purpose. According to most courts, an architect-engineer warrants its work only to the extent that it has used the customary skill of the profession.

Although an architect-engineer firm can be held responsible for lack of care, diligence, or skill, it is not normally considered to be negligent because of errors in judgment. However, there is a definite trend in the courts toward expecting a greater degree of perfection and foresight on the part of architect-engineers and holding these parties responsible for their negligence. In addition, if the architect-engineer represents itself to be a specialist in a certain type of work, it will likely be held responsible for a greater degree of competence and care than would a general practitioner. If the architect-engineer selects and employs consultants such as electrical engineers, mechanical engineers, acoustical engineers, or landscape architects to design or advise concerning specialized portions of the project, the architect-engineer remains responsible to the owner for the overall adequacy of the completed design. In this regard, the design professional has a nondelegable duty to the owner and promises by implication that all services will be properly performed. If a consultant's work should prove to be faulty, the architect-engineer is liable to the owner for any resulting damages. The consultant is, of course, accountable to the architect-engineer for which the special work was done.

If the architect-engineer has an inspection responsibility during construction, it has the duty to ensure that the contractor materially follows the drawings and specifications. It has the responsibility to see that the owner gets substantially the structure called for by the construction contract.

Design professionals sometimes include exculpatory clauses in their design contracts in an attempt to limit their professional liability. Some of these clauses provide that the architect-engineer not be held liable to the owner for damages resulting from the negligence of the designer. Other clauses limit the potential liability of the architect-engineer to a specific amount. However, such agreements apply only to the parties who sign them and cannot limit the architect-engineer's liability to third parties.

With regard to exculpatory language in design contracts, it is to be noted that certain contractual exemptions from liability are declared invalid by state statutes, and that the general enforceability of such contract clauses is uncertain at best. Courts favor strict interpretation of such exculpatory clauses and seek to avoid contract provisions that excuse parties from their own fault or negligence. A new approach, called a contingency reserve, provides that the owner agree to establish a fund of some specific percentage (say, 5 percent) of the construction cost of the project. This reserve is used by the owner to pay the contractor for expenses arising from minor design errors or omissions. The owner agrees not to bring any action against the architect-engineer for additional construction costs within the limits of the contingency reserve, even though they are the fault of the architect-engineer.

4.7 LIABILITY TO THIRD PERSONS

Although it has not always been so, the law now generally recognizes that an architect-engineer can be held liable to a third party if bodily injury or property damage is caused to that party by reason of negligence or failure in duty of the designer. The law of negligence requires a professional to exercise reasonable care to protect third parties during the performance of that party's work. In addition, the architect-engineer may in some instances be held liable to third parties for economic loss suffered as a result of its negligence. In this context, *third party* refers to any party who is a stranger to the design contract between the architect-engineer and owner.

Third-party liability can arise from a project either during its construction or after its acceptance and occupancy by the owner. Under this concept of liability architect-engineers now find themselves subject to damage claims by contractors, subcontractors, construction workers, sureties, suppliers, lenders, and outsiders lawfully on the project premises where the negligent performance of duty by the architect-engineer allegedly caused or contributed to harm or injury.

Recent years have seen the design professional increasingly held liable to general contractors and subcontractors for economic loss caused by the architect-engineer's negligence. Architect-engineers have been held liable for a variety of injuries suffered by workers during construction and by members of the general public after project completion. There have been several cases in which an architect-engineer firm, being responsible in contract with the owner for job inspection, has been judged responsible for the safety of the work and has been made liable for damages when it failed to take corrective measures, on the grounds that it knew or should have known of a dangerous job condition. Negligence suits stemming from injuries suffered on completed projects have been filed by parties whose injuries were caused by alleged improper or inadequate design. An architect-engineer's continuing liability for completed projects is often limited by special statutes of limitations, a subject discussed in the following section.

In any discussion of tort liability resulting from negligence, the matter of strict liability arises. There is an increasing trend in this country toward imposing strict liability, on the basis of implied warranty, for injuries caused to the user or consumer of mass-produced products. Product liability, or strict liability, means liability without proof of fault; that is, liability for damages is not based on a demonstration of negligence on the part of the producer of the goods. Under this theory, the person suffering injury or damages can receive compensation if it can be proven that the product was defective and that this defect caused the injury or other loss. The courts have not yet applied product liability to architect-engineers or construction managers, although it has been applied to the manufacturers of prefabricated buildings and to builder-vendors who produced and sold homes that turned out to be defective. These findings were based on implied warranties of workmanship and habitability.

4.8   STATUTES OF LIMITATIONS

Most states now have special statutes of limitations that apply to the accountability for damages arising from a defective and unsafe condition created as the result of an improvement to real property. These statutes apply to architect-engineers and construction contractors and establish a time beyond which these parties are no longer liable for damages arising out of completed construction projects. Such statutes provide that the time during which the architect-engineer and contractor remain liable starts with the substantial completion of the work or acceptance of the work by the owner. These statutory periods range from 4 to 20 years in the various states, with the average being about 7 years.

In those states without such statutes, the architect-engineer and contractor must rely on the states' general statutes of limitations. The time period within which a given action must be brought under these statutes varies, but a typical statute provides for a three-year period for torts (negligence) and a six-year period for breach of contract. However, a serious question concerning these general statutes of limitations is when the statutory time begins. The usual provision for negligence suits against architect-engineers and contractors is that the time starts when the cause of action accrues. The right of the owner to sue for breach of contract ordinarily starts when the work is accepted. However, there can be exceptions to this standard when the construction defect is concealed. The result of such conditions is that in those states without special statutes of limitations, the architect-engineer and contractor are indefinitely vulnerable to suits charging negligence or breach of contract. It must also be noted that some of these special statutes of limitations have been declared unconstitutional in recent years.

4.9 PROJECT DESCRIPTION

The nature and extent of the construction to be done, the materials to be provided, and the quality of workmanship required are described by the drawings and specifications. Complementing each other very closely, the drawings and specifications present a complete description of the work. The drawings portray graphically the extent and arrangement of the components of the structure. The specifications describe verbally the materials and workmanship required. These documents serve three important functions. First, they serve as a basis for competitive bidding or contract negotiation. Second, they serve as contract administration documents during the construction phase, describing the work to be accomplished and defining the rights and duties of the participants. Third, they are contract documents that constitute the basis for the settlement of claims, disputes, and breaches of contract. It is therefore important that these documents be carefully examined by the contractor during the bidding or negotiation phase.

When lump-sum, unit-price, or guaranteed maximum price contracts are involved, the drawings and specifications must be in a complete and detailed form before the contractors are brought in for bidding or contract negotiation. Contractors can scarcely be expected to obligate themselves to the terms of a fixed-price contract without being able to establish with some precision the nature and extent of the work involved. When a cost-plus type of contract is under consideration, the documents may be more rudimentary. In such cases, detailed drawings and specifications still must be provided, but they can follow as needed during the construction period. However, it is always preferable to have the drawings and specifications as complete and detailed as possible before the owner discusses costs, completion dates, and other such matters with the contractor. When incomplete and preliminary drawings and specifications form the basis for such discussions, subsequent development of the design often creates changes and complicating features that alter the original agreements and understandings.

4.10 OWNERSHIP OF THE DESIGN

Ownership of the drawings, specifications, and associated maps, reports, articles, computer software, and audiovisual tapes, and their possible unauthorized use, can be a matter of considerable importance to the architect-engineer and the owner. Where the owner is a public agency, these documents often become the sole property of the owner as a matter of law. Otherwise, most design contracts between the owner and architect-engineer explicitly designate which of the two parties has legal title to the design documents that are the subject of the agreement. The usual provision is that the design documents are, and remain, the property of the architect-engineer. However, many owners include a specific provision in the design contract that conveys all rights to, and ownership of, the documents to the owner.

Agreements of this type, however, do not bind third parties to the contract. Accordingly, the design contract does not protect the interests of the owner or architect-engineer by preventing third parties from making unauthorized use of the design documents.

To protect against unauthorized usage, a copyright can be obtained in accordance with the U.S. Copyright Law of 1976, as amended. It is not necessary to register a copyright with the U.S. Copyright Office to have protection against unauthorized violation. Copyright protection occurs automatically when the work is created in a tangible form of expression. However, registration establishes a public record of ownership rights and is normally required before any legal action can be taken. To register a copyright, the interested party need only send a properly completed application form, a modest fee, and a deposit copy of the work being registered to the Copyright Office, Library of Congress, Washington, DC 20559. Whether the design work is registered or not, a copyright notice must be shown on the documents. This consists of the symbol © or the abbreviation "Copr.," the year of the first publication, and the name of the owner of the copyright.

Either the architect-engineer or the owner may be able to register the design documents, depending on the arrangements between the two parties. The federal statute gives the copyright protection to the architect-engineer. However, where the architect-engineer is commissioned by the owner to do the design, the owner can obtain the copyright if there is an express agreement to that effect with the designer. Copyright protection extends only to the display, sale, or physical copying of the design documents, not to duplication of the structure or to ideas and processes contained in the documents. Expiration dates for the copyright are 70 years after the creator's death or, if the work was made for hire, 95 years after its publication or 120 years after its creation, whichever comes first.

4.11   THE DRAWINGS

The drawings, or plans, are instrumental in the communication of the architect-engineer's intentions concerning the structure it has conceived and designed. They portray the physical aspects of the structure, showing the arrangement, dimensions, construction details, materials, and other information necessary for estimating and building the project. Drawings are individually prepared for almost every project. A job covered by drawings that are complete, intelligible, accurate, detailed, and well correlated can be priced much more realistically and can also be better constructed than one described by sketchy, poorly drawn, ambiguous, and incomplete documents. When well-prepared documents are provided, disputes and claims for extra payment during construction are minimized, and the owner is likely to get a much better finished product at a lesser cost.

From original drawings prepared by the architect-engineer, booklets or rolls of drawings are reproduced. Custom and usage have evolved more or less standard classifications of drawings and a prescribed order in which they appear in a set. For instance, the drawings for a building typically include subgroups such as a site plan or civil drawings, architectural, structural, plumbing, mechanical, and electrical drawings appearing in that order. The sheets of each category are designated by an identifying capital-letter prefix and are numbered separately and consecutively. The drawing format varies with the type of construction, however. The makeup of the drawings differs considerably among building, industrial, heavy, highway, and utility construction.

4.12   CADD SYSTEMS

Computer production of construction designs and drawings is now an important and growing area. Computer-assisted design and drafting (CADD) systems are now being used by architects, engineers, owners, and contractors to design structures and produce finished, dimensioned working drawings for field construction use, a process that makes their operations much more productive and profitable. CADD systems facilitate the rapid development of the design and production of high-quality construction drawings that convey much more information than their manually created counterparts. Structural and architectural design, electrical circuitry, mechanical flowcharts, piping and instrumentation diagrams, and physical arrangement drawings are all created at a far faster rate. With CADD, the designer can draw, make changes, model, and experiment with speed, efficiency, and precision. Its features include three-dimensional viewing, a variety of display colors, video animations, and pan and zoom capabilities allowing the designer to work with enlarged portions of the drawings. CADD drawings can be continuously updated relative to changes and modifications to plans and elevations. Many of these systems also determine the field quantities of work as the design proceeds, which makes it possible for the designer to prepare a bill of materials for bidding purposes. Also available is software that combines cost estimating with CADD, a development of interest to design-build firms, owner-builders, home builders, and other parties who work with both project design and cost estimating.

4.13   STANDARDIZED DRAWINGS

The use of standardized drawings is common in certain areas of the construction industry. Such drawings are usually standard design details showing materials, dimensions, arrangement, configuration, and other information. Designs for sewer and street work and other municipal construction are common applications. Street and alley paving, manholes, curbs and gutters, drop inlets, catch basins, sewer connections, valve chambers, hydrant and motor settings, and gate wells are examples of municipal work readily susceptible to the standardization of design. Many engineering aspects of bridge and highway construction are also illustrated by standard design details, which may be physically incorporated into the project drawings or included by reference only. The development of such details is of considerable convenience to the construction process and is conducive to economies of fabrication and installation.

4.14   THE SPECIFICATIONS

Specifications are written instructions concerning project requirements. The drawings show what is to be built, and the specifications describe how the project is to be constructed and what results are to be achieved. Historically, the word *specifications* has referred to specific statements concerning technical requirements of a project, such as materials, workmanship, and operating characteristics. However, it has become customary to include the bidding and contract documents together with the technical specifications, the entire aggregation being variously referred to as the project manual, project handbook, construction documents book, or most commonly, simply as the specifications or "specs."

Large projects, or those involving an unusual number of legal or contractual formalities, may have a separate booklet of nontechnical forms and instructions. These document forms are required for bidding and to inform the contractor about the contract terms and provisions. A contractor cannot be expected to enter into a contract with the owner without first having seen the several contract documents. Such nontechnical forms vary considerably with the project and can include such things as an invitation to bid, instructions to bidders, general conditions, supplementary conditions, a proposal form, a bid bond form, contract bond forms, a list of prevailing wage rates, a noncollusion affidavit, a sheet for the listing of subcontractors, and an agreement form. All of these items are discussed in this or the following chapters.

The preparation of the specifications is an important part of an architect-engineer's responsibility and is an arduous chore requiring considerable knowledge and writing ability. Every specification writer must rely on material and equipment manufacturers to furnish technical data and accurate descriptions of their products. The architect-engineer combines this information with its own experience and judgment to specify the appropriate product or material. The volume of manufacturers' literature has reached such enormous proportions that sophisticated systems of storing and retrieving product information have been developed. A number of commercial information retrieval systems are currently available. Many architect-engineers have established computerized systems of specification writing, using word processors and stored master specifications. These methods have proven to be much faster and more economical than the traditional procedures.

4.15   SPECIFICATION DIVISIONS

Specifications are issued by the architect-engineer in duplicated form with stiff paper covers. For large projects, this document can assume the dimensions of a good-sized catalog and may be issued in more than one volume. It is usual practice for the specifications to be segmented into standardized divisions generally used by the construction industry. The initial divisions of the specifications normally contain the nontechnical provisions (mentioned previously) of the contract. The succeeding divisions, which constitute the bulk of the specifications, contain the technical provisions that describe the workmanship and materials for each of the individual construction segments such as site work, concrete, masonry, carpentry, mechanical, and electrical.

The technical divisions that appear within a set of specifications depend on the nature of the project and the general category of construction. Specifications covering the construction of a building, for example, will have only a general resemblance to those for an engineering project such as a highway, tunnel, or marine structure, although both are generally structured in accordance with the same format. Appendix B illustrates a standard structure and format adopted in many construction specifications. This appendix presents the Broadscope Section Titles of MASTERFORMAT as promulgated by the Construction Specifications Institute (CSI) and Construction Specifications Canada (CSC). Although this format has been used primarily for building construction until recently, in 2004 CSI expanded MASTERFORMAT's 16 divisions to 48 divisions and included site and infrastructure, and process equipment, in its scope.

The CSI specification divisions have been widely adopted by many trade, technical, professional, and public groups associated with design and construction. MASTER-

FORMAT is broadly used by architect-engineers in the preparation of specifications for construction projects. Most building materials are now identified by their CSI classification numbers for filing purposes. However, there are other established specification formats in the construction industry. For example, many state highway departments follow the standard format for highway construction specifications sponsored by the American Association of State Highway and Transportation Officials (AASHTO). Some projects, of course, have job specifications whose formats are nonstandard.

4.16   THE GENERAL CONDITIONS

The general conditions, sometimes called general provisions, set forth the manner and procedures whereby the provisions of the contract are to be implemented according to accepted practices in the construction industry. These conditions are intended to govern and regulate the obligations of the formal contract; they exert no effect on any remedy at law that either party to the contract may possess. They are not intended to regulate the internal workings of either party to the agreement, except insofar as the activities of one may affect the contractual rights of the other party or the proper execution of the work. Although the headings and topics included within different sets of general conditions vary, there is a certain similarity of subject matter. The general conditions typically include the following items:

1. Definitions
2. Contract documents
3. Rights and responsibilities of owner
4. Duties and authorities of architect-engineer
5. Rights and responsibilities of contractor
6. Subcontractors
7. Separate contracts
8. Time
9. Payments and completion
10. Changes in the work
11. Protection of persons and property
12. Insurance and bonds
13. Disputes
14. Termination of the contract
15. Miscellaneous provisions

Standardized sets of general conditions have been developed by various segments of the construction industry. The general conditions compiled and published by the American Institute of Architects have found wide acceptance for building construction. These conditions are reproduced in full as Appendix C. The use of standard forms of general conditions can be advantageous from many points of view. They have evolved into a form that has stood the tests of time and experience and have become familiar to contractors, architect-engineers, and owners. Such provisions afford a fair and equitable way of administering construction contracts.

4.17   SUPPLEMENTARY CONDITIONS

Any standard set of general conditions is intended to apply to a relatively broad range of construction and must be amended and/or supplemented at times to conform to the idiosyncrasies of a given project. This is accomplished by a section of the specifications, called the supplementary conditions, which immediately follows the general conditions. Supplementary conditions are occasionally also referred to as special conditions. Common examples of necessary amendments to the general conditions are the number of sets of contract documents to be furnished to the contractor, limitations on surveys to be provided by the owner, special instructions to the contractor when requesting material substitutions, changes in insurance requirements, and special documentations required by the owner as a condition of final payment.

Additional articles must frequently be included that supplement those of a standard set of general conditions. Conditions of project location, order of procedure, times during which the work must proceed, owner-provided materials or equipment, other contracts, unusual contract administration requirements, early occupancy by owner, time of project completion, and liquidated damages are examples of such distinctly individual contract requirements. Appendix D, which contains a representative set of supplementary conditions, illustrates the specialized nature of this division of the specifications.

4.18   THE TECHNICAL SPECIFICATIONS

The technical specifications present verbal descriptions of the technical requirements of the work to be accomplished, with emphasis on the levels of quality to be achieved. These specifications are normally presented in approximately the same general sequence in which the corresponding construction operations actually proceed in the field and are subdivided in accordance with the usual construction craft jurisdictions. It is customary that a separate division of the specifications be devoted to each major type of construction operation that will be involved, such as excavation, concrete, structural steel, piping, insulation, and electrical work.

Many projects—buildings, for example—include elements of such a nature that it is impractical or impossible to verify their adequacy or quality by field tests after completion of the construction. Under such circumstances, the technical specifications prescribe the materials and the workmanship standards required. With other construction elements it is possible to measure, test, or otherwise prove the service performance of the finished product. In such cases, the specifications frequently specify only the desired end result. Such performance, or end-result, specifications are now widely used.

4.19   PERFORMANCE SPECIFICATIONS

A performance specification describes the required performance or service characteristics of the finished product or system without specifying in detail the methods to be used in obtaining the desired end result. This type of specification makes the contractor responsible for obtaining the results expected, and an end product is required that will meet the acceptance tests and standards specified. The selection of construction methods and procedures is left to the contractor. A performance specification obviously allows considerable competition among products and systems. For example, performance specifications are commonly used in conjunction with a project's mechanical and electrical systems, asphaltic and portland cement concrete, and compacted fill.

When service requirements can be established and measured by means of some practical test procedure, the use of end-result specifications can be advantageous. The contracting firm is made responsible for obtaining a satisfactory result. By leaving it free to exercise its ingenuity, skill, and experience to the fullest extent in achieving the desired result, the cost of construction may well be reduced.

4.20   DESIGN SPECIFICATIONS

In a design specification, also referred to as a prescriptive specification or a materials and workmanship specification, the drawings and specifications spell out in explicit terms what must be done to accomplish a desired end result. A design specification describes the kinds and types of materials to be provided, their physical and performance properties, their sizes and dimensions, the standards of installation and workmanship, and the inspection and tests required for verification of quality. It is common practice for a specific brand name or names to be listed with model numbers and other data to establish the standard of quality desired. The use of substitutes by the contractor is discussed in Sections 4.23 and 4.24.

In a design specification the end result may or may not be described, as it is a natural result of the procedure specified. The contractor must perform in accordance with the precepts of quality construction, and the work must conform with the design and dimensions that are spelled out in the technical requirements of the contract. This type of specification is used in connection with structural work where the architect-engineer specifies exact dimensions, methods, materials, weights, sizes, quantities, and procedures. In such a case, the contractor must supply and construct in accordance with the contract and has little or no discretion. Here is a case in which the architect-engineer normally assumes responsibility for the adequacy of the end product.

From the foregoing descriptions of performance and design specifications, it can be seen that many construction specifications involve elements of both. This condition can, at times, be troublesome in assigning responsibility for the adequacy of an end product. There is some case law indicating that where there is conflict between the two, the design specification controls.

4.21   MATERIAL AND PRODUCT STANDARDS

There are many standard material and product specifications that are sponsored by the federal government and a variety of technical societies. These specifications have been devised by specialists in their fields and are accepted as authoritative by the industry. Federal specifications prescribe technical requirements for materials, products, and services procured by the federal government. The American Society for Testing and Materials (ASTM) promulgates large numbers of specifications pertaining to the physical, chemical, electrical, thermal, performance, and acoustic properties of a wide variety of materials, including practically all those used in construction. ASTM standards are widely incorporated into construction specifications by reference as a means of controlling the quality of portland cement, reinforcing steel, asphalt, insulation, and a host of other construction products. The American National Standards Institute (ANSI) publishes widely used material and product standards of great variety, including those used in the construction industry. The American Association of State Highway and Transportation Officials (AASHTO) publishes test standards pertinent to highway and airfield construction.

In addition to these standards, which cover broad ranges of materials, there are many other standard material specifications of a more or less specialized nature. The American Institute of Steel Construction, the American Water Works Association, the American Institute of Timber Construction, the American Concrete Institute, the American Welding Society, the Structural Clay Products Institute, and many others issue standards that apply to the quality and construction application of their product specialties. These standards establish reliable quality criteria for particular work classifications.

4.22   CLOSED SPECIFICATIONS

If a material or process specification is worded such that only one proprietary product will be acceptable, and if no provision is made for substitutions, it is known as a closed specification. In such a proprietary specification, the level of quality is established by naming a specific item or by describing the precise features and characteristics of a given brand-name product. The underlying purpose of a closed specification is to ensure that only a construction product of the precise type and level of quality described will be furnished by the contractor.

A private owner is free to use closed material specifications. This procedure, however, is not generally considered to be in the best interest of the owner because it eliminates the several advantages of competition among suppliers. On public contracts there is a long-established policy against drafting specifications that eliminate or inhibit competition. Hence, a closed material specification is not normally permitted where a public owner is involved.

4.23   OPEN SPECIFICATIONS

For both private and public construction projects, the use of open material specifications is the general rule. Consequently, the architect-engineer must word such specifications so that the products of various manufacturers are acceptable, whether or not they are mentioned by name, provided that such products meet the prescribed operational and quality standards. This type of specification is referred to as an open specification.

To establish a required standard of material quality when using an open specification, two procedures are available. In one case, a generic or prescription specification is used. In this type of specification, brand names or specific products are not listed. Rather, material quality is established by statements concerning performance, physical characteristics, test values, and reference standards. For basic materials such as brick, concrete, steel, wood, and gypsum products, comprehensive reference standards have been developed that define the key properties.

The second procedure to establish a desired level of material quality is the use of a proprietary specification. That is, the quality required is established by listing a specific brand-identified product or products or by describing the characteristics of that specific product or products. When a proprietary product is listed, it is converted into an open specification by suitable wording in the project specifications.

One way to achieve an open specification is to follow a brand name or names by the term "or approved equal." The purpose of an or-equal provision is to enable material or product alternates to be used without adversely affecting quality. This, then, allows an owner to realize the full economic benefit of competitive bidding and contractor expertise. The primary problem with the use of the or-equal provision lies in determining just what constitutes an "equal." It is unusual for comparable products of different manufacturers to be identical in every way. Different brand-name items, equally suitable for a particular application, usually differ in certain respects. The essential concern in determining equality is the end result, not how the end result is achieved.

In the context of an or-equal provision, a substitute need not be identical in every respect to the product specified as a standard of quality. A substitute may be equal but be different in appearance, size, configuration, or design. The equality of an alternative product is established on the basis of the quality and performance of the substitute as compared with the brand-name product specified. The equality of substitutes proposed by the general contractor or a subcontractor is decided by the architect-engineer or owner. For each proposed substitution, the contractor must submit samples, descriptive and technical data, test reports, and other information as a means of demonstrating equality. Even though the or-equal clause is useful in allowing substitutions when cost savings are possible or when the availability of materials specified is uncertain, this provision has been, and continues to be, troublesome for both the architect-engineer and the contractor.

Because it is often not possible to obtain approval for a material substitution before bids are submitted on a competitively bid job, contractors are frequently faced with difficult decisions as to what material prices to use in their proposals. It is not uncommon for the lowest price received by the contractor for a certain item to apply to a brand not listed in the specifications. The contractor is faced with the dilemma of whether to use this low price and gamble that the architect-engineer will subsequently approve the substitution. Alternatively, if the bidding firm states in its proposal that the bid amount is based on stipulated substitutions, its proposal may be rejected by the bidding authority. Although the contractor has the obligation of providing proof that a substitution meets the standards originally specified, the architect-engineer must bear the responsibility for its decisions on material equality. Architect-engineers can be held liable for the inadequacy of approved substitutions and must make such decisions with care. In spite of the troubles encountered, however, the or-equal approach to specification writing is commonly used.

With regard to or-equal clauses, there are several cases on record in which the courts have ruled that when the specifications provide for or-equal substitutions, the contractor is entitled to make a substitution that is equivalent or equal to the product specified. If the owner rejects the substitution of a product less expensive but equal and directs the contractor to install the more expensive product specified, this action has been ruled to constitute a constructive change to the contract. This entitles the contractor to recover the cost difference between the less expensive substitute and the product it was directed to install, plus a reasonable profit.

Another common way to convert a proprietary specification to an open specification is to include a paragraph defining the usage of trade names and describing the applicable substitutions policy. On public projects, there is often a "standard of quality" clause that permits substitution of equal products. In such cases the term "or approved equal" is not used.

4.24 OTHER MATERIAL SPECIFICATION TYPES

As discussed earlier, a closed specification has the advantage of ensuring the desired quality of material but the disadvantage of eliminating competition among suppliers. An open specification can provide competition but, by permitting substitutions, introduces the possibility of materials being used that are inferior to those desired. There have been several specification schemes devised to obtain the advantages and minimize the disadvantages of the open and closed concepts. Unfortunately, a completely satisfactory procedure has yet to be devised.

One of these combination-type specifications is called a base-bid material specification or substitute-bid specification. When this procedure is followed, materials are identified in the technical specifications by a manufacturer's name, model, catalog number, or other specific data. The words "or equal" do not appear. The intent is that only those items listed are to be used in preparing the base bid. However, the bidding contractors may offer alternate items, either on the proposal form or as an attachment to the proposal. These alternate proposals are accompanied by full descriptions and technical data, together with a statement of the cost additional to, or deductive from, the base bid if the substitution is approved. The architect-engineer or owner does not approve or disapprove such alternates before the bid opening, and the submission of material or equipment alternates is voluntary with the contractor. The low bidder is determined on the basis of the base bid. After the bids have been submitted, the owner makes decisions about whether to accept or reject any or all of the alternate proposals suggested by the bidding contractors. If decisions concerning approved substitutions are made prior to execution of the contract, they can be suitably incorporated into the contract documents. Otherwise, approved substitutions can subsequently be incorporated into the contract by change order.

Another scheme sometimes used is for a closed specification to be written, but with a provision that proposals for substitutions can be submitted by a bidder up to a stipulated number of days prior to bid opening. Notice of any approved product substitution is circulated to all bidding contractors by addendum (see Section 5.42). After the deadline for substitute requests has passed, no further substitutions are permitted.

4.25 STANDARD SPECIFICATIONS

Standardized specifications, including both the technical and nontechnical provisions, are used by some segments of the construction industry; they have found considerable application in highway, bridge, and utility construction. These preprinted standard specifications are issued by the contracting agency and may be obtained by any interested party. Although they may not form a physical part of the specification booklet for a specific project, these specifications are made a part thereof by reference. The issued specifications for the individual projects consist merely of the proposal, bond, and agreement forms, together with any necessary modifications and special provisions to the standard specifications. This practice can save considerable time and effort in the preparation of project specifications and is conducive to bidding and construction uniformity.

Similar standard specifications in other fields of construction have been prepared, covering the work of the various trades. Many federal, state, and city agencies utilize these standard trade specifications, augmenting them with modifications in order to make them conform to the peculiarities of the project under consideration. However, to date, the use of standard specifications by the construction industry has met with only limited acceptance, with the notable exceptions of the highway, bridge, and utility categories. Building construction in particular continues to favor the customized specification approach.

QUESTIONS

1. Why do architect and engineering societies encourage owners not to use competitive bidding as a method of selecting professional services?

2. What has been the affect of the Brooks Act on the selection of design professionals on federal government work?

3. Describe the basic services provided to an owner by an architect-engineer firm.

4. When an architect-engineer firm provides field construction services, what are some of the duties?

5. What is the difference between an express warranty and an implied warranty regarding architect/engineer work?

6. If the contract between the owner and the architect-engineer gives ownership of the design to the architect-engineer, why is it necessary to copyright the documents as well?

7. With regard to constructions specifications, why do the technical sections tend to be standardized with CSI MASTERFORMAT and the general conditions standardized with AIA A201?

8. Describe the difference between performance specifications and design specifications. What are the advantages of each?

9. Explain why can "or equal" product specifications be troublesome.

Chapter 5

Cost Estimating and Bidding

5.1 COST ESTIMATING

Construction estimating is the compilation and analysis of the many items that influence and contribute to the cost of a project. Estimating, which is done before the physical performance of the work, requires a detailed study of the bidding documents and the site conditions. It also involves a careful analysis of the results of the study in order to arrive at the most accurate possible assessment of the probable cost, consistent with the bidding time available and the accuracy and completeness of the information submitted.

Construction costs are estimated to serve a variety of purposes, and much of the credit for the success or failure of a contracting enterprise can be ascribed to the skill and astuteness, or lack thereof, of its estimating staff. If the contracting firm obtains its work by competitive bidding, it must be the low bidder on a sufficient number of the projects it bids if it wishes to stay in business. However, the jobs it obtains must not be priced so low that it is impossible to realize a reasonable profit from them. In an atmosphere of intense competition, the preparation of realistic and balanced bids requires the utmost in good judgment and estimating skill.

Although negotiated contracts frequently lack a competitive element, the accurate estimating of construction costs nonetheless constitutes an important aspect of such contracts. The contractor is expected to provide the owner with reliable advance cost information, and its ability to do so determines in large measure its continuing ability to attract owner-clients. In design-construct and construction management contracts, the contractor and the construction manager are called upon to provide expert assistance and advice in regard to costs as the design develops. The advance estimation of costs is a necessary part of any construction operation and is a key element in the conduct of a successful construction contracting business.

It must be understood that construction estimating bears little resemblance to the compilation of industrial "standard costs." By virtue of standardized conditions and close plant control, a manufacturing enterprise can predetermine, almost exactly, the total cost of a unit of production. Construction estimating, in comparison, is a relatively crude process. The absence of any appreciable standardization of conditions from one job to the next, coupled with the inherently complicating factors of weather, materials, labor, transportation, locale, and a myriad of others, makes the advance computation of exact construction costs a matter of accident more than of design. Nevertheless, on the whole, construction estimators do a remarkably good job despite the imponderables involved in any project.

There are probably as many different estimating procedures as there are contractors. In any process involving such a large number of intricate manipulations, innovations and variations naturally result. The form of the worksheets, the order of procedure, the mode of applying costs—all are subject to considerable diversity, with procedures being developed and molded by the individual construction company to suit its own needs. However, in a move designed to eventually introduce a measure of uniformity to construction estimating, two national trade groups representing estimators are working to produce uniform recommended practices and standards for construction estimating—the American Society of Professional Estimators (ASPE), representing estimators who work for contractors, and the Association for the Advancement of Cost Engineering (AACE International), made up of estimators working for large industrial owners and contractors.

In today's market the computer has come to play an indispensable role in the estimation of construction costs. Computers have become a necessity for a competitive contractor, and they are universally used to eliminate errors and produce finished estimates more quickly, while creating a superior product for the contractor's use and application. However, the form and detail of the data produced vary with the software being used, each contractor fashioning its detailed estimating procedures and data format to suit the program being used. Because of such diversity in the computer input and the data generated, this chapter discusses only the basic aspects of construction estimating and the application of the data generated.

5.2 LUMP-SUM ESTIMATES

Two forms of fixed-price estimates are widely used in the construction industry. These are the lump-sum estimate and the unit-price estimate. Cost estimates in the field of building construction are customarily prepared on a lump-sum basis. With this procedure, a fixed price is compiled, for which the contractor agrees to perform a prescribed package of work in full accordance with the drawings and specifications. The contractor agrees to carry out its responsibilities even though the cost may prove to be greater than the stipulated amount. Lump-sum estimates are applicable only when the nature of the work and the quantities involved are well defined by the bidding documents. From the owner's standpoint, such a contract can have many advantages. For example, the contract amount fixes the total project cost, a condition that can be useful when the owner is making the financial arrangements for the project.

Lump-sum estimating requires that a "quantity survey," or "quantity takeoff," be made. This is a complete listing of all the materials and items of work that will be required. Using these work quantities as a basis, the contractor computes the costs of the materials, labor, equipment, subcontracts, overhead, contract bond, and tax. The sum total of these individual items of cost constitutes the anticipated overall cost of the construction. The addition of a markup yields the lump-sum estimate that the contractor submits to the owner as its price for doing the work.

5.3 UNIT-PRICE ESTIMATES

Heavy civil construction projects are generally bid, not on a lump-sum basis, but as a series of unit prices. Unit-price estimates can be compiled when quantities of work items may not be precisely determinable but the nature of the work is well defined. The bidding schedule of Appendix F shows a typical list of bid items. It should be noted that an estimated quantity is shown for each item. These quantity estimates are those of the architect-engineer and are not guaranteed to be accurate. When unit-price proposals are involved, a somewhat different estimating procedure from that described in the previous section must be followed.

A quantity survey is made, much as for a lump-sum estimate, but a separate survey is needed for each bid item. This survey not only serves as a basis for computing costs, but also checks the accuracy of the architect-engineer's estimated quantities. A total project cost, including labor, equipment, materials, subcontracts, overhead, markup, contract bond, and tax, is compiled just as in the case of a lump-sum estimate, but all costs are kept segregated according to the individual bid item to which they apply. More information on this process is presented in Section 5.40.

When computing unit prices, the contractor must keep several important factors in mind. The quantities, as listed in the schedule of bid items, are estimates only. The contractor will be required to complete the work specified in accordance with the contract and at the quoted unit prices, whether quantities greater or smaller than the estimated amounts are involved. This requirement is often modified, however, by contract provisions for the equitable adjustment of a unit price when the actual quantity of a bid item varies more than a stipulated percentage above or below the quantity estimated by the architect-engineer. Values of 15 to 25 percent are often used in this regard. All items of material, labor, supplies, or equipment that are not specifically enumerated for payment as separate items, but which are reasonably required to complete the work as shown on the drawings and as described in the specifications, are considered as incidental to the enumerated bid items. No separate measurement or payment is made for them.

5.4   APPROXIMATE ESTIMATES

The fixed-price costing procedures referred to in Sections 5.2 and 5.3, which are based on a complete quantity survey, furnish the most accurate and reliable estimates possible of what future construction costs will be. The results of these exhaustive and very detailed methods are often referred to as detailed estimates. A detailed estimate of project cost is what the contractor will normally compile for bidding or negotiation purposes when the design documents are finalized and a final working estimate is required.

For a variety of reasons, however, the contractor may wish to determine an approximate or conceptual cost estimate by means of a shortcut method. It may be desirable to make a quick and independent check of a detailed cost estimate. The general contractor may wish to compute an approximate cost of work normally subcontracted, either to serve as a preliminary cost in its bid or to check quotations already received from subcontractors. Negotiated contracts with owners are sometimes consummated while the drawings and specifications are still in a rudimentary stage. In such a case the contractor must compute a target estimate for the owner by some approximating procedure. Approximate estimates are necessarily involved when the contractor is called on to provide order of magnitude, preliminary, and definitive cost estimates as the design develops and progresses.

The making of preliminary estimates is an art quite different from the making of a final, detailed estimate of construction costs. Fundamentally, all approximate price estimates are based on some system of gross unit costs obtained from previous construction work. These unit costs are extrapolated forward in time to reflect current prices, market conditions, and the peculiar character of the job now under consideration. As the design progresses toward finalization, more accurate unit costs can be determined and used. The following methods are commonly used to prepare preliminary estimates:

Cost-per-Function Estimate. An analysis based on the estimated cost per item of use, such as cost per patient, student, seat, car space, or unit of production.

Area Cost Estimate. An approximate cost obtained by using an estimated price for each measurement unit of gross floor area.

Volumetric Cost Estimate. An estimate based on an approximated cost for each measurement unit of the total volume enclosed.

Modular Takeoff Estimate. An analysis based on the estimated cost of a representative module, this cost being extrapolated to the entire structure, plus the estimator's assessment of common central systems.

Partial Takeoff Estimate. An analysis using quantities of composite work items that are priced using estimated unit costs. Preliminary costs of projects can be computed on the basis of estimates of the probable costs of concrete in place, per volumetric unit; structural steel erected, per weight unit; excavation, per bank volumetric unit; hot-mix paving in place, per weight unit; and the like.

Panel Unit Cost Estimate. An analysis based on assumed unit costs per area of floors, perimeter walls, partition walls, ceilings, and roof.

Parameter Cost Estimate. An estimate involving unit costs, called parameter costs, for each of several different building components or systems. The costs of site work, foundations, floors, exterior walls, interior walls, structure, roof, doors, glazed openings, plumbing, heating and ventilating, electrical work, and other items are determined separately by the use of estimated parameter costs. These unit costs can be based on dimensions or quantities of the components themselves or on the common measure of building area.

The unit prices used in conjunction with the foregoing approximate cost methods can be extremely variable, depending on specific contract requirements, geographical location, weather, labor productivity, season, transportation, site conditions, and other factors. There are many sources of such cost information in books, journals, magazines, and the general trade literature. Unit costs are also available commercially from a variety of proprietary sources, as well as derived from the contractor's own past experience. In addition, there are many forms of national price indexes that are useful in updating cost information of past construction projects. When using such costs or cost indexes, care must be taken that the information is adjusted as accurately as possible to conform to local and current project conditions.

5.5 BIDDING PROCEDURES

The procedures, rules, and requirements that pertain to the bidding process followed on public versus private construction projects are quite different. Each is controlled by different aspects of law. It is sometimes difficult to establish whether a given project is public or private. There are many instances when a private owner is involved but the financing is government supported or guaranteed. In such cases, the bidding rules are often mandated by the public agency involved.

Private bidding procedures are normally conducted according to rules and regulations established by the owner, with the advice and assistance of the design professional. As such, the bidding procedures can be modified, altered, or waived at the discretion of the owner.

In contrast, public contracting procedures are prescribed by various procurement statutes. These controlling statutes dictate rules and regulations that must be followed by the public owner. In a general sense, the public authority does not have the right to modify or waive these statutory bidding procedures. Even where a public owner is given discretionary powers, specific procedures must be followed in making any exception to standard directives. Public bidding statutes are designed to protect the public interest, not that of the contractor or the architect-engineer. Their essential purpose is to protect public funds, prevent fraud, collusion, and favoritism, and obtain quality construction at reasonable and fair prices.

5.6 ADVERTISEMENT FOR BIDS (PUBLIC PROJECTS)

In all jurisdictions of the United States, laws regulate and control the awarding of public construction projects. These legal requirements start with the first step in the construction process; that is, notice must be given to interested and qualified members of the construction industry in advance of the bidding on any project financed by public funds. In addition, all bidders must be treated alike and afforded an opportunity to bid under the same terms and conditions.

Public agencies must conform to applicable regulations relating to the dissemination of information pertaining to the bidding of construction projects. The contracting agency may be required to give notice by placing advertisements for bids in newspapers, magazines, trade publications, or other public media. How often and over what periods of time such notices must appear vary from jurisdiction to jurisdiction. Weekly notices for two, three, or four consecutive weeks are common. An advertisement of this type is commonly referred to as a "Notice to Bidders" or an "Invitation to Bid."

The advertisement describes the nature, extent, and location of the work and the authority under which it originates, together with the time, manner, and place in which bids are to be received. It also indicates where bidding documents are available and designates the deposit required, and information is listed concerning the type of contract, bond requirements, dates when the work is to be started and completed, terms of payment, estimate of cost, and the owner's right to reject any or all bids. Statements concerning minimum wage rates, minority participation quotas, and applicable labor statutes are also commonly included. Figure 5.1 is a typical example of such an advertisement.

Rather than advertising as such, agencies of the U.S. government commonly utilize a standard form entitled "Pre-Solicitation Notice (Construction Contract)." This form is posted in public places and is generally distributed to the local construction community as well as to an agency bid list consisting of contractors who have indicated an interest in bidding on work within a given geographical area. It conveys essentially the same information as the advertisement. Figure 5.2 illustrates a typical invitation covering a construction project for the U.S. Army Corps of Engineers. All government Pre-Solicitation

ADVERTISEMENT FOR BIDS

Sealed bids for the construction of a Municipal Airport Terminal Building at Portland, Ohio, will be received by the City of Portland at the City Manager's Office, Portland, Ohio, until 2:30 P.M. (EST), Wednesday, May 19, 20–, and then publicly opened and read aloud. Bids submitted after closing time will be returned unopened. No oral or telephoned proposals or modifications will be considered.

Plans, specifications, and contract documents will be available April 16, 20–, and may be examined without charge in the City Manager's office, in the office of Jones and Smith, Architect-Engineers, 142 Welsh Street, Portland, Ohio, in Plan Services in Cleveland, Akron, Toledo, and Youngstown, Ohio: Pittsburgh, Pennsylvania; Detroit, Michigan; Chicago, Illinois; and Buffalo, New York. General Contractors may procure five sets from the Architect-Engineer upon a deposit of $100.00 per set as a guarantee for the safe return of the plans and specifications within 10 days after receipt of bids. Others my procure sets for the cost of construction.

A cashier's check, certified check, or acceptable bidder's bond payable to the City of Portland in an amount not less than 5% of the largest possible total for the bid submitted including the consideration of additive alternates must accompany each bid as a guarantee that, if awarded the contract, the bidder will promptly enter into a contract and execute such bonds as may be required.

The Architect-Engineer's estimate of cost is $5,600,000.00.

Full compliance with applicable Federal, State and Municipal Wage Laws is required, and not less than the rates of wages legally prescribed or set forth in the Contract, whichever is higher, shall be paid.

Proposals shall be submitted on the forms prescribed, and the Owner reserves the right, as its interest may require, to reject any and all proposals and waive any formalities or technicalities. No bidder may withdraw his proposal after the hour set for the opening thereof, or before award of contract, unless said award is delayed for a period exceeding thirty (30) days.

CITY OF PORTLAND, OHIO
By John Doe, City Manager
April 15, 20–

Figure 5.1   Advertisement for bids.

Notices are posted on the Federal Business Opportunities World Wide Web site managed by the General Services Administration. In October 2001 this site replaced the more conventional *Commerce Business Daily* and now operates as the single point of universal electronic public access on the Internet for government-wide federal procurement opportunities in excess of $25,000. In addition to the invitation for bids, a short synopsis of the proposed project is frequently prepared and sent to trade journals and magazines. Generally, a federal project can be advertised by any means as long as it results in no cost to the government. Where a public agency requires its bidding contractors to be prequalified (see Section 5.11), a bid invitation may be sent only to those contractors that will be permitted to submit proposals.

PRE-SOLICITATION NOTICE *(Construction Contract)*	1. PROJECT NO. ENG 33-9-86	2. DATE OF NOTICE 9 Jul, 20--	3. DATE SOLICITATION DOCUMENTS AVAILABLE *(Approx.)* 10 Jul, 20--	OMB NO.: 9000-0037 Expires: 01/31/93

Public reporting burden for this collection of information is estimated to average 10 minutes per response, including the time for reviewing instructions, searching existing data sources, gathering and maintaining the data needed, and completing and reviewing the collection of information. Send comments regarding this burden estimate or any other aspect of this collection of information, including suggestions for reducing this burden, to the FAR Secretariat (VRS), Office of Federal Acquisition Policy, GSA, Washington, DC 20405; and the Office of Management and Budget, Paperwork Reduction Project (9000-0037), Washington, DC 20503.

NOTE: The project number in Items 1 and 16 may be the same as the Invitation or Proposal Number.

4. OFFERS TO BE RECEIVED BY *(at place specified for receipt of offers)*	A. TIME 2:30 A.M. P.M.	B. DATE *(Month, day, year)* 20 July, 20--	5. TIME FOR COMPLETION *(Calendar days)* 420

6A. ISSUING OFFICE *(Name, address, and ZIP code)* Albuquerque District Corps of Engineers Albuquerque, New Mexico 87101 RETURN NOTICE TO THIS ADDRESS	7. PROJECT TITLE AND LOCATION Taxiways and Aprons Holloman Air Force Base New Mexico

6B. ROOM NUMBER 810	6C. TELEPHONE NO. *(Include area code)* (505) 555-1411	

INSTRUCTIONS: a. Solicitation Documents will be issued upon receipt of your affirmative response to this Pre-Solicitation Notice by the DUE DATE set forth in item 15. b. If a charge is required under Item 8A, your affirmative response must include a check or money order in the applicable amount, made payable to Agency *(shown in Item 9)*. Refund *(when specified in Item 8B)* will be made upon your return of the bid documents in good condition, without marks, notes, or mutilations, within 20 calendar days after the final date for receipt of offers. c. The Issuing Office, at its discretion, may make bid documents available to plan rooms of the Associated General Contractors, Chambers of Commerce, Dodge Reports, and other similar contractors' commercial service facilities. d. Bid guarantee is required with any bid in excess of $100,000. Bid guarantee shall be in the amount of 20 percent of the amount of the bid, or $3,000,000, whichever is less. For bid guarantee purposes, the amount of the bid is the aggregate of the Lump Sum Base Bid, all Alternates *(if any)*, and the product(s) of each unit price *(if any)* multiplied by the applicable number of units shown on the Bid Form. e. NOTICE TO SMALL BUSINESS FIRMS: A program for the purpose of assisting qualified small business concerns in obtaining certain bid, payment, or performance bonds that are otherwise not obtainable is available through the Small Business Administration (SBA). For information concerning SBA's surety bond guarantee assistance, contact your SBA District Office.

8A. CHARGE FOR SOLICITATION DOCUMENTS $ 30.00	8B. IS THIS CHARGE REFUNDABLE? ☐ YES ☒ NO	9. MAKE CHECK PAYABLE TO: Albuquerque District, Corps of Engineers

10. ESTIMATED COST RANGE OF PROJECT		11. OFFERS COVERING THE PROJECT RESTRICTED TO SMALL BUSINESS?	12. SUBCONTRACTING PLAN REQUIRED?
A. FROM $ 7,000,000.00	B. TO $ 10,000,000.00	☐ YES ☒ NO	☐ YES ☒ NO

13. DESCRIPTION OF WORK *(Physical characteristics)*

Work consists of furnishing all labor, equipment, and materials, except as may otherwise be provided herein, to perform the work described in the specifications, schedules, and drawings entitled:
"Taxiways and Aprons, Holloman AFB, Albuquerque, NM".

(If additional space is needed use reverse)

IMPORTANT: FAILURE TO COMPLETE AND RETURN THIS PART OF THE NOTICE TO THE ISSUING OFFICE, AT THE ADDRESS IN ITEM 6A, ON OR BEFORE THE DUE DATE SHOWN IN ITEM 15, MAY RESULT IN YOUR NAME BEING REMOVED FROM OUR MAILING LIST.

14. ACTION REQUESTED *(Check applicable box)*		15. DUE DATE
A. I AM INTERESTED IN BIDDING ON THIS PROJECT AS A: ☐ PRIME CONTRACTOR ☐ PRINCIPAL SUBCONTRACTOR	B. I AM NOT INTERESTED IN BIDDING ON THIS PROJECT. RETAIN MY NAME ON YOUR MAILING LIST.	
NO. OF SET(S) YOU REQUIRE OF SOLICITATION DOCUMENTS	C. REMOVE MY NAME FROM YOUR MAILING LIST.	16. PROJECT NO.

17. NAME, ADDRESS *(City, State, ZIP Code)* **AND TELEPHONE NUMBER OF FIRM**

18A. NAME OF FIRM'S REPRESENTATIVE	19. SIGNATURE OF REPRESENTATIVE	20. DATE SIGNED

NSN 7540-01-148-3531
Previous edition usable

STANDARD FORM 1417 (REV. 8-90)
Prescribd by GSA - FAR (48 CFR) 53.236-1(a)

Figure 5.2 Pre-solication notice.

82

5.7 NOTICE OF PRIVATE PROJECTS

Private owners may select a contractor in any manner they choose. Public advertising is sometimes used on private projects to obtain the advantages of open and free competition. In general, however, private owners do not commonly use public advertising for bids. Negotiated contracts, as opposed to competitive bidding, are now widely used in the private sector. Where competitive bidding is used, private owners frequently use a procedure known as "invitational bidding." In this scheme the owner selects a few prime contractors, each of which is considered to be reputable and qualified, and invites this limited group to bid on the project. This procedure offers the owner the advantage of competition while restricting the bidders to a preselected number.

When private owners wish to put their projects out for open competitive bidding, they need only advise the local construction reporting services (see Section 5.8) and give them the necessary bidding data. These services will immediately pass this information on to their subscribers through a variety of media.

5.8 REPORTING SERVICES

Another source of bidding information is the plan service centers that have been established in larger cities about the country. These centers publish and distribute, on a regular basis, bulletins that describe all projects, public and private, to be bid in the near future within their localities. In addition, these services keep copies of the bidding documents on file for the use of subscribing general contractors, subcontractors, material dealers, and other interested parties.

The plan service centers provide a valuable service in making bidding documents available to a wide circle of potential bidders. Prime contractors use the services as a central source of bidding information, to make a quick check of the bidding documents to determine whether to bid, and as an indication of which subcontractors and material dealers will be bidding a given project. The use of plan centers for actual estimating purposes is generally limited to those subcontractors and material vendors whose takeoff is not extensive. Otherwise, prime contractors and others normally obtain sets of bidding documents for their own use from a plan center's on-line library or directly from the owner or the architect-engineer.

On the national level, *Dodge Reports,* published by the Dodge Division of McGraw-Hill Information Services Company, provides complete coverage of bidding and construction activity within different specified localities. Subscribers receive daily reports concerning jobs to be bid in a particular area, including all known general contractor bidders. This is valuable in that it informs a general contractor about its competition and tells subcontractors and material vendors who the bidding prime contractors are. After each bidding, the subscribers are informed of the low bidders and are given information concerning the contract award. During construction, reports are issued periodically listing the general contractor and the subcontractors. Other services are also available, and subscriptions can be tailored to suit the needs of the contractor.

In recent years, *Dodge Reports* and other plan centers and construction information services have developed well-organized and powerful Internet portals providing subscribers with current information on bidding projects as well as offering instant electronic access to construction plans and contract documents. This trend is quickly replacing the more conventional physical plan room, with hard copy plans, and offers significantly improved service tailoring, access, and navigability.

5.9   THE DECISION TO BID

After learning that proposals are to be taken on a given construction project, the contractor must ascertain whether it is interested in bidding. In a general sense, this depends on the "bidding climate" prevailing at the time. The decision to bid involves a study of many interrelated factors, such as bonding capacity considerations (see Section 7.13), location of the project, severity of contractual terms pertaining to potential contractor responsibilities and liabilities, the owner and its financial status, the architect-engineer, the nature and size of the project as it relates to company experience and equipment, the amount of work presently on hand, the probable competition, labor conditions and supply, and the completion date. It is certainly to a contractor's advantage to make some investigation of a project before it expends the considerable effort, time, and money required to compile and present a bid.

Because bidding a job involves considerable expense, the contractor should exert every effort to be the successful bidder, once a decision to bid is made. The limited data available indicate that estimating expense is on the order of 0.2 to 0.5 percent of the total bid amount.

5.10   SET-ASIDES

It is possible, of course, that a contractor may not be eligible to bid on certain public projects. This can occur on jobs that have been "set aside" for only those contractors that can satisfy specific criteria. This matter is discussed in Section 1.27. For example, under regulations of the U.S. Small Business Administration (SBA), certain federal construction projects are set aside for small contractors. In such an instance, a contractor must conform to certain measures of annual business volume to be eligible to bid on a project. In addition, under the SBA 8(a) program, certain federal construction contracts are removed entirely from competitive bidding and contracts are negotiated with socially and economically disadvantaged small business firms (SDBs) as defined by the Small Business Administration.

Another example is afforded by public construction work at the federal, state, and local levels that is set aside for disadvantaged business enterprises (DBEs). To be able to bid on such projects, a construction company must be owned and controlled by a minority or a woman and meet prescribed certification criteria. There are also some public construction projects that are set aside and reserved for locally based contractors.

5.11   QUALIFICATION

Bidder qualification is based on experience, competence, and financial criteria and is entirely separate from the set-asides discussed in the previous section. Many states have enacted statutes that require a general contracting firm wishing to bid on public work in those states to be qualified before it can be issued bidding documents or before it can submit a proposal. Other states require that the contractor's qualifications be judged after it has submitted a proposal. The first method is called prequalification, and the second is called postqualification. Many other public bodies, including agencies of the federal government, require some form of qualification for contractors bidding on their construction projects. The relative merits and drawbacks of the process have been debated for many years, but qualification in some form has become almost standard practice in the field of public construction. The obvious purpose served by qualification is to eliminate the incompetent, overextended, underfinanced, and inexperienced contractors from consideration.

Prequalification requirements apply almost universally to highway construction, and in certain jurisdictions all construction projects financed with public money require contractor prequalification. In those states in which contractors' licenses are required, contractors must be licensed before applying for prequalification. To prequalify, they must submit detailed information concerning their equipment, experience, finances, current jobs in progress, references, and personnel. Evaluation of these data results in a determination of whether the contractor will be allowed to submit a proposal. Highway contractors usually submit qualification questionnaires at specified intervals and are rated as to their maximum contract capacity. Their construction activities are reflected in their current ratings, with bidding documents being issued only to those qualified to bid on each project. The prequalification certificate may also limit the contractor to certain types of work such as grading, concrete paving, or bridge construction.

In jurisdictions having postqualification, the contractor is called on to furnish certain information along with its bid. The information is much the same as that required for prequalification, but it serves the qualification purpose only for the particular project being bid.

Although the foregoing description of qualification is based entirely on publicly financed projects, the same procedures can also be utilized on private jobs.

A somewhat different procedure, often used in closed bidding, is for the contractor to submit an individualized qualification summary with its bid. This is essentially a sales document and contains information designed to enhance the contractor in the eyes of the owner.

5.12 DEPOSITS

After a general contracting firm has decided it wishes to bid on a given project, it normally obtains the necessary sets of bidding documents from the architect-engineer or the owner. The number of sets needed depends on the size and complexity of the project, the time available for the preparation of the bid, and the distribution the architect-engineer has made to subcontractors, material dealers, and plan services. In general, more sets are required for shorter times of bidding and for more complex projects. It is customary for architect-engineers to require a deposit for each set of documents as a guarantee for its safe return. This deposit, which may range from fifty to several hundred dollars per set, is usually refundable. There are instances, however, when contractors must pay a fee or nonrefundable "deposit" for each set of bidding documents obtained. This procedure is sometimes used on public projects. The objective of what amounts to contractor purchase of the documents is to allow the owner or architect-engineer to recoup some of the printing costs involved.

5.13 AVAILABILITY OF PLANS AND SPECIFICATIONS

Undue limitation by the architect-engineer or owner on the number of sets of plans and specifications made available to the various bidders and plan services can be shortsighted. Any aspect of the bidding process that inhibits competition can only result in a higher cost for the owner. Admittedly, the reproduction of these documents is expensive, and certainly no more sets should be provided than are really necessary. However, there can be no argument that sufficient numbers of bidding documents to achieve a reasonably thorough coverage of the total bidding community produce the best competitive prices, particularly for large and complex projects. Plan services perform a valuable function in this regard by making drawings and specifications readily available to interested bidders.

For takeoff and pricing purposes, subcontractors and material dealers can obtain their own bidding documents from the architect-engineer or owner, or they can use the facilities of local plan services. Yet the prime contractor, in order to obtain certain prices, often finds it necessary to make its own copies of the drawings and specifications available to parties wishing to bid on various aspects of the project. A word of caution to contractors regarding the lending of their drawings and specifications: Care must be taken that all bidding documents in their entirety, including addenda, are made available. This foresight will eliminate possible later difficulties with subcontractors or material suppliers claiming that they bid on the basis of incomplete information and hence did not tender a complete proposal. The general contractor, if possible, should set aside a well-lighted plan room on its own premises for the use of the subcontractors' estimators and material dealers. When such facilities are available, the drawings and specifications need never leave the office and can be made to serve a wider range of bidders in a considerably more efficient manner.

5.14   THE BIDDING PERIOD

A sufficient allocation of time for the bidding process can be an important consideration for the owner. The estimating process takes considerable time and must be dovetailed into the contractors' current operations. Careful study and analysis of the plans and specifications usually result in lower bid prices and substantial savings for the owner. The more complicated and extensive the project, the bigger the dividends a reasonable bidding period will pay to the owner. Unless the work truly entails rapid completion or is an emergency, it is sound economics for the owner to allow sufficient time for the bidding contractors to make a thorough study and examination of the proposed work.

When an insufficient bidding period is allotted, many of the best prices will not have been received by the contractor in time to pass the savings on to the owner. In addition, when an unreasonable bidding deadline is imposed, a certain amount of guesswork is inevitably injected into the estimating process. Contractors understandably compensate for uncertainty by bidding on the high side, which can result in inflated proposals.

From the contractor's point of view, the owner should avoid setting any day following or preceding a weekend or holiday as the date for receiving bids. Morning hours are also undesirable. Such times impose undue hardships on the contractors and other bidders and greatly restrict the availability of prices and other information. If practical, a bidding date should not conflict with other construction bid openings.

During the prescribed bidding period, the various documents are subjected to careful examination. The results of this detailed study give the contractor an overall concept of the total package of work and provide the basis for making the most realistic appraisal of construction costs possible. During the bidding process, it is not uncommon for a contractor to find errors, inconsistencies, and missing information in the bidding documents. It is the responsibility of the contractor to bring such imperfections to the attention of the architect-engineer or the owner so that suitable modifications can be made by the responsible party. Such action by the contractor must be made during the bidding period and before submission of the final proposal.

5.15 INSTRUCTIONS TO BIDDERS

For competitive bidding to be a valid procedure, all competitors must bid under exactly the same conditions for an identical package of work. This requires that a set bidding procedure be established and all bidders be required to conform.

Consequently, when competitive proposals are requested, a considerable amount of information concerning the technicalities of the bidding process must be communicated to the contractors involved. This information usually constitutes one of the first divisions of the specifications and is designated as "Instructions to Bidders." These instructions review the requirements that the owner or contracting authority has set for the form and content of the bids, and prescribe certain procedures with which the bidding contractors are required to conform. This document, for example, states conditions pertaining to the form of the bid, where and when it must be delivered, proposal security required (see Section 5.51), and information concerning late bids and bids submitted by mail or other means. Failure to comply with such instructions can result in a contractor's bid not being accepted.

The instructions commonly include clauses that give the owner the right to reject any or all bids, to postpone the date of bid opening, and to exercise many prerogatives in the selection of the successful bidder. A typical list of instructions to bidders is contained in Appendix G. Standard forms of instructions to bidders are used by many public agencies. *Document A701,* "Instructions to Bidders," of the American Institute of Architects is used for the bidding of private building projects. Standard sets of instructions are designed for general use and do not normally provide all of the information peculiar to a specific project that would be required by prospective bidders. Consequently, supplementary instructions or modifications are customary, or reference is made to certain specific information of interest to bidders already contained elsewhere, such as in the advertisement or invitation to bid.

5.16 PRELIMINARY CONSIDERATIONS

Before starting the detailed estimating procedure, the estimator must become familiar with the instructions to bidders, proposal form, alternates (see Sections 5.49 and 5.50), general and supplementary conditions, drawings, specifications, addenda, and form of contract. Such familiarity is equally important for the quantity takeoff people as it is for the cost estimator. The form and arrangement of the quantity surveys can be dictated by the requirements of the foregoing documents. For example, alternate proposals often require that the quantity surveys of certain portions of the project be compiled separately. When bidding large projects, it is desirable to hold a prebid meeting between the estimators and the persons who will occupy the principal supervisory positions on the project if the bid is successful. The purpose of this meeting is to explore the alternative construction procedures that might be followed and to make tentative decisions regarding methods, equipment, personnel, and time schedules.

Another type of prebid meeting is often held; this one is initiated and conducted by the owner. In this case the owner, early in the bidding period, meets with all the prime contractors and perhaps the major subcontractors who will be submitting proposals. The purpose of this meeting is to give the owner an opportunity to review the project requirements, emphasize certain aspects of the proposed work, clarify or explain difficult points, and answer questions. This kind of meeting can be very useful when the work will be complex or involve unusual procedures.

Attendance at these meetings is critical, as clarifications made by the owner during the meeting are considered an important aspect of the bid documents. Failure to consider these clarifications in the bid may result in under- or overpricing the job or rejection of the bid by the owner.

5.17 JOB SITE VISIT

After preliminary examination of the drawings and specifications, the construction site must be visited. Information is needed concerning a wide variety of site and local conditions, such as the following:

1. Project location
2. Probable weather conditions
3. Availability of electricity, water, telephone, and other services
4. Access to the site
5. Local ordinances and regulations
6. Conditions pertaining to the protection or underpinning of adjacent property
7. Storage and construction operation facilities
8. Surface topography and drainage
9. Subsurface soil, rock, and water conditions
10. Underground obstructions and services
11. Transportation and freight facilities
12. Conditions affecting the hiring, housing, and feeding of workers
13. Material prices and delivery information from local material dealers
14. Rental of construction equipment
15. Local subcontractors
16. Wrecking and site clearing
17. Available borrow or dump sites near the project to accommodate grading operations
18. Capacity of roads and bridges to allow moving oversized and heavy equipment or materials to the site

It is important that an experienced person perform this site inspection, particularly in locations that are relatively new or strange to the contractor's organization. The visitor should become familiar with the project requirements in advance of the visit and take a set of drawings and specifications along for reference. It is desirable that the information developed during the site visit be recorded in a signed report that is made a part of the project estimate. It is usually helpful to have available a camera, tape recorder, measuring tape, and other aids appropriate to the occasion such as an earth auger and hand level. It is not uncommon that a follow-up visit must be made on important projects. When access to the site is restricted for some reason, the site visit may have to be coordinated with other bidding contractors, a representative of the owner and/or the architect-engineer, and the primary subbidders. Site conditions where remodeling or renovation work is at issue can be especially demanding, involving such consideratons as the use of existing elevators, control of noise and dust, matching existing materials and finishes, and maintenance of existing services and operations. Experience shows that the advance preparation of a checklist of items to be investigated during the site visit can be very valuable.

5.18   BID INVITATIONS

Early in the bidding period the contractor usually mails bid invitations to all material dealers and subcontractors who are believed to be interested and whose bids would be desirable. This mailing advises of the project under consideration, the item for which a quotation is requested, the deadline for receipt of proposals by the prime contractor, the place where bids will be accepted, the name of the person to whom a proposal should be directed, the place where bidding documents are available, and any special instructions that may be necessary. Specific reminders concerning the status of addenda, alternates, taxes, and bond are desirable. Contractors customarily maintain a card or computerized file that lists addresses and other pertinent information about material suppliers and subcontractors. These files are maintained both by geographical area and by specialty.

5.19   QUANTITY SURVEYS

A quantity survey, or takeoff, is the detailed compilation of the quantity of each elementary work item that is called for on the project. The quantity survey is the only accurate and dependable procedure for compiling a detailed estimate. Although the bid documents for a unit-price project customarily provide contractors with estimated quantities of each bid item, these are approximate only and the architect-engineer assumes no responsibility for their accuracy or completeness. Architect-engineers' quantities are sometimes computed only within so-called pay lines and may not be representative of work quantities that must actually be done to accomplish the bid item (see Figure 5.3). In addition, more detail is usually required for accurate pricing of the work. Consequently, contractors customarily make their own quantity takeoffs, even on projects for which estimated quantities are given. In making quantity surveys, the estimator must be able to visualize how the work will actually be accomplished in the field. Correspondingly, some appropriate on-the-job experience is normally a necessary qualification for construction estimators.

The quantity survey provides the contractor with valuable information during the construction period as well as for the initial estimation of the project cost. For example, it can provide the contractor with data concerning times to be required in the field for work accomplishment, crew sizes, and equipment needs. The quantity survey provides purchasing information and serves as a reference for various aspects of the field work as it progresses. It is useful in providing the owner with a variety of costs pertaining to the work.

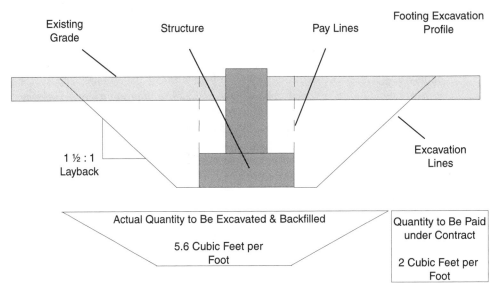

Figure 5.3 Excavation pay lines.

A manual takeoff can be made using pads of standard forms that provide suitable spaces for the entry of dimensions, numbers of units, and extensions, or the process can be computerized. Allowances for the waste of materials such as lumber, concrete, brick, and mortar are made as the takeoff proceeds. In large construction organizations in which there are several takeoff people, it is usual practice for each person to specialize, more or less, along certain lines. For instance, one person may take off only electrical work, whereas another may concentrate on excavation, concrete, and forms. Although this system has many obvious advantages, there is always the possibility of overlooking items or duplicating them. This possibility can be minimized, however, by careful coordination and checking.

Prime contractors do not ordinarily concern themselves with making quantity surveys for all work categories, but limit their takeoff to items that they may reasonably expect to carry out with their own forces. The specialized categories that are done by subcontractors are not usually investigated in detail by prime contractors. However, at times prime contractors may analyze any or all of these categories as a check against the possibility that they may be able to do the work for less money than the best available subcontract bid. In addition, they may have to price a given trade specialty either to check subbids received for reasonableness or when no subcontractor quotations have been submitted. In addition, contractors may not compile the quantities of certain materials, relying on the vendors to do this. Reinforcing steel and structural steel are common examples of materials for which this practice is used.

5.20 PROFESSIONAL QUANTITY SURVEYORS

There are a number of proprietary firms whose principal business is making quantity surveys and cost estimates of selected construction projects. Such a firm may undertake this survey after the design of the project has been completed, or it may act as a cost adviser during the design stage. Owners and architect-engineers usually engage the services of such companies as a means of keeping project costs within budgets and to get the most for the construction dollar. Companies engaged in this form of business are commonly called quantity surveyors.

It is standard practice in the United States for bidding contractors to make their own quantity takeoffs. Thus, there is considerable duplication of effort. Nevertheless, contractors seldom utilize the services of professional quantity surveyors for purposes of preparing proposals. Contractors prefer to prepare their own quantity surveys, in which they have confidence. They maintain a staff of experienced estimators whose reliability and accuracy have been well established. During the takeoff process, the estimator must make many decisions concerning procedure, equipment, sequence of operations, and other such matters. All these decisions must be considered when costs are being applied to the quantity surveys. In addition, work classifications used in quantity surveys must conform to the contractor's established cost accounting system.

It is interesting to note that in Great Britain the quantity surveyor is an established part of the entire construction process. This party, hired by the owner at the inception of the project, makes cost feasibility studies, establishes a construction budget, makes cost checks at all stages of the design process, and prepares a final quantity takeoff. This takeoff is submitted to the contractors bidding the project, who use the information as the basis for their pricing of the work. The quantity surveyor advises the owner on contractual arrangements and compiles certificates of interim and final payment to the contractor doing the work. In some ways, the duties of the British quantity surveyor and the American construction manager are quite similar.

5.21 COST SUMMARIES

On completion of the quantity survey, the total amount of each work classification is obtained and listed. Figure 5.4 presents a form of summary commonly used for a lump-sum project. For this type of contract, a similar summary is prepared for each major work classification. Figure 5.4 contains final quantities of concrete and formwork for the Municipal Airport Terminal Building, a hypothetical project that is used in this book to illustrate selected procedures. Similar summaries are prepared for all of the other principal work types. When the total quantities of work for each classification have been entered, the summary is then used for the pricing of materials and labor. Figure 5.4 does not contain an equipment cost column, although it could. It is common practice on projects that are bid lump-sum, such as buildings, to compute equipment costs for the entire project on a separate sheet rather than on the individual summary sheets.

Figure 5.5 shows a summary for the unit-price project referred to in Appendix F. For this type of bidding, a summary is prepared for each bid item. As illustrated in Figure 5.5, all work types associated with the accomplishment of bid item number 5—Concrete Pavement, 9 inch—are listed on the summary, which often includes several different work categories. With the work quantities obtained from the takeoff, the contractor uses the cost summary to obtain the total direct costs of the bid item. The summary sheets for unit-price projects customarily serve for the computation and recording of material, labor, equipment, and subcontract costs. Figure 5.5 illustrates this point.

SUMMARY SHEET

Project <u>Municipal Airport Terminal Building</u>   Work Items <u>Concrete and Forms</u>

Cost Account Number	Work Item	Unit	Total Quantity	Material Cost Each	Material Cost Total	Labor Cost Each	Labor Cost Total	Total Cost
240	CONCRETE							
.01	Footings	c.y.	1,040	$59.25	$61,620	$6.75	$7,020	$68,640
.05	Grade beams	c.y.	920	$63.75	$58,650	$9.45	$8,694	$67,344
.07	Slab	c.y.	2,772	$59.25	$164,241	$9.00	$24,948	$189,189
.08	Beams	c.y.	508	$63.75	$32,385	$9.45	$4,801	$37,186
.09	Bond beams	c.y.	62	$63.75	$3,953	$13.50	$837	$4,790
.11	Columns	c.y.	102	$63.75	$6,503	$13.05	$1,331	$7,834
.12	Walls	c.y.	466	$63.75	$29,708	$10.13	$4,718	$34,426
.16	Stairs	c.y.	317	$63.75	$20,209	$11.70	$3,709	$23,918
.19	Sidewalks	c.y.	320	$59.25	$18,960	$9.00	$2,880	$21,840
.20	Expansion joint	l.f.	41,060	$0.38	$15,398	$0.23	$9,239	$24,636
.40	Screeds	s.f.	148,000	-	-	$0.18	$26,640	$26,640
.50	Float finish	s.f.	22,000	-	-	$0.23	$4,950	$4,950
.51	Trowel finsih	s.f.	128,000	-	-	$0.38	$48,000	$48,000
.52	Stair finsih	l.f.	2,060	-	-	$1.58	$3,245	$3,245
.60	Rubbing	s.f.	2,530	$0.09	$228	$0.83	$2,087	$2,315
.91	Curing	s.f.	180,000	$0.06	$10,800	$0.05	$8,100	$18,900
					$422,652		$161,198	$583,851
	Indirect Labor Cost					39%	$62,867	$62,867
	Total Concrete				$422,652		$224,065	$646,718
260	FORMS							
.01	Footings	s.f.	1,220	$1.05	$1,281	$1.50	$1,830	$3,111
.05	Grade beams	s.f.	4,840	$1.20	$5,808	$1.50	$7,260	$13,068
.08	Beams	s.f.	1,965	$1.32	$2,594	$1.62	$3,183	$5,777
.11	Columns	s.f.	1,642	$1.62	$1,786	$2.99	$3,267	$5,053
.12	Walls	s.f.	3,240	$1.32	$2,852	$1.50	$4,860	$7,712
.17	Risers	l.f	910	$1.02	$928	$1.92	$1,747	$2,675
.45	Chamfer	l.f.	510	$0.12	$61	$0.12	$61	$122
.60	Oil, ties, nails	s.f.	13,443	$0.09	$1,210	-	-	$1,210
.63	Anchor slot	l.f.	2,800	$0.45	$1,260	$0.09	$252	$1,512
					$17,780		$22,461	$40,241
	Indirect Labor Cost					43%	$6,908	$6,908
	Total Forms				$17,780		$29,369	$47,149

Figure 5.4 Summary sheet, lump-sum bid.

SUMMARY SHEET

Project   Holloman Taxiways and Aprons

Bid Item   No. 5 Concrete Pavement, 9 inch

Cost Account Number	Work Item	Quantity	Unit	Material			Labor			Equipment			Subcontract
				Unit Cost	Total		Unit Cost	Total		Unit Cost	Total		
250.03	Concrete, production	23,625	c.y.	$30.90	$730,013		$22.91	$541,131		$5.16	$121,905		-
254.01	Concrete, hauling	23,625	c.y.	-	-		$0.45	$10,631		$0.53	$12,403		-
258.02	Concrete, lay-down	90,000	s.y.	-	-		$1.59	$143,100		$0.74	$66,150		-
270.08	Reinforcing	50	ton	$825.00	$41,250		$525.00	$26,250		-	-		-
248.20	Grooves	105,000	l.f.	$0.06	$6,300		$0.05	$4,725		$0.02	$1,575		-
240.95	Curing	90,000	s.y.	$0.15	$13,500		$0.06	$5,400		$0.008	$675		-
			Totals		$791,063			$731,237			$202,708		-
						Labor Indirect Cost, 37%		$270,558					
						Labor Cost		$1,001,795					

Figure 5.5   Summary sheet, unit-price bid.

It should be noted that a cost account number identifies each work classification entered on the summary. These are the company's standard cost account numbers that are basic to its cost management system. Each estimate is broken down in accordance with the established cost accounts. Chapter 12 discusses the essentials of a contractor's project cost management system.

The next several sections describe how the many different categories of project cost are evaluated and priced.

5.22   MATERIAL COSTS

For purposes of discussion in this book, the term *materials* is construed to include everything that becomes a part of the finished structure. This includes electrical and mechanical items such as elevators, boilers, escalators, and transformers, as well as the more obvious and traditional items such as lumber, structural steel, concrete, and paint.

Materials required for the accomplishment of a construction project can be and are obtained from a variety of sources. There are times when the contractor produces its own materials. An example is the aggregate that a highway contractor requires for a project. In this case, the contractor must determine its own cost of production, per cubic yard or per ton, by evaluating the costs of drilling, shooting, crushing, screening, and stockpiling the required aggregate.

The usual procedure is to purchase the materials from a standard source such as a dealer, distributor, manufacturer, or manufacturer's representative. Although standard price lists are used occasionally, it is customary for the contractor to solicit and receive special quotations from the suppliers for materials required for the job being bid. General exceptions to such solicitations include stock items like plywood, nails, and lumber; a contractor occasionally purchases these materials in large quantities and maintains a running inventory. Written quotations from the various material sources are desirable so that important considerations such as prices, freight charges, taxes, delivery schedules, and guarantees are explicitly covered. Most suppliers tender their quotations on printed forms that include stipulations pertaining to terms of payment and other considerations. Representatives of material suppliers are a familiar sight in contractors' offices. Even though they are salespersons, they provide valuable services and product information, as well as material prices, to their contractor clients.

As discussed previously in Section 4.23, the material specifications on certain projects require that a specifically named product be furnished but give the bidding contractor the right to propose a less expensive substitute as an alternative in its proposal. When the contractor determines the amount of its bid, the cost of the specified product is used. However, the proposal also identifies the substitute product proposed for use and indicates the difference in price to the owner if the alternative product is approved.

It is not unusual for the bidding documents to provide that certain materials will be supplied by the owner to the contractor at no charge. In such cases, no material costs need be obtained by the contractor nor included in the project cost estimate.

However, all other costs associated with such materials, such as handling, storage, and installation expense, must be included.

Material costs as entered on the summary sheets must all be on a common basis: for example, delivered to the job site and without tax. These costs ordinarily include freight, transportation, storage, and inspection. There are many occasions when the bidding documents designate an "allowance" amount to be included in the bid to cover the cost of a material item. Where this is the case, the contractor does not have to obtain a material price. The designated allowance amount is used to compile the proposal. Allowances are discussed further in Section 5.33.

5.23   DIRECT LABOR COSTS

Direct labor costs are determined from basic wage rates—that is, the hourly rates used for payroll purposes—and do not include indirect labor costs, which are discussed in the next section. To determine the direct labor cost of any given work type, the estimator must know the basic wage rates of the labor categories involved and the production rate that applies to that work type. In addition, the total quantity of the work type must be obtained, as derived from the quantity survey. Direct labor costs are difficult to evaluate and cannot be determined with the same degree of precision as most other elements of construction cost. The pricing of labor is one of the two largest areas of uncertainty in the estimating process. (The costs associated with construction equipment is the other, a subject that is discussed later in this chapter.) To do the best job possible of establishing these two costs, the estimator must make an accurate and detailed quantity takeoff, analyze job conditions carefully, obtain advance decisions concerning the construction methods to be used, and maintain a comprehensive library of costs and production rates from past projects.

Contractors differ widely in how they estimate labor costs. Some choose to include all elements of labor expense into a single hourly rate for each craft. Others evaluate direct labor costs separately from indirect labor costs. Some estimators compute regular and overtime labor costs separately, whereas others combine scheduled overtime with straight time in an average hourly rate. Some evaluate labor costs using production rates directly; others use labor unit costs. There are usually good reasons why a given contractor evaluates labor costs the way it does, and there certainly is no single correct method that must be followed. The procedures described in this chapter compute direct labor costs separately from indirect labor costs, a common practice.

A production rate is normally basic to the determination of the labor cost associated with a given work category. Labor production rates can be expressed in a variety of ways. One of these is the man-hours or crew-hours required to accomplish a unit or prescribed number of units of a given work type. For example, six hours of carpenter time and six hours of carpenter-helper time may be required to assemble, erect, plumb, brace, strip, and clean 100 sq ft of rectangular concrete column forms. To obtain the estimated labor cost for a given quantity of such work, the labor hours required per unit of work are multiplied by the appropriate wage scales per hour and again multiplied by the total number of units of work of this category as obtained from the quantity survey. Other ways to express this type of labor production rate are the number of units of work accomplished per working day by a worker or by a crew, or the total man-hours or total crew-hours to complete a given job.

In estimating the costs of construction labor, a cost per unit of work, or a unit price, can be useful and convenient. For example, if carpenters earn $27.33 per hour and a carpenter-tender's wage scale is $22.50 per hour, the 100 sq ft of rectangular concrete column forms cited previously will have a labor unit cost of $2.99 per square foot. These "labor unit costs" are quick and easy to apply, as illustrated in Figure 5.4, but the cost estimator must make sure that the labor unit costs are kept up-to-date and that they reflect the probable levels of work production and the applicable wage rates.

The most reliable labor productivity information is that which the contractor obtains from its own completed projects. Information on labor productivity and costs is also available from a wide variety of commercial sources. However, although information of this type can be very useful at times, it must be emphasized that labor productivity differs from one geographical location to another and is variable with the season and many other factors. Estimators must also be very circumspect when using labor unit costs that they have not developed themselves. For the same work items, different estimators may include different expense items in their labor unit costs. It is never advisable to use a labor unit cost derived from another source without knowing exactly what it does and does not include.

The pricing of labor in a construction cost estimate can be influenced by local-hire rules and bid advantages offered by some local governments on certain of their projects. For example, hiring quotas or goals are established for local residents, minorities, and women on some public jobs. In other cases, the contractor receives a bidding advantage based on the number of local residents, minorities, and women it promises to employ when bidding a given project.

5.24 INDIRECT LABOR COSTS

Indirect labor costs are those expenses that are additional to the basic hourly rates and are paid by the employer. Indirect labor expense involves various forms of payroll taxes, insurance, and employee fringe benefits of a wide variety. The employer's contribution to Social Security, unemployment insurance, workers' compensation insurance, and contractor's public liability and property damage insurance are all based on payroll. Employers in the construction industry typically provide for various kinds of fringe benefits such as pension plans, health and welfare funds, employee insurance, paid vacations, and apprenticeship programs. The costs of these benefits are customarily based on direct payroll costs. Premiums for workers' compensation insurance and most fringe benefits differ considerably with craft and geographical location.

Indirect labor costs are substantial in amount, usually adding 25 to 50 percent to the direct payroll costs. Indirect labor costs can be included in the cost estimate in different ways. For example, labor unit costs or hourly labor rates that include both the direct and indirect costs of labor can be used. However, this procedure does not always interrelate well with a contractor's cost accounting methods. For this reason direct and indirect labor costs are often computed separately when job costs are being estimated. A commonly used approach is to add the proper percentage allowance for indirect costs to the total direct labor cost, either for the entire project or for each major work category. Because of the appreciable variation of indirect costs from one labor classification to another, it may be preferable to compute indirect labor cost at the same time that direct labor expense is obtained on the summary sheets. The 39 and 43 percent additions to the labor column of Figure 5.4 represent the total indirect labor costs associated with the concrete placing and forming, respectively.

5.25 EQUIPMENT COST ESTIMATING

Unfortunately, the term *equipment* does not have a consistent meaning in the construction industry. A common usage refers to motor graders, power shovels, pile drivers, and other such items used by contractors to accomplish the work. However, equipment is also sometimes used with reference to various kinds of electrical and mechanical furnishings that become part of the completed project. As used herein, *equipment* refers only to a contractor's equipage used for the physical accomplishment of the work.

Equipment costs, like those of labor, are difficult to evaluate with any exactness. Equipment accounts for a substantial proportion of the total cost of most engineering projects, but is less significant for buildings. When the nature of the work requires major items of equipment such as earth-moving machines, concrete plants, and draglines, detailed studies of the associated costs must be made. Costs of minor equipment items such as power tools, concrete vibrators, and power buggies are not normally subjected to detailed study. A standard cost allowance for each such item required is included in the estimate, usually based on the total time it will be needed on the job. The cost of hand tools, wheelbarrows, water hoses, extension cords, and the like can be covered by a lump-sum allowance calculated as a small percentage of the total direct labor cost of the project and included on the job overhead sheet (see Section 5.34).

To estimate the expense of major equipment items as realistically as possible, early management decisions must be made concerning the equipment sizes and types to be required and the manner in which the necessary units will be provided to the project. A scheme sometimes used when the duration of the construction period will be about equal to the service life of the equipment is to purchase all new or renovated equipment for the project and sell it at the end of construction activities. The difference between the purchase price and the estimated salvage value is entered into the job estimate as a lump-sum equipment expense.

Equipment is frequently rented or leased. The equipment expense associated with rental or lease costs is figured into the job by applying the lease or rental rates to the time periods that the equipment will be needed on the project.

The most common manner in which equipment is provided to construction projects is for a contractor who owns the equipment to use it on a succession of projects during its economic lifetime. Rather than a lump-sum equipment purchase expense or cost of rental or lease, ownership expense is involved in this instance. Ownership expense is of a fixed nature and includes depreciation, interest on investment or financing cost, taxes, insurance, and storage.

Regardless of the mode of supplying equipment to a project, equipment operating costs must also be computed and included in the project estimate. Operating costs include such expenses as fuel, oil, grease, filters, repairs and parts, tire replacement and repairs, and maintenance labor and supplies. There is some difference of opinion regarding whether the wages of equipment operators should be included in equipment operating cost. Some contractors prefer to regard the labor associated with equipment operation as a labor expense rather than an equipment cost. Others include the labor cost as a part of the equipment operating expense. For purposes of discussion herein, equipment operators' wages are treated as a labor cost and are not included as part of the equipment expense.

5.26   EQUIPMENT EXPENSE

As described in Section 5.23, the direct labor cost of a given work type is computed by combining a labor production rate with the applicable hourly wage scales and total work quantity. Most equipment costs are calculated in much the same fashion, except, of course, that equipment production rates and hourly equipment costs must be used. The hourly wage rates of various labor categories are immediately determinable from applicable labor contracts, prescribed prevailing wage rates, or established area practice. However, this is not true for equipment. Contractors must establish their own hourly equipment costs as well as their equipment production rates. Ownership, lease, or rental expense is combined with operating costs to derive an estimated total cost per operating hour for most items of production equipment. Power shovels, tractor-scrapers, and ditchers are examples of equipment for which the costs are usually expressed in terms of hourly rates. However, the costs of some classes of production equipment are frequently expressed in terms of expense per unit of product produced. Portland cement concrete mixing plants, asphalt paving plants, and aggregate plants are familiar instances of this expense.

There are some equipment types, normally referred to as support equipment, for which it is more appropriate to express costs in terms of time units other than operating hours. This type of equipment is usually distinguished by the fact that there are no production rates as such involved in their use and application. Prefabricated concrete forms, air compressors, welding machines, cranes, and scaffolding are required at the job site on a continuous basis during particular phases of the work, and operating hours have no real significance in such cases. Costs per week or per month are much more appropriate for such equipment.

Move-in, erection, dismantling, and move-out expenses, also called mobilization and demobilization costs, are independent of equipment operating time and production and therefore are not included in equipment hourly rates. These equipment expenses are separately computed for inclusion in the estimate, often as an item of field overhead.

5.27 EQUIPMENT COST RATES

The determination of the expense rates for the equipment items to be used on a project being estimated is an important matter and requires considerable time and effort. It must be remembered that estimators are always working in the future tense and that the equipment rates they use in their estimates are their best approximations of what such expenses will turn out to be when the project is actually being built. Equipment expense rates are approximations at best and must be regarded as such. The following sections discuss the determination of ownership and operating expense of a piece of contractor-owned production equipment in terms of cost per operating hour. If the equipment is rented or leased, these expenses are reduced to a cost per operating hour and this amount is used in lieu of ownership expense. If support equipment is involved, such as an air compressor or light plant, production rates do not apply, and therefore ownership and operating costs are instead expressed on a weekly or monthly basis.

The standard method of computing future equipment expense is through the use of historical equipment cost records as contained in a contractor's equipment accounting system. The source of ownership and operating cost data for a specific piece of equipment is its ledger account. An important part of a contractor's general accounting system is the equipment accounts, where actual ownership and operating expenses are recorded as they are incurred. A separate account is set up for each major piece of equipment. This account serves to maintain a detailed and cumulative record of the use of, and all expenses chargeable to, the particular piece of equipment. All expenses associated with that piece of equipment, regardless of their nature or the project involved, are charged to that account, including depreciation, investment costs, operating expenses, tire replacement, overhauls, and painting. Also maintained is a cumulative record of that equipment item's use in the field. These data constitute a basic resource for reducing ownership and operating expense to a total equipment cost rate. The sum of ownership and operating expense expressed as a cost per unit of time (operating hour, week, month) is often referred to as the "use rate," "budget rate," or "internal rental rate" of an equipment item. This last term refers to the usual construction accounting practice whereby equipment time on the project during construction is charged against the job at that rate.

The preceding discussion assumes that equipment accounting is done on the basis of the individual machines. However, contractors vary somewhat in how they maintain their equipment accounts, some preferring to keep equipment costs by categories of equipment rather than by the individual unit. These firms use a single account for all equipment items of a given size and type and compute an average budget rate based on the composite experience with all of the units included. Thus, the same expense rate is applied for any unit of a given equipment type regardless of differences in age or condition. In estimating equipment costs on a future project, there may be some logic in using an average hourly cost rate for any given equipment type and capacity because it is often not possible to predict exactly which equipment units will be placed on a particular project.

There are many external sources of information concerning ownership and operating costs for a wide range of construction equipment. Manufacturers, equipment dealers, and a large assortment of publications offer such data. It must be realized, however, that these are typical or average figures and that they must be adjusted to reflect contractor experience and methods and to accommodate the specific circumstances of the project being estimated. Climate, altitude, weather, job location and conditions, operator skill, field supervision, and other factors can and do have a profound influence on equipment costs. When a new piece of equipment is procured for which there is no cost history, equipment expense must be estimated using available sources of information considered to be reliable. For estimating equipment costs, two excellent references are the *Green Guide* and the *Rental Rate Blue Book*, published by EquipmentWatch. Another source of ownership and operating costs of construction equipment are those values predetermined and used by the federal government when paying a contractor for its equipment costs incurred on contract modifications. These are applied when the contractor cannot establish its own hourly costs for equipment from actual accounting records.

5.28   OWNERSHIP AND OPERATING COSTS

Hourly expense rates are computed and periodically updated for each major piece of production equipment. Figure 5.6 illustrates a commonly used procedure for this practice, using a bottom-dump hauler as an example. The average annual usage of this hauler is approximately 1,200 hours, and it is assumed that the useful life of the machine will be a total of 12,000 operating hours, or 10 years.

In Figure 5.6 the original price of the hauler, less tires, is depreciated uniformly over 12,000 hours. The cost of the tires is deducted from the original acquisition price because the tires have a shorter service life than the mechanical part of the equipment. Depreciation is equipment expense caused by wear and obsolescence and allows the recovery of the invested capital over the useful life of the equipment. The method by which the depreciation

ESTIMATE
Hourly Ownership and Operationg Cost for
DP-12 Bottom-Dump Hauler

Ownership Cost

1. Depreciation

Purchase price	=	$	105,450
freight	=	$	3,279
Delivered price	=	$	108,729
Less tires	=	$	17,620
Depreciable value	=	$	91,109

$$\text{Hourly depreciation} = \frac{\$\ 91,109}{12,000} = \$\ 7.59$$

2. Interest, taxes, and storage

$$\frac{\$57,997.50(0.16)}{1,200} = \$\ 7.73$$

Total hourly ownership cost = $ 15.33

Operating Costs

3. Fuel, 4 gallons per hour @ $1.62   = $ 6.48
4. Oil, lubricants, filters (1/3 of fuel cost = 1/3($6.48)   = $ 2.16
5. Repairs, parts, and labor 35% of depreciation = 0.35($7.59)   = $ 2.66
6. Tire replacement = $\dfrac{\$17,620}{3,500}$ = $ 5.03
7. Tire repairs (15% of tire replacement cost) = 0.15(5.03)   = $ 0.76

Total hourly operating cost   = $ 17.09

Total estimated ownership and operating cost   = $ 32.42

Figure 5.6   Estimate of equipment ownership and operating cost.

expense is spread uniformly over the total service life is referred to as straight-line depreciation. Regardless of the mode of equipment depreciation a contractor may use for tax and other accounting purposes, the use of straight-line depreciation for estimating and job cost accounting is usual practice. The straight-line procedure is discussed in Section 9.15.

Taxes, insurance, storage, and interest on investment, together referred to as investment cost, are customarily based on the average annual value of the equipment. This example assumes taxes at 2 percent, insurance and storage at 5 percent, and interest at 9 percent, for a total annual cost of 16 percent of the average yearly value. A standard way of computing the average annual value, based on straight-line depreciation, is by the equation

$$A = \frac{C(n+1) + S(n-1)}{2n}$$

where A = average annual value
    C = delivered cost
    n = number of years of useful life
    S = salvage value

In Figure 5.6, with a salvage value of zero,

$$A = \frac{\$105,450(10+1) + 0(10-1)}{2(10)} = \$57,997.50$$

Operating costs are computed on the basis of the contractor's own experience or by using available national averages. Equipment manufacturers and dealers provide typical ownership and operating costs. It must be realized, however, that these are average figures and that they may not reflect the conditions of the project being estimated. Job estimates must be based on individual job conditions, and the contractor's past experience is the best guide in this regard. Nevertheless, there are times when national averages can be very useful. In the computation of tire replacement cost in Figure 5.6, a new set of tires is assumed to have a life of 3,500 operating hours. Costs of major repair and overhaul are not normally regarded as operating expense, but are capitalized and treated as an element of ownership expense. This calculation is not considered in Figure 5.6.

Figure 5.6 discloses that for the bottom-dump hauler, the ownership cost is $15.33 per operating hour and the estimated operating costs are $17.09 per operating hour. This piece of equipment operating in the field will cost the contractor a total of $32.42 per working hour, not including operating labor.

It is important to note that ownership expense goes on whether or not the equipment is working. If the bottom-dump hauler actually works 1,200 hours or more per year, the owner recovers this cost. If the hauler works less, the contractor does not recover all the annual ownership costs and the deficit comes out of general company profits. Operating costs are, of course, incurred only when the equipment is actually being used. When equipment is leased or rented rather than owned, there is no ownership cost involved. Rather, the lease or rental cost, expressed on an operating hour basis, is used instead. Operating costs apply as before, with the sum of the two again giving the rental rate of the item.

The costing of minor equipment items is done in a somewhat similar, although less exact, manner. A common example is the calculation of the expense of a concrete vibrator whose initial cost is $840. Suppose the service life of this vibrator is approximately four years and about $240 per year must be spent on servicing, repairs, and new parts. With an annual investment cost of approximately $67, this translates into a total cost of $2,068 spread over 48 months, or $43 per month. If concrete pouring on a project will extend over four months, the equipment expense to be added into this job for the concrete vibrator will be $172 plus the estimated fuel cost, if applicable.

5.29   EQUIPMENT PRODUCTION RATES

In addition to equipment cost, equipment production rates are also needed for the computation of equipment expense in a construction estimate. Paralleling the case of labor, applying equipment hourly expenses and production rates to total job quantities enables the estimator to compute the equipment expense for the project. Equipment unit costs, which are equipment costs per unit of production, can also be determined. Equipment production rates, like those of labor, are subject to considerable variation and are influenced by a host of job site conditions. In addition, some equipment production rates must be computed using specific job conditions such as haul distances, grades, and rolling resistance. Estimators must consider and evaluate these factors when pricing a new project.

There are several sources of equipment production information. The most reliable are production records from past projects. Advice from the equipment operators themselves can, at times, be very useful. If a new piece of equipment is involved with which there has been no prior experience, production information provided by the equipment manufacturer or dealer can be of assistance. There are many rules of thumb and published sources of information concerning average equipment production rates. Stopwatch "spot checks," made to ascertain the productivity of specific equipment items, can be of value. In this regard, however, it should be noted that production rates of labor and equipment used for estimating purposes should be average figures taken over a period of time. Daily job production tends to be variable, and this is a hazard of using spot checks. Production records from past projects produce good time average values, whereas a spot check may, by chance, catch production at a high or low point.

Production rates may be applicable to a single piece of equipment or to an equipment spread. A spread of equipment is an aggregation of equipment items working together as a group to accomplish a given aspect of a job. Figure 5.7 illustrates how a production rate and an equipment unit cost are obtained for earth hauled by the bottom-dump haulers. As is true for all such production equipment, the manufacturer of the bottom-dump hauler provides the contractor with much valuable information concerning its capacity and performance characteristics. As illustrated by Figure 5.7, the contractor can use this information, together with operating characteristics of the loading power shovel and the length, grade, and condition of the haul road, to determine the hauling production rate of the bottom-dump unit. Applying the cost figures derived in Figure 5.6 to the production rate in Figure 5.7 discloses that the equipment unit hauling cost is $0.40 per bank cubic yard (b.c.y.) for a fleet of four such units. If an equipment unit cost that includes both loading and hauling is desired, the power shovel cost per bank cubic yard must be obtained and added to the hauling cost. Such equipment unit costs are utilized in the same way as labor unit costs. "Bank" measure refers to soil in its original, undisturbed condition, as compared with "loose" measure, which applies to the soil after excavation. Bank soil is generally denser and has less volume than the same soil after it is excavated.

Equipment production rates given by the manufacturer or observed on the job are usually those that occur when the equipment is turning out its maximum possible output. Obviously, such rates are seldom maintained continuously and allowances must be made for time losses and delays. The "50-minute hour" is commonly used with regard to excavating and other forms of production equipment. This term means that, on the average, the equipment is productive a net time of 50 minutes per hour. The remaining time is consumed

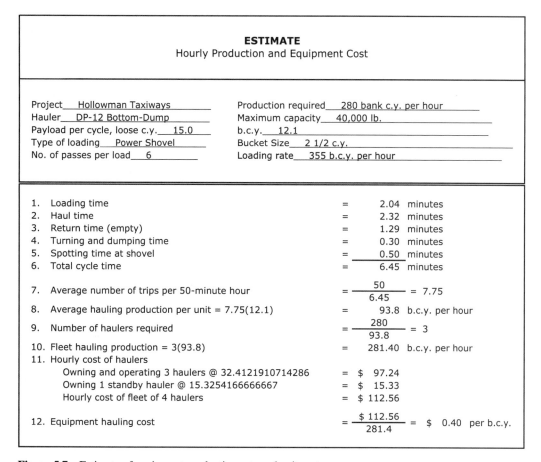

Figure 5.7   Estimate of equipment production rate and unit cost.

by nonproductive actions such as greasing, fueling, minor repairs, maintenance, shifting position, operator delays, and waiting. The application of the 50-minute hour is illustrated in item 7 of Figure 5.7.

5.30   SUBCONTRACTOR BIDS

Quotations from subcontractors can be an important element in the compilation of a project estimate, especially on building construction, where subcontracting is utilized more extensively, on the average, than it is on engineering or industrial construction. Frequently, between 70 and 85 percent of the dollar value of building construction is performed by subcontractors. Unfortunately, many of these bids may not be submitted to the prime contractor until late in the bidding period, a factor that makes it difficult to study and analyze them.

Subcontractors' submission of bids by telephone, together with the practice of making last-minute deductions, can make the final bid compilation difficult during that hectic period immediately before bid opening. The submission of written quotations by subcontractors, with sufficient time to allow the general contractor to analyze the bids, is to be preferred. Stipulations regarding the use of hoisting facilities; the supply of electricity, water, and heat; conditions to be met before starting work; the delivery and storage of materials; and others are commonly attached to and made a part of subcontract proposals. Because subbid stipulations usually require some commitment on the part of the prime contractor, they must receive careful attention during the bidding period so that all costs pertaining to them can be included in the prime contractor's estimate.

Another consideration that makes written subcontractor quotations preferable is the general question of whether oral bids are legally enforceable. The answer depends on the particular facts of the case and the jurisdiction involved. Nevertheless, it must be recognized that there can be serious questions in this regard, and much time and expense can be involved in settling such disputes.

The estimator must be sure as to what each subbid includes and what it does not, a matter that can require considerable checking when many subbids are involved. Frequently, a subcontract quotation provides only for labor and does not include the cost of any materials. The inclusion of taxes and surety bond and the acknowledgment of addenda must either be stipulated by the quotation or determined by the general contractor. On building construction in particular, many items of work must be checked with the bidding subcontractors to be sure of their inclusion. For example, the prime contractor must check with the plumbing and electrical subcontractors to see whether their bids include the cost for connecting temporary water, sanitary, and electrical services. Another consideration is the need to conduct a background check on those subcontractors unknown to the general contractor.

If a subcontract bid is submitted that is substantially lower than any other figure received for the same work, the prime contractor should ask the subcontractor involved to check its proposal before the bid is used to make up the prime contractor's final bid. The reason for this is that if the subcontractor made a mathematical, typographic, or clerical error in compiling the final figure, the courts are apt to permit the subcontractor to withdraw its bid without penalty. In addition, it should be noted that a subcontractor's bid is not binding on the subcontractor if the general contractor has specified a bidding condition through which no contractual obligation is created between the two parties until a formal subcontract is executed.

To ensure time for the evaluation and study of subbids, it may be necessary for the general contractor to establish a deadline beyond which subbids or revisions thereof will no longer be accepted. Two hours before bid time is used in this regard by many contractors. Such an arrangement works best when all of the general contractors in a given area agree to abide by such a deadline. Yet if a contractor establishes this practice unilaterally, it may well miss out on some favorable last-minute figures.

In general, the prime contractor has complete freedom in selecting its subcontractors. On some public projects, however, the prime contractor must award some stipulated percentage of its total contract to minority business enterprises, women-owned businesses, or local firms. In these cases, the contractor must seek out and use the bids of such companies. The contractor's purchase of business services such as contract bonds and insurance from minority-owned businesses can sometimes be counted in meeting subcontracting goals. On private projects, there may occasionally be listed in the instructions to bidders the names of subcontractors acceptable to the owner for major divisions of the work, such as electrical, plumbing, heating and air-conditioning, and elevators. In such cases, the general contractor can subcontract these items of work only to one of those subcontractors listed. At this point, it suffices to say that the low subbid is not necessarily the *best* subbid. The general contractor must consider, in addition to immediate cost, the subcontractor's integrity, organization, financial strength, equipment, and reputation for getting the job done.

Many general contractors have a tendency to award a majority of their subcontracts to the same preferred companies. However, this can be disadvantageous to these general contractors, in that competing subcontractors may not continue to submit their most competitive prices to such firms. It is to the advantage of a prime contractor to establish a good working relationship with several subcontractors associated with each work specialty. When awarding subcontract work, the general contractor must never lose sight of the fact that a subcontractor is reliable only to the extent of its financial capability. The financial soundness of a subcontractor must always be a top criterion in awarding subcontracting work.

5.31 ASSIGNED SUBCONTRACTS

The usual procedure, previously described, of subcontract bids being submitted to the bidding prime contractors for inclusion in their bids, is not always followed in certain states in the bidding of some state-financed projects. In these states, separate bids are requested by the state for general construction and for designated specialty areas such as electrical; plumbing; heating, ventilating, and air-conditioning; and others. The general construction bids cover all work on the project not covered by the designated specialty areas. Bids for work in all of the specified areas are submitted directly to the state agency involved.

After the bidding deadline has passed, the agency identifies the low general construction bidder and the low bidder for each of the specialty areas named in the bidding documents. The state then awards contracts to the low bidders, with the successful general construction bidder now being identified as the project general contractor. The state then assigns its contracts with the specialty contractors to the identified general contractor. From that point on, the specialty contractors are regarded as being subcontractors to the general contractor and the project is constructed according to the usual rules.

An extension of this procedure, called prepurchasing, has emerged on private projects. Here, the developer of a large commercial project breaks the project into several constituent parts. The developer then solicits bids on these separate packages from general contractors, subcontractors, and suppliers. After bids are received, the developer negotiates final prices with the subcontractors and suppliers and assigns them to the low-bidding general contractor.

5.32 BID DEPOSITORIES

The problems associated with the submission of subcontract bids (and certain material prices) were mentioned in preceding sections. The last-minute placing and changing of price quotations can result in a mad scramble as the general contractor attempts to summarize the cost figures and compute its final bid. The process is particularly difficult when there are several alternates or bid items. Serious errors can be made because of the haste with which the final figures must be compiled. The use of bid depositories is one means of circumventing some of the problems associated with the bids of subcontractors and material suppliers. Bid depositories can also reduce bid shopping, a topic discussed in Section 5.55.

Although there are about as many types of bid depositories as there are groups using them, all depositories are designed to accomplish a central purpose. A bid depository is a facility created by a group of contractors or a trade association to collect written bids from subcontractors and material suppliers and convey them to the general contractors. These quotations must be submitted to the depository before an announced deadline; four hours before the time of bid opening is a common requirement. The rules of bid depositories vary, but a common provision is that after a subcontractor bid or material quotation is submitted to the depository, it can be withdrawn before the deadline but cannot be changed except for clarification or correction of obvious errors. After the deadline, the depository transfers these price quotations to the general contractors for their use in compiling final cost figures. Many depositories tabulate the subcontractors' quotations after the bid opening and disseminate this information to the bidders.

Bid depositories are not a new idea and have had a long and sometimes stormy history. Few depositories have experienced unqualified success. It is axiomatic that to be workable the depository must have the wholehearted support of a large majority of the members of the industry within the area served. Some depositories have been found to be in illegal restraint of trade and in violation of antitrust laws. If depositories engage only in the unrestricted collection and distribution of bids, without restraining competition, it seems improbable that such difficulties will be encountered. Rules prohibiting subcontractors who are members of the depository from submitting bids to general contractors who are not members, and forbidding general contractor members from accepting bids from nonmember subcontractors, have led to serious legal difficulties.

5.33   ALLOWANCES

On occasion, the architect-engineer or owner will designate in the bidding documents a fixed sum of money that the contracting firm is directed to include in its estimate to cover the cost of some designated item of work. These sums are called "allowances" and are often used in regard to the costing of designated materials such as finish hardware, face brick, or electric light fixtures. Such an allowance establishes an approximate material cost for bidding purposes in those instances in which the designer does not prepare a detailed material specification. In such cases, the contractor does not solicit material prices for these items, but simply includes the designated allowance amount in the bid. After the contract is let, the owner or architect-engineer makes a final material choice. If the allowance is a material allowance, the contractor must also include the cost of installation in the basic bid. It should be noted that a material allowance is not the same thing as owner-provided materials. With an allowance, the contractor is given a material price to use for bidding purposes. The term *owner-provided materials* refers to those instances in which the owner selects, provides, and directly pays for the designated materials.

The architect-engineer may also state that a given sum is to be included in the estimate to act as a contingency for extras that may be necessary as the work progresses. Such contingency provisions are sometimes utilized for remodeling and other types of work that have many uncertainties associated with them.

Occasionally, allowances are used for subcontracted work. For example, the owner may solicit bids from specialty contractors for specific portions of a project in advance of the date for the opening of the prime proposals. After the low-bidding subcontractors have been determined, the bidding prime contractors are advised of the amounts and nature of the subbids to be included in their estimates. As an alternative, the specifications may merely indicate a sum of money to serve as an allowance for certain subcontracted work that will be let after the award of the prime contract. Landscaping is a common example of such work. In either case, however, the prime contractor must be cognizant of all conditions pertaining to the award of the subcontracts and the exact nature of the work they covered.

Serious operational problems can develop when the general contractor does not select its own subcontractors, as may be the case with subcontract allowances. If the two parties are incompatible or have had previous difficulties, job relationships can deteriorate seriously. This situation can be especially troublesome when one is a union contractor and the other is not.

The inclusion of an allowance in a proposal does not necessarily mean that the contractor is entitled to receive payment in the same amount. The contract documents provide that, after the completion of construction, an adjustment in the contract amount will be made equal to any difference between the specified allowance and the actual cost eventually associated with it.

5.34   PROJECT OVERHEAD

Overhead expenses are costs that do not pertain directly to any given construction work item but are nevertheless necessary for ultimate job completion. *Project overhead*, or *job overhead*, refers to costs of this type that are incurred on the project site. General overhead is discussed in the next section. Job overhead costs can be estimated with reasonable accuracy and are compiled on a separate overhead sheet. This overhead is a significant item of expense and will generally contribute from 5–15 percent of the total project cost, depending somewhat on where certain project costs are included in the cost estimate.

Project overhead expense should be computed by listing and costing each item of overhead individually, rather than by using an arbitrary percentage of project cost. This is true because different projects can and do have widely varying job overhead requirements. The only way to arrive at an accurate estimate of job overhead is to analyze the particular and peculiar needs of each project.

Typical items of job overhead are included in the following list. This list is not represented as being complete, nor would all the items necessarily be applicable to any one project. The project overhead items listed here disclose that the prime contractor normally provides and pays for temporary job utilities and standard site services for use by the entire construction team.

Job mobilization	Storage buildings
Project manager	Sanitary facilities
General superintendent	Field office supplies
Nonworking foremen	Job telephone
Heat	Computer equipment and software
Utilities	Computer networking and Internet connectivity

Small tools

Temporary enclosures

Temporary stairs

Permits and fees

Special insurance

Builder's risk insurance

Photographs

Barricades

Winter operation

Concrete and other tests

Cutting and patching

Drayage

Cleanup

Timekeepers

Security forces

Engineering services

Job signage

Temporary lighting

Drinking water facilities

Badges

Equipment move-in and assembly

Equipment dismantling and move-out

Worker transportation

Worker housing

Legal expenses

Surveys

Field office

Parking areas

Security clearances

Load tests

Temporary roads

Storage area rental

Travel expenses

Protection of adjoining property

First aid

Storage and protection

Temporary partitions

5.35   GENERAL OVERHEAD

General overhead, or office overhead, includes general business expenses such as office rent, office insurance, heat, electricity, office supplies, furniture, telephone and Internet, legal expenses, donations, advertising, travel, association dues, and the salaries of executives and office employees. The total cost of this overhead expense generally ranges from 3 to 10 percent of a contractor's annual business volume. This percentage represents the inescapable costs of doing business, and the contractor must include in the cost estimate of each project an allowance for general overhead expense.

Given the preceding information, it is easy to see that general overhead includes costs that are incurred in support of the overall company construction program and that generally cannot be charged to any specific project. For this reason, general overhead is normally included in an estimate as a percentage of the total estimated job cost. Details vary as to how the contractor does this, but the net effect is the same. A percentage of the total estimated cost of the job can be added as a separate line item, or a suitable markup percentage can be applied that includes general overhead as well as an allowance for profit.

5.36   PROJECT TIME SCHEDULE

When the cost of a project is being estimated, it is important that a general timetable of construction operations be established. Small projects usually require little investigation in this regard, but larger projects need attention to this matter because construction time is of importance when costing such a project. One reason is that the bidding documents normally stipulate the length of time, in calendar days, within which the construction must be completed. In some cases, however, the contractor must indicate on the proposal form its own time requirement for completing the project. In either case, construction time becomes a contract provision, and failure to complete the project on time is a breach of contract that can subject the contractor to damage claims by the owner. Because of the importance of time, the contractor cannot accept at face value the construction period stipulated in the bidding documents, but must make its own independent assessment of the construction time that will actually be required. When the contractor's estimate of construction time is called for on the proposal form, it must be as accurate as possible.

An approximate construction schedule is also needed for pricing purposes, primarily because many items of job overhead expense are almost directly related to the duration of the construction period. In addition, a general calendar schedule of the major job segments provides the estimator with valuable information concerning times available and weather to be anticipated. This offers a basis for decisions regarding equipment and labor productivity, cold weather operations, need for multiple shifts or overtime, and other such considerations. If the contracting firm believes that the stipulated contract time is inadequate for completion of the project using normal construction procedures, it may request the owner and architect-engineer to reconsider and allow a longer construction period. Otherwise, the feasibility of overtime or multiple-shift work must be investigated. If liquidated damages will be assessed against the contractor for failure to complete the project within the designated period, and these damages are less than the costs associated with constructive acceleration, the contractor may decide simply to add the anticipated damages to its job cost estimate. However, the contractor must proceed with caution in this regard. Some construction contracts, in addition to imposing liquidated damages, permit the owner to terminate the contract for default if the contractor fails to complete the work within the time specified. Thus, a liquidated damages clause is for the benefit of the owner and provides an alternative to termination of performance. The contractor cannot regard the presence of a liquidated damages provision as affording a guaranteed means of obtaining additional construction time.

A standard way of devising a general job plan for estimating purposes is to divide the project into major segments, estimate the time necessary to accomplish each of these segments, and establish the sequence in which they will proceed. The result is a bar chart (see Section 11.24) that shows the start and finish dates for the overall project and the approximate calendar times during which the various parts of the job will proceed. A bar chart is a series of bars plotted to a horizontal time scale. Each bar represents the beginning, duration, and completion of some designated segment of the total project. In combination, the bars make up a calendar time schedule for the entire job.

It must be noted that such a bar chart is by no means the equivalent of a detailed network analysis of the project, as discussed subsequently in Chapter 11. However, a complete and detailed time analysis of the project requires considerable time and effort. On a competitively bid job it is not usually possible for contractors to go through the extensive mechanics of a project network time study during the bidding period. When prepared with care, however, the much more approximate bar chart can suffice for estimating purposes.

5.37 MARKUP

On competitively bid projects, a markup or margin is added at the close of the estimating process; this is an allowance for profit plus other items such as general overhead and contingency. The markup, which may vary from 5 to more than 20 percent of the job cost, reflects the contracting firm's considered appraisal of a whole series of imponderables that can influence the probability of its being the low bidder and its chances of making a reasonable profit if it is. Many factors must be considered in deciding a markup figure, and each can have an influence on the value chosen. The size of the project and its complexity, its location, provisions of the contract documents, the contractor's evaluation of the risks and difficulties inherent in the work, the identity of the owner and/or the architect-engineer, and other intangibles can have a bearing on how a contractor marks up a particular job. Many contractors use different markup figures for different parts of a project. For example, contractors commonly apply one markup rate to self-performed work items and a second, smaller markup (perhaps half of the former) to work that it subcontracts.

The contractor is required to bid under a form of contract that has been designed to protect the owner and the architect-engineer. In their attempts to afford these parties protection against liabilities and claims arising from the construction process, the writers of contract documents usually incorporate a great deal of "boilerplate" including disclaimers of one sort or another. A common disclaimer provides that neither the owner nor the architect-engineer assumes any responsibility for the accuracy or completeness of boring logs or other information on subsurface conditions submitted with the bidding documents and that the contractor must assume full responsibility for any and all unknown physical conditions, including subterranean, that may be found at the construction site. Most of the court cases involving contract disclaimers pertaining to subterranean conditions have upheld them in the absence of fraud, warranty of the information by the owner, or deliberate failure to disclose all available information. Generally, the courts have placed the burden of risk for unknown subsurface conditions on the contractor unless there are contractual provisions that give the contractor a right to additional costs (see Section 6.26).

Other disclaimer clauses sometimes force the contractor to assume liability for every conceivable contingency, some of which involve eventualities over which it would have no control whatsoever. For example, provisions are found in some contract documents that absolve the owner from liability in all claims for damages caused by project delay, even delays caused through the owner's own fault or negligence. Provisions are sometimes included that indicate that almost nothing will provide a valid basis for an extension of time. There may be contractual terms providing that the contractor waives all lien rights, provisions stipulating unreasonable change order and claims procedures, or clauses allowing the owner to render binding decisions in case of disputes. When such severe language is used in private contracts, the contractor can sometimes protect itself by inserting protective language on the proposal form or by submitting a qualifying letter with its bid. This is not usually possible on public contracts, however, because of the risk of bid disqualification. When the general tone of the bidding and contract documents assigns the risk of the "unexpected" entirely to the contractor, consideration must be given to the inclusion of a contingency allowance in the markup amount in the bid. It is debatable whether such one-sided and burdensome contract language truly serves the owner's best interests. For one thing, such contract terms can result in owners paying for contingencies that never happen.

The contracting firm's objective when it selects a markup figure is to include the maximum possible profit, at the same time keeping its bid at a competitive level. Stated another way, the firm wants to be the low bidder but with a minimum spread between its low bid and that of the second-low bidder. Procedures have been developed to assist a contractor in selecting markup figures that will maximize its profits over the long term. The two best known and most widely accepted of these bidding strategies are known as the Friedman Model and the Gates Model. The Friedman Model, the simpler of the two, assumes that all bidders are acting independently of one another. This means that the probability of underbidding a group of competitors equals the product of the probabilities of underbidding each competitor separately. The Gates Model assumes that bids from competing contractors are not totally independent of one another, and procedures are applied to take this into account. Appendix H discusses a simple bidding strategy that is based on the Friedman Model using concepts of mathematical expectation and elementary probability.

5.38   THE LUMP-SUM RECAP SHEET

The preparation of a detailed cost estimate results in the determination of a great many separate prices that are scattered throughout a number of worksheets of various kinds. All these costs are finally brought together and entered on a recapitulation, or "recap," sheet. Figure 5.8 illustrates a typical recap sheet for a lump-sum bid. The costs associated with work items the contractor proposes to carry out with its own forces are posted from the summary sheets and entered in the Material and Labor columns. The subcontract bids selected for use are shown under the Subcontract heading. Equipment costs are brought forward from either an equipment sheet or the individual summary sheets. The equipment cost shown in Figure 5.8 was compiled on a separate equipment sheet, as is commonly done with lump-sum bids, and entered in the Material column. Project overhead costs are entered from an overhead sheet, with wages placed under the Labor heading and the total of all other expenses in the Material column. The total of the Material, Labor, and Subcontract columns of the recap sheet must equal the sum of the Total column, which is a useful check on the accuracy of the additions. The bid price is arrived at by the final inclusion of changes, markup, bond, and tax.

Taxes applicable to construction projects vary with locality and can differ between public and private projects. A common tax provision requires the contractor to pay a gross receipts tax or sales tax on its total business volume. Figure 5.8 includes such a tax in the amount of $160,339.

5.39   BID CHANGES

The recap sheet must be totaled and checked at a time when subbids and material prices are still being received and changed. The contractor must have some method for incorporating last-minute price revisions into the final bid amount. Figure 5.9 shows a change sheet that is commonly used by estimators to list and summarize such changes and as a means for adding in items previously forgotten or overlooked.

For an illustration of the procedure, suppose that somewhat in advance of bid time the recap sheet in Figure 5.8 was totaled to obtain the direct cost of $5,046,250. This was based on the best plumbing and heating bid of $813,180. Shortly thereafter a telephone call cut this bid by $69,450. This is entered into Figure 5.9 in the Deduct column. In the brief period before the bid must be finalized and the proposal form filled in, other changes are entered, some additive and some deductive. The final algebraic total of the changes is shown in Figure 5.9 as a deduction of $143,178.

RECAP SHEET

Project Municipal Airport Terminal Building

Work Item	Material	Labor	Subcontract	Total
1 Clearing and grubbing	$1,725	$12,930	-	$14,655
2 Excavation and fill	$21,090	$23,670	-	$44,760
3 Concrete	$ 422,652	$224,065	-	$646,718
4 Forms	$17,780	$29,369	-	$47,149
5 Masonry	-	-	$633,000	$633,000
6 Carpentry	$28,680	$20,460	-	$49,140
7 Millwork	$ 81,090	$30,317	-	$111,407
8 Steel and misc. iron	$377,835	$110,010	-	$487,845
9 Kitchen equipment	-	-	$89,972	$89,972
10 Insulation	-	-	$33,420	$33,420
11 Caulk and weatherstrip	-	-	$7,470	$7,470
12 Lath, plaster, and stucco	-	-	$255,300	$255,300
13 Ceramic tile	-	-	$29,000	$29,000
14 Roofing and sheet metal	-	-	$274,155	$274,155
15 Resilient flooring	-	-	$33,936	$33,936
16 Acoustical tile	-	-	$42,029	$42,029
17 Painting	-	-	$125,010	$125,010
18 Glass and glazing	-	-	$93,479	$93,479
19 Terrazzo	-	-	$126,162	$126,162
20 Miscellaneous metals	$91,842	$16,973	-	$108,815
21 Finish hardware	$65,400	$6,645	-	$72,045
22 Plumbing, heating, and A/C	-	-	$813,180	$813,180
23 Electrical	-	-	$508,107	$508,107
24 Clean glass	-	-	$4,200	$4,200
25 Paving, curb and gutter	-	-	$119,330	$119,330
26 Construction equipment	$84,239	-	-	$84,239
27 Job overhead	$59,325	$132,407	-	$191,732
	$1,251,658	$606,845	$3,187,748	$5,046,250
			Changes	($143,178)
				$4,903,072
			Markup, 8%	$392,246
				$5,295,318
			Bond	$49,317
				$5,344,635
			Sales tax	$160,339
				$5,504,974

Figure 5.8 Recap sheet for lump-sum bid.

112

CHANGE SHEET		
Project Municipal Airport Terminal Building		
	Increase	Deduct
1. Plumbing, heating, air-conditioning	--	$69,450
2. Lath, plaster, and stucco	--	$14,100
3. Roofing and sheet metal	--	$24,720
4. Electrical	--	$33,000
5. Ready-mix concrete	--	$7,875
6. Metal doors	$2,723	--
7. Chain link fence	$3,245	--
8.		
9.		
10.		
Total changes	$5,967	$149,145
		$143,178

Figure 5.9   Bid change sheet.

5.40   THE UNIT-PRICE RECAP SHEET

In a manner similar to that used for a lump-sum bid, all costs associated with a project being bid as unit prices are accumulated and summarized on a final recap sheet. Figure 5.10 illustrates a typical recap sheet for a unit-price bid. To produce a separate bid value for each item designated in the proposal form, this recap sheet must maintain all costs of each bid item separately. The costs associated with each bid item are entered from its summary sheet onto the recap sheet. In Figure 5.10 the total direct cost of the entire quantity of each bid item is obtained by adding its labor, equipment, material, and subcontract costs. The sum of all such bid-item direct costs gives the estimated total direct cost of the entire project ($7,194,789). To this are added the job overhead, markup, bond, and tax, giving the total bid price ($8,918,488). Dividing the total project bid price by the total direct project cost gives a factor of 1.240. By multiplying the total direct cost of each bid item by this factor, the total bid amount for that work item is obtained. Dividing the total bid amount of each work item by the quantity gives the unit price for that bid item.

There are instances in which projects are bid as a combination of lump-sum and unit prices. An example of this is a powerhouse, built under a single contract, the foundation of which was bid on the basis of unit prices and the structure itself was bid lump-sum. Another instance is when the proposal form of a building being bid lump-sum asks for certain unit prices as well. This may be done when some changes to the work during the construction phase appear likely and bid prices are requested on a unit basis for the possible later addition or deletion of work items such as concrete or excavation.

RECAP SHEET

Project Holloman Taxiways and Aprons   Bid Date: August 9, 20–   Estimator: ___GAS___

Item No.	Bid Item	Unit	Estimated Quantity	Labor Cost	Equipment Cost	Material Cost	Subcontract Cost	Direct Cost	Bid Unit	Bid Total
1	Clearing	l.s.	job	$5,139	$11,097	–	–	$16,236	$20,125.76	$20,126
2	Demolition	l.s.	job	$5,726	$7,383	–	–	$13,109	$16,248.98	$16,249
3	Excavation	c.y.	127,000	$75,636	$175,194	–	–	$250,830	$2.45	$310,923
4	Base course	ton	79,500	$352,670	$651,995	$159,479	–	$1,164,143	$18.15	$1,443,043
5	Concrete pavement, 9 in.	s.y.	90,000	$1,001,795	$202,708	$791,063	–	$1,995,565	$27.49	$2,473,655
6	Concrete pavement, 11 in.	s.y.	115,400	$1,572,819	$318,252	$1,241,967	–	$3,133,038	$33.65	$3,883,639
7	Asphalt concrete surface	ton	150	$1,035	$1,304	$2,208	–	$4,547	$37.57	$5,636
8	Concrete pipe, 12 in.	l.f.	1,000	$5,183	$5,439	$15,477	–	$26,099	$32.35	$32,351
9	Concrete pipe, 36 in.	l.f.	300	$2,937	$3,756	$6,809	–	$13,502	$55.79	$16,736
10	Inlet	ea.	2	$111	$164	$801	–	$1,076	$666.58	$1,333
11	Fiber duct, 4-way	l.f.	600	$3,278	$10,367	$5,126	–	$18,770	$38.78	$23,266
12	Fiber duct, 8-way	l.f.	1,200	$9,035	$20,576	$16,095	–	$45,705	$47.21	$56,655
13	Electrical manhole	ea.	6	–	–	–	$9,315	$9,315	$1,924.44	$11,547
14	Underground cable	l.f.	34,000	–	–	–	$255,957	$255,957	$9.33	$317,278
15	Taxiway lights	ea.	120	–	–	–	$45,710	$45,710	$472.17	$56,660
16	Apron lights	ea.	70	–	–	–	$26,313	$26,313	$465.96	$32,617
17	Taxiway marking	l.s.	job	$3,560	$675	$4,683	–	$8,918	$11,053.92	$11,054
18	Fence	l.f.	26,000	$28,506	$7,977	$129,477	–	$165,960	$7.91	$205,720
			Totals	$3,067,427	$1,416,885	$2,373,183	$337,295	$7,194,789		$8,918,488

Job overhead	$272,091
	$7,466,880
Markup, 15%	$1,120,032
	$8,586,912
Bond	$71,815
	$8,658,727
Sales tax	$259,762
Total Project Bid	$8,918,488

$$\text{Factor} = \frac{\$8,918,488}{\$7,194,789} = 1.240$$

Figure 5.10 Recap sheet for unit-price bid.

5.41 COST OF CONTRACT BOND

A large proportion of construction contracts require that the prime contractor provide contract surety bonds to the owner as a guarantee against any default or failure on the part of the contractor during the construction period. The workings and characteristics of contract surety bonds are discussed in Chapter 7. For the present, we are concerned more with the cost of their premiums than with their nature.

The contract documents may require a performance bond only, or both a performance and a payment bond. Many different combinations are used. For example, the contract documents may require a performance bond for 100 percent of the contract price and no payment bond. However, a performance bond of 50 percent and a payment bond of 50 percent are sometimes required. Many of the professional societies recommend a 100 percent performance bond and a 100 percent payment bond. Federal projects require a 100 percent performance bond and a sliding-scale payment bond (see Section 7.6). Generally, however, the contract amount, not the total face value of the bonds, determines the premium the contractor must pay. The explanation for this is that the risk depends on the nature and size of the contract rather than on the bond penalty. Consequently, the bond premium is the same for any of the preceding combinations of performance or performance and payment bonds. Special rules apply when low-percentage performance bonds, or payment bonds only, are required.

The premium the contractor must pay for the required contract bonds is based on the duration of the project, the class of construction, the total contract amount, and the applicable bond rates. These matters are discussed in Section 7.9. Figure 5.8 illustrates the addition of bond premium as the next-to-last step in arriving at the lump-sum bid. The amount of $49,317 is obtained by applying the rates shown in Figure 7.4 for Class B construction (buildings) to the sum of $5,295,318 obtained from the recap sheet in Figure 5.8. The first $100,000 is assessed $25 per $1,000, or at a rate of 2.5 percent. The next $400,000 is charged at the rate of 1.5 percent, the next $2,000,000 at 1.0 percent, the next $2,500,000 at 0.75 percent, and everything over $5,000,000 up to $7,500,000 has 0.7 percent applied to it. The sum of these amounts is $2,500 + $6,000 + $20,000 + $18,750 + 2,067 = $49,317. The bond premium rates given in Figure 7.4 apply for a construction time of up to 12 months or 366 days. For longer periods, the premium is increased by 1 percent per whole month. The estimated time for the Metropolitan Airport Terminal Building is given in Section 5 of Appendix E as 380 calendar days.

5.42 ADDENDA

When a project is being competitively bid, it is occasionally necessary during the bidding period to make changes, modifications, corrections, or additions to the bidding documents. Notice of such revisions is made by means of an addendum issued by the owner or architect-engineer and sent to all bidders of record. On public projects, an addendum is sometimes referred to as an amendment. These addenda serve to notify the bidders of changes in the owner's requirements, detected errors or oversights, clarifications or interpretations of the various provisions of the bidding documents, changes in the design, and similar matters. Changes in the drawings are accomplished by the submittal of revised sheets.

Such corrections, changes, interpretations, and clarifying answers must be in writing.

It is the responsibility of the architect-engineer or owner to see that copies of all addenda promptly reach the parties who hold plans and specifications. Each addendum is made a part of the contract documents and must receive the full attention of all parties who are preparing bids covering any or all portions of the project.

5.43   SCOPE BIDDING

A relatively recent innovation in the competitive bidding of construction projects is the use of "scope" bidding. The usual concept of competitive bidding is that the cost estimate is based principally on a detailed quantity takeoff of work items as they appear on the drawings and are described in the specifications. Where scope bidding is used, the bidding documents notify the bidding contractors that the drawings and specifications do not necessarily indicate or describe all work required for full performance under the contract. The bidders are advised that the successful contractor must furnish all work and materials required for the proper execution and completion of the work, on the basis of the general scope indicated and described. The architect-engineer or owner is responsible for the design and is given the authority to decide what work must be included within the general scope of the contract.

Admittedly, the usual construction plans and specifications do not include every single item of work required. It is usual to require the contractor to provide those items that are a part of its trade and that are installed as a normal and usual part of its work. However, the scope concept can be considerably more demanding and can involve unanticipated work on the basis that it lies within the contemplated scope of the project. In other words, the contractor is required to furnish all work required for the normal functioning of project parts and systems, whether or not shown as such on the bidding documents.

Scope estimating is traditional where performance specifications are used (see Section 4.19) and on negotiated work to establish target estimates or guaranteed maximum contract amounts. However, competitive scope bidding can introduce uncertainty and severe risk for contractors and subcontractors, involving responsibility for major work items that were not anticipated. Should scope bidding be elected as a contracting vehicle, parties are encouraged to review and implement guidelines such as those jointly established by the Associated General Contractors of America, the American Subcontractors Association, and the Associated Specialty Contractors.

5.44   COMPUTER-BASED ESTIMATING

Computers are now widely used by contractors to assist with their cost-estimating functions. Computerized estimating can offer a number of distinct advantages over manual methods, such as increased accuracy and having required cost and production information easily and quickly available. Another important benefit is improved productivity of the estimating staff, thus achieving substantial time savings in producing finalized cost estimates. Computerization also affords the estimator with the necessary information to make the best possible decisions concerning the job being estimated. In a number of ways, the computer now plays a significant role in the preparation of construction cost estimates.

Digitizer pens and other devices enable the estimator to make a quantity takeoff from the drawings, with the measurements and counts being entered directly into the computer as they are determined. The machine converts the takeoff data into pricing units such as square feet and cubic yards. Using database and entered price and productivity information, the computer produces a total estimated cost of the project including overhead, markup, bond, and tax. The computer prices labor and equipment by combining work quantities obtained from the takeoff with unit costs of production or with production rates and hourly costs. These latter values can be taken from the database or can be furnished by the estimator. Computer estimating offers the estimator several different methods of computing labor and equipment costs. With the option of choosing the method of calculating production rates, the estimator can obtain costs that are more realistic. Estimating software allows the computer to store a contractor's cost history in an organized manner for fast and useful access. With computerized estimating the contractor has a tool for storing past experience in a form immediately accessible to company estimators. Stock materials such as concrete and lumber are normally priced using total quantities and unit prices. For custom materials and subcontracts, the estimator enters company names, bid items, and prices as they are received. These prices are then transferred to a summary sheet that displays the contractor's lowest total bid amount as well as the low subcontract and material prices.

There are basically two kinds of cost estimating systems: a server-based system, in which multiple terminals provide more than one estimator access to the same information in preparing separate estimates, and a stand-alone system that generally provides for only one estimate preparation at a time. Estimating practices vary considerably from one contractor to another, making a system that is ideal for one contractor unsatisfactory for another. Experience has shown that there is no single best estimating system. Each contractor must examine its own estimating needs and find the most appropriate solution. For example, the unit cost estimating of highway and heavy projects generally requires a different computer methodology than the lump-sum estimating done by building contractors.

The computer is especially adept at handling last-minute subcontractor quotes and material price changes, and it maintains a running list of low material and subcontract bidders and the contractor's total bid amount. The computer automatically recalculates the estimate whenever additions or changes are made. This is particularly useful in closing out a unit price bid with multiple last-minute pricing changes to a significant number of bid items. The estimating system chooses the optimum cost combination of subcontractors and material suppliers available from the bids submitted. It can make final bid adjustments to reflect the nature of the job and the competitive situation.

In the overall estimating process, the computer relieves the estimator of much of the routine work and allows more time for the study and analysis of job complexities. The end result is a faster and more efficient method of producing cost estimates, a better tool for developing the strategies to win more profitable jobs, and a quick method of assembling the final numbers on bid day with confidence that all costs are included and totaled accurately. Many of today's computerized estimating applications are able to track minority business percentages on subcontractor bids to allow the bidding contractor to see where it stands on minority goals at any time.

It must be recognized that a computer cannot replace the human experience, skill, and judgment so necessary to the estimating process. Any project estimate involves the study and evaluation of that project's unique conditions and characteristics. The estimator must make decisions concerning construction methods, equipment types, and many other variables. The computer assists but does not replace the estimator.

5.45 RANGE ESTIMATING

The estimating of future construction costs always involves many subjective judgments and a considerable degree of uncertainty. Every project estimate is in error. The amount of the error and whether it adds a plus or a minus amount is unknown until completion of the project. There are certain probabilistic procedures that can be introduced in the estimating process that enable the estimator to establish the mathematical probability that the actual cost of the project will not overrun any particular figure. This technique of risk analysis, known as "range estimating," is not appropriate for every project but can be of considerable value to owners, architect-engineers, contractors, and construction managers on large and complex projects.

In an attempt to measure uncertainty, the estimator fixes an upper and lower cost limit, as well as a target value, for each major cost element of the project. Included with these parameters is the confidence level or chance that the actual cost of the project element will be at least as favorable as the target estimate. In effect, this range of possible costs for each critical part of the job brackets the traditional single-value estimate or target cost. The results of these price distributions are fed into a computer, which runs the project through thousands of simulations ranging from conservative to optimistic perspectives of job experience. The computer reports its results in a profile showing different total project costs versus the probability of overrunning each such total cost. Thus, individual uncertainties are combined in such a way that the uncertainty associated with the project's total cost is determined and presented for decision-making purposes. Range estimating can provide management with a means of coping with the problems that major project uncertainties present.

5.46 THE PROPOSAL

A proposal or bid is a written offer, tendered by the contracting firm to the owner, that stipulates the price for which the contractor agrees to perform the work described by the bidding documents. A proposal is also a promise that, on its acceptance by the owner, the bidder will enter into a contract with the owner for the amount of the proposal. Thus, timely acceptance of a proposal by the owner is automatically binding on the bidder.

When open bidding is being used, a prepared proposal form is included with the contract documents and must be used by the contractor to present its bid. Failure to do so will normally result in disqualification. The prepared proposal form is both desirable and necessary so that all bids will be presented and evaluated on the same basis. It facilitates the detection of omissions and other irregularities and makes comparison and analysis of the figures easier for the owner and the architect-engineer.

A typical lump-sum proposal is shown in Appendix I. The usual basis of contract award, assuming that all bidders are considered to be qualified, is the base bid plus or minus any alternates (see Sections 5.49 and 5.50) accepted by the owner.

Appendix F illustrates the usual nature of a unit-price bid form. It should be noted that each bid item in Appendix F is accompanied by an estimate, made by the architect-engineer, of the quantity of that work item which is to be done. The estimated quantities are multiplied by the respective quoted unit prices, and the results are added. This sum represents an approximate total project cost and is the basis for determining the low bidder. In cases of error in multiplication or addition by the contractor in preparing the proposal, there is considerable question as to whether the unit-price amount, the extended line amount, or the total amount of all line items prevails. On public projects, statutes often provide the procedure to be followed. Otherwise, the bid documents specify which amount controls. A common provision in this regard is for the unit price to control and the corrected total sum to govern identification of the low bidder.

Contractors sometimes find that they would very much like to include certain information concerning their bid that is not provided for on the proposal form. This information is often some form of limitation on the bid, either placing certain restrictions on owner acceptance or establishing special conditions pertaining to the conduct of the work. Under such circumstances, the contractor may be tempted to "qualify" its bid. In general, this is not permissible in public bidding, and any bid qualification will make the bid subject to rejection. A private owner generally has more latitude regarding proposal qualifications.

When closed bidding is at issue, the form of the quotation is frequently left up to the individual bidder. In such biddings, the job is not necessarily awarded to the lowest bidder but may be negotiated by the owner with the two or three lowest bidders. Consequently, the contractor designs the wording and content of the proposal to meet circumstances and to put its bid in the best possible light with the owner.

5.47   LIST OF SUBCONTRACTORS

It is a common requirement, especially on public projects, that the prime contractor submit, as a part of its bid, a listing of the subcontractors whose bids were used in the preparation of the prime contractor's proposal. In some cases, the dollar amounts of the subcontractors' bids must also be included. Details of this listing are variable, but the requirement typically includes a listing of the subcontractors doing major categories of work and other categories constituting more than a given percentage of the total bid price. This listing requirement can assist the owner in approving subcontractors, but it is also designed to minimize bid shopping and bid peddling (see Section 5.55). The listing must be submitted at the time of bidding, or the bid will be deemed nonresponsive. Such laws do not require the naming of material suppliers as long as they are not acting as subcontractors or performing job site work. Under some bidding directives of this type, subsequent changes in the named subcontractors can be made only under certain designated circumstances and with permission of the owner. According to others, the general contractor has a prescribed length of time after the bidding within which it can make changes to its subcontractor list. If a contractor violates the provisions of the listing requirement, the owner may either cancel the contract or assess a monetary penalty. In general, a subcontractor listing is favored by subcontractors and is opposed by prime contractors.

The listing of subcontractors can be troublesome on the bidding of projects with alternates. The problem arises because the identity of the low-bidding subcontractor for a given work specialty can depend on which alternates the owner may choose to accept or reject. In such an instance, the prime contractor normally lists its subcontractors based on the base bid, excluding the alternates.

5.48   UNBALANCED BIDS

On a unit-price project, a balanced bid is one in which the unit price for each bid item includes its own direct cost, plus its pro rata share of the project overhead, markup, bond, and tax. The values of the unit prices obtained in Figure 5.10 would constitute a balanced bid. For a variety of reasons, a contractor may occasionally raise the unit prices on certain bid items, and proportionally decrease the prices on others, so that the bid for the total job remains unaffected. This is called an unbalanced bid. It is common practice, for example, for a contractor to increase certain unit prices for items of work that are accomplished early in the course of construction operations and to reduce proportionately the prices for certain bid items that follow later. This process serves the purpose of making the early-progress payments to the contractor of such disproportionate size that a minimum of its own capital is required to finance its initial operations. Such early overpayment is of considerable assistance in helping the contractor to recover heavy move-in and other start-up costs that are involved before the actual start of construction. Unbalancing is unnecessary when a pay item covering job mobilization is provided.

Unbalanced unit-price quotations may also be submitted for other reasons. When the contractor detects what it believes to be a substantial error in the quantities listed in the proposal form, some unbalancing of unit costs is likely to be necessary so that fixed costs such as equipment and overhead will be properly distributed over the true quantities of work. In addition, because of profit motives, the contractor may increase unit prices on items that it believes will substantially exceed the estimated quantities and lower unit costs commensurately on other items. The contractor may simply juggle certain unit costs to disguise the makeup of its prices. When submitting an unbalanced bid, the contractor must be willing to assume the risk of having its proposal declared unacceptable by the contracting authority. Bid rejection because of unbalancing is rare, however, because it is difficult to detect and to substantiate. Some unbalancing is standard practice in most unit-price biddings.

Some of the elements of unbalanced bidding can be illustrated in a numerical example. To understand the rationale behind unbalancing, two characteristics of a unit-price contract must be kept in mind. First, the low bid is determined on the basis of total cost, using the engineer's quantity estimates and the unit prices bid by the contractor. Consequently, the raising of some prices and the corresponding reduction of others is normally done in such a way that the total amount of the bid remains unchanged. Second, the contractor is paid on the basis of its quoted unit prices and the quantities of work actually done.

Suppose, for illustration, we consider two bid items, the estimated quantity of one of which the contractor believes to be substantially in error. The engineer's estimate for ordinary excavation is 150,000 cu yd, but the contractor's takeoff indicates the actual amount will be about 200,000 cu yd. The contractor decides to unbalance its bid as follows:

Bid Item	Engineer's Estimates	Straight Bid		Unbalanced Bid	
		Unit	Total	Unit	Total
Common excavation	150,000 c.y.	$2.50	$375,000	$3.50	$525,000
Selected excavation	100,000 c.y.	$6.10	$610,000	$4.60	$460,000
			$985,000		$985,000

Assume that ordinary excavation actually turns out to be 200,000 cu yd and selected excavation to be 100,000 cu yd. Payment based on actual quantities of work done could vary, as follows, between a straight bid and the unbalanced bid the contractor decided to use:

	Engineer's	Straight Bid		Unbalanced Bid	
Bid Item	Estimates	Unit	Total	Unit	Total
Common excavation	200,000 c.y.	$2.50	$ 500,000	$3.50	$ 700,000
Selected excavation	100,000 c.y.	$6.10	$ 610,000	$4.60	$ 460,000
			$1,110,000		$1,160,000

As can be seen, the contractor's use of unbalanced bidding results in its receiving an increased payment of $50,000.

5.49 OWNER-DESIGNATED ALTERNATES

Proposals on lump-sum biddings are sometimes solicited for two or more alternative ways of accomplishing the same work. This can be done in different ways, but a common procedure involves a request for the costs of additions or deductions, called alternates, to a defined basic structure. Proposal forms frequently request price quotations for alternative methods, materials, or scope of construction. The owner originates and designates these alternates, and the bidding contractor is required to submit a lump-sum price for each alternate named. Alternate proposals may be additive or deductive to the base bid. Appendix I illustrates the use of alternate proposals. Such quotations must be complete within themselves and include all applicable direct costs, job overhead, markup, bond, and tax. Alternates can be of special importance to owners as a means of ensuring that they receive a bid within their limited financing or providing them with an opportunity to make the most judicious selection of a material or process. However, there is no arguing the fact that alternates complicate the bidding process, and a multiplicity of them can work against accurate bidding.

The award of a lump-sum contract with alternates is made to a single contractor. Normally, the low bid is determined from the algebraic sum of the base bid and any alternates accepted by the owner. When there are several alternates, it may be possible for the owner to select or reject certain alternates so that a preferred contractor receives the contract. To combat this possibility, it is usual for the bidding documents to state the order of acceptance of the alternates. Occasionally, the low bidder on a project with alternates is identified on the basis of the low base bid only. This then leaves the owner free to accept or reject any combination of alternates.

Generally, the alternates, as just discussed, are developed by the owner, and the bidding contractor must submit a price for each one in order to be considered responsive. There are some variations from this pattern, however. For example, on bridge construction, the owner may provide a base design and alternate designs. In such a case, the contractor is required to bid only on the alternative design that the contractor believes will result in the least total construction cost.

5.50 CONTRACTOR-DESIGNATED ALTERNATES

Another variant in the use of alternates has appeared in recent years, especially with large bridge projects. In this instance, the owner produces a base bid design. However, contractor-produced design alternates are permitted; that is, a contractor is allowed to make its own design if it believes the cost will be less than the base design proposed by the owner. The contractor normally engages a professional consulting firm to make the design and pays the resulting cost. The contractor then submits its bid based on its design alternate, the owner's base design, or both. The successful contractor is selected on the basis of project cost, owner evaluation of the alternate design, and a study of life-cycle costs.

The entire objective of alternate designs is to produce a quality structure at the least possible cost by allowing competition between design options. The use of contractor design alternates is still relatively new, and questions remain concerning design quality, responsibility of the contractor for the design, economy to the owner, procurement procedures, and selection of the successful bidder. There is no doubt that the procedure has brought forth new and innovative design and construction methods.

On private projects, a bidding contractor may include with its proposal unsolicited alternate prices that it considers potentially attractive to the owner. If a substantial savings is possible for an acceptable substitution, it may win the job for the contractor. Contractors should note, though, that unsolicited alternates are ordinarily unacceptable to public owners.

5.51   BID SECURITY

On practically all public projects and many private ones, a proposal must be accompanied by some form of security as a guarantee that the contractor, on being declared the successful bidder, will enter into a contract with the owner for the amount of its bid and will provide contract bonds as required. Bid bonds, also called proposal bonds, are widely used for the purpose of bid security, although some owners require that each bidding contractor submit a certified check, cashier's check, or other form of negotiable security. When a contractor becomes the successful bidder, the owner retains the security until the contract is signed and satisfactory contract bonds are provided. Ordinarily, the owner returns the bid security of the unsuccessful bidders shortly after the opening of proposals, possibly retaining that of the next lowest bidder or two until after contract signing.

Bid bonds, which are provided by the contractor's surety for a small annual service charge, have the advantage of not immobilizing appreciable sums of the contractor's money. Bid bonds may be executed on the standard form of the surety, or the contract documents may include a specific form of bid bond that the contractor is required to use. On public works the bid bond must be of such form as to satisfy any statutory requirements, and many public agencies require the use of their own standard bid bond forms. A form used by the American Institute of Architects, reproduced in Figure 5.11, illustrates the usual style and content of a bid bond.

The minimum bid security required by the instructions to bidders may be stated as a given percentage of the maximum possible contract amount, including alternates, or as a designated lump sum. Bid security in the amount of 5 or 10 percent of the maximum bid price is a common requirement, although larger percentages are also used. For example, many federal projects require 20 percent bid security. Because bid bonds are prepared in advance of the bidding and before the proposal amount is accurately known, the contractor must arrive at some advance rough estimate of the project cost for the use of its bonding company. If a fixed-sum bid bond is to be written, the amount of which is to be at least a given percentage of the maximum possible contract amount, the preliminary estimate of cost must be made conservatively high to ensure that sufficient bid security is available.

THE AMERICAN INSTITUTE OF ARCHITECTS

AIA Document A310

Bid Bond

KNOW ALL MEN BY THESE PRESENTS, that we

as Principal, hereinafter called the Principal, and

a corporation duly organized under the laws of the State of
as Surety, hereinafter called the Surety, are held and firmly bound unto

as Obligee, hereinafter called the Obligee, in the sum of

Dollars ($),
for the payment of which sum well and truly to be made, the said Principal and the said Surety, bind ourselves, our heirs, executors, administrators, successors and assigns, jointly and severally, firmly by these presents.

WHEREAS, the Principal has submitted a bid for

NOW, THEREFORE, if the Obligee shall accept the bid of the Principal and the Principal shall enter into a Contract with the Obligee in accordance with the terms of such bid, and give such bond or bonds as may be specified in the bidding or Contract Documents with good and sufficient surety for the faithful performance of such Contract and for the prompt payment of labor and material furnished in the prosecution thereof, or in the event of the failure of the Principal to enter such Contract and give such bond or bonds, if the Principal shall pay to the Obligee the difference not to exceed the penalty hereof between the amount specified in said bid and such larger amount for which the Obligee may in good faith contract with another party to perform the Work covered by said bid, then this obligation shall be null and void, otherwise to remain in full force and effect.

Signed and sealed this   day of   19

_____		_____	_____
(Witness)		(Principal)	(Seal)
		(Title)	
_____		_____	_____
(Witness)		(Surety)	(Seal)
		(Title)	

1

Figure 5.11 Bid bond.

INSTRUCTION SHEET
FOR AIA DOCUMENT A310, BID BOND—1970 EDITION

A. GENERAL INFORMATION

1. Purpose

AIA Document A310 establishes the maximum penal amount that may be due the Owner if the Bidder fails to execute the contract and to provide the required performance and payment bonds, if any. It provides assurance that, if a bidder is offered a contract based on its tendered proposal but fails to enter into the contract, then the Owner will be paid the difference in cost to award the contract to the next qualified bidder, so long as the difference does not exceed the maximum penal amount of the bond.

2. Related Documents

The A310 is not incorporated by reference into other AIA documents. For further reference on bonding procedures, see Construction Bonds and Insurance Guide, 2nd Edition, by Bernard B. Rothschild, FAIA, published by the AIA. See also AIA Document A501, Recommended Guide for Competitive Bidding Procedures; AIA Document 701, Instructions to Bidders; AIA Document A771, Instructions to Interiors Bidders; and AIA Document G612, Owner's Instructions Regarding Construction Contract, Insurance and Bonds, and Bidding Procedures.

3. Use of Non-AIA Forms

AIA Document A310 may be used with any appropriate AIA or non-AIA document. CAUTION SHOULD BE EXERCISED BEFORE ITS USE TO VERIFY ITS COMPLIANCE WITH CURRENT LAWS AND REGULATIONS BY CONSULTING WITH AN ATTORNEY OR A BOND SPECIALIST.

B. COMPLETING THE A310 FORM

1. Modifications

Users are encouraged to consult with an attorney or a bond specialist before completing the A310, particularly concerning the effect of federal, state, and local laws on the terms of this document.

2. Identification of the Parties

The Contractor, the Surety, and the Owner should be identified using their respective full names and addresses or legal titles under which the bond is to be executed. The state in which the Surety is incorporated also should be identified in the space provided.

3. Bond Amount

The dollar amount of the bond should be provided in both written and numerical form.

4. Project Description

The proposed project should be described in sufficient detail to identify (1) the official name or title of the facility, (2) the location of the site, and (3) the proposed building type, size, scope, or usage.

C. EXECUTION OF THE BOND

The bond must be signed by both the Contractor and the Surety. The parties executing (signing) the bond should print their title and impress their corporate seal, if any. Where appropriate, attach a copy of the resolution or bylaw authorizing the individual to act on behalf of the firm or entity. As to the Surety, this ususally takes the form of a power of attorney issued by the Surety company to the bond producer (agent) who signs on its behalf.

1/96

INSTRUCTION SHEET FOR AIA DOCUMENT A310 · 1970 EDITION · AIA ® · THE AMERICAN
INSTITUTE OF ARCHITECTS, 1735 NEW YORK AVENUE, N.W., WASHINGTON, D.C. 20006-5292

A310—1970

Figure 5.11 *continued*

124

Should the contractor refuse or be unable to enter into a contract for the amount of its bid or to provide contract bonds as required, the owner can proceed against the bid security as reimbursement for the resulting damages. In this regard, there are two forms of bid bonds in general use. One of these is a "liquidated damages" bond, in which the surety agrees to pay the owner the entire bond amount as damages for default. In this case the bid bond sum is the amount that the owner is entitled to receive, regardless of the actual damages suffered by the owner. In the second type, the bid bond is treated as security. This "difference-in-price" form of bid bond provides that the surety must pay the owner the difference between the defaulted low bid and the price the owner must pay to the next lowest responsible bidder, up to the face value of the bid bond. This is the provision of the bid bond shown in Figure 5.11. However, if the amount of the bid security is not sufficient to pay the owner the additional cost of procuring the work, the contractor can also be liable for the deficit. In any event, the contractor is liable to its surety for any such damages the surety may have to pay to an owner. When the formal application for bond service is signed, the contractor agrees to indemnify the surety company for all claims that may be made against it under the bond.

Upon deciding to bid a project, a contractor should immediately so inform its surety company, particularly if the project is unusually large or different or if the contractor is already burdened with a near-capacity volume of work. When the surety writes the bid bond, it is ordinarily obligated to provide the necessary contract bonds, should the contractor become the successful bidder, or, alternatively, to pay the owner in accordance with the terms of the bid bond. Bidding documents often require certification from the surety providing the bid bond, that the required contract bonds will be provided to the owner if the contractor becomes the successful bidder. Consequently, considerable investigation by the surety may be required before the bid bond can be written, and such investigation can be time-consuming. This matter is discussed more fully in Section 7.12.

With regard to bid bonds, it should be noted that general contractors, on occasion, may require bid bonds from certain of their bidding subcontractors. This is sometimes done when the general contractor is new to the area and has no knowledge of the local subcontractors or their records of performance. Bid bonds may be required of the subcontractors bidding major portions of the work, as a check, of sorts, on their financial stability.

5.52 SUBMISSION OF PROPOSALS

It is the responsibility of the contractor to deliver its bid to the proper place before the deadline time designated. The completed proposal form, together with the bid security and other necessary supplementary information, is sealed in an envelope that is addressed as directed by the instructions to bidders and clearly labeled as a proposal for the project being bid. Bids may be submitted at any time before the deadline scheduled for their acceptance. If feasible, it is usual practice for the contractor to deliver personally the sealed bid shortly before opening time. However, proposals may also be dispatched by letter, fax, E-mail, or messenger service. A common problem associated with the delivery of competitive bids occurs when a bid arrives after the established time. Although there is some disparity in how such cases are handled, most jurisdictions rule that a late bid is not acceptable or, at best, is subject to very restrictive rules in this regard.

Poor scheduling of bid dates, such as close to the Christmas holiday season, occasionally occurs and undoubtedly results in the owner paying more for its project. Studies by the American Society of Professional Estimators (ASPE) indicate that the best time for owners to receive bids is between 2:00 and 4:00 P.M., Tuesday through Thursday. Days before or after holidays and those between Christmas and New Year's Day should be avoided.

Public contract policy dictates that all proposals must be opened publicly and read aloud, a process of open bidding that is also commonly used on private projects. Such bid openings are usually well attended by the bidding general contractors, subcontractors, material vendors, and other interested parties. The ceremony consists of the owner or architect-engineer opening each bidder's sealed envelope, noting the type and amount of bid security, verifying receipt of addenda, and reading the amount of each bid item. Other bid formalities may also be noted for the record. At many bid openings the estimate of cost prepared by the architect-engineer or the owner is also read or distributed to those attending. Bid tabulation forms are made available to those present so they can make a record of the proceedings. It is customary that the bid opening adjourn without an official announcement concerning the identity of the successful bidder. Before the contract award can be made, the bids must be carefully studied and evaluated by the owner and the architect-engineer, a process sometimes referred to as "canvassing."

After the bids have been opened and read, the contractor communicates the results of the bidding to its surety company. This information is incorporated into the surety's permanent file on the contractor and constitutes an important part of the contractor's record of performance.

In closed bidding, the amounts of the bids are not necessarily disclosed. After delivery of the proposals, the owner uses the bids in any way it sees fit to serve its own best interests. It can select any of the bidders it chooses, or reject all bids. The owner often makes a final selection of the successful contractor only after extensive negotiations.

5.53   COMPLIMENTARY BIDS

A complimentary bid is one that a contractor submits to the owner that the contractor itself did not prepare but obtained from another contractor. Complimentary bids can be used for different reasons. For example, there are times when a contractor finds it impossible or undesirable to bid a project when asked to do so by an architect-engineer or owner. To maintain goodwill with this party, the contractor may request and receive a complimentary figure, or courtesy bid in this instance, from a fellow contractor who is bidding the job.

The contractor furnishing the complimentary bid to the other contractor makes it safely larger than its own but sufficiently close to appear to be a genuine proposal. The contractor receiving and submitting the complimentary bid will therefore not be submitting the low bid, nor will it have knowledge of any other contractor's proposal amount, or have entered into any collusive action of any kind. Nevertheless, contractors who provide or use complimentary bids must be circumspect in this regard, especially on publicly financed projects where complimentary bidding may be considered an unlawful form of price-fixing. In addition, many biddings on public projects require that each bidder certify that its bid is noncollusive, that it has compiled its bid independently, that it has not disclosed its bid to any other bidder, and that it has made no attempt to have any other contractor bid or refrain from bidding.

5.54   BID RIGGING

Bid rigging, or price-fixing, involves an arrangement between contractors to control the bid prices of a construction project or to divide up customers or market areas. Bid rigging can occur in a variety of forms, ranging from prearranged agreements among companies to casual, spontaneous actions. It can involve a series of projects over a period of time or can be an isolated incident. Price-fixing can be nationwide in scope or limited to a single locality. In any event, such actions are in restraint of trade and are violations of antitrust laws. Bid rigging is a criminal offense, and a project owner can seek treble damages under the Sherman Act.

An example of bid rigging occurs when a group of contractors decide to divide the available construction contracts among themselves and to fix the amounts of the successful low bids. To accomplish this objective, the participating contractors meet during the bidding period and discuss bids, reach an understanding regarding who the low bidder will be, and decide on the bid amounts to be used. Thus, these firms enter into a continuing understanding in which they allocate the available work, fix the low bids and, all but the selected low bidder, either submit noncompetitive, collusive bids or, perhaps, refrain entirely from bidding. Another instance of price fixing may be a situation in which a contractor, anticipating being the only bidder, persuades other contractors, who do not intend to bid the job, to submit bids higher than that of the initiating contractor. In such cases, the bid winner may later pay designated sums of money to the other participating contractors.

Recent years have seen a large number of bid-rigging cases reach the courts in the United States and in other countries, with dire results for the convicted participants. Some of this nation's largest construction firms and their executives have been found guilty of bid rigging, resulting in severe penalties of fines, imprisonment, and debarment from bidding future work. Projects where bid rigging has occurred include both public and private work involving highway, airport, power plant, and water projects.

5.55   BID ETHICS

How the general contractor handles its subcontract bids and translates them into subcontracts involves ethical considerations that can be very troublesome. Bargaining is the essence of competition, but there are occasions when the propriety of certain procedures is questionable, to say the least. For instance, in an attempt to improve its bidding position, a prime contractor may attempt, during the bidding period, to "chisel down" certain of the subcontract bids that have been received; or the prime contractor may actually reveal low subbids in an attempt to get better prices. This process is often referred to as "bid shopping." The fear that their bids may be disclosed during the bidding period explains, at least partially, the propensity of many subcontractors to submit their final bids at the last minute.

On being designated the successful bidder, the general contractor is then faced with making contract awards to its subcontractors. The conventional concept of this process would normally envisage the awarding of subcontracts to those firms whose bids were used in making up the general contractor's estimate. Thus, in most instances the low-bidding subcontractor for a given portion of the project would automatically get the work, except in those relatively unusual cases in which the low bidder was not considered capable or desirable and its bid was not used. In any case, the subcontract would be for the amount of the subcontractor's bid.

On the other hand, a completely different approach is possible—one in which the contractor regards the procurement of subcontractor services as being basically no different than bargaining for any other commodity on the open market. With this philosophy, the general contractor would feel no obligation to those subcontractors whose prices were incorporated into its own bid. Rather, after being declared the successful bidder, the general contractor may resort to bid shopping in order to obtain lower subcontract prices than those originally used in making up the contractor's bid. At this point, the low-bidding general contractor has a very strong bargaining position, and any lower subcontract prices that can be obtained represent windfall profits. The prime contractor may attempt to persuade an original low-bidding subcontractor to reduce its quotation or to get another subcontractor to underbid the original low bidder. Subcontractors may take the initiative and attempt to better the lowest subcontract prices, an activity sometimes referred to as "bid peddling." These practices are obviously distasteful to subcontractors and often lead them to avoid submitting their bids to those prime contractors who resort to such methods. Another reaction of subcontractors is to inflate their original bids when it appears there is going to be hard bargaining at a later date.

Bid shopping on publicly financed projects has received a great deal of criticism, and considerable effort has been expended to minimize or eliminate this practice. Several states have passed legislation that imposes regulations on the bidding of state-financed projects. For example, in some states, heating, ventilating, and air-conditioning; plumbing; and electrical work on state-financed projects must be awarded as separate contracts and not as portions of the overall prime contract. Another procedure requires general contractors bidding on a public project to name their subcontractors at the time they submit their bids to the owner. A system of bid filing is used on public contracts in a few states. Under this system, subcontractors file their bids with the awarding authority. The authority then furnishes the prime contractors with a list of the subbidders and the amounts of their bids. The general contractors then use these subbids as they choose in the preparation of their own proposals.

5.56 RESPONSIVE BID AND TECHNICALITIES

On construction projects, an acceptable bid must be "responsive" to the invitation for bids and the instructions to bidders. Responsiveness is determined by whether the bid as submitted is an offer to perform, without exception, the exact work as called for by the invitation, and upon acceptance will require the contractor to perform in accordance with all the terms and conditions thereof. A bid is nonresponsive if it contains qualifications or conditions not included in the invitation or if it offers performance that varies from that specified in the invitation. In addition, a bid may be nonresponsive if it does not conform to the technical bidding requirements established in the instructions to bidders. This means that the bid must conform to all the standards, requirements, and conditions listed in the contract solicitation and must not deviate from the criteria in any material way.

The instructions to bidders set forth specific requirements for the submission of the proposal. Any deviation from these technical requirements constitutes a bidding informality that can result in rejection of the proposal. A saving clause is usually present, which allows the owner to waive a minor informality or irregularity and accept the bid as being responsive. However, just what constitutes a minor bidding variation can, at times, be difficult to judge. A private owner has considerable latitude in this regard. Public owners must be much more circumspect in such waivers and generally insist on strict compliance with all bidding technicalities.

5.57   THE ACCEPTANCE PERIOD

The bidding documents normally contain a provision that gives the owner a stated period after the opening of bids to make acceptance. During this period the contractor cannot withdraw or change its bid except under penalty of its bid security. An acceptance period of 30 to 60 days is common, although other periods can be specified. During this waiting period, the subcontractors and material dealers are presumably obligated to the general contractor to stand by their price quotations just as the prime contractor is obligated to the owner. If the owner does not accept one of the bids within the acceptance period, the bidding contractors are free to withdraw or adjust their bids at their discretion.

Contractors are sometimes requested by the owner to grant an extension of the acceptance time. A contractor is generally willing to oblige, but sometimes in its eagerness to get the job, it will agree to such an extension without giving the matter sufficient consideration. Such action means that the completion date of the project will be set back by a length of time equal to the extension of the acceptance period. Raises in labor wages that may occur during the construction period will be in force for the additional time. Moreover, the extension means that the placing of material orders will be delayed, with the ever-present possibility of price advances. Another important aspect of this matter is that a subcontractor or supplier, having submitted what it considers to be an excessively low price, may not be willing to stand by its price quotation past the original acceptance period.

All these factors represent potential elements of additional cost that the contractor may have to assume should it too hastily grant an owner additional acceptance time. When increased costs are anticipated and the contractor does not wish to absorb them, it should quote the required additional amount to the owner in exchange for extending the acceptance period.

5.58   WITHDRAWAL OF BID BY PRIME CONTRACTOR

Under ordinary contract law, a bidder can withdraw or revoke its offer anytime before its acceptance. The situation is considerably different in construction, however. Despite the common-law doctrine of revocability, a prime contractor's bid proposal is normally considered irrevocable after the bid opening and during the acceptance period prescribed by the bidding documents. Such irrevocability results, on public projects, from many and varied statutes and local ordinances. Regarding bid withdrawal on private projects, courts often hold that bids are irrevocable, basing this decision on various legal grounds, although findings on this matter are not entirely consistent. In addition, the bidding documents themselves generally disallow bid withdrawal. It is only realistic, therefore, for a general contractor to regard as normally impossible the withdrawal of its bid after the deadline for the receipt of proposals without forfeiture of the bid security. The courts have long held that a bidder cannot unreasonably refuse to comply with its bid without penalty.

There is, however, a generally accepted legal basis for relieving the prime contractor from its bid—the "doctrine of unilateral mistake." A submitted bid containing a gross mistake may be withdrawn, even though withdrawal is prohibited by provisions of the owner or by rules of law, provided the following conditions are met:

1. The mistake is of such grave consequence that to enforce the contract as offered would be unconscionable.
2. The mistake relates to a material feature of the contract.
3. The mistake has not occurred because of the violation of a positive legal duty or because of culpable negligence.
4. The owner is put in a status quo position to the extent that it suffers no serious prejudice except the loss of its bargain.

When the preceding conditions apply and when the mistake is excusable and one of fact and not caused by a mistake in judgment on the part of the contractor, when the error is of a mechanical or clerical nature, and when the contractor acts promptly to notify the owner of the mistake and to rescind the bid, the courts almost unanimously permit withdrawal of the bid and the return of the contractor's bid security. Errors of a clerical nature include faulty addition, omission, or erroneous entry of certain bid components; misplaced decimals, typographical errors, and transposed figures. There are state statutes and provisions in federal procurement regulations that specifically provide for relief from a unilateral mistake under certain circumstances. The criteria just discussed have served to relieve bidders from their obligations, both before and after the owner has accepted the bid. However, relief for the contractor will likely be denied where it proceeds with the work without first seeking recision of its bid.

In granting relief to the contractor, it must be noted that the courts rarely allow the erring contractor to correct its bid, only to withdraw it without penalty. However, it has been the policy of the federal government and some other public agencies to allow the bid to be corrected and not allow it to be withdrawn in the special case where the bid, as corrected, remains the lowest bid. There has also been at least one case on the bidding of a federal project in which the General Accounting Office (GAO) ruled that the second low bidder, having discovered an error on its bid, was allowed to correct the error and displace the low bidder.

5.59   WITHDRAWAL OF BID BY SUBCONTRACTOR

Occasionally the tables are turned and the general contractor is faced with a request from a subcontractor or material supplier to withdraw its quoted price. The legal question of recision of such a bid offer is a tangled one and is very much a matter of the specific facts of the case. Nevertheless, there are some general concepts that apply. In the following discussion, the question of withdrawal of a subcontract bid is specifically addressed. However, the procedures discussed also generally apply to the withdrawal of a supplier's quotation.

In general, a subcontractor can withdraw its bid any time before the prime contractor's bid is submitted to the owner and the general contractor has not indicated acceptance of the subbid. In many jurisdictions the subcontractor can rescind its bid any time prior to its formal acceptance by the prime contractor. There are some exceptions to this regulation, such as when informal understandings were made between the prime contractor and subcontractor before the award of the general contract. Oral acceptance of a subbid by the general contractor may not be binding because of the applicable state statute of frauds that requires certain contracts to be in writing before they are enforceable. In addition, the courts have sometimes ruled that the general contractor and subcontractor did not intend to bind themselves contractually until a written subcontract was executed. This entire matter is further complicated by the fact that the general contractor is understandably loath to accept a subbid until the owner has made a formal award of the general contract. In light of these complications, it is not surprising that attempts by general contractors to hold low-bidding subcontractors to their bids, based on acceptance before withdrawal, have met with variable results.

When the low-bidding prime contractor bases its proposal on a subcontract bid that is considerably lower than any other submitted, the withdrawal of this low subbid can cause the contractor substantial loss. Because of this condition, the courts have sought rationales by which subcontractors would be required to hold their prices firm for some reasonable time, such as the acceptance period specified in the bidding documents. Probably the most common approach to the matter of holding a subcontractor to its bid is now based on the doctrine of promissory estoppel. This equitable doctrine has been applied to prevent subcontractors from withdrawing their offers even though the offers were not accepted, no contracts had been formed, and the parties had attempted to withdraw their offers. This doctrine avoids entirely the question of whether the subcontractor's bid was formally accepted by the prime contractor. The elements of promissory estoppel can bind a subcontractor to its bid price if the prime contractor can prove (1) that it received a clear and definite offer from the subcontractor, (2) that the subcontractor could expect that the general contractor would rely on the offer, (3) that the contractor actually did rely on the offer and such reliance was reasonable, and (4) that this reliance worked to the general contractor's detriment. Where the general contractor can establish these points, a subcontractor's low bid is generally binding if the detriment can be avoided only by enforcement of the bid.

For the contractor to have the defense of relying on the subbid, it must notify the subcontractor when the bid is so low that it suggests a mistake might have been made, and must request the subcontractor to verify the bid's correctness. In addition, the contractor must accept the subcontractor's bid with reasonable promptness after its own bid is accepted by the owner. In accepting the subbid, the general contractor may not change the terms on which the initial bid was solicited.

There continues to be a question concerning just what constitutes acceptance of a subcontract bid by a general contractor. The law is not entirely consistent on this matter, but there have been recent court decisions to the effect that a general contractor's use of a subcontractor's bid or the listing of the subcontractor's name in the prime contractor's proposal to the owner does not by itself create a binding contract between the two contractors, and the prime contractor is not bound to award the subcontractor the work involved—this is in the absence of any prior agreement or understanding. The subcontractor's bid is an offer and does not ripen into a contract until the prime contractor voluntarily and expressly accepts it. It should be noted, however, that some states and the federal government have enacted statutory regulations concerning the listing of subcontractors. For example, some of these public regulations require a subcontractor listing with the contractor's proposal, and these subcontractors must be used to accomplish the work unless a substitution is permitted by the owner because of special circumstances.

5.60 REJECTION OF PROPOSALS

The contract documents reserve the right for the owner to reject any or all bids. A prime contractor bidder may be rejected because of insufficient finances to handle the project, lack of experience, an unsatisfactory reputation indicating irresponsibility or unreliability, inadequate personnel or equipment, or failure to submit a responsive bid. Private owners are relatively free to ignore technical bidding inadequacies if it is in their interest to do so, but public owners must be careful to observe applicable regulations in this regard. Rejection of proposals on public projects because of bidding informalities is relatively common. Rejection may also be based on irregularities in the bidding procedure, collusion, flagrant unbalancing of bid unit costs, too few bona fide bidders, or unexpectedly high proposals. The lowest proposal sometimes far exceeds the available funds, and the project must then be abandoned or revamped. Public contracts are subject to statutory regulations in this regard. For example, some states have regulations stating that on any bidding where the lowest bid exceeds the estimated cost by more than 10 percent, all bids shall be rejected.

In the event that the owner wishes to reduce the project cost by a minor amount, it is likely to negotiate a final contract with the low-bidding contractor. If major savings are required, it is usual that all proposals be rejected. After redesign or other corrective procedures, the project is readvertised and new bids are taken.

5.61 PREFERENCE STATUTES

Several states have enacted statutes pertaining to the award of state-financed construction projects that give preference to bidding contractors who are domiciled in the particular state or who have previously satisfactorily performed public contracts and have paid specified state taxes for a statutory period. These statutes typically provide that a preferred bidder be awarded the contract for a state public work if its bid does not exceed the lowest submitted proposal by more than a stipulated percentage (5 percent is typical). Such laws do not apply where federal funds are involved. Similar provisions giving a bidding advantage to minority- and women-owned businesses have been applied in some areas.

The motivation behind such protectionist actions is to have the tax expenditures on public works provide a maximum benefit to the local citizenry. Whether this procedure is conducive to obtaining quality construction at the best possible price is debatable, however.

QUESTIONS

1. Explain the dichotomy of cost estimating and competitive bidding on the success or failure of a construction contracting business.
2. What differentiates construction estimating from the compilation of industrial "standard costs"?
3. What initiatives are being taken to help standardize construction estimating practices?

4. What factors should an owner consider when deciding between lump-sum and unit-price contracts? Why are lump-sum contracts typically used on building construction projects while unit-price contracts are typical on heavy civil construction projects?

5. Approximate estimates are developed using a variety of methods and data sources. When developing such estimates, what issues must the contractor consider in order to ensure reasonable accuracy?

6. Public agencies frequently require bidding contractors to be "pre-qualified" whereas private owners make use of "invitational bidding." Discuss the similarities and differences between these two methods.

7. In deciding whether to bid on a project, contractors consider such things as the project location, contract liabilities, and the nature and size of the project. Taken together what are these attributes commonly referred to as?

8. Subcontractors and suppliers frequently share the general contractor's bidding documents in preparing a price. What steps should the general contractor take to ensure that such sharing works effectively?

9. Discuss in detail the possible effects of accelerated bidding periods and limited plan availability on the owner's bidding experience.

10. What is the difference between a construction contract's "instruction to bidders" and its "general conditions"?

11. Good contracting practice suggests that two types of pre-bid meetings be held. One internal to the contracting organization and one called by the project owner. What is the purpose of these meetings, who should attend, which of these meetings should be held before the other and why?

12. Contractors frequently conduct their own quantity take-off before bidding. Considering that these quantities are already provided in the bid documents, why is this a common practice?

13. Discuss the difference between a material allowance and owner furnished materials. How should each be handled in the contractors estimate?

14. Direct labor costs are often developed from labor unit prices. What important issues must the contractor consider when applying these unit prices to an estimate?

15. What is the internal rental rate of a piece of equipment?

16. What sources are available to a contractor when determining equipment costs for estimating future projects?

17. Calculate the ownership costs of a 15 c.y. end dump hauler purchased at auction for $78,290 and shipped to the equipment yard for $1,750. The cost of tires are $15,430. The contractor must also replace the box (the part of the truck where the payload is held) at a cost of $10,590 and anticipates that the machine has a remaining useful life of 7 years with negligible salvage value and an average annual usage of 780 hours. Tax is assumed to be 2.7%, insurance, and storage 4.5%, and interest 10% of the average yearly value.

18. If the hourly operating costs of the end dump hauler above are calculated to be $21.04, what would the cost implications to the contractor's equipment account be if the end dump hauler lasted only 5 years, but operated on average 1,092 hours annually?

19. The contractor has estimated a project using bottom dump haulers as detailed in Figure 5.7, but due lack of equipment availability, is forced to use end dump haulers instead. Assume that the ownership cost of the end dump haulers is $22/hr. and the operating cost is $21.04/hr. Payload, type of loading, number of passes and loading rates are similar for both trucks. In addition, cycle times are identical except that the end dump haulers take 1.75 minute rather than .3 minutes to turn and dump. What is the estimated effect on the contractor's equipment hauling cost?

20. Contractors frequently refer to the room used in coordinating subcontractor quotations before going to bid as "the bull pen." Explain why this aspect of the estimate is frequently so hectic.

21. What is the basic difference between assigned subcontracts and owner designated allowances?

22. Briefly, differentiate between the attributes of an indirect cost and a direct cost.

23. Should a contractor expect to pay significantly more for a 100% performance bond than a 50% performance bond? Explain.

24. Calculate the bond premium for a Class A project with a contract amount of $7,570,320 and a duration of 415 calendar days.

25. Briefly, explain the benefits and risks to the contractor associated with unbalancing a bid.

26. Why might owners prefer difference-in-price bid bonds to liquidated damages bid bonds?

27. On many state projects, plumbing, electrical and HVAC work must be awarded as separate contracts, or general contractors must list subcontractors on their proposals. Why have these initiatives been necessary?

28. Under what conditions is a subcontractor's quote considered to be non-revocable?

Chapter **6**

Construction Contracts

6.1 CONSTRUCTION CONTRACT TYPES

Although there are many different types of construction contracts, they can be divided into two large groups. One group includes those contracts for which the owner selects a contractor based on competitive bidding, and the other includes those in which the owner negotiates a contract directly with a contractor of the owner's choosing. Many public construction contracts, as well as much private work, fall in the first category. Competitive-bid contracts are customarily prepared on a fixed-price basis and consist of two types, lump-sum and unit-price. With a lump-sum contract, the cost amount is a fixed sum that covers all aspects of the work described by the contract documents. A unit-price contract, the second of the two types, is drawn based on estimated quantities of specified work items and a unit price for each item. With regard to competitive-bid contracts, the owner sometimes selects the contractor for a cost-plus-fee contract using contractor credentials, with competitive consideration given to the amount of the fee the contractor proposes.

Negotiated contracts, the second major contracting division, can be drawn on any mutually agreeable basis: lump-sum, unit-price, or cost-plus-fee. Most negotiated contracts use a cost-plus-fee arrangement, whereby the owner agrees to reimburse the contractor for the full amount of the construction cost and pay a stipulated fee for the contractor's services. The major differences between the various types of cost-plus-fee contracts lie in the provisions regarding the compensation of the contractor.

When the owner uses a fixed-price contract or a cost-plus-fee arrangement with a guaranteed maximum, the contractor assumes most of the financial risk associated with the construction process. However, under a cost-plus-fee contract, the owner accepts responsibility for this financial hazard. It must be noted that a single construction contract can and sometimes does contain elements of both contract types. For example, a lump-sum contract may well include unit-price items as well as cost-plus provisions.

Owners, contractors, and engineers can select contracts from a variety of different sources. The two most common sources of construction contracting documents are the American Institute of Architects (AIA) (www.aia.com) and the Construction Management Association of America (CMAA) (www.cmaanet.org). Both organizations publish standard construction contract forms with similar purposes, applications, and clauses. Although certain industry segments give preference to one publisher over another, this chapter, in discussing typical construction contracting themes, refers to the AIA *Document A201* contract. This document is reprinted in this book as Appendix C. Similar terminology can also be found in the CMAA *Document A-3* contract and is reprinted as Appendix D.

6.2   THE LUMP-SUM CONTRACT

In a lump-sum contract, the contractor agrees to perform a stipulated job of work in exchange for a fixed sum of money. The satisfactory completion of the work for the stated amount remains the obligation of the contractor, regardless of the difficulties and troubles that may be experienced in the course of construction activities and even though the total cost of the work may turn out to be greater than the contract price. In the event of construction difficulties and/or excessive costs, the contractor must rely on the remedies contained in the contract for relief, such as clauses providing for changes or changed conditions, or as provided by law for unusual circumstances.

This type of contract is popular with owners for the obvious reason that the total cost of the project is known in advance. However, its use is limited, of necessity, to construction programs that can be accurately and completely described at the time of bidding or negotiation. For this reason, lump-sum contracts are widely used for residential and building construction. If the work is of such a type that its nature and quantity cannot be accurately determined in advance of field operations, the lump-sum type of contract is not suitable.

6.3   THE UNIT-PRICE CONTRACT

A unit-price contract is based on estimated quantities of defined items of work and costs per unit amount of each of these work items. The owner or architect-engineer compiles the estimated quantities, and the unit costs are those bid by the contractor for carrying out the stipulated work in accordance with the contract documents. However, the total sum of money paid to the contractor for each work item remains an indeterminable factor until completion of the project, because payment is made to the contractor based on units of work actually done and measured in the field. Therefore, the owner does not know the exact ultimate cost of the construction until completion of the project. In addition, the owner often must support, either directly or through the architect-engineer, a field force for the measurement and determination of the true quantities of work accomplished.

The contractor is obligated to perform the quantities of work actually required in the field at the quoted unit prices, whether they are greater or less than the architect-engineer's estimates. This obligation is subject to any contract provision for redetermination of unit prices when substantial quantity deviations occur (see Section 5.3). This form of contract has the same requirement for contractor performance, regardless of the difficulties and problems encountered, as described in the previous section for lump-sum contracts.

Unit-price contracts offer the advantages of open competition on projects involving quantities of work that cannot be accurately forecast at the time of bidding or negotiation. Examples of such work include the driving of piles and the excavating of foundations. A price per linear foot of pile or per cubic yard of excavation allows a reasonable variation in the driven length of the individual piles or the actual quantity of excavation because of job conditions that cannot be determined precisely before actual construction operations. However, drawings and specifications that are complete enough for the contractor to assess the overall magnitude of the project and the general nature and complexity of the work must be available for bidding.

6.4   AWARD OF COMPETITIVE-BID CONTRACT

After competitive bids have been submitted and found to be responsive, the owner, after careful study and evaluation of the bids received, must identify the contractor to whom the project will be awarded. Open biddings, involving both private and public owners, customarily award the job to the "lowest responsible bidder." This is normally mandatory on publicly financed projects. Determination of the low-bidding contractor is simple and direct. The matter of contractor responsibility, however, may require some attention.

If the bidding contractors have been prequalified, as is common on public projects, or if the bidding has been limited to an invited list of contractors, a practice occasionally used by private owners, the matter of responsibility has already received attention. In either case, the owner has restricted the bidders to those judged to be qualified and capable, using whatever criteria are believed to be suitable for the purpose. In the absence of bidding discrepancies, the successful bidder is determined on the basis of the lowest total project cost.

Where prequalification or invitational bidding is not applied, the owner must make the determination of responsibility after the bids have been opened. The term *lowest responsible bidder* has been held to mean the lowest bidder whose offer best responds in quality, fitness, and capacity to fulfill the particular requirements of the proposed work and with the necessary qualifications to complete the job in accordance with the terms of the contract. In the case of public contract-awarding bodies, the law gives them discretionary power as to which contractor is the lowest responsible bidder, such discretion not to be interfered with by the courts in the absence of fraud, collusion, or bad faith.

It is difficult to establish practicable criteria to measure and judge "responsibility." The owner must evaluate, by some means, the bidder's capacity, or incapacity, to perform. The owner may request qualification information to be submitted with the bids. Default by the contractor on a previous contract, proof of dishonesty, past difficulties in completion of projects on time, or a reputation for uncooperativeness and cutting corners may be taken as direct evidence of contractor irresponsibility. The determination that a bidder is irresponsible must normally be supported by an investigation and must allow the bidder to present its qualifications. In any event, disqualification of a bidder on the grounds of irresponsibility is a ticklish matter and may lead to litigation.

With regard to the judgment of contractor responsibility, the owner often relies on the contractor's surety company for the answer to this question. Before a surety will provide a contractor with a bid bond and contract bonds (see Chapter 7), it will investigate the contractor's finances, experience record, and other qualifications. In requiring such bonds, the owner may simply assume that if a contractor can "bond" the job, it is a responsible firm.

An owner's estimate of project cost is sometimes a part of bid opening and contract award on lump-sum and unit-price projects. This is especially true on public works. The owner's cost estimate may be prepared by the owner, architect-engineer, or cost surveyor and may or may not be revealed at the time of bid opening. The owner's estimate can be of use to the owner in a number of ways. For example, there are regulations in some public jurisdictions requiring that all bids must be rejected and the job readvertised if the lowest acceptable bid exceeds the owner's cost estimate by more than a stipulated percentage. Some statutes establish this as 10 percent. In addition, the relationship of the owner's cost figure to the contractors' bids can provide the owner with valuable insight into possible bidding errors, bid unbalancing on unit-price jobs, and bidder misunderstandings.

Upon selection of the contractor, the owner advises it in writing that its proposal has been accepted. This acceptance is conveyed by the issuance of a "notice of award," which is forwarded to the contractor together with information concerning arrangements for the signing of the contract. This notice, usually in letterform, sets forth the conditions pertaining to the award. The notice of award does not authorize the start of construction; it is simply formal notification of bid acceptance. The start of work in the field is contingent on the parties executing the formal written agreement and the owner issuing a notice to proceed (see Section 6.35).

In an effort to improve the quality and price of public works, certain agencies have experimented with alternatives to the open competitive bidding process that use the low-bid method of awarding contracts. The prime motivation for competitive bidding is to offer all companies interested in constructing public works an equal opportunity to obtain such contracts. However, abuses of the system by self-serving contractors and incompetent firms with financial problems have prompted such experimentation. So-called "competitive negotiation" has been used, although this procedure works to the disadvantage of many potential participants in public works. "Sole-source negotiation" has been applied by various public agencies, whereby a single contractor is handpicked on the basis of race, ethnic background, or national origin. Such procedures have met with strenuous opposition from contractor organizations.

6.5   COST-PLUS-FEE CONTRACTS

Contracts of the cost-plus-fee (cost-plus) variety are used where, in the judgment of the owner, a fixed-sum contract is undesirable or inappropriate. Cost-plus contracts are normally negotiated between the owner and the contractor. Most cost-plus contracts are open-ended in the sense that the total construction cost to the owner cannot be known until completion of the project. When the drawings and specifications are not complete at the time of contract negotiation, the owner and contractor negotiate what is commonly called a "scope contract." Based on preliminary drawings and outline specifications, the contractor arrives at a project "target estimate." The contract provides that the original contract documents will be subsequently amplified "within the original intent of the preliminary drawings and specifications."

When negotiating contracts of the cost-plus type, the contractor and the owner must pay particular attention to four important considerations:

1. A definite and mutually agreeable subcontract-letting procedure should be arranged. Both parties generally prefer competitively bid subcontracts when they are feasible. If the nature of the work is such that competitive subbids cannot be compiled or are not desirable, a mutually agreeable negotiation procedure will have to be devised.

2. There must be a clearly understood agreement concerning the determination and payment of the contractor's fee. Fees may be determined in many different ways (the most important of these are discussed in subsequent sections). Involved here is not only the amount of the fee, but also the method by which it will be paid to the

contractor during the life of the contract. A statement concerning any variation of fee with changes in the work should be included.

3. A common understanding regarding the accounting methods to be followed is essential. Many problems and controversies can be avoided by working out in advance the details of record keeping, purchasing, and the reimbursement procedure. Some owner-clients have need of accurate and detailed cost information for tax, insurance, and depreciation purposes. Owner requirements of this type, made known at the beginning of the contract, enable the contractor to better serve the owner.

4. A list of job costs to be reimbursable to the contractor should be set forth. Articles 7 and 8 of Appendix K present typical lists of reimbursable and nonreimbursable items for building construction projects.

6.6 SPECIAL REIMBURSABLE COSTS

With respect to reimbursable costs, two categories of expense can be particularly troublesome and merit special attention. One of these is the contractor's general overhead. This expense category involves the costs associated with the preparation of payrolls, purchasing, record keeping, engineering, preparation of working drawings, and similar office functions that are necessary elements of a construction project. If the job is reasonably large, the contractor can establish a project field office directly on the site that performs the necessary support functions. All expenses incurred by the project office, including salaries, are directly assignable to the job and are considered to be legitimate costs for reimbursement. When the project is smaller, such functions are usually carried out in the contractor's main office, and items of general overhead directly assignable to a specific project are difficult to establish. A common procedure is for the owner and the contractor to agree on a percentage of the cost of the work, 7 percent for instance, that the owner will pay to reimburse the contractor for office overhead. Another method is to eliminate general overhead altogether as a reimbursed cost and to increase the contractor's fee by a reasonable amount to provide for that expense.

The second troublesome category of reimbursed expense is that associated with construction equipment. When the equipment is owned or leased by the contractor, it is usual that rental rates are established for the various equipment types, which the owner will pay to the contractor for the time the equipment is required on the job. It is important for specific rates to be established on an hourly, weekly, or monthly basis, and that both parties clearly understand what these rates do and do not include. The costs of move-in, erection, dismantling, and move-out of equipment would ordinarily be separate reimbursable items.

In the event that equipment is to be required that the contractor does not own or lease, stipulation must be made for its provision by the owner or for rental from a third party. When the project is large or when the equipment has a limited useful life, it is common practice for the contractor to purchase the equipment for the owner on a reimbursement basis. At the end of the project, whatever salvage or resale value the equipment might have reverts to the owner. When rental of equipment from third parties is involved, it is best to be as explicit as possible concerning item requirements and rental rates to be paid.

6.7   THE CONTRACTOR'S FEE

With a cost-plus contract, the contractor's fee can be determined in a number of different ways. How this is done is a management responsibility worthy of careful study. The degree of risk and uncertainty associated with the project is clearly a prime consideration. The nature and complexity of the construction operations, geographical location, equipment and manpower requirements, and estimated construction time must also be considered. A project must not be allowed to immobilize company funds, manpower, equipment, and supervision without the contractor's realizing a proper return on its investment. The implication of this consideraton is that the fee designated should be based not only on the size and nature of the project, but also on the contractor resources required and the time they will be needed. In addition, the contractor's fee must often include an allowance for costs not considered reimbursable, such as general overhead.

Although the contractor's fee can be determined in any way mutually acceptable to the two contracting parties, certain procedures are widely used. Some of the commonly used methods are to set the fee amount as a fixed percentage of the cost of the work, a sliding-scale percentage of the cost of the work, a fixed amount, a fixed amount with a guaranteed top price, a fixed amount with bonus, or a fixed amount with an arrangement for sharing any cost savings.

6.8   COST-PLUS-PERCENTAGE-OF-COST CONTRACTS

From the contractor's point of view, one of the most advantageous ways of determining the fee in a cost-plus contract is to set it as a percentage of the cost of construction. This percentage may be a fixed amount or may vary in accordance with a prescribed sliding-scale arrangement. This type of contract is particularly well suited to cover work whose scope and nature are poorly defined at the outset of operations. Work required to meet an emergency may be involved, so that time is not available for the advance preparation of contract documents and for the usual bidding routine. Wars and other periods of extreme urgency afford instances of this type. In other cases, the work entailed may be such that no one can ascertain the difficulties that will be encountered or how great the eventual cost may be. Cleanup and repair of damage resulting from fire, storm, or flood afford many examples of situations where the cost-plus-percentage contract is suitable. Remodeling, expansion of facilities where services must be maintained, underpinning, and certain classes of demolition work are also occasionally done under this form of construction contract.

The cost-plus-percentage-of-cost contract does not provide any direct incentive for the contractor to minimize construction costs. Rather, it seems to work the other way. As a result, this mode of carrying out construction is generally confined to the exceptional and unpredictable described in the preceding paragraph. Public owners are prohibited from negotiating such contracts except under extraordinary circumstances. Contractors must practice strict economy in the interest of their owner-client and be satisfied with a reasonable profit if they are to enjoy continuing success with this type of contract. There is modest usage of this contract form on private construction where the contract includes a maximum, or "upset," price. In this case, if the total construction costs equal a sum less than the upset price, these savings either revert to the owner or are shared by both parties in a predetermined manner.

6.9 COST-PLUS-FIXED-FEE CONTRACTS

A popular type of cost-plus contract is one in which the contractor's fee is established as a fixed sum of money. When this scheme is utilized, the work must be of such a nature that it can be fairly well defined and a reasonably good estimate of cost can be approximated at the time of the negotiations. The contractor computes the amount of the fee on the basis of the size of the project, estimated time of construction, nature and complexity of the work, hazards involved, location of the project, equipment and manpower requirements, and similar considerations.

Under this arrangement, the contractor's fee is fixed and does not fluctuate with the actual cost of the project. Hence, it is to the advantage of the contractor to prosecute the work in as diligent a manner as possible. Expeditious handling of construction operations will minimize construction time and free workers and equipment for other contracts.

6.10 INCENTIVE CONTRACTS

To motivate the contractor to keep the cost of the work and/or the time of construction to a minimum, various forms of bonus and penalty incentives can be applied. Most such provisions are used with cost-plus-fee contracts, although fixed-price contract incentives can also be used. For example, bonus-penalty arrangements can be used in regard to time of contract completion of a fixed-price project. Such provisions provide for a bonus for each day of early completion and a penalty for late completion.

With cost-plus-fee contracts, bonus and penalty provisions are frequently applied to the determination of the fee. Under a cost-incentive-type of contract, the contractor and the owner agree to target estimates of cost and time for the construction. Bonus or penalty arrangements are tied to these target figures. Hence, this type of cost-plus-fee contract must of necessity be applied to work of a fairly definite nature for which drawings and specifications are sufficiently well developed to enable reasonably accurate target values to be determined.

As a stimulant for the contracting firm to minimize costs, a bonus clause providing for shared savings can be written, according to which it will receive, in addition to a base fee, a stated percentage of the amount by which the total actual cost is less than the target estimate. A figure of 25 percent is sometimes used for the contractor's share of the savings. There may also be a provision whereby the contractor's fee is reduced if the construction cost exceeds the target estimate.

When time of completion is of great importance to the owner, the contract can be made to provide that the contractor will receive, in addition to the base fee, a fixed sum of money for each day of beneficial occupancy realized by the owner before the originally agreed-upon completion date. This can be extended to provide that the contractor's fee will be reduced by the same amount for each day completion is delayed. When such a bonus-penalty arrangement is stipulated, the penalty need not be considered as liquidated damages (see Section 6.22), but can be assessed strictly as a penalty. Maximum and minimum fee amounts are sometimes specified when such incentive clauses are applied to cost and time target estimates.

Another form of incentive arrangement is a cost-plus-an-award-fee-type of contract, used principally by public agencies where uncertain conditions prevent fixed-price contracts. In this arrangement, a minimum or base fee is established as a percentage of the target cost. This fee is guaranteed to the contractor for completion of the contract. To the base fee is added an award fee that can range from zero to some maximum value, depending on the contractor's performance as evaluated by the owner. The amount of the award fee is based on the quality of construction work, adherence to schedule, cost efficiency, labor productivity, innovations, and safety performance. This system is advantageous for certain projects in that it provides a substantial contractor motivation, does not require a complete design to commence field operations, and can accommodate major design changes.

6.11   GUARANTEED MAXIMUM COST

An objection of many owners to the cost-plus-type of contract is that an accurate cost of the project is not known until after completion of the construction. A solution to this problem has been to provide for a guaranteed maximum cost to the owner. In this form of contract, the contractor guarantees that the project will be constructed in full accordance with the drawings and specifications and the cost to the owner will not exceed some total upset price. In return for its services, the contractor receives a prescribed fee. If the cost of the work exceeds the assured maximum, the contractor pays for the excess. In this way, a ceiling price is established and the owner is assured it will not be exceeded.

The determination of the upset cost by the contractor must be based on careful estimates made from complete drawings and specifications. Such a contract establishes an ironclad maximum cost above which the contractor will not be reimbursed. Any overage is the contractor's own direct responsibility. An incentive for the contractor to keep costs below the guaranteed maximum is sometimes provided by a bonus clause stating that the contractor and the owner will share any savings.

Contracts providing for a fixed fee with a guaranteed maximum price are sometimes competitively bid in a manner similar to that used for lump-sum contracts. The successful bidder is determined on the combined basis of the quoted maximum price and the fixed fee. The contractor's share of any savings below the guaranteed maximum may also be a criterion in the owner's determination of the successful bid.

Figure 6.1 provides an overview of all major contracting vehicles discussed in this and previous chapters and scales them relative to risk ownership. Although competitive scope bidding is likely to carry the least risk for an owner, it is also expected to carry the largest potential for project claims, disputes, and litigation. Conversely, a cost-plus-percentage-fee contract shifts the risk to the owner, but minimizes the likelihood of contract disputes.

6.12   LETTER OF INTENT

Occasionally an owner may want a contractor to start construction operations before the formalities associated with the signing of the contract can be completed. However, the contractor must proceed with caution in placing material orders, issuing subcontracts, or otherwise obligating itself before it has an executed and signed contract in its possession. In such urgency, it is common practice for the owner to authorize the start of work by a "letter of intent," or "letter contract." This letter is prepared for the signatures of both parties and states their intent to enter into a suitable construction contract at a later date. When signed, the letter is binding on both parties and furnishes the contractor with sufficient authority

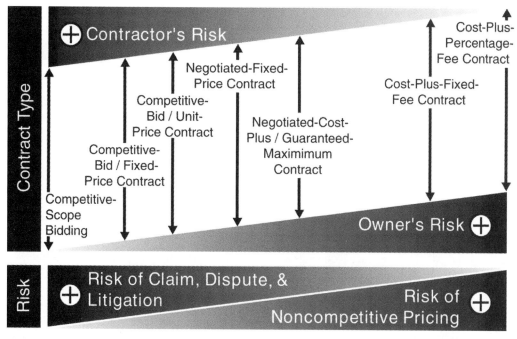

Figure 6.1   Construction risk.

to proceed with construction in the interim before the contract is formally executed. This interim authorization contains explicit information about settlement costs in the event the formal contract is never executed, and often limits the contractor to certain procurement and construction activities. The contractor should have its lawyer examine this document before it is signed.

6.13   THE CONTRACT DOCUMENTS

The construction contract arises from the results of either a competitive bidding procedure or owner-contractor negotiations. It is a universal practice in construction for the contract to be formalized by a written document. The basic purpose of a written contract is to define exactly and explicitly the rights and obligations of each party to the project. The complex nature of construction dictates a form of contract that is relatively lengthy, sacrificing brevity in order to describe precisely the legal, financial, and technical provisions. Construction contracts are substantially different from the usual commercial variety. The commodity concerned is not a standard one, but a structure that is unique in its nature and whose realization involves considerable time, cost, and hazard. Construction contracts involving public bodies are regulated by statute as to content and procedure.

Actually, the usual construction contract consists of a number of different documents. Exactly which documents constitute the construction contract is variable. The following is a listing of the documents that are essential to the bidding, negotiation, and construction process and that, in various combinations, constitute most construction contracts.

1. Invitation to bid
2. Instructions to bidders
3. General conditions
4. Supplementary conditions
5. Technical specifications
6. Drawings
7. Addenda
8. Proposal
9. Bid bond
10. Agreement
11. Performance bond
12. Labor and material payment bond

All of the contract documents are construed together for purposes of contract interpretation, each giving meaning and effect to each other, because it is presumed that everything in the contract has been inserted deliberately and for a purpose. In general, the intention of the contracting parties is determined from the final contract executed by them, rather than from preliminary negotiations and agreements. Contracts are interpreted strictly and to the letter where possible and feasible under the law. It is presumed that those who enter into contracts know what they want, say what they mean, and understand what they have said. It is an accepted rule of law that a person has a duty to read and understand a contract before executing it, and failure to do so is no excuse for not rendering proper performance.

When a conflict exists in a contract, the specific provision prevails over the general provision, the handwritten provision prevails over the typed provision, and the typed provision prevails over the printed provision. In the event that there is an inconsistency where numbers are expressed in words and figures, the words govern. When there are ambiguous terms, they are interpreted against the party who drafted the document. If there is a conflict between the drawings and the specifications, it should be noted that the courts have repeatedly held that the specifications prevail over the drawings. However, the general conditions of the contract may specify the relative order of precedence of the separate contract documents. Should a construction requirement appear only in the specifications and not on the drawings, or vice versa, the contractor must normally provide the requirement just as though it were included in both places (see Subparagraph 1.2.1 in Appendix C and Subparagraph 1.1.19 in Appendix D).

6.14 STANDARD CONTRACT DOCUMENTS

Standard forms of many contract documents are in wide use throughout the construction industry. Such standard forms have the advantage that their record of use has proven them both equitable and workable, and many of the provisions have been tested in the courts. Standardization of contract forms has done much to eliminate areas of disagreement between owners, architect-engineers, and contractors. In addition, they have withstood the test of time and experience and have become familiar to architect-engineers and contractors who clearly understand their meanings and implications. In general, it may be said that contractors prefer contract documents whose arrangement, form, and content are familiar to them.

Standardized versions of several contract forms have been developed by various professional, business, and public organizations. The American Institute of Architects, the Associated General Contractors of America, various branches of the federal, state, and local governments, and others have prepared such document forms. Contract documents of the American Institute of Architects are widely used on private building construction projects. The Engineer's Joint Contract Documents Committee (EJCDC) has prepared forms principally for use on engineering construction. The committee includes the Professional Engineers in Private Practice division of the National Society of Professional Engineers, the American Council of Engineering Companies, the American Society of Civil Engineers, the Associated Contractors of America, and the participation of at least 15 other professional engineering design, construction, owner, legal, and risk management organizations.

6.15 THE AGREEMENT

The agreement is a document specifically designed to formalize the construction contract. It acts as a single instrument that brings together all of the contract segments by reference, and it functions for the formal execution of the contract. It serves the purpose of presenting a condensation of the contract elements, stating the work to be done and the price to be paid for it, and provides suitable spaces for the signatures of the parties. The agreement usually contains a few clauses that are closely akin to the supplementary conditions and serve to amplify them. For example, it is common for the agreement to contain clauses that designate the completion time of the project, liquidated damages, and particulars concerning payments to the contractor, and that list the contract documents. However, practice varies in this regard with such clauses sometimes appearing in the supplementary conditions, at other times in the agreement, and occasionally in both.

Both public and private owners normally use standard agreement forms, although on occasion the agreement may be especially prepared for a given project. Appendix J illustrates the use of *Document A101* of the American Institute of Architects. Private owners often use this agreement form for fixed-price building construction contracts. *Document A101* is designed for use with the AIA "General Conditions of the Contract for Construction," *Document A201,* as contained in Appendix C. Appendix K reproduces *Document A111,* a form of agreement between the contractor and the owner that is used for cost-plus-fee contracts.

6.16 CONTRACT CLAUSES

Construction contracts contain many nontechnical provisions that pertain to the conduct of the work. These clauses, covenants, and agreements constitute the general conditions (Appendix C), supplementary conditions (Appendix E), and provisions of the agreement (Appendixes J and K).

These clauses must be carefully examined and studied by the contractor so that the obligations to be assumed are thoroughly understood. When standard document forms are involved, such as those of the American Institute of Architects, the federal government, or the Construction Management Association of America, there is seldom a need for the contractor to concern itself with identifying "weasel" clauses and "loaded" provisions. These standard contract clauses have well-established records of service and have become familiar to the industry. Many of them have received legal interpretation by the courts.

Nevertheless, construction contracts can, and sometimes do, include provisions pertaining to specified liabilities, waivers, damages, responsibilities, requirements, and other disclaimers that are designed to protect the owner but pose substantial and one-sided risks to the contractor. It is true that there are limits to the enforcement of this type of contract provision by the courts. Such exculpatory language has been invalidated where there was interference by the owner with the contractor's work or where there was a failure on the part of the owner to act in some essential manner, such as the timely approval of submittals. However, the contractor cannot assume that such severe and seemingly unfair contract language will be unenforceable. Obviously, the time for a thorough study and evaluation of all contract documents and provisions is while the project is being priced, not after the contract is signed. After execution of the contract, the contractor is bound by all of its provisions.

Contractors must recognize that they are not lawyers and hence are not competent to appraise the legal implications of contract clauses. Should they have reason to question any aspect of a contract document, they should seek the assistance of their attorneys. Failure to do so can result in serious complications that could have been avoided had legal advice been obtained.

During the bidding period, the contractor must evaluate each clause with regard to its possible or probable contribution to the cost of construction. On becoming the successful bidder, the contractor must again examine the contract, but with a different purpose in mind. Many provisions require specific actions on the part of the contractor during the life of the contract.

It is beyond the scope of this book to attempt a complete discussion of the meanings, implications, and legal record of all contract clauses. Nevertheless, the most important aspects of the principal contract provisions are discussed under appropriate topic headings. The following sections consider contract clauses of special significance that are not treated elsewhere in this book.

6.17   PROGRESS PAYMENTS

It is customary that projects of more than very limited duration require the owner or construction lender to make periodic payments or cost reimbursements to the contractor during the construction period. It is normally neither practicable nor desirable for the contractor to finance the construction from its own resources. Because the contractor often operates on borrowed funds, the terms of payment provided for in the contract are especially important. In general, the contractor must make application for a progress payment a prescribed number of days before it is due, or upon completion of designated phases of the work. In the former case, each payment is based on the value of the work put in place, including that performed by subcontractors during a prescribed period of time. In the latter case, a fixed amount of money or designated percentage of the total contract amount becomes due as each prescribed construction stage is finished.

Fixed-price contracts commonly require the contractor to submit applications for payment at least 10 days before the date established for each progress payment. Depending on the contract terms, a request may be submitted to the architect-engineer or directly to the owner. When it is submitted to the architect-engineer, this party checks the pay request and issues a Certificate of Payment, which is sent on to the owner or lending institution. Many contracts provide that if the owner does not make timely payments to the contractor as required, the contractor has the option of stopping the work.

In lump-sum contracts, the degree of completion of each major work category is usually expressed as a percentage (see Section 9.25). The quantities of work done on unit-price contracts are determined by the field measurement of work put in place. Materials suitably stored on the site or in other agreed-upon locations are customarily taken into account, as well as prefabrication or preassembly work that the contractor may have done at some location other than the job site. It is to be noted that partial payments, limited occupancy by the owner, inspection of the work in the field, or acceptance of the work by the architect-engineer do not constitute an acceptance of the partially completed work by the owner and do not serve as a waiver of any claim the owner may subsequently have against the contractor for defective work or failure to construct in accordance with the contract.

Cost-plus contracts usually provide for the contractor to submit payment vouchers to the owner at specified intervals during the life of the contract. A common contract provision is for weekly or biweekly reimbursement of payrolls and monthly reimbursement of all other costs, including a pro rata share of the contractor's fee.

6.18 RETAINAGE

Many construction contracts, especially those that involve competitive bidding, provide that the owner will retain a certain percentage of the progress payments. In the usual instance, the accumulated retainage remains in the possession of the owner until the project is completed and final payment is made, with the owner paying the contractor no interest on these funds. A retainage of 10 percent for the entire project has been typical, although reduced percentages and other retainage arrangements are now the rule. In any event, retainage on larger projects results in the owner's having custody of appreciable sums of the contractor's funds for extensive periods.

There continues to be considerable discussion and study regarding the need for retainage when the contractor has already provided the owner with performance and payment bonds. However, it must be recognized that contract bonds come into play only on breach of contract by the contractor. Owners look on retainage as further protection against possible eventualities such as contractor failure to remedy defective work, settlement of liens or other claims against the project, collection of damages from the contractor for late completion, payment of damages to others caused by the contractor's performance, and similar claims that the owner may be called upon to settle. Owners also regard retainage as an additional inducement for the contractor to maintain orderly progress of the work, produce quality construction, and keep the work on schedule.

Despite these considerations, however, retainage does have some undesirable aspects for owners, general contractors, and subcontractors alike. Subcontractors are involved because general contractors normally apply retainage to their subcontractors in the same percentage as the owner applies it to the general contractor. Retainage can and does produce real cash flow problems for contractors, resulting in substantial borrowing at hefty interest rates. This results in higher construction costs for owners. In addition, it discourages contractors from bidding some projects, thus reducing competition. Withholding retainage from a subcontractor until completion of a project, even though its work may have been satisfactorily completed long ago, is particularly unfair. Yet the general contractor cannot be expected to remedy this situation from its own funds.

To mitigate the undesirable effects of retainage, a number of changes and innovations have been introduced in recent years. All of these changes have tended to ease the withholding requirement and its impact on the contractors involved. For example, it is now common that when job progress is satisfactory, and with the consent of the surety, percent retainage is withheld only during the first half of the project, with subsequent progress payments being made in full. An alternate to this approach is to apply 5 percent retainage to the entire project. In a more recent development, 10 percent is retained on each work category of the project until that category is 50 percent completed, after which full payment is made if the work is proceeding satisfactorily. On some public projects, 10 percent is withheld for the entire project but the contracting officer may authorize full payment when satisfactory progress is being achieved. The federal government has discontinued the routine use of retainage on direct federal construction that is on or ahead of schedule and otherwise substantially in compliance with contract requirements. A number of public agencies have approved plans for the contractor to substitute certificates of deposit or interest-bearing securities for retention holdbacks. The contractor provides securities, whose value is equal to the required retainage, in escrow to the agency with all interest earned being paid to the contractor. Another scheme used by some public owners is to allow contractors to post surety bonds in lieu of retainage. Section 7.20 discusses several such special uses of surety bonds in the construction industry.

Although retainage is widely used with fixed-price contracts, it is common with cost-plus contracts for the owner to pay all contractor vouchers in full without deducting any percentage as retainage. There is a procedure sometimes used on such contracts that is analogous to retainage. The owner makes full reimbursement to the contractor until some designated percentage (80 percent may be used) of project completion. Further payment by the owner is then withheld until some specified amount of money has been retained. The owner keeps this reserve until final payment is made.

6.19   ACCEPTANCE AND FINAL PAYMENT

The making of periodic payments by the owner as the work progresses does not constitute acceptance of the work. Construction contracts generally provide that acceptance of the contractor's performance is subject to the formal approval of the architect-engineer, the owner, or both. Most contracts provide that mere occupancy and use by the owner do not constitute an acceptance of the work or a waiver of claims. Acceptance of the project and final payment by the owner must proceed in accordance with the terms of the contract. The procedure in this regard is somewhat variable, although inspection and correction of deficiencies are the usual practice.

On building construction, the contractor normally advises the owner or architect-engineer when substantial completion has been achieved. An inspection is held, and a list of items, called a "punch list," requiring completion or correction is compiled. The architect-engineer issues a Certificate of Substantial Completion at that time. This action by the architect-engineer certifies that the owner can now occupy the project for its intended use.

A provision appearing in many contracts is that upon substantial completion, the contractor is entitled to be paid up to a specified percentage of the total contract sum (95 percent is common), less such amounts as the owner may require to cover remedial work indicated by the punch-list items plus unsettled claims and a contingency reserve. The Certificate of Substantial Completion also serves other purposes, including shifting the responsibility for maintenance, heat, utilities, and insurance on the structure from the contractor to the owner. After the contractor has attended to the deficiency list, a final inspection is held and a final Certificate of Payment is issued. A common contract provision is that final payment is due the contractor 30 days after substantial completion. This is applicable, however, only with the consent of the surety and provided all work has been completed satisfactorily and the contractor has provided the owner with all required documentation. Final payment includes all retainage still held by the owner.

A legal question that frequently arises concerning final acceptance and payment involves the degree of contract performance required of the contractor. The modern tendency is to look to the spirit and not the letter of the contract. The vital question is not whether the contractor has complied exactly and literally with the precise terms of the contract, but whether it has done so substantially. Substantial completion may be defined as accomplishment by the contractor of all things essential to fulfillment of the purpose of the contract, although there may be inconsequential deviations from certain terms. Appendix C defines substantial completion as the stage in the progress of the work when the work or designated portion thereof is sufficiently complete in accordance with the contract documents that the owner can occupy or utilize the work for its intended purpose. This is a recent principle, peculiar to construction contracts generally, that has been developed by the courts to mitigate the severity of the rule of exact performance.

A contractor who has in good faith endeavored to perform all that is required by the terms of the contract, and has, in fact, substantially done so, is ordinarily entitled to recover the contract price less proper deductions for minor omissions, deviations, and defects. If defects are easily correctable, the owner is entitled to deduct the cost of correction. Otherwise, the difference in value between the structure specified and that actually built is the usual measure of recovery by the owner. The burden of proving defective performance and the amount of the setoff normally rests on the party asserting the deficiency, the owner in this instance. In evaluating substantial performance, consideration must be given to the contractor's intention, the amount of work performed, and the benefit received by the owner.

Certainly, the substantial-performance concept does not confer on the contractor any right to deviate freely from the contractual undertaking or to substitute materials or procedures that it may consider equivalent to those actually called for by the contract. Only if defects are purely unintentional and not so extensive as to prevent the owner from receiving essentially what was bargained for, does the principle come into play.

6.20 THE WARRANTY PERIOD

Acceptance of the work and payment for it by the owner normally constitutes a waiver of the owner's rights for damages because of defects in the structure if no claim is made within a "reasonable time." Many construction contracts obligate the contractor to make good all defects brought to its attention by the owner during some warranty period after either the time of substantial completion or final acceptance, the point at which the period begins to be variable with the form of contract. One year is a commonly specified warranty period, although periods of up to five years are sometimes required on certain categories of work, such as utility construction. The contractor is required, on notice, to make good, at its own expense, defects detected during this period. In most cases, the prescribed warranty period fixes the reasonable time and releases the contractor from further responsibility after its expiration. The performance bond normally covers warranty periods required by the contract. The warranty requirement does not operate to defer final payment.

An exception to the reasonable-time concept occurs when there is a latent defect caused by the contractor's inadequate performance, which could not have been detected by the owner during ordinary use and maintenance of the structure. In such a case, the owner has a right of action against the contractor beyond the warranty period for such longer period as may be provided for by the contract or as prescribed by law; a term of 10 years is frequently stipulated. For example, the owner would ordinarily have a right of action against the contractor for faulty construction up until the end of the period designated by the applicable statute of limitations for breach of contract or any applicable special statute of limitations (see Section 4.8).

A contractor's warranty does not imply, however, that the contracting firm is liable for the sufficiency of the plans and specifications, unless it prepared them or unless it guaranteed their adequacy. A contractor is required to construct in accordance with the contract documents, and when it does so it cannot ordinarily be held to guarantee that the design will be free from defects or the completed job will accomplish the purpose intended. The contractor is responsible only for improper workmanship, inferior materials, or other faults resulting from its failure to perform in accordance with the contract. Warranties normally exclude contractor responsibility for damage or defect caused by owner abuse, modifications performed by a party other than the contractor, improper or insufficient maintenance, improper operation, or usual wear and tear under normal usage.

6.21   CONTRACT TIME

If the construction contract is silent regarding contract time, the contractor is expected to complete the work within a reasonable time. This condition also applies when the contract requires project completion "at the earliest possible date," "as soon as possible," or "without delay." As one might expect, such imprecise statements regarding contract time can be troublesome. What constitutes a reasonable time is subject to wide differences of opinion and may have to be decided by the courts in case of a dispute.

Most construction contracts are explicit regarding construction time, designating either a completion date or a specific number of calendar days within which the work must be finished. The term "calendar days" includes Saturdays, Sundays, and holidays and is used, rather than "working days," to eliminate possible controversy concerning the effects of weather, overtime, multiple shifts, weekends, and holidays. Despite the seemingly definitive nature of contract time afforded by specifying a completion date or allowable calendar days, experience indicates that a contractor's failure to meet such provisions does not necessarily constitute a material breach of contract. With such contract wording, the courts normally find that time is not, expressly or by implication, a material obligation of the contract, and failure to comply does not permit the owner to rescind the contract and recover damages. For time to be a material feature of the contract, time must be expressly made "of the essence" to the contract by an explicit stipulation to that effect. There are also cases in which the contract provides that the contractor will receive a bonus for completing the contract ahead of schedule. Public agencies sometimes resort to the early-completion bonus system where prompt completion of the project will be of special economic significance to the public.

When the contract declares that time is of the essence, this signifies that the stipulated completion date or time is considered essential to the contract and is an important part of the contractor's obligation. In this instance, failure to complete the project within the time specified is considered a material breach of contract and is actionable by the owner. There have also been instances in which late completion has made the contractor liable for business losses suffered by a tenant when the contractor negligently failed to prosecute the job with due diligence. In such cases, some courts have ruled that the contractor has a "duty of care" to the tenant where the injury was reasonably foreseeable.

When the contract time is stated to be a given number of calendar days, the date on which the time begins is an important matter. Construction contracts usually state that the time will begin on the date the contract is signed or on the date the contractor receives a formal "notice to proceed" (see Section 6.35) from the owner. The contract may establish the finish date as that of substantial completion, final completion, or just "completion" of the project. In general, completion is related to the structure's capability of being utilized for its intended purpose despite small defects to be corrected or minor items not yet accomplished—that is, substantial completion. When the contract stipulates a completion date, the contractor must protect against any delays in getting construction operations under way. If the contract is not promptly executed, a letter of intent may be needed or the contractor may have to inform the owner that an extension of contract time (see Section 6.23) will be claimed.

6.22 LIQUIDATED DAMAGES

Many projects are of such a nature that the owner will incur hardship, expense, or loss of revenue should the contractor fail to complete the work within the time specified by the contract. Where the contract makes time an essential part of the contract, either by the phrase "time is of the essence" or by explicit reference to the stated time requirement, failure to complete the project on time is a breach of contract and can make the contractor liable to the owner for damages. The amount of such damages may be determined by agreement or by litigation. Compensatory damages of this sort are difficult to determine exactly, and in construction contracts it is common practice to provide that the contractor will pay to the owner a fixed sum of money for each calendar day of delay in completion. Most contracts provide that such delay damages run only to the date of substantial completion or beneficial occupancy, although some specify the date that the architect-engineer certifies the project complete.

This assessment against the contractor, known as "liquidated damages," is used in lieu of a determination of the actual damages suffered. The word *liquidated* in this instance merely signifies that the precise amount of the daily damages to the owner has been established by agreement. An advantage to the use of a liquidated damage provision in a construction contract is the possible avoidance of subsequent litigation between owner and contractor. Liquidated damages, when provided for by contract, are enforceable at law provided they are a reasonable measure of the actual damages suffered. The effect of a clause providing for liquidated damages is the ability to use the amount stipulated in lieu of the actual damages that would be caused by late completion, thereby preventing a controversy between the parties over the amount of damages. The courts have ruled that a liquidated damages clause in a contract prevents the owner from recovering its actual damages, even where the actual damages suffered by the owner exceed the amount provided by the liquidated damages provision. The owner deducts any liquidated damages from the sum due the contractor at the time of final payment. Typical values of liquidated damages appearing in construction contracts vary from $500 to $5,000 per calendar day, although much higher values are occasionally stipulated. When a project delay was the fault of both the owner and the contractor, the owner's recovery of liquidated damages is generally precluded. However, in construction, the contractor may still be liable for a portion of the delay. It should be noted that a liquidated damages clause normally works only one way. It conveys no automatic right to the contractor to apply the provision in reverse for owner-caused delay.

It must be emphasized that the courts enforce liquidated damages provisions in construction contracts only when they represent a reasonable forecast of actual damages the owner would be expected to suffer upon breach, and when it is impossible or very difficult to compute or make an accurate estimate of actual damages. When it has been established that the amount was excessive and unreasonable, the courts have ruled that such payment by the contractor to the owner constituted a penalty, and the owner received nothing or its recovery was limited to the actual damages. Punitive damages (those intended to punish) are not ordinarily recoverable for breach of contract. It may be noted that construction contracts specifically provide that the sum named is in the nature of liquidated damages and not a penalty.

A new development relating liquidated damages to retainage has resulted from recent court decisions. As explained in Section 6.18, retainage is withheld by an owner from a general contractor as an incentive for the contractor to complete the work and as security for the owner in the event that the contractor does not complete the work satisfactorily. General contractors often retain funds from their subcontractors for the same reasons. In this regard, recent court decisions have allowed retainage to be forfeited as liquidated damages under certain circumstances.

Courts have ruled that if the amount of retainage withheld is in reasonable proportion to the damage suffered because of breach of contract, the retainage is subject to forfeiture by the breaching party and can be withheld as liquidated damages by the owner (or general contractor). As with liquidated damages, the retainage may be withheld without the injured party having to prove the exact amount of the damages actually suffered.

6.23   EXTENSIONS OF TIME

During the life of a contract, there are often occurrences that cause delay or add to the time necessary to construct the project. Just what kinds of delays will justify an extension of time for the contractor depends on the provisions of the contract. An extension of time relieves the contractor from termination for default or assessment of liquidated damages by the owner because of failure to complete the project on time. In the complete absence of any clause on excusable delay, the contractor can normally expect relief only from delays caused by the law, the owner, the architect-engineer, or an act of God. For this reason construction contracts usually contain an extension-of-time provision that defines which delays are excusable. The provisions of such clauses are quite variable. Because an extension of time constitutes a revision to contract terms, it is formalized by a signed instrument, called a change order (see Section 6.28), which constitutes a binding change to the contract.

The addition of extra work to the contract is a common justification for an extension of time. In this case, the additional contract time made necessary by the broadened scope of the work is negotiated at the same time the cost of the extra work is determined. Delay of the project caused by the owner or its representative is another frequent cause of contract-time extensions.

Many other circumstances arising from unforeseeable causes beyond the control of, and without the fault or negligence of, the contractor can contribute to late project completion. Contracts often list specific causes of delay deemed excusable and beyond the contractor's control. Certain causes are essentially of an undisputed nature, provided the contractor has exercised due care and reasonable foresight. These causes include flood, earthquake, fire, epidemic, war, riots, hurricanes, tornadoes, and similar disasters. Other legitimate but less spectacular causes may or may not be included; some contracts are unduly severe in this regard. Strikes, freight embargoes, acts of the government, project accidents, unusual delays in receiving ordered materials, and owner-caused delays usually constitute acceptable reasons for an extension of contract time. Subparagraph 8.3.1 of Appendix C may be considered a typical contract provision in this regard. The contractor must be alert for unreasonable language relating to time extensions, and give such language proper consideration when bids are being prepared or the contract is being negotiated.

As a rule, claims for extra time are not considered when they are based on delays caused by conditions that existed at the time of bidding and about which the contractor might be reasonably expected to have had full knowledge. Moreover, delays caused by failure of the contractor to anticipate the requirements of the work in regard to materials, labor, or equipment do not constitute reasons for a claim for extra time. Normally, adverse weather does not justify time extensions unless it can be established that such weather was not reasonably foreseeable and was unusually severe at the time of year in which it occurred and in the location of its occurrence.

As discussed earlier, in unit-price contracts actual quantities of work can vary from the architect-engineer's original estimates. Minor variations do not affect the time of completion stated in the contract. However, if substantial variations are encountered, an extension of contract time may very well be justified.

Whenever a delay in construction is encountered over which the contractor has no control, the contractor must bring the condition to the attention of the owner and/or architect-engineer in writing within a designated period after the start of the delay. Ten to twenty calendar days is the period commonly specified in construction contracts. This communication should present specific facts concerning times, dates, and places and include necessary supporting data about the cause of the delay. The contractor's failure to submit a timely written request, with justification, may seriously jeopardize the chances of obtaining an extension of time.

6.24   CHANGES

It is standard practice for a construction contract to give the owner the right to make changes in the work within the general scope of the contract during the construction period. *General scope* in this context may be defined as that which is reasonably within the contemplation of the parties when the contract was made. Depending on the contract and its terms, such changes may involve additions to or deletions from the contract, modifications of the work, changes in the methods or manner of work performance, changes in owner-provided materials or facilities, or even changes in contract time requirements. Changes may have to be made to correct errors in the drawings or specifications. Owner requirements and circumstances sometimes change after the contract award, and changes must be made to meet such conditions. Suggestions by the contractor may even result in the need for changes. How changes are handled depends on the contract provisions as normally contained in a "changes clause" (see Article 7 in Appendix C). Typical language in this regard provides that the owner may make changes in the work and that an equitable adjustment of price and time will be made by a change order to the contract. If the contractor is not satisfied with the owner's or architect-engineer's decisions on cost or time adjustments to the construction contract, the matter normally reverts to the claims and disputes procedure mandated by the agreement. In most cases, the contractor is required to maintain a detailed record of costs invested in executing the change so that they may be used to substantiate its claim.

When the contractor is directed to make a change in the work, a check must be made to ensure that the party making the change is actually authorized to obligate the owner in this regard. In the usual case, neither the architect-engineer, the owner's project representative, nor the job inspector has implied or apparent authority to make work changes or to obligate the owner to pay an increase in the contract amount or to accept an increase in contract time. However, the architect-engineer is normally authorized to make minor changes that do not involve an adjustment of contract cost or time and are consistent with the intent of the contract documents.

If the contractor detects a job condition that it believes will require a contract change, the contractor should so advise the owner or architect-engineer immediately in writing. If the contractor believes the work will require an increase in contract price or an extension of time, it should so state. Many contract documents require that such a notice must be given before work constituting an extra to the contract is performed.

When a change is to be made, the owner or architect-engineer prepares a suitable description including any necessary supplementary drawings and specifications. This information is then presented to the contractor for its action. On a unit-price contract, changes are automatically provided for unless the changed work involves items that were not included in the original contract, unless the changes are so extensive the contractor or owner is authorized by the contract to request adjustments in the unit prices affected, or unless an extension of time will be required. When a change to a publicly financed project is involved, the contractor may be well advised to determine whether the change is within the scope of the changes clause of the contract and that the public agency has followed the procedural requirements of the applicable statute. Where this is not the case, the resulting change order may be declared void, thereby jeopardizing payment to the contractor for the work involved.

When the contract is a lump-sum contract, the contractor determines the cost and time consequences of the change and advises the owner or architect-engineer of them. The adjustment in the contract amount occasioned by the change may be determined as a lump sum, by unit prices, or as the additional cost plus a fee. In this regard, contracts often limit the contractor's markup and indirect cost of the change to some stipulated figure, such as 15 percent for overhead and profit or 10 percent for overhead and 10 percent for profit. Cost-plus contracts present few difficulties in accommodating changes to the work. The contractor's fee may have to be adjusted in accordance with contract terms to reflect the change being made, and target or upset prices must be adjusted by the cost of the change.

Construction contracts universally provide that the contractor is not to proceed with a change until the owner or its representative has authorized it in writing. To proceed without written authorization may make it impossible for the contractor to obtain payment for the additional work. Notwithstanding this requirement, however, the courts, under certain circumstances, have ruled that the lack of written authorization does not automatically invalidate a contractor's claim for extra work, even when authorization in writing is expressly required by the contract. When the owner or its agent orally approved the additional work and promised to pay extra for it, had knowledge that it was being performed without written authorization and did not protest, the courts have ruled either that the work was done under an oral contract separate and distinct from the written one or that the clause requiring written authorization was waived. In either event, the owner was judged to have made an implied promise to pay and the contractor was allowed to recover the cost of the additional work. The courts have sometimes disregarded the requirement for a written order on the grounds that it is unjust enrichment for the owner to enjoy the benefit without having to pay for it.

With regard to changes made to the contract by an action of the owner, "constructive" changes are sometimes involved. A constructive change to the contract is the result of an action, or lack of action, of the owner or its agent that can be construed as a change to the contract even though the owner did not issue a formal, written change order. When the contractor is orally directed to do work different from, in addition to, or in a different manner from that mandated by the contract, a constructive change has been made and the owner is responsible for any additional cost and time. Where this occurs, the contractor must usually give the owner written notice of the application of the constructive change concept.

6.25 ACCELERATION

Acceleration refers to the owner's directing the prime contractor to accelerate its performance to complete the project at an earlier date than the current rate of work advancement will permit. If the job has been delayed through the fault of the contractor, the owner can usually order the contractor to make up the lost time without incurring liability for extra costs of construction. However, if the owner directs acceleration under the changes clause of the contract so that the work will be completed before the required contract date, the contractor can recover under the contract change clause for the increased cost of performance plus a profit.

There is another instance of acceleration, however, in which additional costs may or may not be recoverable by the contractor. This is often referred to as constructive acceleration. Such a case occurs when the owner either fails or refuses to grant the contractor a requested extension of time and still insists that the project be completed by the original contract date. The owner's action in this case compels the contractor to accelerate job progress in order to comply with the original completion date. This obviously increases the contractor's field costs. In such an instance, the contractor must be able to demonstrate certain key elements before being granted relief from these additional field costs:

1. An excusable delay must exist.

2. The contractor must have provided a timely notice of the delay and requested a suitable time extension in accordance with the contract term.

3. The owner must have postponed or denied the contractor's extension request.

4. The owner must have further ordered the contractor to complete the project in accordance with the contractually defined construction period.

5. The contractor must have accelerated its performance in accordance with the owner's directive and thereby incurred additional costs.

6.26   DIFFERING SITE CONDITIONS

A "differing site condition," "changed condition," or "concealed condition" refers to some physical aspect of the project or its site that differs materially from that indicated by the contract documents or that is of an unusual nature and differs materially from the conditions ordinarily encountered. Such conditions include unknown physical conditions of an unusual nature that differ materially from those ordinarily found to exist and are generally recognized as inherent in construction activity of the character provided for in the contract documents. The essential test is whether a given site condition is substantially different from that which the contractor should have reasonably expected and foreseen or could have determined through a reasonable investigation of the site during the bidding phase. Unexpected conditions resulting from hurricanes, floods, abnormal weather, or nonphysical conditions are not normally considered changed conditions.

The construction cost and time of many projects depend heavily on local conditions at the site, conditions that cannot be determined exactly in advance. This applies particularly to underground work, where the nature of the subsoil and groundwater can have great influence on costs. At the time of bidding, the owner and its architect-engineer must make full disclosure of all information available concerning the proposed project and its site. The contractor uses the information as it sees fit and is responsible for its interpretation. The contractor is expected to perform its own inspection of the site and to make a reasonable investigation of local conditions.

Who pays for the "unexpected" depends on the contract provisions. No implied contractual right exists for the contractor to collect for unforeseen conditions. Any ability to collect extra payment and additional time for changed conditions must be provided by contract. An exception is made, however, when the drawings and specifications contain incorrect and misleading information. In such cases the courts generally rule that the owner warrants the adequacy of the contract documents and that the warranty is breached where errors or defects are present.

Many contracts provide for an adjustment of the contract amount and time if actual subsurface or latent physical conditions at the site are found to differ materially from those indicated on the drawings and specifications or from those inherent in the type of work involved. Delay-related expenses resulting from differing site conditions are also considered compensable. This proviso is a "differing site conditions clause," the objective of which is to provide for unexpected physical conditions that become known during the course of construction. When such a contract clause is present, the financial responsibility for unexpected conditions is placed on the owner.

It is the purpose of a changed-conditions clause to reduce the contractor's liability for the unexpected and to mitigate the need for including large contingency sums in the bid to allow for possible serious variations in site conditions from those described and intimated by the contract documents. It must be clearly understood at this point that for a claim to be sustained under this clause, the unknown physical conditions must truly be of an "unusual nature" and must differ "materially" from those ordinarily encountered and generally regarded as inherent in work of the character provided for in the contract. The essential question is whether the conditions are different from those the contractor should have reasonably expected and whether these conditions caused a significant increase in the project time or cost.

Various disclaimers and exculpatory clauses denying liability and responsibility for actual field conditions frequently appear in construction contracts. A common clause provides that neither the owner nor the architect-engineer accepts responsibility for the accuracy or completeness of subsurface data provided, and that all bidders are expected to satisfy themselves as to the character, quantity, and quality of subsurface materials to be encountered. Where a differing-site-conditions clause has been included in a contract, such disclaimers have not acted as a bar to contractor relief under the clause.

When the contract has not included a differing-site-conditions clause and the owner has disclaimed all responsibility for the subsurface information that was given, some courts have held that the contractor could not recover extra costs. The courts, however, seek to avoid contract clauses that exculpate parties from their own fault or negligence, and enforcement by the courts of such provisions in construction contracts is subject to many exceptions. Such exceptions occur when the site conditions were deliberately or negligently misrepresented on the contract documents, when the owner did not reveal all of the relevant information it possessed or of which it had knowledge, or when there was a breach of warranty of the correctness of the bidding information. Disclaimer clauses have also been denied force or effect when they were too strict, when they were in direct contradiction to specific representations, and on the basis that it was unreasonable to expect the contractor to make an extensive investigation of underground conditions during the bidding period.

It must be noted that, regardless of the contract wording, if differing conditions are found at the site and if the contractor intends to claim additional cost or time, it must promptly notify the owner in writing and must also keep detailed and separate cost records of the additional work involved. Failure to do so may make it impossible for the contractor to prevail.

6.27   OWNER-CAUSED DELAY

There are many instances in which the work is delayed by some act of omission or commission by the owner or by someone for whom the owner is responsible, such as the architect-engineer or other contractors. Examples include delays in making the site available to the contractor, failure to deliver owner-provided materials on time, unreasonable delays in the approval of shop drawings, delays caused by another contractor, delays in issuance of change orders, and suspension of the work because of financial or legal difficulties. In addition, work that is unchanged can be prolonged or disrupted because of changes or differing site conditions associated with other portions of the project.

Owner-caused delay is of concern to the contractor for two reasons. First, when the overall completion date of the project is affected, a suitable extension of time must be obtained. As a rule, contracts provide for extensions of time in the case of an owner-caused delay, and this matter is seldom troublesome. The second and more difficult problem for contractors is the ripple effect of such delays, that is, the "consequential damages" or "impact costs" associated with unchanged work that can result from project delay. Examples of impact costs include standby costs of nonproductive workers, supervisors, and equipment, expenses caused by disrupted construction and material delivery schedules, start-up and stopping costs such as those incurred in moving workers and equipment on and off the job, and additional overhead costs. It sometimes happens that work is delayed until cold weather, the rainy season, or spring runoff commences, resulting in greatly increased operating costs. Project delay can defer work until higher wage rates have gone into effect, until material prices have gone up, or until bargains or discounts have been lost. Barring contractual provisions to the contrary, a contractor has the right to recover damages for its increased costs of performance caused by delays attributable to the owner. Such claims are based on the owner's implied warranty that the drawings and specifications provided by it are adequate for construction purposes and its implied promise that it will not disrupt or impede the performance of the construction process. In order to recover damages, the contractor must prove that the owner caused the delays and that the contractor was damaged as a direct result of the delays.

However, construction contracts contain a variety of provisions concerning owner-caused delay. On federal projects the government has assumed responsibility for payment of the impact cost of changes and changed conditions as they affect unchanged work. Federal contracts and many others contain suspension-of-work clauses that provide for the extra cost occasioned by owner-caused delay or suspension. Some contracts contain provisions that expressly limit contractor relief in cases of job delay to time extensions only. Others indicate clearly that the provision for time extension does not preclude delay damages. Many contracts, both public and private, contain "no-damage" clauses whereby the owner is excused from all responsibility for damages resulting from an owner-caused delay.

Where such exculpatory clauses are present, the courts generally give them effect, although they are strictly interpreted and limited to their literal terms. Exceptions that have allowed the contractor to collect increased costs are cases in which the delay was caused by the owner's active interference with the work; the delay was of a type not contemplated by the parties, such as the owner not allowing the work to start on time or the owner acting with malicious intent; the delay resulted from bad faith or misrepresentation by the owner; the delay was caused by the owner's failure to coordinate the work of other contractors on the site; the delay was for an unreasonable period of time; or the delay was caused by the owner's breach of a contractual obligation, such as the nontimely delivery of owner-furnished materials. In general, the courts have granted exception when the owner's actions were intentional, willful, or negligent. Even when consequential damages are awarded, they may reimburse the contractor for extra costs only, with no allowance for profit. It should be noted that a few states have enacted statutes barring no-damage-for-delay clauses in construction contracts, and others have significantly limited their legal application.

A contractor is frequently unable to recover delay damages because of an inability to prove the exact extent of the loss suffered. For this reason, when such a delay occurs, the contractor must keep careful and detailed records if it expects to recover extra costs from the owner. An additional point is that many contracts require the contractor to notify the owner within a designated time after any occurrence that may lead to a claim for additional contract time or extra cost.

6.28   CHANGE ORDERS

When additions, deletions, or changes in the work are made by the owner, a supplement to the contract between the owner and the prime contractor is prepared, which can be on the basis of a lump-sum, unit-price, or cost-plus arrangement. Verbal authorization for such extra work, unless later acknowledged in written form, may not be enforceable in a court of law. However, the legal requirement for a written change directive can be waived by the words, acts, or conduct of the parties. The contract supplement, called a "change order" (see Paragraph 7.2 in Appendix C), is consummated by a written document that describes the modification to be made, the change in the contract amount, and any authorized extension of contract time. The change order form identifies the change being made as a modification to the original construction contract and normally bears the acceptance signatures of the owner and the prime contractor. If it is not the owner who signs, the contractor must be sure that the party executing the change order is an authorized agent of the owner with the authority to obligate the owner and to make a binding change to the contract. For example, by virtue of its position alone, the architect-engineer has no authority to order changes and must be authorized in some way by the owner to act on the owner's behalf. Many architect-engineer firms have developed their own individual change order forms, an example of which is shown in Figure 6.2. The previous contract amount shown in Figure 6.2 is obtained from the owner's acceptance of a deductive alternate (see Section 5.49) for $66,723. With a base bid of $5,504,974 (Figure 5.8) and a deductive alternate of $66,723 (Appendix I), the original contract amount in Figure 6.2 was $5,438,251. A change order, as a modification of the contract, means that the parties have entered into a new contract. It is presumed that such a contract modification has taken into account all prior negotiations and understandings leading up to its signing and that the terms of each change order reflect proper consideration of these negotiations.

As mentioned earlier in this text, a usual contract requirement is that the contractor is not to proceed with any contract change before the owner or its agent has authorized it in writing. However, there are occasional instances in which a contract change is necessary and the contractor must proceed with the work involved before an agreement as to the effect of the change on contract price and time can be reached with the owner. Many construction companies utilize informal "field orders" when a change will not have a material effect on contract provisions. When the change is of significance, a formal "construction change directive" can be used (see Paragraph 7.3 in Appendix C). Such a directive is an order by the owner to the contractor to perform work differently from the manner specified by the contract documents. This order is an instruction by the owner to the contractor and serves to change the work but does not modify the contract. When both parties subsequently sign a change order effecting their agreement on the impact of the change, this constitutes a formal modification of the contract.

6.29   VALUE ENGINEERING

Understandably, construction contractors are well informed on construction costs and time requirements, and owners are increasingly approaching the design and construction of their projects as a team effort. Experience shows that the project owner, designer, and contractor,

---

JONES & SMITH
ARCHITECT-ENGINEERS
PORTLAND, OHIO

<u>CONTRACT CHANGE ORDER</u>

Project: Municipal Airport Terminal Building   Change Order No.   1
For: City of Portland, Ohio   Date: July 30, 20–

To: The Blank Construction Co., Inc.
           1938 Cranbrook Lane
           Portland, Ohio

Revised Contract Amount

Previous contract amount	$5,438,251
Amount of this order	
(~~decrease~~) (increase)	5,240
Revised contract amount	$5,443,491

An (~~increase~~) (~~decrease~~) (no change) of ___ days in the contract time is hereby authorized.

This order covers the contract modification hereunder described:

   Providing and installing folding partitions and ornamental screens as shown and described by supplemental Drawing X-1 attached hereto. This change includes all grounds, nailer blocks, and other provisions required for the satisfactory installation of said partitions and screens.

The work covered by this order shall be performed under the same terms and conditions as included in the original construction contract.

Changes approved   Jones & Smith, Architect-Engineers

_____   by _____
              (Owner)

by _____

_____
              (Contractor)

by _____

Figure 6.2   Contract change order.

working cooperatively, can appreciably reduce the required construction time and life-cycle cost of a project without sacrificing the design. This process, referred to as "value engineering," can be described as finding a better way to accomplish a construction project at less cost.

   Many construction contracts now include value engineering incentive clauses. In this context, value engineering applies after the contract is awarded and is concerned with the elimination or modification of any contract provision that adds cost to a project but is not necessary to the structure's required performance, safety, appearance, or maintenance. The basic objective of value engineering is often stated to be the elimination of "gold plating." A clause of this type provides an opportunity for the contractor to suggest changes in the plans or specifications and to share in the resulting savings. These changes may involve substitutions of materials, change or improvement of the design, reductions in quantities, or procedures other than those set forth and required by the contract documents. Value engineering is not designed to enable the contractor to second-guess the designer, but to take advantage of the contractor's special knowledge and to cut the cost of a project to the lowest practicable level without compromising its function or sacrificing quality or reliability. In short, the contractor is encouraged to develop and submit to the owner cost-reducing proposals leading to changes in the plans or specifications. If the owner accepts them, a change order is processed and the owner and the contractor usually share the savings equally. Many value engineering clauses require the prime contractor to include similar clauses in its subcontracts and provide that it will share in any cost savings proposed by a subcontractor.

It should to be noted that value engineering refers to making a change in the drawings, specifications, or other contract provision and requires the approval of the owner. Savings that can be realized without a contract change properly belong to the contractor.

Value engineering is applied to the planning and design of construction projects as well as to their construction. For example, value engineering during the planning and design phases has been an in-house activity of the U.S. Army Corps of Engineers for several years. Some public agencies are now including a requirement for value engineering in many of their professional service contracts for architect-engineer and construction management services. This is a contract requirement beyond the normal design considerations of cost and function. It calls for a systematized, formal effort directed at isolation and elimination of unnecessary costs. It may be described as a "second look" by a qualified team of specialists whose purpose is to challenge the basic system and the choice of materials in the design, with the objective of eliminating excess costs without diluting the quality or function of the various segments of the project. Value engineering at the design stage is geared to long-term cost rather than just to original construction expense. The Society of American Value Engineers (SAVE) (www.value-eng.org) is a professional group concerned with value engineering and its application.

6.30   RIGHTS AND RESPONSIBILITIES OF THE OWNER

Construction contracts typically reserve several rights to the owner. Depending on the type of contract and its specific wording, the owner may be authorized to award other contracts in connection with the work, to require contract bonds from the contractor, to approve the surety proposed, to retain a specified portion of the contractor's periodic payments, to make changes in the work, to carry out portions of the work in case of contractor default or neglect, to withhold payments from the contractor for adequate reason, and to terminate the contract for cause. The rights of the owner to inspect the work as it proceeds, to direct the contractor to expedite the work, to use completed portions of the project before contract termination, and to make payment deductions for uncompleted or faulty work are common construction contract provisions. The owner is normally free to perform construction or operations at the site with its own forces or with separate contractors.

Similarly, the contract between owner and contractor imposes certain responsibilities on the owner. For example, construction contracts make the owner responsible for furnishing property surveys that describe and locate the project, granting the contractor timely access to the work site, securing and paying for necessary easements, providing certain insurance, and making periodic payments to the contractor. The owner is required to make extra payments and grant extensions of time in the event of certain eventualities provided for in the contract. When there are two or more prime contractors on a project, the owner may have a duty to coordinate them and synchronize their field operations.

It is important to note that the owner cannot intrude on the direction and control of the contractor's operations. By the terms of the usual construction contract, the contractor is known at law as an "independent contractor." Even though the owner enjoys certain rights with respect to the conduct of the work, it cannot issue direct instructions as to method or procedure, unreasonably interfere with construction operations, or otherwise unduly assume the functions of directing and controlling the work. By so doing, the owner can relieve the contractor from many of the latter's rightful legal and contractual responsibilities. If the owner oversteps its rights, it may not only assume responsibility for the accomplished work but may also become liable for negligent acts committed by the contractor in the course of construction operations.

Under the laws of most states, the owner is responsible to the contractor for the adequacy of the design. If the drawings and specifications are defective or insufficient, the contractor can recover the resulting costs and extensions of time from the owner. Because the contractor must construct in accordance with the contract documents, the owner implicitly warrants their accuracy and sufficiency for the intended purpose. The contractor is not responsible for the results and is not liable for the consequences of design defects. The responsibility of the owner for the design is not overcome by the usual bidding clauses requiring bidders to visit the site, to check the drawings and specifications, and to inform themselves concerning the requirements of the work. It must be noted, however, that the owner's responsibility for the design does not protect the contractor who does not fully comply with the contract. The implied warranty that the contract drawings are adequate for the purpose intended is limited to those cases where the contractor has followed the drawings and specifications.

6.31 DUTIES AND AUTHORITIES OF THE ARCHITECT-ENGINEER

Other than cases in which the same contracting party performs both design and construction or in which the owner has an in-house design capability, the architect-engineer firm is not a party to the construction contract, and no contractual relationship exists between it and the contractor. The architect-engineer is a third party that derives its rights and authorities over the construction process from the general contract between the owner and the prime contractor. When the owner uses private design professionals, the effect on the construction contract is to substitute the architect-engineer for the owner in many important respects. However, the jurisdiction of the architect-engineer to make determinations and render decisions is limited to, and circumscribed by, the terms of the construction contract. The architect-engineer represents the owner in the administration of the contract and acts for the owner during the day-to-day construction operations. The architect-engineer advises and consults with the owner, and communications between owner and contractor are made through the architect-engineer. Paragraph 4.2 in Appendix C contains typical provisions regarding the architect-engineer's role in contract administration.

Construction contracts often impose many duties and bestow considerable authority on the design firm. All construction operations are conducted under its surveillance, and it generally oversees the progress of the work. It has a direct responsibility to see that the workmanship and materials fulfill the requirements of the drawings and specifications. To ensure this fulfillment, the architect-engineer can exercise the right of job inspection and approval of materials. Also involved may be the privilege of approving the contractor's general program of field procedure and even the construction equipment the contractor proposes to use. Should the work be lagging behind schedule, the architect-engineer may reasonably instruct the contractor to speed up its activities. Many contracts bestow upon the architect-engineer the right to stop the project, or any part thereof, to correct unsatisfactory work or conditions. The specific language in this regard is widely variable, ranging from a broad to a limited right.

The foregoing paragraph does not mean that the architect-engineer assumes responsibility for the contractor's methods merely because it retains the privilege of approval. The rights of the architect-engineer are essentially concerned with verifying that the contractor is proceeding in accordance with the contract documents. It should be pointed out, however, that the architect-engineer cannot unreasonably interfere with the conduct of the work or dictate the contractor's procedures. Here again, if the direction and control of the construction are taken out of the hands of the contractor, the construction firm is effectively relieved of many of its legal and contractual obligations.

The contract documents often authorize the architect-engineer to interpret the requirements of the contract and to judge the acceptability of the work performed by the contractor. The architect-engineer occupies a position of trust and confidence in this regard and must act in good faith throughout the construction process. When this party acts as the official interpreter of contract conditions and as a judge of the contractor's performance, it must exercise impartial judgment and cannot favor either the owner or the contractor. Some contracts provide that the architect-engineer's decisions are not final and that the owner and contractor can exercise their rights to arbitration or the courts, providing the architect-engineer has rendered a first-level decision. Other contracts state that the architect-engineer's decisions are final and binding on both parties with regard to artistic effect only. Still other contracts give the architect-engineer broad authority to make final decisions concerning the quality and fitness of the work and to interpret the contract documents.

Where the architect-engineer is given final and binding authority, this is necessarily restricted to questions of fact. In the absence of fraud, bad faith, and gross mistake, the decision of the architect-engineer can be considered as final, provided the subject matter falls within the proper scope of authority given to the design firm by the construction contract. With respect to disputed questions of law, however, the architect-engineer has no jurisdiction. It cannot deny the right of a contractor to due process of law, and the contractor has the right to submit a dispute concerning a legal aspect of the contract to arbitration or to the courts, as may be provided by the contract. Matters pertaining to time of completion, extensions of time, liquidated damages, and claims for extra payment usually involve points of law.

Architect-engineers can be liable to the owner and third parties for damages resulting from their negligence or lack of care. Where the contractor has followed the plans and specifications prepared by the architect-engineer and these documents prove to be defective or insufficient, the architect-engineer will be responsible for any loss or damage resulting solely from the design defect. This rule is subject to the absence of any negligence on the part of the contractor, or any express warranty by the contractor that the drawings and specifications were sufficient or free from defects.

6.32 RIGHTS AND RESPONSIBILITIES OF THE CONTRACTOR

As one might expect from a document prepared especially for the owner, the contractor has few rights and many obligations under the contract. Its major responsibility, of course, is to construct the project in conformance with the contract documents. Despite all the troubles, delays, adversities, accidents, and mischances that may occur, the contractor is expected to "deliver the goods" and finish the work in the prescribed manner. Although some difficulties justify allowing more construction time, only severe contingencies, such as impossibility of performance, can serve to relieve the contractor from its obligations under the contract. In an era of rapidly rising costs of fuel, materials, and labor, the contract may provide a degree of financial protection by the inclusion of an escalation clause. Such a clause transfers some of the risk of inflation from the contractor to the owner by providing that the contract amount will be adjusted according to how designated key cost factors change during the contract period.

The contractor is expected to give its personal attention to the conduct of the work, and a responsible company representative must be on the job site at all times during working hours. The contractor is required to conform to laws and ordinances concerning job safety, licensing, employment of labor, sanitation, insurance, traffic and pedestrian control, explosives, and other aspects of the work. Many contracts now include tough rules designed to decrease air and noise pollution on construction projects and imposing regulations and restrictions concerning trash disposal, pile driving, riveting, demolition, fences, dust, and housekeeping.

The contractor is required to follow the drawings and specifications and cannot be held to guarantee that the completed project will be free of defects or will accomplish the purpose intended. However, if the contractor deviates from the design documents without consent of the owner, the contractor does so at its own peril and assumes the risk of the variation. Even if the construction contract does not expressly warrant that the contractor will comply with the design documents, such a warranty is implied. The contractor has the right to rely on the accuracy and adequacy of the contract documents. Unless the contract expressly requires such action, the contractor has no duty to review or verify the design. The contractor is responsible for, and warrants, all materials and workmanship, whether put in place by its own forces or by those of its subcontractors. Contracts typically provide that the contractor will be responsible for the preservation of the work until its final acceptance, although engineering projects commonly include an "act of God" exclusion (see Section 8.8). Even though the contractor has no direct responsibility for the adequacy of the plans and specifications, it can incur a contingent liability for proceeding with faulty work whose defects are apparent. The contractor must call all obvious design discrepancies to the attention of the owner or architect-engineer. Otherwise, the contractor proceeds at its own risk, because a party who builds according to patently defective drawings and specifications can be held responsible for any resulting liability. Should an instance occur in which the contractor is directed to do something it feels is not proper and not in accordance with good construction practice, it should protect itself by writing a letter of protest to the owner and the architect-engineer stating its position before proceeding with the matter in dispute.

Insurance coverage is an important contractual responsibility of the contractor, both as to the type of insurance and the policy limits. The contractor is required to provide insurance not only for its own direct and contingent liability, but frequently also for the owner's protection. The contractor is expected to exercise every reasonable safeguard for the protection of persons and property in, on, and adjacent to the construction site. Where the owner requires contract surety bonds, the contractor is responsible for furnishing them in the proper amounts before field operations commence.

The most important contractor rights concern progress payments, recourse should the owner fail to make payments, termination of the contract for sufficient cause, right to extra payment and extensions of time, and appeals from decisions of the owner or architect-engineer. Subject to contractual requirements and limitations therein, the contractor is free to subcontract portions of the contract, purchase materials where it chooses, and process the work in any way and in any order it pleases.

6.33 INDEMNIFICATION

The vulnerability of architect-engineers to third-party suits is discussed earlier in this book (see Section 4.7). In like manner, owners are subject to claims by third parties to the construction contract for damages arising out of construction operations. The rule stating that one who employs an independent contractor is not liable for the negligence or misconduct of the contractor or its employees, is subject to many exceptions. To protect the owner and the architect-engineer from third-person liability, many construction contracts now include indemnification or "hold-harmless" clauses. Indemnification requires that one party compensate a second party for a loss that the second party would otherwise bear.

Hold-harmless clauses typically require that the contractor indemnify and hold harmless the owner and the architect-engineer, and their agents and employees, from all loss or expense by reason of liability imposed by law on them for damages because of bodily injury or damage to property arising out of, or as a consequence of, the work. The present tendency is for parties injured by construction operations to sue practically everyone associated with the construction. In addition, the courts show a growing inclination to ascribe liability, in whole or in part, for such damages to the owner or architect-engineer. For example, owners and architect-engineers have been made responsible for damages caused by construction accidents where adequate safety measures were not taken on the job.

Hold-harmless clauses are not uniform in their wording. However, these clauses can be grouped into three main categories:

Limited-form Indemnification. The limited form holds the owner and the architect-engineer harmless against claims caused solely by the negligence of the prime contractor or a subcontractor.

Intermediate-form Indemnification. The intermediate form includes not only claims caused by the contractor or its subcontractors but also those in which the owner and/or architect-engineer may be jointly responsible, regardless of degree of fault. This is the most prevalent type of indemnification clause, an example of which can be seen in Paragraph 3.18 in Appendix C.

Broad-form Indemnification. The broad form indemnifies the owner and/or architect-engineer against all losses, even when the party indemnified is solely responsible for the loss.

When broad-form hold-harmless clauses are used, contractual liability coverage (see Section 8.25) is expensive and some provisions may not be insurable. The effect is compounded when the general contractor logically requires the same indemnification from its subcontractors. A majority of states have enacted anti-indemnification statutes prohibiting "an indemnitor from indemnifying an indemnitee for the indemnitee's sole negligence" or "an indemnitor from indemnifying an indemnitee for the indemnitee's sole or concurrent negligence." Such statutes make broad-form and intermediate-form indemnities difficult to enforce in these states. These statutes have largely been enacted to preserve the integrity of insurance regulations, to support a legislative policy for preventing the transfer of liability from a party at fault to one without fault, and to promote construction workplace safety. For example, subcontractors are frequently required by contract to sign a broad-form indemnity in support of the general contractor. Should the general contractor fail to maintain safe working conditions on-site, and this negligence results in the injury or death of one of the subcontractor's employees through no fault of the subcontractor, the indemnity would protect the general contractor from any legal action taken by the employee or any other party and assign full liability to the subcontractor. Such clauses may therefore serve to upset the balance of responsibility in contracts and protect the guilty while burdening the innocent.

Indemnification clauses are also routinely included in subcontracts, under which the subcontractor agrees to hold harmless the general contractor, the owner, and perhaps the architect-engineer. In the presence of such language, these indemnified parties are protected against liability that may devolve to them out of, or as a consequence of, the performance of the subcontract. Just as the general contract can relieve the owner and/or architect-engineer from liability even when they are at fault, a subcontract indemnity clause can be worded so that it acts to relieve the general contractor, owner, and architect-engineer from liability even when they may be at fault.

It is a principle of long standing that a party may protect itself from losses resulting from liability for negligence by means of an agreement to indemnify. However, the rule is restricted to the extent that contractual indemnity provisions will not be construed to indemnify a party against its own negligence unless such intention is expressed in unequivocal terms. When such a clause is to be included in subcontracts, it must explicitly provide that the subcontractor indemnifies the designated parties whether or not they, solely, may have caused the liability.

6.34   TERMINATION OF THE CONTRACT

Construction contracts may be ended in a variety of ways, full and satisfactory performance by both parties being the usual manner of contract termination. However, there are other means of ending a contract that are of interest and importance to the construction industry.

Material breach of contract by either party is occasionally a cause for contract termination. Failing to make prescribed payments to the contractor or causing unreasonable delay of the project are probably the most common breaches by the owner. Other possible breaches can be failure or delay in performance, failure to coordinate work, failure to provide access to the project site, or financial insolvency of owner. In such circumstances, the contractor is entitled to damages caused by the owner's failure to carry out its responsibilities under the contract.

Default and failure to perform under the contract are the usual breaches committed by contractors. Nonperformance, faulty performance, failure to show reasonable progress, failure to meet financial obligations, persistent disregard of applicable laws or instructions of the owner or architect-engineer—these are examples of material default by contractors that convey to the owner a right to terminate the contract. When the owner terminates the contract because of a breach by the contractor, the owner is entitled to take possession of all job materials and to make reasonable arrangements for completion of the work. It is worthy of note that when failure to complete a project within the contract time is the breach involved, the owner probably will not be awarded liquidated damages if it terminates the contract and does not allow the contractor to finish the job late when the latter is making a genuine effort to complete the work. The record is not completely clear on this point, however.

A third way in which a contract can be terminated is by mutual agreement of both parties. This is not common in the construction industry, although there have been such instances. For example, it sometimes happens that the contractor faces unanticipated contingencies, such as financial reverses, labor troubles, or loss of key personnel, that make proper performance under the contract a matter of considerable doubt. Under such circumstances, the owner and the contractor may agree to rescind the contract and to engage another contractor. When little or no work has been done, termination of the contract by mutual consent can sometimes be attractive to both parties.

Construction contracts, particularly those publicly financed, normally provide that the owner can terminate the contract at any time it may be in its best interests to do so, by giving the contractor written notice to this effect. When such a provision is present, the contractor has agreed to termination at the prerogative of the owner. However, in such an event, the contractor is entitled to payment for all work done up to that time, including a reasonable profit, plus such expenses as may be incurred in canceling subcontracts and material orders and in demobilizing the work. If the owner terminates the contract capriciously, it may become liable for the full amount of the contractor's anticipated profit plus costs.

A contract may be rescinded because of the impossibility of performance under circumstances beyond the control of either party. Impossibility of performance is not the case when one party finds it an economic burden to continue. To demonstrate grounds for termination of a contract, it must indeed be impossible or impracticable to proceed. For instance, unexpected site conditions may be found that make it impossible to carry out the construction described by the contract. Operation of law may render the contract impossible to fulfill. However, the doctrine of impossibility does not demand a showing of actual or literal impossibility. Impossibility of performance has been applied to cases in which, even if performance were technically possible, the costs of performance would have been so disproportionate to that cost reasonably contemplated by the contracting parties as to make the contract totally impracticable in a commercial sense. The deciding factor was that an unanticipated circumstance made performance of the contract vitally different from that which was reasonably to be expected.

Other unusual but possible causes of contract termination can occur, such as destruction of subject matter. For example, there may be termination of a contract to remodel an existing building if that building is destroyed by fire before the renovation can be started. Frustration of purpose also can occur. For example, suppose a contract was let to construct a boat pier on a lake, but before the pier could be built, the lake was permanently drained.

6.35   THE NOTICE TO PROCEED

The beginning of contract time is often established by a written notice to proceed, which the owner dispatches to the contractor. The date of its receipt is normally considered to mark the formal start of operations. This notice, in the form of a letter advising the contractor that it may enter the site immediately, directs the contractor to start work. If by some chance the owner does not yet have clear title to the property, the notice will relieve the contractor of liability caused by its construction operations. Contracts usually require the contractor to commence operations within some specified period, such as 10 days after receipt of the notice to proceed.

6.36   SUBCONTRACTS

A subcontract is an agreement between a prime contractor and a subcontractor under which the subcontractor agrees to perform a certain specialized part of the work. A subcontract does not establish any contractual relationship between the subcontractor and the owner, and neither is liable to the other in contract. A subcontract binds only the parties to the agreement: the prime contractor and the subcontractor. Nevertheless, construction contracts frequently stipulate that the owner or the architect-engineer must approve all subcontractors. The bidding documents may require that a list of subcontractors be submitted with the proposal or that a similar list be submitted by the low-bidding prime contractor for approval by the owner after the proposal has been accepted (see Subparagraph 5.2.1 of Appendix C). When the owner is given the right of approval, the general contractor is not relieved of any of its responsibilities by the owner's exercise of its prerogative.

If approval of subcontractors is required, such approval must be obtained before the general contractor enters into agreements with its subcontractors. There have been cases in which the owner refused to accept a subcontractor with which the prime contractor had already signed an agreement. It is possible in such a predicament that the general contractor must abrogate the existing subcontract and thereby render itself liable for damages to the subcontractor. Actually, the disapproval of a subcontractor is not a common occurrence. This is particularly true on public projects, where disapproval of a subcontractor may result in litigation and can be difficult to sustain by the public owner.

Although informal letters of proposal and acceptance may suffice for subcontracted work of small consequence, it is preferable that the prime contractor formalize all of its subcontracts with a written document that sets forth in detail the rights and responsibilities of each party to the contract. A well-prepared subcontract document can eliminate many potential disputes concerning the conduct of subcontracted work. The prime contractor may use a standard subcontract form, or it may develop its own special form to suit its particular requirements. The Associated General Contractors of America has prepared several subcontract forms for building construction, depending on who assumes the risk for owner payment and whether the contract is being written on privately or federally funded work. A sample of the short-form agreement between contractor and subcontractor in which the two share the risk of owner payment is presented in Appendix M. Subcontract forms must be prepared with extreme care and with the advice of an experienced lawyer.

A frequent area of disagreement between the general contractor and a subcontractor is the form of subcontract used and the terms and conditions that the form includes. It is standard practice that a formal written contract be used between general contractor and subcontractor. The subcontractor knows this when it tenders its bid to the general contractor. When a subcontractor has previously done work for a general contractor, the subcontractor tacitly assumes that the standard subcontract form of the general contractor will again be used. Consequently, the subcontractor's bid will normally include, without specific reference, the standard contract terms and conditions used by that general contractor. Problems arise when the subcontractor submits a bid to a general contractor for the first time or when the general contractor makes some major revisions to its company subcontract form. Here, the contract terms contemplated by the two parties can be quite different. A possible complication is that if the general contractor presents a subcontract containing significant conditions not included in the bidding documents or instructions, the subcontractor can consider it as a counteroffer and the bid of the subcontractor is not binding. To avoid such difficulties, many general contractors require that subcontract bids be made on a standard bid quotation form that binds the parties to the use of a particular subcontract.

Change orders to construction contracts often involve modifications to subcontracted work. When this is the case, the general contractor and the subcontractor must execute a suitable change order to the subcontract affected. The prime contractor normally uses a standard form for this purpose, similar to that shown in Figure 6.2.

6.37   SUBCONTRACT PROVISIONS

Subcontracts, in many respects, are similar in both form and content to the prime construction contract. Two parties contract for a specific job of work that is to be performed in accordance with a prescribed set of contract documents. On some projects, the owner reserves the right to approve the form of subcontract to be used by the general contractor. Normally, the general contractor will include in its subcontracts express provisions that the subcontractor is bound to the prime contractor to the same extent that the prime contractor is bound to the owner by the terms of the prime contract. In addition, the subcontract must provide that the subcontractor assumes all obligations to the prime contractor that the prime contractor assumes to the owner. An illustration of this requirement is afforded by Subparagraph 5.3.1 in Appendix C. With such wording, provisions of the general contract such as changes, changed conditions, prevailing wages, warranty period, compliance with applicable laws, approval of shop drawings, and responsibility for safety extend to the subcontractors. Other owners require the general contractor to include in its subcontracts certain provisions or designated clauses that appear in the general construction contract.

Project specifications normally include a description of those work items for which warranties and guarantees are required. A common area of dispute between the general contractor and its subcontractors is whether these warranty and guarantee periods begin when the specific work types are completed or when the entire project is finalized. A common procedure in this regard is for all subcontract agreements to clearly state that the effective beginning dates of warranty and guarantee periods will coincide with the completion and acceptance of the total project.

In addition to the provisions of the general construction contract, there are many relationships applicable to the conduct of the subcontracted work itself for which provision must be made. Many of these are in the nature of general conditions to the subcontract, which are made an integral part of the printed subcontract form. Clauses pertaining to temporary site facilities to be furnished, insurance and surety bonds to be provided by the subcontractor, arbitration of disputes, extensions of time, and indemnification of the general contractor by the subcontractor are typical instances. A detailed description of the work to be accomplished by the subcontractor must be included, which normally includes references to page numbers of drawings and sections of the specifications. A statement concerning the starting time and schedule of the subcontracted work, including any liquidated damages, is important. Subcontracts often contain a "no damage for delay" clause. This clause provides that in the case of delay caused by the general contractor, the subcontractor's sole remedy will be an extension of time to complete its work. Special conditions must often be included, such as security clearances for workers, job site storage, work to be accomplished after hours, and other requirements that arise from the peculiarities of a given project. Some subcontracts require that the subcontractor waive its right to file a mechanic's lien (see Section 9.34) against the project property site. However, some states have enacted statutes voiding such waivers as being against public policy.

Because the subcontractor has no contract with the owner, the subcontractor normally cannot sue the owner directly as a means of settling disputes or claims. Rather, the subcontractor must either initiate a claim in the name of the prime contractor or have the prime contractor prosecute the claim for the subcontractor. In addition, the subcontractor cannot submit a claim against the owner after the general contractor has finalized the project and released the owner by acceptance of final payment. Consequently, subcontracts commonly include a clause obligating the prime contractor to pursue any subcontractor claims against the owner or allowing the subcontractor to prosecute them in the name of the prime contractor. It should be noted that owners are generally precluded from suing a subcontractor for a breach of the subcontract, not only because these two parties are not in contract, but also because the law generally regards the owner as an incidental beneficiary to the subcontract and not as a third-party beneficiary.

Terms of payment to the subcontractor and retainage by the general contractor are, of course, especially significant. Payment to the subcontractor can be established on a lump-sum, unit-price, or cost-plus basis. A matter of continuing concern to all parties is payment to the subcontractor by the general contractor when, for some reason, the general contractor has not been paid by the owner. Different subcontracts make quite different provisions in this regard. All subcontracts stipulate that the general contractor shall pay its subcontractors promptly after payment is actually made by the owner, although some obligate the contractor to pay the subcontractors *only* after owner payment. Subparagraph 10.2 in Appendix M provides that payment to the subcontractor is conditioned on the prime contractor receiving payment from the owner.

Such contract provisions have raised many questions concerning the responsibility of the prime contractor to pay the subcontractor and have resulted in much litigation. Most courts now hold that the subcontract must be explicit concerning the assumption of risk by the subcontractor in regard to delay of payment by the owner, for the contractor to be authorized to withhold payment from the subcontractor. If such contractual wording is not present, the subcontractor must normally be paid within a reasonable time for completed and acceptable work. In addition, the courts note that delay in payment by the owner generally involves a dispute between the owner and the general contractor. If this dispute does not involve a particular subcontractor, the courts have ruled that the uninvolved subcontractor is entitled to payment within a reasonable time, even if the prime contractor has not been paid. It is to be noted that some subcontract forms do not contain a contingent payment provision but do provide that subcontractor payments are contingent on payment certificates being issued by the architect-engineer.

Progress payments to subcontractors by the prime contractor are normally subject to the same retainage provisions that apply to payments made by the owner to the prime contractor. The subcontract instrument provides for such retainage.

A matter that requires close attention by the general contractor is the possible lack of coordination between provisions of the prime contract and those of the subcontract form used on a given project. Noncoordinated contract terms can lead to serious legal difficulties for the prime contractor. For example, this may happen when the dispute resolution procedures in the subcontract fail to correspond with such clauses in the prime contract.

6.38 PURCHASE ORDERS

A purchase order is a written document that defines and prescribes the conditions pertaining to a purchase of materials, supplies, equipment, machinery, and similar goods. When both the buyer (contractor) and the seller (vendor) sign a purchase agreement, it becomes a purchase contract between the two parties. Although there can be oral purchase agreements, such contracts are not enforceable in most states if the price of the goods is $500 or more. This regulation is in accordance with the statute of frauds section of the Uniform Commercial Code. This minimum has been amended to $5,000 in some states. Oral supply contracts, just as oral subcontracts, often lead to disagreements concerning the terms of the agreement and are to be avoided. Contracts for the sale of materials, like other contracts, contain conditions and terms applying to the purchase. Purchase order forms used in the construction industry normally include characteristic clauses of terms pertaining to the materials being ordered and their delivery.

Normally, the contractor, using its own standard printed form, prepares the purchase order. Any terms and conditions included are therefore directed primarily toward protecting the interests of the contractor and ensuring that the materials and their delivery conform to the requirements of the construction contract and the contractor. It is not unusual, however, for vendors to sell materials only on the basis of their own terms and conditions. Under such circumstances the contractor must sign a purchase order form prepared by the seller. By so doing, the contractor accepts all terms and conditions of the resulting purchase contract.

Purchase orders must be prepared with care and attention to detail. It is of great importance that they contain all of the agreements, promises, and provisions pertaining to the procurement at hand. For instance, purchase orders not only describe and specify the materials and quantities to be provided, but also define the conditions pertaining to taxes, freight, shipping and packaging, insurance, time of delivery, discounts, credit terms, liquidated damages, warranties, prepayment terms, and other such important matters. A properly prepared purchase order can be of great importance to the contractor in the event of improper shipment, loss, damage, late delivery, shortage, or other procurement problems. As it does in developing its subcontracts, the prime contractor often incorporates into its material purchase orders, by reference, the provisions of the prime contract, including the general conditions.

QUESTIONS

1. Explain the issue or risk assumption for contractor and owner when selecting negotiated contracting methods.

2. Lump-sum contracts are beneficial to owners because the total cost of a project can be determined in advance. Yet, owners are not always able to use this contracting method. Explain what obstacles might prevent its use by an owner.

3. Most competitively bid projects are awarded to the "lowest responsible bidder". Explain the meaning and importance of "responsible" in this context.

4. Why does construction equipment demand special attention on a cost reimbursable contract?

5. Describe the difference between a cost-plus-award-fee contract and a cost-incentive-contract.

6. A variety of contracting methods exist. At one end of the spectrum, risk is transferred to owners, at the other, risk is accepted by the contractor. Besides risk, what other issues should owners consider when selecting a contracting method?

7. What advantage does contract standardization hold for owners?

8. Once a progress payment for a given scope of work has been approved, can the owner file a claim against the contractor for faulty workmanship of this scope? Explain.

9. Why do many owners require retainage on contracts even though the contractor has provided the owner with a performance bond?

10. Substantial completion has varied in its meaning and interpretation over the years. In recent years, what major quality must be met to attain substantial completion?

11. If, during the course of construction, the cost to the owner of delayed completion of the project increases significantly beyond the estimate used in the liquidated damages clause in the contract, can the owner re-assess its claim against the contractor based on actual losses? Explain. If actual losses are significantly less than the liquidated damages in the contract, can the owner still enforce the clause?

12. Explain the special circumstances surrounding a constructive change?

13. How does constructive acceleration differ from the more standard form of owner directed acceleration?

14. Give a good example scenario of consequential damages associated with owner-caused delay.

15. Construction contractors are known at law as "independent contractors." What implications does this legal term have on owner rights during project construction?

16. What consequences may occur if the subcontract form delivered by the general contractor to the subcontractor contains significant conditions not included in the original bidding documents?

Chapter 7

Contract Surety Bonds

7.1 SURETY BONDS IN CONSTRUCTION

The use of contract surety bonds in the construction industry is but one of many applications of surety bonds in business and commerce. In law, a surety is a party that assumes liability for the debt, default, or failure in duty of another. A surety bond is the contract that describes the conditions and obligations pertaining to such an agreement. A surety bond is not an insurance policy. Rather, it serves as an extension of credit by the surety, not in the sense of a financial loan, but as an endorsement. Insurance protects a party from the risk of loss, whereas suretyship guarantees the performance of a defined contractual duty.

By the terms of a construction contract bond the surety agrees to indemnify the owner, called the obligee, against any default or failure in duty of the prime contractor, called the principal. Contract bonds are three-party agreements that guarantee that the work will be completed in accordance with the contract documents and that all construction costs will be paid. When the contractor properly discharges its obligations and after any warranty period covered by the bond expires, the bond agreement is discharged and no longer has any force or effect. Regardless of the reason, if the prime contractor fails to fulfill its contractual obligations, the surety must assume the obligations of the contractor and see that the contract is completed, paying all costs up to the face amount of the bond. It should be noted that dual obligee bonds are sometimes used. Such bonds protect both the owner and the lending institution that advances construction funds to the owner. The need for surety bonds is a consequence of the fact that construction is a very risky business. Past experience suggests that approximately one-half of all construction contracting firms now in operation will not be in business seven years from now.

A contract bond form is a simple document that makes no attempt to describe in detail the specific liabilities of the surety. The bond guarantees the construction contract in all of its provisions, and the obligations of the bond are identical with the provisions of the contract. The bond does not impose on the surety any obligations that are separate from, or in addition to, those assumed by the general contractor. By the same token, by being bonded, the contractor assumes no additional obligation to the owner that it has not already assumed by contract or by operation of law. The bond can be invoked by the owner only if the contractor is in breach of contract. The statutes of frauds of the various states require that contracts of suretyship must be in writing to be enforceable. For this reason, contract bonds are always written documents.

Contract bonds are not required of the contractor on all construction projects. Although such bonds are required by law on public jobs, not all private construction is bonded. This is especially true when invitational bidding or negotiation is used. With regard to contract bonds, the owner must keep one important fact in mind. Contract bonds cannot be viewed as a satisfactory substitute for an able, honest, adequately financed contractor. Unfortunately, owners must sometimes look to the surety for job completion and make the best of a bad situation.

7.2   FORMS OF CONTRACT BONDS

By the terms of the construction contract with the owner, the prime contractor accepts two principal responsibilities: to perform the objective of the contract and to pay all costs associated with the work. Both of these obligations can be included within a single bond instrument, and combined performance and payment bonds are written on a few projects, almost all of which are privately financed. However, it is usual practice for construction contracts to require two separate contract bonds, one bond covering performance of the contract and the other guaranteeing payment for labor and materials. The separate forms bear the endorsement of the American Institute of Architects, the Associated General Contractors, and others, and virtually all statutory bonds on public work are on separate forms.

Under the single type of bond, there is a potential conflict of interest between the owner and the persons furnishing labor and materials. Because the owner has priority, the face value of the bond can be entirely consumed in satisfying its claims. Thus, in many instances the single bond form has afforded little or no protection for material dealers, workers, and subcontractors. In addition, there have been serious problems with the priority of rights of the persons covered. The double form of bond covers separately the interest of the owner and that of subcontractors, material suppliers, and workers. The premium cost of the bond protection is not increased by furnishing two separate bonds rather than one.

7.3   PERFORMANCE BONDS

The owner is entitled to receive what it contracted for or the equivalent. A performance bond acts primarily for the protection of the owner. It guarantees that the contract will be performed and that the owner will receive its structure, built in substantial accordance with the terms of the contract. A performance bond incorporates, by reference, the terms of the contract, and the responsibility of the contractor is the measure of the surety's obligation. On the contractor's default, the burden of contract performance is that of the surety. A performance bond customarily covers any warranty period that may be required by the contract, the usual bond premium including one year of such coverage. All performance bonds, as well as payment bonds, discussed in the following section, have a face value that acts as an upper limit of the expense the surety will incur in finishing the contract, should that action become necessary. This face value is expressed as a fixed sum of money, the amount of which is usually derived from some percentage, 100 percent for instance, of the total contract price.

Figure 7.1 reproduces AIA *Document A312,* "Performance Bond," of the American Institute of Architects and illustrates the provisions of a typical performance bond used on private projects. The standard form used by the federal government provides that the bond will apply to all contract modifications and that the life of the bond must include all extensions of time and any guarantee period required.

THE AMERICAN INSTITUTE OF ARCHITECTS

AIA Document A312

Performance Bond

Any singular reference to Contractor, Surety, Owner or other party shall be considered plural where applicable.

CONTRACTOR (Name and Address):   SURETY (Name and Principal Place of Business):

OWNER (Name and Address):

CONSTRUCTION CONTRACT
  Date:
  Amount:
  Description (Name and Location):

BOND
  Date (Not earlier than Construction Contract Date):
  Amount:
  Modifications to this Bond: ☐ None   ☐ See Page 3

CONTRACTOR AS PRINCIPAL   SURETY
Company: (Corporate Seal)   Company: (Corporate Seal)

Signature: _____   Signature: _____
Name and Title: Name and Title:

(Any additional signatures appear on page 3)

(FOR INFORMATION ONLY—Name, Address and Telephone)
AGENT or BROKER: OWNER'S REPRESENTATIVE (Architect, Engineer or
                                         other party):

AIA DOCUMENT A312 · PERFORMANCE BOND AND PAYMENT BOND · DECEMBER 1984 ED. · AIA ®
THE AMERICAN INSTITUTE OF ARCHITECTS, 1735 NEW YORK AVE., N.W., WASHINGTON, D.C. 20006   **A312-1984   1**
THIRD PRINTING · MARCH 1987

Figure 7.1   Performance bond.

1 The Contractor and the Surety, jointly and severally, bind themselves, their heirs, executors, administrators, successors and assigns to the Owner for the performance of the Construction Contract, which is incorporated herein by reference.

2 If the Contractor performs the Construction Contract, the Surety and the Contractor shall have no obligation under this Bond, except to participate in conferences as provided in Subparagraph 3.1.

3 If there is no Owner Default, the Surety's obligation under this Bond shall arise after:

3.1 The Owner has notified the Contractor and the Surety at its address described in Paragraph 10 below that the Owner is considering declaring a Contractor Default and has requested and attempted to arrange a conference with the Contractor and the Surety to be held not later than fifteen days after receipt of such notice to discuss methods of performing the Construction Contract. If the Owner, the Contractor and the Surety agree, the Contractor shall be allowed a reasonable time to perform the Construction Contract, but such an agreement shall not waive the Owner's right, if any, subsequently to declare a Contractor Default; and

3.2 The Owner has declared a Contractor Default and formally terminated the Contractor's right to complete the contract. Such Contractor Default shall not be declared earlier than twenty days after the Contractor and the Surety have received notice as provided in Subparagraph 3.1; and

3.3 The Owner has agreed to pay the Balance of the Contract Price to the Surety in accordance with the terms of the Construction Contract or to a contractor selected to perform the Construction Contract in accordance with the terms of the contract with the Owner.

4 When the Owner has satisfied the conditions of Paragraph 3, the Surety shall promptly and at the Surety's expense take one of the following actions:

4.1 Arrange for the Contractor, with consent of the Owner, to perform and complete the Construction Contract; or

4.2 Undertake to perform and complete the Construction Contract itself, through its agents or through independent contractors; or

4.3 Obtain bids or negotiated proposals from qualified contractors acceptable to the Owner for a contract for performance and completion of the Construction Contract, arrange for a contract to be prepared for execution by the Owner and the contractor selected with the Owner's concurrence, to be secured with performance and payment bonds executed by a qualified surety equivalent to the bonds issued on the Construction Contract, and pay to the Owner the amount of damages as described in Paragraph 6 in excess of the Balance of the Contract Price incurred by the Owner resulting from the Contractor's default; or

4.4 Waive its right to perform and complete, arrange for completion, or obtain a new contractor and with reasonable promptness under the circumstances:

.1 After investigation, determine the amount for

which it may be liable to the Owner and, as soon as practicable after the amount is determined, tender payment therefor to the Owner; or

.2 Deny liability in whole or in part and notify the Owner citing reasons therefor.

5 If the Surety does not proceed as provided in Paragraph 4 with reasonable promptness, the Surety shall be deemed to be in default on this Bond fifteen days after receipt of an additional written notice from the Owner to the Surety demanding that the Surety perform its obligations under this Bond, and the Owner shall be entitled to enforce any remedy available to the Owner. If the Surety proceeds as provided in Subparagraph 4.4, and the Owner refuses the payment tendered or the Surety has denied liability, in whole or in part, without further notice the Owner shall be entitled to enforce any remedy available to the Owner.

6 After the Owner has terminated the Contractor's right to complete the Construction Contract, and if the Surety elects to act under Subparagraph 4.1, 4.2, or 4.3 above, then the responsibilities of the Surety to the Owner shall not be greater than those of the Contractor under the Construction Contract, and the responsibilities of the Owner to the Surety shall not be greater than those of the Owner under the Construction Contract. To the limit of the amount of this Bond, but subject to commitment by the Owner of the Balance of the Contract Price to mitigation of costs and damages on the Construction Contract, the Surety is obligated without duplication for:

6.1 The responsibilities of the Contractor for correction of defective work and completion of the Construction Contract;

6.2 Additional legal, design professional and delay costs resulting from the Contractor's Default, and resulting from the actions or failure to act of the Surety under Paragraph 4; and

6.3 Liquidated damages, or if no liquidated damages are specified in the Construction Contract, actual damages caused by delayed performance or non-performance of the Contractor.

7 The Surety shall not be liable to the Owner or others for obligations of the Contractor that are unrelated to the Construction Contract, and the Balance of the Contract Price shall not be reduced or set off on account of any such unrelated obligations. No right of action shall accrue on this Bond to any person or entity other than the Owner or its heirs, executors, administrators or successors.

8 The Surety hereby waives notice of any change, including changes of time, to the Construction Contract or to related subcontracts, purchase orders and other obligations.

9 Any proceeding, legal or equitable, under this Bond may be instituted in any court of competent jurisdiction in the location in which the work or part of the work is located and shall be instituted within two years after Contractor Default or within two years after the Contractor ceased working or within two years after the Surety refuses or fails to perform its obligations under this Bond, whichever occurs first. If the provisions of this Paragraph are void or prohibited by law, the minimum period of limitation avail-

AIA DOCUMENT A312 • PERFORMANCE BOND AND PAYMENT BOND • DECEMBER 1984 ED. • AIA ®
THE AMERICAN INSTITUTE OF ARCHITECTS, 1735 NEW YORK AVE., N.W., WASHINGTON, D.C. 20006
THIRD PRINTING • MARCH 1987

A312-1984   2

Figure 7.1 *continued*

176

able to sureties as a defense in the jurisdiction of the suit shall be applicable.

10 Notice to the Surety, the Owner or the Contractor shall be mailed or delivered to the address shown on the signature page.

11 When this Bond has been furnished to comply with a statutory or other legal requirement in the location where the construction was to be performed, any provision in this Bond conflicting with said statutory or legal requirement shall be deemed deleted herefrom and provisions conforming to such statutory or other legal requirement shall be deemed incorporated herein. The intent is that this Bond shall be construed as a statutory bond and not as a common law bond.

12 DEFINITIONS

12.1 Balance of the Contract Price: The total amount payable by the Owner to the Contractor under the Construction Contract after all proper adjustments have been made, including allowance to the Con-

tractor of any amounts received or to be received by the Owner in settlement of insurance or other claims for damages to which the Contractor is entitled, reduced by all valid and proper payments made to or on behalf of the Contractor under the Construction Contract.

12.2 Construction Contract: The agreement between the Owner and the Contractor identified on the signature page, including all Contract Documents and changes thereto.

12.3 Contractor Default: Failure of the Contractor, which has neither been remedied nor waived, to perform or otherwise to comply with the terms of the Construction Contract.

12.4 Owner Default: Failure of the Owner, which has neither been remedied nor waived, to pay the Contractor as required by the Construction Contract or to perform and complete or comply with the other terms thereof.

MODIFICATIONS TO THIS BOND ARE AS FOLLOWS:

(Space is provided below for additional signatures of added parties, other than those appearing on the cover page.)

CONTRACTOR AS PRINCIPAL		SURETY	
Company:	(Corporate Seal)	Company:	(Corporate Seal)
Signature: _____		Signature: _____	
Name and Title:		Name and Title:	
Address:		Address:	

Figure 7.1 *continued*

7.4 PAYMENT BONDS

A payment bond acts primarily for the protection of third parties to the contract and guarantees payment for labor and materials used or supplied in the performance of the construction. The private owner and the financial institutions providing construction and permanent financing are thereby protected against liens (see Section 9.34) that can be filed on the project by subcontractors, material vendors, workers, and other unpaid parties to the work. Although a private owner ordinarily decides for itself whether to require contract bonds from the contractor, it is worthy of note that there are a few state statutes that require payment bonds on privately financed work. Figure 7.2 presents AIA *Document A312,* "Labor and Material Payment Bond," of the American Institute of Architects. This bond, typical of common-law payment bonds employed by corporate sureties for privately financed projects, provides the following:

1. The claimant must have had a direct contract with either the general contractor or a subcontractor.

2. Labor and materials include water, gas, power, light, heat, oil, gasoline, telephone, and rental of equipment directly applicable to the contract.

3. Written notice must be given by the claimant, other than one having a direct contract with the general contractor, to any two of these: general contractor, owner, or surety, within 90 days after claimant performed its last work or furnished the last of the materials.

4. The owner is exempted from any liabilities in connection with such claims.

5. Claims must be filed in the appropriate court.

6. No claims shall be commenced after the expiration of one year following the date on which the general contractor stopped work, barring a statute to the contrary.

Payment bonds exclude from their coverage parties who are remote from the general contractor. It should be noted that the bond form in Figure 7.2 includes only those claimants who have a direct contract with the prime contractor or one of the subcontractors. Whether specific instances of labor, material suppliers, or sub-subcontractors on public projects are protected by statutory payment bonds depends on the language of the related statute. Because liens cannot be filed against public property, the payment bond may well be the only protection vendors, workers, and subcontractors have for payment on public projects.

It is worthy of note that a subcontractor's supplier may have the right to recover payment under a general contractor's payment bond, even though the general contractor has made payment in full to the subcontractor. In many areas, the prime contractor is, under its bond, subject to double payment obligations if a subcontractor has not paid its suppliers. When the general contractor does not require payment bonds from its subcontractors, it must use special procedures when paying these parties (see Section 9.29).

7.5 STATUTORY AND COMMON-LAW BONDS

Payment bonds are either statutory or common-law, and there are important differences between the two. The bonding requirements on public projects are prescribed by law, and a statutory bond, at least by reference, contains the provisions of the statute that

THE AMERICAN INSTITUTE OF ARCHITECTS

AIA Document A312

Payment Bond

Any singular reference to Contractor, Surety, Owner or other party shall be considered plural where applicable.

CONTRACTOR (Name and Address):   SURETY (Name and Principal Place of Business):

OWNER (Name and Address):

CONSTRUCTION CONTRACT
   Date:
   Amount:
   Description (Name and Location):

BOND
   Date (Not earlier than Construction Contract Date):
   Amount:
   Modifications to this Bond: ☐ None   ☐ See Page 6

CONTRACTOR AS PRINCIPAL   SURETY
Company: (Corporate Seal)   Company: (Corporate Seal)

Signature: _____   Signature: _____
Name and Title: Name and Title:

(Any additional signatures appear on page 6)

(FOR INFORMATION ONLY—Name, Address and Telephone)
AGENT or BROKER: OWNER'S REPRESENTATIVE (Architect, Engineer or
                                           other party):

AIA DOCUMENT A312 • PERFORMANCE BOND AND PAYMENT BOND • DECEMBER 1984 ED. • AIA ®
THE AMERICAN INSTITUTE OF ARCHITECTS, 1735 NEW YORK AVE., N.W., WASHINGTON, D.C. 20006   **A312-1984 4**
THIRD PRINTING • MARCH 1987

Figure 7.2 Labor and material payment bond.

1 The Contractor and the Surety, jointly and severally, bind themselves, their heirs, executors, administrators, successors and assigns to the Owner to pay for labor, materials and equipment furnished for use in the performance of the Construction Contract, which is incorporated herein by reference.

2 With respect to the Owner, this obligation shall be null and void if the Contractor:

2.1 Promptly makes payment, directly or indirectly, for all sums due Claimants, and

2.2 Defends, indemnifies and holds harmless the Owner from claims, demands, liens or suits by any person or entity whose claim, demand, lien or suit is for the payment for labor, materials or equipment furnished for use in the performance of the Construction Contract, provided the Owner has promptly notified the Contractor and the Surety (at the address described in Paragraph 12) of any claims, demands, liens or suits and tendered defense of such claims, demands, liens or suits to the Contractor and the Surety, and provided there is no Owner Default.

3 With respect to Claimants, this obligation shall be null and void if the Contractor promptly makes payment, directly or indirectly, for all sums due.

4 The Surety shall have no obligation to Claimants under this Bond until:

4.1 Claimants who are employed by or have a direct contract with the Contractor have given notice to the Surety (at the address described in Paragraph 12) and sent a copy, or notice thereof, to the Owner, stating that a claim is being made under this Bond and, with substantial accuracy, the amount of the claim.

4.2 Claimants who do not have a direct contract with the Contractor:

.1 Have furnished written notice to the Contractor and sent a copy, or notice thereof, to the Owner, within 90 days after having last performed labor or last furnished materials or equipment included in the claim stating, with substantial accuracy, the amount of the claim and the name of the party to whom the materials were furnished or supplied or for whom the labor was done or performed; and

.2 Have either received a rejection in whole or in part from the Contractor, or not received within 30 days of furnishing the above notice any communication from the Contractor by which the Contractor has indicated the claim will be paid directly or indirectly; and

.3 Not having been paid within the above 30 days, have sent a written notice to the Surety (at the address described in Paragraph 12) and sent a copy, or notice thereof, to the Owner, stating that a claim is being made under this Bond and enclosing a copy of the previous written notice furnished to the Contractor.

5 If a notice required by Paragraph 4 is given by the Owner to the Contractor or to the Surety, that is sufficient compliance.

6 When the Claimant has satisfied the conditions of Paragraph 4, the Surety shall promptly and at the Surety's expense take the following actions:

6.1 Send an answer to the Claimant, with a copy to the Owner, within 45 days after receipt of the claim, stating the amounts that are undisputed and the basis for challenging any amounts that are disputed.

6.2 Pay or arrange for payment of any undisputed amounts.

7 The Surety's total obligation shall not exceed the amount of this Bond, and the amount of this Bond shall be credited for any payments made in good faith by the Surety.

8 Amounts owed by the Owner to the Contractor under the Construction Contract shall be used for the performance of the Construction Contract and to satisfy claims, if any, under any Construction Performance Bond. By the Contractor furnishing and the Owner accepting this Bond, they agree that all funds earned by the Contractor in the performance of the Construction Contract are dedicated to satisfy obligations of the Contractor and the Surety under this Bond, subject to the Owner's priority to use the funds for the completion of the work.

9 The Surety shall not be liable to the Owner, Claimants or others for obligations of the Contractor that are unrelated to the Construction Contract. The Owner shall not be liable for payment of any costs or expenses of any Claimant under this Bond, and shall have under this Bond no obligations to make payments to, give notices on behalf of, or otherwise have obligations to Claimants under this Bond.

10 The Surety hereby waives notice of any change, including changes of time, to the Construction Contract or to related subcontracts, purchase orders and other obligations.

11 No suit or action shall be commenced by a Claimant under this Bond other than in a court of competent jurisdiction in the location in which the work or part of the work is located or after the expiration of one year from the date (1) on which the Claimant gave the notice required by Subparagraph 4.1 or Clause 4.2.3, or (2) on which the last labor or service was performed by anyone or the last materials or equipment were furnished by anyone under the Construction Contract, whichever of (1) or (2) first occurs. If the provisions of this Paragraph are void or prohibited by law, the minimum period of limitation available to sureties as a defense in the jurisdiction of the suit shall be applicable.

12 Notice to the Surety, the Owner or the Contractor shall be mailed or delivered to the address shown on the signature page. Actual receipt of notice by Surety, the Owner or the Contractor, however accomplished, shall be sufficient compliance as of the date received at the address shown on the signature page.

13 When this Bond has been furnished to comply with a statutory or other legal requirement in the location where the construction was to be performed, any provision in this Bond conflicting with said statutory or legal requirement shall be deemed deleted herefrom and provisions conforming to such statutory or other legal requirement shall be deemed incorporated herein. The intent is that this

AIA DOCUMENT A312 • PERFORMANCE BOND AND PAYMENT BOND • DECEMBER 1984 ED. • AIA®
THE AMERICAN INSTITUTE OF ARCHITECTS, 1735 NEW YORK AVE., N.W., WASHINGTON, D.C. 20006
THIRD PRINTING • MARCH 1987

A312-1984   5

Figure 7.2 *continued*

Bond shall be construed as a statutory bond and not as a common law bond.

14 Upon request by any person or entity appearing to be a potential beneficiary of this Bond, the Contractor shall promptly furnish a copy of this Bond or shall permit a copy to be made.

15 DEFINITIONS

15.1 Claimant: An individual or entity having a direct contract with the Contractor or with a subcontractor of the Contractor to furnish labor, materials or equipment for use in the performance of the Contract. The intent of this Bond shall be to include without limitation in the terms "labor, materials or equipment" that part of water, gas, power, light, heat, oil, gasoline, telephone service or rental equipment used in the Construction Contract, architectural and engineering services required for performance of the work of the Contractor and the Contractor's subcontractors, and all other items for which a mechanic's lien may be asserted in the jurisdiction where the labor, materials or equipment were furnished.

15.2 Construction Contract: The agreement between the Owner and the Contractor identified on the signature page, including all Contract Documents and changes thereto.

15.3 Owner Default: Failure of the Owner, which has neither been remedied nor waived, to pay the Contractor as required by the Construction Contract or to perform and complete or comply with the other terms thereof.

MODIFICATIONS TO THIS BOND ARE AS FOLLOWS:

(Space is provided below for additional signatures of added parties, other than those appearing on the cover page.)

CONTRACTOR AS PRINCIPAL		SURETY	
Company:	(Corporate Seal)	Company:	(Corporate Seal)

Signature: _____   Signature: _____
Name and Title: Name and Title:
Address: Address:

AIA DOCUMENT A312 · PERFORMANCE BOND AND PAYMENT BOND · DECEMBER 1984 ED. · AIA ®
THE AMERICAN INSTITUTE OF ARCHITECTS, 1735 NEW YORK AVE., N.W., WASHINGTON, D.C. 20006
THIRD PRINTING • MARCH 1987

A312-1984 6

Figure 7.2 *continued*

make the bond a requirement. Private projects use common-law bonds whose coverage and functionings stand entirely on the provisions contained in the bond instrument itself.

The distinction between statutory and common-law bonds is an important matter to the parties for whose protection the payment bond is written. On public projects the action of claimants to obtain protection under the bond must be in accordance with the applicable statute. This applies whether or not the statutory requirements are contained in the language of the bond itself. When payment bonds are required by statute on public projects, the right to recover on them is limited by the conditions of the statute to the same extent as though the provisions of the statute were fully incorporated into the bond instrument. If a claimant fails to comply with the statutory requirements applying to enforcement of rights under the bond, it will not be permitted to recover.

A common-law bond is used when there are no statutory requirements. It is a contract that stands by its own language and that is enforced in the usual manner for contracts. In this case a claimant must proceed as described on the face of the bond.

On private projects the use of the standard common-law payment bond used by the various surety companies is the usual practice. This form is standardized nationally and is approved by professional groups such as the American Institute of Architects. When statutory bonds are required, most public agencies having substantial building programs have developed standard bond forms that conform with the applicable statute. Because the laws pertaining to bonding requirements differ somewhat from one jurisdiction to another, bond forms for public contracts are not standardized nationally. The federal government and many states and municipalities use their own bond forms.

The standard payment bond used by the federal government is written to comply with the provisions of the Miller Act (see Section 7.6). This statutory bond protects laborers, material vendors, and subcontractors who perform work or supply materials for the project, although the extent of this protection depends on how far removed the unpaid party is from the general contractor.

7.6 THE MILLER ACT

The Miller Act prescribes the requirements of performance and of payment bonds used in conjunction with federal construction projects. Enacted in 1935 and subsequently amended, this statute provides that on all federal construction contracts of more than $100,000, the contractor shall furnish a performance bond for the protection of the United States and a payment bond for the protection of persons supplying labor and materials in the prosecution of the work.

The Act provides that the performance bond must be written in such amount that, in the opinion of the contracting officer, the interests of the United States are adequately protected. Under issued regulations of the comptroller general, federal agencies customarily require a performance bond in the amount of 100 percent of the contract amount.

The Construction Industry Payment Protection Act of 1999 makes a number of changes to the Miller Act. Payment bond amounts are now equal to the amount of the performance bond, affording contractors and subcontractors more protection.

To ensure that bonds are supplied to the U.S. government by a financially responsible surety, the U.S. Department of the Treasury maintains a list of surety companies acceptable for federal bonds. This list is updated regularly and can be found at the following website: *www.fms.treas.gov/c570/c570.html.*

The Miller Act gives workers, subcontractors, and material vendors who deal directly with the prime contractor the right to sue on the prime contractor's payment bond if payment is not received in full within 90 days after the date on which the last of the labor was done or the last of the materials were furnished. The law further provides that any person having a direct contractual relationship with a subcontractor, but no contractual relationship with the prime contractor, will have a right of action on the prime contractor's payment bond, provided the claimant gives written notice to the prime contractor within a 90-day period. There is no requirement for notice in the case of a party who deals directly with the prime contractor.

Under the Miller Act, first-tier subcontractors and material suppliers and second-tier subcontractors and material suppliers are protected, but the payment protection of this federal statute extends no further. In addition, second-tier parties must deal with first-tier subcontractors. A subcontractor under the Miller Act has been held to mean one that performs for the prime contractor a specific part of the labor or material requirement of the project. Thus, the term *subcontractor* has been construed by the courts to include a party who supplies custom-made materials but does not install them. An unpaid bond claimant cannot sue on the payment bond until 90 days after the last of the labor was performed or the last of the materials was delivered. However, suit must be brought within one year after last work or delivery. Suits authorized by law are brought in the name of the United States, for the use of the party suing, in the appropriate district court. Suit is brought and prosecuted by the unpaid party's own attorney.

Since passage of the Miller Act, all 50 states have followed with the enactment of their own "Little Miller Acts." These statutes apply to projects financed by the states and establish contract bond requirements similar to those imposed by the Miller Act. These state bonding statutes do differ, however, with regard to which parties can recover under the payment bond.

7.7 CLAIMS FOR PAYMENT

An unpaid party to the construction process who looks to the prime contractor's payment bond for compensation must process its claim in accordance with the terms of the bond instrument on private projects or the governing statute on public work. The processing of such a claim is a technical procedure requiring the services of an attorney. However, attorney's fees cannot normally be collected by a Miller Act claimant. To perfect a claim, the claimant must be generally aware of the notice and time requirements involved. On a private project, the unpaid party can obtain a copy of the payment bond from the owner, surety, or architect-engineer. Public owners or surety companies can provide information concerning statutory claim requirements.

The party seeking compensation may be required to give written notice of the outstanding debt. However, the notice requirement varies with the type of owner involved and the form of payment bond used. On most private projects, no notice is required from an unpaid party that is in contract with the principal (general contractor) on the payment bond. Notice to any two of the bond principal, owner, or surety is needed if the unpaid party has no contract with the principal. Under the Miller Act, which prescribes bonding requirements on federal government construction, one in privity of contract with the prime contractor has no notice requirement. Otherwise, notice must be given to the prime contractor.

Where written notice of an unpaid debt is required, this must be given within a specified period of time. Under the Miller Act and the usual bond forms used on private work, a claimant that was not in contract with the prime contractor must give notice within 90 days after furnishing the last work or material for which the claim is made. An unpaid party must bring suit on the payment bond within the time limits specified by the bonding statute on public works or by the bond instrument itself on private jobs.

7.8   CONTRACT CHANGES

Construction contracts typically give the owner the right to make changes in the work. Because the contract comes before the bond and the bond guarantees the contract, it is commonly assumed that extension of the contract bond to include changes in the contract is provided for automatically. However, the construction contract is between the owner and the contractor, and in the event of changes, the surety is put in the position of being obligated by the terms of a contract to which it is not a party. Common law does not allow two contracting parties to bind a third without its consent. For this reason it is always advisable for the owner to obtain the prior written consent of the surety to any change or modification of the contract. For instance, it is advisable that an executed consent-of-surety form be attached to each project change order. Approval of a contract change in writing is needed because of the aforementioned application of statutes of frauds to contracts of suretyship.

One aspect of the matter of contract modification is that it is possible for the surety to be exonerated from its original obligation, regardless of any provision in the construction contract that changes to the contract do not release the surety under any bond previously provided. Each such case is decided on the facts of the particular situation, and the court record is not entirely clear on this matter. For instance, changes in the contract may constitute a material departure from the original contract or substantially change the manner of payment, the way the contract is to be performed, or the time for performance so as to make the contract significantly more difficult or costly to complete. When this has occurred, the courts have relieved the surety from its obligation on the basis that the contract was not the one the surety originally underwrote and agreed to be bound by. For relief in such cases it is necessary to show that the contract change was made without the consent of the surety, and proof is required that the change in the contract substantially increased the risks of the surety and of the contractor. As a general rule of thumb, changes in the contract that increase the amount of the contract by more than 10 percent are considered to be significant.

A second consideration in obtaining written consent of the surety for a contract change is that the surety is not obligated to provide a bond for any additional or modified work unless the surety has expressly waived the right of notice. In this regard the performance bond in Figure 7.1 provides that the surety waives notice of any alteration or extension of time made by the owner. The payment bond in Figure 7.2, however, contains no such waiver clause. Some bond forms used by the federal government stipulate that the surety waives notice of all extensions or modifications to the contract. However, a number of these now also include a provision that limits the value of changes that can be made in a bonded contract without the consent of the surety to 10 percent of the amount of the contract.

7.9 BOND PREMIUMS

For the purpose of computing contract bond premiums, construction contracts are divided into four classifications: A-I, A, B, and miscellaneous. Figure 7.3 contains a representative listing of these contract classifications.

The individual states must approve advisory rates submitted by surety companies and associations to state insurance commissions. These rates can vary somewhat with the surety company, and there are many deviated rates that are used. Such rates are adjusted up or down periodically to reflect loss experience. It is up to the contractor to determine from its surety firm the bond rates applicable to the specific project being priced. The premium rates shown in Figure 7.4 are for illustrative purposes only and serve to demonstrate representative contract bond rates and how premium costs for a given construction project are determined. The rates in Figure 7.4 are those used in conjunction with lump-sum and unit-price contracts where performance or performance and payment bonds are required. These rates include a warranty period of one year if required by the contract. If a longer period is required, an additional premium charge is made. The rates shown in Figure 7.4 apply to the total contract amount and include all subcontracted work. An example showing the application of these rates is given in Section 5.41.

When the work can reasonably be assigned to more than one classification, the classification requiring the higher premium rate controls. Neither the classification of the contract nor the premium rate can be altered by segmenting the work or by parties other than the contractor furnishing the materials. All separate contracts are assigned the same classification as the general contract. Bond premiums for cost-plus contracts are subject to some variation from the rates of fixed-sum contracts as contained in Figure 7.4. Premium rates for cost-plus contracts with guaranteed maximum amounts are the same as for fixed-sum contracts. Cost-plus-percentage-of-cost contracts have a premium charge of 60 percent of the fixed-sum rates, excluding the contractor's fee from the final contract amount. Cost-plus-fixed-fee contracts are assessed a premium of 30 percent of the fixed-sum rates, excluding the contractor's fee.

The premiums for contract bonds are payable in advance, and the bonds are customarily delivered to the owner at the time the contract is signed. This premium payment is subject to later adjustment based on the ultimate contract amount, reflecting final work quantities on unit-price contracts and including all change orders and contract adjustments.

7.10 THE SURETY

Essentially all contractors utilize the services of national corporate surety companies whose specialties are the writing of bid bonds and contract bonds for contractors. These firms are subject to public regulation in the same manner as are insurance companies. They operate under charters and file their schedules of premium rates with designated public authorities. Because the true worth of the bond is no greater than the surety's ability to pay, the owner reserves the right to approve the surety company and the form of bond. The federal government requires that all corporate sureties proposed for use on government projects be approved by the U.S. Treasury Department. This list of surety companies approved for federal projects can be a valuable reference for private owners when faced with the approval of a contractor-proposed surety. The Treasury list can be found

Classification			
A-1	A	B	Miscellaneous Contracts
Ash conveyors	Airfield grading	Air-conditioning	Ash removal
Boiler repairs	Airfield surfacing	Airport buildings	Bridges
Conveyors	Airfield runways	Aqueducts	Buildings,
Doors	Aluminum siding	Buildings	prefabricated
Fire alarms	Athletic fields	Canals	Culverts
Fire escapes	Beacons	Dams	Demolition
Flagpoles	Ceilings, metal or	Dikes	Draying
Floors, wood and	acoustical tile	Docks	Dredging
composition	Coal storage	Electrical work	Garbage removal
Gas tanks	Curb and gutter	Excavation	Grade eliminations
Generators	Curtain walls	Foundations	Hauling
Guardrails	Ducts, underground	Gas piping	Highways
Ironwork,	Elevators	Grain elevators	Maintenance
ornamental	Floodlights	Heating systems	Overpasses
Kitchen equipment	Glazing	Incinerators	Roads
Lock gates	Greenhouses	Jetties	Shoring
Metal windows	Machinery	Locks	Street paving
Parking meters	Millwork	Masonry	Structural iron and
Pipelines, oil or gas	Murals	Piers	steel
Police alarms	Parking areas	Piling	Test borings
Radio towers	Parks	Pipelines, water	Timber cutting
Refrigerating plants	Piping, high-pressure	Plants, power	Underpasses
Scaffolding	Playgrounds	Plants, sewage-	Viaducts
Sidewalks	Riverbank protection	disposal	
Signal systems,	Road medians	Plastering	
railroad	Roofing	Plumbing	
Signs	Ski lifts	Seawalls	
Stack rooms	Sprinkler systems	Sewers	
Standpipes	Stone, furnishing	Stone setting	
Street lighting	Storage tanks, metal	Subways	
Tanks, gas	Tennis courts	Tunnels	
Thermostat equip-	Waterproofing	Waterworks	
ment	Wind tunnels	Wells	
Towers, water		Wharves	
Track laying			
Traffic control			
systems			
Weatherstripping			
Window cleaning			

Figure 7.3 Construction contract classification for contract surety bond premium rate.

186

Contract Price		Premium Rate per $1,000 of Contract Price for First 12 Months* (Subject to change without notice)							
		Class A-1		Class A		Class B		Misc.	
First	$100,000	$	9.40	$	15.00	$	25.00	$	19.50
Next	$400,000	$	7.20	$	10.00	$	15.00	$	15.00
Next	$2,000,000	$	6.00	$	7.00	$	10.00	$	12.00
Next	$2,500,000	$	5.00	$	5.50	$	7.50	$	10.00
Next	$2,500,000	$	4.50	$	5.00	$	7.00	$	9.00
Over	$7,500,000	$	4.00	$	6.75	$	6.50	$	8.00

*For construction time in excess of 12 months or 366 calendar days, compute basic premium at above rates and increase this computation by 1% per month for each month over 12 months (disregarding a fraction of a month).

Figure 7.4 Premium rates for performance or performance and payment bonds. Lump-sum or unit-price contracts.

at *www.fms.treas.gov/c570/c570.html*. Another source of information in this regard is *A. M. Best Insurance Reports*, which provide financial ratings for insurance and surety companies.

Occasionally, on private work, the contract documents require that the contract bonds be obtained from a particular surety company. This requirement of specifying a particular surety usually means that the contractor must do business with an unfamiliar company. In this situation, obtaining the bond may turn out to be a lengthy and laborious process, requiring the submission of financial reports, a list of jobs in progress, an experience record, and other data of voluminous proportions that may be needed to establish the contractor's record and financial standing. The professional associations of contractors and architect-engineers oppose the practice of having the surety designated by the owner and support a policy of leaving the contractor free to obtain surety bonds from a company of its choice. However, the private owner is at liberty to follow whatever practice it wishes in this regard, and the contractor must either comply with its instructions or not bid. In this regard, there are a few states with statutes that prevent an owner from requiring a contractor to obtain contract bonds from a designated surety.

On large contracts, a single surety may seek protection for itself by enlisting other sureties to underwrite a portion of the contract. This is very much like reinsurance. The original surety remains completely responsible for guaranteeing the proper performance of the contract. If the bond is invoked, it is up to the original surety company to get its underwriting sureties to stand behind it in the completion of the contract.

In some instances the owner requires that the contract bonds be provided by cosureties, which means that two or more sureties divide the total contract obligation among themselves. On very large contracts this practice spreads the risk over the participating cosureties and correspondingly reduces the magnitude of the risk to which any one of them is exposed. This procedure also affords the owner a measurable degree of protection against possible financial default by a single surety. Occasionally it is necessary to have cosureties on large federal contracts because of limits established by the U.S. Treasury Department on the maximum amounts of single contract bonds that a given surety is authorized to execute.

7.11 INDEMNITY OF SURETY

A contract bond is not insurance for the general contractor and does not function for its protection. Under the bond, the surety indemnifies the owner against default by the contractor. However, the contractor in turn must indemnify the surety against any claim that may be brought against the surety because of the contractor's failure to perform in the prescribed manner. Legal fees incurred by the surety because of claims under the bond are also recoverable from the contractor. Before the surety will provide bond service, the contractor must first sign a formal application form or General Indemnity Agreement. This form is lengthy and contains a great deal of fine print. The net result, however, is that the contractor agrees to indemnify the surety and hold it harmless from expenses of every nature that the surety may sustain by reason of the invocation of the bond.

When the application is signed by an individual contractor or a partnership, each principal is obligated to the entire extent of his personal fortune. If a corporation makes application, only the corporation's assets are pledged. However, the corporate officers and company shareholders normally submit their personal contracts of indemnity to the surety in order to increase the firm's bonding capacity. Another means of providing needed financial capacity to an otherwise qualified firm is to obtain the personal guarantee or indemnity of a third-party financial backer. This is a case in which a person or entity having substantial financial resources contributes credit to the construction firm in return for a share of the profits.

Many contractors believe that once the surety has provided a bid bond, it is then obligated to issue payment and performance bonds for the same project if the contractor submits the successful bid. However, many indemnity agreements stipulate that the surety has no such duty. The agreement may provide that the surety can refuse to execute such bonds without incurring any liability and without affecting the mutual obligations of the parties with respect to other contract bonds issued under the agreement.

7.12 INVESTIGATION BY SURETY

Before a surety will furnish an unknown contracting firm with a bid bond or contract bond, a thorough program of investigation is carried out to establish the past record and current commitments of the company. To establish a bond relationship between a contractor and a surety is a time-consuming, costly, and laborious process for the contractor. The experience, character, reputation, financial standing, equipment, integrity, personal habits, and professional ability of the firm's owners and key personnel are carefully examined. A track record of satisfactorily completed projects, owner satisfaction, and prompt payment of financial obligations is essential. The surety checks to see that the contractor is well managed, has a history of meeting its financial obligations promptly, is reliable, deals fairly, and performs its work in a timely fashion. An attempt is made to identify the key employees, evaluate their experience, and determine whether the company has adequate estimating, construction, and administrative experience to accomplish the proposed construction work. The surety also checks the adequacy of the contractor's financial assets, construction equipment, and physical facilities. Company financial statements, both for the current period and for some years past, are subject to study and analysis. The firm's bank credit is verified, together with its relations with its sources of credit and supply. Information must be given about the company's cost management and accounting systems. In short, a surety will issue bonds to a contractor only when the traditional "three C's" requirement is met: character, capacity, and capital. The information outlined here assists the underwriter in evaluating the ability of the contractor to complete work undertaken in accordance with contract terms and to meet the resulting financial obligations.

The providing of bonding services to a contractor is a highly individual matter. An exception to this general rule, however, can be found in the surety bond guarantee program that has been used by the Small Business Administration to assist qualified small businesses and minority-owned businesses to obtain construction bonds that they could not otherwise obtain. Under this program, the U.S. government guarantees to repay the participating sureties 90 percent of any loss caused by the default of a covered contractor on bonds up to $100,000 and 80 percent on bonds above that amount. The Small Business Administration (SBA) will guarantee bonds only for firms with three-year average annual revenues under $6 million and will bond projects worth up to $2 million. This program is called SBA Plan A.

Once a contractor has firmly established relations with a bonding company, the contractor's bonding capacity becomes reasonably well established, and future investigations by the surety underwriter are concerned with keeping the contractor's records current and investigating the individual bond requests as they are submitted. If the contractor's workload is well below its limit and contracts of the usual variety are proposed, a bond application is generally approved without delay. However, when the maximum bonding capacity is to be approached, or when an unusually large or completely new type of construction project is proposed, approval of the bond application may require a considerably longer period of time or may not be forthcoming at all.

When the contractor makes application for a bond for a new project, it will find that the surety is interested in many aspects of the work that is proposed. The following are the usual and most important subjects of investigation:

1. The essential characteristics of the project under consideration, including its size, type, and nature. Evaluation of the hazards of construction cannot proceed until the surety is apprised of the work. Included here are the identity of the owner and its ability to pay for the construction as it proceeds. The contractor must have adequate equipment, expertise, experience, and resources to do work of the type and size proposed.

2. The total amount of uncompleted work the contractor presently has on hand, of both the bonded and the unbonded variety. This must, of necessity, include work that has not yet been awarded. The obvious point of concern here is to prevent the contractor from becoming overextended with regard to working capital, equipment, and organization.

3. The adequacy of working capital and the availability of credit. The contractor can assist its own cause by keeping the surety fully informed as to its activities and supplied with up-to-date financial reports. This phase of investigation can often protect the contractor against taking on a project that is too big for it to handle.

4. The amount of money the contractor "left on the table," that is, the spread between the low bid and the next highest bid. Competitive conditions in the construction

industry are such that a spread of more than 5 or 6 percent between the two lowest bidders can be the cause of some concern. The surety wishes to ensure that the contractor's estimating and bidding procedures are sound.

5. The largest contract amount of similar work the contractor has successfully completed in the past. Inexperience in a new field of construction has caused many contractor failures. The surety would like the contractor to stay with the kind of work in which it is most experienced. If the contractor wishes to change to another type, the surety will urge that the first steps be small ones until the contractor acquires the necessary experience. If the contractor is not properly equipped for the new work, it must be demonstrated to the surety how the equipment problems will be solved.

6. Terms of the contract and bonds to be required, details of how payment will be made to the contractor, retainage, time for completion, liquidated damages, and the nature of job warranties—these all influence the surety's appraisal of the contractor's ability to do the work.

7. The amount of work subcontracted and the qualifications of the subcontractors. The surety's concern here is that the prospective subcontractors possess the necessary organization, experience, and financial resources to carry out their portions of the work.

After each bonded project is completed, the surety sends to the owner a request for a final report on the contractor's performance. The owner is asked to submit to the surety a statement concerning the contractor's handling of the job, changes that were made in the work, and the final total contract amount. This figure is used as a basis for any terminal adjustment of the bond premium.

7.13   BONDING CAPACITY

A useful concept widely used by the construction industry is that of "bonding capacity" or "bonding line." These terms have no precise definition but refer to the maximum value of uncompleted work the surety will allow the contractor to have on hand at any one time. A contractor's bonding capacity is a function of its net worth and cash liquidity and can vary according to the volume of work on hand, accumulated retainage on current jobs, type of work involved, time durations of outstanding contracts, and other considerations. Bonding capacity, or the amount of surety credit that will be extended to a given contractor, is commonly obtained as a multiple of the contractor's net quick worth, perhaps augmented by the amount of money the contractor could realize by liquidating its fixed assets such as real estate and equipment. Net quick worth is obtained as quick assets minus current liabilities (see Section 9.10). Quick assets are those that are immediately convertible to cash. The multiple applied to net quick worth can vary substantially with the individual contractor and the field of construction involved. On building construction, where 75 to 80 percent of the contract is normally sublet, the factor may be 10 to 20 or more. With heavy construction, where few subcontracts are involved, large amounts of expensive equipment are required, and the work is often in isolated areas, the multiple is likely to be somewhat smaller. New contractors will have a smaller value than older and more experienced firms. When a surety grants a line of credit to the contractor, it may require that no one contract shall exceed a certain percentage of the total work on hand.

The difference between a contractor's bonding capacity and its current total of uncompleted work, both bonded and unbonded, is a measure of the additional work for which the surety will provide a bond. There are, of course, other factors involved, such as the type of new work being considered and the size of the project. In computing the current value of a contractor's uncompleted work, the surety may decide to deduct at least part of the value of the subcontracts that are bonded.

7.14   THE SURETY AGENT

The local representative of a surety company is the surety agent. This is the person with whom the contractor must deal directly in all matters concerning bid and contract bonds. There are many advantages to be gained by the contractor in selecting a thoroughly qualified and experienced agent who has bona fide surety company affiliations that enable him to act on behalf of the surety company he represents and to provide prompt bonding service. The surety agent is a very important member of the construction circle, who can contribute significantly to the success or failure of a contractor's business. The agent is a trained observer of the construction industry who has a detached point of view and whose advice is therefore particularly valuable to the contractor.

Most bonding agents are sincere and astute representatives of large corporate sureties whose businesses are solidly based on long experience and competent service. Given the discussion in the foregoing sections, it is easy to understand how the contractor may sometimes get the impression that the surety representative is unduly meddling in its affairs or is overly limiting its volume of work. In all fairness, however, the contractor should feel fortunate that the surety is interested and alert enough to be of assistance in helping avoid the many pitfalls associated with the management of a construction firm. The contractor must realize that both are working toward the common goal of a prosperous and successful contracting business. Some surety agents are, of course, more conservative than others. It is up to the contractor to select a bonding representative who is responsive to its needs within the limits of responsible and competent service.

The bid bonds and contract bonds that are provided to the contractor seldom, if ever, originate directly from the home office of the corporate surety. Rather, these documents are prepared by the local agent. In order to verify that the local representative is indeed an agent of the surety and is authorized to execute bond instruments that are binding on the surety company, it is customary that each construction surety bond include an appropriate power-of-attorney form, which is attached to the bond itself.

7.15   DEFAULT BY THE CONTRACTOR

Should the contractor default, the surety is required to perform in accordance with the terms of the bond. On contractor default, the surety may assume charge of the project and complete the contract or allow the owner to contract with another contractor. If the surety completes the contract, it is responsible for the total cost of completing the work, less the unpaid balance of the contract, even if the net cost exceeds the face amount of the bond. In the second case, the surety makes available to the owner funds sufficient to pay the cost of construction, less the balance of the contract price, but not to exceed the face amount of the bond. How the surety company elects to complete the contract is usually a matter for it to decide. There are several alternative courses of action, but a common one is for the original contractor to finish the job with the financial help of the surety and under the eye of a construction consultant hired by the surety. The surety may decide, however, to put the remaining work out for competitive bids and select a new contractor to take over, or a new contractor may be engaged on a cost-plus basis. The new contractor may contract with the surety or the owner. It should be noted here that upon default by the contractor, the surety can disclaim liability under the bond on the grounds that the owner acted improperly in terminating the contractor.

Contract bonds can offer the contractor a genuine advantage when its ability to proceed has been temporarily curtailed by legal or financial difficulties. If the financial condition of the contracting firm is basically sound, the surety may choose to help it get back on its feet and in business again. It may elect to advance the contractor credit in sufficient amount for the contractor to proceed with its work. If claims have tied up the contractor's capital, the surety may furnish bonds to discharge these claims, thereby enabling the contractor to proceed on its own. Such arrangements are made privately and occur more often than might be thought.

When the contractor defaults and the surety undertakes to complete the work, the surety becomes entitled to all of the remedies the owner has against the contractor under the contract. In addition, the surety is entitled to receive from the owner the balance of the contract price, which is defined in Figure 7.1 as "the total amount payable by Owner to Contractor under the Contract and any amendments thereto, less the amount properly paid by Owner to Contractor." Provisions of the governing statute apply to this matter when a statutory bond is used. The surety may also press any claims against the owner that the defaulting prime contractor may have had. Regardless of the bond wording, the surety often finds that it must compete with other claimants for the retainage withheld by the owner.

The surety also has another right in its relationship with the owner. If the owner defaults, for example, by not making progress payments, the contractor is released from liability under the contract. Because the responsibility of the surety to the owner is the same as that of the contractor to the owner, any act of the owner that releases the contractor from its contractual obligation also releases the surety.

An interesting situation regarding the protection afforded by contract bonds can arise on those construction projects where the owner has awarded work on its project to more than one prime contractor (see Section 1.19). On such projects, each prime contractor is normally required to provide the owner with a performance and a payment bond. When one prime contractor fails to perform in accordance with its contract, there is a question as to whether the defaulting contractor's surety is liable to the other prime contractors for interference or disruption to their work. The majority rule in this regard is that the surety of the defaulting contractor is not liable to the other prime contractors for additional costs that they may have suffered by reason of the disruption. Most jurisdictions deny that one prime contractor is a third-party beneficiary of the bonds provided by another prime contractor.

7.16   CONTRACT BONDS AND TYPE OF CONTRACT

The preceding sections have presented the basic workings of contract surety bonds. The face values of these performance and payment bonds are dictated by the construction contract (private work) or by governing statutes (public work), with 100 percent of the contract price now being the maximum possible value for each bond type. The details of such bonding are quite variable with the type of contract involved and can become very complex in some instances. A particular aspect of the construction process should be noted at this point. When the owner contracts with an architect-engineer for project design, the owner receives a professional service and no contract surety bond is involved. The owner's interests are protected, not by a surety bond, but by the architect-engineer's professional liability insurance (see Section 8.31).

The general bonding procedures normally utilized with the usual construction contract types are as follows.

Single Prime Contractor, Fixed-Price Contract. With this traditional contract arrangement, the single prime contractor provides the owner with the required performance and payment bonds. In the case of a lump-sum contract, the face values of the bonds are determined from the original contract amount. The premium paid is adjusted at project completion to reflect any change in contract price caused by change orders. With a unit-price contract, the same procedure is followed, with the bond amounts being based on the estimated total project cost. Here again, the bond premiums are adjusted at contract completion to reflect the final contract price.

Multiple Prime Contractors, Fixed-Price Contract. In this instance, each prime contractor provides the owner with bonds covering that contractor's site work and whose face values are based on that individual contractor's contract amount. In the usual instance, each of the prime contractors is free to use the services of its customary surety company.

Cost-Plus-Fee Contract. Here the contractor provides the owner with contract bonds whose face values are determined from the initial target price established for the work to be done by that contractor. As in other cases, the bond premium amount is finalized when the final contract price of the project has been determined.

Joint Venture. In a joint venture agreement, each participating contractor normally undertakes a specific amount or percentage of the total contract. Accordingly, each coventurer bonds its proportionate share of the contract price. The normal procedure is for the usual sureties of the various contractors to jointly underwrite the project and sign bonds as cosureties. Each surety provides its contractor's share of the contract bonds provided to the owner. In many cases, a working fund of 5 to 10 percent of the contract amount must be established by the joint venturers before the bonds are executed.

Construction Management. The construction manager (CM), as an agent of the owner, manages the project from its inception through the construction process. The CM is paid a professional fee of perhaps 3 or 4 percent of the total cost of construction for its managerial services. In so doing, the CM carries out many duties ordinarily performed by the architect-engineer, especially during the construction process. These functions of the CM are in the area of professional services, where contract bonds are not used. The CM also performs many management duties usually done by the general contractor. In the usual construction management arrangement, the owner contracts directly with a series of prime trade contractors that do the actual construction. The CM coordinates, schedules, and controls the work of these contractors.

In the standard agency form of CM, the construction manager does not provide contract surety bonds to the owner. Rather, these bonds are provided to the owner by each of the prime trade contractors. Protection for the owner from CM negligence is provided by professional liability insurance carried by the CM. The CM can be bonded, but the protection so afforded to the owner is small because the bond face amount is based on the CM fee contract, which is only a small percentage of the value of the entire project.

Design-Construct. It should be remembered that when the design-construct concept is used, the owner receives professional design services and project construction services from a single contracting party. Where design services are a part of the contract requirements, a surety firm will issue a bond for the entire project but may require a bond surcharge, for example, 120 percent of the normal bond premium. A complication here is that the cost of construction is often not accurately known at the time the design-construct contract is consummated and the bonds are procured. As a result, the initial face amounts of the contract bonds may be based on an initial, approximate cost of construction. The amounts of the bonds and their cost are adjusted at a later date when the design is complete and the final construction cost is determined.

7.17   SUBCONTRACT BONDS

An earlier discussion pointed out that the general contractor is responsible for the job performance of the subcontractors. In addition, it can also be held liable if a subcontractor does not pay for materials, labor, or sub-subcontracts pertaining to the project. Consequently, the general contractor, to protect itself against subcontractor debt and default, may require certain or all of them to provide performance and payment bonds. In this instance, the subcontractor is the principal and the prime contractor is the obligee. General contractors have found, to their dismay, however, that not every subcontract bond form nor every surety company provides them with the kind of financial protection desired. It is for this reason that many prime contractors require the use of their own subcontract bond forms. Otherwise, the contractor must reserve the right to approve the surety and bond form proposed for use by a subcontractor.

A factor of major importance to be emphasized at this point is that the bond serves in no way to replace honesty, integrity, and competence on the part of the subcontractor. Bond or no bond, an inferior subcontractor means trouble. Although the bond will afford the general contractor some measure of protection against financial loss directly attributable to a particular subcontract, it does not and cannot cover expenses caused by work stoppages, delays, and disruptions of the overall construction program that inevitably result from subcontractor default.

7.18   MAINTENANCE BONDS

Construction contracts provide, expressly or by implication, that the general contractor must warrant the work against faulty workmanship and materials for some period of time after project completion. A warranty period of one year is widely used, this being the period of time during which the contractor is required to remedy construction defects for which it is responsible at no cost to the owner. In case the contractor does not honor this responsibility, the owner is normally protected by the usual performance bond, which continues in effect during the one-year warranty period. If the contractor does not remedy defects, then the surety is required to do so.

However, one year may be too short a period for some parts of the work, and the owner may require positive protection against certain defects in the work beyond the usual one-year period. For example, the contract may require the contractor to guarantee the paving for 5 years or the roofing for 10 to 40 years. Manufacturer's or applicator's warranties are also sometimes required to guarantee the performance characteristics of machinery, processes, or materials.

The true value of a warranty is quite obviously only as good as the integrity of the firm standing behind it. In the event of breach of warranty, the warrantor becomes liable to the owner for damages. However, whether damages can actually be collected may be an entirely different matter. To protect the owner from breach of warranty by the contractor, subcontractor, or supplier, a surety bond called a maintenance bond or warranty bond is sometimes required by the construction contract. A maintenance bond, provided by the responsible party to the owner, binds the surety, if necessary, to correct defined defects in the contracted work that appear within a specified time after job completion. A common example of a maintenance bond is a roof bond, the terms of which guarantee the roof against defects of workmanship and materials for some specified period of time. A bond of this type is executed by a corporate surety, with the manufacturer of the roofing materials named as the principal. The owner must be careful in this regard because some roofing "bonds" are not bonds at all, but merely the unsupported warranty of the roofing contractor or manufacturer without surety.

When a maintenance bond covering some aspect of the work is provided to the owner, the owner must first look to the principal if defects materialize. If the principal refuses or is unable to remedy the situation, the owner can invoke the bond, and the surety must make good the warranty as guaranteed by the bond.

7.19   CONTRACT BOND ALTERNATIVES

If a prime contractor is not able to obtain the customary contract bonds from a commercial surety company, alternatives are occasionally used. Under the Miller Act, such bonds on federal construction projects can also be provided by personal or individual sureties or they can be in the form of U.S. bonds or notes, certified or cashier's checks, bank drafts, postal service money orders, or cash. A personal surety can be a company principal or stockholder who pledges personal assets or any other individual or group that agrees to answer for the debt or default of the contracting party. Such personal sureties must provide evidence of having a specified net worth or collateral sufficient to back up their surety obligation. The bonding statutes of many states and some private owners also permit prime contractors, and some prime contractors permit subcontractors, to use such alternatives to the usual commercial surety bonds.

The Office of Federal Procurement Policy now has a rule in effect that allows contractors to use an irrevocable letter of credit in lieu of payment and performance bonds on federal construction contracts. This policy is designed to give more flexibility and contracting opportunities to smaller firms that may not be able to secure bonds. Such letters of credit are called guaranteed letters of credit, as distinguished from the usual commercial letters of credit used in sales transactions, and are backed by the contractor's funds, which are held by the bank. It should be noted that certain legal aspects of letters of credit differ greatly from the usual surety bonds.

With letters of credit, for example, the bank's obligation to pay is not conditioned on the mutual performance of construction contract terms by both the owner and contractor. As such, the bank is obligated to pay the owner if the owner presents the proper documentation, even if the owner has failed to perform as required by the construction contract. If the owner presents the documentation specified by the letter of credit to the issuing bank and if these documents conform to the letter of credit requirements, the bank will pay the owner. The bank will then seek reimbursement from its customer, the contractor, for the amount paid. However, a contractor who disputes whether it is in default can file suit to enjoin the bank from paying the owner on the letter of credit.

Letters of credit, unlike surety bonds, place the financial burden of a default dispute on the contractor rather than the owner. With a surety bond, if there is a dispute about whether a contractor is in default, the owner must wait until the dispute is resolved in its favor if it is to get its money from the surety. With a letter of credit, the owner gets its money immediately and the contractor must then sue the owner if it disputes the default. Contractors must be aware of the risks involved with letters of credit before using them.

7.20   MISCELLANEOUS SURETY BONDS

In addition to bid, performance, payment, subcontract, and maintenance bonds, the contractor occasionally finds it necessary or desirable to furnish or accept several other forms of surety bonds. The most important of these are the following:

Bonds to Release Retainage. Where a construction contract provides for the owner's retainage of a given percentage of the prime contractor's progress payments, the owner will sometimes make full payment to the contractor if the contractor will post a surety bond with the owner as obligee for the amount of the accumulated retainage involved.

Bonds to Discharge Liens or Claims. Persons who have not received payment for labor or materials supplied to a construction project are entitled to file a mechanic's lien against the property of a private owner or against moneys due and payable to the general contractor in the case of a public contract. Such actions can freeze capital needed by the contractor to conduct its operations. A surety bond in an amount fixed by an order of the court can be used to discharge a mechanic's lien. The bond functions as a financial guarantee to the owner and releases money that has been withheld from the contractor.

Bonds to Indemnify Owner against Liens. The contractor may be called upon to post a bond in advance that indemnifies the owner against any impairment of title or other damage that may be suffered by reason of liens or claims filed on its property. In this situation, a bond is required before any such liens are filed, rather than being used to discharge liens after they are filed in the way described in the previous paragraph.

Bonds to Protect Owners of Rented Equipment and Leased Property. During construction operations, the contractor may find it desirable to rent or lease equipment, parking lots, access roads, storage installations, and similar facilities. The owner of such property often requires the contractor to post a bond that guarantees proper maintenance and payment of rental charges, and indemnifies the owner against loss, damage, or excessive wear on the property.

Fidelity Bonds. The usual fidelity bond protects the employer against dishonest acts of an employee such as theft, forgery, or embezzlement. The employee covered by the bond is the principal, and the employer is the obligee. If the employer suffers a loss proven to have been caused by the employee, the surety will reimburse the employer up to the face value of the bond. This type of surety bond is more completely discussed in Section 8.18.

Judicial or Court Bonds. When the contractor is the plaintiff in a legal action, it sometimes is required to furnish security for court costs, possible judgments, and similar financial eventualities. Such security is often provided in the form of judicial or court bonds. This legal requirement is commonly applied when the contractor institutes court proceedings in a state or jurisdiction other than its own.

License Bond. Also known as a permit bond, this is a bond required by state law or municipal ordinance as a condition precedent to the granting of a contractor's license or permit. License bonds guarantee compliance with statutes or ordinances and make provision for payment to the obligee or members of the general public in the event that the licensee violates its legal or financial obligations. This topic is also discussed in Section 1.30.

Termite Bond. This form of bond is given by manufacturers or applicators of substances intended to prevent damage caused by termites for a specific period of time. The bond protects the property owner in the event that termite damage occurs after treatment is applied.

Subdivision Bond. This bond, given by a developer to a public body, guarantees construction of all necessary improvements and utilities.

Self-Insurers' Workers' Compensation Bond. This bond, given by a self-insured contractor to the state, guarantees payment of all statutory benefits to injured employees.

Union Wage Bond. This bond, given by a contractor to a union, guarantees that the contractor will pay union wages and will make proper payment of fringe benefits required by union contract.

QUESTIONS

1. What is the difference between a performance bond and a payment bond?

2. What is the difference between a statutory bond and a common law bond?

3. The Miller Act requires performance and payment bonds on all federal projects. How far down the subcontractor chain does the payment bond apply?

4. Why should the surety be notified of contract changes?

5. What are some of the factors that affect a contractor's bonding capacity?

6. What is the surety's obligation to the owner if the owner breaches the contract and the contractor stops work?

7. How are performance and payment bonds handled on joint venture construction projects?

8. If a performance bond covers the typical one year warranty period, why do some owners require a warranty or maintenance bond as well?

9. How does surety differ from insurance?

Chapter 8

Construction Insurance

8.1 RISK MANAGEMENT

Risk management may be defined as a comprehensive approach to handling exposures to loss. Any peril that can cause financial impairment to a business enterprise is subject to risk management. The following are five steps that a contracting firm can follow in applying risk management to its business.

1. Recognize and identify the various risks that apply to the construction process. These may arise as a consequence of contract provisions, the nature of the work, site conditions, or the operation of law.

2. Measure the degree of exposure presented by the risks identified. This involves establishing the frequency of losses and the potential severity of the losses that may occur.

3. Decide how to protect against those risks that have been identified. If a risk cannot be eliminated by an alternative procedure or by contractual transfer to another party, such as hold harmless/indemnity provisions or waivers of subrogation, the choice may be to purchase commercial insurance, to use deductibles, to self-insure, or to assume the risk.

4. Conduct a company-wide program of loss control and prevention. The project safety programs discussed in Chapter 15 constitute an important part of this procedure.

5. Monitor the results.

In devising the best risk-handling program, commercial insurance is the keystone to adequate financial protection. Insurance does not eliminate the risks involved in construction contracting, but it does shift most of the financial threat to a professional risk bearer. In addition to paying losses, insurance companies offer valuable services to the contractor in the areas of safety, loss prevention, educational and training programs, site inspections, and others. Where self-insurance is involved, it is not usually the sole risk-bearing element but works in conjunction with a commercial insurance program. Insurance is also involved with a form of risk assumption in the use of deductible amounts. The primary thrust of this chapter is a comprehensive discussion of standard commercial insurance coverages used in the construction industry.

8.2 CONSTRUCTION RISKS

By nature, construction work is hazardous, and accidents are frequent and often severe. The annual toll of deaths, bodily injuries, and property damage in the construction industry is extremely high. The potential severity of accidents and the frequency with which they occur require that the contractor protect itself with a variety of complex and expensive insurance coverages. Without adequate insurance protection, the contractor would be constantly faced with the possibility of serious or even ruinous financial loss.

As discussed previously in this book, construction projects usually have in force several simultaneous contractual arrangements: those between the owner and the architect-engineer, between the owner and the general contractor, and between the general contractor and its several subcontractors. Contracts that provide for design-construct and construction management services, and the use of separate prime contracts, introduce additional features. Construed as a whole, these contracts establish a complicated structure of responsibilities for damages arising out of construction operations. Liability for accidents can devolve to the owner or architect-engineer, as well as to the prime contractor and subcontractors whose equipment and employees perform the actual work. Construction contracts typically require the contractor to assume the owner's and architect-engineer's legal liability for construction accidents or to provide insurance for the owner's direct protection. Consequently, a contractor's insurance program normally includes coverages to protect parties other than itself and to protect itself from liabilities not legally its own.

The matter of risk and insurance for the construction contractor is rendered even more difficult on large projects by confusion as to which party is responsible for a given loss or liability. The matter of how the responsibility should be divided among owner, architect-engineer, fabricators, general contractor, subcontractors, and construction manager has become a tangled and extremely complex legal matter. Modern conditions of fast-tracking, alternate designs, shop drawings, design-build, and construction management, have blurred the lines that divide the multiple participants involved in today's construction projects. The division of responsibility in such cases can be very imprecise.

8.3 THE INSURANCE POLICY

An insurance policy is a contract under which the insurer promises, for a consideration, to assume financial responsibility for a specified loss or liability. Under the insurance contract the insurer has a duty to defend the contractor in an action brought against it if the complaint alleges facts that are potentially within the coverage of the policy, and to indemnify the contractor from loss covered by the policy. The policy contains many provisions pertaining to the loss against which it affords protection. Fundamentally, the law of insurance is identical to the law of contracts. However, because of its intimate association with the public welfare, the insurance field is controlled and regulated by federal and state statutes. Each state has an insurance regulatory agency that administers that state's insurance code, a set of statutory provisions that imposes regulations on insurance companies concerning investments, reserves, annual financial statements, and periodic examinations. The organizational structure, financial affairs, and business methods of insurance companies are controlled. In most states insurance policies must conform to statutory requirements regarding form and content.

A loss suffered by a contracting firm as a result of its own deliberate action cannot be recovered under an insurance policy. However, negligence or oversight on the part of the contractor will not generally invalidate the insurance contract. The contractor must pay a premium as the consideration for the insurance company's promise of protection against a designated loss. Most types of insurance require that the premium be paid before the policy becomes effective. In the event of a loss covered by an insurance policy, the contractor cannot recover more than the loss; that is, a profit cannot be made at the expense of the insurance company.

The premiums of many insurance types are adjusted up or down according to the contractor's loss experience record. When annual premium payments reach a given level, the contractor becomes eligible for an experience rating. In this process, credits or debits determined for the individual contractor are applied to the manual rates in accordance with how the premiums paid in compare with the losses paid out over a period of time. A manual rate is a standard premium charge based on a probable loss experience for a given class of risks. Such rates are published and are used for general insurance cost information. A contractor whose losses are low enjoys a considerable savings on the costs of its experience-rated coverages. Thus, there is wisdom in a contractor's continuing to do business with an insurance company with which it has established a considerable background of favorable experience.

Insurance companies can be organized as stock companies or as mutual companies. Stock companies are organized in a manner similar to that of a bank, and ownership is vested in stockholders. The owner of an insurance policy has no ownership in the company and assumes no risk of assessment if the insurance company encounters financial reversals.

A mutual company is one in which the policyholders constitute the members of the insuring company or association. Every policyholder of the mutual company is, at the same time, an insurer and an insured. If it happens that the premiums collected are in excess of the losses, the excess is returned to the policyholders as dividends. By the same token, if losses outweigh income, assessments may be levied against the policyholders. State laws permit mutual companies that satisfy certain tests to limit or eliminate the assessment that can be levied against the members. Consequently, the policies of many mutual companies are nonassessable. This varies considerably with the bylaws and policies of the individual mutual companies.

In property and casualty insurance, a field of insurance especially important to contractors, several mutuals are among the largest companies. In life insurance, probably a majority of the largest companies are mutuals.

8.4   CONTRACT REQUIREMENTS

With the many hazards that confront a construction business and the plethora of insurance types that can be purchased, one may wonder how a contractor decides just what insurance is really needed. In reality, however, the contractor quite often has no choice. For example, it is standard practice for construction contracts to require the contractor to provide certain insurance coverages.

Construction contracts typically make the contractor responsible for obtaining coverages such as workers' compensation insurance, employer's liability insurance, and comprehensive general liability insurance. Property insurance to protect the project and liability insurance to protect the owner may be made the responsibility of either the owner or the contractor, depending on the contract. Article 11 of AIA *Document A201* (Appendix C) makes the owner responsible for obtaining both of these insurance coverages. There are, of course, many examples of special insurance being required by contract when the construction involves unusual risks or conditions. When the contract delegates specific responsibility to the contracting firm for obtaining certain insurance, it is customary that insurance certificates (see Section 8.51) must be submitted to the owner or the architect-engineer as proof that the coverage stipulated has, in fact, been provided.

As discussed in Section 6.33, construction contracts frequently require the contractor to hold the owner and the architect-engineer harmless by accepting any liability that either of them may incur because of operations performed under the contract. Most contract documents that contain such indemnity clauses are explicit in requiring the contractor to procure appropriate contractual liability insurance (see Section 8.25). Subparagraph 11.1.1.8 of Appendix C illustrates this point.

With regard to project insurance requirements, it is always good practice for a contractor to submit a copy of the contract documents to its insurance company before construction operations commence. The contracting firm is not an insurance expert and is not usually competent to evaluate the risks and liabilities placed on it by the contract. Its insurance agents or brokers are qualified to comb the documents and advise the firm concerning the insurance needs dictated by the language and requirements of a given construction contract.

8.5   LEGAL REQUIREMENTS

Certain kinds of insurance are required by law, and the contractor must provide them whether or not they are called for by the contract. Workers' compensation, automobile, unemployment, and Social Security are examples of coverages required by statute. It can be argued that unemployment and Social Security payments made by the contractor are more in the nature of taxes than of insurance premiums in the usual sense. Nevertheless, both unemployment and Social Security are treated as forms of insurance for the purpose of discussion in this chapter.

The law makes the independent contractor liable for damages caused by its own acts of omission or commission. In addition, the prime contractor has a contingent liability for the actions of its subcontractors. Therefore, whether or not the law is specific concerning certain types of insurance, the contractor as a practical matter must procure several different categories of liability insurance to protect itself from liability for damages caused by its own construction operations as well as those of its subcontractors.

8.6   ANALYSIS OF INSURABLE RISKS

Aside from coverages required by law and the construction contract, it is the contractor's prerogative to decide what insurance it will carry. Such elective coverages pertain principally to the contractor's own property or to property for which it is responsible. It is not economically possible for the contractor to carry all the insurance coverages available to it. If it purchased insurance protection against every risk that is insurable, the cost of the resulting premiums would impose an impossible financial burden on its business. The extent and magnitude of a contractor's insurance program can be decided only after careful study and consideration. If a risk is insurable, the cost of the premiums must be balanced against the possible loss, the probability of its occurrence, and the contractor's ability to withstand the loss. There are, of course, risks that are not insurable or for which insurance is not economically practical. Associated losses must be regarded simply as ordinary business expenses.

At times, careful planning and meticulous construction procedures can minimize a risk at less cost than the premium of a covering insurance policy. Thus, the contractor may choose to assume a calculated risk rather than pay a high insurance premium. A common example of assuming such a risk involves construction that is to be erected immediately adjacent to an existing structure. If the nature of the new construction is such that the existing structure may be endangered by settlement or collapse, there are two courses of action open to the contractor. As one alternative, the contractor can include in its estimate the premium for a collapse policy. Such protection is high in cost and is generally available only with substantial deductible amounts. Instead, the contractor can assume the risk without insurance protection, choosing to rely on skill and extraordinary precautions in construction procedures to get the job done without mishap. Many analogous cases can be cited with respect to pile driving, blasting, water damage, and other potential hazards.

8.7 CONSTRUCTION INSURANCE CHECKLIST

Insurance coverages are complex, and each new construction contract presents its own problems in this regard. The contractor should select a competent insurance agent or broker who is experienced in construction work and familiar with contractor's insurance problems. Without competent advice, the contractor may either incur the needless expense of overlapping protection or expose itself to the danger of vital gaps in insurance coverage. The contractor can often reduce insurance costs by keeping its agent or broker advised in detail as to the nature and conduct of its construction operations.

In the long list of possible construction insurance coverages, not every policy is applicable to a given firm's operations. The following checklist is not represented as being complete, but it does include insurance coverages typical of the construction industry.

A. Property Insurance on Project during Construction

   1. All-risk builder's risk insurance. This insurance protects against all risks of direct physical loss or damage to the project or to associated materials caused by any external effect, with noted exclusions.

   2. Named-peril builder's risk insurance. The basic policy provides protection for the project, including stored materials, against direct loss by fire or lightning. A number of separate endorsements to this policy are available that add coverage for specific losses.

     a. Extended coverage endorsement. This covers property against all direct loss caused by windstorm, hail, explosion, riot, civil commotion, aircraft, vehicles, and smoke.

     b. Vandalism and malicious mischief endorsement.

     c. Water damage endorsement. Insurance of this type indemnifies for loss or damage caused by accidental discharge, leakage, or overflow of water or steam.

Included are defective pipes, roofs, and water tanks. This does not include damage caused by sprinkler leakage, floods, or high water.

    d. Sprinkler leakage endorsement. This provides protection against all direct loss to a building project as a result of leakage, freezing, or breaking of sprinkler installations.

3. **Earthquake insurance**. This coverage may be provided by an endorsement to the builder's risk policy in some states. Elsewhere a separate policy must be issued.

4. **Bridge insurance**. This insurance is of the inland marine type and is often termed the "bridge builder's risk policy." It affords protection during construction against damage that may be caused by fire, lightning, flood, ice, collision, explosion, riot, vandalism, wind, tornado, and earthquake.

5. **Steam boiler and machinery insurance**. A contractor or owner may purchase this form of insurance when the boiler of a building under construction is being tested and balanced or when being used to heat the structure for plastering, floor laying, or other purposes. Unlike other property insurances listed here, this type includes some liability coverage. This policy covers any injury or damage that may occur to, or be caused by, the boiler during its use by the contractor.

6. **Installation floater policy**. Insurance of this type provides named-peril or all-risk protection for property of various kinds, such as project machinery (heating and air-conditioning systems, for example), from the time it leaves the place of shipment until it is installed on the project and tested. Coverage terminates when the insured's interest in the property ceases, when the property is accepted, or when it is taken over by the owner.

B. Property Insurance on Contractor's Own Property

1. **Property insurance on contractor's own buildings**. This coverage affords protection for offices, sheds, warehouses, and contained personal property. Several different forms of this insurance are available.

2. **Contractor's equipment insurance**. This type of policy, often termed a "floater," insures a contractor's construction equipment regardless of its location.

3. **Motor truck cargo policy**. This insurance covers loss by named hazards to materials or supplies carried on the contractor's own trucks from supplier to warehouse or building site.

4. **Transportation floater**. Insurance of this type provides coverage against damage to property belonging to the contractor or others while it is being transported by a public carrier. It may be obtained on a per-trip, project, or annual basis.

5. **Fidelity bond**. This surety bond affords the contractor protection against loss caused by dishonesty of its own employees.

6. **Crime insurance**. This form of insurance protects the contractor against the loss of money, securities, office equipment, and similar valuables through burglary, theft, robbery, destruction, disappearance, or wrongful abstraction. It insures against loss of valuables caused by safe deposit burglary and forgery.

7. **Valuable papers destruction insurance**. This policy protects the contractor against the loss, damage, or destruction of valuable papers such as books, records, maps, drawings, abstracts, deeds, mortgages, contracts, and documents. It does not cover loss by misplacement, unexplained disappearance, wear and tear, deterioration, vermin, or war.

C. **Liability Insurance**

1. **Contractor's public liability and property damage insurance**. This insurance protects the contractor from its legal liability for injuries to persons not in its employ and for damage to the property of others, if the property is not in the contractor's care, custody, or control when such injuries or damage arise out of the operations of the contractor.

2. **Contractor's protective public and property damage liability insurance**. This protects the contractor against liability imposed by law arising out of acts or omissions of its subcontractors.

3. **Completed-operations liability insurance**. This form of insurance protects the contractor from damage claims stemming from its faulty performance on projects already completed and handed over to the owner. This form of insurance is often required because the usual forms of liability insurance provide protection only while the contractor is performing the work and not after it has been completed and accepted by the owner.

4. **Contractual liability insurance**. This form of insurance is required when one party to a contract, by terms of that contract, assumes certain legal liabilities of the other party. The usual forms of liability insurance do not afford this coverage.

5. **Professional liability insurance**. This insurance protects the contractor against damage claims arising out of design and other professional services rendered by the contractor to the owner.

6. **Workers' compensation insurance**. This insurance provides all benefits required by law to employees killed or injured in the course of their employment.

7. **Employer's liability insurance**. This insurance is customarily written in combination with workers' compensation insurance. It affords the contractor broad coverage for the bodily injury or death of an employee in the course of his employment, but outside of, and distinct from, any claims under workers' compensation laws.

8. **Owner's protective liability insurance**. This insurance protects the owner from its contingent liability for damages arising from the operations of the prime contractor or its subcontractors.

D. **Employee Insurance**

1. **Employee benefit insurance**. This coverage provides employees with designated fringe-benefit insurance such as medical, hospital, surgical, life, and similar coverages.

2. **Social Security**. This all-federal insurance system operated by the U.S. government provides retirement benefits to an insured worker, survivor's benefits to his family

when the worker dies, disability benefits, hospitalization benefits, and medical insurance.

   3. Unemployment insurance. This federal-state insurance plan provides qualified workers with a weekly income during periods of unemployment between jobs.

   4. Disability insurance. This insurance, required by some states, provides benefits to employees for disabilities caused by nonoccupational accidents and disease.

E. **Automobile Insurance**. Various forms of insurance are available in connection with the ownership and use of motor vehicles. Liability coverages protect the contractor against third-party claims of bodily injury or property damage involving the contractor's vehicles or nonowned vehicles that are used in its interest. Physical damage coverage indemnifies the contractor for damage to its own vehicles.

F. **Business, Accident, and Life Insurance**

   1. Business interruption insurance. This insurance is designed to reimburse the insured for losses suffered because of an interruption of its business.

   2. Sole proprietorship insurance. A policy of this type provides cash to assist heirs in continuing or disposing of the business without sacrifice in the event of death of the owner.

   3. Key person life insurance. This insurance reimburses the business for financial loss resulting from the death of a key person in the business. It also builds up a sinking fund to be available upon retirement.

   4. Corporate continuity insurance. In the event of a stockholder's death, this insurance furnishes cash for the purchase of his corporate stock. This provides liquidity for the decedent's estate and prevents corporate stock from falling into undesirable hands.

The next several sections present a detailed discussion of the insurance types of major importance to the construction industry.

8.8   PROJECT PROPERTY INSURANCE

Construction contracts make the general contractor responsible for a project during construction until it is accepted by the owner. Consequently, it is up to the contractor to take all reasonable steps necessary to protect the project from loss or damage and to see that suitable insurance is provided for this purpose. Most construction contracts require that project insurance be purchased and maintained on the entire project to the full insurable value thereof, including all subcontracted work. The possibilities of property damage to the project itself and to stored materials can depend considerably on the nature of the work, its geographical location, and the season during which the work will be performed. On building construction, the loss potential is usually large and diverse and can include fire, smoke, explosion, collapse, vandalism, water, wind, freezing, and physical damage of a wide variety. Marine structures can be especially vulnerable to wind, ice, wave action, tides, water currents, collision, and collapse. In a general sense, the highway contractor faces comparatively few hazards to project property, the elements of nature probably constituting the source of greatest risk.

Depending on the terms of the contract, the project insurance may be provided by the owner or the prime contractor. Regardless of who buys this insurance, however, the intent is to protect the interests of the owner, the prime contractor, the subcontractors, and perhaps the lending institution in the event a loss occurs during construction and to provide funds for repairs or rebuilding.

In a very real sense, project insurance must be procured that is tailored to meet the specific risks intrinsic to the work. Builder's risk policies are standard on building construction. An all-risk installation floater policy is commonly used on projects such as water and sewer jobs, where there is little or no exposure to fire or hazards typically covered by extended coverage insurance. Policies designed for the specific construction hazards of bridges, tunnels, radio and television towers, and other special construction types are available. Highway, reclamation, and other engineering contractors often carry no project insurance, choosing to self-insure the small risks and relying on "acts of God" contract clauses to protect them from major damage to the project. A typical clause of this type provides that the contractor will make good all damage to any portion of the work, except those damages due to unforeseeable causes beyond the control of, and without the fault of, the contractor, including acts of God or extraordinary action of the elements. An act of God has been defined as a natural occurrence of extraordinary and unprecedented proportions, whose magnitude and destructiveness could not have been protected against by the exercise of ordinary foresight.

8.9   BUILDER'S RISK INSURANCE

It is quite impractical to attempt herein any comprehensive treatment of the many forms and provisions of project insurance. Builder's risk insurance, a discussion of which follows, has been selected because of its broad applicability and similarity to other types. This type of insurance provides property coverage only and does not include any form of liability coverage.

On building construction, where builder's risk insurance normally constitutes the basic project policy, it is customary to deduct the costs associated with land preparation, landscaping, excavation, underground utilities, and foundations below the lowest basement floor in computing the insurable value of the structure. The rationale is that these portions of the project are not directly susceptible to loss by fire or other usual hazards. Premium rates for builder's risk insurance can vary considerably with the type of construction and the availability of fire-fighting facilities. Premium rates are higher for unprotected areas and for the more combustible classes of construction.

There are two types of builder's risk policies; the all-risk form is the one most generally used. The named-peril form of insurance with endorsements is available but seldom utilized. The all-risk form is much broader in scope than the named-peril form and provides insurance protection for all losses not specifically excluded in the policy. Both types of builder's risk policies normally provide that the structure must not be occupied or used by the owner in the course of construction without obtaining the consent of the insurance carrier. Builder's risk policies routinely contain coverage exclusions of one kind or another, a matter that requires careful study before construction operations commence. A usual exclusion is coverage for the cost of correcting faulty workmanship, materials, construction, or design, as well as loss or damage caused by error, omission, or deficiency in design, specifications, workmanship, or materials. Some policies exclude loss caused by the testing of project machinery, although this exclusion can normally be removed for an additional premium. Loss caused by floods or earthquake are similar examples. The policy should be checked to see that job materials are covered, not only while stored at the job site but also while in transit to the job site or stored at an off-site location. As a general rule, any loss of specifications, drawings, records, documents, accounts, deeds, currency, notes, securities, or designs is specifically excluded.

Builder's risk policies can be written to reimburse the insured on an actual cash value or a replacement cost basis. Replacement cost is the usual valuation used. Builder's risk insurance can be purchased on a specific or blanket basis. With the specific form, the contractor purchases a separate policy for each project as it comes along. Under blanket coverage, all projects acquired during the policy period are included under the same policy. The blanket form is usually preferable where the contractor has an appreciable volume of business.

Builder's risk policies are very flexible and can include many kinds of protective provisions other than merely covering the direct expense of physical loss or damage to a project. For example, should there be fire, storm, or other damage to the project, the repair of which delays project completion or interferes with the use of the premises by the owner or tenant, the repair cost is covered by the project property insurance. However, consequential loss coverage is also available under the policy to protect the parties from the consequences of business interruption, loss of rents, or extra expense. Very broad coverage under builder's risk policies is available for certain classes of projects. Project property insurance can now include performance and process guarantees to ensure that a project will meet specified output or service requirements. Completion guarantees are available that will cover debt service and penalties after the designated finish date in the event of a covered loss.

8.10   ALL-RISK BUILDER'S RISK INSURANCE

An all-risk builder's risk insurance policy, widely used for building construction projects, covers the project proper as well as temporary structures at the job site. Materials and supplies pertaining to the construction are protected while held temporarily in storage prior to delivery, while in transit to the job site, and after their delivery and while awaiting installation.

The policy insures against all direct physical loss or damage from any external cause to the property covered, except for stated exclusions. There are many such exclusions, however, including certain damage due to freezing, explosion of steam boilers or pipes, glass breakage, subsidence and settling, artificially generated electrical currents, rain and snow, earthquake, floods, nuclear radiation, and several others. Many of these exclusions can be removed or modified for an additional premium. The policy exclusions and format vary somewhat from one insurance company to another.

The all-risk policy is very flexible and can be varied substantially. The contractor is generally able to work out specific coverage to suit its needs.

8.11   NAMED-PERIL BUILDER'S RISK INSURANCE

Named-peril builder's risk insurance, unlike the all-risk form that protects the project against all losses except those excluded, affords coverage for only those risks specifically listed. The basic policy of this form protects building projects only against direct loss caused by fire or lightning. Such insurance ordinarily covers the cost of facilities and materials connected or adjacent to the structure insured, including temporary structures, materials, machinery, and supplies of all kinds incidental to the construction of the structure. When not otherwise covered by insurance, this policy also protects construction equipment and tools owned by the contractor or similar property of others for which it is legally responsible. All property is protected that is a part of or is contained in the structure, in temporary structures, or on vehicles, or is stored on the premises adjacent to the project.

There are, of course, many possible causes of physical loss or damage to a construction project besides fire and lightning. Other significant risks can be insured against by purchasing various endorsements to the basic named-peril policy. An extended coverage endorsement provides protection against damage or loss caused by windstorm, hail, explosion, riot, riot attending a strike, civil commotion, aircraft, vehicles, and smoke. Another common endorsement covers vandalism and malicious mischief. It should be noted that each endorsement protects against only the hazards named and that each additional endorsement is obtained by the payment of an extra premium. Many other special endorsements to builder's risk are available, including protection against water damage, sprinkler leakage, early occupancy by the owner, and earthquake.

8.12   BUILDER'S RISK PREMIUMS

There are two ways in which the premiums of builder's risk insurance can be paid, the coverage being the same in either case. One payment mode is the reporting form, which establishes the insurable value of the structure, and hence the face value of the policy, in accordance with periodic progress reports submitted to the insurance company by the contractor. This form requires the contractor to make a monthly report of the insurable value of the work in place, including all materials stored on the site. The premium for this form of policy is paid in monthly installments, the amount of each monthly premium payment being computed on the basis of the last progress report. Premium payments are therefore initially low but increase progressively each month as the job advances toward completion. An important consideration with the reporting form is keeping the reports accurate and up-to-date. Should a report be undervalued or overlooked, the insured stands the risk of assuming a part of the liability if a severe loss occurs. Because of the complications of the reporting form, contractors tend to use the second payment mode, the completed-value form.

For the completed-value form, the policy is written for the full amount of the project value, which is the contract price less the cost of the foundations and other excluded work. The coverage must be written by the time the foundations are completed, and the premium is payable as a lump sum in advance. This form requires no monthly report. The premium rate for the completed-value form is about 50 percent of the full-term rate, on the assumption that the insurable value starts from zero and increases approximately linearly during the construction period.

In general, the insured is free to choose the form of premium payment to be used. At times the choice between them may be made on the basis of the least probable premium cost. The premium for the completed-value form is based on a linear increase of insurable value with time. If a project's insurable value initially increases slowly but later rises rapidly, the total of the monthly premiums paid with the reporting form will be less than the single completed-value premium. Such a project is normally large and requires considerable time for excavating, pile driving, foundation construction, and other initial work, which is exempted from the insurable value.

As a rule, the completed-value form of builder's risk insurance is somewhat cheaper for short-term jobs. In addition, it does not require monthly progress reports. However, the preparation of these reports does not usually involve any substantial amount of additional time or expense because the essential information is already available from the monthly project pay requests.

8.13   PROVIDING OF BUILDER'S RISK INSURANCE BY THE OWNER

The standard documents of the American Institute of Architects stipulate that, unless otherwise provided, the owner shall purchase and maintain the builder's risk policy (see Subparagraph 11.3.1 in Appendix C). With other contract forms, however, it is common practice for the general contractor to provide this coverage, especially on public projects. Most contractors prefer to provide builder's risk insurance themselves because owners, often unfamiliar with this type of insurance, may procure insurance policies that provide less coverage than is necessary or coverage with deductible amounts.

There are, however, certain types of projects that do lend themselves well to the provision of builder's risk by the owner. One such example is a project that involves several prime contractors. Other good examples are a remodeling job and an addition to an existing building. If the contractor is to provide the builder's risk insurance, there is always considerable uncertainty about protection for the existing structure. In such circumstances the contractor may make an arrangement with the owner to add the additional coverage to the owner's existing policy, although this matter can become involved from an insurance standpoint. It is much simpler and perhaps cheaper for the owner in such cases to obtain the insurance in the first place. However, there is a contractor's-interest form of builder's risk insurance that can be purchased for alteration or remodeling projects to protect the contractor's interest only. If the owner does purchase the builder's risk insurance, the contractor must determine whether the policy provides the customary construction coverage. If the coverage is not adequate, the contractor can request that the owner obtain broader coverage. Otherwise, the contractor can usually arrange additional coverage on its own behalf in the form of "difference-in-conditions" coverage.

8.14   SUBROGATION

An important aspect of builder's risk insurance is the subrogation clause, a standard provision in many forms of insurance, including builder's risk. The workings of subrogation may be illustrated by the following example. If the owner of insured property should sustain a loss to this property, the insurance company will pay the insured for the damage suffered, up to the face amount of the policy. However, by the terms of the subrogation clause in the policy, the insurance company acquires the right of the insured to recover from the party whose negligence caused the loss. This process of subrogation gives the insurance company the right to sue the offending party in the insured's name for recovery of its loss. In the case where the owner provides builder's risk insurance and the general contractor's operations cause or contribute to project damage, the contractor may be exposed to action by the builder's risk carrier for recovery of its loss under the policy. Alternatively, if a subcontractor or sub-subcontractor causes or contributes to a loss on the project, this party may be subject to suit by the insurance company. If the general contractor buys the builder's risk insurance, subrogation applies to its subcontractors or sub-subcontractors if they cause a loss on the project. It is easy to see that application of the subrogation clause by the insurance company could defeat the entire purpose of the project property insurance.

One means of at least partially avoiding the undesirable effects of subrogation is to make the owner, prime contractor, and all the subcontractors and sub-subcontractors named insureds under the policy. Subrogation cannot usually be employed against parties who are insured under the policy. However, this exemption may apply only to damage that a contractor does to its own work and not to damage caused to the work of others where the contractor has no insurable interest. Another approach is for the owner and all of the contractors to waive all rights against each other for damages caused by fire or other perils. Subparagraph 11.3.2 in Appendix C illustrates this point. However, there are some builder's risk forms that provide that policyholders cannot waive such rights unless written permission of the insurance company has been received. Thus, both the naming of all parties as insureds and the waiver of rights among all parties are desirable when either the owner or the general contractor provides the builder's risk insurance.

8.15   TERMINATION OF BUILDER'S RISK INSURANCE

Builder's risk policies may be canceled on a pro rata basis at any time requested by the policyholder. If the contractor provides this insurance, the time at which it can terminate the policy is an important matter. On the one hand, the premiums are expensive and the contractor naturally wishes the expenditure to cease at the earliest possible moment. On the other hand, the contractor cannot dispense with this protection until such time as the owner is legally responsible for the project. When the contract has been silent in this regard, the courts have repeatedly found that the contractor remains responsible for the project until the owner has made formal acceptance.

8.16   CONTRACTOR'S EQUIPMENT FLOATER POLICY

The insurance that a contractor procures to protect its construction equipment from loss or damage is commonly referred to as an equipment floater policy. This insurance is very flexible and can be designed to fit the specific needs of the contractor. A floater policy provides protection against physical loss or damage by external means that may occur to the contractor's portable equipment while on the job, in transit, or on the contractor's own premises. As a general rule, a contractor's floater policy does not provide any liability coverage for loss of, or damage to, leased or rented equipment. However, an endorsement is usually available to remove this liability exposure. There are no standard premium rates for an equipment floater policy. Each construction company is rated according to the type of work performed, past loss experience, company reputation, and dispersion of risk.

Equipment insurance can be obtained on a named-peril or all-risk basis. With named-peril coverage, the policy provides coverage only for equipment losses that are specifically named in the policy. Such coverage can be made to include such hazards as transportation, upset, theft, collision, overturn, fire, explosion, tornado, windstorm, landslide, and flood.

The all-risk form protects equipment against all losses other than those specifically excluded in the policy. Such exclusions can include loss of equipment lent to others, equipment overload, loss resulting from maintenance or repair, and waterborne equipment. Deductible amounts often apply to most or all losses. The cost of the all-risk form is normally higher than that for the named-peril, and the size of the deductible amounts can have a substantial effect on the premiums.

The equipment floater policy is available to the contractor on either a schedule or blanket coverage basis. Under the schedule form, each item of equipment must be listed in the policy to be covered and a specified value must be given. For contractors with large equipment spreads, maintaining an up-to-date schedule is difficult, and coverage is usually maintained on a blanket basis. With blanket coverage, the contractor submits a listing of all owned equipment and the value of each item at the beginning of the policy period. A similar listing is presented at the end of the policy period. In this way, policy coverage is automatically adjusted to reflect the acquisition or sale of equipment units. Equipment is normally insured for its actual cash value, which is defined as its replacement cost less reasonable depreciation.

To cover a key piece of the contractor's equipment that is damaged or destroyed, a rental cost reimbursement endorsement is normally available. This endorsement will pay the cost of renting replacement equipment on a temporary basis until repairs can be made or a new unit obtained.

Equipment insurance, like many other forms of property insurance, is usually sold on a 100 percent co-insurance basis. This means the contractor is required to carry insurance to the full value of the equipment. If this is not done, the contractor becomes a co-insurer with the insurance company on any loss that may occur. For example, if a contractor is carrying insurance coverage at only 80 percent of the actual value of the equipment, compensation by the insurer will be for only 80 percent of any loss, subject to overall policy limits. The contractor will have to make up the remaining 20 percent.

8.17   PROPERTY INSURANCE

Property insurance is purchased by the contractor to protect its offices, warehouses, and other buildings from physical damage. Such a policy is designed to cover buildings, business personal property of the contractor, and the personal property of others in the contractor's care, custody, or control. This insurance includes protection for building contents and can be made to provide coverage for a wide variety of loss possibilities. The basic form includes coverage for direct loss by fire, lightning, vandalism, sprinkler leakage, hail, smoke, aircraft, vehicles, riot, explosion, windstorm, and other named hazards. This policy can be expanded to a broad form of coverage that includes several other named loss possibilities. A third option is the special form of property insurance that covers all risks of physical loss except those perils specifically excluded in the policy.

The property policy can be a specific policy in which the dollar value of coverage for each building and the value of personal property covered at each location are specified. An alternative to this is blanket coverage in which a single policy amount covers the entire schedule of property. In either case, a co-insurance provision is usual, which requires the insured to either maintain insurance equal to at least a specified percentage of the property's value (80 percent, for example) or to act as a co-insurer with the insurance company. In the latter event, the contractor will receive a reduced payment in case of a loss.

A property insurance policy can be written to reimburse the contractor for property loss on an actual cash-value basis or on a replacement-cost basis. The property policy normally covers only direct loss to the insured property caused by covered perils. However, it can be made to cover indirect losses such as business interruption, rental value, and extra expense. The cost of this insurance is obtained from the values of the property covered.

8.18 FIDELITY BONDS

A contracting firm, like other businesses, is subject to loss caused by a variety of forms of criminal action. A common source of such loss is the dishonesty of the firm's own employees. To provide the contractor with protection against such loss, surety bonds called fidelity bonds are normally used rather than insurance as such. A fidelity bond indemnifies the contractor for loss of money or other property caused by dishonest acts of its bonded employees. This includes losses sustained through larceny, theft, embezzlement, forgery, and other fraudulent or dishonest acts. Under this form of surety bond, the employer is the obligee and the employee is the principal. Although a fidelity bond is not insurance, it is hardly distinguishable from insurance insofar as the employer is concerned. A fidelity bond in physical form resembles an insurance policy and is sometimes referred to as employee dishonesty insurance.

There are a number of ways in which fidelity bond protection can be written. The most important are described as follows:

Named Schedule Form. Employees named in the bond are covered.

Position Schedule Form. The occupants of listed company positions are covered.

Commercial Blanket Form. This form covers all employees of the firm without naming them or listing positions. This is an aggregate limit form, meaning that regardless of the number of employees involved in the loss, there is an aggregate and total limit of liability.

As indicated, the contractor can purchase protection on specific persons or on positions, such as a bookkeeper and other selected personnel. However, records almost invariably show that it is impossible to designate in advance the person who will be dishonest. On this basis a blanket form of bond may be preferable, inasmuch as it covers all employees. In the event of loss under this form of policy, the contractor need only prove that it was caused by an employee or employees unknown. Under the named schedule or position schedule form, however, the dishonest act must be shown to have been that of an individual whose name is listed or who occupied a listed position. In addition, schedule forms require continuous checking as names and positions change. Other key advantages of the blanket form are that automatic additions and deletions can be made without notice to the bonding company and that no premium adjustment must be made during a premium year for additional employees.

8.19 CRIME INSURANCE

Crime insurance is available in many forms to cover a wide variety of losses a contractor may suffer. About every type of crime peril and property subject to crime can be covered under a separate insurance policy. However, contractors usually obtain package coverage plans that include various forms of combination crime coverage, including fidelity bonds to protect against employee dishonesty as discussed in the previous section. Under the package coverage plan, the contractor can purchase an assortment of crime coverages. These include employee dishonesty; theft, disappearance, and destruction; premises burglary; computer fraud; forgery and alteration; safe deposit box burglary and robbery; robbery and safe burglary; and other criminal acts.

8.20 LIABILITY INSURANCE

Liability is an obligation imposed by law. In the course of conducting business, a contractor may incur liability for damages in any one of the following ways:

1. Direct responsibility for injury to persons (not employees) or damage to property of third parties caused by an act of omission or comission by the contractor.

2. Contingent liability, which involves the indirect liability of the general contractor for the acts of parties for whom it is responsible, such as subcontractors

3. Liability that arises out of a project after the work has been completed and the structure has been accepted by the owner

4. Contractual liability, whereby the contractor has assumed the legal liability of the owner, or other party, by the terms of a contract

5. Liability that may devolve to the contractor as a result of the operation of its motor vehicles

6. Liability that arises from design and associated professional services rendered by the contractor to the owner

7. Liability to injured employees, both those covered and those not covered by worker's compensation laws

Liability insurance, also called defense coverage, serves no purpose other than to protect the contractor against claims brought against it by third parties. Insurance of this type pays the costs of the contractor's legal defense as well as paying judgments for which the contractor becomes legally liable, up to the face value of the policy. In the settlement of liability claims against the contractor, the insurance company has the right to settle as it sees fit without the approval or consent of the contractor. Liability insurance provides no protection to the contractor for loss of, or damage to, its own property. It is important to note that many forms of liability insurance customarily include subrogation clauses that give the insurance company the right to file suit to recover losses.

Liability insurance for firms involved with design-construct and construction management has become a very technical and complex matter. It is not possible to generalize on this subject because of the many variables inherent in such contractual arrangements. The liability coverage needed is a matter for expert advice and depends on the specific type of company organization, the particular role being played by the insured, and the provisions of the contractor with the owner. For example, there have been several instances in which a construction manager has been held liable for bodily injury and property damage resulting from unsafe or defective job site conditions, even though the construction manager did not perform any of the actual work. Such liability depends on the construction manager's authority and degree of control of the project.

It should also be noted that when contractors affiliate to form a joint venture, which is a legal entity separate from its constituent members, the liability coverages of the individual contractors will not apply to claims arising from a joint venture project. Separate policies must be obtained, with the joint venture indicated as the named insured.

For a variety of good reasons, a contractor should consider obtaining all of its liability insurance from one insurance company. This arrangement often offers the possibility of substantial premium discounts because of the larger total premium amounts involved, a better overall loss rating, and other factors. In addition, a single liability insurance carrier helps to avoid gaps and overlaps in coverage and eliminates squabbles between insurance companies about loss responsibility.

8.21 COMMERCIAL GENERAL LIABILITY INSURANCE

Because most contractors require the same basic liability coverages, certain of these are packaged together into a single commercial general liability policy. This policy protects contractors against third party liability claims arising from the contractor's operations as well as those of independent contractors, completed operations, contractual liability, personal injury, and certain other hazards. The liability coverages named here are discussed in the following sections.

Commercial general liability policies are written to cover bodily injury and property damage caused by "occurrences." An occurrence, in this context, is defined as an accident including continuous or repeated exposure to the same harmful conditions. Thus, coverage is not limited to a single event and the occurrence need not be sudden but may be produced over a period of time. Consequently, in construction, bodily injury and property damage caused by pile driving, water leakage and seepage, dust, settlement, inadequate shoring, and other instances of injurious exposure are normally covered by the policy. An occurrence must be unforeseen, unexpected, and not intended by the insured. Thus, if a loss occurs, failure to correct the condition that caused the loss may permit the insurer to deny coverage for a second event on the grounds that it could logically have been expected.

The term *occurrence* is also used in a different way with regard to general liability insurance, this having to do with the event that must happen during the policy period in order to "trigger" the insurance coverage. With liability insurance written on an "occurrence" basis, the insurance covers bodily injury and property damage that occurs during the policy period even though the claim may not be made until after the policy expires. Under the "claims-made" form, the injury or damage is covered by the policy in effect at the time the claim is made, even though the loss may have occurred prior to the present policy. If a claim is made when no policy is in effect, the contractor has no insurance protection. However, with claims-made coverage, this restriction can be altered somewhat by retroactive coverage, extended reporting period coverage (tail coverage), and special exclusion endorsements.

8.22 BODILY INJURY AND PROPERTY DAMAGE LIABILITY INSURANCE

The basic form of general liability coverage, often simply called public liability insurance or premises-operations insurance, protects the contractor against its legal liability to third persons for bodily injury and property damage arising from its own operations. Excluded from coverage are injuries to the contractor's own employees as well as all forms of automobile liability. The contractor's buildings and premises, owned or leased, are included as well as its construction operations in progress anywhere in the United States. Normally included in the property damage liability coverage, either as a part of the basic policy or by broad-form property damage endorsement, is damage to the property of other contractors (including subcontractors). Consequently, if the operations of Contractor A cause damage to the work of another contractor on a construction project, the public liability insurance of Contractor A will pay the cost. Not covered, however, would be any damage that Contractor A did to its *own* work. Also normally included in this insurance is an elevator liability clause that covers the insured's legal liability for bodily injury or property damage arising from ownership, maintenance, or use of elevators owned, controlled, or operated by the contractor. The term *elevators* in this context includes material and personnel hoists. Contractors must ordinarily carry very high limits of coverage (bodily injury and property damage) for this type of insurance, either as required by contract or simply to be adequately protected. This is often done through the use of umbrella excess liability coverage (see Section 8.43).

Standard forms of this insurance include personal coverage for individual proprietors, partners, and executive officers, directors, and stockholders of corporations while acting within the scope of their duties. For an additional premium, the policy may be endorsed to extend coverage to company employees as additional insureds while carrying out their company duties. This affords protection to employees who are named as individuals in lawsuits together with their employers. Premiums for public liability insurance are based on the contractor's payroll.

With respect to a contractor's liability for injury to third parties, the concept of "attractive nuisance" is important. The doctrine of attractive nuisance establishes that a possessor of land has a special obligation to children who trespass on the property. When a contractor is in possession of land, such as a job site, its operations frequently attract children. Because a child may not realize that he is entering on the land of another or may not appreciate the hazards involved, the contractor has a responsibility to protect trespassing children. The standards of due care, however, apply only to artificial conditions on the land, not to natural features such as trees, cliffs, or bodies of water. Courts usually hold that a trespassing adult is not entitled to damages for injury not willfully or negligently inflicted by the contractor. However, the situation is different with respect to children. The courts hold that a contractor who maintains a dangerous appliance or hazardous premises is expected to exercise reasonable care to eliminate the danger or otherwise to protect children by the use of security guards, fences, or other protective measures.

8.23   CONTRACTOR'S PROTECTIVE PUBLIC AND PROPERTY DAMAGE LIABILITY INSURANCE

Contractor's protective public and property damage liability insurance, often called contractor's contingent liability insurance, protects the contractor from its contingent liability imposed by law because of injuries to persons or damage to property of others arising from the acts or omissions of independent contractors (e.g., subcontractors). This policy includes protection for the general contractor if the subcontractor's insurance is inadequate or nonexistent. A claimant alleging damages caused by a subcontractor may sue not only the subcontractor, but the prime contractor as well. This situation arises from the fact that the prime contractor exercises general supervision over the work and is responsible for the conduct of construction operations, including those of the subcontractors. Contingent liability insurance not only covers accidents arising from operations performed for the contractor by an independent subcontractor, but also protects the contractor from liability it may incur because of any supervisory act it performs in connection with a subcontractor's work. If a subcontractor further subcontracts portions of its work, it will also need this form of protection.

The premiums for contractor's contingent liability insurance are derived from the subcontract amounts and do not generally vary with the work classifications. Insurance companies writing this form of insurance may require that the prime contractor obtain from its subcontractors certificates of insurance verifying that they have purchased public liability and property damage insurance to cover their own direct liability.

8.24 COMPLETED-OPERATIONS LIABILITY INSURANCE

Completed-operations liability insurance protects the contractor from liabilities arising out of projects that have been completed or abandoned. In a limited sense, the contractor is not liable for damages suffered by a third party by reason of the condition of the work after the project has been completed and accepted by the owner. According to the "completed and accepted" rule, once the owner accepts the work it becomes the party responsible for any defects in the work. When a dangerous condition exists that is obvious or readily discoverable upon inspection, and the owner allows this condition to continue, the owner is substituted for the contractor as the party answerable for damage to a third party. This is true even though the dangerous condition was originally created by the contractor. Nor is the contracting firm liable if it merely carries out plans, specifications, and directions given to it by another, unless the information is so obviously faulty that no reasonable person would follow it. However, there are many exceptions to these rules and the courts now increasingly hold contractors liable to anyone who might be injured because of their negligence. A contractor who creates a dangerous condition on the property of another may be held responsible to parties injured thereby, even after owner acceptance, if the parties so injured could reasonably have been expected to come in contact with the dangerous condition and provided the contractor knew or should have known of the hazard and did not exercise reasonable care to warn of the dangers.

The rule of owner responsibility does not apply to dangerous conditions of such a latent nature that their existence would not be discovered by the owner during the exercise of ordinary care and usage. Indeed, courts have recently imposed liability on contractors without requiring a showing of fault in cases involving latent defects in completed projects. This is an extension to the construction of the doctrine of manufacturers' or product liability whereby builders of mass-produced homes and other contractors have been held subject to strict liability for damages to third parties, even in the absence of contractor negligence. This follows from the legal theory that a producer of goods is strictly liable for injuries or damages resulting from a product defect. Consequently, a contractor's responsibility does not end with project completion. Rather, its liability for completed operations continues for the full period of the applicable state statutes of limitations.

When contractors obtain completed-operations liability insurance for the first time, they are automatically protected against loss for damage to the persons or property of others arising from any completed project, regardless of its completion date. This is true, of course, in the absence of any specific exclusion to the contrary. The protection under this policy starts with the effective date of the insurance and continues as long as the insurance is kept in force. The date of injury or damage, however, must fall within the policy period. Protection under the policy extends only to liability to damaged parties and does not apply to damage to the work that was performed by the contractor. This exclusion can be relieved to some extent by a broad-form property damage endorsement applied to the completed-operations policy. The premiums for completed-operations insurance are based on the contractor's payroll.

8.25 CONTRACTUAL LIABILITY INSURANCE

The liability coverages previously discussed protect the contractor only with respect to its liability as imposed by law. In many instances in construction, however, the contractor, by terms of a construction contract, purchase order, or other form of agreement, assumes the legal liability of another. This form of liability, called contractual liability, refers to the contractor's acceptance by contract of another party's legal responsibility. The contractor is protected from such assumed liability by contractual liability insurance. This coverage provides protection from the tort liability of another for bodily injury or property damage caused to a third party, such liability being assumed by the contractor through a business contract. The premium for contractual liability insurance is based on the contract amounts involved.

Routine purchase agreements can impose serious contractual liability obligations on the contractor. For example, the purchase orders used by some transit-mix concrete companies contain a clause whereby the contractor agrees to hold harmless the concrete dealer from all liability that may arise out of any accident involving the transit-mix trucks or their drivers while on a job site. On signing such a purchase order, the contractor assumes a legal responsibility normally belonging to the dealer. In the absence of contractual liability coverage, the contractor will be unprotected should such a loss occur.

As previously discussed in Section 6.33, the growing incidence of third-party suits against owners and architect-engineers has led to the widespread inclusion of indemnity clauses in construction contracts. By the terms of such hold-harmless agreements, the contractor assumes a liability that is not legally its own but that of another. This is contractual liability, and the contractor is not protected unless blanket contractual liability coverage is procured or each such clause is individually insured. For the purpose of rating hold-harmless agreements, the contractor must usually provide its insurance company with a copy of the contract section under which the hazard is assumed.

Contractual liability policies often have a number of important exclusions. A common exclusion pertains to bodily injury or property damage occurring within 50 ft of railroad property involving a railroad bridge, trestle, tracks, road beds, tunnel, underpass, or crossing. Third-party beneficiary losses, explosion, underground damage, collapse, and other such losses may be exempted from the policy. Also excluded is contractual liability associated with the operation, ownership, maintenance, or use of automobiles. Such coverage must be provided by automobile insurance.

8.26 THIRD-PARTY BENEFICIARY CLAUSES

A third-party beneficiary contract is a contract in which two parties enter into an agreement whereby one of the parties is to perform an obligation for a third party. In such a contract, if the obligated party does not perform as promised for the benefit of the third party, the third party can enforce its rights under the contract even though it is not a party to the contract. For example, contracts with public agencies may require that the contractor be responsible for all damage to property, however caused by the construction operations. Such clauses often refer specifically to blasting. This is the natural result of the desire of governmental bodies to protect adjoining property owners from the hazards of construction operations. However, when the contract wording makes the contractor assume direct liability to third parties, a citizen property owner can assert rights as a third-party beneficiary even though he is not a party to the contract and even though the contractor was not negligent in its operations. In other words, an owner whose property has been damaged by the contractor's operations can sue for damages on the basis of breach of contract without having to prove contractor negligence. This contingency is not covered under usual contractual liability coverage. Third-party beneficiary coverage can be added for an additional premium, usually on a specific exposure basis.

It is of interest to note that special provisions are now being included in some public construction contracts to limit the risk of third-party beneficiary suits. These "no third-party liability" clauses, as they are known, provide that the contracting parties do not intend to make the public or any member thereof a third-party beneficiary under the contract, or to authorize anyone not a party to the contract to maintain a suit for bodily injury or property damage pursuant to the terms or provisions of the contract. The intent of such a clause is, of course, to remove the construction contract as a vehicle for third-party suits and to ensure that the duties, obligations, and responsibilities of the parties to the contract remain as imposed by law.

8.27   PERSONAL INJURY

Bodily injury, as used in an insurance context, refers to physical injury to the body, including shock, mental anguish, mental injury, sickness, disease, or death. *Personal injury* refers to intangible harm, and the usual commercial general policy covers personal injury liability. There are a number of reasons why this type of insurance coverage can be important to the contractor.

Personal injury liability insurance protects the contractor from liability it may incur for (1) false arrest, malicious prosecution, willful detention, or false imprisonment, (2) libel, slander, or defamation of character, and (3) wrongful eviction, invasion of privacy, or wrongful entry. Such protection may be needed, for example, if the contractor causes the arrest of someone it suspected of theft or damage to its property and is sued for false arrest. This could happen, for example, if a contractor's security guard detained someone in the course of his duties. Personal injury liability coverage is normally written subject to the exclusion of the contractor's own employees from coverage, but the policy can be made to include them.

8.28   EXCLUSIONS FROM COMMERCIAL GENERAL LIABILITY POLICY

Those liability coverages usually contained in a contractor's commercial general liability insurance are discussed in the previous sections. However, despite its name this policy is not all-inclusive in its coverage, and there are many exclusions. The following are important examples of these exclusions:

Property damage liability

Automobile, watercraft, and aircraft liability

Professional liability

Liability for injury to the contractor's own employees

Liability arising from pollution

Most of the items listed here are insurable at extra cost, some as endorsements to the commercial general liability policy and some as separate policies. Except for watercraft and aircraft, the first four of the listed exclusions are discussed in the following sections.

8.29   PROPERTY DAMAGE LIABILITY EXCLUSIONS

A contractor's commercial general liability policy excludes coverage for liability arising from damage to:

1. Property owned by, leased to, or rented to the contractor
2. Personal property in the contractor's care, custody, or control
3. Premises sold, abandoned, or given away by the contractor
4. Property lent to the contractor

During the course of construction, damage to the following property is excluded:

5. That particular part of real property on which the contractor, or subcontractor working on the contractor's behalf, is performing operations, if property damage arises out of these operations
6. That particular part of any property that must be restored, repaired, or replaced because of faulty workmanship

These exclusions mean that the commercial general policy does not protect the contractor from liability for damage done to another person's property while the contractor is working on it or with it, and that it does not cover damage to the insured's own property. This provision is consonant with the general rule that liability insurance is not intended to cover damage to one's own property. The most troublesome of these exclusions is the "care, custody, or control" provision and how it applies to construction projects.

The effect of these exclusions can be eased somewhat by obtaining the appropriate form of property insurance or by the use of a broad form of property damage endorsement that, to some extent, restricts application of the exclusions. Actually, this endorsement uses more explicit language with respect to the exclusions, thereby covering more damage situations and providing somewhat broader protection for the contractor. Although it does not completely eliminate the property damage exclusions, it does cover more claims.

It can be readily seen that much of the property damage exempted in the commercial general policy is normally insured by other standard property policies. Builder's risk, equipment floater, installation floater, and other types of policies cover much of the property excluded in the liability coverage.

8.30 AUTOMOBILE INSURANCE

The operation of automobiles exposes the contractor to two broad categories of risk. One form of risk is loss or damage to the contractor's own vehicles caused by collision, fire, theft, vandalism, and similar hazards. The other form of risk is liability for bodily injury to others or damage to the property of others caused in some way by the operation of the contractor's automobiles. Many states have statutory requirements concerning the purchase of liability insurance by owners of motor vehicles. In an insurance context, automobiles or vehicles are construed to include passenger cars, trucks, truck-type tractors, trailers, semi-trailers, and similar land motor vehicles designed for travel on public roads. The contractor is free to choose which vehicles will be insured for automobile liability and which for physical damage coverage.

Automobile liability insurance provides financial protection when the contractor is legally obligated to pay for bodily injury or property damage arising from the ownership, maintenance, or use of a covered vehicle. This includes liability coverage for over-the-road hazards for self-propelled motor vehicles used for the sole purpose of providing mobility to construction equipment such as pumps, power cranes, air compressors, electric generators, welders, and drills. The insurer also provides the contractor with legal defense against liability actions. Automobile liability coverage has several exclusions, such as injury to an employee, property owned or transported by the contractor, and others. The insurance carrier will pay all damages resulting from an accident up to the liability limits specified in the policy. In view of recent large awards in cases of bodily injury and the claims consciousness of the public in general, the contractor must maintain high limits of motor vehicle liability coverage. The contractor's umbrella liability insurance (see Section 8.43) normally provides liability limits above those in an automobile policy.

Physical damage coverage is available in three major forms: collision, specified perils, and comprehensive coverage. Collision insurance pays for loss to a covered vehicle resulting from a collision with another object or the vehicle's overturning. If only a few motor vehicles are involved, collision coverage is often purchased. However, many contractors with large fleets of vehicles prefer to self-insure the collision exposure because of the high premium costs involved. Collision insurance customarily provides for a specified deductible for each collision loss occurrence. This deductible amount can be made to vary, correspondingly changing the premium rate charged.

Specified perils coverage pays only for losses caused by hazards listed in the policy, such as theft, fire, earthquake, explosion, flood, vandalism, wind, hail, and others, not including collision or overturning. Comprehensive coverage pays for damage to, or the loss of, a covered automobile for any cause except collision and overturning. The specified coverage is normally less costly than the comprehensive form, and many contractors choose to obtain the specified perils coverage to reduce premium expense. It is normally possible to put some vehicles under comprehensive coverage and others under specified perils coverage.

There are endorsements that the contractor should consider when purchasing its motor vehicle insurance. One provides protection when an employee uses his private automobile in doing the contractor's business. In this regard, the contractor can purchase an "employees as insured" endorsement. When this endorsement is added to the contractor's policy, it will provide coverage if an employee is found liable for an accident that occurs while he is using a private vehicle on company business. Other endorsements are available to provide the contractor protection when using leased or hired units. Medical payments coverage for persons occupying an insured vehicle and uninsured motorists coverage endorsements are also available.

8.31 PROFESSIONAL LIABILITY INSURANCE

Professional liability insurance provides protection from liability arising from errors, omissions, or negligent acts of the insured in performing design and other professional services. When the contractor's responsibilities to the owner include professional as well as construction duties, professional liability insurance must usually be obtained. Although this risk applies primarily to design-construct contractors and construction managers, other contractors can have varying exposure in this area. Professional liability may devolve to the design-build contractor in two ways. Direct responsibility may arise from a design performed by in-house architect-engineers. Contingent professional liability is possible where professional design services are subcontracted to an outside architect-engineer firm or under a design-construct contract to a subcontractor. Liability exposure of a construction manager arises out of its involvement in the feasibility and design phases of the project and the supervision of workforces other than those of the construction manager itself. Most professional liability policies are written on a claims-made basis, a topic discussed in Section 8.21. In addition, such policies generally provide for relatively large deductible amounts. Professional liability coverage can be obtained on a project or blanket basis.

When professional liability insurance is obtained, it should include contingent professional liability coverage where design work is subcontracted to outside professionals or subcontractors. The total design liability exposure of the contractor is actually covered by a combination of professional liability insurance, the comprehensive general liability policy, the umbrella excess policy, and the builder's risk policy. However, this matter can be complicated by the fact that the usual commercial general liability, umbrella excess, and builder's risk policies specifically exclude professional liability protection. Insurance protection for contractors in the area of design liability is a very complex matter, and the contractor should seek expert advice in this area.

8.32 THE PRINCIPLES OF WORKERS' COMPENSATION

Before the present era of workers' compensation laws, an employer was obligated to protect its employees only to the extent of providing a safe place to work. To obtain redress at the common law, an injured employee had to file suit against the employer and prove that the injury was due to the latter's negligence. Available to the employer were the accepted common-law defenses of contributory negligence of the injured employee, assumption of risk by the injured employee, and negligent acts of fellow employees. Under the contributory negligence doctrine, the employee could not obtain redress if he were negligent to any degree, regardless of the employer's negligence. The assumption-of-risk doctrine denied recovery if the worker knew or should have known about the inherent risks. The fellow-servant doctrine held the employer liable for its own actions but not for those of a fellow worker of the injured party. This trinity of common-law defenses made it difficult for a disabled worker to prove employer responsibility and negligence. The process was at best a slow, costly, and uncertain one for the employee.

The social and economic consequences of this problem were instrumental in the development of workers' compensation laws. New York enacted the first state workers' compensation law, which became effective in 1910. Since that time, compensation legislation has been passed by the federal government, every state and territory of the United States, and each dominion of Canada. Workers' compensation statutes apply to most private employment and much public employment (federal civil service and the military are notable exceptions).

The underlying economic principle of workers' compensation is that the costs associated with an on-the-job injury or death of an employee, regardless of fault, is an expense of production and should be borne by the industry. The expense incurred by employers in providing suitable protection for their employees is considered as another cost of doing business and, as such, is presumably reflected in the selling price of their products or services. The fundamental objective of the compensation statutes is to ensure that an injured worker receives prompt medical attention and monetary assistance. Such support is provided by the employer and involves a minimum of legal formality.

Another basic principle is the strict liability of the employer, regardless of any fault of the injured employee. Contributory negligence of the employee, such as failure to wear a company-provided safety helmet or to conform with posted safety regulations, will not usually affect the employer's liability. As a matter of fact, about the only exceptions to the payment of compensation benefits occur when the worker deliberately inflicts the injury on himself, the injury occurs when the worker is intoxicated, the injury is sustained in the course of committing a felony or misdemeanor, or the injury is intentionally caused by a coemployee or third person for reasons not associated with the employment.

Secondary objectives of compensation laws are to free the courts from the tremendous volume of personal injury litigation, to eliminate the expense and time involved in court trials, and to serve as an instigating agent in the development of effective safety programs.

8.33 WORKERS' COMPENSATION LAWS

Although all of the various workers' compensation laws embody the same general principles, they differ considerably in their working details, and no two of them are exactly alike. The contractor must be especially careful to buy proper insurance and conform with the legal requirements of the workers' compensation laws of each state in which it works. Every law makes certain exclusions to its coverage. For example, most of the compensation acts exclude domestic servants, farm labor, casual employees, independent contractors, and workers in religious or charitable organizations. Businesses that employ fewer than a specified number of employees are exempted in some of the states. Interstate railway workers and maritime employees are not covered by the compensation acts but are protected by separate federal legislation.*

Some compensation laws are designated as compulsory and others as elective. Every employer whose employees are covered by a compulsory law must accept the act and

---

* Contractors that own watercraft, operate wharf facilities, engage in marine construction, conduct operations on or near navigable waters, or are directly engaged in maritime construction at adjacent buildings, piers, or other facilities may employ workers covered by the Longshoremen's and Harbor Workers' Compensation Act. Workers covered by this federal statute are not ordinarily protected by workers' compensation statutes, and the usual workers' compensation insurance does not apply. Separate insurance or an endorsement to the regular workers' compensation policy is needed in such cases.

provide for the benefits specified. Failure to comply with the prescribed provisions can result in severe penalties, the payment of damages to injured workers, and possible imprisonment. In areas where the law is elective, employers have the option of either accepting or rejecting it. However, if an employer elects to remain outside the legislation, it risks an injured worker's civil suit for damages and simultaneously loses the three common-law defenses: assumption of risk, negligence of fellow employees, and contributory negligence. In effect, this means that all compensation laws are compulsory. In most states the presumed-acceptance principle applies, whereby in the absence of specific notice to the contrary, the employer is presumed to have accepted coverage by the compensation act. Workers in excepted or excluded employments may usually be brought within the act through voluntary action by their employers.

An injured employee who files a claim for, and accepts assistance under, a workers' compensation statute ordinarily forfeits the right to sue his employer for damages. Compensation statutes usually provide that a workers' compensation claim is the only remedy the employee has against his employer. According to law, the employer is strictly liable and the injured employee is guaranteed benefits as prescribed by law. In return, the employer avoids the danger of unpredictably large jury verdicts for bodily injury caused by employer negligence. However, during recent years there has been a gradual erosion of the employer's no-fault immunity from suit when an injured employee accepts workers' compensation benefits. In many states, injured workers can now sue their employers for tort damages in addition to receiving state compensation benefits. For an injured worker to have this right, there usually must be a subjective realization on the part of the employer that injury or death could occur because of an unsafe condition. The usual basis for suit by the employee is some form of gross negligence on the part of the employer.

Workers' compensation statutes normally protect the general contractor from tort liability to employees of its subcontractors. Such statutes also protect project supervisors and managers from lawsuits for negligence or tort where they are acting within the scope of their employment. States vary as to whether an injured employee or the workers' compensation insurance company can sue third parties who may have caused the injury. In most states, an injured construction worker can recover from the employer's compensation insurance and then sue a third party who caused the injury. In some areas, the worker cannot collect further compensation from "fellow employees." This term is considered to include other contractors and other workers on the project. However, several states do allow coemployee suits, some of these only when the injury involved an intentional act, intoxication, a reckless act, willful and malicious conduct, unprovoked aggression, or gross negligence or when the accident involved a motor vehicle. In most states, however, an injured employee has considerable latitude in suing third persons for negligence. This contributes appreciably to the number of third-party suits filed against owners and architect-engineers. To be successful, of course, the injured employee must demonstrate that the party sued was negligent and that this negligence caused or contributed to the injury. In some states, however, statutes now provide that an injured worker cannot proceed against the architect-engineer after receiving state workers' compensation benefits. Exempted from this protection, however, are claims based on negligent design.

All of the states, territories, and the District of Columbia have enacted child labor laws that regulate the conditions under which minors may be employed. Workers' compensation laws cover legally employed minors. In some jurisdictions, double compensation or added penalties are provided in cases of injury to illegally employed minors. Minors also enjoy special benefit provisions.

8.34   ADMINISTRATION OF WORKERS' COMPENSATION LAWS

Workers' compensation laws are generally administered by commissions or boards created by law. A few states provide for court administration. Statutory provisions relating to administration vary somewhat from state to state, but each of the laws contains certain regulations pertaining to its implementation. Notice to the employer of the injury by the injured worker is required within stipulated time limits unless the contractor or its agent had actual knowledge of the injury. A claim must be filed by the injured worker within a statutory period. Such claims are normally settled by agreement subject to approval of the administrative body.

Review and appeal of the compensation awarded to the injured worker, as well as regulation of attorney's fees, are provided for by the various acts. Requirements vary concerning the keeping of accident records by the employer, but all states require that the employer report injuries to a designated authority. Failure to report in accordance with the applicable statute can result in fines and, in some jurisdictions, even imprisonment. Except for preliminary reports, the contractor's insurance company usually makes the formal reports required by law except in states with monopolistic funds (see Section 8.37).

Benefits to an injured worker are only those provided by the law and approved by the administrative authority. Should the injured party not be satisfied with the award as specified by law, he can appeal to the court designated by the act. Such appeal must be initiated within a statutory period. In some jurisdictions the employee can file suit against his employer under the compensation act for alleged failure to provide safety devices.

8.35   WORKERS' COMPENSATION BENEFITS

All workers' compensation laws provide various forms of benefits for the disabled employee or his dependents for job-connected injury, death, or sickness regardless of how the disability or death may have been caused. The monetary value of such benefits varies substantially from one state to another. These benefits include medical treatment, hospitalization, and income payments for the worker during his disability. Also provided are funeral expense and survivor death benefits. Many jurisdictions include special benefits such as a lump-sum payment for disfigurement, rehabilitation services, and extra benefits for minors injured while illegally employed. Most statutes now stipulate that medical benefits shall be provided without limitation as to maximum total cost and that wage compensation shall be paid for as long as the worker is disabled. In those states where benefits are limited by statute, either in terms of dollars or period of time, the contractor may purchase an endorsement to its workers' compensation insurance policy that provides "extralegal" or additional benefits for its injured employees. Most states require a waiting period before a worker can collect certain benefits, seven days being typical. However, these states normally allow a worker who has been off the job for a specified length of time to receive benefits for the waiting period retroactively.

Four classifications of work injuries are used in conjunction with workers' compensation benefits: (1) temporary-total, (2) permanent-partial, (3) permanent-total, and (4) death.

Other than medical assistance, the usual benefits provided by workers' compensation laws are quite modest, especially as compared with the worker's usual earning capacity. The idea is, of course, to give the worker sufficient monetary help so that he will not become a burden to others.

The great majority of compensation cases involve injuries of the temporary-total classification, meaning that the worker is temporarily unable to work but ultimately recovers fully from his injuries and returns to employment. Income benefits payable during the period of convalescence are determined as a percentage of the worker's average wages. Most states limit the minimum and maximum benefits payable weekly as well as the number of weeks and the total dollar amount of benefit eligibility.

Permanent-partial disability connotes a continuing partial disability, although the worker is eventually able to return to work. This form of disability is classified either as a schedule injury, meaning the loss or loss of use of a finger, eye, leg, or other part of the body, or a nonschedule injury, which is of a more general nature. Compensation acts provide additional benefits for schedule injuries.

A permanent-total disability injury prevents the worker from future employment in his former craft. Most of the state workers' compensation laws provide that specified benefit payments shall be made for life in cases of permanent-total disability. The other statutes limit the benefits as to time, amount, or both. Rehabilitation services are provided to train disabled people in new skills or occupations.

In the event of the accidental death of a worker, all states provide for the payment of death benefits to his family or other dependents. A few acts provide for payment of benefits to the widow or widower for life or until remarriage, and to children until they reach a prescribed age. Most states place a limitation on the time period or total amount of such payments.

8.36   ADDITIONAL PROVISIONS OF WORKERS' COMPENSATION LAWS

Workers' compensation statutes now include occupational diseases within their coverage. Provisions vary, but compensation benefits are generally the same as for other forms of disability. Many states have provided for extended periods of time during which claims may be filed as a result of certain latent, slowly developing occupational diseases.

The identification and compensation of occupational diseases is presently in a state of transition and is becoming a disruptive element of major proportions in the area of workers' compensation. Claims under the name of occupational disease have increased greatly, stemming from certain practices and the use of many substances common in construction. An example of this trend is the growing number of claims made because of the effects of asbestos. Claims are often for very large sums of money, reflecting the long periods of time between the first exposure and discovery of the disease, as well as the attendant disability. New products are constantly coming into use, and the effects of different substances in combination are only now being discovered. There are serious questions concerning who should pay, how much should be paid, and who is eligible to receive such payments.

In addition to the increasing incidence of occupational disease, the definition of what constitutes an occupational disease is being broadened. Benefits are now being awarded for illnesses caused by job mental stress and for cumulative injury such as loss of hearing and cardiovascular problems. More and more claims under workers' compensation are coming from older workers and as the result of more diseases being found to be occupational and compensable. Occupational diseases are beginning to put a severe strain on the various state workers' compensation plans.

Second-injury or special-disability funds have been developed to meet problems arising when an employee, previously injured, suffers a second injury that, together with the first, results in a combined disability much more severe than that caused by the second injury alone. Under ordinary circumstances the employer in whose employ the second injury took place must provide benefits as dictated by the total resulting disability. This condition has sometimes made employers reluctant to hire previously injured persons. All states now provide that the employer is responsible for payments required by the second injury only. Additional compensation to the injured employee, as called for by the combined effects of his injuries, comes from second-injury funds, which were created at the time the second-injury provisions were enacted into law.

8.37   WORKERS' COMPENSATION INSURANCE

A workers' compensation statute requires the contractor to provide its injured workers with all benefits as may be required by that state's law. The usual way in which a contractor does this is through workers' compensation insurance, although almost all states allow the contractor to carry its own risk as a self-insurer if it can provide satisfactory evidence of its financial ability to do so. Monopolistic state funds have been established in six states and the provinces of Canada. Under the laws of these jurisdictions, employers whose operations are covered by their compensation laws are required to insure in that state's fund, although in some instances employers can qualify as self-insurers. Another twelve states have competitive state funds whereby the employer may purchase compensation insurance either from a private insurance carrier or a state fund. In the remaining states, private insurance companies provide all workers' compensation insurance. Workers' compensation insurance provides benefits only as prescribed by state laws. Claims made under federal compensation laws can be covered by an endorsement to the contractor's workers' compensation policy.

Workers' compensation insurance is unlimited in the policy and will pay the medical costs and provide the benefits required by law. This insurance also provides legal defense for the insured and pays any court awards in jurisdictions in which an injured worker can bring suit against his employer under the compensation act. If injury to sole proprietors, partners, or corporate officers is to be covered, a specific policy endorsement to that effect is required in some states. In other states, those parties are automatically included.

The contractor must have workers' compensation insurance in force for each state in which its employees may be located. The workers' compensation policy applies only to obligations imposed on the insured by compensation laws of those states listed in the policy. If a claim against the contractor originates in a state not listed, there is no coverage. It is possible for the policy to list some or all of the states except the monopolistic fund states that do not permit private workers' compensation insurance. At present, these states are Nevada, North Dakota, Ohio, Washington, West Virginia, and Wyoming.

8.38   WORKERS' COMPENSATION INSURANCE RATES

The manual rates for workers' compensation insurance are set by rating bureaus of the various states and are adjusted annually. In almost all states, premium costs are computed by multiplying each employee's total wages, exclusive of overtime pay, by the rate specified for his classification. In a few states, there is a $300-per-week limitation. In these states, any excess of wages over $300 per week is not subject to assessment. The insurance company conducts periodic audits of the contractor's payroll records to verify that sufficient premiums have been paid.

Premium rates for compensation insurance vary considerably from state to state and with the classifications of work involved. Because a higher premium rate is charged for the types of work that entail greater risks of injury, it pays the contractor to show accurately the work classifications its employees are actually performing. For example, the following is an instance of the base rates for compensation insurance per $100 of wages:

Masonry, general	$ 32.44
Carpentry, general	24.96
Roofing	49.90
Painting	17.12

Workers' compensation insurance is an expensive coverage whose premiums can constitute one-half or more of the total cost of a contractor's insurance program. Noticeable recent increases in premiums have been due to the rising cost of health care. This often leads construction businesses to achieve loss reduction through more selective hiring, substance abuse programs, strict safety regulations, employee training, and postloss claims management.

8.39   WORKERS' COMPENSATION INSURANCE RATING PLANS

There are two ways in which the annual premium the contractor pays for its workers' compensation insurance is determined. Referred to as "rating plans," one format is the traditional guaranteed-cost plan and the second is the retrospective plan. The former is more commonly used and is discussed first.

Workers' compensation insurance companies have become active participants in the field of construction safety and have established merit systems whereby contractors are rated according to their accident experience. Contractors are experience-rated by various rating bureaus, according to which those contractors with good safety records are rewarded by reductions in their compensation rates, and those with poor records are penalized through rate increases. Such experience modification operates so that an employer's past loss experience is a determining factor in its present and future workers' compensation costs. The higher the ratio of chargeable claim costs paid by the insurance company to payments received from a given employer, the higher its premium rates. Under this plan, workers' compensation insurance premiums are determined by multiplying the payroll by the appropriate manual rates and then applying the contractor's individual experience modifier. This rating modifier is the means by which the contractor's insurance cost is adjusted up or down according to its loss experience. Values of the experience modification rating can vary from approximately 0.5 to more than 2.0. A volume discount is applied regardless of insurance losses. This discount rate increases as the amount of the annual premium increases. The experience modifier is often used by a private owner when prequalifying a contractor for a project. A common procedure is to refuse to deal with a contractor whose current modifier is greater than 1.00, especially if this value has been on the rise during recent years.

Workers' compensation insurance, as well as other liability coverages, is also sold on a retrospective basis. When the retrospective rating plan is used, the contractor pays manual premium rates during the life of the policy. The insurance carrier periodically evaluates the contractor's losses under the policy. The contractor receives a rebate if its loss experience has been good and is required to pay an additional premium if its loss experience has been unfavorable. Compensation insurance sold on a retrospective basis stipulates both a minimum and a maximum premium rate under the policy. Several different retrospective plans are available, each plan specifying a different set of minimum and maximum premium rates and providing for different possible ranges of final premium adjustments. Retrospective-rated insurance is truly a cost-plus form of insurance subject to a guaranteed maximum price but is not suitable for all contractors. Expert advice is highly desirable. As a usual rule, larger contractors with good safety programs stand to gain most from this arrangement.

8.40   WORKERS' COMPENSATION DEDUCTIBLE PLAN

In some states, a plan called the Workers' Compensation Deductible Plan is offered by insurance companies. Under this plan, if the contractor assumes personal responsibility for paying a prescribed portion of a workers' compensation claim, the premiums for the contractor's workers' compensation insurance coverage are reduced. For example, if the per accident deductible credit applied to insurance company payments to a given claimant is established as $5,000, the contractor's workers' compensation insurance premium payments are reduced by, say, 14 percent. Under this plan, the insurance company pays the entire claim and then looks for reimbursement from the contractor, generally through monthly or quarterly billings. Because the contractor holds the funds until the insurance company actually pays the claim, the insurance company usually requires financial protection from the contractor in the form of a letter of credit, surety bond, or premium trust fund.

8.41   WORKERS' COMPENSATION SELF-INSURANCE

A large majority of the compensation acts allow an employer to act as its own insurer, provided that it can satisfy certain minimum financial requirements as stipulated by the various state insurance departments. Self-insurance is limited in a practical sense to big companies having such a large spread of risks that they can assume their own liability on workers' compensation to their financial advantage. For complete self-insurance, the contracting firm must establish its own services of claim adjustment, claim investigation, safety engineering, and others similar to those furnished by insurance companies. To qualify as a self-insurer with state officials, a contractor may be required to furnish a surety bond in an amount fixed by law or the administrative agency.

Self-insurance programs have been devised between employers and insurance companies whereby the employer is self-insured up to certain maximum amounts, payments in excess of which are guaranteed by excess insurance purchased from the insurance company. Under such plans, the employer deposits a percentage (75 percent, for example) of the usual workers' compensation insurance premiums in a bank. This establishes a fund for the payment of workers' compensation benefits. The rest of the premium (25 percent, in our example) is paid to the insurance company. This payment goes in part to purchase the excess insurance coverage. The rest of the payment to the insurance company provides the contractor with the usual insurance company services pertaining to claims, safety inspections, auditing, accident reports, and medical and legal services.

8.42   EMPLOYER'S LIABILITY INSURANCE

Employer's liability insurance is written in conjunction with workers' compensation insurance and affords the contractor broad coverage for bodily injury or death of an employee arising from, or occurring in conjunction with, his employment but not covered under workers' compensation law. Employer's liability insurance applies to the contractor's operations only in those states listed in the contractor's workers' compensation policy and to those operations in other states that are necessary and incidental to operations in the listed states.

There are instances when an injury to an employee can fall outside the coverage of workers' compensation. For example, if an employee is injured through the failure of the contractor to provide safety appliances or working conditions required by state law, the employee may elect not to receive workers' compensation benefits but to sue the contractor for damages under common law. The contractor may have an employee who is injured on a minor operation in another state where the contractor does not have workers' compensation insurance in effect. The injured employee of the contractor may collect workers' compensation benefits from the contractor's insurance company and then file suit against a third party such as the owner or a subcontractor. In this event, it is usual for the owner or subcontractor to make the contractor a party to the action. In each of the instances just cited, workers' compensation insurance provides no direct protection for the contractor. However, employer's liability insurance will pay the legal costs incurred in the contractor's defense as well as any judgment, up to the face amount of the policy.

As previously discussed, a contractor doing business in a monopolistic fund state must purchase workers' compensation insurance from that state fund. These state funds, however, do not include employer's liability insurance with the workers' compensation coverage. Here, the contractor must obtain a "stopgap endorsement," which will provide employer's liability insurance in the monopolistic states. This endorsement can apply to the commercial general liability coverage or the workers' compensation policy.

The standard employer's liability insurance coverage, as provided by the present workers' compensation rate structure, is $100,000 for each accident to one or more employees. The limit for bodily injury by disease is $100,000 per employee, with a $500,000 policy aggregate. The premium for this basic amount is included in the regular workers' compensation insurance premiums. However, higher limits can be obtained by the payment of an additional premium.

8.43   UMBRELLA EXCESS LIABILITY INSURANCE

Because of the substantial hazards that its operations present and the very high awards now occasionally made in bodily injury cases, the contractor may sometimes be concerned about the adequacy of the liability insurance it carries. A common way to eliminate this question is to purchase umbrella excess liability insurance that provides excess coverage above the underlying commercial general liability, automobile liability, and employer's liability policies.

Although the coverage of a policy of this type is not at all standard, it serves a three-fold purpose. One purpose is to cover liability claims that might otherwise fall between the coverages carried under the separate underlying liability policies. A second purpose is to raise the policy limits of the contractor's existing liability insurance to values high enough to provide protection against any foreseeable loss or combination of losses. The umbrella coverage does not eliminate the existing policies; it merely provides substantially greater protection for the same hazards insured against under the primary coverages. If the contractor incurs a liability beyond the limits of a primary policy, then the umbrella coverage is invoked. Umbrella coverage is sold on either an occurrence or claims-made basis.

The third purpose of an umbrella policy is to provide some coverages that are not provided by the underlying liability policies. Under this coverage, losses are normally subject to a $10,000 to $50,000 deductible per occurrence, however. To illustrate how an umbrella policy works in this instance, suppose a contractor has an umbrella policy with a $25,000 self-insured deductible. If the contractor's operations should cause damage to adjoining property, a loss not covered by its underlying insurance, the contractor pays the first $25,000 of the loss and the umbrella coverage pays the remainder up to the face amount of the policy.

Umbrella policies do contain a number of exclusions. Among these are damage to property owned or controlled by the insured, damage to insured's work, contractual liability, explosion, collapse, and others.

8.44   WRAP-UP INSURANCE

On large construction projects, all work is often performed under a coordinated or wrap-up insurance program provided by the owner. In such cases, the owner elects to furnish certain insurance coverages for the entire construction team of owner, architect-engineer, construction manager, prime contractors, and all subcontractors. On such projects, the contractors are required to accept the coverages that are provided by the owner. The usual procedure is for a single insurance company to provide all of the workers' compensation and employer's liability, general liability, umbrella, and builder's risk insurance for the entire project. When the wrap-up scheme is used, the owner buys and pays for the insurance stipulated. The general contractors and the subcontractors bid on the project without including the cost of the insurance in their proposals.

The prime attraction of the wrap-up form of insurance coverage is the reduced cost to the owner, both in the direct cost of the premiums and in the saving of the contractor's and subcontractors' markup on the cost of the premiums when the owner buys and pays for the insurance. Lower premium rates are effected through volume purchase from a single carrier. Wrap-up insurance also eliminates much of the administrative detail involved when each contractor and subcontractor on a project buys its own insurance. With such individual purchases, layer upon layer of hold-harmless clauses can increase the cost of the public liability insurance. On large projects, there seems to be little question that the direct cost of insurance can be reduced through use of the wrap-up insurance concept. In addition, a single carrier can provide concentrated safety inspections and quick claims service. Gaps and overlaps in coverage can be minimized, and disputes between insurance carriers eliminated.

It should be noted, however, that wrap-up plans upset the normal business arrangements that contractors have with their usual agents and insurance companies. The contractors must normally deal with a different set of safety engineers, underwriters, claims people, and auditors for each different wrap-up insurer. Repeated audits of the contractor's accounting records by different insurance companies can seriously disrupt a contractor's normal office routine and add to administrative costs.

8.45 OWNER'S LIABILITY INSURANCE

The owner is responsible for procuring its own general liability insurance that applies to its normal operations and protects it from liability that may arise because of its own negligent acts or those of its employees. It may also choose to obtain owner's protective liability coverage, either as a part of its general liability policy or by having the protective coverage furnished by the general contractor.

Despite the fact that the general contractor and the subcontractors are directly and legally responsible for liabilities arising from their operations, the owner is often made a party to legal actions arising from acts or omissions connected with their construction activities. Protective liability insurance is available, which protects the owner from injury or damage claims caused by the operations of the general contractor or any of the subcontractors. This insurance, with the owner as the named insured, covers its contingent liability for personal injury, including death or property damage, that may occur during the construction operations, as well as any liability it may incur as a result of its direction or supervisory acts in connection with the work being performed or through the omission of a duty that cannot be lawfully delegated to an independent contractor.

Where the contractor is made responsible for obtaining protective liability insurance for the owner, this insurance is normally obtained as a separate policy, although in some instances the owner can be listed as an additional insured on the contractor's commercial general liability policy. When a construction contract requires that the contractor provide protective liability insurance for both the owner and the architect-engineer, both parties can be named insureds on the same protective policy for an extra premium. If the construction contract contains an indemnity clause in favor of the owner and also requires that the contractor provide protective liability coverage, there is some duplication of coverage.

8.46 NON-OCCUPATIONAL DISABILITY INSURANCE

Disability benefit laws in a few states require most employers to provide insurance protection for their employees for disabilities arising from accidents or diseases not attributable to their occupation. Disability insurance pays a weekly benefit when an eligible wage earner is disabled by an off-the-job injury or illness. Workers' compensation insurance does not apply in such instances and does not provide any benefits. Usually, a waiting period of seven days must elapse before benefits begin, thus eliminating claims for minor disabilities. All state plans provide for maximum weekly benefits and a maximum number of weeks for which benefits are paid.

In some states, non-occupational disability insurance may be obtained through either a private or a state plan. In others, all disability insurance required by law must be acquired through the state. State-administered plans are supported by payroll taxes levied against both the employer and the employee. Private plans are acceptable, particularly if underwritten by a reputable insurance carrier. Benefits provided by private plans must at least equal those in the state plan, whereas contributions from the employee must be no greater. The cost of disability insurance is usually shared by employer and employee. Some employers, however, regard these benefits as essential to their employee relations programs and underwrite the entire cost.

In some areas the contractor's legal obligation for providing disability benefit insurance to its employees can be satisfied by contributions to union welfare funds that provide disability benefits substantially equivalent to those required by state law. Protection for nonunion employees and for union members who are not covered by such union welfare funds must be acquired through other sources.

8.47   INSURANCE CLAIMS

Every loss or liability for which the contractor's insurance may be responsible must be brought to the attention of the insurance carrier, and perhaps other parties, through the medium of a written notice or report. State workers' compensation statutes require that accident reports be submitted covering all compensable accidents to employees. In the event of accidental injury to a person who is not an employee, the contractor should, for its own protection, submit a complete report of the matter to the company carrying its commercial general liability insurance. Motor vehicle accidents must usually be reported both to the insurance company and to law enforcement agencies on special forms provided. Property damage is reported in a proof-of-loss affidavit, although final settlement is usually made on the basis of a detailed schedule of costs.

To ensure that claims are properly made and tracked, it is wise for the contractor to designate one person in its organization to assume complete responsibility for matters pertaining to insurance. In addition to the filing of reports and claims, this individual must be familiar with all aspects of the various insurance policies of the contractor and of the subcontractors. This person should keep a checklist of all coverages, have all new or potential construction contracts examined for insurance requirements, make necessary cancellations and renewals, keep a file of subcontractors' insurance certificates, and generally oversee the contractor's insurance program.

8.48   SUBCONTRACTORS' INSURANCE

By the terms of its subcontracts, the prime contractor normally requires each subcontractor to provide and maintain certain insurance coverages. Article 3 in Appendix N illustrates this requirement. For the most part, prime contractors look to their subcontractors for the same coverages and limits as are required by the construction contract with the owner. The insurance carried by its subcontractors is a serious matter to the prime contractor. The subcontractors' coverages play an important role in the total insurance protection of the prime contractor and often, either by itself or in conjunction with the prime contractor's own insurance, provide direct protection for the prime contractor. Many prime contractors require their subcontractors to name the prime contractor as an additional insured party in the subcontractors' employer's liability, commercial general liability, and automobile liability policies.

In cases where the liability insurance coverage of a subcontractor proves to be faulty or inadequate, responsibility can devolve from the subcontractor to the prime contractor. In addition, when the subcontractor does not provide insurance coverage required by law, such as workers' compensation insurance, the general contractor may be made responsible. For instance, most of the states have what are called "subcontractor-under" provisions in their workers' compensation statutes. The effect of such a provision is that a prime contractor (or subcontractor who further subcontracts) is considered to be the statutory employer of the employees of any subcontractor that does not procure the required workers' compensation insurance. In such an instance an injured employee of the subcontractor may be able to recover under the insurance of the prime contractor. Construction subcontracts may provide that if the subcontractor does not procure the insurance required, the prime contractor has the right to obtain such insurance for the subcontractor and to charge the account of the latter with the premium cost involved or to terminate the subcontract.

8.49   GROUP INSURANCE PLANS

In the usual case, the contractor procures its insurance individually from commercial companies of its choice. However, in today's market, it is not uncommon for contractor groups or members of trade associations to join together for the purpose of obtaining certain insurance coverages from a common source. The impetus for such an association lies in possible premium savings, readily available coverage to members in a tight insurance market, and possible investment income for member contractors in some cases. Many organizational schemes have been devised to provide contractor members with a variety of insurance types, some of which are discussed in the following paragraphs.

A common group insurance arrangement occurs when member contractors form a group purchase pool, called a risk retention group (RRG). This pool serves as a co-op in the purchase of designated insurance types from a selected commercial carrier, for its contractor members. Such a pool assumes no risk but uses the group's purchase volume to negotiate better coverages at reduced cost.

A more ambitious arrangement occurs when the members of a construction group establish themselves as a self-insured pool. Members of the pool invest capital into the pool and make premium payments into a trust fund. The plan is implemented by the group forming an insurance subsidiary that provides only member contractors with designated insurance coverages. Such subsidiaries are often referred to as group-owned captive insurance companies. These programs are often used when a given type of insurance becomes very expensive or difficult to obtain. For example, several local chapters of the Associated General Contractors of America, Inc. now have self-funded workers' compensation insurance programs. In such a plan, the contractor members pool their funds and insure themselves. This enables the plan participants to save on this line item of operating cost and become more cost-competitive. The groups' self-insured trusts are directed and controlled by their own trustees and pay losses covered by the insurance provided. Some of these companies prepare their own policies and handle their own claims and underwriting. Such companies function as small, self-contained insurance firms.

It is a more common practice with captive companies to engage a commercial insurance company to perform the normal duties and activities of a typical insurance firm. Participating contractors pay premiums into a trust fund, and this fund pays policy losses. The trust distributes any dividends back to the member contractors after paying claims and maintaining a reserve to meet future claims and expenses. The trust purchases reinsurance to protect itself against catastrophic loss.

There are also so-called offshore captive insurance companies. In this case the group-owned company is located in an offshore location, such as Bermuda, where taxes and other conditions are favorable to private business.

8.50   EMPLOYEE BENEFIT INSURANCE

Construction firms, like other businesses, often provide their employees with certain forms of insurance as fringe benefits. Major medical, hospital, surgical, life, accidental death, disability, and weekly income coverages are examples. Group coverage can be arranged between a construction company and an insurance carrier where certain benefits are made available to all employees of that company or to all employees in a specified class or category. As a result of the seasonal nature and labor mobility in the construction industry, an employee must normally serve a prescribed minimum time and work at least a specified number of hours per week to be eligible for company group insurance coverage. Most group insurance is sold on a yearly term basis with the rates being redetermined annually.

There are also group insurance plans that provide such benefits for all company members of a trade association or other body rather than for an individual employer. This is a broad-scale arrangement whereby all eligible firms can obtain selected group coverages from the same source. Such benefits are often provided by a multiple-employer insurance trust underwritten by a large national insurance company.

Group insurance is a useful management device in attracting and keeping company personnel. The group insurance package can be a valuable recruiting device and is a vital part of employee compensation. This is very important to company personnel, because the contractor ordinarily pays all or a substantial part of the premium cost of such insurance coverages.

8.51   CERTIFICATES OF INSURANCE

An insurance certificate is a printed form executed by an insurance company certifying that a named insured has in force the insurance designated by the certificate. Such certificates are addressed to the party requiring the evidence of insurance and list the types and amounts of insurance the insured has purchased. These certificates note the expiration dates of the policies and contain a statement to the effect that the party in whose favor the certificate is drawn will be informed in the event of cancellation or change of the insurance described.

The general contractor is often required to submit certificates of its insurance to owners and other parties. Construction contracts usually contain the provision that the contractor shall submit suitable insurance certificates to the owner or the architect-engineer at the time the contract is signed or, at least, before field operations are begun (see Subparagraph 11.1.3 in Appendix C). Many building departments require proof that certain insurance is in effect before they will issue building permits.

It is standard practice for a general contractor to require suitable and up-to-date insurance certificates from all of its subcontractors. A properly maintained file of such certificates will ensure that each subcontractor provides and maintains the insurance coverages and amounts required by law and subcontract. Many general contractors have a company policy stating that subcontractors cannot start their field operations until suitable certificates of insurance have been filed with the general contractor.

8.52 SOCIAL SECURITY

Social Security, operated by the federal government through the Social Security Administration, provides four basic types of benefits for the workers covered. Monthly cash payments are provided after a wage earner reaches a certain age and retires. When a qualified worker reaches the age of 65, hospitalization benefits are provided and supplementary medical insurance is available on application and payment of a small monthly premium. Survivor benefits are provided for the dependents of a worker when that worker dies, regardless of age. Benefits are made available to a worker who suffers a disability that renders him unable, for a period of 12 calendar months or longer, to do any substantial gainful work for which he is qualified by age, experience, training, and education. The Social Security Administration also administers the Supplemental Security Income program, which provides financial assistance to people who are blind or disabled, or who are 65 years of age or older, and who can prove a genuine financial need. Since the passage of the original Social Security Act in 1935, benefits have been liberalized and the number of workers covered has grown substantially.

Employees and employers share the cost of Social Security by paying special taxes into a fund in the U.S. Treasury, out of which benefits are paid. For each employee, the employer must contribute the same amount that is deducted from the employee's pay. The tax rates paid by employer and employee are statutory, as is the annual amount of wages subject to the tax. Both the tax rates and taxable earnings have been increased several times by congressional action.

The Social Security Administration keeps a record of each worker's wages received while in employment covered by the act. This record is maintained as a separate account for each worker, under his name and identifying number. Each individual must have his own Social Security number, which he can obtain from any local Social Security office. Any worker can check on his Social Security account by writing to the Social Security Administration and asking for a statement of his wage credits.

8.53 UNEMPLOYMENT INSURANCE

Unemployment insurance is a hybrid coverage, involving both federal and state laws, that provides weekly benefit payments to a worker whose employment is terminated through no fault of his own. The cost of the unemployment compensation system is paid by employers through federal and state taxes. Each state has some form of unemployment compensation law that works in conjunction with the Federal Unemployment Tax Act. In addition, each state has established its own administrative agency, which works in partnership with the Bureau of Employment Security, an agency of the U.S. Department of Labor. The federal government sets minimum benefit standards for the states. Each state specifies its own qualifications, the amount of the benefit, the duration of payments, and the employment covered. Excluded from most state laws are railroad workers; domestic workers; federal, state, and municipal workers; workers in nonprofit educational, religious, or charitable organizations; agricultural workers; casual labor; and those who are self-employed.

Unemployment insurance is intended to provide workers with a weekly income to assist them during periods of unemployment. There is no intent to benefit those who cannot or will not work. Only persons who have been working for a specified period of time on jobs covered by their state unemployment compensation law, who are able and willing to work, and who are unemployed through no fault of their own are eligible to receive benefits. However, some states do provide unemployment benefits under certain circumstances to workers on strike.

All employers that come under the provisions of the state unemployment insurance laws must pay taxes based on their payrolls up to a prescribed amount per calendar year for each employee. In a few states, unemployment tax is also assessed against employees. The tax applies to a contractor's entire workforce, including both field and office employees. The employer pays a part of such payroll taxes to the federal government and a larger share to the state in which the employment takes place. Federal law imposes on all covered employment a basic payroll tax whose rate is variable from year to year, depending on the financial condition of the national fund. With certain restrictions, the employer can take credit against its federal unemployment tax for its payments to the state fund and for amounts that it is excused from paying to the state because of its favorable claim experience.

Each state has an established tax rate that is adjusted up or down for a given employer, depending on its experience. A separate account is set up by the state for each employer covered. Each account is credited with tax payments made and charged with benefits paid to former employees. The extent to which the tax credits exceed or are less than the benefits charged determines by how much the employer's rate is adjusted above or below the basic tax rate. A contractor who finds it possible to offer a maximum amount of steady employment to its workers can enjoy a substantial reduction in its state unemployment tax rate. Conversely, if it has a large turnover, its tax rate can be raised. The state tax rate can also vary with the solvency of the state unemployment benefit fund.

When a worker files a claim for benefits, the employee's last employer is sent a notice. If the employer replies that the worker was separated from the firm for a reason other than lack of work, the benefits may not be chargeable against that employer's account. In addition, an unemployed worker may disqualify himself from unemployment benefits by voluntarily quitting his last job without good cause, being discharged for misconduct, being directly engaged in a strike or other labor dispute, failing to apply for or accept an offer of suitable work, and other causes. Under the existing compensation program, the states pay jobless benefits for up to 26 weeks. If the state unemployment level reaches a defined "trigger" point, the states and federal government split the cost of additional benefits for up to a total of 39 weeks. Federal supplemental compensation can provide benefits in some areas beyond 39 weeks at federal expense during periods of high national unemployment. The amount of the unemployment payments a worker can receive varies from state to state.

QUESTIONS

1. Who typically pays for an all risk builder's risk insurance policy, who is the beneficiary, and what is covered?

  2. What is the difference between a mutual insurance company and a stock insurance company?

  3. The five major types of contractor's insurance are:

  Builder's Risk

  Property

  Liability

  Employee

  Business, Accident, Life

      a. Which of these are primarily direct job costs and which are home office overhead costs?

      b. Which of these are typically required by the construction contract?

      c. Which of these basically protect the assets of the contractor?

      d. In which category above would an equipment floater policy belong?

  4. What is subrogation and how does it affect builder's risk coverage?

  5. What kind of contractor would be most apt to have an equipment floater policy and why?

  6. What is a fidelity bond and how does it work?

  7. What are some of the ways in which a contractor may incur liability?

  8. Why is it necessary for a contractor to have completed operations insurance?

  9. How does general liability insurance differ from professional liability insurance?

  10. What are the variables that affect the premium rates on worker's compensation insurance?

  11. What factors would a contractor consider when thinking of self insurance for worker's compensation?

  12. What is wrap-up insurance and why is it used on some projects?

  13. What are certificates of insurance and how are they used?

Chapter 9

Business Methods

9.1 THE CONSTRUCTION BUSINESS

This chapter discusses selected accounting and commercial procedures that are of particular importance to the conduct of a construction business. Although the popular conception of contracting involves pouring concrete and driving nails, there is much more to such a business than the construction itself. The management of a construction enterprise is vitally concerned with its field operations. However, the necessary support functions performed by the central office are an indispensable part of the total picture.

9.2 FINANCIAL RECORDS

The downfall of many otherwise well-run construction companies is the lack of accurate, detailed, and current information concerning all aspects of their financial affairs. Financial records must be maintained to serve a variety of business and management purposes. One of the important reasons why construction contractors, like other businesspeople, must keep accurate records is that the law requires it. An assortment of governmental agencies require data pertaining to taxes, payrolls, and other company information, together with a system of accounts adequate to serve the purpose. In addition, the contractor's records must provide the source material for obtaining indispensable support services. Financial statements and reports are required for submittal to bankers, sureties, owners, insurance companies, public agencies, lending firms, and others.

Beyond this area of legal and practical compulsion lies the realm of record keeping to serve the purposes of company management. The functions of a construction company's accounting system are not limited to keeping and producing the records. Although such information is basic and essential to the conduct of company operations, it is necessary to analyze and summarize such data so that it can be used to best effect. Conscious direction of company affairs through the continuous generation of carefully selected and summarized intelligence can lead to a company's success. The accounting system must provide information to assist management in controlling company operations and in utilizing its available capital to the greatest possible advantage. Without proper records it is impossible for a contractor to estimate construction costs accurately, control costs on ongoing projects, keep the company in a fluid cash position, make sound judgments concerning the acquisition of equipment, and perform the many other functions associated with the financial management of the business.

9.3   ACCOUNTING METHODS

Although the details of record keeping vary among firms, a pattern of basic accounting procedures exists that is common to the construction industry. The main focus of a contractor's accounting system centers on the determination of income and expense from each of its construction projects; that is, each contract is treated as a separate profit center. In an accounting sense, a profit center is any group of associated activities whose profit or loss performance is separately measured and analyzed. The original estimate of costs pertaining to each contract serves as a budget for that project. As costs are incurred, they are charged, directly or indirectly, against the job to which they pertain. Customarily, the keeping of a contractor's business accounts is concentrated in the home office. However, on large projects and cost-plus contracts, a subsidiary set of accounts is frequently maintained in a field office where all record keeping pertaining to the project is done. The controlling accounts of such projects are incorporated into a master set of accounting records maintained in the contractor's central office.

The cash method and the accrual method are two basic accounting procedures. Under the cash method, income is taken into account only when cash is actually received, and expense is taken into account only when cash is actually expended. Depreciation of equipment and other capital assets is, of course, a noncash expense. The cash method is a simple and straightforward form of income recognition, and no attempt is made to match revenues with the accompanying expenses. This accounting method is used principally by professional people, small businesses, and nonprofit organizations, as well as for personal records. Many small construction contractors use the cash method because of its simplicity and the fact that tax liability for contract profits is recognized only when payment has been received. If a construction company does only small jobs, maintains little or no materials inventory, and owns no equipment of consequence, the cash method is simple and adequate. However, for most construction companies the cash method does not work to the advantage of the contractor. Income earned but not received and expenses incurred but not paid as of the end of a reporting period are not recorded or reflected in the company records. An income statement (see Section 9.8) prepared on such a basis will not present a realistic indication of true profit or loss. Similarly, a balance sheet (see Section 9.9) cannot reflect accurately the company's financial condition. Such financial reports are entirely inadequate for credit and other purposes.

The accrual method is the second basic accounting procedure. Under this method, income is taken into account in the fiscal period during which it is earned, regardless of whether payment is actually received. Similarly, items of expense are entered as they are incurred, whether actually paid out during the period or not. The accrual method of income recognition is more complex because a connection is maintained between the revenues earned and the associated expenses, and it requires a more elaborate system of accounting records.

9.4   ACCOUNTING FOR LONG-TERM CONTRACTS

A long-term contract is one that is not completed within the taxable year in which such contract was entered into. The Tax Reform Act of 1986 mandates that one of three methods of accounting must be applied by contractors to their long-term construction contracts:

1. Percentage of completion method
2. Percentage of completion–capitalized cost method
3. Completed contract method

These three methods differ in the manner in which costs and expenses are matched against project revenue and in the identification of the period in which the income from a job is taken into account. The workings of these three procedures are discussed in the following sections.

9.5 PERCENTAGE OF COMPLETION METHOD

The percentage of completion method recognizes job income from long-term contracts as the work advances. Thus, the profit is distributed and taxes are paid over the fiscal years during which the construction is under way. This method has the advantage of recognizing project income periodically on a current basis, rather than irregularly as contracts are completed. Under this method, gross project income is recognized according to the percentage of the contract completed during a given fiscal year. The percentage used in this regard is the ratio of project costs incurred during the fiscal year to the total estimated contract cost, including any revised amounts to complete the work. Applying this percentage of completion to the total contract price and deducting the applicable project costs yields the net project income for the fiscal year involved. The major weakness of this procedure is its dependence on estimates of costs to complete the work, a value that can, at times, be subject to considerable uncertainty.

An alternative to the percentage of completion method of reporting income from long-term construction projects is the hybrid percentage of completion–capitalized cost procedure. Using this method, 90 percent of the revenues and expenses under the contract are accounted for using the percentage of completion method. The remaining 10 percent of the project costs are taken into account using the completed contract method. This procedure is discussed in the next section.

9.6 COMPLETED CONTRACT METHOD

The completed contract method of accounting recognizes project income only when the contract is completed. What actually constitutes completion of a contract for income-reporting purposes has been subject to varying interpretations by the courts. Present practice tends largely to establish completion only after work finalization and acceptance by the owner, although the Internal Revenue Service (IRS) takes the position that a contract is complete when it is "substantially" finished. Contractors often use one standard for their own internal accounting and reporting and another for tax reporting to the IRS. The contractor may not delay completion of a contract if its principal purpose is to defer federal income tax.

Under the completed contract method, project costs are accumulated during construction using the IRS's long-term project cost capitalization method. The scope of contract costs that must be capitalized when this method is used includes not only the costs directly related to the contract but also indirect costs attributable to the project. Costs such as general and administrative expenses and interest expense related to the performance of the contract are included, whereas overhead not directly associated with the long-term project, such as marketing, sales, preparation of unsuccessful bids and proposals, are not.

The Tax Reform Act of 1986 severely restricted the application of the completed contract method of accounting for long-term construction projects. At present, this procedure can be applied to an entire contract only by contractors whose average annual business revenue over the last three years is less than $10 million, and can be applied only to projects that require less than two years to complete.

The completed contract method can be advantageous when, on long-term contracts, the contractor cannot accurately predict the economic results of its future contract performance. This applies when project uncertainties preclude the making of accurate estimates of profits earned. Another advantage of the completed contract method is the deferral of income tax until the end of the contract and the collection of final payment. Similarly, however, loss recognition is also deferred. A disadvantage of the method is that it does not reflect current performance for long-term contracts and may result in irregular recognition of income and hence, in some situations, in greater income tax liabilities. Another potential disadvantage of the completed contract method is that the year of completion of a project can be subject to considerable uncertainty if the end of the project coincides with the end of the accounting period or if there are disputed claims unsettled at the end of that year.

9.7   FINANCIAL STATEMENTS

At intervals of time, called accounting periods, it is necessary to make a determination of the financial condition of any business. How often this is done is a matter of company management requirements and the need of outside agencies to be informed concerning the company's financial strength. Accounting periods can be of variable length, such as a month, a quarter, or a year. Accounts are closed at the end of each accounting period, and relevant financial statements are prepared. Several forms of financial statements can be derived from the company books of account. It is the purpose of such statements to group together significant facts in a way that will enable the person reading it to form an accurate judgment concerning some aspect of the company operation, such as the overall financial condition of the organization or the profit-loss results of its operations. As may be expected, company management itself makes much use of a company's financial statements. These reports are invaluable for the advance planning of operations. They reflect the company's borrowing and bonding capacities and yield much information concerning its policies with respect to purchasing, equipment ownership, and office overhead.

In addition, these financial statements serve many important functions with respect to external agencies. Bankers, surety and insurance companies, equipment dealers, credit-reporting agencies, and clients are all concerned with the contractor's financial status and profit experience. Stockholders, partners, and others with a proprietary interest use the statements to obtain information concerning the company's financial condition and the status of their investments. Two financial statements of particular importance are the income statement and the balance sheet.

9.8 THE INCOME STATEMENT

The income statement is an abstract of the nature and amounts of the company's income and expenses for a given time period, usually a quarter or full fiscal year.[1] This statement shows the profit or loss as the difference between the income received and the expenses paid out during the period. Figure 9.1 is an income statement of the Blank Construction Company, Inc. for the fiscal year ending December 31, 20–. The following paragraphs, keyed to the lowercase letters in parentheses as they appear in Figure 9.1, explain the various items:

(a) This statement has been prepared on a completed-contract basis. Project income is the total contract value of all projects completed during the period covered by the statement. The total is obtained from a supporting schedule that shows the income figures for each completed project.

(b) Project costs include all materials, labor, equipment, subcontracts, job overhead, and other expenses that have been charged to the completed projects. A supporting schedule of job costs lists these expenses for the individual projects. Included with job costs is the general overhead expense that has been allocated to the completed projects. General overhead is periodically distributed among the various projects in proportion to the costs incurred by them during the period.

(c) Subtracting project costs from project income yields net project income for the statement period.

(d) Other income lists the net income received from other sources.

(e) Total net income before taxes is the amount realized from all operations during the statement period.

(f) This sum represents federal and state income taxes paid or due for the statement period.

(g) Net income after taxes is the amount available for company expansion or for distribution to stockholders.

(h) This sum represents the earnings accumulated as of the start of the statement period. A corporation can retain its earnings for business purposes or can distribute them as dividends to the stockholders. However, there is an accumulated-earnings tax, which is a penalty tax applicable to corporate earnings accumulated for the purpose of avoiding income tax payable by its shareholders by permitting profits to accumulate in the corporation. The Internal Revenue Service currently allows corporations to accumulate up to $250,000 generally, and $150,000 for businesses serving fields such as architecture, consulting, and engineering.

(i) Dividends paid represent distribution of earnings made to stockholders in the form of dividends on common stock during the statement period. This construction company has 4,610 shares of common stock outstanding. A dividend of $8 per share was declared and paid during the fiscal year just ended.

(j) The balance represents the total retained earnings of the corporation through the end of the report period.

---

[1] The income statement is also known by other names: profit and loss statement, statement of earnings, statement of loss and gain, income sheet, summary of income and expense, profit and loss summary, statement of operations, and operating statement.

THE BLANK CONSTRUCTION COMPANY, INC.

PORTLAND, OHIO

INCOME STATEMENT
For the year ending December 31, 20-

ITEM		TOTAL
(a) PROJECT INCOME:	$8,859,138.39	
(b) Less Project Costs, including office overhead		
expense of $239,757.04	$8,705,820.15	
(c) Net Project Income	$153,318.24	$153,318.24
(d) OTHER INCOME:		
Discounts Earned	$23,064.93	
Equipment Rentals	$23,758.93	
Miscellaneous	$12,882.64	
Total Other Income	$59,706.50	$59,706.50
(e) NET INCOME BEFORE TAXES ON INCOME		$213,024.74
(f) Less Federal and State Taxes on Income		$97,616.66
(g) NET INCOME AFTER TAXES ON INCOME		$115,408.08
RETAINED EARNINGS:		
(h) Balance, January 1, 20-	$106,127.24	
(i) Dividends Paid	$36,880.00	
Total Retained Earnings	$69,247.24	$69,247.24
(j) BALANCE, December 31, 20-		$184,655.32
(k) EARNINGS PER SHARE ON NET INCOME		$40.06

Figure 9.1   Income statement.

(k) In a corporate form of business, it is usual for the income statement to show net earnings for the year (after taxes) per share of outstanding stock.

It must be realized that an income statement has several limitations. Concentrating as it does on past events, the income statement reports only the company's profit or loss experience during the reporting period and does not show the present overall financial condition of the firm. The amounts shown are, at best, only approximations. The precision implied by the figures exists in appearance only, not in fact. However, assuming that reasonable care is exercised in compiling the data, the results can be of great value when used with judgment. In the highly competitive and unpredictable construction industry, the profit reported does not always indicate the management quality of the company.

THE BLANK CONSTRUCTION COMPANY, INC.

PORTLAND, OHIO

BALANCE SHEET
December 31, 20-

ASSETS		
(a) CURRENT ASSETS:		
Cash on hand and on deposit	$389,927.04	
Notes receivable, current	$16,629.39	
Accounts receivable, including retainage of $265,686.39	$1,222,346.26	
Deposits and miscellaneous receivables	$15,867.80	
Inventory	$26,530.14	
Prepaid expenses	$8,490.68	
TOTAL CURRENT ASSETS		$1,679,791.31
(b) NOTES RECEIVABLE, NONCURRENT		$12,777.97
(c) PROPERTY:		
Buildings	$55,244.50	
Construction equipment	$388,289.80	
Motor vehicles	$97,576.04	
Office furniture and equipment	$23,596.18	
TOTAL PROPERTY	$564,706.52	
(d) Less accumulated depreciation	$422,722.51	
NET PROPERTY		$141,984.01
(e) TOTAL ASSETS		$1,834,553.29

LIABILITIES		
(f) CURRENT LIABILITIES:		
Accounts payable	$306,820.29	
Due subcontractors	$713,991.66	
Accrued expenses and taxes	$50,559.69	
Equipment contracts, current	$2,838.60	
Provision for income taxes	$97,616.66	
Total		$1,171,826.90
(g) DEFERRED CREDITS:		
Income billed on jobs in progress on December 31, 20-	$2,728,331.36	
Costs incurred to December 31, 20- on completed jobs	$2,718,738.01	
Deferred credits		$9,593.35
TOTAL CURRENT LIABILITIES		$1,181,420.25
EQUIPMENT CONTRACTS, NONCURRENT		$7,477.72
(h) TOTAL LIABILITIES		$1,188,897.97
NET WORTH:		
(i) Common stock, 4,610 shares	$461,000.00	
Retained earnings	$184,655.32	
(j) TOTAL NET WORTH		$645,655.32
(k) TOTAL LIABILITIES AND NET WORTH		$1,834,553.29

Figure 9.2 Balance sheet.

9.9 THE BALANCE SHEET

A balance sheet presents a summary of the assets, liabilities, and net worth of a company at a particular time, for example, at the close of business on the final day of a fiscal year.[2] Balance sheets are universally used to describe the financial condition of business concerns. A representative example of a contractor's balance sheet is depicted in Figure 9.2, showing assets on the left side of the balance sheet and liabilities and net worth on the right side.

The basic balance sheet equation may be stated as follows: Assets = liabilities + net worth. The balance sheet presents in analytical form all company-owned property, or interests in property, and the balancing claims of stockholders or others against this property. The foregoing equation expresses the equality of assets to the claims against these assets. Assets are defined as anything of value, tangible or intangible. Liabilities involve obligations to pay assets or to render services to other parties. Net worth is obtained as the excess of assets over liabilities and represents the contractor's equity in the business. Here, "the contractor" is used in the broad sense meaning proprietor, partners, or stockholders.

The following paragraphs consider the major headings in Figure 9.2.

(a) Current assets include cash, materials, and other resources that may reasonably be expected to be sold, consumed, or realized in cash during the normal operating cycle of the business. When the business has no clearly defined cycle, or when several operating cycles occur within a year, current assets (and current liabilities) are construed on a 12-month basis. This one-year rule applies to most contractors. Prepaid expenses represent goods or services for which payment has already been made and which will be consumed in the future course of operations.

(b) Noncurrent notes receivable are in the nature of deferred assets, representing the value of notes that become receivable at some future date or dates.

(c) Property represents the fixed assets of the business. These assets are more or less permanent in nature and cannot readily be converted into cash, at least not in amounts commensurate with their true values to the contractor. These assets generally have a useful life of several years, although assets such as buildings, equipment, vehicles, and furnishings do gradually wear out. These assets are capitalized at their purchase prices or in accordance with appropriate cost appraisals.

(d) Accumulated depreciation represents the total decrease in value of the property due to age, wear, and obsolescence. Depreciation is more fully discussed in Section 9.14.

(e) Total assets are the sum of everything of value that is in the possession of, or is controlled by, the company.

(f) Current liabilities are debts that become payable within a normal operating cycle of the business [see item (a)]. Presumably, payment will be made from current assets.

(g) In our example, the Blank Construction Company, Inc. is using the completed-contract method of reporting income. Deferred credits represent the excess of project billings over related costs on current contracts that have not reached completion by the end of the reporting period. This amount is treated as a current liability, because the company is required to render services in the future and payment has already been made for these

---

[2] The balance sheet may also be called by other names: financial statement, statement of financial condition, statement of worth, or statement of assets and liabilities.

services. If the contractor had underbilled, that is, if the billings were less than related costs, the resulting amount would be treated as a current asset.

(h) Total liabilities are the sum of every debt and financial obligation of the company.

(i) This is the capital stock account, showing the classes and amounts of stock that have been actually issued and paid for by the stockholders. In our example, the owners have purchased 4,610 shares of $100-par-value common stock.

(j) This represents the net ownership interest the corporation "owes" to the stockholders. In our example, it is obtained as the sum of the common stock investments plus retained earnings (also called earned surplus). The book value of the common stock as of December 31, 20–, is obtained by dividing $645,655.32 by 4,610, which gives $140.06 per common share. Book value excludes intangibles of all kinds that would have no value upon liquidation. The market value of this common stock, that is, the price for which the stock can be sold, may depart substantially from its book value. What shares are truly worth in a closely held corporation can be very difficult to determine. Yet such a value may be needed for estate, gift, or income tax purposes. The Internal Revenue Service has issued guides to assist in valuing closely held shares.

(k) The equality of (e) to (k) illustrates that total assets = total liabilities + net worth.

The balance sheet has considerable analytical value for those who wish to determine the financial condition of a firm. It discloses the nature and composition of a company's assets and shows how these assets are financed. The sources of funds tell a great deal about the quality of management and the stability of a contractor. The balance sheet gives a good idea of the liquidity of a firm, that is, its ability to meet its short-term financial obligations. A comparison of balance sheet values over time discloses trends in the company's management policies and financial position. There are, of course, some inadequacies associated with balance sheets. The information presented applies as of a specific date and may not be representative of the company's normal financial condition. Many asset values are only approximate, and the accounting method chosen for the preparation of a balance sheet can appreciably influence the data presented.

9.10   FINANCIAL RATIOS

Ratios of various kinds are frequently employed as quantitative guides for the assessment of a company's financial and earning position. For example, ratios can be used to analyze income statements and balance sheets. The financial analyst is not as concerned with the absolute size of the amounts in this document, but rather the relationships, or ratios, between the different values. By comparing the same ratios over a series of financial reports, very significant information can be obtained regarding a company's financial performance over the years. When such ratios are compared with similar figures of other contractors, a comparative financial picture of the business is obtained.

Many different ratios can be used to provide company management with guidance concerning the financial condition of the firm and areas that need attention. Four types of ratios are in wide use:

Liquidity Ratios. These measures reveal a company's ability to meet its financial obligations.

Activity Ratios. These ratios indicate the level of investment turnover or how well the company is using its working capital and other assets.

Profitability Ratios. These values relate company profits to various parameters such as contract volume or total assets.

Leverage Ratios. These ratios compare company debt with other financial measures such as total assets or net worth.

A number of financial ratios are commonly used by construction contractors. The value of each ratio is computed for the Blank Construction Company, Inc., using data from the company's income statement (Figure 9.1) and balance sheet (Figure 9.2). This company is represented to be a building contractor whose overall financial condition is relatively good. In brackets next to each ratio obtained for the Blank Construction Company, Inc. are shown typical national median ratios for commercial building contractors. Such median values change from year to year and vary with the size of the contractor and type of construction involved, so the values given are for illustrative purposes only.

The following data are taken from, or calculated based on values in, Figures 9.1 and 9.2:

Current assets	=	$1,679,791.31
Inventory	=	26,530.14
Quick assets = current assets minus inventory	=	1,653,261.17
Current liabilities	=	1,181,420.25
Total assets	=	1,834,553.29
Total liabilities	=	1,188,897.97
Fixed assets	=	141,984.01
Net worth	=	645,655.32
Net working capital = current assets minus current liabilities	=	498,371.06
Net project income (before taxes)	=	153,318.24
Project income	=	8,859,138.39

1. Quick assets to current liabilities (quick ratio) $= \dfrac{\$1,653,261.17}{\$1,181,420.25} = 1.40\ [1.2]$

2. Current assets to current liabilities (current ratio) $= \dfrac{\$1,679,791.31}{\$1,181,420.25} = 1.42\ [1.3]$

3. Total liabilities to net worth $= \dfrac{\$1,188,897.97}{\$645,655.32} = 1.84\ [2.0]$

4. Project income to net working capital $= \dfrac{\$8,859,138.39}{\$498,371.06} = 17.8\ [17.9]$

5. Project income to net worth $= \dfrac{\$8,859,138.39}{\$645,655.32} = 13.7\ [9.9]$

6. Fixed assets to net worth $= \dfrac{\$141,984.01}{\$645,655.32} = 0.22\ [0.3]$

7. Percent net project income (before taxes) to project income $= \dfrac{\$153,318.24\ (100)}{\$8,859,138.39} = 1.73\ [1.6]$

8. Percent net project income (before taxes) to net worth $= \dfrac{\$153{,}318.24\,(100)}{\$645{,}655.32} = 23.7\,[19.1]$

9. Percent net project income (before taxes) to total assets $= \dfrac{\$153{,}318.24\,(100)}{\$1{,}834{,}553.29} = 8.4\,[5.5]$

The ratios cited here are illustrative only of contractors' financial experience in general, and the ratios for an individual contractor can, and often do, vary substantially from these median values. Nonetheless, to the trained eye such financial ratios tell a great deal about the business, its financial condition, and the quality of its management.

9.11 SIGNIFICANCE OF RATIOS

Each financial ratio conveys its own message. In the following paragraphs, the numbers in parentheses correspond to the ratio equations shown in the preceding section.

(1) The ratio of quick assets to current liabilities (quick ratio) shows the number of dollars immediately available to cover current debt. This ratio does not consider the value of inventory, as inventory may be hard to convert into cash in an emergency. When this ratio is 1.0 or larger, the business is said to be liquid and the ratio is usually considered adequate. This ratio is sometimes referred to as the acid test for liquidity.

(2) The ratio of current assets to current liabilities (current ratio) is probably the most significant short-term financial measure. It is a test of the firm's ability to meet current obligations. The higher the ratio, the greater the assurance that short-term debts can be paid. In the construction industry, a current ratio of at least 1.5 is generally regarded as being favorable.

(3) Total liabilities to net worth (debt to equity ratio) indicates the relative amounts that creditors and owners have invested in the business. Values of 1.0 to 2.0 for this ratio are usually considered acceptable in the construction industry.

(4) Project income to net working capital and (5) project income to net worth convey the same general message. These ratios measure the rate of capital turnover, showing how actively the firm's capital is being put to work. If capital is turned over too rapidly, liabilities can build up at an excessive rate. If the ratios are too low, funds become stagnant and profitability suffers. In construction, these ratios are higher than in most other industries. These high values partially explain why entry into the construction business is so easy. A construction contractor can do more work per dollar of invested capital than is possible for most other industries.

(6) Fixed assets to net worth can vary greatly from one construction type to another, being much higher for engineering than for building construction. In general, a higher value is undesirable because heavy investment in fixed assets (principally equipment) indicates that the firm may have low net working capital or has incurred substantial funded debt to supplement working capital.

(7) Percent net project income (before taxes) to project income reveals the average profit margin realized on the fieldwork. This figure is typically small in the construction industry and reveals why the rate of contractor business failures is so high. It does not require much to raise the cost of a project by a percent or two and change it from a winner to a loser.

(8) Percent net project income (before taxes) to net worth is perhaps the most important of the long-term ratios, because it reflects the efficiency with which invested capital is employed. Considering the risk, the contractor should expect to earn more on its investment than it could realize from current market dividends or interest rates.

(9) Percent net project income (before taxes) to total assets relates company profits to the level of assets available to earn a profit. Smaller construction corporations may have somewhat low profitability ratios, which can, at least partially, be explained by the respectable salaries of company officers. A significant proportion of company profits is often taken in the form of personal salaries.

9.12 CONSTRUCTION EQUIPMENT ACQUISITION

For contractors whose business operations require substantial spreads of construction equipment, the manner in which the necessary units are acquired is an important matter. Construction equipment is very expensive and operates in an environment that requires major support in the areas of maintenance, repairs, and parts. The historical pattern of equipment acquisition in the construction industry has been for the contractor to purchase and own the required units. However, contractors are coming to the realization that construction equipment is a very costly business aspect that must be carefully managed. Most construction firms with a relatively stable volume of work within a limited geographical area continue to find it desirable and economical to own their own fleets. However, contractors increasingly find that renting or leasing their construction equipment is economically preferable to ownership. Equipment management now requires that the contractor treat his equipment as an independent profit center and carefully evaluate every aspect of equipment acquisition.

Although the direct cost of equipment rental can be substantially higher than the cost of either leasing or ownership, rental can have its advantages. For example, it can be an advantageous way in which to keep the work progressing when there is an equipment breakdown on the job or to meet specialty job requirements. Rental can be very efficient for low-percentage utilization and for short-term peak or seasonal use. Rentals may also be considered for the purpose of evaluating machines for eventual purchase or to fill in for units that are being repaired. Equipment is not rented for a guaranteed period and can be returned at any time. Rental can be a valuable option when the job site is far removed geographically from the contractor's other operations, or for providing specialized items. Equipment maintenance and repair are the responsibility of the dealer or rental yard. Rental can conserve company capital and give a better ratio of assets to liabilities. Many equipment centers now specialize in the rental of construction equipment. Rental rates normally vary with the length of time a unit is needed, with short-term rates being higher. It is often possible to arrange rental with an option to purchase.

Leasing is a common and widely used means of acquiring construction equipment for a term of one year or longer and has several potential financial advantages for the contractor. Leasing is a useful step between renting and buying. This form of financing is considered an operating expense, not a liability, as with a bank loan. Leasing can improve a contractor's working capital position by avoiding having funds tied up in fixed assets. Thus, contractors can acquire equipment without adversely affecting their budgets. Leasing can require no initial cash outlay, because it can provide complete financing of the equipment. Certain tax advantages are possible with leasing, and options can be added as necessary to meet a given contractor's needs. Under certain circumstances, lease payments compare favorably with ownership costs. Many leases provide that, at the expiration of the lease period, the contractor has a purchase option if there is a continuing need for the machine and if it is worth the additional payment. Lease agreements for construction equipment normally extend for periods of one year or more, whereas renting is for shorter terms.

Purchase of new or used equipment is the most common method of acquiring units that receive substantial usage in the contractor's field operations. The purchase of construction equipment is a major management decision and is one deserving of careful study. Considerations such as present and future need, cost, unit size and capacity, model reliability and ruggedness, risk of obsolescence, and how the unit will fit in with present equipment are all important aspects of selection. A major aspect of the decision is how the acquisition is to be financed. When purchasing equipment, the contractor is well advised to investigate the source of that equipment. Equipment obtained from full-service, factory-authorized distributors usually costs more but can offer added value to the customer in the crucial areas of service and parts support, product warranties, purchase options, and manufacturer/distributor financing plans.

9.13 EQUIPMENT MANAGEMENT

Many contractors own and use extensive spreads of equipment to accomplish their construction projects. Equipment is very expensive, and the contractor's investment in "iron" can easily run into the tens of millions of dollars. Equipment makes money for the contractor, but it must be carefully managed to produce the expected income. Consequently, equipment must be managed just as is any other capital investment. In the case of construction equipment, management involves making informed judgments about equipment acquisition and financing, establishing a comprehensive preventive maintenance program, maintaining accurate and current records of equipment income, expense, and usage, and establishing an appropriate company policy concerning equipment replacement.

A rigorous preventive maintenance program is essential to profitable equipment utilization and management. Equipment downtime can have a serious effect on project costs and schedules. A contractor's preventive maintenance program must be tailored to its own equipment specifications and the job site conditions where the equipment operates. Special attention must be given to those key machines whose downtime would have a particularly severe effect on production and costs.

Once equipment is obtained, the contractor attempts to recover the acquisition costs by using it and realizing residual values upon its disposition. The accounting procedures used to charge construction projects with equipment costs vary substantially from one contractor to another. However, a common procedure is to establish an internal rental rate for each piece of equipment, a topic previously discussed in Sections 5.26–5.28. During the progress of work in the field, a charge equal to the internal rental rate times the number of equipment hours, weeks, or months is made against the project. At the same time, an equal but opposite credit is made to the ledger account of that equipment unit. This is the same equipment account previously discussed in Section 5.27, where all items of expense, exclusive of operating labor and hours of usage, are continuously maintained for that equipment item.

In essence, what the contractor is doing is establishing, at least on paper, a separate company that owns, services, and maintains all of its major equipment and rents it to the contractor at predetermined rates. This equipment accounting procedure provides a cumulative record of expense and earnings for each major equipment unit. How these figures compare provides company management with invaluable information concerning equipment utilization, maintenance, costs, and replacement. In short, the company knows which machines are paying their way and which are not. Many construction firms have a company policy stating that when the annual cost of an equipment item exceeds its annual earnings, it is either replaced or sold. A large construction company is apt to treat its equipment division as a separate profit center. This just says that the equipment investment is treated as a separate business, with the return on the investment required to be commensurate with the risk. Whether a contractor expects a break-even performance or a profit on its equipment investment depends on company policy.

The replacement of a large equipment item is a major decision that requires considerable study. Deciding on the replacement life presents a problem in timing and is done by contractors in different ways. However, it is well established that the replacement life is not the age at which the machine is no longer able to produce. Rather, replacement life is the equipment age that optimizes its advantage to the owner. Methods are available to determine the point in time when the production costs of a machine are at a minimum. Another form of study establishes when profits earned by the equipment item are at a maximum. Various forms of discounted cash flow models are available. Every construction firm needs some formal system of analysis to indicate the most propitious time to replace a given major piece of equipment. In a general sense, when the average time rate of expense for a given machine exceeds the historical rate of similar machines, it needs to be replaced or rebuilt.

Associated with the replacement of certain equipment types is the possibility of rebuilding a presently owned unit rather than replacing it. The cost of rebuilding a machine is normally one-half or less of the price of new equipment. This can be an efficient way to extend the life of older machinery and restore its original productivity. This procedure can be an attractive alternative for contractors who may have trouble supporting the financing of new equipment models.

9.14 EQUIPMENT DEPRECIATION

From the day of its acquisition, most business property steadily declines in value because of age, wear, and obsolescence. This reduction in value, called depreciation, is a cost of doing business and represents the physical depletion of assets such as offices, warehouses, vehicles, computers, furniture, and construction equipment of all kinds. Because of the continued decline in the value of such property, its cost must be amortized over its useful life so that the contractor recovers its investment and has capital available when replacement of the property becomes necessary. Because depreciation is a cost of doing business, it is treated as an expense for tax purposes. As such, the Internal Revenue Service dictates how depreciation is to be calculated for tax purposes.

Depreciation is an especially important matter for contractors that own large fleets of equipment. The average contractor specializing in heavy and highway work may well have an equipment investment of $25 million or more, 20 times that of a building contractor with a comparable contract volume. Depreciation charges for equipment-oriented contractors account for an appreciable portion of their annual operation expense. Basically, depreciation systematically reduces the value of a piece of equipment on an annual basis. The sum of these reductions at any time is a depreciation reserve which, when subtracted from the initial cost of the equipment, gives its current book value. The initial cost of an asset, for depreciation purposes, includes not only the sale price but also the costs of taxes, freight, unloading, assembly, delivery, and setup expenses. The book value represents the portion of the original cost that has not yet been amortized. As previously stated, depreciation charges are a cost of doing business and correspondingly reduce company earnings. At the same time, the depreciation reserve represents capital retained in the business, ostensibly for the ultimate replacement of the capital assets being depreciated. For purposes of illustration and comparison of depreciation methods, let us consider a $57,000 trencher with a service life of five years.

Figure 9.3 presents the results of depreciating this trencher by three different procedures. The depreciation values shown in Figure 9.3 for years 1 and 6 reflect a half-year convention rule. This is a procedure mandated by the Internal Revenue Service whereby property placed in service at any time during a given tax year is treated as having been placed in service at the midpoint of that tax year. Although this rule may seem to encourage contractors to purchase all their equipment as late in the year as possible, the Internal Revenue Service has limited application of this rule to contractors who make less than 40 percent (by dollar amount) of their annual equipment purchases in the final quarter of the year. When a contractor exceeds this limit, the midquarter convention comes into effect. This convention requires the contractor to begin depreciation for each piece of equipment midway through the quarter in which it was put in use. For the purpose of this example, the midyear convention applies. The trencher was obtained during year 1 and, with a service life of five years, will be completely charged off in year 6. One-half year of depreciation is charged for year 1 and for year 6 in Figure 9.3. The sections following discuss methods of depreciating assets for income tax purposes, using the trencher for purposes of illustration.

9.15   STRAIGHT-LINE DEPRECIATION

The straight-line method is the simplest depreciation procedure. It writes off the depreciable value of an asset at a uniform rate throughout its service life. It assumes, for record-keeping purposes, that the amount of value loss is constant year by year. There are sometimes objections to the use of the straight-line method, because the value decline of the capital item never actually follows such a course. It can be argued that most assets decrease in value more rapidly during the early years of their life.

In reality, few would contend that the straight-line procedure measures accurately the actual decline in asset value. In this regard, however, it must be recognized that asset appraisement is not the objective of depreciation accounting. This accounting method is concerned with spreading the cost of an asset over its useful life in a systematic and rational manner, rather than attempting to gauge depreciation charges to parallel physical decline. Basically, it must be viewed as a process of allocation and not one of valuation. Contractors widely apply straight-line depreciation to their equipment for internal purposes such as estimating equipment costs and equipment cost control on ongoing projects. The third part of Figure 9.3 illustrates the workings of the straight-line method using a salvage value of zero.

TRENCHER DEPRECIATION
5-year property

Original Cost = $57,000   Depreciable Value = $57,000

Placed in Service During	End of Year	MACRS Annual Depreciation Calculation	Accumulated Depreciation	Book Value
	2005	20.00% ($57,000) = $11,400	$11,400	$45,600
	2006	32.00% ($57,000) = $18,240	$29,640	$27,360
2005	2007	19.20% ($57,000) = $10,944	$40,584	$16,416
	2008	11.52% ($57,000) = $6,566	$47,150	$9,850
	2009	11.52% ($57,000) = $6,566	$53,717	$3,283
	2010	5.76% ($57,000) = $3,283	$57,000	$0

Placed in Service During	End of Year	MACRS Special Depreciation Allowance (SDA) Annual Depreciation Calculation	Accumulated Depreciation	Book Value
	SDA	30.00% ($57,000) = $17,100	$17,100	$39,900
	2005	20.00% ($39,900) = $7,980	$25,080	$31,920
	2006	32.00% ($39,900) = $12,768	$37,848	$19,152
2005	2007	19.20% ($39,900) = $7,661	$45,509	$11,491
	2008	11.52% ($39,900) = $4,596	$50,105	$6,895
	2009	11.52% ($39,900) = $4,596	$54,702	$2,298
	2010	5.76% ($39,900) = $2,298	$57,000	$0

Placed in Service During	End of Year	Straight-Line Alternative Annual Depreciation Calculation	Accumulated Depreciation	Book Value
	2005	10.00% ($57,000) = $5,700	$5,700	$51,300
	2006	20.00% ($57,000) = $11,400	$17,100	$39,900
2005	2007	20.00% ($57,000) = $11,400	$28,500	$28,500
	2008	20.00% ($57,000) = $11,400	$39,900	$17,100
	2009	20.00% ($57,000) = $11,400	$51,300	$5,700
	2010	10.00% ($57,000) = $5,700	$57,000	$0

Declining Book Values for Various Depreciation Methods

Figure 9.3 Annual depreciation of trencher.

254

9.16   ACCELERATED DEPRECIATION

Accelerated depreciation gives a faster write-off of the cost of an asset during the first years of its life than in the later years. There is considerable adherence to the opinion that a better matching of revenue and expense is achieved by applying accelerated depreciation to construction equipment, because new equipment is generally more productive than old. The use of procedures that afford larger depreciation deductions during the earlier years of asset ownership may also have other important advantages for a contractor. For instance, rapid initial depreciation causes cash to be retained in the business. Larger depreciation charges against regular income cause a reduction of income taxes for that year.

Accelerated methods provide for faster recovery of equipment value, consequently offering the contractor some measure of protection against unanticipated contingencies later in the life of the equipment, such as obsolescence, excessive maintenance, rapid wear, or need for equipment of a different size or capacity. Rapid depreciation amortizes the contractor's equipment investment more quickly and simultaneously decreases the book value at the same rate. Speedy reduction of book value leaves the contractor much freer to dispose of unsatisfactory equipment and, to some degree, can also assist in combating the problem of replacing equipment during periods of inflation. In the usual case, contractors favor using accelerated depreciation of their equipment and other business assets for purposes of income tax reporting. At present, the only accelerated method of depreciating business property approved by the Internal Revenue Service is the Modified Accelerated Cost Recovery System.

9.17   MODIFIED ACCELERATED COST RECOVERY SYSTEM

The Economic Recovery Tax Act of 1981 established a detailed depreciation system for business property, which has gone through several subsequent revisions. It is currently known as the Modified Accelerated Cost Recovery System (MACRS). This method is required for business income tax reporting and is, basically, an accelerated depreciation procedure, with the provision that taxpayers can elect straight-line depreciation if they wish. This system is relatively simple and easy to use. Equipment life and salvage value are a matter of law rather than negotiation. This system makes it possible to depreciate construction equipment to zero salvage value over periods much shorter than their typical useful lives.

MACRS identifies several classes of property and defines the allowable depreciation amounts that are calculated as a stipulated percentage of an asset's adjusted cost. The personal property classes of interest to contractors are three- and five-year assets, the year classification depending on the property type involved. For example, three-year property includes automobiles and light-duty trucks, and five-year property includes essentially all other contractor equipment.

Using the MACRS procedure, the value of each piece of equipment is decreased annually by the depreciation charge calculated. The sum of these charges at any point in time is a depreciation reserve which, when subtracted from the initial cost of the equipment, gives its current book value.

9.18   MACRS DEPRECIATION PROCEDURE

The Modified Accelerated Cost Recovery System has been gradually phased in over the years. All assets placed in service before 1981 continue to be depreciated using the same methods the contractor had been using: the straight-line, declining-balance, sum-of-the-years, or production procedure. The cost recovery laws instituted over the past 25 years prescribe annual recovery percentages for property placed in service in the different years.

For tax purposes, a contractor wants to write off construction equipment as rapidly as possible. Using the prescribed MACRS deduction rates accomplishes this. The 2003 Jobs and Growth Act further accelerated the deduction rate for MACRS depreciation. This Act has allowed, on a limited basis, additional first-year depreciation of certain assets of 30 percent and 50 percent of the depreciable basis (depending on the asset type). This allowance is referred to as the Special Depreciation Allowance (SDA). The center portion of Figure 9.3 illustrates the effect of this SDA. Although contractors generally prefer rapid depreciation on assets, taxpayers have the option of using either MACRS accelerated rates or straight-line depreciation for three- or five-year property. In addition, there are provisions allowing the taxpayer to depreciate three- and five-year property over longer periods using straight-line depreciation. These MACRS options seemingly would rarely work to the advantage of a construction firm for income tax reporting.

It is permissible by Internal Revenue Service regulations to employ different depreciation methods for internal financial reporting and tax reporting purposes. Contractors may use straight-line (or any other conventional procedure) for the preparation of their usual financial statements and the MACRS accelerated rates for income tax purposes.

Figure 9.3 illustrates how recovery values are obtained using equipment put in service after May 5, 2003. The new $57,000 ditcher is used for purposes of illustration. Under MACRS, the ditcher is classified as five-year property and uses the one-half-year convention rule, as discussed in Section 9.14. There are numerous other rules and regulations associated with MACRS. The contractor should seek advice on matters pertaining to real property, disposition of property, partial expensing, and other depreciation tax matters.

9.19   PROCUREMENT

Procurement is a function of paramount importance in a contractor's organization. This activity includes a number of different actions, each playing an indispensable role in obtaining the goods and services needed for company operations. Procurement is a centralized company effort, often operating as a separate department. Gathering together all such activities is economical in terms of both time and money. It concentrates expertise and affords protection against improper and unnecessary expenditures of many kinds. Procurement personnel occupy a position of great responsibility within a contractor's organization and perform many duties. Procurement involves the preparation and use of a number of standard document forms such as requisitions, purchase orders, and subcontracts. The principal procurement functions are discussed in the following paragraphs.

Purchasing. Essentially, purchasing is the obtaining of equipment, tools, materials, stores, supplies, fuel, parts, motor vehicles, and services of every sort that are needed by the office organization, the storage yard and warehouse, and the field projects. This involves the processing of requisitions, obtaining and analysis of bids, and preparation and issuance of purchase orders. Whereas most project-related purchases are governed by owner specifications, nonproject purchases frequently require the development of purchasing specifications. This task is an essential component of the contractor's purchasing function. When a general contract change order affects an outstanding purchase order, the contractor's purchasing agent must make a suitable change.

Expediting and Receiving. After an order has been placed, expediting contact must be maintained with the vendor to ensure timely delivery. The procurement department must remain in regular contact with the supplier until orders are received. This is especially true for project materials. The designation of a required delivery date in a purchase order is no guarantee that the vendor will deliver on schedule. Energetic follow-up action is a necessity if the most favorable delivery schedule is to be realized. When the goods are shipped, adequate arrangements must be made for their receipt, unloading, and storage.

Inspection. Upon receipt of the goods, they must be inspected immediately to verify quantity, quality, and other essential characteristics. In case of shortage, loss, damage in transit, or shipment of the wrong goods, immediate action is required. If quality-control verification tests are called for, samples are taken for that purpose. A written receiving report is prepared and filed for each delivery.

Shipping. The contractor's purchase order, in many instances, designates the method of shipping the goods. This is an important aspect of procurement and requires detailed knowledge concerning the various shipping alternatives and the time and cost implications of each. In the case of strikes or other disruptive events affecting the transportation industry, last-minute changes in shipping arrangements are sometimes needed. Shipments lost or misplaced in transit must be traced, and suitable claims made where goods are damaged or lost.

Subcontracts. This procurement function includes the preparation and processing of project subcontracts. The essential information needed to do this comes from the project estimate. When a general contract change order affects subcontractors, suitable changes to the subcontracts affected are prepared and processed.

9.20   CASH DISCOUNTS

Discounts are frequently offered to the contractor by material dealers and others as an inducement for early payment of bills. These are not trade discounts, such as those provided by adjusting a catalog price or making a concession for quantity buying. Rather, cash discounts are in the nature of a premium given in exchange for payment of an invoice before it becomes due, and the buyer is entitled to the discount only when payment is made within the time specified.

Many cash discount terms are in commercial usage; three are discussed here. The expression "2/10 net 30" means that if the contractor makes payment within 10 days of the invoice date, 2 percent can be deducted from the face amount of the invoice. The bill in its full amount is due and payable 30 days after the date of the invoice. Material dealers normally date their invoices the day the goods are shipped. If the customer's location is close by, there is time to check the goods before paying within the discount period. A customer who is far removed geographically, however, may have to pay before the goods are actually received in order to take advantage of the discount. To overcome this disadvantage to distant customers, some vendors, upon request, will mark their invoices "ROG" or "AOG" to indicate that the discount period begins on "receipt of goods" or on "arrival of goods."

A cash discount provision similar to that just discussed uses the term "prox," which means the discount period is expressed in terms of a specified date in the next month after a shipment is made or billed. The expression "2/10 prox net 30" means a 2 percent cash discount is allowed if the invoice is paid not later than the 10th day of the month following the purchase. The net due date of the account is 30 days from the first of that month.

Another commonly used discount is expressed as "2/10 EOM," which indicates a 2 percent discount can be taken if the invoice is paid by the 10th day of the month after the purchase is shipped. If the buyer does not make payment within this period, the bill is considered to be due net thereafter.

Cash discounts are treated as income and appear as "discounts earned" on the income statement (Figure 9.1). Based on a single invoice, 2 percent may appear to be inconsequential. However, if a contractor is able to take this discount on a million dollars' worth of materials in a year's time, it has increased its earnings by $20,000. When a contractor fails to discount its bills, its sources of credit generally view this with some agitation, because it may be symptomatic of cash flow problems (see Section 9.32) and, perhaps, of more serious financial ills.

The status of cash discounts is a matter of sufficient import to be clearly spelled out in cost-plus contracts. It is standard practice in such agreements that all cash discounts accrue to the contractor, except when the owner has advanced the contractor money from which to make payments. Documents of the American Institute of Architects reflect this policy, as shown by Article 9.1 in Appendix K. However, some forms of cost-plus contracts stipulate that the reimbursable cost of materials shall be decreased by the amount of any cash discounts taken, regardless of who provides the funds to make payment.

9.21   TITLE OF PURCHASES

Most contractors regularly purchase appreciable quantities of construction materials. During the process of transport, delivery, unloading, and storage of these materials, a variety of losses or damage can and do occur. When such difficulties arise, the Uniform Commercial Code largely controls the mutual rights and responsibilities of the buyer and the seller. Normally, the risk of loss or damage as related to personal property rests with the person holding title. Responsibility for loss and the rights of each party as buyer or seller depend on whether title of the goods has passed from seller to buyer at the time the issue is raised. For this reason, the time at which the title of a purchase passes from the vendor to the contractor is a matter of considerable importance.

Fundamentally, title to personal property that is the basis of a sale passes from seller to buyer when the parties intend it to pass. If the sales contract either states or clearly implies at what time title is to pass, the terms of the agreement govern. Seldom, however, do the parties to a sales agreement insert specific provisions concerning the passage of title. Consequently, the Uniform Commercial Code has established a standard set of rules in this matter. In the absence of any expressed intention to the contrary, title passes in accordance with the following general rules:

Cash Sale. Agreements that call for delivery and payment to take place concurrently are called cash sales. Title to the goods passes when the goods are paid for and delivery takes place.

On-Approval Sale. When goods are delivered to the buyer on approval or trial, title changes when the buyer indicates acceptance, when the goods are retained beyond the time fixed for their return, or when the goods are retained beyond a reasonable time.

Sale or Return. Title passes when goods are delivered to the buyer. However, the buyer has the option to return the goods within a fixed period.

Delivery by Vendor. If the purchase order requires the seller to deliver the goods to the buyer's destination and the seller itself makes delivery, the seller retains title until the goods are delivered.

Shipment by Common Carrier. There is a general rule that when the goods are shipped by common carrier to the buyer, title passes to the buyer when the seller delivers the goods to the carrier for transportation. However, the following exceptions apply:

1. When the seller fails to follow shipping instructions given by the buyer, such as when the buyer names a particular carrier and the seller ships by another.

2. When the seller is required to deliver at a particular place, such as the buyer's dock or railroad siding.

3. When the seller is required to pay freight up to a given point, as in FOB agreements.

4. When the seller is required by purchase order or custom to make arrangements with the carrier to protect the buyer, as by declaring the value of the shipment or as in CIF agreements, and fails to do so.

5. When the goods shipped do not correspond in both quality and quantity to those ordered.

6. When the seller reserves title by retaining the bill of lading.

FOB. Standing for "free on board," FOB indicates that the seller must put the goods on board the common carrier free of expense to the buyer, with freight paid to the FOB point designated. For example, contractors' purchase orders frequently specify delivery as FOB job site or FOB storage yard. Under an FOB agreement, title goes to the buyer when the carrier delivers the goods at the place indicated. FAS, free aboard ship," signifies the same arrangement where delivery is made by ship.

CIF. Standing for "cost, insurance, freight," CIF indicates that the purchase-order price includes the cost of the goods, customary insurance, and freight to the buyer's destination. Title passes when the seller delivers the merchandise to the carrier and forwards to the buyer the bill of lading, insurance policy, and receipt showing payment of freight. C&F, "cost and freight," indicates the same shipping arrangement except that the vendor need not obtain insurance.

COD. Meaning "collect on delivery," COD indicates that title passes to the buyer, if the buyer is to pay the transportation, at the time the goods are received by the carrier. However, the seller reserves the right to receive payment before surrendering possession to the buyer.

9.22 CHECK ON PROJECT FINANCING

An important aspect of a contractor's business is to obtain advance information concerning a private owner's finances and credit rating. Just as the owner is concerned with the capacity and capabilities of the contractor, so the contractor must be assured that the owner has the necessary financial means to carry out its end of the bargain. Contractors too often pay little or no attention to the matter of what arrangements a private owner has made for project funding. This matter cannot be safely ignored, because the contractor must usually invest a substantial amount of its own money in a project before receiving covering payments from the owner. A default by the owner in meeting its payment obligations under the contract can have a very serious impact on the contractor's business. The appreciable risks assumed by the contractor need not be increased by the chance of a default in owner payments.

Before obligating itself by bid or contract to a private owner, a contractor should make some investigation concerning the financial integrity of the owner and the source of financing for the project. In this regard, many construction contract forms provide that the contractor can request and receive reasonable evidence from the owner that suitable financial arrangements have been made (see Subparagraph 2.2.1 in Appendix C). However, certain inquiries should be made regardless of whether there is an express right to determine the source of project financing. Where the owner is completely unknown to the contractor, additional credit information can be obtained from sources such as financial and credit rating services, industry trade groups, other contractors who may have previously dealt with the owner, or with the assistance of the contractor's banker.

There are cases in which the contractor must make certain with whom it is dealing. An example is an instance in which the owner is a shell, dummy, or subsidiary corporation especially formed for the sole purpose of getting the project constructed and minimizing liability of the principals or parent company. Such an owner corporation would have very limited assets and would be a poor credit risk. Until any new corporation has had time to establish a record of financial stability, the principals or parent company must demonstrate project funding or personally guarantee to make good on contractual debts. When dealing with a substantial corporation, the contractor must know that it is dealing with that corporation and not an undercapitalized subsidiary. In addition, when the contractor is dealing with an agent of the owner, the contractor must be given written assurance that the agent has authority to bind the principal to the contract.

Where the owner is a public agency, the contractor is normally assured that funds required to pay for public works will be available. However, there can be exceptions to this rule when state or local governments create quasi-governmental "corporations" to perform various construction functions. Such entities may not possess the full faith and credit backing of the state. Once construction monies have been exhausted, the contractor can find itself without recourse for payment. The contractor should perform some analysis of funding for these types of agencies before bidding such projects.

On public contracts, statutory requirements must have been observed by the responsible public officials for a valid contract to exist. In most jurisdictions, the omission of any required procedural step by a public owner can leave the contractor without remedy to obtain payment for work performed. If a government body enters into a contract without complying with the statutory requirements pertaining to the bidding and awarding of public construction contracts, the contract is considered beyond the power of the public agency and may be declared void. There is a growing body of legal precedent, however, indicating that when an imperfect public contract was entered into in good faith and is devoid of fraud or collusion, the contractor is entitled to relief based on the equitable doctrine of unjust enrichment. Notwithstanding this precedent, however, the public may become the unintended beneficiary of the work without paying for it. One who does business with a public entity must be aware of the laws governing its administration and the limitations on the powers and authorities of the public officers involved.

9.23   PERIODIC PAYMENTS

Construction contracts typically provide that partial payments of the contract amount shall be made to the prime contractor as the work progresses. The exact procedure followed can depend substantially on the position of the owner and the type of construction involved. For example, in housing and some other types of construction, specified payments are made at the completion of certain stages of construction. In housing, payments are due at defined points of progress, such as at completion of excavation and foundations, at completion of framing, when the structure is enclosed and mechanical and electrical rough-in is installed, at completion of the interior finish, and at full completion. The owner or mortgagee makes each of these payments to the contractor after job inspection and receipt of the necessary documentation from the contractor.

Payment to the contractor at monthly intervals is the more usual contract proviso. The contractor, the architect-engineer, or the owner may prepare the pay requests. The general conditions or supplementary conditions of the contract normally stipulate which party is to have the responsibility and authority for compiling these requests. Such periodic payments made by the owner to the prime contractor are subject to the retainage provisions contained in the construction contract (see Section 6.18).

Each periodic pay request is based on the work accomplished since the last payment was made. Consequently, field progress must be measured or estimated at intervals, and a payment form used that clearly identifies the units of work that have been accomplished and on which the pay request is based. The owner or lending agency may require the use of a pre-scribed payment requisition form, or the contractor may be free to devise its own. When the pay request has been compiled, it is transmitted to the owner or lending agency, commonly through the architect-engineer or contracting officer who approves the request and certifies it for payment. With variations between public and private projects, the contractor must normally include with each payment request a waiver of lien and an affidavit of payment of subcontractors and material suppliers. This is a reasonable requirement, but some owner payment procedures go beyond this and require release by the contractor of any and all claims against the owner. In such cases, the contractor must be very circumspect about sub-mitting such releases if it wishes to protect any claim rights it may have against the owner.

Approval of a contractor's pay requests can be an important responsibility of the architect-engineer. In protecting the interests of its client, the owner, it is incumbent upon the architect-engineer to see that the periodic payments made to the contractor are reason-able measures of the work actually accomplished and that the work has been performed properly. In addition, recent trends indicate that an architect-engineer may be held responsible if its negligence enables a contractor to divert funds. An example of this situation involved a major contract on which the contractor went bankrupt. The bonding company completed the project. The dollar value of the work in place when the bonding company took over was substantially less than the payments the contractor had received. The bonding company sued the architect-engineer for negligence in permitting overpayment to the contractor, and the architect-engineer was found liable for the overpayments.

Slow payment by the owner to the contractor continues to be a serious problem for the construction industry. Some owners contract for a project without having completed arrangements for funding. Promoters may not have enough equity capital or permanent mortgage commitments for the full contract amount. At times, owners deliberately delay payments to the prime contractor, even though funds are available, in order to have the use of the funds themselves. Owners may sometimes delay payments to force the contractor into expediting the work or to exact more work than is called for by the contract documents. An associated problem is the slow processing of contract change orders, resulting in long delays in paying the contractor for extra work the owner itself initiated and approved. In the event of payment disputes or an owner's failure to make payments in accordance with the contract, the contractor must use great care to fully comply with contract provisions in this regard before stopping work or terminating the contract for nonpayment.

Federal and state statutes now discourage slow payment to construction contractors on public projects. The Prompt Payment Act of 1982 provides an incentive for the federal government to pay its bills on time, including progress payments to contractors. This Act mandates that governmental agencies must pay interest penalties for tardy payment under contracts with private businesses. Most of the states have now enacted similar prompt-pay legislation applicable to state contracts. Payment provisions vary considerably from state to state.

Because of the several problems associated with the disbursement of construction funds, including late payment by the general contractor to its subcontractors and suppliers, there has been some movement toward having third parties be responsible for making payments on construction contracts. There are now private firms whose services to lending agencies include ensuring that the money moves quickly down the chain of command from lender to owner, then to general contractor, subcontractor, and supplier. In this way, funds are not diverted and all parties to the project are paid on time.

In 1988 an amendment to the Prompt Payment Act provided that the government is required to make payments to prime contractors within 14 days of receiving a progress payment application or pay the contractor interest at the Prompt Payment Act interest rate established by the Contract Disputes Act. The prime contractor, in turn, is required to pay its subcontractors within 7 days of receiving payment from the government or pay interest. The prime contractor is allowed to withhold payment for cause, such as deficient performance. The prime contractor must also certify to the government that all payments from the government have been earned and that subcontractors and suppliers have been paid out of previous payments.

In 1999 the federal government updated the Prompt Payment Act through CFR 1315 to reflect requirements established by the Debt Collection Improvement Act of 1996 and the increased use of electronic commerce in government and industry. This new rule requires all contractors submitting invoices on federal projects to include electronic funds transfer information and their Federal Tax Identification Numbers on all invoices. Inclusion of this information benefits the contractor by allowing for faster payment. Yet invoices that fail to include this information will be rejected and the contractor will loose all rights to late payment interest penalties. This revision further entitles the government to make early payment to the contractor if it is deemed to be in the government's best interest and if it promotes the use of electronic payments.

In 2000 the federal government implemented an interim rule making cost-reimbursement service contracts subject to the payment and interest requirements of the Prompt Payment Act as well.

9.24 PROJECT COST BREAKDOWNS

Construction contracts for lump-sum projects usually require that the contractor submit some form of cost breakdown of the project to the owner or architect-engineer for approval before any payment request by the contractor is presented. This cost breakdown, also called a "schedule of values," is intended to serve as the basis for the subsequent monthly pay requests. The breakdown, which is actually a schedule of costs of the various components of the structure, including all work done by subcontractors, is prepared in sufficient detail for the architect-engineer or owner to readily check the contractor's pay requests. Figure 9.4, which presents a typical monthly pay request, illustrates in column 1 a cost breakdown for the Municipal Airport Terminal Building. Occasionally, to satisfy an owner requirement, the specifications stipulate the individual items for which the contractor is to present cost figures. In the absence of such instructions, it is usual to prepare the cost schedule by using the same general items as they appear in the specifications and on the final recap sheet of the estimate. This practice minimizes the time and effort necessary to prepare the breakdown figures and ensures a maximum of accuracy in the results.

Inspection of Figure 9.4 shows that the cost figures given for the individual work classifications do not match those indicated on the recap sheet (Figure 5.8). The explanation for this discrepancy is that construction equipment, job overhead, markup, bond, and tax (often referred to collectively as job burden) are not shown on the cost breakdown as line items but have been incorporated into the listed job costs. The sum of these five amounts has not been distributed on a completely pro rata basis, however. Rather, items of work completed early in the construction period, such as excavation, concrete, reinforcing steel, and forms, have been increased proportionately. This procedure, called "front-end loading," serves the purpose of helping to reimburse the contractor for the initial costs of moving in, setting up, and commencing operations, for which a specific pay item is seldom provided. Such moderate unbalancing of cost figures assists the contractor financially and minimizes its initial investment in the owner's project. In addition, since the project was estimated in Chapter 5, the owner has elected to include bid alternate three (shown in Appendix J) in the contract, reducing the total contract amount by $66,723. All of these elements taken together account for the apparent discrepancy between Figure 9.4 and Figure 5.8.

It should be pointed out that a cost breakdown compiled for payment purposes cannot be used for the pricing of either extra work to, or deductions from, the contract. Items of general expense have been prorated and added to the various pay items whether or not they apply directly to a given item of work. In addition, as discussed earlier, these cost breakdowns are usually unbalanced to some degree.

		Total Cost (1)	Completed to Date (2)	Cost to Complete (3)	Percent Complete (4)

THE BLANK CONSTRUCTION COMPANY, INC.
1938 Cranbrook Lane
Portland, Ohio

PERIODIC ESTIMATE FOR PARTIAL PAYMENT

Project Municipal Airport Terminal Building   Location   Portland, Ohio

Periodic Estimate No.   4   For Period   September 1, 20- to September 30, 20-

Item No.	Item Description	Total Cost (1)	Completed to Date (2)	Cost to Complete (3)	Percent Complete (4)
1	Clearing and Grubbing	$22,508	$22,508	$0	100
2	Excavation and Fill	$67,559	$57,425	$10,134	85
3	Concrete and Forms				
	Footings	$108,572	$77,086	$31,486	71
	Grade Beams	$108,898	$63,161	$45,737	58
	Beams	$65,960	$6,596	$59,364	10
	Columns	$21,306	$6,392	$14,914	30
	Slabs	$383,093	$45,971	$337,122	12
	Walls	$59,566	$21,444	$38,122	36
	Stairs	$43,380	$0	$43,380	0
	Sidewalks	$35,662	$0	$35,662	0
4	Masonry	$734,584	$22,038	$712,546	3
5	Carpentry	$57,025	$0	$57,025	0
6	Millwork	$129,284	$0	$129,284	0
7	Steel and Misc. Iron				
	Reinforcing Steel	$129,218	$67,193	$62,025	52
	Mesh	$28,382	$4,257	$24,125	15
	Joist	$140,334	$0	$140,334	0
	Structural	$269,908	$45,884	$224,024	17
8	Insulation	$38,782	$0	$38,782	0
9	Caulk and Weatherstrip	$8,667	$0	$8,667	0
10	Lath, Planter, and Stucco	$296,270	$0	$296,270	0
11	Ceramic Tile	$33,651	$0	$33,651	0
12	Roofing and Sheet Metal	$312,079	$12,483	$299,596	4
13	Resilient Flooring	$39,381	$0	$39,381	0
14	Acoustical Tile	$48,771	$0	$48,771	0
15	Painting	$145,071	$0	$145,071	0
16	Glass and Glazing	$108,478	$0	$108,478	0
17	Terrazzo	$138,859	$0	$138,859	0
18	Miscellaneous Metals	$120,238	$0	$120,238	0
19	Finish Hardware	$83,304	$0	$83,304	0
20	Plumbing, Heating, A/C	$939,150	$84,524	$854,627	9
21	Electrical	$592,057	$29,603	$562,454	5
22	Clean Glass	$4,873	$0	$4,873	0
23	Paving, Curb, and Gutter	$123,381	$0	$123,381	0
		$5,438,251	$566,565	$4,871,686	
24	Change Order No. 1	$5,240	$0	$5,240	0
	Total Contract Amount	$5,443,491	$566,565	$4,876,926	10.4%
A	Cost of Work Performed to Date	$566,565			
B	Materials Stored On-site (Schedule Attached)	$102,207			
C	Total Work Performed and Materials Stored	$668,772			
D	Less 10% Retainage	$66,877			
E	Net Work Performed and Materials Stored	$601,895			
F	Less Amount of Previous Payments	$272,309			
G	Balance Due This Payment	$329,586			

Figure 9.4 Periodic estimate for partial payment.

9.25   PAYMENT REQUESTS FOR LUMP-SUM CONTRACTS

Figure 9.4 illustrates the form of a typical pay request for a lump-sum building contract. Usually prepared by the contractor, it includes all subcontracted work as well as that done by the contractor's own forces. For each work classification that it does itself, the contractor estimates the percentage completed and in place. From invoices submitted by the subcontractors, suitable percentage figures are entered for all subcontracted work. These percentages are shown in column 4 of Figure 9.4. The total value of each work classification is multiplied by its percent completion, and these figures are shown in column 2. To the total completed work is added the value of all materials stored on the site but not yet incorporated into the work. The cost of stored materials includes that of the subcontractors' materials and is customarily set forth in a supporting schedule. From the total of work in place and materials stored on-site is subtracted the retainage. This gives the total amount of money due the contractor up to the date of the pay request. From this is subtracted the amount of progress payments already made. The resulting figure gives the net amount now payable to the prime contractor.

Although the pay request procedure for lump-sum contracts has been in general use for many years, it has one serious defect. As shown in Figure 9.4, the project is divided for payment purposes into relatively few work classifications, most of which are extensive and often extend over appreciable portions of the construction period. This situation can make it difficult to estimate accurately the percentages of the various work categories completed. Actual measurement of the work quantities accomplished to date is the key to accurate percentage figures, but this can become very laborious. Therefore, most of the percentages are established by visual appraisal or other approximating procedures. Contractors want these estimates to be fair representations of the actual work achieved, but, understandably, they do not want them to be too low. Hence, most of their percentage estimates are apt to be on the generous side.

This circumstance continues to produce vexing problems for both the contractor and the owner. If it is difficult for the contractor to estimate the completion percentages accurately, it is at least equally difficult for the architect-engineer or owner to check these reported values. This presents the architect-engineer with a difficult problem because, in the interest of its client, it must make an honest effort to see that the monthly payments made to the contractor are reasonably representative of the actual progress of the job. Architect-engineers are at times casual about the processing of a pay request, feeling that a delay in payment will offset the generous completion percentages provided by the contractor. Although effective, this approach is often at odds with contract provisions regarding payment. It also ignores the fact that the progress payments are being reduced by the prescribed retainage.

One way to obtain more accurate payment information is to require the contractor to submit a detailed critical path schedule with costs attached to each activity prior to the first payment. With the use of this schedule, costs for each completed activity are charged and a pro rata percentage of each activity in progress is estimated. The result is a payment request that is accurate and easy for the architect-engineer to certify.

9.26   PAYMENT REQUESTS FOR UNIT-PRICE CONTRACTS

Payment requests under unit-price contracts are based on actual quantities of each bid item completed to date. The determination of quantities accomplished in the field is done in several different ways, depending on the nature of the particular bid item. When volume of aggregate, weight of asphaltic concrete, or bags of portland cement are set up as bid items, these quantities are usually measured as they are delivered to the work site. Delivery tickets or fabricator's certificates are used to establish the weight of reinforcing or structural steel. Other work classifications, such as volume of excavation, length of pipe, or volume of concrete, are measured in place or computed from field dimensions. Survey crews of the owner and of the contractor often make their measurements independently of one another and adjust any differences.

Many owners use their own standard forms for monthly pay estimates. On unit-price contracts, the owner or the architect-engineer often prepares the pay request and sends it to the contractor for checking and approval before payment is made. On unit-price contracts that involve a substantial number of bid items, each monthly pay request is a sizable document consisting of many pages. In essence, the total amount of work accomplished to date on each bid item is multiplied by its corresponding contract unit price. All of the bid items are totaled, and the value of materials stored on the site is then added. From this total is subtracted the prescribed retainage. The resulting figure represents the entire amount due the contractor for its work to date. The sum of all prior progress payments that have already been paid is then subtracted, thus yielding the net amount of money payable to the contractor for its work that month.

9.27 PAYMENT REQUESTS FOR COST-PLUS CONTRACTS

Negotiated contracts of the cost-plus variety provide numerous methods for making payments to contractors. In many cases, contractors furnish their own capital, receiving periodic reimbursements from owners for costs incurred. Other contracts provide that owners will advance money to the contractors for the purpose of meeting their payrolls and paying other expenses associated with the work. In one scheme, the contractor prepares estimates of expenses for the coming month and receives the money in advance. Then, at month's end, the contractor prepares an accounting of the actual expenses. Any difference between the estimated expenses and actual expenses is adjusted with the next monthly estimate. Other contracts provide for constant-balance bank accounts, whereby the contractor writes checks and the owner furnishes funds. In any event, the contractor must make periodic accountings to the owner of the cost of the work, either to receive direct payment from the owner or to obtain further advances of funds.

A periodic payment to the contractor under a cost-plus type of contract is not usually based on quantities of work performed but on reimbursement of expenses incurred by the contractor during the preceding pay period. Consequently, such pay requests consist primarily of the submission of cost records. Copies of invoices, payrolls, vouchers, and receipts are submitted in substantiation of the contractor's claims. In addition to cost records of payments made by the contractor to third parties, the periodic pay requests customarily include equipment expense and a pro rata share of the negotiated fee. If the owner furnishes funds to the contractor to pay for construction costs, the owner is credited with all cash discounts.

Because of the sensitive nature of cost reimbursement, it is common practice to maintain a separate set of accounting records for each cost-plus project. When the project size warrants it, a field office is established where all matters pertaining to payroll, purchasing, disbursements, and record keeping for the project are performed. Project financial records are either routed through the owner's representative or are available for inspection at any time. This procedure does much to eliminate misunderstandings and facilitates the final audit. Cost-plus contracts with public agencies customarily impose special conditions on the contractor pertaining to form of payment, payment application, affidavits, and preservation of project records.

9.28   FINAL PAYMENT

The procedural steps leading up to acceptance of the project and final payment by the owner vary with the nature of the work and the specific provisions of the contract. In building construction, the process typically commences when the contractor, having achieved substantial completion, requests a preliminary inspection. The owner or its authorized representative, in company with general contractor and subcontractor personnel, inspects the work. A list of deficiencies to be completed or corrected is prepared, and the architect-engineer issues a certificate of substantial completion. After the deficiencies have been remedied, a final inspection is held and the contractor presents its application for final payment. If the work is determined to be complete and acceptable, the architect-engineer issues a certificate for final payment that includes retainage still withheld by the owner. Under a lump-sum form of contract, the final payment is the total contract price less the sum of all progress payments previously made. With a unit-price contract, the final total quantities of all payment items are obtained, and the final contract amount is determined. Final payment is again equal to the total contract price less the sum of all previous payment installments made by the owner.

Construction contracts typically require that the general contractor's request for final payment be accompanied by a number of documents. For example, releases or waivers of lien executed by the general contractor, all subcontractors, and material suppliers are common requirements on privately financed jobs. The owner may also require an affidavit from the contractor stating that all parties who might be entitled to place a lien on the project have furnished the required releases and waivers of that right. Other contracts call for an affidavit certifying that all payrolls, bills for materials, payments to subcontractors, and other indebtedness connected with the work have been paid or otherwise satisfied. Claims and disputes still unresolved at the time of final payment should be expressly identified in writing by the contractor as unsettled and reserved in the final payment application and in the final lien waivers and releases. Construction contracts frequently require the contractor to provide the owner with as-built drawings, various forms of written warranties, maintenance bonds, and literature pertaining to the operation and maintenance of job machinery. Consent of surety to final payment is a common prerequisite. Subparagraph 9.10.2 in Appendix C is an example of a contractual provision concerning documentation required for final payment.

If there are unpaid construction costs still outstanding at the time of request for final payment, the contractor must report the parties to whom money is owed and the amounts due. When the owner has been informed of sums payable on the project, either by affidavit from the contractor or by direct notice from the unpaid party(ies), the owner may become liable for these debts if sufficient funds are not withheld from the general contractor to cover them. If the general contractor certifies payment of all bills and the owner has not received any notification of debt, the owner is generally entitled to rely on the general contractor's statement and make final payment. There can be exceptions to this rule, however, in those states whose lien statutes impose unlimited liability on the owner's property. In such cases, the owner must require releases of lien from all possible claimants.

9.29 PAYMENTS TO SUBCONTRACTORS

When the owner makes payment to the general contractor, it is expected that the general contractor will then make prompt payment to its subcontractors. Payment to subcontractors, although usually a routine process for the general contractor, is often a troublesome one and requires attention and care. Indeed, the wise general contractor will apply many of the same precautions used by the owner in paying the general contractor to its payment of subcontractors. It does seem clear, however, that the safeguards applied will depend considerably on the particular subcontractor and the contractor's experience with that subcontractor. Joint checks by owners and general contractors to pay subcontractors and their suppliers have become commonplace. Joint checks generally protect owners and general contractors from liens and payment bond claims.

The general contractor must check each monthly pay request from a subcontractor to ensure that it is a fair measure of work actually performed. The prime contractor does not wish to allow its subcontractors to be overpaid any more than the owner wants to overpay the general contractor. To assist in this matter, the prime contractor frequently requires the subcontractors to submit cost breakdowns for use in checking their requisitions. The prime contractor will normally apply the same retainage rate to its subcontractor payments that the owner applies to payments made to the prime contractor.

When the general contractor makes payment to a subcontractor, it is understandable that the general contractor wants assurance that the subcontractor is meeting its financial obligations. In this regard, the contractor should use certain precautions to protect itself from the hazard of a subcontractor's failure to pay its bills and meet its payrolls. When the prime contractor furnishes the owner with a payment bond, it has a liability to those unpaid parties within the coverage of the bond who comply with the notice provisions of that bond. Even though the contractor has paid a subcontractor in full, an unpaid creditor of that subcontractor can still file a claim on the payment bond. Contractors have been required to pay twice for the same item of work because of their payment bond liability. As a consequence of this possibility, the contractor must take steps to obtain appropriate releases of lien, receipts, or affidavits of payment when payments are made to subcontractors. Another action in this regard is to require that subcontractors furnish payment bonds to the prime contractor. Subcontracts frequently provide that the general contractor can withhold payments from a subcontractor if it appears that the subcontractor is not paying its bills, although such a clause may make the contractor liable to third persons as third-party beneficiaries unless care is used in the subcontract language.

The general contractor frequently makes payment to a subcontractor without specification as to how the payment will be applied by the recipient. Some caution may be in order here if the two parties have dealings on more than one job. In many jurisdictions, if the general contractor does not designate how the payment is to be applied, the subcontractor may apply the payment to a job other than the one for which payment is being made. This follows from a generally recognized rule of law that when a debtor who owes two or more separate debts makes a payment, it has the right to designate which will be credited. If it fails to do so, the creditor can choose. If proper application of payment is important, the general contractor can indicate on the check the application of proceeds desired and include a letter of transmittal that specifically states the purpose for which the payment is being made. If a general contractor makes payment to a subcontractor's supplier by a check payable to both the supplier and the subcontractor, specific instructions on how the payment should be credited by the payees should be provided.

There is another side of the coin, however. Frequently, payment to a subcontractor is delayed because the prime contractor diverts the funds to other areas of its own business. This may follow from the general contractor's attempt to perform a volume of work in excess of its financial capability, resulting in payments from the newer jobs being used to pay the bills of older work. The subcontractor is often reticent to press for payment, not wanting to antagonize the general contractor. In the final analysis, however, the subcontractor can seek relief in different ways, such as through the general contractor's payment bond, a suit for breach of contract, and the filing of a mechanic's lien. There is a fourth avenue for unpaid subcontractors and material dealers in those states with construction trust-fund statutes. A construction trust-fund statute declares that the funds paid to a prime contractor by the owner are trust funds held by the contractor as trustee for the benefit of its subcontractors and material suppliers. Such statutes afford these parties an additional avenue of recovery from the general contractor.

Slow payment by general contractors to their subcontractors is the source of considerable friction between the two parties. Subcontractors maintain that the working relationships between the two could be much more amicable and productive if general contractors would pay their subcontractors more promptly. Lower subcontractor bids and more efficient field production are suggested as possible benefits to the general contractor. A few states now have statutes that provide for fines when contractors on public or private projects fail to pay their subcontractors on time. Many subcontract forms now give the subcontractor the right to request, directly from the architect-engineer or owner, information about percentages of completion or the payment amount to the prime contractor certified by the architect-engineer or owner for work done by the subcontractor. In this regard, the American Subcontractors Association has established a national reporting network to provide member subcontractors with information about the business practices of specific general contractors. This includes data concerning periodic and final payments.

It occasionally happens that a general contractor will incur or assume an expense that is chargeable to a subcontractor. The general contractor may pay some of the subcontractor's bills; temporarily provide the subcontractor with labor, equipment, materials, or facilities; or perform cleanup or hauling services on behalf of the subcontractor. The usual way for the general contractor to recoup these expenses is to subtract them from payments made to the subcontractor. Such a deduction, called a "backcharge," may lead to a dispute between the general contractor and the subcontractor who is backcharged, unless the amount involved is discussed and, preferably, agreed to in advance.

9.30   PAYMENTS TO MATERIAL SUPPLIERS

Payments by a contractor to its material suppliers are made in accordance with the terms of the applicable purchase order or usual commercial terms. Payment for materials is not normally dependent on any disbursements made by the owner to the general contractor, but is due and payable in full 30 days after invoice date, upon receipt of materials, at the end of the month in which delivery was made, or according to other stipulations. Such payment terms frequently include provisions for cash discounts to encourage early payment by the contractor. Disbursements to material dealers are made in full because retainage does not apply to purchase order payments.

With reference to payment for materials, most contractors have established some system of checking and control. For example, in a typical procedure, a purchase order is issued for all project materials of any consequence. Such a purchase order not only specifies the cost of the materials but also serves as an internal control for the contractor. As deliveries are made, the materials are inspected, quantities are verified, and any quality-control or acceptance tests are performed. After inspection and testing, a receiving report is sent to the disbursement section of the contractor's office listing the materials received, identifying the vendor, and describing any shortages, damage, or variations from specified quality. No payment of the vendor's invoice is made until a receiving report has been received from the project or from the contractor's warehouse, attesting to proper delivery, and until the invoice amount has been proved against the purchase order. Backcharges, as discussed in the preceding section in regard to subcontractors, can also apply to material vendors.

9.31   DIRECT PAYMENT

There has been considerable effort by the American Subcontractors Association (ASA) to change the method by which payments are made to subcontractors. The ASA maintains that the traditional procedure of the owner paying the general contractor and the general contractor then paying its subcontractors exposes subcontractors to slow payment, cash flow problems, and other abuses. For this reason, considerable effort has been expended to change the procedure. One method suggested to speed up payments is to have a third party act as paymaster under a common agreement of the owner, lender, and general contractor. Owners make periodic payments to an escrow account, from which subcontractors are paid directly in amounts approved by the general contractor. It is claimed that such direct disbursement will result in lower prices, better performance, and faster project completion.

The concept of direct payment to subcontractors has been stoutly resisted by general contractors on the basis that it would be detrimental to their overall control of the project. They maintain that the present system is not defective if the subcontractors act in a responsible, prudent, and businesslike manner. The basic position of general contractors is that the advantages and benefits to the owner of the single-contract method of construction would be adversely affected by a direct payment system. The direct pay procedure is now used in this country to a modest extent on both public and private projects.

There are times when the owner and general contractor may agree to have the owner make progress payments and/or final payment directly to the project subcontractors and material suppliers. In such cases, the owner deducts the amounts involved from payments made to the general contractor. There are various reasons for doing this, a common rationale being the owner's desire that the subcontractors and suppliers be paid promptly and in full for their contributions to the work. This procedure can protect the owner from claims and liens. Alternately, if the general contractor–subcontractor relationship is strained, direct payment by the owner may improve subcontractor performance. Although such a direct payment arrangement is normally not desirable, it can at times ameliorate problems on the job.

9.32 CASH FLOW

Many contractors have gotten into deep financial trouble when they have suddenly and unexpectedly run out of cash to pay their bills. Although it can happen to any business, negative cash flow is one of the major causes of failure for small construction firms. *Cash flow* refers to a contractor's income and outgo of cash. The net cash flow is the difference between disbursements and income over a period of time. A positive cash flow indicates that cash income is exceeding disbursements, and a negative cash flow signifies just the opposite. Cash is the fuel that runs the business, and a contractor must maintain a cash balance sufficient to meet payrolls, pay for materials, make equipment payments, and satisfy other financial obligations as they become due.

In itself, however, cash is not a productive asset; it must be invested in some way to make it productive. Therefore, management wants to have enough funds readily available for needs as they occur but to keep excess cash suitably invested. Similarly, if a contractor must borrow funds on which to operate, it wishes to recognize this need as early as possible and does not want to borrow any more than is necessary. The timing of cash flow is especially important in determining a firm's working capital requirements, and the future cash position of a construction firm is of great importance to its management.

9.33 CASH FORECASTS

A cash forecast is a schedule that summarizes the estimated cash receipts, estimated disbursements, and available cash balances for some period into the future. The forecast, starting with a known beginning cash balance, estimates cash income and disbursements, either on a weekly or a monthly basis, yielding an estimated cash balance at the beginning of each period. For the most part, only short-term future cash predictions are optimally useful for contractors because of the unpredictability of new contract acquisition, as well as other financial matters generally. These figures indicate when the available cash balance will be below minimum needs and when it will be above. This, in turn, gives advance warning that additional funds must be obtained by borrowing or that excess funds will be available for investment or company growth.

The preparation of a cash forecast begins with the collection of detailed information regarding future cash income and expenditures. For a construction contractor this must usually first be done for each individual project, based on its proposed progress schedule. The resulting cash flow figures (the net result of cumulative cash income and outgo on a time basis) are then combined with the company's general and administrative disbursements to develop a total cash flow forecast for the given planning period. The computational process is complicated somewhat by the fact that both project earnings and expenses occur on discrete bases, and these are normally not on the same time frequency. For example, project income is received in some regular and periodic way such as in monthly payments. Project expenses are extremely variable in their payment patterns. Payrolls are on a weekly basis, whereas disbursements to material dealers and subcontractors are on a monthly basis. Payment of taxes, insurance premiums, and certain equipment expenses are largely uncoordinated with the physical progress of the work. When done properly, the preparation of a cash forecast is a complex and time-consuming operation, but the resulting management information can make the expense and effort well worthwhile. After the forecast estimates are made, they must be reviewed periodically to determine their continued validity. When material differences develop, new estimates are prepared and the forecast revised.

9.34 THE MECHANIC'S LIEN

A mechanic's lien is a right created by law to secure payment for work performed and materials furnished in the improvement of land. This statutory right attaches to the land itself in much the same way as a mortgage. The purpose of a lien statute is to permit a claim on premises where the value or condition of real property has been increased or improved and where the owner has not made suitable payment. Following the lead of Maryland in 1791, every state has enacted some form of mechanic's lien law. These laws are similar in general import but differ considerably in detail. The courts construe lien laws strictly, and full compliance with all provisions of the local statute is mandatory.

Lien laws are based on the theory of unjust enrichment and are designed to protect workers, material suppliers, and, under certain conditions, general contractors and subcontractors. The rights accruing to general contractors and subcontractors are the same, but there are differences in the notices required of the two parties. For a general contractor to obtain a lien, a usual requirement is that the owner had previously agreed to have the work done and to pay for it. In some states, a written contract must exist. In case of default by a private owner on a construction contract, the general contractor actually has two remedies available. It can file suit against the owner for breach of contract, or it can exercise its right of lien. In some cases, the contractor takes both courses of action.

Architect-engineers are given lien rights by some state statutes. However, for such a lien right to be available, work usually has to have commenced on the property. Design service, in and of itself, may not be sufficient unless construction has actually proceeded. In some states, only services that directly benefit the land, such as supervision, apply. Design itself is not included. Because of wide variations in state laws, generalizations about the lien rights of design professionals can be badly misleading and the intricacies of individual state statutes must be investigated.

Public property is not subject to a statutory lien. However, many states allow unpaid workers, subcontractors, and material dealers to establish lien claims against contract funds that are still held by a contracting public agency. This type of claim, referred to as a municipal mechanic's lien, applies to contract funds, not to the real property. Under such liens, when a payment claim is filed, disbursements to the prime contractor are stopped and the unpaid funds are preserved for payment to the claimant. The right to recover from a public owner is limited to contract funds still in control of the agency.

Although lien laws are beneficial, there are many drawbacks associated with them. For the claimant to thread its way through the intricacies of the law requires a number of time- and money-consuming measures. Recovery is technical, cumbersome, and often unsatisfactory for all concerned.

The lien rights of parties dealing with a general contractor depend on whether the state in which the work was performed has a lien statute based on the New York system or the Pennsylvania system. Under the New York system, the amount such a party can collect is limited to the amount due the general contractor from the owner. If nothing is due when the lien is filed, the lienor may not look to the owner of the premises for payment. However, in states that abide by the Pennsylvania system, the owner is not so protected. The unpaid party has a right to file a lien even if the entire contract amount has been paid to the prime contractor. Correspondingly, the owner may be forced to pay for some of the contract work twice, should the general contractor fail to pay its subcontractors or material suppliers. Withholding payment from the general contractor in an amount sufficient to cover open accounts and waivers of lien afford the owner protection against such possible double payment.

The lien statutes of certain states restrict the lien rights of unpaid parties where a private owner has required the general contractor to furnish a payment bond. In those states, on receipt of the bond, the owner is exempt from liabilities for mechanic's liens filed by unpaid claimants other than the general contractor. The payment bond takes the place of the owner's real property as security for the lien. The bond exempts the owner's property from the possibility of lien foreclosure.

9.35 FILING A LIEN CLAIM

For any party who contracts directly with the owner to be able to obtain a lien, the statutes typically require that party to record a notarized claim for public record with a county authority within the statutorily prescribed time. In most jurisdictions, this lien claim is considered sufficient if it names the owner, describes the project, contains appropriate allegations as to the work performed or materials furnished, and states the amount and from whom it is due. The time for filing differs from state to state, but general contractors typically have 60 days after the owner files a notice of project completion or, if no notice is filed, 90 days after the physical completion of the work that is the subject of the lien. Many states require, in addition to the filing of a lien claim, that written notice also be given to the owner or its agent within a specified time. In some cases, notice to the owner must precede the filing of the claim. The advice of a local attorney can be invaluable when filing a claim of lien.

Laborers, subcontractors, or material dealers who contract with the general contractor rather than directly with the owner are also entitled to liens, but the statutory requirements for these parties are often different from those for the general contractor. They must not only file a notice of lien for the public record, but are also usually required to give notice in writing to the owner or its agent. The time limit for such filing is prescribed by statute and is frequently different from that required of parties contracting directly with the owner. In addition, the statutory time for filing a claim may begin with the date when the last labor or materials were furnished, rather than with project completion.

9.36 FORECLOSURE OF LIEN

Once a lien claim is filed and no payment has been made, proceedings must be brought to enforce the lien, usually within 90 days after filing. The court procedure by which the claim is judicially determined is known as a foreclosure action; it varies from state to state but is highly technical and requires the services of a lawyer. If the evidence substantiates the claim under the mechanic's lien and the court finds that there is a sum of money due and owing, the court can order the property to be sold and the proceeds used to satisfy the indebtedness. The sale of the property is made at auction by the sheriff, and a certificate of sale is executed to the successful bidder. The holder of the lien is paid from the proceeds of the sale. Most lien statutes give the original owner a prescribed period of time to redeem its property upon payment of the judgment, interest, and costs.

The priority of claims is a matter treated by the pertinent state statute. A mortgage subsequent to the construction contract is usually inferior to a mechanic's lien that arises out of work done under the contract. Mortgages outstanding at the time of the contract may or may not be inferior, depending on the law. Generally, different liens stand on an equal footing with one another regardless of the order in which they are filed. An exception is the preferred status usually given to workers and material dealers.

9.37   RELEASE OF LIEN

The right to place a lien may be released or waived in a number of ways, depending on the particular statute. Construction contracts with private owners sometimes include a clause whereby the general contractor agrees not to file or place any liens against the owner's premises and waives its right in this regard. After the contract has been signed, this clause is binding on the contractor. The courts have long held that a contractor, by the terms of a contract, may release its right of lien. However, some states have enacted legislation that makes the waiver of a mechanic's lien void as against public policy, and hence unenforceable.

A broader form of release of lien sometimes used in construction contracts provides that the general contractor or any subcontractor or material dealer cannot file liens. This proviso can be made binding on the subcontractors and material dealers, provided that it is permitted by the statute involved and that the general contractor takes certain actions, these actions depending on the requirements of the law. For example, the contractor may be required to give timely notice of the waiver before the purchase orders and subcontracts are signed, either by direct notification in writing to the subcontractors and material dealers or by making the release agreement a matter of public record. In addition, the subcontracts and purchase orders may have to include a clause that expressly provides for the release of lien by the subcontractors and material dealers. This is a consequence of the fact that lien statutes often provide that the right of lien may be waived only by an express agreement in writing specifically to that effect.

Rather than the contractor releasing its right of lien through the medium of a contract clause to that effect, it may be required to submit a signed release form to the owner as a condition of receiving final payment, or perhaps each time it requests a progress payment. Similar releases are often also required from the subcontractors and material vendors. Standard waiver-of-lien forms are available from suppliers of commercial and business forms and are widely used for the purpose.

9.38   ASSIGNMENTS

Practically all rights arising out of contracts are assignable. A common contractual right is that of receiving payment in exchange for the performance of a stipulated contractual duty.

Assignment means the transfer of such a right from the party to whom the right belongs by contract, to a third party. The ability of a party to assign to others the right to receive funds by contract is subject to certain limitations, but these seldom apply to construction contracts. To provide security for a loan or to pay off an impatient creditor, a contractor may assign funds due, or to become due, under a construction contract, to a lending institution or the creditor. After notice of the assignment has been given to the owner, it must make payments to the assignee. Basically, the assignee acquires the same but no greater rights than its assignor had, and the owner is required only to make payments as required by contract. The owner cannot be placed in a worse position than it would otherwise have been in if the assignment had not been made. For example, payment by the owner to the assignee can be excused by the contractor's failure to perform properly.

A party to a contract can assign its rights under the contract without the consent of, or notice to, the other contracting party (the debtor). Freedom of assignment, however, can be regulated by the terms of the contract itself. Construction contracts usually contain provisions that expressly forbid the owner or the general contractor to assign the contract as a whole to third parties without the written consent of the other (see Subparagraph 13.2.1 in Appendix C). With regard to this reference in Appendix C, a contract clause banning assignments is usually interpreted by the courts as preventing delegation of performance. To prevent the assignment of a contract right, such as to receive moneys due, specific provision to this effect must normally be made.

Subcontract forms used by general contractors customarily include a restriction of assignment (see Article 7 in Appendix M). The general contractor is often requested by subcontractors to approve an assignment of moneys due or to become due under a subcontract. By assignment of such funds to a bank, for example, the subcontractor can receive a loan from the bank to finance its operations. It is important to note that the failure of the general contractor to honor a subcontractor's assignment of funds has resulted in liability of the general contractor to the assignee. On receipt of notice of an assignment, the general contractor generally has the duty to pay the assigned whether it has accepted the assignment or not. This follows from the Uniform Commercial Code, which provides that if an assignment is made as security for a loan, consent of the other party is not required despite any contract clause to the contrary.

9.39 COMPUTER USAGE

Computers are invaluable and indispensable in the conduct of a successful construction business. They are a necessity for efficient and profitable operation, and contractors of all kinds utilize computers as a usual and ordinary part of their daily business operations. Experience has proven conclusively that computers, properly selected and used, can enhance the decision-making capability through greater access to more current information and thereby improve the effectiveness of company management and increase overall productivity. With the aid of computers, the contractor can reduce costs, improve internal operating procedures, and maintain a competitive position in the current market. Computer applications can be useful to any construction business, large or small, because of their speed, accuracy, and ability to store, analyze, and retrieve large volumes of information. Every day, new hardware and software emerge, placing new and powerful capabilities in the hands of construction personnel.

The functions listed here are areas of computer usage that are representative of current practice in the construction industry:

Financial Accounting. The computer performs basic accounting functions in areas such as accounts receivable, accounts payable, general ledger, inventory, cash forecasts, financial statements, and subcontractor control.

Estimating. Computer programs used in estimating range from spreadsheet applications to complex estimating software packages. Some estimating programs use digitizers to assist with quantity take off. These quantities are matched to estimated unit cost databases. These programs make it possible for the contractor to prepare complete, accurate, and detailed cost estimates in less time.

Equipment Accounting. The computer maintains records of equipment depreciation, ownership and operating costs, hours of operation, maintenance, spare parts, production rates, and unit costs.

Payroll. Using time card input, the computer prepares payroll checks, periodic and special payroll and tax reports, and the payroll register and updates the employee master files.

Project Management. The computer can be used to assist with project budgeting and scheduling, progress reports, schedule updating, labor and equipment cost reports, job status reports, project cost forecasts, progress payments, and quality control.

Purchasing. The computer participates in the tabulation and handling of bids, purchase order preparation, expediting, and shop drawing development.

Word Processing. The computer is very efficient in handling company paperwork and correspondence. Standard letters, purchase orders, memoranda, contracts, and other documents can be maintained, revised, and restructured as needed.

CAD and GIS. Most architects and engineers now produce design drawings using computer-based design applications. These applications fall into two major categories. The first, computer aided design (CAD), is used for designing buildings, industrial facilities, and civil engineering construction projects contained within bounded project sites. The second application, called a Geographic Information System (GIS), is frequently used for the design of municipal utility systems such as a city storm drain system or a telecommunications network. Construction contractors often find that these electronic design files can be very useful during the construction phase itself. Modern survey equipment is able to reference the electronic design files in order to partially automate construction surveying and other layout functions. The contractor can also use electronic designs in planning the movement of critical equipment on the project site. For example, contractors use CAD files to determine, in advance, the best route within a refinery project for the mobilization and demobilization of a tower crane.

9.40   MARKETING

One of the most neglected aspects of a construction contracting enterprise is the development of a company marketing plan, even though such a plan can sometimes spell the difference between success and failure. Marketing efforts can serve the contractor well when there is a need to keep the present organization busy or when there is a desire to expand the firm's operations either geographically, to include new work types, or to do larger jobs. Basically, marketing involves identifying a target audience that needs construction services and developing materials and a strategy for reaching and influencing this group. This can be done by identifying market needs and directions and establishing the types of work the company can do most profitably, based on its strengths and capabilities. From these determinations, the contractor can identity its company's market, remembering that the construction customer is very sensitive to price, time, and quality of product. The company then communicates its capabilities to its potential clients as a means of attracting additional contracts. A formal marketing program is an investment in long-range success, and it strives to establish a positive image of the firm's capabilities in the minds of those in a position to influence the procurement of construction services.

When a construction company makes a decision to undertake an active marketing program, a detailed plan of action must be developed to determine what is to be done, how it is to be done, and who is to do it. Initially, a study must be made of company strengths and weaknesses, what markets should be pursued, and a finding made as to how the firm is perceived by outsiders. The marketing initiative must instill potential clients with a sense of trust in the company, define company goals and objectives, broadcast the message that the company wants a certain type of business, and stress that the contractor is the most qualified firm available and can do top-quality work at a lower cost. One of the best sources of new business can be previous customers. If a contractor serves owners well and provides excellent service in everything it does, those owners will continue to bring their business to that contractor. They will also recommend the contractor to others. Certainly, the users of construction services rank trust as one of the most important criteria when selecting a construction contractor. When this research has been done, the company plan may involve all or some of the following actions:

Project Signs. The contractor should erect suitably placed signs on its construction projects, showing its name, address, and telephone number. This is an obvious and effective public relations device and serves to keep the name of the company before the public.

Company Brochure. This publication is a basic marketing tool and can include information and pictures concerning completed projects, personnel, equipment, company history, type of work, specialties, testimonials from former clients, and facilities. Such a brochure can be sent to a select mailing list and should emphasize company competence, resourcefulness, and credibility.

Web Page. Corporate web pages are a good and inexpensive means of providing potential clients with current marketing information and recent project success stories. To be effective, web pages must be professional in appearance and should be regularly updated with new information.

Advertising. General advertising is usually a waste of money. To be effective, modest advertising can be placed in trade publications and other business resources such as those of chambers of commerce.

Newsletter. This is a company house organ that includes news of employees, job site stories, company developments, project completions, new jobs, and similar topics. Copies of this publication can be distributed to a select mailing list. Apart from its role in marketing, a company newsletter can decidedly improve a firm's general operations. It provides a personal touch to internal relations, fosters team spirit, and is a valuable mode of increased internal communication.

Publicity. Information concerning project completions; company and employee news; anniversaries, new contracts, special events, and other newsworthy items can be provided to the local news media. Distinctive company hard hats and an eye-catching logo can be useful in promoting company pride and public recognition.

Public Affairs. The involvement of company personnel in various public affairs can be very valuable to the creation of a favorable company image. Participation in trade shows, service clubs, and seminars, membership in public bodies, committee work, and other public participation can be beneficial to the company's marketing objectives.

Satisfying construction owners requires that the contractor relieve their concerns and misgivings during the construction period. When service to a customer requires an extensive duration, as it does in construction, the anxieties of an owner mount. Easing these anxieties is a fundamental aspect of marketing. An effective way of doing this is through telephone calls, regular visits, and invitations to visit the job site. Project progress reports can also be very effective. Relieving an owner's concerns and providing reassurance can be a valuable and effective customer service.

The total marketing program should project an image of professionalism and integrity and emphasize company skills and specialties. It should convey the impression that the company will efficiently execute a construction contract with a personal touch of concern for the client's interests. Construction is an evolving and changing area of business, so the marketing program must occasionally be reevaluated and revised to reflect such changes. There are private consulting firms that specialize in providing construction marketing services to contractors, and contractor trade associations can provide substantial assistance in this regard to their members.

9.41   SUBSTANCE ABUSE PROGRAMS

Alcohol and drug abuse has grown to epidemic proportions—a national problem that is shared by the construction industry. The U.S. Department of Labor reports that the construction industry currently has some of the highest rates of drug and alcohol abuse in the country. Substance abuse poses a severe threat to the safety, productivity, and image of construction. It presents the contractor with major problems of decreased productivity, absenteeism, job accidents, damage and destruction of property, increased insurance claims, lower quality of workmanship, employee turnover, increased cost of insurance, low morale, and reduced efficiency of company operations. To combat this threat, contractors establish company substance abuse programs and rigorous preemployment drug and alcohol screening policies.

The basis for such action must be a written company policy that is effectively communicated to all employees and with written guidelines to supervisory personnel. There is normally a preliminary orientation program explaining the reasons for such actions and the workings of the policy. Employees sign consent forms indicating that they are aware of, and intend to comply with, the company substance abuse policy as a condition of employment or continued employment. The everyday functionings of the company procedure may involve undercover investigations on job sites, urine-screening tests for prospective and present employees, supervisor training, searches, counseling programs, and internal employee assistance programs geared to rehabilitation. A common requirement is that any employee involved in an accident or suffering an injury on a work site must be given a urine test. Failure of an employee to conform to company policy can result in termination.

The preparation and implementation of such a company policy is a difficult task and requires competent legal advice. Contractor trade organizations and the U.S. Department of Labor (www.dol.gov) can provide invaluable assistance in this regard. There are now reasonably safe legal guidelines that allow a contractor to establish substance abuse control programs based on legitimate business interests. With regard to union contractors establishing programs that include drug or alcohol testing for current employees and job applicants, the National Labor Relations Board has ruled that this is a mandatory subject of collective bargaining and is not a management prerogative. The contractor must notify the union of its intent to initiate such testing and, upon request, to bargain to an agreement or good-faith impasse before implementing such a program.

Private owners have added considerable impetus to the establishment and use of such control efforts by contractors. It is a common contractual requirement that the contractor must demonstrate to the owner that a program for controlling drug and alcohol abuse has been established and is in effect on the project. Some private owners require that all contractor employees be tested for drugs before they enter the job site. The prime contractor passes this requirement down the line by including a similar provision in all of its subcontracts. Construction labor unions have also adopted substance abuse programs and provide drug and counseling services to their members.

9.42   EMPLOYEE MOTIVATION

An important aspect of the operation of a successful and cost-effective construction firm is the motivation of company employees. The ability of a contractor to motivate its workforce largely determines its success in constructing projects on time and within budget. Unfortunately, motivation is a nebulous concept, and it is difficult to know just what will stir individuals to put forth their best efforts. Nevertheless, there are some guides, in this regard, that can be very effective.

Fundamental to motivation is giving the worker the idea that the company is a good place to work and a feeling of involvement, in the sense of being a member of a team. Involving employees in the company effort through communication with managers and supervisors can do this. The more an employee knows about the company, how it functions, and the role that he plays in the scheme of things, the greater will be that employee's interest in contributing to the company's success. An effort by management in getting to know the workers and soliciting their ideas and suggestions can create feelings of belonging and generate interest in working to achieve a common goal. There needs to be an established procedure to air their grievances and to instill the feeling that they are being treated fairly. Open lines of communication between supervisor and worker and an established complaint system can be very effective in improving worker attitudes.

The opportunity for growth within a company, as provided for by company personnel policies and formal training programs at all levels, can be a powerful stimulant. A variety of incentives can be used to foster team spirit and to reward employees for creative thinking and work well done. Profit sharing, bonuses, and public recognition can be effective motivators. Company newsletters and individual awards can be used to recognize outstanding performance. Money is important, but it is not the only motivator to productive performance. A company's ability to communicate with its workers, to understand their problems, and to work as a team toward a common goal can be the firm's greatest asset.

9.43   EMPLOYEE TRAINING PROGRAMS

Many construction contractors find it necessary to provide their employees with various forms of specialized training. The company expense associated with such programs is clearly justified by the increased skill, efficiency, and productive capacity of the construction team. Open-shop contractors, not having the construction unions as a ready source of skilled craftsmen, must often train their own field forces. All contractors are concerned with the need to train their supervisory and middle management personnel in how to perform their job responsibilities in the most capable fashion.

There are a wide variety of programs which do not follow formal apprenticeship practices, available to instruct construction craftsmen (see Section 13.36). There are public plans at both the federal and local levels designed primarily to train minorities, women, and unemployed workers. These procedures concentrate on hands-on pre-apprenticeship preparation for minority workers and women and the upgrading of skills for the unemployed. In addition, many vocational-technical schools provide construction craft training.

Privately supported and conducted trade skills programs are largely concentrated in the open-shop sector of the construction industry. Some contractors implement their own in-house training plans. Many local chapters of contractor associations sponsor and conduct programs to instruct field workers. These efforts include all types of training in schemes of varying style and scope. Many involve specialized instruction to develop craftsmen highly skilled in a particular class of work, including diverse forms of training geared to the needs of specific jobs. Classroom instruction, on-the-job training, and home study are all used to varying degrees by different training projects. Many of these programs train people as helpers. These on-the-job procedures prepare workers to fill jobs in support of skilled craftsmen. Many schemes offer both specific-task and cross-craft instruction. These training plans stress the efficiency of competency-based, task-oriented, on-the-job training, and inter-craft mobility for workers who can advance to more highly skilled jobs.

Training for supervisory and middle management personnel is no less important than that for the trade skills. Contractor associations, professional groups, and consulting firms are sources of customized programs for construction firms. Such training is available in the form of manuals, resource materials, teleconferences, and audiotapes. Many large contractors have devised their own in-house programs for training their supervisory and management employees.

9.44   JOB SITE CRIME

Job site theft and vandalism have become a major source of financial loss to construction contractors across the country. Job site crime takes many forms, ranging from malicious damage done by trespassers, to the pilfering of small tools by employees, to the theft of heavy equipment by organized criminals. The Insurance Services Office reports that theft of heavy equipment has increased at an average rate of 22 percent per year since 1996, costing construction companies nearly a billion dollars a year in property losses. Only 10 to 15 percent of stolen equipment is recovered. In addition, losses from job site theft and vandalism can cost the contractor considerably more than just the value of what is actually stolen or vandalized. When theft or damage occurs at a job site, the work is often slowed or brought to a temporary halt, and these delays are very expensive in effect and cause serious interruption in the construction time schedule. A further result is that company insurance rates are often increased. There are no foolproof procedures that can be followed, but there are effective steps that a contractor can take to combat this problem. Certainly, crime prevention is an important facet of the contractor's business.

To minimize losses caused by job site crime, company management must establish and commit itself to a crime prevention program. The details of such a company activity will vary with company management, the type of equipment owned, the type of work normally performed, and the general location of company projects. The cost and time devoted to designing and implementing such a crime prevention program can be recovered many times over by reducing the costs resulting from theft and vandalism. Company employees should be made to understand that such a program is implemented for the reduction of losses and is essential to protect the company's and their own interests.

In recent years wireless video surveillance systems and global positioning systems tracking technology have become commonly available at affordable prices. Such systems can assist considerably in the recovery of stolen equipment and materials and the successful prosecution of the wrongdoers.

QUESTIONS

1. Explain the difference between cash based and accrual based accounting methods.

2. Describe why both an income statement and balance sheet are necessary when determining the financial status of a company.

3. What can you infer from a company with a quick ratio of 0.72 and a current ratio of 1.52? What strategic change should the company consider making?

4. What effect might the outside rental of equipment have on a company's ratio of assets to liabilities?

5. Why do many contractors elect straight-line depreciation for internal purposes and accelerated depreciation for external reporting?

6. Describe the salient differences between payment requests developed for lump-sum contracts, unit-price contracts and cost-plus contracts.

7. Explain the difference between the New York and Pennsylvania systems of lien rights.

Chapter 10

Project Management
and Administration

10.1 THE NEED FOR PROJECT MANAGEMENT

The essential focus of a construction company is its field projects. These are, after all, the heart and soul of any such business enterprise. If a project is to meet its established time schedule, cost budget, and quality requirements, close management control of field operations is a necessity. Project conditions such as technical complexity, importance of timely completion, resource limitations, and substantial costs put great emphasis on the planning, scheduling, and control of construction operations. Unfortunately, the construction process, once it is set in motion, is not a self-regulating mechanism but requires expert guidance if events are to conform to plans.

It must be remembered that projects are one-time and largely unique efforts of limited time duration that involve work of a nonstandardized and variable nature. Field construction work can be profoundly affected by events that are difficult, if not impossible, to anticipate. Under such uncertain and shifting conditions, field construction costs and time requirements are constantly changing and can escalate substantially with little or no advance warning. Skilled and unremitting management effort is not only desirable, but is imperative for a satisfactory result.

10.2 PROJECT ORGANIZATION

All construction projects require some field organization, and large jobs will require considerably greater organization than smaller jobs. Terminology differs somewhat from one construction firm to another and organizational patterns vary, but the following description is more or less representative of current trade practice.

The management of field construction is customarily conducted on an individual project basis, with a project manager being made responsible for all aspects of the project. Project management cuts across functional lines of the parent organization, and the central office acts in a service role to support the field operations. Working relations with a variety of outside organizations, including architect-engineers, owners or owner representatives, subcontractors, material and equipment dealers, regulatory agencies, and possibly labor unions, are an important part of guiding a job through to its conclusion. Project management is directed toward pulling together all the diverse elements involved into a going venture with the common objective of project completion.

The form and extent of a project's organization depend on the nature of the work, size of the project, and type of construction contract. A firm whose jobs are not particularly extensive will have essentially all office functions, such as accounting, payroll, and purchasing, concentrated in its main or area office. Only larger projects can justify the additional overhead necessary to carry out the required office tasks in a field office on the job site. Extensive projects frequently support a substantial field management team, the extent of which depending on the nature of the work, its geographical location, and the type of contract. For example, a large cost-plus contract may well have all associated office functions performed at the project site. A project management staff is customarily developed along much the same lines as the contractor's main operating organization.

10.3   THE PROJECT MANAGER

The project manager organizes, plans, schedules, and controls the fieldwork and is responsible for getting the project completed within the time and cost limitations. He attends to change orders, project budgets, records, payment requests, progress schedules, shop drawings, and samples and acts as the principal liaison between the contractor and the owner and its representatives. The project manager provides a focal point for all facets of the project and brings together the efforts of those organizations having input into the construction process. He coordinates matters relevant to the project and expedites project operations by dealing directly with the individuals and organizations involved. In any such situation where events progress rapidly and decisions must be consistent and informed, the specific leadership of one person is needed. Because he has overall responsibility, the project manager must have broad authority over all elements of the project. The nature of construction is such that he must often take action quickly on his own initiative, and it is necessary that he be empowered to do so. To be effective, the project manager must have full control of the job and be the one voice that speaks for the project. Project management is a function of executive leadership and provides the cohesive force that binds the several diverse elements into a team effort for project completion.

When small contracts are involved, a single individual may act as project manager for several jobs simultaneously. Figure 10.1 illustrates such an organizational plan. Larger projects normally have a full-time project manager who reports to a senior executive of the company. The manager may have a project team to assist him, or he may be supported by a central office functional group. Figure 10.2 shows such an arrangement.

The project manager must have expertise and experience in the application of specialized management techniques for the planning, scheduling, and cost control of construction operations. These procedures, developed specifically for application to construction projects, are discussed in Chapters 11 and 12. Because much of the project management system is usually computer based, the project manager must have access to adequate computer support services and a good working knowledge of the software applications being used.

10.4   THE PROJECT SUPERINTENDENT

Project management and field supervision are quite different responsibilities. A site supervisor or project superintendent handles the day-to-day direction of project operations. His duties involve supervising and directing the trades, coordinating the subcontractors,

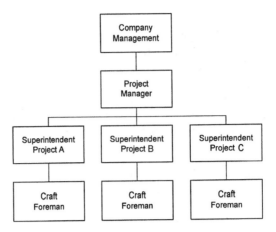

Figure 10.1   Project organization, small projects.

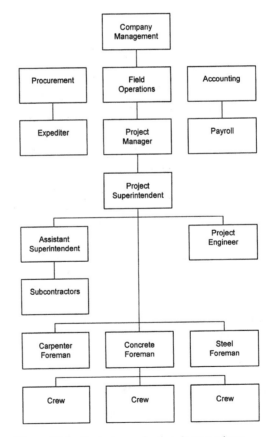

Figure 10.2   Project organization, large projects.

working closely with the owner's or architect-engineer's field representative, checking daily production, and keeping the work progressing smoothly and on schedule. He is responsible for material receiving and storage, equipment scheduling and maintenance, project safety, and job records and reports. The superintendent is normally authorized to make small purchases directly from the field, using a petty cash fund or a simplified field purchase order form.

Centralized authority is necessary for the proper conduct of a construction project, and the project manager is the central figure in that respect. Nevertheless, some freedom of action by the field superintendent is required in field construction work. In practice, construction project authority is wielded much as a partnership effort, with the project manager and the project superintendent functioning much as allied equals. Notwithstanding such practical functioning, however, top company management must make a clear delegation of authority to the project manager and must also make sure the duties and responsibilities of each are explicitly understood.

10.5   THE JOB ENGINEER

On large projects, the field staff normally includes a field engineer who reports either to the project manager or to the project superintendent. The field engineer is generally assigned such responsibilities as project scheduling, progress measurement and reporting, progress billing, keeping job records and reports, cost studies, testing, job engineering and surveys, safety and first aid, and payrolls. He may also represent the project manager in certain duties such as holding project coordination meetings or negotiating change orders with owners. Job engineers are frequently assigned technical responsibilities such as reviewing and approving shop drawings, materials, and technical submittals. Sometimes the title "Job Engineer" or "Project Engineer" is used to describe this position. This is generally an entry-level position in the contractor's organization for degreed civil/construction engineers. Although professional registration is generally not a prerequisite for this position, a degree of care must be exercised when "engineer" is used in a job title. Many states reserve the term *engineer* in a job title exclusively for professionally registered engineers.

10.6   OWNER'S PROJECT REPRESENTATIVE

On large projects and those where the construction is of a specialized or highly technical nature, a full-time representative of the owner is normally part of the project team. This party may be referred to as the owner representative, resident project representative, resident engineer, resident architect, contracting officer, or by another title. Although this person may be an employee of the owner or the architect-engineer, his full-time responsibility is to protect and ensure the contractual rights of the owner during the construction process.

The duties of the owner's representative can be highly variable, depending on the form of contract, type of construction, and wishes of the owner. This person spends full time on the job site, working closely with contractor personnel. Depending on the size and nature of the job, he may handle the assigned responsibilities personally or may be in charge of a project team consisting of several persons. The duties of an owner's representative typically include such things as inspection and quality control, checking shop drawings, keeping job records, checking pay requests, attending project meetings, making location surveys, materials and acceptance testing, processing of change orders, measurement of work quantities, performance testing, preparing as-built drawings, providing status reports, and similar functions.

10.7   JOB SITE COMPUTERS

Computers are now commonly used on the job site itself. They are frequently used for word processing; organizing and controlling documentation; storing, transmitting, and generating job information pertaining to time schedules, labor and equipment costs, time status reports, payroll data, estimating, materials, and subcontractors; handling on-site accounting functions and performing schedule updates and revisions. Contractors often establish network connections between their home offices and each of their field offices. This is particularly effective when some of the project functions are supported out of a home office. There can be a substantial advantage in having such computer resources located on larger jobs, principally because it affords project management personnel the capability to develop needed information and to collect project data promptly and easily. With the proper software, job site computers can substantially increase the productivity and effectiveness of field forces by providing the project manager and superintendent with accurate and current job information.

10.8   ASPECTS OF PROJECT MANAGEMENT

In general terms, project management may be described as the judicious allocation and efficient use of resources to achieve timely completion of a project within the established construction budget. The resources required are money, manpower, equipment, materials, and time. Considerable management effort is required if the contractor is to meet its construction objectives. The achievement of a favorable time-cost balance by the careful scheduling and coordination of labor, equipment, and subcontractors, and the maintaining of a material supply to sustain this schedule, requires effort and skill. The project organization must blend the quantitative techniques of scientific management together with the subjective ingredients of experienced judgment and intuition into an effective and efficient operating procedure. Astute project management requires at least as much art as science—as much skill in human relations as management technique.

The details of a job management system depend greatly on the contractual arrangements with the owner. Basic to any contractor's project management system, however, is the control of project time and cost during the construction period. Before field operations begin, a detailed time schedule of operations and a comprehensive construction budget are prepared. These constitute the accepted time and cost goals that will be used as a "flight plan" during the actual construction process. After the project has been started, monitoring systems are established that measure the actual costs and progress of the work at periodic intervals. The reporting system provides progress information that is measured against the programmed targets. A comparison of field costs and progress with the established plan quickly detects exceptions that must receive prompt management attention. Data from the system can be used to make corrected forecasts of the costs and time necessary to complete the work.

10.9 FIELD PRODUCTIVITY

Project management is vitally concerned with field productivity, because this is the measure that determines whether the project will be completed within the established cost budget and time schedule. In the early 1980s, the Business Roundtable's Construction Industry Cost Effectiveness Project (CICE) produced a 24-report study that marked a milestone for the industry. According to the study, field productivity had declined seriously during the past several years. In part, these reports precipitated a renewed dedication by the industry to motivate its workforce and find new means to increase worker productivity. This became a primary focus of project management.

The CICE study disclosed that an appreciable proportion of a construction worker's time is completely nonproductive, and at least one-half of the time wasted is caused by poor job management. Such productivity loss is attributed to demotivated workers, lack of material management, poorly trained supervisors, failure to utilize modern time and cost management systems, poor communications, and lack of teamwork. To overcome these negative factors, there must be a commitment by company management to establish a detailed program of action.

Many things can demotivate a craftsperson, but the CICE study reported that some of the most common complaints were materials unavailable, unsafe working conditions, redoing work already completed, unavailability of tools or equipment, lack of communication, and disrespectful treatment by supervisors. These are all factors that can be minimized or eliminated by proper management action. Associated with workers being "turned off" are absenteeism and job turnover, both factors in poor productivity.

It is a surprising fact that few construction companies make any attempt to give their foremen and superintendents supervisory training. There are now many supervisory training programs available through contractors associations, consultants, and technical schools. In this regard, the Associated General Contractors of America sponsors the Supervisory Training Program (STP) and the Associated Builders and Contractors has several educational programs aimed at foremen, superintendents, site managers, and project managers. These programs include such topics as communications, work planning, motivation, leadership, cost-effectiveness, human relations, safety, contract documents, and problem solving.

The CICE study also disclosed that scheduled overtime can be counterproductive and inefficient. Scheduled overtime does not mean the occasional periods when more than the usual 40 hours per week are worked. *Scheduled overtime* refers to excess hours being made a part of the usual workweek for extended periods. As such overtime increases, productivity actually decreases.

10.10 PROJECT ADMINISTRATION

Project administration refers to those actions that are required to achieve the established project goals. These involve duties that may be imposed by the construction contract or that are required by good construction and business practice. The efficient handling, control, and disposition of contractual and administrative matters in a timely fashion are of paramount importance for a smooth-running job.

Specifically, project administration means those practices and procedures, usually routine, that keep the project progressing in the desired fashion. Whatever is required to efficiently provide the project in a timely manner with the materials, labor, equipment, and services required lies within the general jurisdiction of project administration. The rest of this chapter describes a number of essential elements of project administration.

10.11 PROJECT MEETINGS

On larger projects, after the principal subcontracts have been awarded and before the beginning of actual construction, it is common practice for the general contractor to call a pre-construction meeting or series of meetings between representatives of the owner, architect-engineer, subcontractors, and prime contractor. This get-together serves to introduce members of the construction team to each other, and to establish ground rules of the construction process. Matters of common concern are discussed, which generally include the following:

Shop drawings	Job site security
Project time schedules	Owner-furnished materials
Storage and hoisting facilities	Site surveys
Project site offices	Quality control
Temporary job services and utilities	

The architect-engineer may also use this meeting to discuss prerequisite items needed prior to the start of field construction activities, such as the following:

Insurance certificates	Construction schedule
Required permits	Schedule of owner payments
Cost breakdowns	Project safety plans approval
List of subcontractors	Contractor's quality control plan approval

In addition, the preconstruction meeting should address important contractual provisions such as:

Completion date	Site access provisions
Liquidated damages	Environmental impact initiatives
Progress payments	Time extensions
Bonus clauses	

The preconstruction meeting gives those attending a chance to raise questions, reach agreements, and clear up misunderstandings.

Once construction has started, regular job site meetings are standard practice on larger projects. The project manager runs these meetings with representatives of the owner, architect-engineer, material vendors, and subcontractors attending. Minutes of these meetings are kept, and copies are distributed to interested parties. At such meetings job progress is discussed; trouble spots are identified and corrective action is planned. These regular meetings are very valuable, in that all parties concerned are kept fully informed concerning the current job status and are made to realize the importance of meeting their own obligations and commitments. Such face-to-face exchanges are invaluable in resolving misunderstandings and quickly identifying the real sources of problems and difficulties on the project.

10.12   SCHEDULE OF OWNER PAYMENTS

A common contract provision requires the contractor to provide the owner with an estimated schedule of monthly payments that will become due during the construction period. The owner needs this information so that cash will be available as needed to make the necessary periodic payments to the contractor. Because the owner must often sell bonds or other forms of securities to obtain funds with which to pay the contractor, it is important that the schedule of anticipated payments be an accurate forecast.

The most accurate basis for determining such payment information is the project time schedule (see Section 11.9) that is established by the contractor either before field operations start or very early in the construction period. By establishing the total costs associated with each scheduled segment of the project and by making some reasonable assumptions concerning how construction costs vary with time (a linear relationship is often assumed), a reasonably accurate prediction of the value of construction in place at the end of each month can be made. By considering retainage, these data can be reduced to an estimated schedule of owner's monthly payments. These values may subsequently have to be revised as the project progresses.

Many project scheduling software applications currently available on the market support cost-loaded schedules and allow the contractor to quickly and easily develop and update the schedule of owner payments throughout the project.

10.13   SHOP DRAWINGS

The working drawings and specifications prepared by the architect-engineer, although adequate for job pricing and general construction purposes, are not suitable for the fabrication and production of many required construction products. Manufacture of the necessary job materials and machinery often requires that the contract drawings be amplified by detailed shop drawings that supplement, enlarge, and clarify the contract design.* Such descriptive technical submissions are prepared by the producers or fabricators of the materials and are submitted to the general contractor and thence to the architect-engineer for approval before the items are supplied. This procedure also applies to materials provided by subcontractors, in which case shop drawings are submitted to the general contractor through the subcontractor. Shop drawings are required for almost every product that is fabricated away from the building site. For instance, in building construction shop drawings must be prepared for everything from reinforcing steel and metal door frames to millwork and finish hardware. The general contractor must sometimes prepare shop drawings covering work items or appurtenances it has designed and will fabricate.

10.14   APPROVAL OF THE SHOP DRAWINGS

When shop drawings are first received from a supplier, the contractor is responsible to the owner for checking them carefully to verify materials, field measurements, and field

---

* The term *shop drawings* refers to fabrication, erection, and setting drawings; manufacturer's standard drawings or catalog cuts; performance and test data; wiring and control diagrams; schedules; samples; and descriptive data pertaining to material, machinery, and methods of construction as may be necessary to carry out the intent of the contract drawings and specifications.

☐ Reviewed   ☐ Furnish as Corrected

☐ Rejected   ☐ Revise and Resubmit

☐ Submit Specific Item

This review is only for general conformance with the design
concept of the project and general compliance with the information
given in the Contract Documents. Corrections or comments made
on the shop drawings during this review do not relieve the
contractor from compliance with the requirements of the plans
and specifications. Approval of a specific item shall not include
approval of an assembly of which the item is a component.
Contractor is responsible for: dimensions to be confirmed and
correlated at the job site; information that pertains solely to
the techniques, sequences, and procedures of construction;
coordination of his or her work with that of all other trades,
and for performing all work in a safe and satisfactory manner.

Jones and Smith, Architect-Engineers

Date: _____   By: _____

Figure 10.3   Typical architect-engineer's submittal review stamp.

construction criteria and to check and coordinate submittal information with the work and
the contract documents. The shop drawings are then forwarded to the architect-engineer so
that they may be checked for conformance with information given and the design concept
expressed in the contract documents.

In general, most construction contracts put responsibility on both the architect-engineer
and the contractor to verify that submittals conform to the contract documents, and in
addition make the contractor responsible for ensuring that the submittals comply with
field conditions and constructability requirements. Paragraph 3.12 and its subparagraphs
in Appendix C describes the workings of a typical approval procedure. Although approval
of shop drawings can possibly put legal responsibility on the architect-engineer or general
contactor if an error in the drawings is later found to be the proximate cause of damages,
responsibility for their correctness belongs to the subcontractor or supplier who prepared the
submittal. In recent years architect-engineers have largely stopped using the word *approved*
when indicating that shop drawings have been found satisfactory in order to limit their
liability. Rather, *reviewed, no exceptions taken, accepted*, and *examined* are now frequently
used. Figure 10.3 provides commonly used language provided by the architect-engineer on
submittal review stamps.

A sufficient number of copies of each drawing must be provided for distribution to all
interested parties. Such requirements are generally spelled out in the contract documents.
After the architect-engineer has returned the approved drawings, the contractor notes the
nature of any comments or corrections and routes copies as necessary, returning at least one

SUBMITTAL AND SHOP DRAWING LOG

PROJECT NAME _____

PROJECT NUMBER _____

| DATE RECEIVED | SPECIFICATION SECTION NO. / SHOP DRAWING NO. / SUBMITTAL NO. | CONTRACTOR, SUBCONTRACTOR OR SUPPLIER | COPIES RECEIVED | REFERRED | | | | ACTION | | | | COPIES TO | | | |
				TO	DATE SENT	COPIES SENT	DATE RETURNED	APPROVED	APPROVED AS NOTED	REVISE & RESUBMIT	NOT APPROVED	SUBCONTRACTOR	SUPPLIER	FIELD	FILE

Figure 10.4 Submittal and shop drawing log.

copy to the supplier. Occasionally, the shop drawings must be redone and resubmitted to the architect-engineer. Because the material supplier must receive a copy of the approved shop drawings before putting the order into production, it is important that the submittal-approval-return process be expedited. Otherwise, materials may not be delivered to the project when they are needed. It is usual for the contractor to establish some form of logging and control system with regard to shop drawings to guard against oversight or delay in the approval process. A sample of a typical submittal and shop drawing log is provided as Figure 10.4.

It is important to note that only a qualified approval is given to shop drawings by architect-engineers. Such approval relates only to conformance with the design concept and overall compliance to the contract drawings and specifications. Checking does not include quantities, dimensions, fabrication methods, or construction techniques. Approval of shop drawings by the architect-engineer does not relieve the contractor of its responsibility for errors or inadequacies in the shop drawings or for any failure to perform the requirements and intent of the contract documents. Approval of shop drawings does not authorize any deviation from the contract unless the contractor gives specific notice of the variance and receives express permission to proceed accordingly. However, as long as there is no explicit disagreement between the shop drawings and the construction contract, approval is usually binding in the event of a subsequent dispute over design requirements. So long as the work covered conforms to approved shop drawings, the contractor is considered to have complied with its contractual obligation. For obvious reasons, it is important that contractors carefully check their project shop drawings. They cannot act merely as go-betweens with suppliers and architect-engineers. An interesting aspect of approved shop drawings is that they are not usually considered to be contract documents, and they do not modify or extend the obligation of either party to the construction contract.

10.15 QUALITY CONTROL

Quality control during field construction is concerned with ensuring that the work is accomplished in accordance with the requirements specified in the contract. The architect-engineer establishes the criteria for construction, and the quality control program checks contractor compliance with those standards. A field quality control program involves inspection, testing, and documentation for the control of the quality of materials, workmanship, and methods. On a given project, the architect-engineer, owner, consultants, prime contractor, or construction manager may administer the quality control program. The inspector has no authority to give directions, render interpretations, or change the contract requirements. The inspector is not present to manage the job, direct the work, or to relieve the contractor from any of its obligations. He cannot tell the contractor what to do or how to do it, nor interfere in field operations unless it is to prevent something from being done improperly. The function of the inspector is to observe the construction process and to ensure compliance with contract requirements by the contractor.

The architect-engineer's providing of field inspection services can significantly increase his vulnerability to a wide range of potential liabilities. For instance, the responsibility for field inspection has led to the liability of architect-engineers for construction defects caused by the contractor's failure to follow contract requirements. Design professionals have been made responsible for contractors' construction methods and have been held liable to workers and members of the public injured by construction operations. As a result, there has been a trend for architect-engineers to withdraw as much as possible from any responsibility for site operations. Extensive revisions have been made to the contract language that defines their field inspection responsibilities. The word *supervision* has proven to be especially troublesome for architect-engineers because it suggests that the designer has some degree of control over, and thus responsibility for, the contractor's day-to-day operations. By the terms of most design contracts between owner and architect-engineer, the designer accepts only very limited responsibility for field operations. For example, Subparagraph 2.6.2.1 in Appendix A provides that the designer will make periodic site visits, but will not assume any responsibility for continuous on-site inspections unless provided for under a supplementary agreement. If the architect-engineer does agree to administer the project quality control program, it will assign a full-time employee to the project to provide administrative and surveillance services. Laboratory and field testing may be done by this person, or a commercial testing laboratory may be engaged to perform this specialized work.

Many public agencies and corporate owners establish their own internal quality control programs to monitor their ongoing construction projects. Where there is a continuous construction program or where very large and complex projects are involved, the owner may establish a functional department within its overall organization that acquires trained personnel and develops standards for quality control application in the field.

Owners sometimes hire specialized consultants to provide field quality control services. In this context, a consultant is a technical firm such as a testing laboratory, a specialized consulting engineer (not the project designer), or a construction management firm that provides quality control services. Such consultants are frequently employed on extremely complex construction or on projects that have specialized and highly technical quality control requirements. These firms are entirely independent of the architect-engineer and contractor and are usually hired before the design is completed to provide advance quality assurance input and advice.

During recent years, some public and private organizations have required prime contractors to take a more proactive role in the control of project quality by having them develop and manage their own quality control programs. Construction contracts with these owners require the contractor to maintain a job surveillance system of its own and to perform inspections that will ensure that the work performed conforms to contract requirements. The contractor is required to maintain and make available adequate records of such inspections. Owner representatives monitor the contractor's quality control plan and make spot-check inspections during the construction process. This owner oversight is frequently referred to as quality assurance. Under such contracts, the contractor is required to provide significant and specific inspection and documentation to satisfy both itself and the owner that the work being performed meets the contract requirements. In the usual case, the contractor is required to report, on a daily basis, the construction progress, problems encountered, and corrective action taken and to certify that the completed work conforms to the drawings and specifications. The owner is responsible for final inspection and may inspect at any other time deemed necessary to ensure strict compliance with the contract provisions.

Many state and federal owners now require the contractor to submit a formal quality control plan for approval before commencing construction or, in some cases, before being invited to submit a bid or proposal. The contractor's quality control plan must generally cover all aspects of the construction process—from raw material sourcing through processing, delivery, storage, installation, and finishing. Therefore, the contractor must work in close partnership with its subcontractors and suppliers to prepare a comprehensive plan for tracking and controlling product quality at each successive stage of development. Some owners require that the contractor also list the qualifications of the craftspersons participating in the construction process and establish a standard for their training and certification. Process planning, standardization, and control are vital to the contractor's quality plan. The contractor must make certain that systems are instituted at the inception of an activity that ensure that quality standards will be met or exceeded in the finished product, as failures often have costly consequences. The Construction Industry Institute (www.construction-institute.org) produces several publications designed to assist contractors with developing successful quality control programs.

The last aspect of quality control includes the final inspection, field acceptance testing, and start-up of the facility. The prime contractor, appropriate subcontractors, and manufacturers' representatives start up the project equipment and systems, with the owner checking all control and instrument operations. This process includes simulation of both normal operating and emergency conditions. The facility is then turned over to the owner with a complete set of job files, shop drawings, maintenance and operating manuals, and as-built drawings.

Recent years have witnessed what can only be described as increasing neglect of construction quality control. Modern procedures involve so many players on the construction team that responsibility for the end product has become splintered and diffused. The withdrawal of the architect-engineer has been mentioned earlier. Inspection by the contractor is criticized on the basis that it is unrealistic to expect profit-oriented contractors to rigorously inspect their own work. As a result, there is a sentiment in the industry that construction quality control should be a budgeted part of the total construction process and that this responsibility should be contracted to third-party professional firms.

10.16 MATERIALS MANAGEMENT

The management of materials is an essential element in the conduct of any construction project. Because the expense of materials and the cost of labor to install them account for more than one-half the total cost of many projects, the supply of materials commands management attention. The process begins with the materials take-off and continues through requisitioning, purchasing, receiving, storage, and distribution. If the owner is providing materials, this party is a necessary part of the management process and there must be clear communication between owner and prime contractor coordinating the procurement and delivery process. This will also apply if the owner or prime contractor is supplying materials to a subcontractor.

Computerizing the contractor's material procurement and management system is now common with larger firms, a move that can yield substantial labor cost savings and reduce the overall costs of material procurement and handling. These are computerized procedures to track and control the flow of construction materials to work crews in the field, starting with quantity take-off and continuing through purchasing and allocation. The principal aspects of the management system after the material purchase order has been issued are discussed in the following sections.

10.17 EXPEDITING

The nature of a construction enterprise is such that the timely delivery of project materials is of extreme importance. If required items are not available when needed, the contractor can experience major difficulties because of the disruption of the construction schedule. Such delays are expensive, awkward, and inconvenient, and every effort must be made to avoid them. When purchase orders are written, delivery dates are designated which, if met, will ensure that the materials will be available when needed. These dates are established on the basis of the project progress schedule and must necessarily make allowance for the approval of shop drawings.

Unfortunately, the contractor cannot assume that the designation of delivery dates in its purchase orders or the securing of delivery promises from the sellers will automatically ensure that the materials will appear on schedule. To obtain the best service possible, a series of follow-up actions, referred to as expediting, are taken after each material order is placed so as to keep the supplier constantly reminded of the importance of timely delivery. Expediting may be a job site function, or the construction firm may provide all of its construction projects with a centralized expediting service. A full-time expediter is sometimes required on a large project. When the owner is especially concerned with completion of a job, or when certain material deliveries are crucial, the owner often participates with the contractor in cooperative expediting efforts.

A necessary adjunct to the expediting function is the maintaining of a check-off system or log where the many steps in the material delivery process are recorded. Starting with the issuance of the purchase order, a record is kept of the dates of receipt of shop drawings, their submittal to the architect-engineer, receipt of approved copies, return of the approved drawings to the vendor, and delivery of the materials. Because shop drawings from subcontractors are submitted for approval through the general contractor, the check-off system should also include materials being provided by the subcontractors. This is desirable because project delay can be caused by any late material delivery, regardless of who provides the material. This same documentation procedure is followed for samples, mill certificates, concrete-mix designs, and other submittal information required. General contractors sometimes find it necessary, in regard to critical material items, to determine the manufacturer's production calendar, testing schedule if required, method of transportation to the site, and data concerning the carrier and shipment routing. This kind of information is especially helpful in working the production and transportation around strikes and other delays.

Each step in the approval, manufacture, and delivery process is recorded, and the status of all materials is checked frequently. At intervals, a material status report is forwarded to the project manager for his information. This system enables job management to stay current on material supply information and serves as an early-warning device when slippages in delivery dates seem likely to occur.

The intensity with which the delivery status of materials is monitored depends on the nature of the materials concerned. Routine materials such as sand, gravel, brick, and lumber usually require little follow-up. Critical made-to-order items, whose late delivery would badly cripple construction operations, must be closely monitored. In such cases, the first follow-up action should be taken weeks or months in advance of the scheduled delivery date. This action, perhaps via a letter showing order number, date of order, and delivery promise, requests specific information on the anticipated date of shipment. Return answers to such inquiries can be very helpful. If a delay appears likely, strong and immediate action is necessary. Letters, telegrams, telephone calls, and personal visits, in that order, may be required to keep the order progressing on schedule.

10.18   DELIVERIES

In addition to working for the timely delivery of materials, the expediter is also usually responsible for their receipt, unloading, and storage. In general, deliveries are made directly to the projects to minimize handling, storage, insurance, and transportation costs. However, there are often instances when it is preferable or necessary to store materials temporarily at off-site locations until they are needed on the job. A common example of when this is done is in the construction of buildings in crowded urban settings where storage space is extremely limited.

When notice of a material delivery is received, suitable receiving arrangements must be made. Advance notice of shipments is provided directly by the vendors or through bills of lading or other shipping papers. If a shipment is due at a job site, notice is given to the project superintendent. If suitable unloading equipment is not available on the site, such equipment must be scheduled or the project superintendent must be authorized to obtain whatever may be required. If a shipment is to be made to the contractor's storage yard or warehouse, the person in charge must be advised and unloading equipment must be scheduled, if required.

The scheduling of material deliveries to the job site can be especially important on some projects. For example, consider the delivery of structural steel to a building project in a downtown city area. On projects of this type, storage space is extremely limited and deliveries must be carefully scheduled to arrive in the order needed and at a rate commensurate with the advancement of the structure. There must also be close cooperation between the contractor and the steel supplier. An additional factor is the routing of the trucks through the city streets, often at off-hours, and arranging for the direction of traffic around the vehicles during the delivery and unloading operations. Arrangements for necessary permits, police escorts, labor, and unloading equipment must be made in advance. On such projects, many material deliveries are not made directly to the site but to temporary storage facilities owned or rented by the contractor. When such off-site storage is used, deliveries to the job site are made in accordance with short-term job needs.

10.19   RECEIVING

Job materials are usually delivered directly to the job site. There are times, however, when it is either undesirable or impossible to accept shipments at the project. Construction in congested urban areas is an instance already mentioned. Another example is early delivery of items that would be susceptible to damage, loss, or theft if stored on the job for extended periods. Whenever possible, such materials are stored in the contractor's yard or warehouse until they are needed.

Truck shipments may be made by common carriers or the vendor's own vehicles. In either case, the material must be checked for damage as it is being unloaded, and quantities checked against the freight bill or vendor's delivery slip. Observed damage must always be noted on all copies of the freight bill and be witnessed by the truck driver's signature. The receiver should not sign the delivery slip or freight bill until the quantity delivered has been checked against that indicated.

When shipment is made by rail car, the contractor advises the carrier as to where it desires the car to be spotted as soon as the contractor is advised of the car number. The shipment should be checked after the car is placed for unloading and any visible damage reported to the railroad claim agent. In case of damage, unloading must be deferred until the shipment has been inspected and proper notations made on the bill of lading. A claim for damage or loss is submitted to the freight claim agent on the carrier's standard form. This claim must be accompanied by the original bill of lading, the receipted original freight bill, the original or a certified copy of the vendor's invoice, and other information in substantiation of the claim. Should damage be such that it is not visible and cannot be detected until the goods are unpacked, the contractor must make its claim at that time on the carrier's special form that is used for concealed damage. Rail shipments of less-than-carload (LCL) may be such that the contractor must pick up the material at the freight depot, or it may be delivered to the project or yard by truck. Delivery depends on the FOB point designated by the purchase order.

The party who receives a shipment on behalf of the contractor should immediately transmit the covering delivery ticket, freight bill, or bill of lading to the contractor's office. Information pertaining to damage or shortage and the location of material storage should be included.

Although most purchase orders include freight charges in their face amounts, material vendors do not always prepay freight charges. As a result, materials often arrive at the contractor's location with freight charges to be collected. The contractor usually has an account with the carrier that allows it to receive the goods without having to pay the transportation charges at the time of delivery. However, in the case of common carriers, Interstate Commerce Commission regulations require that freight charges be paid within a short time after delivery. Therefore, it is important that freight bills for collect shipments be transmitted immediately to the contractor's office for payment. Where the purchase order amount includes freight, it is usual for the contractor to pay the freight charges and backcharge the account of the vendor.

10.20   INSPECTION OF MATERIALS

It is preferable that inspection of delivered goods for quantity and quality be done concurrently with their unloading and storage. This is not always possible, however, and there may often be an objectionable delay in the unloading and release of transporting equipment. The checking of the package count as shown by the freight bill or delivery ticket should always be done, with any variations being indicated on the bill or ticket. However, the quantities of different items and their quality often must be verified at the first opportunity after receipt. To do this, the party making the inspection obtains copies of the covering purchase order and approved shop drawings. A thorough check of the delivered items is made to verify both item quantities and quality. This verification can become quite laborious in some cases, but can pay big dividends in minimizing later job delays caused by missing, faulty, or erroneous materials. The project inspector frequently participates and assists in this inspection process. Inspection of material deliveries must be done with reasonable promptness so that there is time to take any corrective measures necessary.

MATERIAL RECEIVING REPORT

1. CONTRACT NUMBER		ORDER NO.	6. INVOICE NO./DATE		7. PAGE NO.	8. ACCEPTANCE POINT
2. SHIPMENT NO.	3. DATE SHIPPED	4. BILL OF LADING	5. DISCOUNT TERMS			
9. GENERAL CONTRACTOR			10. ADMINISTERED BY			
11. SHIPPED FROM ADDRESS	FOB:		12. PAYMENT WILL BE MADE BY			
13. SHIPPED TO ADDRESS			14. MARKED FOR			

15. ITEM NO.	16. STOCK/PART NO.	17. QUANTITY SHIPPED/RECEIVED	18. UNIT	19. UNIT PRICE	20. AMOUNT

21. CERTIFICATION: I hereby certify that I have received and checked the items listed in sections 15, 16, 17, 18, 19, and 20 and found them to comply with the contract documents. The quality of the items have been checked and found to be:

☐ Acceptable   ☐ Accepted as Noted   ☐ Rejected

SIGNATURE	DATE

Figure 10.5   Material receiving report.

After the delivered materials have been reconciled with the shop drawings and purchase orders, the inspector normally files a receiving report with the company's procurement section. A sample of such a report is provided in Figure 10.5. This report shows the purchase order number, date of inspection, material, location of storage, quantity, remarks, and signature of the inspector. The receiving report verifying receipt of the proper count and quality clears the order and authorizes payment to the vendor. With partial shipments, several receiving reports may be required to clear the entire order.

Inspection duties at times involve the sampling of various kinds of construction materials. Construction contracts may require the laboratory testing of certain materials as proof of quality. Thus, inspectors must be acquainted with the standard methods of sampling sand, gravel, bulk cement, asphalt, reinforcing steel, and other construction commodities. Another aspect of materials inspection is the obtaining of certification of quality or the results of laboratory control tests from the manufacturer or producer. Submittal of the manufacturer's certification of quality may enable the contractor to avoid duplicate acceptance testing.

Commonly associated with the inspection of construction materials is the process of grouping materials together by task. As the materials are checked, they are separated into groups, each of which is associated with a specific future work type or activity. This simplifies the matter of later identifying the materials needed by a given labor crew. When it is done some time before the need arises, such grouping can identify incorrect or missing items while there is yet time to take corrective action.

10.21   SUBCONTRACTOR SCHEDULING

As important members of the field construction team, subcontractors have an obligation to pursue their work in accordance with the project schedule established by the prime contractor. Failure of a subcontractor to commence its operations when required or to pursue its share of the work diligently can be a serious matter for the general contractor. Consequently, the scheduling of subcontractors deserves and must receive appropriate action by the general contractor. Subcontractors should be provided a copy of the job schedule showing the date their work is scheduled and the time allotted for its accomplishment as soon as the schedule is available. A practice followed by many contractors is to notify each subcontractor by letter two weeks or more before the subcontractor is expected to move onto the project and commence operations. Subcontractors must be given adequate time to plan their work and make the necessary arrangements to start their operations. Follow-up telephone calls are made if needed. In the interest of good subcontractor relations, the project manager should not schedule a subcontractor to appear on the site until the job is ready and the subcontracted work can proceed unimpeded. After a subcontractor is on the job, its progress must be monitored to ensure that its operations are keeping pace with the overall project time schedule. If its work falls behind, the project manager may reasonably instruct the subcontractor to respond appropriately to accelerate its progress.

With regard to the general matter of subcontractor scheduling, the form and content of the subcontract can be very important. A carefully written document with specific provisions regarding conformance with time schedule, material orders, and shop drawing submittals can strengthen the project manager's hand in keeping all aspects of the project on schedule. In this context, a common problem is the failure of a subcontractor to order major materials in ample time to meet the construction schedule. General contractors occasionally find it advisable to monitor their subcontractors' material purchases. This can be accomplished by including a subcontract requirement that the subcontractor submit unpriced copies of its purchase orders to the general contractor within 10 days after execution of the subcontract. In this way, the general contractor can oversee the expediting of key materials provided by subcontractors along with its own.

10.22   RECORD DRAWINGS

A common general contract requirement is that the contractor must maintain and prepare one set of full-size contract drawings marked to show various kinds of "as-built" information. These drawings show the actual manner, location, and dimensions of all work as actually performed. This involves marking a set of drawings to show details of work items that were not performed exactly as they were originally shown, such as changed work, changed site conditions, and variations in alignment or location. In addition, details and exact dimensions are given for those work items that were not precisely located on the original contract drawings. Depths, locations, and routings of electrical service and underground piping and utilities are examples of this requirement. The set of record drawings is prepared by the contractor as the work progresses and is turned over to the architect-engineer or owner at the end of the project.

10.23   DISBURSEMENT CONTROLS

To coordinate the actions of the company accounting office with the project, it is necessary to implement a system of disbursement controls. These controls are directed toward controlling payments made to vendors and subcontractors and require that no such payments be made without proper approval from the field. The basic purpose of disbursement control is twofold: first, to ensure that payment is made up to the value of the goods and services received to date and, second, to see that total payment does not exceed the amount established by the purchase order or subcontract.

Payments made for materials are based on the terms and conditions of the covering purchase orders. Copies of all job purchase orders are provided for the project manager's use and information. Purchase order disbursement by the accounting office is conditioned on the receipt of a signed delivery ticket or receiving report from the job site. Suitable internal controls are established to ensure that total payments do not exceed the purchase order amount. Any change in purchase order amount, terms, or conditions is in the form of a formal written modification, with copies sent to the job site.

Disbursements to subcontractors follow a similar pattern. Because there are no delivery tickets or receiving reports for subcontractors, all subcontractor invoices are routed for approval through the project manager, who has copies of all the subcontracts. The project manager determines whether the invoice reflects actual job progress and either approves the invoice or makes appropriate changes. General contractors normally withhold the same percentage from their subcontractors that the owners retain from them. If the subcontractor bills for materials stored on-site, a common requirement is that copies of invoices be submitted to substantiate the amounts billed. Any change to a subcontract is accomplished by a formal change order.

10.24   JOB RECORDS

To serve a variety of purposes, a documentation system is needed on each project that will produce a comprehensive record of events that transpired during the construction period. The extent to which this is done and the job records that are maintained are very much functions of the provisions of the construction contract and the size, complexity, and risks inherent in the work. It is up to the contractor to establish which records are appropriate for a given project and to see that they are properly kept and filed.

The original estimating file and the contract documents are basic job records. During the construction phase, periodic progress reports, cost reports, a job log, correspondence, minutes of job meetings, time schedules, subcontracts, purchase orders, field surveys, test reports, progress photographs, shop drawings, and change orders are routinely maintained as a permanent project record. On larger projects additional records on manpower, equipment, operating tests, and back charges; pile driving and welding records; progress evaluation studies; and other records are kept.

10.25   THE DAILY JOB LOG

A job log (or diary) is a historical record of the daily events that take place on the job site. The information to be included is a matter of personal judgment, but the log should include matters relevant to the work and its performance. The date, weather conditions, job accidents, numbers of workers, and amounts of equipment should always be noted. It is advisable to indicate the numbers of workers by craft and to list the equipment items by type. A general discussion of daily progress, including a description of the activities completed and started and an assessment of the work accomplished, is important. Where possible and appropriate, the quantities of work put in place can be included. The job log should be maintained in a hardcover, bound booklet or journal. Pages should be numbered consecutively in ink, with no numbers being skipped. Every day should be reported and all entries made on the same day as they occur. There should be no erasures, and each diary entry should be signed immediately under the last line of the day's entry.

The diary should list the subcontractors who worked on the site, as well as the workers and equipment provided. The performance of subcontractors and how well they are conforming to the project time schedule should be noted. Material deliveries received must also be noted, along with any shortages or damage incurred.

It is especially important to note when material delivery dates are not met and to record the effect of such delays on job progress and costs.

The diary should include the names of visitors to the site and facts pertinent thereto. Visits by owner representatives, the architect-engineer, safety inspectors, union representatives, and people from utilities and government agencies should be documented and described. Meetings of various groups at the job site should be recorded, including the names of people in attendance, problems discussed, and conclusions reached. An important part of the job log is on-site photographs that are taken at regular intervals. Such a photographic record can serve a variety of purposes by providing a clear and indisputable time record of job progress. Consistent pictorial documentation of the construction process can keep the owner constantly updated on the progress of the project and can clearly demonstrate responsibility for accidents on the job and delays, thus avoiding much potential costly litigation. Such a photographic record preserves the working details of a construction project long after it has been completed.

Complete diary information is occasionally necessary to substantiate payment for extra work and is always needed for any work that may involve a claim. The daily diary should always include a description of problems on the job and what steps are being taken to correct them. The job log is an especially important document where disputes result in arbitration or litigation. To be accepted by the courts as evidence, the job diary must meet several criteria. The entries in the log must be original entries made on the dates shown. The entries must have been made in the regular course of business and must constitute a regular business record. The entries must be original entries, made contemporaneously with the events being recorded and based on the personal knowledge of the person making them. Where these criteria have been met, the courts have generally ruled that the diary itself can be entered as evidence, even if its author is not available to testify.

10.26   CLAIMS AND DISPUTES

A claim can be defined as a formal demand for compensation made by one party to a contract to another in accordance with the contract document. Construction contracts typically require the contractor to advise the owner in writing, within a prescribed period, concerning any event that will result in a delay of the project and/or additional cost. The actual claim against the owner follows and includes a detailed description of the job condition underlying the claim, identification of the contract provisions or legal basis under which the claim is made, details of how the condition has caused the extra cost and/or delay, and a summary of the increased costs and extension of time requested. A claim usually involves both questions of entitlement, which refers to the merit of the claim, and the cost and time extension involved. Claims stem from a wide variety of conditions, including (1) late payments, (2) changes, (3) constructive changes, (4) changed conditions, (5) delay or interference, (6) acceleration, (7) errors or omissions in design, (8) suspension of the work, (9) variations in bid-item quantities, or (10) rejection of or-equal substitutions. Almost any extra cost or time required of the contractor by the action or inaction of the owner or the owner's agent can be a valid basis for a claim against the owner. Refusal by the owner to recognize the claim does not ordinarily authorize the contractor to refuse to continue its field operations. However, the contractor should proceed with the work in dispute only after filing a written protest with the owner.

Although their provisions vary, construction contracts typically require the contractor to formally advise the owner within a specified time period when a situation arises that could lead to a claim. In the event of a dispute concerning a potential claim, the contractor usually cannot refuse to proceed with the work without committing breach of contract. Although the disputed work may be performed under protest, the contractor must continue field operations with diligence, relying on remedies in the contract to settle the questions of compensation and extension of time. The requirement that the contractor must proceed with the disputed work does not apply, however, if the change or modification involved is beyond the scope of the contract. The distinction between what is and what is not within the scope of the contract is often difficult to establish, however.

Contracts normally stipulate that claims and disputes be first submitted to the owner or its representative. If the claim is denied, the resulting dispute can then be submitted to various levels of appeal, such as appeals boards, arbitration, or the courts. The remedies available depend considerably on whether the dispute involves questions of fact or of law. Most public owners have statutory or administrative procedures established for the settlement of contract disputes. Such disputes involving the federal government are settled in accordance with the Contract Disputes Act of 1978. According to this statute, if there is a dispute on a federal construction project, the contractor must first present the matter to the contracting officer and request a final decision. If this decision is not acceptable to the contractor, an appeal can then be made to the appropriate agency board of contract appeals or the U.S. Claims Court.

In the final analysis, the successful settlement of a disputed claim depends largely on painstaking documentation. Preparation of a claim may be based on the routine project records compiled during the construction process. However, these records may not contain the detailed information needed to substantiate such a demand. For this reason, the contractor may be well advised to maintain a special set of records that pertains specifically to the matter in dispute. The standard dictum "put everything in writing" applies here. It should be noted in this regard that, to be effective, documentation must be prepared during the construction process. Records created after the fact will not generally receive consideration. Project photographs, dated and identified, can be very effective in working with claims. When the project manager becomes aware of an issue that has the potential of becoming a contract dispute, a copy of all documentation pertaining to the issue should be maintained in a separate file. If the matter is later resolved without incident, the document archiving for the issue can be discontinued.

Construction contracts frequently contain a provision stating that acceptance of final payment by the contractor shall be considered as a general release in full of all unsettled claims against the owner arising out of, or in consequence of, the work. Other contracts provide for a conditional release, the contractor being permitted to maintain other causes of action. Subparagraph 9.10.5 in Appendix C provides that the acceptance of final payment shall constitute a waiver of all claims by the contractor except those previously made in writing and still unsettled. This can be an important point because, for one reason or another, contractors often do not or cannot file claims until after the work has been completed. Where such contract provisions appear, the courts give them effect.

10.27   CLAIMS IN THE CONSTRUCTION INDUSTRY

In the construction industry, claims by one party against another are very common, breach of contract between the two parties to the contract often being the basis of such claims. However, many members of a construction team—architect-engineer, owner, prime contractor, subcontractors, material dealers, construction managers, and others—are not bound to each other by contract. The actions, or lack thereof, of one party can, nevertheless, seriously affect the rights and responsibilities of others during the construction process. Contract privity generally establishes that only parties bound to each other by contract may file claims against one another. Under certain circumstances, many courts have abandoned the contract privity requirement and allow court actions "in negligence" where no contract exists. This occurs when the court finds that the party sued owed a duty to the claimant. For example, under special circumstances, contractors and subcontractors can sue the architect-engineer and contractors can sue project managers. The party suing must be of a class whose reliance on the other party was clearly foreseeable. For example, the defendant party must have had a duty of due care; a duty to manage, supervise, or inspect the construction; a duty to review drawings and specifications and to identify design defects; or a similar responsibility.

A common action by the owner is to make a notation on the check used for final payment, stating, in effect, "By endorsement, this check is accepted in full payment of the account indicated." Ordinarily, if the contractor cashes the check, it is barred from suing to collect an additional sum. This follows from a generally recognized rule of law that when there is a dispute between debtor and creditor over the amount due, acceptance of such a check amounts to an agreement to accept it as final payment. In most states, however, if there is no reasonable basis for a debtor's denial that it owes the sum claimed, acceptance of a check for a smaller sum does not constitute a binding agreement for final payment, even if there is such a notation on the check. The contractor should consult with its attorney when such matters are at issue.

Claims by contractors against owners stating that they are entitled to additional compensation as a result of encountering site conditions more difficult than anticipated, are very common. However, it must be kept in mind that there is a long history of court decisions to the effect that a contractor, in an express contract, does not have the right to collect additional payments from the owner just because the work was more expensive than expected. The courts have long held that the contract prevails over the contractor's right to collect the reasonable value of the work. Much more must be proven than merely to demonstrate that the work was more troublesome than was originally contemplated. The courts are reluctant to alter a construction contract simply because one party underestimated the risk.

Construction contracts sometimes include a "no-damage-for-delay" provision, which indicates that the owner will not be liable to the prime contractor or any subcontractor for monetary claims arising out of work delays. The contractor's sole remedy in such contracts is an extension of contract time. Experience with such contract clauses indicates that exceptions are rare and that these clauses are normally valid and enforceable.

Recent years have witnessed a tremendous increase in the number of claims and disputes arising from construction operations. The teamwork that has been traditional in the industry has degraded to a process of faultfinding and defensiveness. The tenor of the industry has become increasingly adversarial, and costs have soared because of excessive lawsuits. Litigation, as the standard means of settling construction disputes, has reached epic proportions in the United States. Recent years have witnessed a pronounced increase in the use of onerous construction documents and contracts focused on punitive measures to enforce performance by the contractor.

The resulting swell of litigation has been expensive and counterproductive, inhibiting efforts to produce quality projects on time and within the budget. The construction industry is attempting to avoid litigation as a cost-effective way of resolving disputes. Experience shows that the process is slow, expensive, and often does not result in fair play. In an effort to provide more speedy and satisfactory alternatives to the clogged judicial system, the construction industry is developing and promoting innovative methods of claims prevention, such as "partnering," and alternative methods of dispute resolution (ADR).

10.28   PARTNERING

Partnering can be described as the conducting of a cooperative enterprise between two or more parties dedicated to achieving a common goal for their mutual benefit with a minimum of dispute and conflict. In the construction industry, partnering involves all members of the construction team practicing the philosophy of cooperation, communication, and accepting full responsibility for their actions. It works to minimize risk by establishing a cooperative atmosphere in which the project takes priority. Experience with partnering indicates that an adversarial relationship between contractor and owner can be avoided with the cooperation and goodwill of all members of the building team. There are many variations in the concept of partnering, but the practice is generally proving to be a positive development that results in a departure from the litigation and contention that has come to characterize the construction industry. Disputes still occur, but partnering has considerably reduced the need to resort to the courts.

Partnering is a growing practice among owners, design firms, contractors, major subcontractors, and suppliers. It is presently being applied to both public- and private-sector projects and focuses on achieving common objectives and creating an atmosphere conducive to enhancing communication between parties and minimizing disputes. Benefits include improved efficiency and cost-effectiveness, improved communications, timely problem identification and resolution, improved scheduling, better subcontractor relations, increased opportunity for innovation, and improvement of products and services. The hallmarks of this relationship are trust, cooperation, teamwork, shared vision, and the development and attainment of mutual objectives. In a partnering relationship, the parties seek to transform the traditional adversarial relationship between the owner and contractor to a more collaborative and productive atmosphere. It is a change of attitude rather than a formally structured contractual agreement that characterizes partnering.

It must be clearly understood that partnering and the partnering agreement are not considered part of the contract documents and cannot modify, eliminate or extend any part of the construction contract, the drawings, or the specifications. Therefore, agreements made during a partnering session that contradict the contract or contract documents are invalid and cannot be relied upon by any party. Such modifications always require a formal contract change order.

10.29 TOTAL QUALITY MANAGEMENT

A relatively recent and innovative approach to the management of private business is now becoming widely practiced throughout American industry. More businesses are finding that they are being required to conduct their affairs differently if they are to survive in today's highly competitive market. A new, smarter management style, often referred to as Total Quality Management (TQM), is spreading across the American business scene. Commensurately, American construction companies are now finding that they must organize their operations differently if they are to prosper in today's marketplace. As such, construction project managers are examining issues of quality in every phase of the construction process. Workers are being encouraged to contribute their experience and know-how to the improvement of field construction methods. Contractors are striving for long-term strength, not just short-term profit. The movement toward improved product quality now pervades management practices in a wide range of industries, including construction. More and more construction contractors are turning to TQM as a competitive necessity.

In the construction industry, the major barriers to the successful implementation of TQM have been identified as the following: (1) lack of a properly trained workforce, (2) intensive competition and low quality standards, (3) poor drawings and specifications, (4) inferior management practice and training, and (5) lack of competent field managers. The financial success of a contractor is now heavily influenced by how effectively it can overcome these problems.

In the construction industry, TQM can be the price of admission for winning new contracts. Today's owners may now require that a contractor demonstrate how it will provide high quality in its services and in the product it builds. Some large owners have now implemented programs to evaluate contractor performance, in which proper job planning, skilled supervision, material supply, construction procedures, and tool and equipment selection are key factors.

Although many contractors have found it difficult to implement a formal TQM initiative within their companies, the introduction of several of TQM's basic elements has proven to have a positive impact on performance. These basic elements include the following:

1. Top management commitment to quality
2. A company-wide dedication to continuous improvement
3. Customer focus
4. A team structure with company-wide participation

5. The use of tools and techniques to base actions on facts, data, and analysis

6. Training and education of the entire project team

A critical factor in the successful implementation of TQM and other such quality programs is the contractor's ability to quantitatively measure performance. Not only must the contractor implement a process of capturing and recording data on a variety of indicators, but a system must be established to distill the data for proper analysis and benchmarking. Pareto diagrams, control charts, graphs, histograms, scatter diagrams, and cause-and-effect diagrams are often useful in capturing trends and highlighting effects.

10.30 ALTERNATIVE DISPUTE RESOLUTION (ADR)

As mentioned earlier, over the past several years the construction industry has become excessively combative and unduly concerned with disputes, dispute resolution, and litigation. Unfortunately, nobody really wins in such a process, the net result being a reduction of profitability, productivity, and quality. In an era when litigation costs can literally put a contractor out of business, construction industry leaders are increasingly seeking alternative methods to prevent and settle disputes.

As an alternative to the traditional judicial process, Alternative Dispute Resolution (ADR), is a process for resolving disputes without resorting to construction litigation. Included in ADR are arbitration, mediation, mini-trials, dispute review boards, and other means of dispute resolution that do not involve litigation. Each of these procedures is designed to facilitate the achievement of a private settlement or resolution of a dispute by the parties themselves with the aid of a person or process that assists them in reaching a solution.

10.31 NATIONAL CONSTRUCTION DISPUTE RESOLUTION COMMITTEE (NCDRC)

In 1991, in an effort to provide a remedy for excessive litigation arising out of construction projects, leaders of the industry established a task force to promote awareness, understanding, and the use of private dispute-resolution techniques and to encourage their use as standard practice in the industry. The American Arbitration Association's National Construction Dispute Resolution Committee (NCDRC) (www.adr.com) represents all segments of the construction industry: public and private owners, architects, engineers, contractors, subcontractors, insurers, sureties, lenders, and others. All share a common goal—to declare war on unnecessary disputes and litigation. The NCDRC group has emerged to provide leadership to the industry in moving toward the best utilization of ADR for the benefit of the construction process.

10.32 ARBITRATION

Customarily, when a contractual dispute arises between a contractor and an owner, the matter is first referred to the owner or its representative. However, as has been pointed out previously, almost all contracts provide for certain appeals from such first-level decisions. In the construction industry, arbitration is frequently the next, and last, step in the settlement of such controversies between the owner and the prime contractor. Arbitration is the referral of a dispute to one or more impartial persons for final and binding determination.

Court action can impose delay, expense, and inconvenience on both the contractor and the owner. For this reason, many construction contracts provide for the arbitration of disputes. Actually, no contract clause is necessary for arbitration, because a dispute can be arbitrated at any time by mutual consent of the parties. Arbitration implies a common consent by the disputants to have their differences settled. It offers the advantages of a settlement that is prompt, informal, private, convenient, and economical, and that has been decided upon by experts in the field. Arbitration is not a replacement for the law, but rather an adjunct to it.

Arbitration makes it possible for a construction dispute to be judged by professionals experienced in the construction industry. Although arbitration is an orderly proceeding governed by rules of procedure and standards of conduct, it is informal and need not conform to the adversary rules of conduct that the courts require. Finality is an important reason for the use of arbitration. Court decisions are open to lengthy appeals, resulting in long and costly delays in settling many cases. The award in an arbitration hearing cannot be changed without both parties agreeing to reopen the case.

Arbitration clauses in contracts can, and sometimes do, limit the scope of contractual disagreements that must be referred to arbitration. However, construction contracts normally contain a broad-form arbitration provision that covers all claims, disputes, and other matters arising out of, or in relation to, the contract or the breach thereof. Hence, contract clauses that provide for arbitration are normally phrased so that the parties to the contract agree to submit to arbitration any future disputes that may arise during the course of construction operations. Subparagraph 4.6.1 in Appendix C illustrates this point. However, although all arbitration statutes make agreements to arbitrate existing disputes irrevocable and enforceable, agreements to arbitrate unknown future disputes are not enforceable in all states. Where certain state arbitration statutes apply, either party can refuse to submit to arbitration even though that party may have promised to do so by the terms of a contract. However, most construction involves interstate elements and is thereby covered by the Federal Arbitration Act that applies to interstate commerce. The federal statute makes agreements to arbitrate future disputes binding. Historically, arbitration was not commonly provided for in public contracts. However, recent years have seen the use of the arbitration process become an accepted means of resolving commercial disputes involving governmental units. This has been especially true in the case of states, state agencies, counties, municipal corporations, and other state political subdivisions. Legal authority for these public owners to agree to arbitration in their construction contracts comes from specific statutory authorization or is inferred from their capacity to enter into contracts. The use of arbitration to resolve contractual disputes does not apply to the federal government, however. The Contract Disputes Act of 1978 mandates a process that must be followed to resolve contractual claims and disputes. According to the Act, the dispute is first referred by the contractor to the contracting officer. Appeals to the decision can be made to the agency boards of contract appeals or the U.S. Claims Court. At present, the federal government does not accept arbitration as a means of settling its construction disputes.

Most construction contracts that provide for arbitration stipulate that it shall be conducted under the Construction Industry Arbitration Rules as administered by the American Arbitration Association (AAA). These rules are reproduced in full in Appendix L. The

AAA neither gives legal advice nor arbitrates disputes, but it does provide assistance in obtaining arbitrators, furnishing rules of procedure, and obtaining other help. In exchange for its assistance, the AAA charges a nominal fee. Where contract bonds or insurance may be involved in an arbitration, the contractor should give its surety or insurance company advance notification. A reason for this requirement is that an arbitration award is not binding on a party that did not agree to submit its rights to arbitration. An insurer, for example, that has not specifically agreed to be bound by an arbitration award or to indemnify the contractor for any liability imposed by arbitration, may claim to be a nonconsenting third party and thus would not be bound by the arbitration award. Some insurance policies specifically provide that the insurer will defend arbitration proceedings as well as lawsuits and will pay arbitration awards as well as court judgments. The contractor should check this matter before signing an agreement containing an arbitration clause.

General contractors commonly include arbitration clauses in their subcontracts (see Article 13.9, Appendix M). Such clauses provide that disputes between a subcontractor and a general contractor that cannot be settled by mutual agreement or mediation shall be referred for arbitration or litigation. When a contract with the owner provides for arbitration, the prime contractor may assume that the arbitration requirement extends to the subcontractors. Experience indicates, however, that when a general contractor wishes arbitration to be used, the inclusion of a specific arbitration provision in its subcontracts will help to ensure the primacy of arbitration with respect to disputes between the general contractor and its subcontractors.

10.33   ARBITRATION PROCEDURE

Whether or not conducted strictly under the rules of the American Arbitration Association, the general arbitration procedure is well established. The party wishing to initiate arbitration makes a written demand of the other side stating the nature of the dispute, the amount involved, and the remedy sought, and requests that the matter be submitted to arbitration. If the other party agrees to arbitrate, a board of arbitration, consisting of one or three persons, is then selected. Arbitrators are picked not only for their impartiality and disinterest in the subject at arbitration but also for their experience in, and knowledge of, the construction field. No arbitrator should have a family, business, or financial relationship with either party to the controversy. A point to be stressed is that the arbitrators' authority to hear and decide exists only by virtue of the agreement of the parties. They are endowed with only such authority as these parties may confer upon them.

After the board has been selected, a hearing is conducted during which each side is free to call witnesses and to present such evidence as it wishes and the arbitrators consider admissible. Each party can be represented by counsel and is entitled to question the other party and its witnesses. Arbitration is a less formal process than litigation in a court of law. The parties, having elected to resolve their controversy by arbitration rather than in a lawsuit, have themselves agreed not to be bound by strict rules of evidence. The principal legal requirement is that a fair and full hearing for both sides be held.

Where the matter at arbitration is very complex and involves large sums of money, a prehearing conference may be held. In such a case, an informal meeting of the parties or their attorneys is called prior to the formal hearing to exchange information, to stipulate uncontested facts, and to resolve certain collateral issues. A prehearing conference serves some of the same purposes that discovery serves for cases that go to the courts.

After the hearing has been completed, an award is made within a reasonable period of time. In the usual case, arbitrators award compensation for damages suffered as a result of breach of contract. Until recently, punitive damages could not be awarded by arbitration. However, there is now a noticeable trend in both state and federal arbitration law to permit punitive damage claims where a broadly worded arbitration agreement is involved. A written copy of the findings and award, signed by the arbitrators, is sent to each of the parties. The arbitrators are not required to explain the rationale of their findings, only to decide all of the questions that were submitted by the disputants. It is usual practice that the arbitrators stipulate how the fees and costs shall be apportioned between the parties. Once the disputants have submitted to arbitration and an award is handed down, the parties are bound to it. The decision can be submitted to the appropriate court for an order confirming the award. Once the award has been confirmed, it has the same legal force as any other court judgment and can be enforced, if necessary, in the usual fashion. No appeal can be made against the arbitrators' findings, although the award can be challenged if a party thinks that the award was not within the submission; that there was bias, collusion, or prejudice on the part of the board; or that a full and fair hearing was not held. Reversal of awards on such grounds is seldom requested and almost never found justified. Courts give every reasonable presumption in favor of the award and of the arbitrators' proceedings. The burden rests on the party attacking the award to produce evidence sufficient to invalidate it.

10.34 MEDIATION

In recent years another method of dealing with disputes in the construction industry has been introduced, that of mediation. Mediation is a less formal procedure than arbitration and is an alternative course of action in the early stages of a dispute. Mediation involves a mutually agreed-upon and impartial third party who attempts to assist the disputing parties in reaching an agreeable settlement, but lacking any power to impose a decision. Mediation has long been associated with labor contract disputes. It is now being applied to situations concerning construction contracts in which disputants seek outside assistance in settling their differences. Mediation is a completely voluntary process. The mediator cannot impose a settlement but can only seek to assist the parties in making a direct settlement between themselves. Mediation can be provided for by contractual agreement or resorted to by mutual consent. A common contractual clause in recent years requires that the parties first mediate disputes and escalate their action to arbitration only in the event that resolution cannot be found through direct discussion or mediation (see Article 13.9 in Appendix M and Subparagraph 4.6.2 in Appendix C).

When mediation is at issue, the parties to a dispute submit to the mediator a summary of their positions, plus pertinent documentation. The mediator meets with the two sides, together and separately. The mediator tells each party about the flaws and strengths of its case, without revealing anything about the other's position. Subsequently, the mediator suggests a settlement to both sides, which they can accept, reject, or negotiate.

10.35 MINI-TRIALS

A mini-trial is an abbreviated trial. Each party to a dispute is given an opportunity to present its position using witnesses or oral presentations. The process is presided over by a so-called judge or referee who is selected by both sides to the dispute. Depending on the nature of the dispute, the individual may be a lawyer or a construction professional. After both sides have made their presentations, the judge conveys his findings and attempts to accomplish a settlement.

10.36   DISPUTE REVIEW BOARDS

In addition to arbitration and mediation, a further method of dispute resolution has recently appeared. Dispute review boards (DRBs) have proven to be relatively successful and cost-effective; this technique, however, has mostly been utilized by public agencies and has found little use among private owners.

After the construction contract is signed, a DRB is organized before any construction work begins. One board member is selected by the contractor, subject to approval by the owner. A second board member is chosen by the owner with acceptance by the contractor. These two board members now choose a third, who serves as chairman of the board and is subject to approval by both the owner and the contractor. Once the board is established, regular site meetings are held. The recommendations of the board are not binding and often constitute only an intermediate step in the final dispute resolution. The board is not designed to replace the courts or arbitration. However, DRB findings can provide a reliable, contemporaneous record of the facts as well as expert opinions from a disinterested, impartial panel of experts who observed the events as they occurred. Experience shows that DRBs put strong emphasis on the avoidance of legal action.

QUESTIONS

1. Describe the difference in duties between a project manager and field supervisor.

2. Project meetings, development of payment schedules, shop drawings, quality control, materials management, expediting, deliveries, receiving, material inspections, subcontractor scheduling, the creation of record drawings, disbursement controls, and daily job logs are all part of what project function?

3. Owner project financing often demands the sale of bonds and other securities. How does the owner time the sale of these assets to liquidate capital for contractor payments? Explain the process.

4. Why are shop drawings frequently the source of project dispute between the supplier, contractor, and architect-engineer?

5. Explain the difference between quality control and quality assurance and the relationship they have to one another.

6. The on-time delivery of project materials is rarely an accidental occurrence. What tools and processes aid the proper management of the materials procurement process?

7. What two basic purposes do disbursement controls serve?

8. What procedures should a contractor follow should a claim submitted to the owner be rejected?

9. In protecting his interests, what two vital checks must a contractor make before accepting final payment from the owner?

10. What elements of the contract documents can a partnering agreement modify?

11. What role does physical performance measurement play in TQM and other similar quality programs?

12. How does arbitration differ from litigation? What advantages does arbitration offer over litigation in the settlement of construction disputes?

13. What purpose does the AAA serve in the arbitration processes?

14. How does mediation differ from arbitration? What advantages does mediation offer over arbitration? Are these dispute resolution methods mutually exclusive?

Chapter 11

Project Time Management

11.1 INTRODUCTION

Time is an important aspect of job management. If a construction project is to proceed efficiently and be completed within the contract time, the work must be carefully planned and scheduled in advance. Construction projects are complex, and a large job can involve literally thousands of separate operations. If these tasks were to follow one another in consecutive order, job planning and scheduling would be relatively simple, but this is not the case. Each operation has its own time requirement, and its start depends on the completion of certain preceding operations. At the same time, many tasks are independent of one another and can be carried out simultaneously. Thus, a typical construction project involves many mutually dependent and interrelated operations that, in total combination, constitute a tangled web of individual time and sequential relationships. When individual task requirements for materials, equipment, and labor are superimposed, it becomes obvious that project planning and scheduling is a very complicated and difficult management function.

11.2 THE CRITICAL PATH METHOD

The critical path method (CPM) is a procedure developed especially for the time management of construction projects. CPM involves the analysis of the sequential and time characteristics of projects by the use of networks. It is a widely used procedure for construction time control, and contractors are now frequently required by contract to apply network methods to the planning and scheduling of their fieldwork. Complete and comprehensive time management systems have been developed based on the CPM procedure. The reader is referred to books that describe these systems in detail.[*] This chapter confines itself to a discussion of the basics.

CPM is a project management system that offers a basis for informed decision making on projects of any size. It provides information necessary for the time scheduling of a construction project, guides the contractor in selecting the best way to shorten the project duration, and predicts future manpower and equipment requirements.

The procedure starts with project planning. This phase consists of (1) identifying the elementary items of work necessary to achieve job completion, (2) establishing the order in which these work items will be done, and (3) preparing a graphical display of this planning information in the form of a network. The procedure just described may suggest that project

---

[*] See, for instance, R. H. Clough, G. A. Sears, and S. K. Sears, *Construction Project Management,* 4th ed. (New York: Wiley-Interscience, 2000).

planning must follow a definite step-by-step order of development. In actual practice this is not the case; the three planning steps generally proceed more or less simultaneously. However, for purposes of discussion, the three steps described are treated separately in the order mentioned.

The scheduling phase that follows requires an estimate of the time required to accomplish each of the work items identified. With the use of the network, computations are then made that provide information concerning the time schedule characteristics of each work item and the total time necessary to achieve project completion. The computations associated with project scheduling are simple additions and subtractions. Manual computation is easy and logical but can become tedious and very time-consuming on larger projects. For this reason contractors use computers to produce their project schedules and to update them periodically during construction in the field. For a full understanding of the methods used and a thorough appreciation of the data generated, however, it is necessary that the practitioner be familiar with how the calculations are made. Consequently, the following discussion of planning and scheduling is based on manual procedures.

11.3   GENERAL CONSIDERATIONS

Although some preliminary study of a project may occur during the estimating or negotiation process, it is usual that the detailed planning be started immediately after the contractor has been awarded the construction contract. A characteristic of CPM is that its effectiveness depends on the accuracy of the input information and the skill and judgment with which the generated data are used. Consequently, the development and application of a project plan and schedule should be made the responsibility of people who are experienced in, and familiar with, the type of field construction involved. Because the prime purpose of CPM is to produce a coordinated project plan, key subcontractors are often included in the planning. Their input can be vital to the development of a workable construction schedule. Normally, the prime contractor sets the general timing reference for the project. The individual subcontractors then review the portions of the plan relevant to their work, and needed adjustments are made.

It is usual to develop the network diagram in rough form as the job is dissected into its basic elements and the sequential order of construction operations is established. It is often helpful to list the major segments of the project and to use them to develop a preliminary diagram. This diagram can serve as a basis for discussion and as a basic framework for the subsequent development of a fully detailed network. It is important that the plan prepared be the one the contractor actually expects to follow. This means, therefore, that those preparing the plan must have authority to make decisions concerning methods, procedures, equipment, and labor.

11.4   PROJECT PLANNING

Planning is the devising of a workable scheme of operations to accomplish an established objective when put into action. Besides being the most time-consuming and difficult aspect of the job time management system, planning is also the most important. It requires an intimate knowledge of construction methods combined with the ability to visualize discrete work elements and to establish their interdependencies. If planning were to be the only job analysis made, the time would be well spent. It involves a depth and thoroughness of study that gives the construction team an invaluable understanding and appreciation of job requirements.

For purposes of planning, the project must first be broken down into elemental time-consuming activities. An activity is a single discrete work step in the total project. The extent to which the project is subdivided into activities depends on a number of practical considerations, and the following factors should be taken into account:

1. Different areas of responsibility, such as subcontracted work, that are distinct and separate from work being done directly by the prime contractor
2. Different categories of work as distinguished by craft or crew requirements
3. Different categories of work as distinguished by equipment requirements
4. Different categories of work as distinguished by materials such as concrete, timber, or steel
5. Distinct and identifiable subdivisions of structural work such as walls, slabs, beams, and columns
6. Location of the work within the project, the performance of which necessitates different times or different crews
7. Owner's breakdown for bidding or payment purposes
8. Contractor's breakdown for estimating purposes

The activities chosen may represent relatively large segments of the project or may be limited to small steps. For example, a concrete slab may represent a single activity, or it may be broken down into the erection of forms, placing of reinforcing steel, pouring of concrete, finishing, curing, and stripping of forms. If the activities are too large, the job plan developed will not yield information in sufficient detail to be optimally useful. However, if the subdivision of the work is too elemental, the excessive detail tends to obscure the truly significant planning factors. Basically, the extent to which the work is broken down into activities is determined by the party who will be using the information. If the owner, architect-engineer, or project manager is to be the recipient, only moderate detail is needed because their use of the data is normally limited to overall monitoring of job progress. However, if the information is being prepared for field supervisors who are concerned with day-to-day direction, a considerable degree of subdivision of the work is required.

As the separate activities are identified and defined, the sequential relationships between them are determined. This process is referred to as "job logic" and consists of the established order of construction operations. When the time sequence of activities is being determined, restraints must be recognized and taken into consideration. Restraints are practical limitations of one sort or another that can influence or control the start of certain activities. For example, an activity that involves the placing of reinforcing steel obviously cannot start until the steel is on the site. Hence, the start of this activity is restrained by the time required to prepare and approve the necessary shop drawings, fabricate the steel, and deliver it to the job. In like manner, the start of an activity may depend on the availability of labor, subcontractors, equipment, completed construction drawings, owner-provided materials, and other required inputs. Failure to consider such constraints can be a serious failing of a job plan and schedule.

Some restraints are shown as time-consuming activities. For example, the preparation of shop drawings and the fabrication and delivery of job materials are material restraints, that require time to accomplish and are depicted as activities on project networks. Restraints are also shown in the form of dependencies between activities. If the same crane is required for two activities, an equipment restraint is imposed by having the start of one activity depend on the finish of the other.

11.5   PRECEDENCE NOTATION

As the identification of activities proceeds and their logic is established, the resulting job plan is depicted graphically in the form of a network. There are two commonly used symbolic conventions used to portray network activities. One, called "precedence notation," depicts each activity as a rectangular box. The other shows each activity as an arrow and is called "arrow notation." Both kinds of diagrams are widely used, but precedence diagrams are considered to have some important advantages. For this reason, precedence notation is used here.

In drawing a precedence network, each time-consuming activity is portrayed by a rectangular figure. The dependencies between activities are indicated by dependency or sequence lines going from one activity to another. The identity of the activity and a considerable amount of other information pertaining to it are entered into its rectangular box. This matter is developed further as the discussion progresses.

11.6   THE PRECEDENCE DIAGRAM

The preparation of a realistic precedence diagram requires time, work, and experience with the type of construction involved. The management data extracted from the network can be no better than the diagram itself. The diagram is the key to the entire time control process. When the network is first being developed, the planner must concentrate on job logic. The only consideration at this stage is to establish a complete and accurate picture of activity dependencies and interrelationships. Restraints that can be recognized at this point should be included. The time durations of the individual activities are not of concern during the planning stage.

Each activity in the network must be preceded either by the start of the project or by the completion of a previous activity. Each path through the network must be continuous, with no gaps, discontinuities, or dangling activities. Consequently, all activites must have at least one activity following, except the activity or activities that terminate the project. With the usual project network, it is not possible to have the finish of one activity overlap beyond the start of a succeeding activity. When such a condition exists, the work must be further subdivided. It follows, therefore, that a given activity cannot start until *all* those activities immediately preceding it have been completed.

Each activity is given a unique numerical designation, with the numbering proceeding generally from project start to finish. The usual practice is that numbering is not done until after the network has been completed. Leaving gaps in the activity numbers is desirable so spare numbers are available for subsequent refinements and revisions. In this book, activities are numbered by multiples of ten. It is standard practice, as well as being a requirement of many computer programs, that precedence diagrams start with a single opening activity and conclude with a single closing activity.

11.7   EXAMPLE PROBLEM 1

To illustrate how the three planning steps of activities, job logic, and network go together, consider a simple project consisting of a few steps, such as the construction of a heavy reinforced concrete slab on grade. This will be called Example Problem 1. Let us assume the job is subdivided into eight activities as listed in Figure 11.1. The activity "Procure reinforcing steel" will be recognized as a job restraint.

The job logic or the time-sequence relationships among the activities of Example Problem 1 must now be determined. The sequence of operations will be the following.

The procurement of reinforcing steel, the excavation, and the building of forms are all opening activities that can proceed independently of one another. Fine grading will follow excavation, but forms cannot be set until both the excavation and form building have been completed. The placing of reinforcing steel cannot start until fine grading, form setting, and steel procurement have all been carried to completion. Concrete will be poured after the steel has been placed, and finishing will be the terminal operation, proceeding after the concrete pour.

As has already been mentioned, the diagram is normally drawn concurrently with the development of the job plan, and the operational sequence is not otherwise recorded or listed. There are occasions, however, when it is useful to list the job logic. A common way of expressing the necessary activity sequence is to list for each activity those preceding activities that must be completed before the given activity is started. This is a sufficient system for enumerating job logic and will be used here. Figure 11.1 presents the previously discussed job logic of Example Problem 1. The precedence diagram describing the prescribed sequence of activities is shown in Figure 11.2

EXAMPLE PROBLEM 1: REINFORCED CONCRETE SLAB ON GRADE		
Activity	Activity Number	Activities Immediately Preceding
Start	10	—
Excavate	20	10
Build Forms	30	10
Procure Reinforcing Steel	40	10
Fine Grade	50	20
Set Forms	60	20, 30
Place Reinforcing Steel	70	40, 50, 60
Pour Concrete	80	70
Finish Concrete	90	80

Figure 11.1   Example Project 1, job logic.

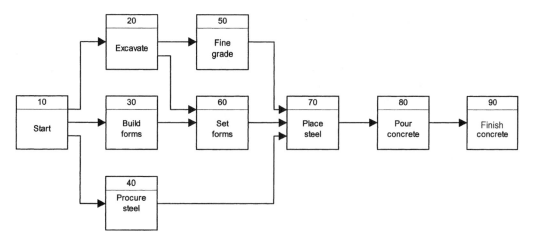

Figure 11.2   Example Project 1, precedence diagram.

Figure 11.2 is a complete project plan in graphical form. It shows what is to be done and in what order it will be accomplished. The diagram serves as the construction flight plan and points the way on a daily basis. A precedence network of the project is an extraordinarily efficient medium for communication among owner, architect-engineer, contractor, subcontractors, and other members of the construction team.

11.8   THE NETWORK FORMAT

A horizontal diagram format has become standard in the construction industry. The general synthesis of a network is from start to finish, from project beginning on the left to project completion on the right. The sequential relationship of one activity to another is shown by the dependency lines between them, and the essence of the network is the manner in which the constituent activities are joined into a total operational pattern. In the usual precedence diagram, the lengths of the lines between activities have no significance; the lines indicate only the dependency of one activity on another. Arrowheads are not always shown on the dependency lines because of the obvious left-to-right flow of time. However, arrowheads are shown here for additional clarity.

In the initial stages of developing a diagram, the network is sketched emphasizing activity relationships rather than the appearance or style of the diagram. The first version is apt to be a freewheeling conglomeration of sweeping curves, random direction lines, and many sequence-line crossovers. Corrections, revisions, and erasures are plentiful. The rough appearance of the first version of the diagram is of no concern, but its completeness and accuracy are important. The finished diagram can always be put in a tidier form at a later date.

The activity numbers appearing in Figures 11.1 and 11.2 are used only as a convenience in depicting the job logic. The use of symbols on the precedence network in lieu of writing out an abbreviated description of each activity is not recommended. Even mnemonic codes make a diagram difficult to read, and the user must spend a considerable amount of time consulting the symbol listing to check the identity of activities. Networks are much more intelligible and useful when each activity is clearly identified on the diagram.

When a working precedence diagram is being prepared, the scale and spacing of the activities deserve attention. If the scale is too large and activities are widely spaced, the resulting network is likely to become so large that it is unmanageable. Yet a small scale and an unduly compact makeup render the diagram difficult to read and inhibit corrections and modifications. With experience and observation of the work of others, practitioners will soon learn to adjust the scale and structure of their diagrams to the scope and complexity of the project involved. It must not be forgotten that the network is intended to be an everyday tool, used and consulted by a variety of people. The emphasis here is not on drafting elegance but on producing the most realistic and intelligible form of network possible.

Dependency lines that go backward from one activity to another should not be used. Backward, in this context, means going from right to left on the diagram, which is against the established direction of time flow. Backward sequence lines are confusing and increase the chances of unintentional logical loops being included in the network. A logical loop involves the impossible requirement that an activity be followed by an activity that has already been accomplished. Logical loops are, of course, completely illogical, but they can be inadvertently included in large and complex networks if backwardly directed sequence lines are permitted. Crossovers occur when one dependency line must cross over another to satisfy job logic. Careful layout can minimize the number of crossovers, but there are usually some that cannot be avoided. Any convenient symbol can be used to indicate a crossover.

11.9 PROJECT SCHEDULING

In the discussion thus far, the vital concern has been with the identification of activities and their sequential constraints. Before the procedure can be carried further, attention must be given to time requirements. This is called the "scheduling phase" of CPM. The essential input information for project scheduling is the network diagram and a time duration estimate for each activity.

Project scheduling is accomplished by carrying out a series of simple computations that yield valuable project control information. An overall project completion time is first established. Times or dates within which each activity must start and finish if the established project completion date is to be met, are then determined.

11.10 ACTIVITY DURATIONS

The first step in the scheduling process is to estimate the time necessary to carry out each activity. These durations are generally expressed in working days and do not include nonproductive periods such as holidays and weekends. Other units of time may be used (e.g., hours, weeks, or shifts). The only requirement is that the same unit of time be used throughout. The process of estimating activity durations usually results in a certain amount of network refinement and redefinition of activities.

Much of the value of the planning and scheduling process depends on the accuracy of the estimated activity durations. The following rules constitute the basis for this important step:

1. It is important that each activity be evaluated independently of all the others. For a given activity, assume that materials, labor, equipment, and other needs will be available when required. If there is reason to believe that this will not be so, then the use of a suitable preceding restraint may be in order.

2. For each activity, assume a normal level of manpower and/or equipment. Exactly what "normal" is in this context is difficult to define. Based on experience, customary and relatively standard crew sizes and equipment spreads have emerged as being efficient and economical. In short, a normal level is intrinsic to the situation and is optimum insofar as expedient completion and minimum costs are concerned. At times, what is normal may be dictated by the availability of labor or equipment. If shortages are anticipated, this factor must be taken into account.

3. A normal workday or workweek is assumed. Overtime and multiple shifts are not considered unless this is standard procedure or they are part of a normal work period. Around-the-clock operations are customary in many tunnel jobs, for example, and overtime is routinely utilized on highway projects during the summer months because of a typically short working season.

4. Activity durations must be estimated without regard to any predetermined contract completion date. Otherwise, there is apt to be, consciously or unconsciously, an effort made to fit the activities within the total time available. The only consideration pertinent to estimating an activity time is the amount of time required to accomplish *that* activity, and that activity only.

5. Use consistent time units throughout. It must be remembered, for example, that when working days are used, weekends and holidays are not included. Certain job activities, such as concrete curing or plaster drying, carry through nonworking days, and their times must be adjusted accordingly (computer programs typically enable the use of different calendars for different types of activities). Material delivery times are invariably given in terms of calendar days. A delivery time of 30 calendar days translates into approximately 21 working days.

Quantitative information is available to assist in the estimation of activity durations; for example, the estimating sheets compiled when the job was priced will yield the man-hours or equipment-hours for each activity. By making assumptions concerning the size of the work crew or the number of equipment units, activity durations can be computed. Time estimates of good accuracy can often be made informally. Off-the-cuff time estimates made by experienced construction supervisors usually prove to be surprisingly accurate. In any event, it is important that someone experienced in, and familiar with, the type of work involved estimate the activity times. Up to a point, the accuracy of duration estimates can be improved by subdividing the work into smaller units.

11.11 TIME CONTINGENCY

When activity times are being estimated, it is assumed that the work will progress reasonably well. However, Murphy's Law is especially applicable to construction, and a safety factor must be included in time estimates to allow for possible disruptions of many kinds. Practice varies in regard to adding contingency allowances to the estimated durations of activities. There is general agreement, however, that a contingency allowance should not be applied to individual activities to allow for unforeseen project delays caused by fires, accidents, equipment breakdowns, strikes, late material deliveries, difficult site conditions, floods, legal delays, and the like. In such cases, it is generally impossible to predict which activities will be affected and by how much. A suitable contingency is better included in the time required for overall project completion (see Section 11.15).

The method of providing allowances for time lost because of inclement weather depends on the type of work involved. When projects, such as most highway, heavy, and utility work, are affected by the weather, the entire project is normally shut down. As a result, it is usual for an estimated number of days lost because of weather to be added to the overall project duration.

On buildings and other work that can be protected from the weather, allowances for time lost are commonly added to the durations of those activities or groups of activities susceptible to weather delay. It is relatively easy to establish average time losses for particular activity types during the seasons when they will be under way. After a job of this type passes a certain stage, it is not often completely shut down by bad weather. Although some parts of the project may be at a standstill, others can still proceed. Consequently, a better overall job of scheduling will possibly result if allowances for weather are included in the durations for those activities likely to be involved rather than adding a weather contingency to the entire project.

11.12 EXAMPLE PROJECT 2

To illustrate the subsequent discussion, a somewhat more involved construction project is considered, Example Project 2. Although much simplified from an actual construction project, it serves to illustrate the essential workings of the scheduling procedure.

The project to be discussed is a natural gas compressor station involving a prefabricated metal building, compressor foundation, standby butane fuel system, and compressor appurtenances such as piping, electrical services, and controls. Figure 11.3 presents the job logic and the estimated activity durations, and Figure 11.4 shows the completed precedence diagram. For the sample activity in Figure 11.4, each activity duration in terms of working days is shown in the lower central part of the activity box. The identifying number of each activity is located in the upper central part of the activity box. The other numerical values shown with the activities and the contingency activity at the right terminus of Figure 11.4 are discussed subsequently.

11.13 NETWORK COMPUTATIONS

After a time duration has been estimated for each activity, some simple and step-by-step computations are performed. The purpose of these calculations is to determine (1) the overall project completion time and (2) the time brackets within which each activity must be accomplished if this completion time is to be met. The network calculations involve only additions and subtractions and can be made in different ways, although the data produced are comparable in all cases. The usual procedure is to calculate what are referred to as "activity times."

EXAMPLE PROBLEM 2: COMPRESSOR STATION			
Activity	Activity Number	Duration (working days)	Activities Immediately Preceding
Start	10	0	—
Grading	20	7	10
Granular fill	30	21	10
Electric run-in	40	42	10
Procure metal building	50	28	10
Excavate butane	60	17	20
Compressor foundation	70	8	20
Electrical to compressor	80	20	40
Butane storage	90	34	60
Connect cooling	100	11	30,70
Bolts and plates	110	6	30,70
Slab and footings	120	14	30,70
Piping system	130	15	90,100
Set compressor	140	13	110,120
Erect building	150	29	50,120
Set control panel	160	14	80,140
Wire building	170	17	150
Connect piping	180	10	130,160
Test and clean	190	4	170,180
Contingency	200	5	190

Figure 11.3 Example Project 2, job logic and activity times.

The calculation of activity times involves the determination of four limiting times for each network activity. The "early start" or "earliest start" (ES) of an activity is the earliest time the activity can possibly start, allowing for the times required to complete the preceding activities. The "early finish" or "earliest finish" (EF) of an activity is the earliest possible time by which it can be completed, which is determined by adding that activity's duration to its early start time. The "late finish" or "latest finish" (LF) of an activity is the very latest it can finish and allow the entire project to be completed by a designated time or date. The "late start" or "latest start" (LS) of an activity is the latest possible time it can be started if the project target completion date is to be met and is obtained by subtracting the activity's duration from its latest finish time.

The computation of activity times can be performed manually or by computer. When calculations are made by hand, they are normally performed directly on the network itself. When making the initial computations, it is usual for the activity times to be expressed in terms of *expired* working days. Commensurately, the start of the project is customarily taken to be at time zero.

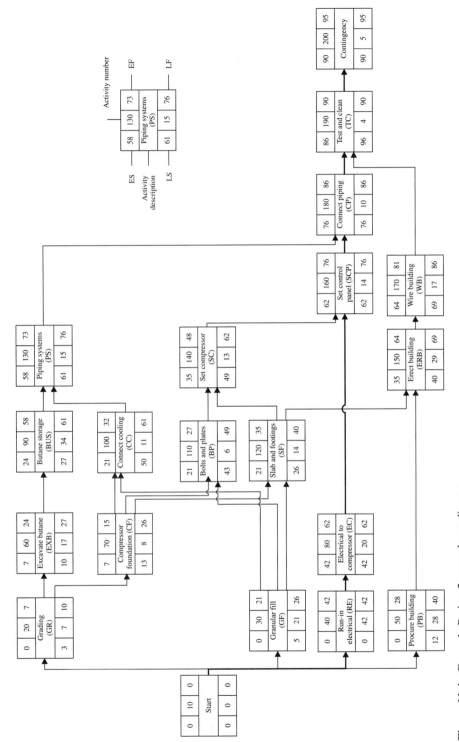

Figure 11.4 Example Project 2, precedence diagram.

11.14   EARLY ACTIVITY TIMES

The compressor station network shown in Figure 11.4 is used here to describe how the manual calculation of activity times is performed directly on a precedence diagram. Such networks are exceptionally convenient for the manual calculation of activity times and afford an excellent basis for describing how such calculations are made. The computation of the early start and early finish times is treated in this section. The determination of the late activity times is described later. The early time computations proceed from project start to project finish, from left to right in Figure 11.4; this process is referred to as the "forward pass." The basic assumption for the computation of early activity times is that every activity will start as early as possible. That is, each activity will start just as soon as the last of its predecessors is finished.

The ES value of each activity is determined first, with the EF time then being obtained. Figure 11.4 shows that activity 10 is the initial activity. Its earliest possible start is, therefore, zero elapsed time. As explained by the sample activity shown in Figure 11.4, the ES of each activity is entered in the upper left of its activity box. The value of zero is commensurately entered at the upper left of activity 10 in Figure 11.4. The EF of an activity is obtained by adding the activity duration to its ES value. Activity 10 has a duration of zero. Hence the EF of activity 10 is its ES of zero added to its duration of zero, or a value of zero. EF values are entered into the upper right of the activity boxes, and Figure 11.4 shows the EF of activity 10 to be zero. Activity 10 calculations are trivial, but the use of a single opening activity is customary in precedence diagrams and is required by some computer programs.

Figure 11.4 shows that activities 20, 30, 40, and 50 can all start after activity 10 has been completed. In going forward through the network, the earliest that these four activities can start is controlled by the EF of the preceding activity. Because activity 10 has an EF equal to zero, then each of the following four activities can start as early as zero. Consequently, zero values appear in the upper left of activities 20, 30, 40, and 50. The EF value of each of these four is obtained by adding its ES of zero to its respective duration. For example, the EF value of activity 20 is zero plus 7, or a value of 7. Thus, a 7 appears in the upper right of activity 20.

Continuing into the network, neither activity 60 or 70 can start until 20 has been completed. The earliest that 20 can be finished is at the end of the seventh day, so the earliest that activity 60 or 70 can be started is after the expiration of 7 working days (or at a time of 7). The ES of activities 60 and 70 are indicated to be 7 in Figure 11.4. The EF of activity 60 will be its ES value of 7 plus its duration of 17, or a value of 24. In like fashion, activity 70 will have an EF value of 15.

The start of activities 100, 110, and 120 must all await the completion of *both* activities 30 and 70. The earliest that activity 70 can be finished is time 15, but activity 30 cannot be completed until time 21. Because activities 100, 110, and 120 cannot start until *both* activities 30 and 70 are finished, the earliest possible start times for each of the three activities is time 21. Activities 100, 110, and 120 are examples of what are called merge activities. This is an activity whose start depends on the completion of two or more antecedent activities. The computation rule for a merge activity on the forward pass is that its earliest possible start time is equal to the latest (or largest) of the EF values of the immediately preceding activities.

The forward-pass calculations consist of only repeated applications of the few simple rules just discussed. Working methodically in step-by-step fashion, the computations in Figure 11.4 proceed from activity to activity until the end of the project, activity 190, is reached.

11.15   PROJECT DURATION

Figure 11.4 shows that the early finish time for the last work activity (190) is 90 expired working days. For the job logic established and the activity durations estimated, it will require 90 working days to complete the work. What this says is that if a competent job of planning has been done, if activity durations have been estimated with reasonable accuracy, and if everything goes reasonably well in the field, project completion can be anticipated in 90 working days or about $(7/5) \times 90 = 126$ calendar days.

The matter of contingency must again be considered at this point. Although allowances for lost time caused by inclement weather can be assigned to the affected activities in some cases, an overall weather contingency is more appropriate in other cases (see Section 11.11). In addition, some provision must be made for general project delays caused by a variety of troubles, mischances, oversights, difficulties, and job casualties. Many contractors, at this stage of the project scheduling, plan on a time overrun of 5 to 10 percent and add this to the overall projected time requirement for the entire work. The percentage actually added must be based on the contractor's judgment and experience. In Figure 11.4, a contingency of five working days has been added to the diagram in the form of the final contingency activity 200. Adding an overall contingency of five working days gives a probable job duration of 95 working days. In the contractor's way of thinking, 95 working days represents a more realistic estimate of actual project duration than the value of 90. If everything goes as planned, the contractor will probably finish the job in about 90 working days. However, if the usual difficulties arise, it has allowed for a 95 working-day construction period.

Whether the contractor chooses to add in a contingency allowance or not, now is the time to compare the computed project duration with any established project time requirement. Continuing with Example Project 2 and assuming that a contingency of 5 working days will be used, the figure of $(7/5) \times 95 = 133$ calendar days, plus any holidays that occur during that period of time, is compared with the required completion date. If the compressor station must be completed in 120 calendar days, then the contractor will have to consider how to shorten the time duration. If a construction period of 140 calendar days is permissible, the contractor can feel reasonably confident that this requirement can be met and no action to shorten the work is required. How to go about decreasing a project's duration is the subject of Section 11.20.

The probable project duration of 133 calendar days is a valuable piece of information. For the first time, the contractor has an estimate of overall project duration that can be relied upon with considerable trust.

11.16   LATE ACTIVITY TIMES

Project calculations now "turn around" on the value of 95 working days, and a second series of calculations is performed to find the late start (LS) and the late finish (LF) times for each activity. These calculations, called the "backward pass," start at the project end and proceed backward through the network, going from right to left in Figure 11.4. The late activity times now to be computed are the latest times at which the several activities can be started and finished with project completion still achievable in 95 working days. The supposition during the backward pass is that each activity finishes as late as possible without delaying project completion. The LF value of each activity is obtained first and is entered into the lower right portion of the activity box. The LS in each case is obtained by subtracting the activity duration from the LF value. The late start time is then shown at the lower left.

The backward pass through Figure 11.4 is begun by giving activity 200 an LF time of 95, as shown in the lower right of the activity box. The LS of an activity is obtained by subtracting the activity duration from its LF value. Activity 200 has a duration of 5. Hence, the LS of activity 200 is its LF of 95 minus its duration of 5, or a value of 90. This value of 90 is entered at the lower left of activity 200 in Figure 11.4.

Figure 11.4 shows that activity 190 immediately precedes activity 200. In working backward through the network, the latest that activity 190 can finish is obviously controlled by the LS of its succeeding activity, 200. If activity 200 must start no later than day 90, then activity 190 must finish no later than that same time. Consequently, activity 190 has an LF time equal to the LS of the activity following (200), or a value of 90. Activity 190, with a duration of 4, has an LS value of 90 minus 4, or 86. These values are shown on activity 190 in Figure 11.4.

Figure 11.4 shows that activity 190 is preceded by two activities: 170 and 180. The LF of each of the two antecedent activities is set equal to the LS of activity 190, or day 86. Subtracting the activity durations from the LF values yields the LS times. The LS times for activities 170 and 180 are, correspondingly, equal to 69 and 76.

Some explanation is needed to clarify the calculation when the backward pass reaches a burst activity, which is one that has more than one activity immediately following it. In Figure 11.4, activity 120 is the first such activity reached during the backward pass, this activity being followed immediately by activities 140 and 150. To obtain the late finish of activity 120, the late start of the immediately succeeding activities are noted. These are obtained from Figure 11.4 as 49 for activity 140 and 40 for activity 150. Keeping in mind that activity 120 must be finished before either activity 140 or 150 can begin, it is obvious that activity 120 must be finished by no later than day 40. If it is finished any later than this, the entire project will be delayed by the same number of days. The rule for this and other burst activities is that the LF value for such an activity is equal to the earliest (or smallest) of the LS times of the activities following.

The backward-pass computations proceed from activity to activity until the start of the project is reached. All that is involved are repetitions of the rules just discussed.

11.17 TOTAL FLOAT

Examination of the activity times appearing in Figure 11.4 discloses that the early and late start times (also early and late finish times) are the same for certain activities but not for others. The significance of this finding is that there is time leeway in the scheduling of some activities and none at all in the scheduling of others. This leeway is a measure of the time available for a given activity above and beyond its estimated duration. This extra time is called *float*, two classifications of which are in general usage: total float and free float.

The total float of an activity is obtained by subtracting its ES time from its LS time. Subtracting the EF from the LF gives the same result. Once the activity times have been computed on the precedence diagram, values of total float are easily computed and may be noted on the network if desired, although this has not been done in Figure 11.4. As in Figure 11.4, the total float for a given activity is found as the difference between the two times at the left of the activity box or between the two at the right. An activity with zero total float has no spare time and is, therefore, one of the operations that controls project completion time. For this reason, activities with zero total float are called "critical" activities. The second of the common float types, free float, is discussed in Section 11.19.

11.18   THE CRITICAL PATH

In a precedence diagram, a critical activity is quickly identified as one whose two start times at the left of the activity box are equal. Also equal are the two finish times at the right of the activity box. Inspection of the activities in Figure 11.4 discloses that there are seven activities (10, 40, 80, 160, 180, 190, and 200) that have total float values of zero. Plotting these on the figure shows that these seven activities form a continuous path from project beginning to project end, and this chain of critical activities is called the "critical path." The critical path is normally indicated on the diagram in some distinctive way, such as with colors, heavy lines, or double lines. Heavy lines are used in Figure 11.4.

Inspection of the network diagram in Figure 11.4 shows that there are numerous paths between the start and finish of the diagram. These paths do not represent alternate choices through the network. Rather, each of these paths must be traversed during the actual construction process. If the time durations of the activities forming a continuous path were to be added for each of the many possible routes through the network, a number of different totals would be obtained. The largest of these totals is the critical, or minimum, time for overall project completion. Each path must be traveled, so the longest of these paths determines the length of time necessary to complete *all* of the activities in accordance with the established project logic.

If the total times for all of the network paths in Figure 11.4 were to be obtained, it would be found that the longest path is the critical path already identified using zero total floats and that its total time duration is 95 days. Consequently, it is possible to locate the critical path of any network by merely determining the longest path. However, this is not usually a practical procedure. The critical path is normally found by means of zero total float values. Although there is only one critical path in Figure 11.4, more than one such path is always a possibility in network diagrams. One path can branch out into a number of paths, or several paths can combine into one. In any event, the critical path or paths must consist of an unbroken chain of activities from start to finish of the diagram. There must be at least one such critical path, and it cannot be intermittent. A break in the path indicates an error in the computations. On the compressor station, five of the eighteen activities (exclusive of start and contingency), or about 28 percent, are critical. This is considerably higher than for most construction networks because of the small size of the project. In larger diagrams, critical activities generally constitute 20 percent or less of the total.

Any delay in a critical activity automatically lengthens the critical path. Because the length of the critical path determines project duration, any delay in the finish date of a critical activity, for whatever reason, automatically prolongs overall project completion by the same number of days. As a consequence, identification of the critical activities is an important aspect of job scheduling because it pinpoints those job areas that must be closely monitored at all times if the project is to be kept on schedule.

11.19 FREE FLOAT

Free float is another category of spare time. The free float of an activity is found by subtracting its early finish time from the earliest start time of the activities directly following. To illustrate how free floats are computed, consider activity 20. Figure 11.4 shows the earliest finish time of activity 20 to be 7. The activities immediately following are 60 and 70, and the earliest start time for both is 7. The difference between the earliest finish time of activity 20 (day 7) and the earliest start time of the activities immediately following (day 7) is zero. Hence, activity 20 has a free float of zero. Another example is activity 70. Figure 11.4 shows the earliest finish time of activity 70 to be at time 15. The activities immediately following are 100, 110, and 120, and their earliest start time is 21. Thus, activity 70 has a free float of $21 - 15 = 6$ days. Alternately, the free float of an activity can be obtained by subtracting its EF (upper right) from the smallest of the ES values (upper left) of those activities immediately following.

11.20 LEAST-COST PROJECT SHORTENING

If the 95-working-day construction time is not satisfactory, the contractor must reexamine the operational plan. It may be, for example, that the construction contract for Example Project 2 stipulates a completion time of less than 95 working days. In the absence of some analytical procedure such as CPM, there is no way in which those activities that truly control total project time can be identified. Lacking such a procedure, the usual impulse is to accelerate most, if not all, of the job operations when the construction time for a project has to be reduced or when a project is falling behind schedule. This procedure is neither necessary nor economical, because there is seldom any need to speed up any but the critical job activities. On large construction networks, only 10 to 20 percent of the activities normally prove to be critical. Consequently, most job activities are floaters; that is, more time is available for them than is required by their estimated durations. There is nothing to be gained by diminishing even further the durations of such noncritical activities.

Because the critical path determines the overall job duration, the only way in which job duration can be shortened is to reduce the length of the critical path, either by revising the job logic or by shortening some of the individual critical activities, or perhaps both. The first action is to determine the possibility of performing some of the critical activities in parallel with one another rather than in series. Sometimes a localized reworking of job logic will make it possible to shorten the critical path. A caution is needed at this point: Any revision of the network diagram may result in new or additional critical paths. Thus, when the critical path is being shortened, the network must be continuously reexamined to ensure that all critical operations have been identified.

If the time reduction achieved by reworking the job logic is not sufficient, the durations of the critical activities themselves must be examined with the intent of decreasing the duration of certain ones. This shortening of critical activities cannot be done arbitrarily. It must be both practical and feasible in the field.

Assume that it has become necessary to shorten the job duration of Example Project 2 by means of reducing the times of the critical activities. For this purpose, there are only five critical activities that are susceptible to possible shortening. Activity 10, Start, is already at zero duration, and activity 200, Contingency, must remain at 5 days. Consequently, each of the five critical activities (40, 80, 160, 180, and 190) must first be studied to ascertain whether it can be shortened and, if so, what the additional direct cost will be. When necessary, most activities can be completed in less than the normal time. Speedup actions increase the direct cost of the activity, however, because they involve the additional expense of overtime work, multiple shifts, more equipment, payment of premiums for quick delivery of materials, larger but less efficient crews, and similar speedup factors. There is a point, of course, beyond which the duration of an activity cannot be compressed further.

Suppose the contractor on the compressor station of Example Project 2 ascertains that it can shorten activity 40 (Run-in electrical) by 3 days by sending in another pole-setting truck crew. The additional expense of moving the extra truck and crew to and from the job site will be $2,000. Activity 80 (Electrical to compressor) can be diminished by 1 day if the electricians work overtime. The additional cost of premium time will amount to $400. Activity 160 (Set control panel) can be shortened by 1 day if the manufacturer's installation engineer does part of his work at night. This will require special lighting that will cost about $600 to install and remove. There does not seem to be any feasible way in which activity 180 (Connect piping) can be shortened significantly. Activity 190 (Test and clean) is already scheduled as a 24-hour-a-day operation and cannot be shortened.

The contractor's analysis discloses, first of all, that the original critical path can be shortened by a maximum of 5 days. This means that the probable job duration can be decreased from 95 to 90 days, assuming any new critical paths created during the process can also be shortened. The project duration cannot be reduced by more than 5 days, however, because the original critical path remains a critical path during the entire shortening process.

The critical activities to be shortened are selected on the basis of least cost. The exact activities selected depend on the number of days the project is to be reduced. In Example Project 2, the time-shortening pattern and cost for each of the first 3 days of reduction are as follows:

Step	Project Duration (Working Days)	Critical Activities Shortened	Total Direct Cost
0	95	—	$0
1	94	Activity 80 — 1 day ($400)	$400
2	93	Activity 80 — 1 day ($400)	—
		Activity 160 — 1 day ($600)	$1,000
3	92	Activity 40 — 3 days ($2,000)	$2,000

This action now reduces the original critical path to 92 days. Consequently, a path consisting of activities 10, 20, 60, 90, 130, 180, 190, and 200, whose total length is also 92 days, now becomes a second critical path. To shorten the project further, it is now necessary to compress both critical paths simultaneously. This could be done by diminishing either activity 180 or 190, because they are common to both critical paths. However, it has already been established that neither of these can be reduced. Further project shortening, therefore, depends on reducing concurrently the two parallel branches of the critical path. Suppose decreasing activity 90 is the cheapest way to shorten the new critical loop, and it can be decreased by as much as 3 days at an extra cost of $300 per day. The pattern and cost for the next 2 days' reduction is as follows:

Step	Project Duration (working days)	Critical Activities Shortened	Total Direct Cost
4	91	Activity 40 — 3 days ($2,000)	—
		Activity 80 — 1 day ($400)	—
		Activity 90 — 1 day ($300)	$2,700
5	90	Activity 40 — 3 days ($2,000)	—
		Activity 90 — 2 days ($600)	—
		Activity 80 — 1 day ($400)	—
		Activity 160 — 1 day ($600)	$3,600

This decrease now reduces the probable project duration to 90 working days, and this is the most that can be achieved. When there is more than one critical path (this example now has three), all critical paths must be shortened simultaneously to achieve any overall project reduction. Because the original critical path has now been decreased to an irreducible minimum, no further project shortening is possible.

At this point, the contractor must make a management decision concerning how much project shortening it is willing to buy. To answer this question, the consequences of running over the original contract time must be weighed against the costs of shortening the project.

11.21   TIME-SCALED NETWORKS

The project network shown in Figure 11.4 shows activities located in the general order of their accomplishment but is not plotted to a time scale. Project diagrams can be, and frequently are, drawn to a horizontal time scale to serve a variety of useful purposes. When drawing a time-scaled diagram, either of two time scales can be used: one in terms of expired working days and the other as calendar dates. One scale is, of course, convertible to the other. Figure 11.5 is the time-scaled plot of Example Project 2, plotted to a working-day scale with the calendar months also indicated. This is the same network diagram shown in Figure 11.4, and the two portray the same logic and convey the same scheduling information.

Figure 11.5 is obtained by plotting, for each activity, its ES and EF values. The horizontal distance between is equal to the estimated time duration of the particular activity. Although arrowheads are shown in this figure, they are not actually needed and do not appear in most time-scaled diagrams. Vertical solid lines indicate sequential dependence of one activity on another. When an activity has an early finish time that precedes the earliest start of activities following, the time interval between the two is, by definition, the free float of the activity. Free floats are shown as horizontal dashed lines in Figure 11.5, and a time-scaled plot of this type automatically yields to scale the free float of each activity. When an activity has no free float, no dashed extension to the right of that activity appears. The horizontal dashed lines also represent total float for groups or strings of activities, a topic that is discussed in Section 11.22. The circles shown in Figure 11.5 represent the ES times of the activities that immediately follow. The numerical values in these circles are the ES times in terms of expired working days.

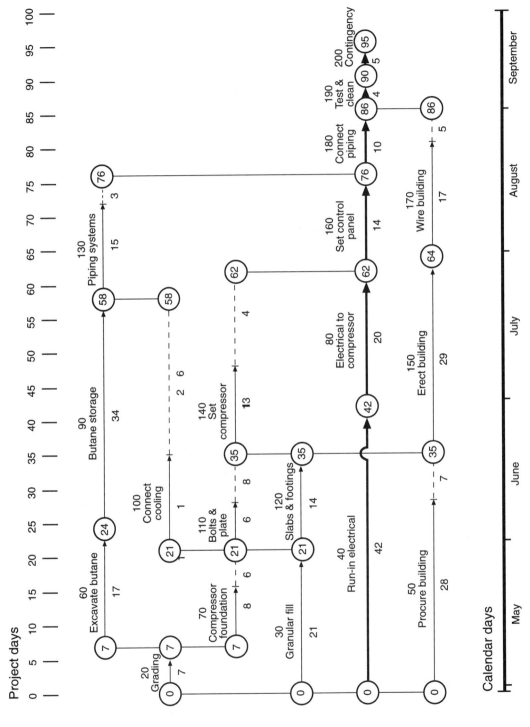

Figure 11.5 Example Project 2, time-scaled network.

The time-scaled diagram has some distinct advantages over a regular network for certain applications because it provides a graphical portrayal of the time interrelationships between activities as well as their sequential order. Such a plot enables one to determine immediately which activities are scheduled to be in process at any point in time and to detect quickly where time and resource problems exist. These networks are very convenient devices for checking daily project needs of labor and equipment and for the advance detection of conflicting demands among activities for the same resource. Using such a diagram is also an effective way to record and analyze the progress of the work in the field. As of any given date, progress can be plotted for each activity in process according to its percentage of completion. The resulting plot shows at a glance which activities are on, ahead of, or behind schedule.

11.22 SIGNIFICANCE OF FLOATS

The float of an activity represents potential scheduling leeway. When an activity has float time, this extra time may be utilized to serve a variety of scheduling purposes. When float is available, the early start of an activity can be delayed, its duration can be extended, or a combination of both can occur. To do a proper job of scheduling noncritical activities, it is important that the practitioner understand the workings of float times.

The total float of an activity is the maximum time that its actual completion date can extend beyond its earliest finish time without delaying the entire project. It is the time leeway available for that activity if the activities preceding it are started as early as possible and those following it are started as late as possible. If all the total float of any one activity is utilized, a new critical path is created. Correspondingly, the free float of an activity is the maximum time by which its actual completion date can exceed its earliest finish date without affecting either overall project completion or the times of any subsequent activities. If an operation is delayed to the extent of its free float, the activities following are not affected and they can still start at their earliest start times.

A time-scaled network illustrates and clarifies the nature of total and free float. In Figure 11.5, the critical path can be visualized as a rigid time spine extending through the diagram. The horizontal dashed lines (float) can be regarded as compressible connections between activities. Activities, singly or in groups, can move to the right (be delayed), provided that the correct dependency relationships are maintained and there is flexibility (float) present.

To understand the nature of total float, consider the rigid subassembly in Figure 11.5 that consists of activities 20, 60, 90, and 130. Figure 11.5 shows that not only is the free float of activity 130 equal to 3 days but also that the total float of each of the activities 20, 60, 90, and 130 is the same 3 days. The diagram also shows that if any one of these four activities is allowed to consume an additional 3 days, the succeeding activities are pushed along to the right and all the float in that path is gone. Consequently, it becomes a second critical path. The result would be the same if 3 additional working days were to be consumed by any combination of activities 20, 60, 90, and 130. Figure 11.5 clearly illustrates that total float is usually shared by a string or aggregation of activities, and the additional time it represents is associated with the group.

The significance of free float is equally obvious from a study of Figure 11.5. For example, consider activity 100, which has a duration of 11 days and a free float of 26 days. Figure 11.5 shows that there is a total of 37 days in which to accomplish activity 100. Within its two boundaries, activity 100 can be delayed in starting, have its duration increased, or a combination of the two without disturbing any other activity. For all practical purposes, activity 100 can be treated as 11 abacus beads that can be moved back and forth on a wire 37 units long. The free float of an activity, therefore, is extra time associated with that activity alone and that can be used or consumed without affecting the early start time of any succeeding activity.

11.23   THE EARLY-START SCHEDULE

If the network calculations are made by computer, the printout will give activity start and finish times as calendar dates. However, if manual computations are made directly on the network, as in Figure 11.4, the numerical data obtained must be translated into calendar dates at which the activities are scheduled to start and finish. This calendar date schedule is customarily based on the early start and finish times of each activity. If modifications to the computed early times have resulted from project shortening, this can be incorporated into the schedule.

Thus far, project times have been expressed only in terms of expired working days. Conversion of expired working days into calendar dates is easily done with the aid of a calendar on which the working days are numbered consecutively, starting with number 1 on the anticipated start date and skipping all weekends, holidays, and vacation periods. It is assumed that the start date of Example Project 2 will be the morning of Monday, April 29.

When making up a job calendar, the true meaning of expired days must be kept in mind. To illustrate, the early start of activity 90 in Figure 11.4 is shown to be 24. This means that this activity can start *after the expiration* of 24 working days, so the *starting time* of activity 90 will be the morning of calendar date numbered 25 (June 3). There is no such adjustment for early finish dates. In the case of activity 90, its early finish time derived from Figure 11.4 is 58, which means that it is finished by the end of the 58th working day. This is indicated by the calendar date numbered 58 (July 19).

Figure 11.6 is the early-start schedule for Example Project 2, with the activities listed in the order of their starting dates. Also indicated are critical activities and free float values, information that is very useful to field supervisors. Not only is this form of operational schedule useful to the contractor, but it can also satisfy the usual contract requirement of providing the owner and/or architect-engineer with a projected timetable of construction operations.

11.24   BAR CHARTS

A conventional bar chart is used as a simple graphical picture of the job schedule, showing the early start and early finish times of various project segments plotted to a horizontal time scale. The unsurpassed visual clarity makes the bar chart a very valuable medium for displaying schedule information. It is immediately intelligible to people who have no knowledge of CPM or network diagrams. It affords an easy and convenient means of monitoring job progress, checking delivery of materials, scheduling equipment and subcontractors, and recording project advancement. Bar charts continue to be widely used on a project site to inform one and all concerning the job time schedule and progress.

	EXAMPLE PROBLEM 2: COMPRESSOR STATION Early Start Schedule			
Activity	Schedule Duration (working days)	Free Float	Scheduled Starting Date (early start) AM	Scheduled Finish Date (early start) PM
20	7	0	April 29, 20–	May 7, 20–
30	21	0	April 29, 20–	May 28, 20–
*40	42	0	April 29, 20–	June 26, 20–
50	28	7	April 29, 20–	June 6, 20–
70	8	6	May 8, 20–	May 17, 20–
60	17	0	May 8, 20–	May 31, 20–
100	11	26	May 29, 20–	June 12, 20–
120	14	0	May 29, 20–	June 17, 20–
110	6	8	May 29, 20–	June 5, 20–
90	34	0	June 3, 20–	July 19, 20–
140	13	14	June 18, 20–	July 5, 20–
150	29	0	June 18, 20–	July 29, 20–
*80	20	0	June 27, 20–	July 25, 20–
130	15	3	July 22, 20–	August 9, 20–
*160	14	0	July 26, 20–	August 14, 20–
170	17	5	July 30, 20–	August 21, 20–
*180	10	0	August 15, 20–	August 28, 20–
*190	4	0	August 29, 20–	Setember 4, 20–

*Critical activities

Figure 11.6   Example Project 2, early start schedule.

It is easy to recognize that a time-scaled network *is* a comprehensive form of bar chart. Commensurately, bar charts can be made to show certain job information in addition to the early start and finish dates. Free float and total float can be plotted, as well as the sequential interdependencies between the activities. (Bar charts of this latter type are sometimes referred to as fence bar charts.) Conventional bar charts can be quickly derived from the network diagram or project schedule. Activities may not always be the most desirable basis for preparing a bar chart. Simpler diagrams with fewer bars and showing more comprehensive segments of the work may be more suitable for some job applications. In such cases, it is an easy matter to combine strings or groups of associated activities into a single bar chart item.

11.25   RESOURCE SCHEDULING

If a project schedule is to be workable, it must be sustained by adequate resources as they are needed on the project. In a construction context, "resources" include materials, subcontractors, labor crews, and equipment. The calendar-date job schedule provides reasonably accurate information concerning when such resources will be required. It is then the responsibility of job management to follow through and see that they are available when needed.

From the job schedule now established, the contractor can determine the calendar dates by which the various materials must be delivered to the job site. Allowing some amount of time as a safety factor, the purchase orders can specify reasonable delivery times. In a similar manner, a calendar schedule of subcontractor operations can be abstracted. Subcontractors can be notified well in advance of when their crews must be on the job. Advance information regarding materials and subcontracted work can be of inestimable value to the contractor's procurement people and expediters in their efforts to provide timely and adequate support to the project.

The job schedule can serve a somewhat different purpose with respect to labor crews and equipment. Not only does the contractor now know when the resources will be required on the job, but conflicting demands for the same resource can also be detected. For example, from the calendar-date schedule for the job activities it is possible to determine total job labor and equipment requirements on a day-to-day basis. This information can be valuable in that it can disclose hitherto unsuspected peaks and valleys in labor demand and conflicts between activities for the same equipment item. It is often possible to smooth manpower demands or remove equipment conflicts by utilizing the floats of noncritical activities.

As a simple illustration, suppose the start of activity 140 in Example Project 2 required the use of the same heavy-duty crane already engaged in activity 90. Two such cranes on the project simultaneously would be difficult and expensive to arrange. Figure 11.5 shows that the start of activity 140 can be deferred by as much as 14 days. Rescheduling the start of activity 140 will undoubtedly remove the equipment conflict. As can be seen, a time-scaled plot of the network diagram can be valuable and convenient for the scheduling of construction equipment.

The procedure just discussed can also be applied to labor demands. The main objective here is to detect in advance where several activities may concurrently require workers of the same craft, thereby causing pronounced peaks in labor demand. Sharp fluctuations in labor needs on a project are undesirable and often impossible to arrange. The judicious rescheduling of activities by the use of float times can do much to alleviate impractical fluctuations in labor demand. In addition, it assists management in recognizing manpower needs well ahead of time where shortages of certain crafts or highly specialized labor may be a factor.

11.26 PROGRESS MONITORING

After the operational plan and calendar schedule have been prepared, the work can now proceed with the assurance that the entire project has been thoroughly studied and analyzed. A plan and schedule have been devised that will provide specific guidance for the conduct of the work in the field. However, it is axiomatic that no plan is ever perfect, nor can the planner possibly anticipate every future job circumstance and contingency. Problems arise every day that could not have been foreseen. Adverse weather, material delivery delays, labor disputes, equipment breakdowns, job accidents, and other conditions can and do disrupt the original plan and schedule. Thus, after construction operations commence, there must be continual evaluation of field performance as compared with the established schedule. Considerable time and effort are expended to check and analyze the progress of the job and to take whatever action may be required, either to bring the work back on schedule or to modify the schedule to reflect changed job conditions.

How often field progress should be measured and evaluated depends on the degree of time control that is considered feasible for the particular work involved. Fast-paced projects using multiple shifts may demand daily progress reports. Large-scale jobs, such as earth dams, that involve only a limited range of work classifications may use a reporting frequency of once a month. However, for most jobs, progress reporting is normally done on a weekly basis.

11.27   PROGRESS ANALYSIS

The analysis of job progress is concerned primarily with determining the effect that reported job progress has on the established project completion date and any intermediate time goals that have been identified. When a progress report is received from the field, the time status of the activities, completed and in progress, is compared with the project schedule. Of particular importance in this regard are the critical activities. If a critical activity has been started late, a setback in project completion will occur unless the delay can be somehow made up by the time the activity is completed. If a critical activity has been finished late, the overall completion date is delayed commensurately. The analysis of job progress data establishes the current time status of the project and where the work is ahead of or behind schedule.

11.28   SCHEDULE UPDATING

Updating is primarily concerned with the effect that schedule deviations and changes in the job plan have on the portions of the project yet to be constructed. One of the nice features of CPM is its ability to incorporate changes as the project goes along. No project plan is ever perfect, and deviations from the original program are inevitable. Consequently, occasional updating of the plan is necessary during the construction period. Updating involves revising the schedule of the work yet to be accomplished to reflect the effect of changes that have occurred as work on the project has progressed. The objective is to determine corrected information regarding activity start and finish dates, floats, and critical activities and to establish a revised project completion date.

How often the project schedule should be updated depends on how often it is needed. If the progress of the work has followed the plan closely and the schedule has held up reasonably well, there is really no need for a network updating. If there have been a number of significant deviations from the plan, then an updating is probably in order. Attempting to make the project fit an obsolete schedule can undo much of the good realized from formalized planning and scheduling. When the accumulated changes start to render the network and its schedule ineffective, it is time to revise them.

11.29   FAST-TRACKING

CPM is used as a planning and scheduling device for a broad range of industrial applications. Included among these are the planning and design of construction projects. Many state highway departments, for instance, use CPM methods for this purpose. When the usual linear construction mode is involved, the design is accomplished and finalized before field construction starts. As such, the planning and scheduling of the design and of the construction phases are done essentially independently of one another. However, the two are closely coordinated under a design-construct or construction management type of contract in which phased construction or fast-tracking is normally utilized. This topic is also discussed in Section 1.14.

Fast-tracking refers to the planned coordination and overlapping achievement of the design and construction of a project, as opposed to the traditional sequential completion of each step before the next is initiated. For the process to be truly effective, however, the best possible cooperative efforts of the architect-engineer and the contractor or construction manager are required. It is also necessary that the owner have a well-thought-out program of requirements and be able to make immediate design decisions. CPM is an especially useful device to achieve a truly integrated and practicable fast-track schedule. No new principles are involved, but the planning, design, and construction of the project must be merged into a single concerted course of action.

11.30   COMPUTERS AND TIME MANAGEMENT

Personal computers have made effective time management of construction jobs available to contractors for a wide range of field projects. With inexpensive hardware and software, the contractor can perform substantially all of the usual time control functions. Computers are now used to create activity networks, make network computations, derive calendar-date schedules, print bar charts, make update calculations, and maintain an as-built network showing the actual activity starts, finishes, and durations. Project supervisory personnel are thus constantly aware of the current time status of the work, the identity of critical activities, and the values of free float. These computer operations are now commonly located directly on the job site.

11.31   DELAY CLAIMS

Delays in construction can have a serious effect on field costs and project time. Contractors sometimes find it necessary to make formal claims for delay damages and extensions of time caused by the owner, architect-engineer, another prime contractor, subcontractor, or material vendor. In such cases, the contractor must be able to establish the cause of the delay and its total impact on individual activities and on the project as a whole. Proving the direct effect of a project delay or delays can often be a very complex and difficult matter.

In attempting to ascertain the effect of a delay on the performance of a contract, a common complication is that the delay affects only certain of the individual job activities. Experience shows that the detailed planning network used for time management can be an excellent tool for analyzing delay effects because it clearly shows the time interrelationships between the various project segments. A detailed planning network affords an opportunity to demonstrate the true effects of project delays. This can help establish the compensable damages and time extensions due the contractor. Courts and other adjudicative bodies now utilize project time networks in apportioning responsibility for job delays. To demonstrate the effect of a given delay, either on a given activity, group of activities, or the total project, a detailed comparison can be made between the as-planned and the actual as-built networks.

Such a comparison can clearly reveal extended durations, start and completion delays, and required logic changes. In this way, the effects of individual or multiple delays can be analyzed, a process that is of great assistance in establishing compensable damages and contract time extensions. The planning and scheduling network has an invaluable ability to reveal the true effect of a given delay, as can be seen by considering a simple example: A 7-day delay of a critical activity obviously has a commensurate impact on the overall project completion date. However, a 7-day delay of an activity with 20 days of free float would likely have no effect.

QUESTIONS

1. Why is the planning phase of CPM the most difficult and important?

2. When breaking a project down into activities, what are some of the considerations that have to be taken into account?

3. Why is time contingency allowances placed at discrete places in a network rather than added to specific activities that are subject to unanticipated delays?

4. What is the significance of the difference between early start times and late start times?

5. Using the network diagram in Figure 11.4, alter the durations of the different activities and recalculate the project duration.

6. Using the data from the previous problem, is there more than one critical path and if so, what is the significance of multiple critical paths?

7. What is the significance of a non-critical path that has one day of float?

8. Total float is sometimes referred to as shared float. Why is the float considered shared?

9. When shortening a project's duration using the least cost procedure, what things limit the amount of time any individual activity can be shortened?

10. What are the attributes of a time scaled network diagram?

Chapter 12

Project Cost Management

12.1 THE IMPORTANCE OF COST CONTROL

Project cost and its control are important management responsibilities. After construction begins, the project cost management system retrieves costs, labor and equipment hours, and production quantities from the job site as the work progresses. This information is used in two important ways. One is to develop production and cost data for labor and equipment in a form suitable for estimating the cost of future work. The other application of cost and production information is to assist in keeping the construction costs of an ongoing project within its established budget. Regardless of the type of contract with the owner, it is important that the contractor exercise the maximum control possible over field costs during the construction period. A functioning and reliable cost system plays a vital role in the proper management of a construction project.

How the costs of an ongoing construction project are obtained and used depends on its size and character. A large, complex job requires a detailed reporting and information system to serve project management needs. Simpler and less elaborate procedures are sufficient for smaller and less complex projects. In any event, the only justification for the expense of a project cost system is the value of the management data it provides. If the information produced is not used or if it is not supplied in a usable form or timely fashion, then the system has no real value and its cost cannot be justified. Properly designed and implemented, project cost management is an investment rather than an expense.

This chapter discusses the methods involved in project cost control and the estimating feedback process. Although the details of how these actions are actually accomplished vary substantially from one construction firm to another, the process discussed here can be regarded as being reasonably typical of present practice within the construction industry.

12.2 PROJECT COST CONTROL

Project cost control is a company information system designed to assist the project manager in controlling construction costs. It is a monitoring process that provides feedback to the manager concerning project expenses and how they compare with the established budget. Project cost control actually begins with the preparation of the original cost estimate. For each project, the contractor prepares a detailed estimate of prices, which serves as the basis of its proposal if competitive bidding is involved, or a target estimate if the contract is negotiated. This estimate serves as a budget for cost control during the construction process.

338

The cost control system identifies project expenses as they occur and charges them against the project elements to which they apply. The costs, as they actually occur, are continuously compared with the budget. Keeping within the budget and knowing when and where job expenses are excessive are key factors of profitable operation. The field costs are obtained in substantial detail because this is the way jobs are originally estimated and because excessive costs in the field can be corrected only if the exact causes can be isolated. To learn that construction expenses are going over the budget is not helpful if it is impossible to determine where the trouble is occurring.

Summary cost reports are prepared at regular time intervals. These reports are designed to serve as management-by-exception devices, making it possible for the project manager to determine the cost status of the project and to pinpoint those work classifications showing excessive expenses. In this way, management attention is quickly focused on those job areas that need it. Timely information is required if effective action against cost overruns is to be taken. Detecting excessive costs only after completion of the work leaves the contractor with no possibility of taking corrective action.

12.3   DATA FOR ESTIMATING

As discussed in Chapter 5, when the cost of a project is being estimated, many elements of cost must be evaluated. Labor and equipment expenses, in particular, are priced in the light of past experience. In essence, historical production records are the only reliable source of information available for estimating these two categories of job expense. The company cost system provides a reliable and systematic means of accumulating labor and equipment productivity and costs for use in estimating future jobs.

With regard to feedback information for cost estimating purposes, there is some variation in practice about the form of production data that are recovered. It can be argued that production rates are fundamental to the estimating of labor and equipment costs. However, as seen in Chapter 5, costs per unit of production, or "unit costs," are widely used for estimating labor and equipment expense because of the convenience of their application. Such unit costs are, of course, determined from production rates and hourly costs of labor and equipment. Unit costs can be kept up-to-date by being adjusted for changes in hourly rates and production efficiencies. Information generated for company estimating purposes can, therefore, be in terms of labor and equipment production rates, unit costs, or both. For estimating purposes, the feedback system must be designed to produce information in whatever form or forms are compatible with company needs and procedures.

Figure 12.1 illustrates the construction cost feedback loop established by a properly functioning project cost management system. The project estimate is developed to support the contractor's bid and establishes the total price the contractor will receive for accomplishing the project scope. Therefore, by necessity, the project estimate establishes the project control budget. Once construction operations begin, cost data are generated. These data are captured in the contractor's cost reports and labor and equipment production reports. Both unit cost and productivity information can then be extracted from these reports for use in preparing better cost estimates of future work. This feedback loop ensures that cost estimates of future projects are grounded by the contractor's historical achievements in the field. This link is vital to long-term success in the construction contracting business.

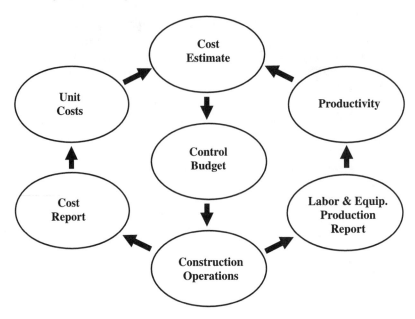

Figure 12.1   Construction cost feedback loop.

12.4   ACCOUNTING CODES

It is customary that an identifying code designation be assigned to each individual account of a contractor's accounting system. The coding systems used by contractors are not standardized, although a number of schemes have been proposed by various technical and professional groups. Standard account systems have been developed by organizations such as the Associated General Contractors of America, American Home Builders Association, American Road and Transportation Builders Association, and the Construction Specifications Institute. Contractors often use their own individual coding systems that they have tailored to suit their particular operations. Alphabetic, decimal, and mixed cost codes are in use. The decimal code used in this text may be considered typical of construction accounting practice. An advantage of a decimal system is that it is expandable to admit any level of detail desired.

Appendix O contains an abbreviated list of typical ledger accounts in common use by contractors. The asset accounts, as included in the general ledger, are identified by whole numbers from 10 through 39. General ledger liability accounts are designated by 40 through 49, net worth by 50 through 69, income accounts by 70 through 79, and expense accounts by 80 through 99. Subaccounts are assigned a distinctive decimal number, the first part of which identifies the general ledger or control account under which the subaccount exists. For example, consider the account 14.105. The whole number 14 indicates a Property, Plant, and Equipment account. The first decimal number, 0.1, indicates that it is associated with Real Estate and Improvements. The last two numbers identify the specific piece of property to which this account applies.

MASTER LIST OF PROJECT COST ACCOUNTS			
Subaccounts of General Ledger Acount 80.000			
PROJECT EXPENSE			
Project Work Accounts .100–.699		Project Overhead Accounts .700–.999	
100	Clearing and Grubbing	700	Project administration
101	Demolition	.01	project manager
102	Underpinning	.02	office engineer
103	Earth excavation	701	Construction supervision
104	Rock excavation	.01	superintendent
105	Backfill	.02	carpenter forman
115	Wood structural piles	.03	concrete foreman
116	Steel structural piles	702	Project office
117	Concrete structural piles	.01	move-in and move-out
121	Steel sheet piling	.02	furniture
240	Concrete, poured	.03	supplies
.01	footings	703	Timekeeping and security
.05	grade beams	.01	timekeeper
.07	slab on grade	.02	watchmen
.08	bcarns	.03	gaurds
.10	slab on forms	705	Utilities and services
.11	columns	.01	water
.12	walls	.02	gas
.16	stairs	.03	electricity
.20	expansion joint	.04	telephone
.40	screeds	710	Storage facilities
.50	float finish	711	Temporary fences
.51	trowel finish	712	Temporary bulkheads
.60	rubbing	715	Storage area rental
.90	curing	717	Job sign
245	Procast concrete	720	Drinking water
260	Concrete forms	721	Sanitary facilities
.01	footings	722	First-aid facilities
.05	grade beams	725	Temporary lighting
.07	slab on grade	726	Temporary stairs
.08	beams	730	Load tests
.10	slab	740	Small tools
.11	columns	750	Permits and fees
.12	walls	755	Concrete tests
270	Reinforcing steel	756	Compaction tests
.01	footings	760	Photographs
.12	walls	761	Surveys
280	Structural steel	765	Cutting and patching
350	Masonry	770	Winter operation
.01	8 in. block	780	Drayage
.02	12 in. block	785	Parking
.06	common brick	790	Protection of adjoining
.20	face brick		property
.60	glazed tile	795	Drawings
400	Carpentry	796	Engineering
440	Millwork	800	Worker transportation
500	Miscellaneous metals	805	Worker housing
.01	metal door frames	810	Worker feeding
.20	window sash	780	General cleanup
.50	toilet partitions	950	Equipment
560	Finish hardware	.01	move-in
620	Paving	.02	setup
680	Allowances	.03	dismantling
685	Fencing	.04	move-out

Figure 12.2 Master list of project cost accounts.

In Appendix O, Project Expense is designated by account number 80.000. This is where every item of expense chargeable to a particular job is recorded. This major category of Project Expense is often subdivided into two major subdivisions: Project Work Accounts and Project Overhead Accounts. Each of these major subdivisions has an extensive internal breakdown into detailed items of cost. An abridged list of cost accounts for each of these is shown in Figure 12.2. The project cost breakdown shown is not intended to be complete nor to apply to all categories of construction. The cost accounts shown in Figure 12.2 would best apply to building construction. Once established, the code number for a given cost account remains the same. It is used consistently throughout the company and does not vary from one project to another. The use of "general" or "miscellaneous" cost accounts is avoided.

12.5 JOB COST ACCOUNTS

The starting point for project cost management is the company financial accounting system. A basic accounting principle for construction contractors is that project costs are recorded by job. Cost keeping must of necessity be a function of each project, and profit or loss is evaluated at the individual job level. However, project cost data obtained directly from the usual accounting records have only very limited value for the project cost system. Although the normal accounting routine periodically reports project costs associated with labor, materials, subcontracts, job overhead, and equipment, these are total costs. Details concerning their makeup, such as the expenses associated with basic work elements like excavation, concrete work, or carpentry, are not available. In addition, the information provided is entirely in terms of costs and includes no data concerning work quantities put in place. As a result, job management cannot determine on which work items costs are within the budget or over the budget. What is required for cost management is more detail concerning the makeup of project costs and the work quantities accomplished. This requires the collection of additional data not provided by the basic financial accounting system.

The required cost breakdown is achieved by maintaining a detailed set of separate cost records for each project. A separate account is set up for each cost item that pertains to a given project. These accounts are subsidiary to the general ledger accounts. Each job expense is then posted to the appropriate cost account. Expenditures for labor, materials, equipment, subcontracts, and field overhead are charged to the appropriate cost accounts throughout the life of the project. Accounts that are active and relevant to the project under consideration are chosen from the chart of cost accounts. The details of the job cost accounts can vary from one job to another. The project may be divided into only a few cost accounts or into a great many line items. This depends on the size of the project and the type of work involved. Because the cost system is directed toward making comparisons between actual and budgeted costs, cost accounts chosen for the project will parallel the work breakdown used for estimating that job. Normally, the accounts used for the project cost system are identical with those used for the original cost estimate.

12.6 MONTHLY COST REPORT

After project commencement, the information contained in the job cost accounts is periodically summarized to prepare various forms of cost reports. These routine reports are designed to inform project management in sufficient detail and in a timely manner concerning

MONTHLY COST REPORT

Project_____   Project No._____
Period Ending_____   Prepared by_____

| Cost Code (1) | Description (2) | Budget (3) | MATERIALS | | | |
			Cost to Date (4)	Estimated Cost to Complete (5)	Estimated Final Cost (6)	Variance (7)
103	Excavation					
240	Concrete					
260	Concrete forms					
270	Reinforcing steel					
280	Structural steel					

Figure 12.3 Monthy material cost report. (Similar forms are prepared for labor, equipment, subcontracts, and total project.)

the cost status of the work. The information conveyed and the frequency of such reporting are chosen to meet the specific needs of job management. It is very important that each cost report be designed to convey its particular message in the most succinct and usable form possible.

The formats used by contractors in preparing their periodic cost reports vary widely, as do the nature of the cost information presented and the frequency of issue. A common cost report is a monthly reporting of the overall cost status of the project. This report can also forecast the eventual total cost. Figure 12.3 presents an example of this type of job cost summary. Actually, Figure 12.3 shows only one of five parts that make up the complete report. The job cost information presented in Figure 12.3 for Materials is also compiled for Labor, Equipment, Subcontracts, and Total Project. Values for Total Project are simply the respective totals of the four constituent cost categories.

For each of the five work items represented in Figure 12.3, budget values in column 3 are obtained directly from the original job cost estimate. The job cost accounts provide the values of costs to date, column 4. The estimated costs to complete, column 5, are based on the original budget amended to reflect job experience to date. The estimated final cost, column 6, is obtained as the sum of columns 4 and 5. The variance, column 7, is column 3 subtracted from column 6. Thus, the variance is the difference between the anticipated actual cost and the budget amount. A positive value of variance is the amount of the anticipated cost overrun; a negative value is, of course, just the opposite. The algebraic sum of the total variance values is the amount by which it now appears that the actual total project cost will be more or less than the budgeted amount. For work not yet started, estimates of cost to finish are normally taken as the budgeted amounts. For completed items of work, the actual final cost is used for the estimated final cost. Although not highly accurate, the resulting figures for the total project clearly reveal the current job cost status.

<table>
<tr><td colspan="7" align="center">MONTHLY OVERHEAD COST REPORT</td></tr>
</table>

MONTHLY OVERHEAD COST REPORT						
Project _____ Project No. _____						
Period Ending _____ Prepared by _____						
Cost Code (1)	Description (2)	Budget (3)	Cost to Date (4)	Estimated Cost to Complete (5)	Estimated Final Cost (6)	Variance (7)
700	Project manager					
701	Supervisors					
702	Project office					
705	Utilities					
740	Small tools					

Figure 12.4   Monthly overhead cost report.

12.7 PROJECT OVERHEAD

The overall project cost report just discussed does not include a reporting of job overhead, because such expense is general in nature and not associated with specific work items. For this reason, project indirect expense is normally reported separately. Figure 12.4 shows a typical form used for monthly reporting of job overhead expense. Values for the various items in Figure 12.4 are obtained in much the same fashion as for the preceding monthly cost report, except that the costs now pertain to job overhead cost accounts rather than work classifications.

12.8 LABOR AND EQUIPMENT COSTS

Job costs associated with materials, subcontracts, and project overhead are of a reasonably fixed nature. Barring an estimating oversight or mistake, these costs are determined with reasonable accuracy when the job is estimated, and such costs do not ordinarily vary a great deal from their budgeted amounts. For this reason, costs of this type are maintained as project expenses but are not usually subjected to detailed analysis. The cost information available from the monthly project cost reportings is normally detailed and timely enough for purposes of project cost management.

Labor and equipment costs, however, are subject to considerable variability. These two categories of job expense can vary substantially during the construction process. The inherent variability of labor and equipment costs explains the fact that contractors' cost management systems usually concentrate on these two items of job expense. The monthly cost reports previously discussed, although useful, do not provide the needed information concerning these expense types, in either suitable detail or sufficient time. Accordingly, labor and equipment expenses are customarily subjected to another level of analysis and reporting. The process of determining, at short intervals, how much work is being put in place in relation to the costs of labor and equipment being supplied is described in the following sections.

12.9   COST ACCOUNTING

To accumulate and analyze labor and equipment cost and production information from an ongoing project, special records and detailed accounts are required. The maintenance of such detailed job records is referred to, in the aggregate, as "cost accounting." Cost accounting is not independent of the contractor's job cost accounts. Rather, it is in the nature of an elaboration of the basic project expense accounts. Cost accounting involves the continuous determination of labor and equipment costs, together with the work quantities produced, the analysis of these data, and the presentation of the results in summary form.

It can thus be seen that project cost accounting differs from the usual accounting routines in that the information gathered, recorded, and analyzed is not entirely in terms of dollars and cents. Construction cost accounting is necessarily concerned not only with costs, but also with labor-hours, equipment-hours, and the amounts of work accomplished. The systematic and regular checking of costs is a necessary part of obtaining reliable, time-average production information. A system that evaluates field performance only periodically in the form of occasional "spot checks" does not provide trustworthy feedback information, either for cost control or for estimating purposes.

Project cost accounting must strike a workable balance between too little and too much detail. A system that is too general will not produce the detailed costs necessary for meaningful management control. Excessive detail will result in the objectives of the cost system being obliterated by masses of data and paperwork, as well as needlessly increasing the time lag in making the information available. The company cost system must be tailored to suit the contractor's own particular mode of operation. The detail used in this book is reasonably typical of actual practice in the industry.

A project cost accounting system supplements field supervision; it does not replace it. In the final analysis, the best cost control system a contractor can have is skilled, experienced, and energetic field supervision. It is important that field supervisors realize that project cost accounting is meant to assist them by the early detection of troublesome areas. Trade and site supervisors are key members of the team, and without their support and cooperation, the cost accounting system cannot and will not perform satisfactorily.

For purposes of both estimating and cost accounting it is necessary that the project be broken down into the same elementary work classifications and be identified by the same cost code numbers used in previous and similar projects. Figures 5.4 and 5.5 show that the estimator identifies each work item by its code designation when he sets down the results of his quantity takeoff. Throughout the construction process, a continuous record is kept of the actual costs of production of these same work items. From estimate to project completion, the same work breakdown and cost code numbers apply.

12.10 LABOR AND EQUIPMENT BUDGET

For purposes of cost accounting, a labor and equipment budget must be prepared before the beginning of field operations. This budget information may be in the form of labor and equipment hours, unit costs, or both. For the discussions in this chapter, it is assumed that the budget is prepared in terms of the total estimated work quantity, unit cost of labor and/or equipment, and total labor and/or equipment cost for each work cost code involved on the job. This budget then serves as the yardstick against which the actual field costs are measured during the progress of the work.

Contractors vary as to whether the budget costs include or exclude indirect labor costs (see Section 5.24). Practice in this regard depends on the detailed estimating and cost accounting procedures being used. What is important is that the budget and field costs be expressed on exactly the same basis and include the same items of cost. In this way, valid comparisons can be made between estimated and actual costs of production. In this text, the project cost budget and labor cost reports are all in terms of direct labor costs only and do not include indirect labor expense.

12.11 COST ACCOUNTING REPORTS

Summary labor and equipment cost reports must be compiled sufficiently often to detect excessive project costs while there is still time to remedy the situation. Cost report intervals are very much a function of project size, the nature of the work, and the type of construction contract involved. A balance must be struck between the cost of generating such reports and the value of the management information received. Daily cost reports are sometimes prepared on complex projects involving multiple shifts, but it is unusual for a job to profit from such frequent cost reporting. Some very large projects involving relatively uncomplicated work classifications may find intervals of a month to be satisfactory. For most construction projects, however, labor and equipment cost reports are needed more frequently.

It is generally agreed that weekly labor and equipment cost reports are about optimum for most construction operations, and this is the basis for discussion here. The cutoff time can be any desired day of the week, although a contractor often matches its cost system to its usual payroll periods and to its time schedule monitoring system. The procedures involved with labor cost accounting are discussed first, followed by an examination of equipment cost accounting methods.

12.12 LABOR TIME CARDS

The source document for both payroll purposes and labor cost accounting is the labor time card. It is used to report the hours of labor time for each worker and the work categories to which the labor applies. Figure 12.5 shows a typical daily labor time card, and Figure 12.6, a weekly labor time card. Which card is used depends on company preference and policy.

When daily time cards are used, a time card is filled out each day. The head of each card provides for entry of the project name and number, date, weather conditions, and the name of the person preparing the card. The body of the time card provides for the name, badge number, and craft of each worker covered. Hours are reported as regular time (RT) or overtime (OT), as the case may be. For each worker listed, several slots are provided for the distribution of the hours to those cost codes involved during that day. Absolute perfection

THE BLANK CONSTRUCTION COMPANY, INC.
DAILY LABOR TIME CARD

Project   Municipal Airport Terminal Building   Project Number   1286   Weather   Cloudy-windy

Date   August 18, 20-   Prepared by   R. D. Jones

Badge Number	Name	Craft	Time Classif.	Hourly Rate	Cost Code							Total Hours	Gross Amount
					701.03	240.01	240.05	240.07	240.08	240.51	240.91		
316	Jones, Richard D.	CF	RT	$ 29.25	4		2		2			8	$ 234.00
			OT										
109	Adams, Claude	CM	RT	$ 27.75						6	2	8	$ 222.00
			OT										
422	Chavez, S. C.	CM	RT	$ 27.75				4		4		8	$ 222.00
			OT										
461	Womac, C. T.	CM	RT	$ 27.75		4	2			2		8	$ 222.00
			OT										
247	Sears, A. M.	L	RT	$ 21.45		4	4					8	$ 171.60
			OT										
356	Johnson, Clyde	L	RT	$ 19.95				4	4			8	$ 159.60
			OT										
393	Newman, Stan	L	RT	$ 19.95		4	4					8	$ 159.60
			OT										
211	Pong, Glace	L	RT	$ 19.95				4	4			8	$ 159.60
			OT										
	Total Labor Cost				$117.00	$276.60	$279.60	$270.60	$218.10	$333.00	$ 55.50		$1,550.40

Figure 12.5 Daily labor time card.

347

THE BLANK CONSTRUCTION COMPANY, INC.
WEEKLY LABOR TIME CARD

Name: Womac, C. T.
Week Ending: August 22, 20-
Craft: CM
Project: Municipal Airport Terminal Building
Project Number: 1286
Prepared by: R. D. Jones

Cost Code	Time Classif.	Hourly Rate	Monday	Tuesday	Wednesday	Thursday	Friday	Saturday	Sunday	Total Hours	Total Cost
	RT	$ 27.75	4	8	8					20	$ 555.00
240.01	OT										
	RT	$ 27.75	2				4			6	$ 166.50
240.05	OT										
	RT	$ 27.75				4				4	$ 111.00
240.07	OT										
	RT	$ 27.75				4				4	$ 111.00
240.08	OT										
	RT	$ 27.75	2				4			6	$ 166.50
240.51	OT										
	RT										
	OT										
	RT										
	OT										
Total Hours	RT		8	8	8	8	8	0	0	40	
	OT		0	0	0	0	0	0	0		
Gross Amount			$222.00	$222.00	$222.00	$222.00	$222.00				$ 1,110.00
Weather			Sunny	Rain	Rain	Clouds	Wind				

Figure 12.6 Weekly labor time card.

in time distribution is not possible; nevertheless, the need for care and reasonable accuracy cannot be overemphasized.

In Figure 12.5 it can be noted that the foreman's time is charged to account 701.03, which is an overhead account, and to accounts 240.05 and 240.08, which are work accounts. This is illustrative of the usual practice that a foreman's time spent on supervisory duties is charged to an overhead account and time spent working with his tools is charged to the appropriate work cost codes. As shown in Figure 12.5, the foreman spent four hours supervising his crew, two hours pouring grade beams, and two hours pouring beams.

If weekly time cards are used, a separate card for each individual worker is prepared to record the hours worked that week. Although the arrangement of the weekly time card is different from that of the daily time card, it presents the same information concerning the individual worker. The hourly rates shown on the time cards in Figures 12.5 and 12.6 are base wage rates only and do not include any indirect labor costs such as payroll taxes, insurance, or fringe benefits. Where the contractor is using computer-based payroll and cost accounting procedures, the computer can preprint all standard information on the cards or on a daily time sheet. The person making the time report need add only the hours and cost codes.

12.13   TIME CARD PREPARATION

The foreman normally does the distribution of each worker's time to the proper cost accounts because he is in the best position to know how each person's time is actually spent. Usually this information is first recorded in the foreman's pocket time book, with the nature of the work performed by the individuals of his crew often being described by words rather than cost code numbers. The foreman identifies the craft and position of each worker by using any simple letter or numerical code the contractor may decide on. For example, in Figure 12.5, CF indicates concrete foreman, CM means cement mason, and L indicates laborer. Ordinarily, the foreman enters the hourly rate for each worker because it is not unusual for an individual during a week, or even in a single day, to be employed on work requiring different rates of pay. However, others may extend the hours and wage rates into totals.

The importance of accurate and honest time reporting cannot be overemphasized. Based on the allocation of labor (and equipment) time to the various account numbers, cost and production information is generated. If this information is inaccurate or distorted, it not only is erroneous but can be seriously misleading when used for estimating or cost control purposes. Loss items must be identified as such without any attempt at cover-up by charging time to other cost accounts.

It may well be that the formal time card is filled in by someone other than the foreman, such as the field engineer, field superintendent, or project manager. The field engineer, for example, may collect the foremen's time books at some convenient time each day and fill out the individual time cards, adding the necessary information for payroll and cost accounting, such as employee number and work cost codes, and perhaps make the extensions.

Even if a weekly time card is used, it is preferable that the information be entered on a daily basis. Setting down the time and its cost distribution as the work progresses lends a precision that is lost if the foreman enters the information from memory at the end of the week. It is probable that completing the labor record on a daily basis can improve the accuracy of the distribution. It is for this and other reasons that many contractors favor the use of daily time cards.

12.14   MEASUREMENT OF WORK QUANTITIES

To determine labor production rates or unit costs, it is necessary to obtain not only the hours and costs involved but also the amount of each elementary work classification that has been accomplished. On some types of work, it may be feasible and convenient for the foreman to include with his daily time cards the work quantities accomplished for that day. It is more common, however, for work measurements to be made at the end of each weekly payroll period, which is the basis for discussion here. The items of work measured must be identical to the standard cost code work classifications. Note that the weekly measurement of work quantities includes all work items performed, whether accomplished by labor, by equipment, or by their combination. Consequently, the same weekly work quantity determination serves for both labor and equipment cost accounting.

Work quantities can be obtained in a variety of ways, depending on the nature of the work involved and the company's management method. Direct field measurement on the job, estimation of percentages completed, computation from the drawings, and the obtaining of quantities from the estimating sheets or CPM activities are all used. Direct measurement in the field is common, either in terms of total units achieved to date or units accomplished that week. This procedure is easily applied to projects that involve few cost code classifications. Many heavy, highway, and utility contracts are of this type. There are often instances in which quantities can be obtained with reasonable accuracy by applying estimated percentages of completion to the work amount totals. Although not as accurate as direct measurement of actual quantities, this procedure can yield useful information as long as accurate determinations are occasionally made, such as for monthly pay requests. Total work quantities of all types should occasionally be checked against the estimating sheets as a verification of the overall accuracy of quantity reporting.

The field measurement of work quantities on projects that involve many cost code classifications can become a substantial chore. Most building and industrial projects entail substantial numbers of different cost classifications. One convenient procedure in such cases is to mark off and dimension the work advancement in colored pencil on a set of project drawings reserved for that purpose. The extent of work put in place can be indicated as of the end of each day or each week, as desired. By using different colors and dating successive stages of progress, work quantities can be determined from the drawings or estimating sheets as of any date desired.

The field measurement of work quantities can be, and often is, done by the field supervisors. However, on large projects and especially those with many work codes, it may be desirable that the field engineer or project manager carry out this function because of the time and effort required. Weekly reports of work done are submitted on standard forms, such as that in Figure 12.7.

12.15   FORMS OF LABOR REPORTS

Weekly labor reports can be prepared on either a labor-hour or cost basis. That is, labor productivity can be monitored in terms of either labor-hours per unit of work (production

THE BLANK CONSTRUCTION COMPANY, INC.
WEEKLY QUANTITY REPORT

Project _Municipal Airport Terminal Building_   Project No. ___1286___

Week Ending _____August 22, 20-_____   Prepared by _R. D. Jones_

Cost Code	Work Description	Unit	Total Last Report	Total This Week	Total to Date
240.01	Concrete, footings	c.y.	479	196	675
240.05	Concrete, grade beams	c.y.	208	208	416
240.07	Concrete, slab	c.y.	595	65	660
240.08	Concrete, beams	c.y.	0	60	60
240.51	Concrete, trowel finish	s.f.	50,595	2,865	53,460
240.91	Concrete, curing, slab	s.f.	50,595	2,865	53,460

Figure 12.7   Weekly quantity report.

rates) or cost per unit of work (unit prices). Which of these is used is a function of project size, type of work involved, and project management procedures. Where labor-hour control is used, a budget of labor-hours per unit of work is prepared. In this regard, total labor-hours are usually used with no attempt to subdivide labor time by trade specialty, such as so many hours of carpenter time and so many of ironworker time. Total labor-hour estimates are based on an "average" crew mix for each work type.

During field operations, actual labor-hours and work quantities are obtained. This makes possible a direct comparison of actual to budgeted productivity. Such an approach, of course, reflects productivity but not cost. It is simple to implement and avoids many problems associated with labor cost analysis. One such problem occurs on projects that require a long time to complete. Wage rates may be increased several times during the life of such a job. When the project cost is first being estimated, educated guesses project the likely wage increases. What this means is that labor costs on long-term jobs are often estimated without exact knowledge of the wage rates that will actually apply during the construction process. For this reason, the actual labor wage rates during the work period may turn out to be different from the rates used in preparing the original control budget. Thus, labor costs produced by the project cost system are not directly comparable to the budget amounts. To make valid cost comparisons, it is necessary to either adjust these costs to a common wage-rate basis or work in terms of labor-hours rather than labor cost.

In the problem just discussed, labor-hours can serve as a very effective basis for labor cost control. However, for most construction applications, labor cost analysis is more widely applied than labor-hour analysis. For this reason, the labor cost reports discussed here are all based on costs.

12.16   WEEKLY LABOR COST REPORTS

Once a week, labor costs obtained from the time cards are matched to the work quantities produced. The results of this analysis are summarized in a weekly labor cost report, two different forms of which are illustrated in Figures 12.8 and 12.9. These labor reports classify and summarize all labor costs incurred on the project up through the effective date of the report (August 22). The labor costs in these two reports are direct labor costs only and do not include indirect labor costs. The objective of these reports is to provide job management with detailed information concerning the current status of labor costs and to indicate how these costs compare with those estimated. Both of the labor report forms are designed to identify immediately those work classifications having excessive labor costs and to give an indication of how serious these overruns are. Labor cost reports vary considerably in format and content from one construction company to another, although all such report forms are designed to convey much the same kind of management information.

Figure 12.8 is a weekly cost report for concrete placement that summarizes labor costs as budgeted, for the week being reported, and to date. Not all cost report forms include costs for the week being reported. These values can be of significance, however, in indicating downward or upward trends in labor costs. The labor report form in Figure 12.8 involves work quantities, as well as labor expense, and yields unit costs for each work type. Unit prices, obtained by dividing the total labor cost in each work category by the respective total quantity, enable direct comparisons to be made between the actual costs and the costs as budgeted. In Figure 12.8, the budgeted total quantity, budgeted total labor cost, and budgeted unit cost for each work type are taken from the project budget. The other quantities and labor costs are actual values, either for the week reported or to date.

When the total quantity of a given work item has been completed, its to-date and projected savings or loss figures are obtained merely by subtracting its actual total labor cost from its estimated total cost. When a work item has been only partially accomplished, the to-date savings or loss of that work item is obtained by multiplying the quantity in place to date by the underrun or overrun of the unit price. The projected savings or loss for each work type can be obtained in different ways. In Figure 12.8, it is determined by assuming that the unit cost to date will continue to completion of that work type. Multiplying the total estimated quantity by the underrun or overrun of the unit price to date yields the projected savings or loss figure.

The projected savings and loss figures shown in Figure 12.8 afford a quick, informative summary of how the project is doing as far as labor cost is concerned. Those work types with labor overruns are identified, together with the financial consequences if nothing changes. Some labor cost reports indicate the trend for each cost code—that is, whether the unit cost involved has been increasing or decreasing. This information can be helpful in assessing whether a given cost overrun is improving or worsening and in evaluating the efficacy of cost reduction efforts.

Figure 12.9 is an alternative form of a weekly labor cost report that presents, in somewhat different form, the same weekly cost information for concrete placement. This figure

THE BLANK CONSTRUCTION COMPANY, INC.

WEEKLY LABOR COST REPORT

Project _Municipal Airport Terminal Building_ Project No. _1286_

Week ending _August 22, 20-_ Prepared by _W. W. Smith_

Cost Code	Work Description	Unit	Quantity			Direct Labor Cost			Unit Labor Cost			To Date		Projected	
			Budget	This Week	To Date	Budget	This Week	To Date	Budget	This Week	To Date	Savings	Loss	Savings	Loss
240.01	Concrete, footings	c.y.	$1,040	$196	$675	$7,020	$1,391	$4,865	$6.75	$7.09	$7.21		$308		$475
240.05	Concrete, grade beams	c.y.	$920	$208	$416	$8,694	$1,871	$3,779	$9.45	$8.99	$9.08	$153		$338	
240.07	Concrete, slab	c.y.	$2,772	$65	$660	$24,948	$675	$6,710	$9.00	$10.38	$10.17		$770		$3,232
240.08	Concrete, beams	c.y.	$508	$60	$60	$4,800	$572	$572	$9.45	$9.53	$9.53		$5		$39
240.51	Concrete, trowel finish	s.f.	$128,000	$2,865	$53,460	$48,000	$1,076	$18,716	$0.38	$0.38	$0.35	$1,332		$3,189	
240.91	Concrete, curing, slab	s.f.	$180,000	$2,865	$53,460	$8,100	$129	$1,929	$0.05	$0.05	$0.036	$477		$1,605	

Figure 12.8 Weekly labor cost report (#1).

THE BLANK CONSTRUCTION COMPANY, INC.

WEEKLY LABOR COST REPORT

Project Municipal Airport Terminal Building   Project No.   1286

Week ending   August 22, 20-   Prepared by   W. W. Smith

Cost Code (1)	Work Description (2)	Unit (3)	Total Quantity Budgeted (4)	Total Quantity to Date (5)	Percent Complete (6)	Budgeted Direct Labor (7)	Budgeted Labor to Date (8)	Actual Labor to Date (9)	Cost Difference (10)	Deviation (11)
240.01	Concrete, footings	c.y.	1,040	675	64.9	$7,020	$4,556	$4,865	($308)	1.07
240.05	Concrete, grade beams	c.y.	920	416	45.2	$8,694	$3,931	$3,779	$153	0.96
240.07	Concrete, slab	c.y.	2,772	660	23.8	$24,948	$5,940	$6,710	($770)	1.13
240.08	Concrete, beams	c.y.	508	60	11.8	$4,800	$567	$572	($5)	1.01
240.51	Concrete, trowel finish	s.f.	128,000	53,460	41.8	$48,000	$20,048	$18,716	$1,332	0.93
240.91	Concrete, curing, slab	s.f.	180,000	53,460	29.7	$8,100	$2,406	$1,929	$477	0.80
	Totals to Date						$37,448	$36,569	$879	0.98

Figure 12.9   Weekly labor cost report (#2).

shows actual and budgeted total labor costs to date for each cost classification. The budgeted total quantities and total labor cost for each cost code are obtained from the project budget. The actual work quantities and labor costs to date are cumulative totals for each work classification obtained from the time cards and weekly quantity reports. Column 10 of Figure 12.9 shows the cost difference as column 8 minus column 9, with a positive difference indicating that the cost as estimated exceeds the actual cost to date. Hence, in column 10, a positive number is desirable; a negative number, undesirable. The deviation is the actual cost to date, column 9, divided by the budgeted cost to date, column 8. A deviation of less than 1 indicates that labor costs are within the budget, whereas a deviation of more than 1 indicates a cost overrun.

Although column 10 does indicate the magnitude of the labor cost variation for each cost code, it does not indicate the relative seriousness of the cost overruns. The deviation is of value in this regard because it shows the relative magnitude of the labor cost variance. For those work types not yet completed, the cost differences listed in column 10 of Figure 12.9 do not always check exactly with the to-date savings and loss values of Figure 12.8. These small variations are caused by the rounding off of numbers and are not important.

12.17 EQUIPMENT EXPENSE

Cost keeping for equipment, especially on engineering construction, is an important cost accounting application. Equipment constitutes a substantial proportion of the cost of such projects, and the need to analyze equipment costs parallels the need to determine labor costs. The costs associated with large pieces of construction equipment are substantial, inherently variable, and deserving of a comprehensive record-keeping system. The objectives of equipment cost accounting are the same as those discussed for labor costs. Management requires timely information for effective project cost control, and estimators need data to use in future bids. Only major equipment items merit detailed cost study, however. Lesser equipment such as power saws, concrete vibrators, and hand-operated soil compactors are normally charged to a project on a flat rate or lump-sum basis and do not require detailed cost analysis.

Labor wage rates are, in almost all instances, fixed by, or related to, local labor agreements. No such determination exists for equipment costs, however, and it is up to the contractor to establish its own equipment expense rates. In the case of rental or leased equipment, the rental or lease rates are known, but the contractor must still establish the field operating costs. In the case of contractor-owned equipment, both ownership and operating costs must be determined. As discussed in Section 5.25, ownership expense includes depreciation and investment costs. Investment expense includes the costs of interest, insurance, taxes, and storage. Operating costs are on-the-job expenses such as tire replacement and repairs; mechanical repairs and parts; fuel, oil, and grease; and, possibly, operating labor. Some contractors prefer to regard the labor associated with equipment operation as a labor expense rather than an equipment cost. Others include the labor cost as a part of equipment-operating expense. There are some cost accounting advantages in treating equipment-operating labor in the same way that other labor cost is treated, and not lumping it together with equipment costs proper. This practice is the basis used in the discussions in this text.

When contractors estimate new work, they must obtain the most accurate values possible of the ownership and operating expenses of the various equipment types that will be required. Figure 5.6 illustrates a widely used procedure for estimating such costs. For most items of production equipment, ownership, lease, or rental expense is combined with operating costs in a total cost per operating hour. When a project is under construction, it is the purpose of the cost accounting system to determine the number of hours each equipment type is in operation and the project work accounts to which these hours apply. With the use of the hourly equipment rates previously established for estimating purposes, equipment costs are periodically determined for each work type. When these costs are matched with work quantities accomplished, equipment unit costs of production are obtained. Thus, once the hourly equipment rates are determined, equipment cost accounting proceeds very much like labor cost accounting and similar kinds of cost reports are produced.

Generally, equipment expense is directly chargeable to a single project work account. However, there are occasions when this is not true, and equipment costs must be accumulated in a suspense account until they can be distributed to the proper cost accounts. An example of this procedure is the treatment of a central concrete-mixing plant, consisting of many separate equipment items, that is producing concrete for several different cost accounts. Equipment expenses of this type must be put into a suspense account and periodically distributed equitably to the appropriate cost accounts, based on the quantities involved.

Another special concern regarding cost accounting is support equipment, equipment items that serve many different operations. Examples include cranes, hoists, air compressors, and electric generators. The allocation of time for such machines to specific work codes can be very difficult, if not impossible. A common approach is to establish special cost codes for this equipment, often in field overhead. All equipment time is charged to this account, with no effort made to distribute the time to the various work accounts involved.

The internal rental rates used to charge equipment time to projects are based on time-average ownership and operating expenses that actually vary over the service life of the equipment. For example, investment costs decrease and repair costs increase with equipment age. However, the use of lifetime average costs is the only way to have each project bear its proper share of the ultimate total expense associated with any particular equipment item. When equipment rental rates are charged to a job, this is an all-inclusive charge. Correspondingly, the costs of fuel, lubrication, maintenance, repairs, and other such equipment expenses are not charged to the job on which they are actually incurred, but to the applicable equipment accounts.

12.18   EQUIPMENT TIME CARDS

Because equipment costs are expressed as a time rate of expense, time reporting is the starting point for equipment cost accounting. Equipment time is kept in much the same way as labor time. Where major items of equipment are involved, a common procedure is to have the equipment supervisor make out daily or weekly equipment time cards that include each equipment item on the job. Equipment time cards are separate from, and in addition to, the operators' time cards. This procedure has merit because, by using different time cards, separate reportings are available for payroll and labor cost accounting purposes and for equipment cost accounting. In addition, equipment items such as pumps and air compressors may not have full-time operators and may otherwise be overlooked. Figure 12.10 is a typical daily equipment time card.

An equipment time card performs the same cost accounting function as a labor time

THE EXCELLO COMPANY, INC.

DAILY EQUIPMENT TIME CARD

Project ___ Holloman Taxiways ___ Weather ___ Warm-clear ___ Project No. ___ 8608 ___

Date ___ October 16, 20- ___ Prepared by ___ J. Brown ___

| Machine No. | Machine | Rate per Hour | Cost Code | | | | Total Hours | | | Total Cost |
			101.05	103.07			W	R	I	
16	Bottom dump hauler	$32.41	8				7	1	0	$259.30
12	Bottom dump hauler	$32.41	8				8	0	0	$259.30
17	Bottom dump hauler	$32.41	8				8	0	0	$259.30
48	2 c.y. shovel	$61.80		8			8	0	0	$494.40
7	HG-11 tractor	$29.76	4	4			7	0	1	$238.08
21	Air compressor	$10.88		8			8	0	0	$87.00
	Total Cost		$896.93	$700.44						$1,597.37

Figure 12.10 Daily equipment time card.

card. By allocating equipment times to the proper cost codes, it is possible to determine the equipment costs chargeable to the various work types. Accuracy of time allocation to cost codes is just as important for equipment as it is for labor if reliable information is to be obtained for purposes of cost control and estimating. Even if a weekly time card is used, it is preferable that the time information be entered on a daily basis. Just as with labor time, distributing the equipment time daily is conducive to better accuracy. Someone other than the field supervisor, such as the field engineer, may fill out the time card and enter the budget rates of the individual equipment items reported and make the cost extensions.

Figure 12.10 records equipment time as working time (W), repair time (R), and idle time (I). Excessive equipment idle time (it may be difficult to have this item reported honestly) may indicate field management problems, such as too much equipment on the job, lack of operator skill, improper balance of the equipment spread, or poor field supervision. Appreciable repair time can indicate inadequate equipment maintenance, worn-out equipment, severe working conditions, or operator abuse. A substantial amount of unproductive time can be caused by job accidents, inclement weather, unanticipated job problems, or unfavorable site conditions.

12.19   EQUIPMENT COST REPORTS

Once each week, equipment costs are matched with the corresponding quantities of work produced. Work quantities are derived from the weekly quantity report, discussed in Section 12.14. By following the same process described for labor, a weekly equipment cost report is prepared. An equipment cost report summarizes all equipment costs incurred on the project up through the effective date of the report. Either of the two cost report forms used for labor (Figures 12.7 and 12.8) can be used for equipment. Figure 12.11 is a frequently used format for weekly equipment cost reports. As can be seen, this figure is very similar to the weekly labor cost report form shown in Figure 12.8. Figure 12.11 serves to inform project management in a quick and concise manner regarding both current equipment costs and those work items that are over budget. The equipment report form in Figure 12.11 presents both work quantities and equipment expense and yields actual unit costs for each work type. Comparing the estimated with the actual equipment unit costs discloses where equipment costs are overrunning the project budget. The budgeted total quantity, budgeted total equipment cost, and budgeted unit cost for each work type are taken from the project budget. The other quantities and equipment costs are actual values, either for the week reported or to date. The to-date and projected savings and loss values capsulate the equipment cost experience on the project up through the report date of October 18.

12.20   COST INFORMATION AND FIELD SUPERVISORS

There is considerable difference of opinion over whether detailed cost information should be divulged to lower-level field supervisors. It has been suggested, for example, that craft foremen can be tempted to charge labor or equipment time incurred on operations showing losses to other cost codes for operations in which performance has been good. It is also possible that confidential cost information may be compromised or that a foreman may tend to relax when he knows that his costs are within the estimate. There is some truth to all of these concerns.

THE EXCELLO COMPANY, INC.
WEEKLY EQUIPMENT COST REPORT

Project _____ Holloman Taxiways _____
Week ending _____ October 18, 20- _____

Project No. _____ 8608 _____
Prepared by _____ J. Brown _____

Cost Code	Work Description	Unit	Quantity			Equipment Cost			Equipment Unit Cost			To Date		Projected	
			Budget	This Week	To Date	Budget	This Week	To Date	Budget	This Week	To Date	Savings	Loss	Savings	Loss
101.05	Excavation, hauling	c.y.	$127,000	$12,200	$113,680	$102,870	$9,515	$90,383	$0.81	$0.78	$0.80	$1,698		$1,897	
103.07	Excavation, common	c.y.	$127,000	$12,200	$113,680	$72,390	$8,232	$81,846	$0.57	$0.67	$0.72		$17,048		$19,046
145.11	Base course, spreading	ton	$79,500	$7,360	$77,420	$65,588	$6,515	$65,039	$0.83	$0.89	$0.84		$1,167		$1,198
250.03	Concrete, production	c.y.	$23,625	$2,090	$4,139	$121,905	$10,376	$21,105	$5.16	$4.96	$5.10	$252		$1,440	
254.01	Concrete, hauling	c.y.	$23,625	$2,090	$4,139	$12,404	$1,193	$2,484	$0.53	$0.57	$0.60		$311		$1,775
258.02	Concrete, lay-down	s.y.	$90,000	$8,360	$16,556	$66,150	$5,769	$10,923	$0.74	$0.69	$0.66	$1,246		$6,772	

Figure 12.11　Weekly equipment cost report.

Yet it is well recognized that the only way a cost system can succeed is with the support and cooperation of the field supervisors. They try to achieve the best possible performance and expect to receive credit, and perhaps a bonus, if they beat the estimated costs. Field costs are very much involved when companies enter into profit-sharing or incentive plans with their supervisors. Probably the best answer is to provide craft foremen with the total amounts of cost overruns without specifying the estimated or actual unit costs.

12.21   COST CONTROL

Weekly labor and equipment cost reports make it possible for company management to assess quickly the cost status of the project and to pinpoint the work areas where such expenses are proving to be excessive. In this way, management attention is quickly focused on those work classifications that are in need. If the project expense information is developed promptly, it may be possible to bring the offending costs back in line. In fact, of course, project cost control starts when the job is first priced, because this is when the control budget is actually established. No amount of management expertise or corrective action can salvage a project that was initially priced too low. Thus, there are always likely to be some work classifications whose actual costs will exceed those estimated. The project manager is primarily responsible for getting the total project built for the estimated cost. If some costs go over budget, then, with luck, savings in other areas may counterbalance these overruns.

Having identified excessive production costs, project management must now decide how to proceed. The hourly rates for labor and equipment are not controllable by management. The only real opportunity for cost control lies in improving production rates. Skilled field supervision, astute job management, energetic resource expediting, and the improved makeup of labor crews and selection of equipment can, to a degree, favorably influence this element of work performance.

Any efforts to improve field production must be based on detailed knowledge of the pertinent facts. If the cause of excessive costs cannot be specifically identified, then a satisfactory solution is not likely to be found. It is impossible to generalize on this particular matter, but certainly, the solution must be gauged to suit the issue. In the usual case, full cooperation between the field supervisors and project management is needed before any real cost improvement can be realized. Field supervisors play a key role in implementing corrective procedures. There are no precise guidelines for reducing excessive project costs. The effectiveness of corrective procedures depends largely on the ingenuity, resourcefulness, and energy of the people involved.

Production costs are frequently high early in the construction process, but tend to become lower as the work progresses. This is a "learning curve" phenomenon whereby costs decrease as experience and familiarity with the job are gained. In a general way, production costs usually tend to decrease as the job goes along, because crew members learn how to work as a team, become familiar with the job, and find out what their foremen expect of them.

12.22   INFORMATION FOR ESTIMATING

Estimating requires production rates and unit costs that are a balanced time average of good days and bad days, high production and low production. For this reason, information for estimating is normally not recovered from the cost accounting system until after project completion, or at least not until all the work type being reported has been finished. It is at this point that the best possible time-average rates can be obtained. Permanent files of cost and productivity information are maintained, providing the estimator with immediate access to data accumulated from prior projects.

Both production rates and unit costs are available from the project cost accounting system. To be of maximum value in the future, however, it is important that such productivity data be accompanied by a description of the project work conditions that applied while the work was being done. Knowledge of the work methods, equipment types, weather, problems, and other job circumstances will make the basic cost and productivity information much more useful to an estimator. Such a written narrative becomes a part of the total historical record of each cost account.

12.23   COMPUTER APPLICATION

Computers are widely used by construction contractors in conjunction with their project cost systems. Because cost accounting can become laborious and time-consuming, even for relatively small operations, the use of a computer is recommended, as it has an advantage over manual methods, offering greater economy, speed, and accuracy. In addition, a computer provides a cost system with flexibility and depth that manual systems cannot match. This does not mean that job costs cannot be developed satisfactorily by hand. Many small contractors have completely adequate manual cost systems. Experience indicates, however, that few contractors of substantial size are able to manually generate field cost reports in time to serve a genuine cost-control purpose. Contractors often find that manual methods serve well as generators of estimating information, but not for cost control. With the use of manual methods, a common experience is that the project is finished before the contractor knows the profit status of the work. It is only being realistic to recognize that most contractors find computer support to be a necessary part of their project cost accounting system.

Supporting computer software is now very flexible, in the sense that just about any cost report or information can be generated that project management may desire. Several current computer programs produce cost reports in the same general formats as those presented in this chapter. The programs commonly used by contractors actually perform a whole series of cost accounting and financial accounting functions. After the input of cost and production information, the computer generates payroll checks, keeps payroll records, maintains the equipment accounts, and performs other functions, as well as producing a variety of productivity and cost reports and project cost forecasts. By computerizing job cost accounting, the contractor can obtain accurate and timely project financial information that makes it possible to control costs, manage cash flow, improve cost estimates, and increase profitability.

12.24   EQUIPMENT CHARGES TO PROJECTS

The usual procedure for charging equipment costs to construction projects is discussed earlier in Section 9.13. There are, however, some aspects of equipment charges that require special arrangements. For example, some equipment expenses are not included in the usual hourly rental rates. The costs of move-in, erection, dismantling, and move-out are fixed costs that cannot be incorporated into time rates of expense. Such costs are normally charged to appropriate job overhead accounts and are not included in the hourly or monthly equipment rates.

In regard to equipment charged to the project at an hourly rate, how idle time and repair time are handled is a matter of company policy. Several different procedures are followed. Probably the most common approach is to charge the project at the established rates for the full working day for each piece of equipment on the job. However, credit is given for repair time and for idle time caused by weather and other uncontrollable causes. The significance of this procedure is that the job is charged for all equipment on the site whether it is used or not. This policy can materially assist in controlling underusage of equipment. Where standby equipment units are purposely kept on a job to handle emergencies, the project is usually charged only for the ownership expense involved.

Where the project is charged for extensive periods of equipment idle time, it is probably best to charge the individual cost accounts for net operating hours, plus ordinary or usual idle time. The excessive idle time can be charged to a special overhead account. The reason for this accounting maneuver is that the equipment cost charged to a given work item is used to compute summary information for purposes of cost control and estimating. The use of the special account avoids having excessive idle time distort the reported equipment unit costs.

QUESTIONS

1. In what ways are jobsite costs, labor and equipment hours and production quantities used by a contractor?

2. Why are summary cost reports considered "management-by-exception" tools? Why is this aspect of the report useful?

3. Explain why planning and continuity are vital to a successful cost coding system.

4. What five cost categories are needed for a complete and comprehensive monthly construction cost report?

5. Most contractors only report on materials, subcontracts, and overhead monthly. If more frequent reports are produced by the contractor, why are these items rarely included?

6. How does cost accounting differ from the more usual accounting routines?

7. Equipment costs developed for monthly cost reports generally exclude equipment move-in, erection, dismantling, and move-out. Why are these costs excluded and how are they generally handled?

Chapter **13**

Labor Law

13.1 THE IMPORTANCE OF LABOR LAW

The employment of labor by a construction firm is subject to the provisions of an imposing array of both federal and state statutes. These laws have such an important bearing on the conduct of a contracting business that the contractor must have at least a general grasp of their workings and implications. This chapter discusses the important features of the principal federal statutes that apply to the employment of construction workers. Federal laws are discussed because of their wide applicability and because most state labor statutes are patterned after federal law. Greater emphasis is placed on the broad implications of these laws than on the intricacies of case studies.

For purposes of discussion in this chapter, the statutes pertaining to labor-management relations are considered first (Sections 13.2 through 13.25). Following are the federal laws pertinent to equal employment opportunity (Sections 13.26 through 13.28). The last sections of the chapter discuss labor-standards legislation and other topics.

13.2 HISTORY OF THE LAWS OF LABOR RELATIONS

In the early days of this nation, the right of working people to associate for their mutual aid and protection was severely restricted. Unions were strictly curtailed and were sometimes referred to as "unlawful conspiracies." Up until the 1930s the laws governing labor relations were created principally by the courts. In the almost complete absence of applicable statute law, employer and union complaints were adjudicated primarily in accordance with common law. In general, the courts tended to grant employers relief from unionizing activities, but refused to assist unions against employers because there was no precedent for this in the common law. Court injunctions were widely used to negate the usual union weapons—strikes, picketing, and boycotts. At the same time, there was no comparable judicial instrument available to the unions to assist their organizing efforts.

This state of affairs continued until 1932, although the Sherman Anti-Trust Act of 1890 did provide the beginnings of a statutory basis for labor-management policy. This Act made statutory provisions against the restraint of trade. It was primarily intended to limit the growth of business cartels, and it is debatable whether it was ever intended to apply to labor unions. However, the U.S. Supreme Court ruled in 1908 that the provisions of the Act covered labor organizations. The Sherman Act provided a broad new basis for the use of court injunctions against unions and placed another effective weapon of unions, the boycott, in jeopardy. The Supreme Court ruled that a union could be sued for damages suffered as a result of a boycott and that the union and its members were individually liable. In summary, it can be reported that the law was discriminatory against labor unions until the advent of the New Deal in the early 1930s.

In 1914 a brief but unsuccessful effort was made to alleviate the discriminatory effects of the Sherman Act on organized labor through the passage of the Clayton Act. This Act amended the Sherman Act by declaring, among other things, "The labor of a human being is not a commodity or article of commerce." However, the U.S. Supreme Court later ruled that the Clayton Act applied only to the peaceful activities of unions, and labor disputes were not considered peaceful activities. Thus, the only practical advantage the Clayton Act offered unions was that they were no longer considered "conspiracies" in restraint of trade.

During the intervening years, Congress has passed a series of major federal labor statutes that contain positive and detailed statements of national labor policy. Under current laws the right of workers to form or join unions and to take concerted action to improve their economic condition is guaranteed, and the exercise of that right is protected. National labor policy today is the sum of the policies and provisions contained in the major federal labor relations statutes: the Norris-LaGuardia Act (1932), the National Labor Relations Act (also known as the Wagner Act) (1935), the Labor Management Relations Act (also known as the Taft-Hartley Act) (1947), and the Labor-Management Reporting and Disclosure Act (also known as the Landrum-Griffin Act) (1959). Although there appear to be a great many labor acts, most are technically amendments to the National Labor relations Act.

The following sections discuss these major pieces of federal labor-management legislation. The discussion of these laws concentrates on those provisions of the law that pertain especially to the construction industry.

13.3   THE NORRIS-LAGUARDIA ACT

In 1932, Congress enacted the Norris-LaGuardia Act, which strictly limits the power of the federal courts to issue injunctions against union activities in labor disputes and protects the right of workers to strike and picket peaceably. Also called the Anti-Injunction Act, this statute makes it very difficult for an employer to secure injunctions in a federal court against union activities in labor disputes. Although the Act itself pertains only to federal courts, many states have enacted similar injunction-control legislation.

Although it is difficult for private parties to obtain federal court injunctions against peacefully conducted labor action, injunctions are available to certain government agents under modern labor relations statutes. In this regard, however, injunctions are issued against only those union activities that are in violation of the law or that imperil national health or safety. In addition, the U.S. Supreme Court has decreed that an employer can obtain a federal court injunction against a striking union that is violating the no-strike provision in a dispute subject to arbitration.

The Norris-LaGuardia Act also expressly prohibits "yellow-dog contracts" and makes them unenforceable in federal courts. Designed to discourage union membership, such employment contracts provide that a job applicant will not be hired until he promises not to join a union during his tenure of employment and to renounce any existing membership. Such contracts were widely used by many industrial employers before the passage of the Norris-LaGuardia Act.

13.4   THE NATIONAL LABOR RELATIONS ACT

Congress passed the National Labor Relations Act (NLRA), also known as the Wagner Act, in 1935. Enacted in an atmosphere of depressed business conditions and extensive unemployment, the central purpose of the Wagner Act was to protect union-organizing activity and to foster collective bargaining. Employers were required to bargain in good faith with the legally determined representatives of their workers. Employers were forbidden to practice discrimination against their employees for labor activities or to influence their membership in any labor organization. The Wagner Act defined these and other unfair labor practices as they pertained to employers. However, no such restrictions were applied to employees or unions in their relations with employers. Enforcement of the law was vested in a National Labor Relations Board, which the Wagner Act created. Under the shelter of this piece of legislation, union strength and membership increased enormously between 1937 and 1945. A number of state acts followed, which were more or less patterned after the federal law.

It was almost inevitable that the sudden removal of the traditional restraints would result in union excesses. Starting at about 1938, public opinion concerning organized labor became increasingly antagonistic with the mounting incidence of union restrictive practices, wartime strikes, and criminal activities by some labor leaders. Congressional resentment against organized labor's high-handed actions resulted in the War Labor Disputes Act (Smith-Connally Act) of 1943. However, the provisions of this Act proved to be largely ineffective. If nothing else, however, the Act reflected the mounting popular sentiment for the enactment of positive union-control legislation. During this period several state legislatures passed statutes that regulated and curbed union activities. By 1947, 37 states had passed some form of labor-control legislation.

13.5   THE LABOR MANAGEMENT RELATIONS ACT

In 1947, Congress passed the Labor Management Relations Act, commonly known as the Taft-Hartley Act. This was the first federal statute that imposed comprehensive controls on the activities of organized labor. It amended the earlier National Labor Relations Act (Wagner Act) in several important respects and added new provisions of its own. The National Labor Relations Board was reconstituted, and its authority was redefined. Section 7 of the Taft-Hartley Act established the basic right of every worker to participate in union activities or to refrain from them, subject to authorized Union Security agreements requiring membership in a union as a condition of employment. To protect such rights, Section 8 of the Taft-Hartley Act defined unfair labor practices for both employers and labor organizations. The Act established the Federal Mediation and Conciliation Service, gave the president of the United States certain powers regarding labor disputes imperiling national health or safety, and restricted political contributions by labor organizations and business corporations (see Section 13.25).

In contrast to the Wagner Act, the Taft-Hartley Act was designed to curtail the freedom of action of unions in several different and important ways. Although the Act reiterated a national labor policy of encouraging and assisting collective bargaining, it provided that the public interest must prevail in the conduct of labor affairs. The provisions of the Taft-Hartley Act have had far-reaching effects on the labor-management scene.

Experience with the provisions of the Act quickly revealed several imperfections and shortcomings. Some features of labor employment peculiar to the construction industry proved to be inadequately covered. For example, because of the transience of construction labor and the limited duration of most construction projects, organizing and bargaining for union representation under the Taft-Hartley Act could not reasonably be applied (this is further discussed in Section 13.15). However, the Act's extremely controversial nature and the appreciable strengths of both its backers and opponents made revision of the law a very touchy and difficult matter.

13.6   THE LABOR-MANAGEMENT REPORTING AND DISCLOSURE ACT

In 1959, Congress passed the Labor-Management Reporting and Disclosure Act, also known as the Landrum-Griffin Act. This Act established a code of conduct for unions, union officers, employers, and labor relations consultants. In addition, it guaranteed certain rights to rank-and-file union members and imposed stringent controls on union internal affairs. The principal thrust of this law was to safeguard the rights of the individual union member, to ensure democratic elections in unions, to combat corruption and racketeering in unions, and to protect the public and innocent parties against unscrupulous union tactics. Under the Act, reports pertaining to union organization, finances, activities, and policies are required from unions, union officials, and employees; and of employers, labor relations consultants, and union trusteeships. It was made illegal for an employer to pay or lend money to any labor representative or union of its employees (see Section 13.24).

In addition, the Landrum-Griffin Act amended the National Labor Relations Act and the Taft-Hartley Act. The 1959 law enumerated additional unfair union labor practices and remedied several inadequacies of the Taft-Hartley Act with respect to pressures that unions and their agents can legally apply to employers and their employees. So-called hot-cargo labor agreements (see Section 13.10, item 6) were forbidden, with an exception made for the construction industry. Limitations were applied to organizational picketing of employers by labor organizations. Most of the restrictions on union-security agreements in the construction industry were removed, and union hiring halls (see Section 13.16) were made lawful.

13.7   COVERAGE OF THE NATIONAL LABOR RELATIONS ACT

The National Labor Relations Act, as amended by the Taft-Hartley Act and the Landrum-Griffin Act, plays a dominant role in national labor relations policy. Its declared purpose is to state the recognized rights of employees, employers, and labor unions in their relations with one another and with the public, and to provide the machinery to prevent or remedy any interference by one with the legitimate rights of another. In particular, it protects employees in the free exercise of their right to join or not to join a union, to bargain collectively through representatives of their own choosing, and to act together with other employees for mutual aid and protection. Any violation of these rights, whether by management or by labor representatives, is declared an unfair labor practice.

The National Labor Relations Act applies to employers and employees engaged in interstate commerce or the production of goods for such commerce. Interpretation by the courts as to what constitutes interstate commerce in the construction industry has been so broad that the Act's authority extends to almost all construction work of any consequence. The Act specifically excludes the following employers and employees from its coverage:

1. Exempted Employers

    a. The United States government

    b. State governments and their political subdivisions

    c. Wholly owned government corporations

    d. Federal reserve banks

    e. Employers subject to the Railway Labor Act

    f. Labor organizations (when not acting as employers)

    g. Officers or agents of labor organizations

2. Exempted Employees

    a. Employees of exempted employers as listed in item 1

    b. Agricultural laborers

    c. Domestic servants

    d. Individuals employed by their parents or spouses

    e. Independent contractors

    f. Supervisors

Certain selected portions of the National Labor Relations Act, as amended, have been selected for discussion. These are presented in Sections 13.8 through 13.25.

13.8   THE NATIONAL LABOR RELATIONS BOARD

Administration of the National Labor Relations Act is the responsibility of the National Labor Relations Board (NLRB), which is composed of five members, and the general counsel of the board. The president of the United States, with the consent of the Senate, appoints members of the NLRB for terms of five years. The president, with the consent of the Senate, appoints the general counsel for a term of four years. The NLRB has two primary functions: (1) to establish, usually by secret-ballot elections, whether groups of employees wish to be represented by designated labor organizations for collective bargaining purposes and (2) to prevent and remedy unfair labor practices.

Much of the day-to-day work of investigating and processing charges of unfair labor practices and handling representation proceedings has been delegated by the board to the various NLRB regional offices located in major cities throughout the nation. The board has given its regional directors final authority in election cases, subject to limited review. In unfair labor practice cases, the board acts much like an appellate court to determine whether an unfair labor practice actually exists and how such practices should be remedied. The board does not ordinarily become involved until an investigation has been conducted and recommendations have been made by a regional office. The general counsel has largely independent authority in the prosecution of unfair labor practices and determines which cases are put before the board. Responsible for general supervision over the regional NLRB offices and for most of the administrative routine of the agency, the general counsel has broad and direct authority to seek injunctions against unfair labor practices.

By statute, the NLRB exercises its powers over all enterprises whose operations affect interstate commerce. It does not act, however, on every case over which it could exercise jurisdiction. Rather, the board restricts its attention to a caseload it can handle expeditiously and within its budgetary limitations. The Landrum-Griffin Act authorized the NLRB to limit its cases to those whose effect on commerce is, in the board's opinion, substantial. As a guide to when it will exercise its power, the board has established minimum measures of the annual volume of business that must be involved before the NLRB will accept the case. These standards are expressed in terms of the gross dollar volume of the employer's sales and purchases that cross state lines, and they vary for different areas of industry. The Landrum-Griffin Act further provides that state and territorial courts and agencies can assume jurisdiction over labor disputes the NLRB declines to hear.

13.9   REPRESENTATION ELECTIONS

Section 9(a) of the National Labor Relations Act requires that an employer bargain with the representative selected by a majority of its employees, but does not stipulate a selection procedure. The only requirement is that the representative clearly be the choice of the majority. The representative may be an individual or a labor union but cannot be a supervisor or other representative of the employer.

For the employees to select a majority representative, it is usual for the nearest regional office of the NLRB to conduct representation elections. However, such an election can be held only when a petition has been filed by the employees, by an individual or a labor organization acting on their behalf, or by an employer who has been confronted with a claim of representation from an individual or labor organization. The present rule is that a union may secure an election to determine the wishes of the employees if it can show that at least 30 percent of the eligible employees have indicated they desire such representation. The board, through its regional office, supervises every step in the election procedure.

In a representation election, the employees are given a choice of one or more bargaining representatives or no representative at all. To be chosen, a labor organization must receive a majority of the valid votes cast. The NLRB will certify the choice of the majority of employees for a bargaining representative only after a secret-ballot election.

Section 7 of the Taft-Hartley Act has a free speech provision that establishes the employee's right to hear the arguments of both labor and management. The expressing or disseminating of any views, arguments, or opinions by either side does not constitute an unfair labor practice as long as it contains no threat of reprisal or force or promise of benefit. Within these limitations, an employer that wants to stay nonunion can state its opinions to its employees.

Because the election process is lengthy and the duration of most construction projects does not allow adequate time to conduct an election in accordance with Section 9(a) of the National Labor Relations Act, this provision is rarely used in the construction industry. Instead, prehire agreements provide the more common mechanism for the election of union representation. This concept is discussed further in Section 13.15.

13.10   EMPLOYER UNFAIR LABOR PRACTICES

Under the National Labor Relations Act, as amended, an employer commits an unfair labor practice if it:

1. Interferes with, restrains, or coerces employees in the exercise of rights protected by the Act, such as their right of self-organization for the purpose of collective bargaining or other mutual assistance [Section 8(a)(1)].

2. Dominates or interferes with any labor organization in either its formation or its administration, or contributes financial or other support to it [Section 8(a)(2)]. Thus, "company" unions that are dominated by the employer are prohibited, and employers may not unlawfully assist any union financially or otherwise.

3. Discriminates against an employee in order to encourage or discourage union membership [Section 8(a)(3)]. It is illegal for an employer to discharge or demote an employee or to single him out in any other discriminatory manner simply because he is or is not a member of a union. In this regard, however, it is not unlawful for employers and unions to enter into compulsory union-membership agreements permitted by the National Labor Relations Act. This is subject to applicable state laws prohibiting compulsory unionism.

4. Discharges or otherwise discriminates against an employee because he has filed charges or given testimony under the Act [Section 8(a)(4)]. This provision protects the employee from retaliation if he seeks help in enforcing his rights under the Act.

5. Refuses to bargain in good faith about wages, hours, and other conditions of employment with the properly chosen representative of its employees [Section 8(a)(5)]. Matters concerning rates of pay, wages, hours, and other conditions of employment are called mandatory subjects, about which the employer and the union must bargain in good faith, although the law does not require either party to agree to a proposal or to make concessions.

6. Enters into a hot-cargo agreement with a union [Section 8(e)]. Under a hot-cargo agreement, the employer promises not to do business with, or not to handle, use, transport, sell, or otherwise deal in the products of, another person or employer. This unfair labor practice can be committed only by an employer and a labor organization acting together. A limited exception to this ban on hot-cargo clauses is made for the garment industry and the construction industry.

13.11   UNION UNFAIR LABOR PRACTICES

Under the National Labor Relations Act, as amended, it is an unfair labor practice for a labor organization or its agents:

1. **a.** To restrain or coerce employees in the exercise of their rights guaranteed in Section 7 of the Taft-Hartley Act [Section 8(b)(1)(A)]. In essence, Section 7 gives an employee the right to join a union, to assist in the promotion of a labor organization, or to refrain from such activities. This section further provides that it is not intended to impair the right of a union to prescribe its own rules concerning membership.

1. **b.** To restrain or coerce an employer in its selection of a representative for collective bargaining purposes [Section 8(b)(1)(B)].

2. To cause an employer to discriminate against an employee with regard to wages, hours, or other conditions of employment for the purpose of encouraging or discouraging membership in a labor organization [Section 8(b)(2)]. This section includes employer discrimination against an employee whose membership in the union has been denied or terminated for cause other than failure to pay customary dues or initiation fees. Contracts or informal arrangements with a union under which an employer gives preferential treatment to union members are violations of this section. It is not unlawful, however, for an employer and a union to enter into an agreement whereby the employer will hire new employees exclusively through a union hiring hall as long as there is no discrimination against nonunion members. This section also permits union-security agreements that require employees to become members of the union after they are hired.

3. To refuse to bargain in good faith with an employer about wages, hours, and other conditions of employment if the union is the representative of its employees [Section 8(b)(3)]. This section imposes on labor organizations the same duty to bargain in good faith that is imposed on employers.

4. To engage in, or to induce or encourage others to engage in, strike or boycott activities, or to threaten or coerce any person, if in either case an object thereof is:[1]

   a. To force or require any employer or self-employed person to join any labor or employer organization, or to enter a hot-cargo agreement that is prohibited by Section 8(e) [Section 8(b)(4)(A)].

   b. To force or require any person to cease using or dealing in the products of any other producer or to cease doing business with any other person [Section 8(b)(4)(B)]. This is a prohibition against secondary boycotts, a subject discussed further in Section 13.17 of this chapter. This section of the National Labor Relations Act further provides that, when not otherwise unlawful, a primary strike or primary picketing is a permissible union activity.

   c. To force or require any employer to recognize or bargain with a particular labor organization, as the representative of its employees, that has not been certified as the representative of such employees [Section 8(b)(4)(C)].

   d. To force or require any employer to assign certain work to the employees of a particular labor organization or craft rather than to employees in another labor organization or craft, unless the employer is failing to conform with an order or certification of the NLRB I Section 8(b)(4)(D)]. This provision is directed against jurisdictional disputes, a topic discussed in Section 13.21 of this chapter.

5. To require of employees covered by a valid union shop, membership fees that the NLRB finds to be excessive or discriminatory [Section 8(b)(5)].

---

[1] Section 8(b)(4) permits "publicity," other than picketing, provided such action does not have the effect of inducing a work stoppage by neutral employees. Such publicity can notify the public, including the customers of a neutral employer, that a labor dispute exists concerning a primary employer's products being distributed by the neutral.

6. To cause or attempt to cause an employer to pay or agree to pay for services that are not performed or not to be performed [Section 8(b)(6)]. This section forbids practices commonly known as featherbedding.

7. To picket or threaten to picket any employer to force it to recognize or bargain with a union:

    a. When the employees of the employer are already lawfully represented by another union [Section 8(b)(7)(A)].

    b. When a valid election has been held within the past 12 months [Section 8(b)(7)(B)].

    c. When no petition for an NLRB election has been filed within a reasonable period of time, not to exceed 30 days from the commencement of such picketing [Section 8(b)(7)(C)].[2]

The National Labor Relations Board has ruled that discrimination by a labor union because of race is an unfair practice under the Taft-Hartley Act. Discrimination based on race in determining eligibility for full and equal union membership and segregation based on race has been found unlawful by the NLRB. This has made possible the filing of unfair labor practice charges against a union because of alleged racial discrimination. A union found guilty of such practices faces cease-and-desist orders, as well as possible rescission of its right to continue as the authorized employee representative.

13.12 CHARGES OF UNFAIR LABOR PRACTICES

A contractor, a union, or an individual worker can file charges of unfair labor practices. These charges must usually be filed with the NLRB regional office that serves the area in which the case arose within six months from the date of the alleged unfair activity. After charges are filed, field examiners investigate the circumstances, and a formal complaint is issued if the charges are found to be well grounded and the case cannot be settled by informal adjustment.

When a complaint is issued, a public hearing is held before a trial examiner whose findings and recommendations are served on the parties and are sent to the NLRB in Washington, D.C. If no exceptions are filed by either party within a statutory period, the examiner's judgment takes the full effect of an order by the NLRB. If exceptions are taken, the NLRB reviews the case and makes a decision. This process is illustrated in Figure 13.1.

If a contractor or a union fails to comply with an order of the NLRB, the board has no statutory power of enforcement of its own but can petition the appropriate United States Court of Appeals for a decree enforcing the order. If the court issues such a decree, failure to comply may be punishable by fine or imprisonment for contempt of court. Parties aggrieved by the order may seek judicial review.

---

[2] Subparagraph (c) does not apply to picketing or other publicity for the purpose of truthfully advising the public that an employer does not have a union contract or employ union labor, unless it has the effect of inducing employees of persons doing business with the picketed employer not to pick up, deliver, or transport goods or not to perform services. If the purpose is purely informational, it can be continued indefinitely. If the purpose is organizational, it cannot be continued more than 30 days without a petition for an election.

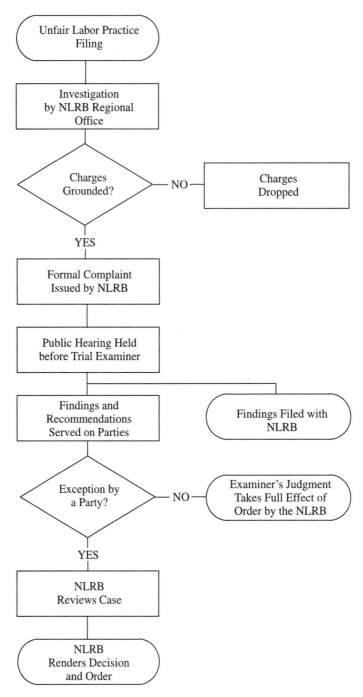

Figure 13.1 Process for filing unfair labor practice charge.

13.13   REMEDIES

When the NLRB finds that a contractor or a union has engaged in an unfair labor practice, it is empowered to issue a cease-and-desist order and to take such affirmative action as is deemed necessary to remove the effects of the unfair practice found to have been committed. The purpose of the board's orders is remedial, and it has broad discretion in fashioning remedies for unfair labor practices. Typical affirmative actions ordered by the NLRB include reinstatement of persons discharged, reimbursement of wages lost, or refund of dues or fees illegally collected.

The law provides that whenever a charge is filed alleging certain unfair labor practices relating to secondary boycotts, hot-cargo clauses, or organization or recognition picketing, the preliminary investigation of the charge must be given first priority. The board or the general counsel is authorized to petition the appropriate federal district court for an injunction to stop any conduct alleged to constitute an unfair labor practice. If the preliminary investigation of a first-priority case reveals reasonable cause to believe the charge is true, the law requires that the general counsel seek such injunctive relief or temporary restraining order as seems appropriate under the circumstances.

In addition to filing charges of unfair labor practices when faced with illegal union activity, the contractor has access to other remedies. In the case of a union's illegal use of various economic weapons such as secondary boycotts, unlawful picketing, and illegal strikes, there are corrective actions available to the contractor. A court injunction directing that the activity cease is probably the most powerful weapon for prompt resolution of the matter. This method also prevents serious economic damage. An action for damages is usually possible after the illegal conduct has caused harm to the contractor. Discipline or discharge of the striking employees and arbitration are also possible remedies.

13.14   UNION-SHOP AGREEMENTS

The National Labor Relations Act outlaws the closed shop but permits the establishment of a union shop. A closed shop requires that a worker be a member of the appropriate union at the time he is hired. In the case of a union shop, a new employee need not be a union member at the time of employment, but must join within a stipulated period to retain his job. Therefore, a union security agreement (an agreement providing for compulsory union membership) cannot require that, to be hired, applicants for employment be members of the union, but can stipulate that all employees covered by the agreement must become members of the union within a certain period of time. This grace period cannot be fewer than 30 days after hiring, except in the building and construction industry, in which a shorter grace period of 7 days is permissible.

Union-shop agreements often provide for the checkoff of union dues, an arrangement whereby the employers deduct dues from their employees' wages and pay the withheld money to the union. Under the NLRA, such checkoff is permitted only by written assignment of each employee, and such assignment is not to remain in effect for a period of more than one year or beyond the end of the current collective agreement, whichever occurs first. A mandatory checkoff is illegal and is an unfair labor practice on the part of both the employer and the union.

Section 14(b) of the Taft-Hartley Act provides that the individual states have the right to forbid negotiated labor agreements that require union membership as a condition of employment. In other words, any state or territory of the United States may, if it chooses, pass a law making a union-shop labor agreement illegal. This is called the "right-to-work" section of the Act, and such state laws are termed right-to-work statutes. At present, 22 states have such laws in force.[3] It is interesting to note that most of these state right-to-work laws go beyond the mere issue of compulsory unionism inherent in the union shop. Most of them outlaw the agency shop under which workers, in lieu of joining a union, must pay as a condition of continued employment, the same initiation fees, dues, and assessments as union members. Some of these laws explicitly forbid unions to strike over the issue of employment of nonunion workers.

Basically, state right-to-work laws prohibit discrimination in employment on the basis of membership or nonmembership in a labor union. It is interesting to note that in some right-to-work states, their statutes have been interpreted to protect subcontractors as well as workers. In these states, the courts have ruled that a general contractor cannot discriminate against a subcontractor based on whether the subcontractor is union or nonunion.

13.15   PREHIRE AGREEMENTS

Section 8(f) of the National Labor Relations Act allows an employer engaged primarily in the building and construction industry to sign a labor agreement with a union prior to the hiring of any workers or before the union can show that it represents a majority of the employees involved. Such labor contracts are called prehire or Section 8(f) agreements and may apply to only one specific project or to a designated geographical area. Such prehire arrangements are permitted only in the construction industry, which is unique because of the transience of its workers and the relatively short durations of its projects. Contractors enter such agreements primarily as a means of access to established labor costs, stable labor relations, and a ready-made source of skilled manpower. By its 1987 rulings in the *Deklewa* case, the National Labor Relations Board established the rules now governing construction industry prehire agreements. The general workings of these rules are discussed in the following paragraphs.

During the life of a prehire agreement between a union and a contractor, the union bargaining unit's employees may hold an NLRB representation election. Should the majority of employees vote in favor of union representation, the prehire agreement currently in force does not automatically convert to a conventional labor contract. Rather, the contractor must first formally recognize the union. In addition, the employees, a rival union, or the contractor can petition for a union decertification election at any time during the life of the 8(f) arrangement and the prehire agreement cannot bar such an election.

If a union receives formal recognition from the contractor after establishing majority support, the union then becomes the exclusive bargaining agent and the obligation of the contractor to bargain in good faith for a new labor contract will apply. If the union loses the election, the prehire agreement is terminated and the parties are prohibited from entering into another prehire pact for a period of one year.

---

[3] Alabama, Arizona, Arkansas, Florida, Georgia, Idaho, Iowa, Kansas, Louisiana, Mississippi, Nebraska, Nevada, North Carolina, North Dakota, South Carolina, South Dakota, Tennessee, Texas, Utah, Virginia, Wyoming, and Oklahoma now have right-to-work legislation in effect.

When an employer has entered into a prehire contract with a union and there is no representation election, the contractor cannot repudiate the agreement during its tenure. The contractor is bound by the terms of the contract for all work done on any job site within the area covered unless the agreement is limited to a particular job site. Under NLRB rules, parties to prehire agreements are bound by them unless the workers vote to decertify or change the union in an NLRB election. During its term, a prehire pact is just as enforceable as a labor contract negotiated with the union involved. However, when the 8(f) arrangement expires, there is no presumption that the union represents a majority of the workforce covered and either party can end the relationship at that time. The employer must be left free from coercive union efforts to compel the negotiation or adoption of a successor agreement. By removing the presumption of the union's majority status when the prehire arrangement expires, neither employer nor employees are locked into a union relationship and the contractor can walk away without bargaining.

13.16 UNION HIRING HALLS

The NLRB provides that, in the construction industry, a labor agreement can require the contractor to acquire its workers only through a designated local union. Such an arrangement, referred to as a "union hiring hall," requires the contractor to notify the union of employment opportunities and to give the union an opportunity to refer qualified applicants. The agreement may specify minimum training or experience qualifications for employment or provide for priority in job referrals based on length of service with the employer, in the industry, or in the particular geographical area. However, hiring-hall agreements that give priority to employees who previously worked for employers subject to collective bargaining agreements with the union have been found illegal under the National Labor Relations Act.

Contracts or informal arrangements with a union under which an employer gives preferential treatment to union members are illegal. It is not unlawful, however, for an employer and a union to enter into an agreement whereby the employer will hire new employees exclusively through a union hiring hall, as long as there is no discrimination against nonunion members in favor of union members. Both the agreement and the actual operation of the hiring hall must be nondiscriminatory; job referrals must be made without reference to race, color, religion, sex, national origin, or union membership. The employer must not discriminate against a nonunion employee if union membership is not available to that employee under the usual terms or if membership is denied to that employee for a reason other than nonpayment of union dues and fees. Hiring-hall provisions in labor contracts usually give the contractor the right to reject any applicant and the right to obtain employees from other sources when the union is unable to supply a sufficient number of qualified people.

The courts have held that a lawful hiring hall is a mandatory subject of bargaining between employers and unions and that a union can strike and picket to press its demands for an exclusive nondiscriminatory hiring-hall referral system. Once obtained in a collective bargaining agreement, a hiring-hall arrangement is enforceable in the federal courts or before the NLRB. Contractor responsibility for discriminatory hiring-hall practices was a thorny issue for many years. However, by its ruling in 1982, the U.S. Supreme Court judged that contractors that are not guilty of intentional discrimination are not liable if the union-run hiring halls they use are operated on a biased basis. The courts have also ruled that union hiring halls are permissible in states with right-to-work laws as long as the hiring-hall agreement expressly states that union membership is not to be considered in job referrals.

Construction hiring halls or referral systems that discriminate against minorities are illegal under civil rights statutes, even though they may appear to be legal under the National Labor Relations Act. The matter of contractor responsibility has become particularly troublesome in regard to the hiring of workers from minority groups. In this regard, the NLRB has held that hiring is a management responsibility and cannot be delegated to a union. The contractor, not the union, must be the judge of a worker's competence, and a worker's access to a construction job cannot be conditioned on his ability to pass a union examination. If a contractor hires or retains a worker the union will not accept, the union is liable for the consequences if it strikes to force that worker off the job.

13.17   SECONDARY BOYCOTTS

A primary boycott arises when a union, engaged in a dispute with an employer, exhorts that firm's customers and the public to refrain from all dealings with that employer. A secondary boycott occurs if a union has a dispute with Company A and attempts to exert pressure on it by causing the employees of Company B to stop handling or using the products of Company A or otherwise forces Company B to stop doing business with Company A. The primary employer, in this case Company A, is the employer with whom the union has the dispute. Company B is the neutral secondary employer; hence the name "secondary boycott."

Secondary boycotts have a long and turbulent legal history. Illegal at common law, secondary boycotts were ruled to have been forbidden by the Sherman Act in a decision by the U.S. Supreme Court. The Norris-LaGuardia Act and the Wagner Act, which together gave unions almost complete immunity from liability for damages arising out of secondary boycotts, reversed this prohibition. The pendulum has since returned almost to its original position, with the National Labor Relations Act forbidding secondary boycotts.

Secondary boycotts in construction can assume many forms, and the dividing line between a legal primary boycott and an illegal secondary boycott is sometimes hazy and difficult to establish. The wording of the law is strictly construed in determining the legality of a boycott action. A form of secondary boycott that is of extraordinary importance to the construction industry is common situs picketing.

13.18   COMMON SITUS PICKETING

A common situs is a given location, such as an industrial plant or a construction project, at which several different employers are simultaneously engaged in their individual business activities. A dispute between one of these employers and a union is likely to involve the other, neutral employers, especially if picketing is involved. Decisions of the NLRB and of the courts have evolved some rules for establishing whether such common situs picketing constitutes an illegal secondary boycott action.

In an attempt to give effect to both the union's right to picket the primary employer and the right of the secondary employer to be free from disputes that are not its own, the NLRB established in 1950 the Moore Dry Dock tests, which determine when a union may picket a common site without committing an illegal secondary boycott. Its rules are as follows:

1. That the picketing be limited to times when the employees of the primary employer are working on the premises

2. That the picketing be limited to times when the primary employer is carrying on its normal business there

3. That the picket signs clearly indicate the identity of the primary employer with whom the union is having the dispute

4. That the picketing be carried on reasonably close to where the employees of the primary employer are working

Common situs picketing occurs frequently in the construction industry, often as the result of a union contractor awarding work to a nonunion subcontractor. The resulting picketing causes the employees of the union contractors on the project to refuse to cross the picket line. In 1951, the U.S. Supreme Court decided the *Denver Building and Construction Trades Council* case. This dispute involved a general contractor whose employees were union members and a nonunion subcontractor. The project was picketed and shut down by the construction unions, which demanded that the subcontractor be discharged. The U.S. Supreme Court found that the unions were guilty of an illegal secondary boycott because the object of the picketing was to force the general contractor (secondary employer) to quit doing business with the nonunion subcontractor (primary employer).

Subsequent to the *Denver Building Trades* case, the "separate gate" doctrine was developed. On multiemployer construction sites one gate is reserved and marked for the primary contractor involved in the labor dispute and another gate for the neutral contractors not involved. On the basis of the Moore Dry Dock standards traditionally applied to common situs picketing and the *Denver Building Trades* case, the NLRB and the courts now hold that picketing of the gate reserved for the contractor directly involved in the dispute is permissible, but that the unions cannot picket separate gates used by employees of neutral secondary contractors if the effect is to keep them off the project. In essence, this decision says that construction contractors at the site are separate and distinct employers and union picketing must be limited to the primary contractor involved in the dispute. However, a recent court decision ruled that common situs picketing is permissible under certain circumstances. The court decreed that a union can engage in common situs picketing when the gate reserved for the open-shop employer is so located that the union cannot communicate its picketing message to the public. Recent years have seen a sustained attempt by organized labor to prevail upon Congress to amend the National Labor Relations Act to permit unrestricted picketing of construction sites. To date, however, these efforts have failed.

The U.S. Supreme Court, in the *General Electric* case (1961), decided that the matter of situs picketing is different, however, when a construction contractor is doing construction work at an industrial plant. The Court ruled that picketing by plant strikers of gates reserved exclusively for contractor personnel can be banned only if there is a separate, marked gate set apart for the contractor, if the work being done by the contractor is unrelated to the normal operations of the industrial company, and if the work is of a kind that will not curtail normal plant operations. The courts have ruled that the *General Electric* decision does not apply to a prime contractor nor to subcontractors at the usual construction site. Although the past 15 years has seen numerous court decisions further defining common situs picketing and secondary boycott practices by unions, the laws established by the NLRB have consistently been upheld by the courts.

13.19 SUBCONTRACTOR AGREEMENTS

The Landrum-Griffin Act made it an unfair labor practice for an employer and a union to enter into an agreement whereby the employer refrains from handling the products of another employer or ceases doing business with any other person. As pointed out earlier, such a contract provision is called a hot-cargo clause. However, the construction industry was exempted from the ban under certain circumstances. Under the construction industry proviso to Section 8(e) of the National Labor Relations Act, contractors can agree to restrictions on subcontracting or can agree not to handle certain products as long as the restrictions relate to the contracting or subcontracting of work to be done at the site. This construction industry exemption has led to the widespread use of two forms of hot-cargo clauses in construction labor contracts. One of these is the subcontractor agreement, a subject discussed the next paragraph. The other is the prefabrication clause, which is discussed in the following section.

Subcontractor agreements typically require the general contractor to award work only to those subcontractors who are signatory to a specific union labor contract or who are under agreement with the appropriate union. Such agreements do not extend to supplies or other products produced or manufactured elsewhere and delivered to the construction site. The NLRB and the courts have ruled that construction unions may strike to obtain subcontractor clauses in their labor contracts if no secondary boycott is involved. For example, picketing to induce a general contractor to accept a subcontractor clause is legal, but picketing is illegal as a secondary boycott if it is designed to force a neutral general contractor to stop doing business with an existing and identified nonunion subcontractor. The courts have held that unions cannot enforce subcontractor clauses through threats, coercion, strikes, or picketing, but that violations of such labor contract provisions can be submitted to arbitration or a civil action can be taken under Section 301 of the Taft-Hartley Act. Self-enforcing clauses have been ruled illegal and unenforceable. These are clauses in which the contracting firm agrees that if it violates the subcontract agreement, the union can take action against it such as picketing, refusing to provide workers, or canceling the labor contract between them.

For many years there has been considerable uncertainty as to just what constitutes a legal subcontracting clause. As a result of U.S. Supreme Court decisions in the *Connell* case (1975) and the *Woelke & Romero* case (1982), the following guidelines have emerged suggesting that subcontractor clauses are not in violation of federal labor or antitrust statutes. These guidelines state that (1) the agreement with a union containing subcontracting restrictions must pertain only to work performed at a construction site; there is no requirement, however, that the pact be limited to the subcontracting of work to be performed at a particular project, (2) the workers represented by the union that makes the agreement restricting the contractor's right to subcontract must have an employer-employee relationship with that contractor, and (3) the subcontracting restrictions must be provided for in a collective bargaining agreement. The courts have determined that two types of subcontractor clauses are legal: a "signatory" clause, requiring the subcontractor to be signed to the same agreement as the general contractor, and a "terms and conditions" clause, whereby the subcontractor agrees to be bound by the same terms and conditions as the general contractor, but does not necessarily sign an individual agreement.

13.20   PREFABRICATION CLAUSES

The prefabrication clause is the second form of hot-cargo provision ruled to be permissible under the construction industry exemption proviso of the National Labor Relations Act. In accordance with a U.S. Supreme Court decision made in 1967 concerning the installation of precut doors in Philadelphia, construction unions may legally obtain labor agreements to bar the use of prefabricated products in construction in order to preserve their customary on-site work and can enforce such clauses by strikes and picketing. These provisions ban the use of prefabricated construction products manufactured off the site, products that eliminate work normally done on the project itself. Examples of these construction products are precut and prefitted wooden doors, precut pipe insulation, prefabricated trusses, and prepackaged boilers. Such product boycotts have been construed to fall within the construction industry exemption from the ban on hot-cargo clauses and can be legal provisions in a labor agreement if the prefabricated products replace work customarily and traditionally performed on the site by union members.

The NLRB applies the "right-to-control test" as one factor in determining the legality of prefabrication clauses. Under this test, if a prefabricated product is specified by the architect-engineer and/or owner, then the contractor is required by the terms of the construction contract to provide the materials as specified and has no control over product selection. In such a case, the union cannot refuse to handle and install the product, whether or not a prefabrication clause exists. However, if the construction contract does not specify the use of a prefabricated product, but the contractor, on its own volition, decides to use such materials, then the union is entitled to enforce the clause and to strike and picket the offending contractor to block the use of the prefabricated materials. Application of the right-to-control test to determine when a prefabrication dispute becomes an illegal secondary boycott was upheld by the U.S. Supreme Court in the *Enterprise* case (1977).

13.21   JURISDICTIONAL DISPUTES

A jurisdictional dispute can arise when more than one union claims jurisdiction over a given item of work on a job site. The dispute is between unions, and the prime contractor or subcontractor responsible for the work is caught in the middle. The unionized segment of the construction industry has many disputes of this kind, because each of the many craft unions regards its type of work as a proprietary right and jealously guards against any encroachment on its traditional sphere by other unions. Craft jurisdiction is of great economic and personal importance to the unionized worker and has been a continual source of strife between construction unions.

Lines of demarcation between the various jurisdictions are sometimes indistinct, and the development of new products and methods often brings with it jurisdictional clashes between unions whose members claim an exclusive right to the work assignment.

Realizing that jurisdictional conflict is an ever-present possibility on union projects, the prudent project manager may take any of a number of commonsense precautions to head off such disputes before they occur. One such action is to hold a meeting, sometimes referred to as a prejob conference, of the craft foremen before work starts in the field. At this meeting, items of work are discussed and work assignments are made by the prime contractor. This allows potential disputes to surface and gives the conflicting unions time to resolve their differences. However, if agreement between the unions is not forthcoming, the contractor makes the final assignment. The craft stewards can informally settle many jurisdictional differences that arise during the construction period. Despite these precautions, however, jurisdictional disputes do occur and the contractor is faced with the need to get such matters resolved.

When disputed work is at issue, the contractor has the authority and responsibility to assign the work to one of the unions involved. Although strikes, picketing, and other coercive action by a union to gain a work assignment is an unfair labor practice [Section 8(b)(4)(D)] and most labor contracts contain provisions forbidding such union actions, the union not receiving the assignment may well resort to a slowdown, walkoff, or other disruptive actions. In any event, the contractor now has a jurisdictional dispute on its hands.

If informal efforts to avert a jurisdictional dispute are to no avail, the contractor responsible for the work must now make an assignment to one of the disputing unions. Although contractors' decision criteria tend to vary, time efficiency and economy of operation are normally important aspects of the choice. If the jurisdictional dispute persists, the contractor can take the matter to the National Labor Relations Board or utilize a voluntary plan for the resolution of such matters.

13.22   NLRB JURISDICTIONAL SETTLEMENT

Section 10(k) of the Taft-Hartley Act provides that the NLRB will hear and determine a jurisdictional dispute if the parties thereto have not agreed to a voluntary procedure. After an employer has made the work assignment and the offended union begins or threatens picketing or other coercive action, the contractor can initiate a proceedings with the NLRB. To do this, the contractor must file an unfair labor practice charge with the regional NLRB office. The regional office conducts a hearing on the matter, and the results are sent to the NLRB in Washington, D.C., for its ruling. In case of union noncompliance with this decision, the board can petition the appropriate U.S. Circuit Court of Appeals to enforce its order. The NLRB can obtain an injunction to halt any project picketing, strikes, or other union activities associated with the dispute.

Many contractors utilize the NLRB to resolve their jurisdictional disputes. Work awards made by the NLRB are limited in scope to the disputes from which they arise. The Taft-Hartley Act gives private parties damaged by a jurisdictional strike, picketing, or other coercion the right to sue the union or unions involved. The NLRB has consistently given priority to economy and efficiency of operation when making work awards. Because the employer's assignment is based on these same considerations, NLRB awards uphold the contractor's original assignment in more than 90 percent of the cases. Unions do not favor the use of 10(k) proceedings for the aforementioned reasons.

In the *Texas Tile* case (1971), the U.S. Supreme Court held that the contractor is a party to a jurisdictional dispute as well as the rival unions. The effect of this ruling is that any private arrangement between unions providing for the settlement of their jurisdictional differences must include the employer if the employer is to be bound by it. Otherwise, such an agreement between the unions does not prevent a contractor from referring the matter to the NLRB and will not bar the NLRB from hearing the dispute and assigning the contested work.

13.23   VOLUNTARY JURISDICTIONAL SETTLEMENT PLANS

The Taft-Hartley Act does not require the NLRB to rule on jurisdictional matters when the disputants have agreed to voluntary methods of settlement. Accordingly, the construction industry has set up a number of voluntary plans, one national in scope and many strictly local in coverage.

The present Plan for the Settlement of Jurisdictional Disputes in the Construction Industry is a national procedure for the resolution of disputes that are referred to it. The AFL-CIO Building and Construction Trades Department and several contractor associations established this joint labor-management mechanism in 1984. The five contractor associations that participate in this plan are the National Erectors Association, the Northern American Contractors Association, the Sheet Metal and Air Conditioning Contractors Association, the National Electrical Contractors Association, and the Mechanical Contractors Association. This limited participation in the national plan represents only a relatively small segment of the total construction industry and does not include large industry groups such as the Associated General Contractors of America. Under this plan, a Joint Administrative Committee oversees administration of the plan and produces a list of arbitrators from which one is selected to decide a given jurisdictional dispute. Rulings of arbitrators under the plan are enforceable in federal court. Other regional and state plans have a broader makeup and are sometimes elected by participants because the national plan has not worked well for them.

The criteria used by the arbitrator in reaching a decision are prescribed by the settlement plan. The arbitrator must first determine whether there has been a previous decision or national agreement between the unions concerning the matter at issue. This involves checking jurisdictional agreements and decisions of record such as those contained in the "Green Book" and the "Gray Book."[4] If there is no such established precedent, the arbitrator must consider relevant agreements between the crafts. If no such agreement exists, consideration must be given to established and prevailing trade practice in the locality. In so doing, the arbitrator cannot ignore the interests of the consumer or the past practices of the contractor.

Although the national plan procedure can be faster than that of the NLRB, many contractors are not in accord with its decision criteria. The national plan follows a procedure, favored by the construction unions, that relies on past practice. Contractors want more emphasis placed on efficiency and good business precepts. To date, the national plan has not received wide acceptance from contractor ranks.

In addition to the national plan just discussed, many local areas have set up machinery for the voluntary settlement of jurisdictional disputes. Several large metropolitan areas maintain their own boards for settling jurisdictional strikes. Local plans and procedures of a wide variety are used throughout the country. Without doubt, a large proportion of the jurisdictional disputes in the construction industry are settled at the local level.

---

[4] The "Green Book," approved by the Building and Construction Trades Department, AFL-CIO, and the "Gray Book," published by the Associated General Contractors of America, contain permanent jurisdictional agreements reached between unions that are pertinent to the construction industry.

13.24   PAYMENTS TO EMPLOYEE REPRESENTATIVES

Section 302 of the Taft-Hartley Act prohibits any employer or association of employers from paying, lending, or delivering money or other things of value to its employees or their representatives if the purpose of the payment is to influence the right of employees to organize and bargain collectively. This includes labor unions or officers thereof and any employee or group of employees. Specifically permitted by this section, however, are various fringe benefits such as health, welfare, pension, vacation, holiday, and annuity payments, as well as apprenticeship plans and prepaid legal services for which employer contributions are permissible and over which unions are given some control. Such contributions must be paid to a trust fund, and the trustees must consist of an equal number of labor and management representatives.

13.25   POLITICAL CONTRIBUTIONS

The Taft-Hartley Act makes it unlawful for certain organizations, including business corporations, labor organizations, and trade associations, to make a contribution or expenditure in connection with the election of federal officials. A labor organization is defined as any organization in which employees participate and which exists for the purpose of dealing with employers concerning grievances, labor disputes, wages, rates of pay, hours of employment, or conditions of work.

Despite these restrictions, however, business corporations, labor unions, and trade associations are very active on the American political scene through the medium of political action committees (PACs). Under federal campaign laws and Federal Election Commission regulations, corporations, labor unions, and trade associations are permitted to establish PACs at the national, state, and local levels. Within limitations established by law, these organizations are authorized to solicit and receive personal contributions from individuals and disburse these to selected candidates for political office. As of this time, an individual may give a maximum of $2,000 per election to a federal candidate or the candidate's campaign committee. The individual can give up to $5,000 per calendar year to a PAC, and $10,000 to a state party committee that supports federal candidates. An individual can give a maximum of $25,000 per calendar year to a national party committee. The biennial limit placed on an individual's total political contributions is $95,000 ($37,000 to all candidates and $57,500 to all PACs and parties). PACs established by corporate businesses and trade associations use contributed funds to help elect candidates philosophically in tune with management viewpoints. However, PACs associated with labor unions contribute large sums toward the election of candidates who support the causes of organized labor. PACs now account for a large percentage of the campaign funds raised by public office seekers, with business PACs providing considerably more money than labor PACs.

In addition, labor unions maintain local "education" funds through the AFL-CIO Committee on Political Education (COPE). This falls under a federal law that allows labor unions to spend unlimited amounts of their own monies on communications to members and their families on any subject. These funds support get-out-the-vote efforts through the publication and distribution of union newsletters that make specific candidate endorsements. They also finance meet-the-candidates sessions where union-endorsed candidates may speak and solicit funds and volunteer labor. Although the campaign laws allow business corporations to conduct similar education activities with their management employees and stockholders, this is not commonly done.

13.26   THE CIVIL RIGHTS ACT OF 1964

In passing the Civil Rights Act of 1964, Congress confirmed and established certain basic individual rights pertaining to voting; access to public accommodations, public facilities, and public education; participation in federally assisted programs; and opportunities for employment. Title VII of this Act, Equal Employment Opportunity, prohibits discrimination in employment or union membership. It is an unlawful practice for an employer (1) to refuse to hire or to discharge any individual or otherwise discriminate against him or her regarding conditions of employment because of race, color, religion, sex, or national origin or (2) to limit, segregate, or classify employees in any way that would deprive the individual of an employment opportunity or adversely affect his or her status as an employee because of race, color, religion, sex, or national origin.

Administration and enforcement of the Civil Rights Act is made the responsibility of the Equal Employment Opportunity Commission (EEOC), which the Act created. The responsibility of the commission is to ensure that consideration for hiring and promotion is based on ability and qualifications, without discrimination. Title VII prohibits discriminatory practices on the part of employers, employment agencies, labor organizations, and apprenticeship or training programs. The Civil Rights Act applies to interstate commerce and covers employers and labor organizations. There are several exemptions from the Act, some of which are local, state, and federal agencies, government-owned corporations, American Indian tribes, and religious organizations. The law requires that employers, labor unions, employment agencies, and joint labor-management apprenticeship committees keep such records and submit such reports as the EEOC may require. Special rules apply in states that have their own enforceable fair employment practice laws.

The Equal Employment Opportunity Act of 1972 amended the Civil Rights Act of 1964 in several important respects and substantially expanded its coverage. It authorized the EEOC for the first time to go directly to court for temporary restraining orders and for permanent injunctions against unlawful discrimination. This is in addition to other remedies such as reinstatement or hiring with back pay and appropriate affirmative action directives. The Civil Rights Act of 1964 now covers joint labor-management committees for apprenticeship and other training programs. The coverage of the Act was expanded to include employers of 15 or more employees and unions that operate hiring halls or referral systems or that have 15 or more members (25 in the original Civil Rights Act of 1964). Also created was the position of general counsel to act along much the same lines as the general counsel under the NLRB. The general counsel of the EEOC has authority to bring civil court actions against patterns and practices of employment discrimination in interstate commerce.

It is interesting to note that all of the rules, regulations, laws, and court cases relating to sexual harassment have their origins and enabling authority in the Civil Rights Act of 1964. This Act, although initially designed to promote racial integration through the prohibition of discrimination in voting, education, and the use of public facilities, now plays a vital role in modern employment policy, office ethics, and federal contracting guidelines.

13.27   EXECUTIVE ORDER 11246

Issued in 1965, Executive Order 11246 applies to contracts and subcontracts exceeding $10,000 on federal and federally assisted construction projects. Federal and federally assisted construction is not normally construed to include projects involving federal assistance that is in the nature of a loan guarantee or insurance. Contractors are prohibited from discriminating against any employee or applicant for employment because of race, color, religion, or national origin. The contractor must take positive action to ensure that applicants are employed, and that employees are treated during employment without discrimination. Affirmative action must be taken by contractors on projects covered by the Executive Order to increase the level of minority representation in their workforces. Actions pertaining to employment, promotion, transfer, recruitment, layoffs, rates of pay, training, and apprenticeship must not be discriminatory. Executive Order 11375 (1968), as an extension of Executive Order 11246, applies to federal and federal-aid contracts and prohibits discrimination against any employee because of sex.

Executive Order 11246, administered by the Office of Federal Contract Compliance Programs (OFCCP), U.S. Department of Labor, states that each federal contracting agency shall be primarily responsible for obtaining compliance with the provisions of the order. In addition, each administering agency is made responsible for compliance by the recipients of federal financial assistance. Federal agencies have compliance officers whose duties are to ensure adherence to the objectives of the order, which include conducting compliance reviews. A compliance review is a procedure used to check an ongoing contract. The contractor is required to give information to show that it is complying with the nondiscriminatory requirements of its contract, including affirmative action.

In the event of noncompliance with OFCCP rules, the contract may be canceled or suspended, and the contractor can be declared ineligible for further government or federally assisted construction contracts. In addition, the OFCCP has the authority to withhold progress payments from contractors that are in violation of Executive Order 11246. This authority stems from the government's right to suspend payment when a contractor fails to comply with any requirement of the contract. Compliance reports from contractors are required, and the general contractor must include suitable provisions concerning compliance with the order in its subcontracts and purchase orders.

13.28   THE AGE DISCRIMINATION IN EMPLOYMENT ACT

The Age Discrimination in Employment Act of 1967 prohibits arbitrary age discrimination in employment. This Act protects individuals 40 years of age or older from age discrimination by employers of 20 or more persons in an industry involving interstate commerce. Employment agencies, labor organizations, and most employees of federal, state, and local governments are also covered by the Act.

By the terms of the Act it is against the law for an employer to:

1. Fail or refuse to hire, to discharge, or to otherwise discriminate against any individual as to conditions of employment because of age

2. Limit, segregate, or classify its employees so as to deprive any individual of

employment opportunities or to adversely affect his status as an employee because of age

3. Reduce the wage rate of any employee in order to comply with the Act

The prohibitions against discrimination because of age do not apply when age is a bona fide occupational qualification, when differentiation is based on reasonable factors other than age, when the differentiation is caused by the terms of a bona fide seniority system or employee benefit plan, or when the discharge or discipline of the individual is for good cause.

Employers must post an approved notice of the Age Discrimination in Employment Act in a prominent place where employees can see it, and must maintain records as required. The Act is enforced by the Equal Employment Opportunity Commission (EEOC), which can make investigations, issue procedural rules, and enforce its provisions through the courts.

13.29   THE DAVIS-BACON ACT

The Davis-Bacon Act (1931), as subsequently amended, is a federal law that determines the wage rates, including fringe benefits, that must be paid workers on all federal construction projects and on a host of federally assisted jobs. The law applies to contracts in excess of $2,000 and states that the wages of workers shall not be less than the wage rates specified in the schedule of prevailing wages as determined by the secretary of labor for comparable work on similar projects in the vicinity in which the work is to be performed. The contractor must pay overtime at the rate of time and one-half for all work done by an employee in excess of 40 hours per week. General contractors and subcontractors are required to pay at least once a week all workers employed directly on the site of the work at wage rates no lower than those prescribed. The prime contractor is responsible for its subcontractors' compliance with prevailing wage requirements, a matter that may require some checking to be sure that the required compensation package is being paid. Full payment must be made, with the exception of such payroll deductions as are permitted by the Copeland Act (see Section 13.31). The law's purpose is to protect the local wage rates and local economies of each community and presumably to put union and nonunion contractors on a more nearly equal competitive footing in the bidding of federal and federally assisted projects.

For purposes of defining the coverage of prevailing wages prescribed for a given project, the work site is defined as being limited to the physical place or places where the construction called for will remain and to other adjacent or nearby property used by the contractor or subcontractors that can reasonably be included in the "site" because of proximity. Fabrication plants, batch plants, borrow pits, tool yards, and the like are considered to be a part of the work site, provided that they are dedicated exclusively, or nearly so, to performance of the contract and are so located in relation to the actual construction location that it would be reasonable to include them. Exempt from the work site definition, and therefore from Davis-Bacon wage rates, are permanent offices, branch plants, or fabrication plants whose locations and continuance are governed by the contractor's or subcontractors' general business operations.

The Davis-Bacon Act is administered by the U.S. Department of Labor. Contractors whose projects are covered must keep certain records, file periodic reports, and comply with various regulations with respect to the use of apprentices. Violation of Davis-Bacon requirements constitutes a breach of contract and exposes the contractor and subcontractors to government compliance action. The prevailing legal opinion is that workers cannot sue employers who fail to pay prevailing wages required under the Davis-Bacon Act. Only the federal government can enforce the law. Restitution is secured for workers found to have been underpaid, and penalties are assessed for violations of the overtime requirements. Violators can be denied the right to bid on other federal or federal-aid projects.

The Act does not provide for judicial review of Labor Department wage determinations, but a Wage Appeals Board operating under delegated authority from the Secretary of Labor has been established to hear appeals from findings or decisions of the Davis-Bacon Division. Contracting agencies have primary responsibility for Davis-Bacon enforcement, because obligations under the Act become a part of the construction contract. Most of the states and many cities have some type of prevailing wage requirement covering state and locally funded construction work.

13.30   DAVIS-BACON ADMINISTRATION

Recent years have seen the Davis-Bacon Act come under attack from a number of different quarters. The General Accounting Office of the federal government has recommended that the Act be repealed, and many responsible parties, including committees of the U.S. Congress, have demanded its reform. Much of the criticism has been directed at the administration of the Act and the procedures that have been followed in determining prevailing wages. There have been many allegations that prevailing wages as determined were usually too high, resulting in excessive construction costs. In many instances, prevailing wage rates have been merely equated to local union pay scales. It has been stated that the Davis-Bacon Act eliminates competition between union and nonunion contractors and imposes costly and burdensome reporting requirements on contractors.

The Act is now being applied to more and more construction projects as a result of the Department of Labor rendering more expansive interpretations of "site of the work" and "expenditure of federal funds." However, in 1984 the following new regulations were put in effect:

1. A prevailing wage is determined as a weighted average when a single wage rate does not apply to a majority of workers in a given area.

2. Wage data from prior Davis-Bacon projects are excluded from wage surveys in setting prevailing wages for building and residential construction work. (This rule does not apply to heavy and highway projects where there is little nonfederal construction.)

3. Urban wage data cannot be used in the determination of rural prevailing wages, and vice versa.

The Davis-Bacon Act requires that workers on covered projects receive at least the hourly wages and fringe benefits prevailing in the project locality. Payment by the contractor must equal or exceed this total hourly sum. However, the law does not limit the portion of the total amount that can be paid in the form of fringe benefits. Contractors could conceivably abuse this allowance by reducing a worker's cash wages and paying more of the required compensation into fringe benefit trust funds, thus reducing the employer payments for Social Security, workers' compensation insurance, and unemployment tax. To guard against this possibility, the Internal Revenue Service and the Department of Labor limit the contractor's contribution for fringe benefits to 25 percent of an employee's annual compensation.

The craft classification of "helpers" is now allowed on Davis-Bacon projects located in those areas where the use of such a labor classification is a prevailing practice. Helpers are defined as semiskilled workers who perform their jobs under the direction of, and with the assistance of, a journeyman tradesman. A maximum ratio of two helpers for every three journeymen in a given craft is normally permitted, although variances allowing a larger ratio can be granted under certain circumstances.

Another recent action regarding Davis-Bacon is based on the Freedom of Information Act. The courts have ruled that a contractor on a Davis-Bacon project must, on request, provide unions with copies of its certified payroll reports showing the workers' names, job classifications, pay scales, and fringe benefits. Given this information, unions can check on the enforcement of the prevailing wage law.

13.31   THE COPELAND ACT

As passed in 1934, and since amended, the Copeland Act makes it a punishable offense for an employer to deprive those employed on federal construction work or work financed in whole or in part by federal funds, of any portion of the compensation to which they are entitled. Other than deductions provided by law, the employer may not induce "kickbacks" from its employees by force, intimidation, threat of dismissal, or any other means whatsoever. This portion of the Copeland Act is commonly known as the Anti-Kickback Law, violation of which may be punished by fine, imprisonment, or both. Regulations issued by the secretary of labor allow the contractor or subcontractor to make additional deductions from wages by showing that the proposed deductions are proper, provided that the prior approval of the Department of Labor is obtained. The employer may deduct union dues if such holdback is consented to by the employee and is provided for in a collective bargaining agreement.

The law stipulates that payroll records shall be maintained and reports submitted by contractors as the Department of Labor may require. The Copeland Act covers all construction projects on which Davis-Bacon prevailing wages apply. The contracting agency is responsible for enforcing compliance with the Act.

13.32   THE FAIR LABOR STANDARDS ACT

First enacted by Congress in 1939 and since amended several times, the Fair Labor Standards Act, also known as the Wage and Hour Law, contains minimum wage, maximum hours, overtime pay, equal pay, and child-labor standards. Workers whose employment is related to interstate commerce or consists of producing goods for interstate commerce are covered without regard to the dollar volume of business.

The Fair Labor Standards Act provides for a minimum wage for all employees covered, this minimum wage having been steadily increased over the years. Also required is payment of an overtime rate of one and one-half times the regular hourly rate of pay for all hours worked in excess of 40 hours in any workweek. However, payment of overtime is not required for more than 8 hours per day, nor is there a limit set on the number of hours that may be worked in any one day or during any one week. The law does not require premium pay for Saturday, Sunday, or holiday work, or vacation or severance pay.

An employer who violates the wage and hour requirements is liable to its employees for double the unpaid minimum wages or overtime compensation plus associated court costs and attorney's fees. Willful violation of the law is made a criminal act, and the errant employer can be prosecuted. Several classes of employees are exempted from coverage under the Act, such as bona fide executive, administrative, and professional employees who meet certain tests established for exemption.

The Fair Labor Standards Act, as amended by the Equal Pay Act of 1963, provides that an employer must not discriminate on the basis of sex by paying employees of one sex wages at rates lower than it pays employees of the other sex for doing equal work on jobs requiring comparable skill, effort, and responsibility and performed under similar working conditions. Pay differentials can be justified by a seniority system, merit system, piecework pay system, or other system based on factors other than sex.

The basic minimum age for employment covered by the Act is 16 years, except for occupations declared hazardous by the secretary of labor, to which an 18-year minimum age applies. Construction per se is not designated as hazardous, but specified work assignments such as truck driving, wrecking and demolition, roofing, and power tool operation are so designated.

13.33 THE CONTRACT WORK HOURS AND SAFETY STANDARDS ACT

In 1962, Congress passed the Contract Work Hours and Safety Standards Act, also known as the Work Hours Act of 1962. This Act, as subsequently amended, applies to federal construction projects and to projects financed in whole or in part by the federal government. The main requirement of this law is that every worker shall be paid at a rate not less than one and one-half times the basic rate of pay for all hours worked in excess of 8 hours per day or 40 hours per week. In the event of violation, the contractor or subcontractor responsible is liable for unpaid wages to the employees affected and for liquidated damages to the federal government. Willful violation of the Work Hours Act is punishable by fine, imprisonment, or both. The enforcement of the law and the withholding of funds from the contractor to secure compliance with the Act are made the responsibility of the government agency for which the work is being done.

An earlier congressional enactment related to overtime pay was the Walsh-Healey Public Contracts Act (1936), which required contractors performing federal or federally assisted work to pay overtime wages after 8 working hours per day. However, amendments made in 1986 to the Walsh-Healey Public Contracts Act and the Contract Work Hours and Safety Standards Act eliminated the requirement that contractors pay workers overtime when they work more than 8 hours per day. Employees of contractors on federally funded construction can now work flexible hours up to 40 per week before they must be paid overtime rates. For example, a contractor can schedule four days of 10-hour shifts in one week without being required to pay overtime wages. Contractors have long supported such a change, maintaining that it would clear the way for more flexible and compressed schedules, resulting in savings of time and cost to the federal government. This change does not affect any obligation to pay overtime after 8 hours per day contained in state laws, local laws, collective bargaining agreements, or employment contracts.

13.34   THE HOBBS ACT

Also known as the Anti-Racketeering Act, the Hobbs Act, enacted in 1946, makes it a felony to obstruct, delay, or affect interstate commerce by robbery or extortion. To attempt or conspire to do so is also made a felony. *Robbery* is defined as the unlawful taking or obtaining of personal property from a person against that person's will by means of actual or threatened force or violence. *Extortion* is defined as the obtaining of property from another, with that person's consent, induced by the wrongful use of actual or threatened force, violence, or fear. The underlying motive of the Hobbs Act was to put an end to the use of threats, force, or violence by union officials to obtain payment from employers under the guise of recompense for services rendered. Prosecution of violators is placed in the hands of the U.S. Department of Justice.

Extortion by unions and union officials has been practiced in many guises against contractors, with payments being required as a condition of avoiding "labor trouble." These payments have been concealed behind many subterfuges such as "gifts," "commissions," "equipment rentals," and "services." The courts have held that, under the Hobbs Act, extortion extends to attempts by a union or union officials to obtain money from an employer in the form of wages for imposed, unwanted, and superfluous services. Violence or threats of violence need not be involved if such attempts involve fear of economic loss, injury to employees, or damage to equipment. However, by its 1973 decision in the *Enmons* case, the U.S. Supreme Court held that the Hobbs Act does not extend to extortion connected with legitimate labor disputes such as acts of violence committed on a picket line set up to obtain economic bargaining demands. However, it is possible to prosecute a union for its extortionate demands for superfluous and unnecessary employees. Such demands are not a legitimate labor objective under the law.

13.35   IMMIGRATION REFORM AND CONTROL ACT

The Immigration Reform and Control Act of 1986 contains civil and criminal penalties for employers who knowingly hire aliens not authorized to work in the United States. The law requires every employer to establish both the identity and the eligibility for employment of each person hired. This Act requires all employers to verify the employment status and keep records on each new employee. Under the law, employers are subject to sanction if they knowingly hire unauthorized aliens or if they discriminate against prospective employees based on national origin or citizenship status. One of the principal goals of this Act was to stem the flood of illegal aliens into the United States by making it difficult for them to find work. It should be noted that the construction industry has been among the top employers of illegal aliens nationally. The Act requires employers to verify the status of all their new hires and to examine certain documents to ensure identity and work authorization. The law forbids employers from hiring any aliens known to be unauthorized to work in the United States or to continue to employ any of these aliens. Employers face stiff fines for not requiring proper documentation and knowingly hiring an illegal alien. Federal agents can seek criminal penalties for a pattern or practice of violating the provisions of this Act.

Under the Act, employers must check to see that employees hired have documents that establish both identity and employment authorization. Documents such as a driver's license, certificate of naturalization, Social Security card, U.S. Military card, certificate of citizenship, passport, birth certificate, or alien registration card must be provided by the prospective employee. The employer is required to have each newly hired employee complete a sworn statement on Immigration and Naturalization Service (INS) Form I-9, which attests that the person is either a citizen of the United States or is authorized to work in the United States. The employer provides a sworn statement on the same Form I-9 that the employer has examined the documents provided showing both the individual's identity and employment authorization. The specific documents examined must be listed, including their identification numbers and expiration dates. The employers must retain Form I-9 for three years after the date of hire or one year after employment is terminated, whichever is later.

The INS has ruled that contractors do not need to complete verification forms for temporary workers each time they are rehired. Employers must prepare the forms only once a year for such employees. Employers may, but are not required to, retain copies of the documents used to complete the verification procedures. Contractors may, under certain circumstances, rely on unions, central clearing houses, or state employment agencies to complete the verification process.

13.36   THE NATIONAL APPRENTICESHIP ACT

In 1937, Congress passed the National Apprenticeship Act, authorizing the establishment of the Bureau of Apprenticeship and Training (BAT) of the U.S. Department of Labor. The bureau has the responsibility to encourage the establishment of apprenticeship programs and to help improve existing ones, but it does not conduct them itself. One of its prime objectives is to promote the development of such programs. Since its establishment, the bureau has provided technical assistance in developing and improving apprenticeship and other industrial training programs and has set minimum standards for the registration of local apprenticeship programs. The bureau works closely with state apprenticeship agencies, trade and industrial education institutions, and management and labor. Through its field staff, it cooperates with local employers and unions in developing apprenticeship programs to meet specific needs.

To implement apprentice training on a local level, many states have passed laws that provide for the establishment of state apprenticeship councils. These state agencies, which function cooperatively with the Bureau of Apprenticeship and Training, are made up of labor, management, and public representatives, often with the addition of members from the state labor departments and others. Using the standards recommended by the bureau as a guide, the councils have established detailed standards and procedures to which apprenticeship and training programs in the state are expected to conform. The state council becomes a part of the national apprenticeship program by securing recognition of its standards and procedures by the Bureau of Apprenticeship and Training. To date, the Bureau of Apprenticeship and Training has displayed a reluctance to register construction apprenticeship programs for new or nontraditional crafts or a combination of crafts.

13.37   THE DRUG-FREE WORKPLACE ACT

In 1988, the Drug-Free Workplace Act was passed, which requires federal government contractors and employers that receive federal contracts and grants to maintain drug-free workplaces. Construction contractors performing federal work valued at $100,000 or more are required by law to take various steps to discourage drug use by their employees. Specifically, the contractor must (1) publish and distribute to each employee a statement prohibiting drugs in the workplace and specifying punitive actions, (2) establish a drug-free awareness program, (3) notify employees that as a condition of employment they must abide by the terms of the policy and notify the employer (contractor) within five days if they are convicted of a drug violation in the workplace, (4) notify the contracting agency (owner) of a workplace drug conviction within ten days, (5) discipline any employee convicted of a criminal drug offense at the workplace or require the employee to participate in an approved drug abuse treatment program, and (6) make a good-faith effort to maintain a drug-free workplace.

Contracting agencies can suspend or terminate contracts and can bar contractors from federal work for up to five years if they make a false certification, fail to carry out the Act's requirements, or have such a number of employees convicted of drug offenses as to indicate that the contractor failed to make a good-faith effort to provide a drug-free workplace.

Drug testing has become a common employment requirement on both union and open-shop construction projects. Many unions have entered into labor contracts that require prehire drug testing as a part of the contractors' antidrug programs. Many open-shop firms now reserve the right to spot-check specific employees on their job sites. Employees of subcontractors are also subject to such examination. Both union and contractor testing policies call for the firing of employees who test positive.

13.38   AMERICANS WITH DISABILITIES ACT (ADA)

Enacted in 1990, the Americans with Disabilities Act (ADA) was the first federal statute to extend civil rights protections to disabled persons. This Act extends substantial legal protections and remedies to persons with disabilities and those defined as being disabled. The Act applies to both public- and private-sector employers with 15 or more employees. The stated purpose of this Act is "to provide a national mandate to end discrimination against disabled individuals and to bring them into the economic and social mainstream of life, to provide enforceable standards addressing discrimination against disabled individuals, and to ensure that the federal government plays a central role in enforcing the standards on behalf of disabled individuals." The ADA prohibits discrimination based on a disability in both the private and public sectors in areas of employment, public accommodations, public service, transportation, and telecommunications. The Act sets forth prohibitions against disability-based discrimination by employers, employment agencies, labor organizations, or joint labor-management committees with respect to hiring and all terms, conditions, and privileges of employment.

With respect to an individual, the term *disabled* means (1) having a physical or mental impairment that substantially limits one or more of the major life activities of such an individual, (2) having a record of such an impairment, or (3) being regarded as having such an impairment. Under the ADA, employers may not discriminate against any qualified individual with a disability in any aspect of hiring or employment. Only disabled individuals who are otherwise "qualified" can be protected against employment discrimination. A "qualified individual" is a person with a disability who, with or without reasonable accommodation, can perform the essential functions of the job. The phrase "essential functions" means job tasks that are fundamental and not marginal.

In addition to the ADA's general prohibitions against employment discrimination, this law contains a number of specific provisions, such as those that prohibit limiting, segregating, or classifying a job applicant or employee in a manner that adversely affects his opportunities or status; using standards, criteria, or methods of administration in a manner that results in or perpetuates discrimination; imposing or applying tests and other selection criteria that screen out disabled individuals unless the test or selection criteria are job-related or consistent with business necessity; or denying employment opportunities because a qualified individual with a disability needs reasonable accommodation.

13.39   THE FAMILY AND MEDICAL LEAVE ACT

In 1993, the Family and Medical Leave Act (FMLA) was enacted to help employees better balance their work and family lives by providing them with unpaid leave for certain reasons. The Act is administered by the U.S. Department of Labor's Employment Standards Administration, Wage and Hour Division, and requires most employers to provide eligible employees with up to 12 weeks of unpaid leave in a 12-month period for specified family and medical reasons.

All public agencies, including state, local, and federal employers, local education agencies and private-sector employers, who employed 50 or more employees in 20 or more workweeks in the current or preceding year and who are engaged in commerce or in any industry or activity affecting commerce, are bound by the Family and Medical Leave Act. The requirements of the Act override any collective bargaining agreements in place between employer and employee. In order for an employee to qualify for leave under the Act, he must:

1. Work for a covered employer
2. Have worked for the employer for a total of 12 months
3. Have worked at least 1,250 hours over the previous 12 months
4. Work at a location in the United States where at least 50 employees are employed by the employer within a 75-mile radius

Generally, employers must permit a qualifying employee to take leave in order to care for a newborn child, during the placement of an adopted child or foster child in the employee's immediate family, to care for an immediate family member with a serious health condition, or to take medical leave when the employee is unable to work because of a serious health condition. During this period employers are required to protect the employee's job and provide uninterrupted health benefits (if the employee was receiving those benefits prior to taking leave), and upon returning from leave, the employee must be restored to his original job (or equivalent) with equivalent pay, benefits, and other terms and conditions of employment.

Although the Family and Medical Leave Act went into effect on August 5, 1993, if a collective bargaining agreement between employer and employee was in effect on that date, the Act became effective on the expiration of the collective bargaining agreement or February 5, 1994, whichever was earlier.

QUESTIONS

1. What vital change in legal policy did the Norris-LaGuardia Act of 1932 mark for organized labor?

2. In recent years, the President of the United States has intervened to halt strikes and force workers back to work. Which labor act provides the President with this power and authority? Under what conditions is such Presidential intervention allowed?

3. The Taft-Hartley Act first established rules governing the way unions could organize and bargain in 1947. Briefly, describe the evolution of these rules over the next several decades and the problems encountered as they apply to the construction industry.

4. The Wagner Act and its amendments are designed to stop unfair labor practices. What basic labor rights do these acts serve to protect?

5. How are unfair labor practice cases distributed and processed by the NLRB itself, its regional offices, and territorial courts and agencies?

6. At what point can a union secure a representation election to determine whether a body of employees desires union representation? What percentage of votes must the union receive to be declared the representative?

7. What is a "hot-cargo" agreement and how does the law on this type of agreement affect the construction industry?

8. In a jurisdictional dispute, can a union strike to force an employer to assign certain work to the employees of one union over another? Explain your answer.

9. Explain the difference between a secondary boycott and a hot-cargo subcontractor agreement.

10. What purpose does the Davis-Bacon Act serve?

11. Sexual harassment law and policy have their foundation in which legislation?

Chapter 14

Labor Relations

14.1 THE CONSTRUCTION WORKER

All construction projects are of limited duration. When a job is completed, the field crews move on to another project or are laid off until more work becomes available. The average construction worker has no fixed relationship with any one contractor, and his tenure of employment with a given employer is normally indefinite and temporal. He is tightly bound to his occupation and is only loosely associated with any given construction company. The worker may work for several different employers over a period and is known more as a carpenter or cement mason than as the employee of any particular firm. Although these generalities are much less true for workers in the specialty trades, such as electricians and plumbers, than they are for carpenters, ironworkers, and others in the basic trades, they still typify employment in the construction industry. The construction worker is in the atypical position of being a skilled artisan with no permanent place of employment. This circumstance is a manifestation of the fact that the projects of a typical construction company are relatively short-term, variable as to location, and demanding of different combinations of trade skills. All contractors experience fluctuating requirements for manpower as new jobs are started and existing projects are completed. Economical operation dictates that most of a contractor's workforce be drawn from a local pool of manpower as needs dictate.

Although the popular conception of a construction worker is that of a "migratory bird," the geographical mobility of such people is variable with the craft and the individual circumstances of employment. Crews on large industrial construction projects and on highways, pipelines, bridges, transmission lines, tunnels, and other engineering construction are necessarily mobile. Millwrights, boilermakers, pipe fitters, pile drivers, and structural ironworkers often follow their specialties over wide geographical areas, moving from project to project. Building construction tradesmen are more apt to find continuous or relatively continuous employment within a given locality. A number of considerations, such as home ownership, family ties, schools, pensions, and other factors, make many construction workers reluctant to move their places of residence. For this reason, among others, there is a certain amount of movement into and out of the construction industry itself. Many tradesmen find jobs elsewhere in the economy when local job opportunities in construction for their particular craft skills are limited.

In the construction industry, more or less standard terms are applied to construction workers and are used to designate the skill level of the individual worker. The term *journeyman* (sometimes *craftsman*) is used to designate a skilled worker who is fully qualified in a given craft or trade. An *apprentice* is a person who is enrolled in a craft apprenticeship program certified by the Bureau of Apprenticeship and Training. *Helper, trainee, preapprentice*, and *subjourneyman* all refer to unskilled and semiskilled craft workers who are below the level of journeyman and are not laborers or apprentices.

14.2 EMPLOYMENT IN THE CONSTRUCTION INDUSTRY

The construction industry, as this nation's largest single economic effort, provides about 5 percent of the total national civil employment. Currently, employment in this vital industry totals approximately 6.9 million workers. Construction employees are predominantly male, with about 12 percent of workers being women. Construction relies more on younger workers than other industries, and many employed in this industry are younger than 30 years of age. This circumstance, however, results in a higher attrition rate among new entrants.

Recent years have seen a reduced influx of new workers into the construction industry. The recruitment of young workers by organized labor has also been hampered by the growth of high-tech industries. With an estimated need for more than 200,000 new tradesmen per year, the construction industry is exerting every effort to attract additional workers by improving the industry's image, actively recruiting young workers, improving retention, establishing better training programs, and providing more on-the-job experience.

14.3 EMPLOYEE BENEFITS

Construction contractors now provide their employees with an imposing array of pay augmentations known collectively as fringe benefits. Employee benefits can be defined as programs, other than direct wages, that compensate employees in some way. The nature of these benefits varies somewhat between union-shop and open-shop contractors insofar as the details of operation and administration are concerned, but their provision is now a common feature throughout the construction industry. Some of these benefits are mandated by law, such as Social Security, Medicare, and worker's compensation. Voluntary benefits include such things as pension plans, profit sharing, health and welfare funds, life insurance, paid vacations, paid holidays, employee education, legal aid funds, annuities, sick leave, bonuses, supplemental unemployment payment plans, retirement, apprenticeship programs, employee stock ownership plans (ESOPs), employee-sponsored 401(k) retirement savings programs, and other types of benefits. Such benefits are made available to construction workers through private company plans and/or those of contractor associations, unions, and public organizations. National figures indicate that employer contributions to such fringe benefits for construction workers now average about 32 percent of direct payroll costs.

The motivation for such generous employer-provided benefits is obvious. These programs are designed to meet employee needs and to attract and retain skilled and productive tradesmen, improve employee morale, and provide a high level of job satisfaction. To receive the full advantage from such program expense, contractors exert every effort to communicate information concerning the nature and value of such benefits to their employees.

14.4 THE UNION CONTRACTOR

In the conduct of their labor relations, some contractors work under union contracts, others operate open shop. The construction unions are considered first, followed by a discussion of the open shop.

A union contractor is one whose labor relations are largely determined by labor contract. Although the contractor may or may not have been directly involved in negotiating the union agreement, the contractor is undeniably bound by the terms of the contract to which it becomes a signatory. The workers of such employers are required to belong to, or to join, the appropriate union and often must be hired through the local union hiring hall. The union contractor has the real advantage of having access to a ready-made pool of skilled labor, but wages, hours, fringe benefits, and other conditions of employment are largely dictated by contract and are not matters for company determination.

Generally, relations between the contractor and its unionized employees are casual and impersonal. The tradesman is referred to a project by his union, and his job performance is judged largely by his foreman, a fellow union member. There is little direct contact between the construction worker and the contractor-employer. Personal relations are almost nonexistent, because rates of pay, holidays, overtime, and other conditions of employment are not negotiated with the employee directly, but with his union, and are dictated by the labor contract.

When he reports to a project, the tradesman is assumed to know his trade and is expected to perform any task that is assignable to his craft. He receives the same rate of pay as his fellow craftsmen, regardless of their relative skills and abilities to produce. Largely, his loyalties are to his union, whose well being he closely associates with his own. He is not especially "company minded," because he has no fixed relationship with any given contracting firm.

Because of these circumstances pertaining to the employment of unionized tradesmen, a contractor's labor relations are conducted almost entirely with the craft union locals whose members the contractor employs, and not with the workers themselves. Labor relations for a union-shop contractor consist primarily, therefore, not of interpersonal relationships but of labor contract negotiation and administration.

14.5   THE ROLE OF THE UNIONS

Despite divided opinion in the industry, there is no denying the fact that unions make an important contribution to the operation of the construction industry. The unions have a stabilizing influence on a basically unstable business, an influence that, from the point of view of the contractor, has both advantages and disadvantages. Through the medium of negotiated labor contracts, fixed wage rates are established and much uncertainty associated with the employment of labor is removed. The unions provide a pool of skilled and experienced workers from which the contractor can draw as its needs dictate.

Belonging to a union can offer many compensations to the working person. A prime consideration is the leading role organized labor takes in raising wages and establishing some form of job security for its members. Although the bargaining power of the individual worker is weak, that of an organization of workers can be very strong. The workers are secure in the belief that their union will protect them from unfair treatment and will exert every effort to improve their situation. They enjoy a sense of belonging to a group with the common purpose of mutual help, and through their elected union representatives, they have a voice in the determination of their wages and conditions of employment.

It is true that some union members belong, not of their own volition, but as a matter of necessity to keep their jobs. However, available evidence incontrovertibly indicates that most union members are not unwilling captives of organized labor but belong because it is their desire. Of course, not all members are enthusiastic unionists. Many of them do not participate actively in union affairs nor even attend meetings regularly. However, this passive attitude cannot be interpreted as a lack of union loyalty.

14.6   UNION HISTORY

Unions exist worldwide and originated at least 150 years ago. However, organized labor has become a stable, responsible element in America only within the past century. In 1886 the first enduring union association was founded after a long era of repeated failures. The American Federation of Labor (AFL) was organized in that year, with Samuel Gompers as its first president. Since the time of its inception, the AFL has traditionally been identified with the skilled craft worker. The AFL is a confederation of many sovereign national unions, each of which remains free to manage its own internal affairs. The construction trades were charter members of the AFL.

At the time the AFL was founded, semiskilled and unskilled factory workers were largely unorganized. The Knights of Labor, an organization of diverse membership, had made some progress in that direction but was waning rapidly. Nevertheless, the general sentiment of the AFL was not concordant with organization of the industrial worker. In 1905, the Industrial Workers of the World (IWW or "Wobblies") was organized to fill the void left by the AFL's failure to act. The IWW advocated the elimination of capitalism, engaged frequently in violent strikes, and declined rapidly in the post–World War I era. Its form of radicalism was never acceptable to the American public. Once again, there was no central force for the organization of the mass-production worker, although several independent industrial unions were leading somewhat precarious existences.

Enactment of the Norris-LaGuardia and Wagner Acts in the 1930s encouraged and assisted union activity. The AFL began to extend a lukewarm welcome to some industrial unions, but relegated them to a second-class position known as "federal locals." This scheme enjoyed no great success, because the AFL craft unions, apprehensive that they would be obliterated in a huge mass of industrial workers, pressed demands on the federal locals for jurisdiction over members who were engaged in craft occupations. In 1935 the disgruntled leaders of eight industrial unions associated with the AFL formed the Committee for Industrial Organization for the avowed purpose of organizing the mass-production industries. This committee was organized within the AFL and sought to induce it to assume a dual personality, representing both craft and industry. This was nothing short of treason in the eyes of the AFL hierarchy, and the committee was summarily ordered to disband or be expelled from the federation. In 1936, the AFL executive council suspended the Committee for Industrial Organization. In 1938, after two fruitless years of attempts at reunification, the committee became the Congress of Industrial Organizations (CIO), an association of autonomous industrial unions, with John L. Lewis of the United Mine Workers as its first president.

In the following years, the AFL and the CIO engaged in a bitter struggle for the leadership of American labor, and both sides came to recognize the need for reconciliation. However, personal animosities and conflicts of interest proved difficult to resolve, and the merger was delayed for many years. Not until 1955 were the two groups able to patch up their differences and reassociate.

14.7   AFL-CIO CONSTRUCTION UNIONS

A large proportion of the organized construction workers belong to one of the 14 international unions that form the Building and Construction Trades Department of the AFL-CIO, including the Teamsters, who have reaffiliated after an absence of many years. These unions are listed in Figure 14.1. This very large AFL-CIO union includes construction truck drivers as a part of its total membership. As used here, the term *international* refers to the fact that such unions have jurisdiction over some members in Canada and some effort has been made by the Laborers International to establish a presence in Mexico. The Building and Construction Trades Department reports that the total membership of the 14 building trades unions listed in Figure 14.1 is more than 3 million. At present, approximately 16 percent of the construction industry workforce is unionized.

At this point it may be helpful to consider the two major types of labor organizations, the craft union and the industrial union. The craft union follows jurisdictional lines and is made up exclusively of workers in a single craft, trade, or occupation. Its members may be, and often are, employed in different industries. All the building and construction trades unions are of this type. An industrial union includes everyone who works in a particular industry, regardless of what his individual job responsibilities may be. The United Auto Workers (UAW) is a good example of this type of union.

14.8   OTHER CONSTRUCTION UNIONS

The vast majority of unionized construction tradesmen belong to one of the 14 AFL-CIO construction trade unions and the International Brotherhood of Teamsters. However,

---

AFL-CIO CONSTRUCTION UNIONS

1. International Association of Heat and Frost Insulators and Asbestos Workers
2. International Brotherhood of Boilermakers, Iron Ship Builders, BlackSmiths, Forgers and Helpers
3. International Union of Bricklayers and Allied Craftworkers
4. International Brotherhood of Electrical Workers
5. International Union of Elevator Constructors
6. International Association of Bridge, Structural, Ornamental and Reinforcing Iron Workers
7. Laborers' International Union of North America
8. International Union of Operating Engineers
9. Operative Plasterers' and Cement Masons' International Association of the United States and Canada
10. International Union of Painters and Allied Trades
11. United Union of Roofers, Waterproofers and Allied Workers
12. Sheet Metal Workers' International Association
13. National Brotherhood of Teamsters
14. United Association of Journeymen and Apprentices of the Plumbing and Pipe Fitting Industry of the United States and Canada

Figure 14.1   International unions affiliated with the Building and Construction Trades Department, AFL-CIO.

other unions represent significant numbers of construction workers. There are a number of small independent unions that represent limited numbers of construction workers in a few localities. "Independent," in this context, means that the unions involved are not affiliated with the AFL-CIO. Most of these unions are organized along industrial lines without trade jurisdictions.

In 2000, the United Brotherhood of Carpenters and Joiners of America, one of the largest AFL-CIO affiliates, with more than 500,000 members, elected to become independent. The Carpenter's Union, and other independent unions like it, often elect their autonomy from the AFL-CIO in order to have greater internal management flexibility, control local finances, resolve jurisdictional issues, and remain competitive with nonunion labor. In terms of total numbers, however, the AFL-CIO craft unions represent a very large majority of the organized construction workers. For this reason the discussion in the following sections pertains principally to these unions.

14.9   THE LOCAL UNION

The basic unit of a construction union is the "local," each local exercising jurisdiction over an assigned geographical area such as a borough, city, county, or state. The local serves as a headquarters and is responsible for all union activities of that craft within its boundaries. The geographical area assigned to a given local can vary widely from craft to craft. For example, there may be an appreciable number of laborer's locals within a given state but only a few locals of the plumber's union. Throughout the continental United States there are approximately 8,000 locals of the AFL-CIO building and construction trades unions.

Each local union elects its own officers, these usually consisting of a president, vice president, secretary-treasurer, and sergeant at arms. These officials may or may not be salaried; a typical arrangement is that they are paid only for time actually spent on union business. An executive board or some equivalent body may also be established, which is concerned primarily with the admission of new members, discipline, financial matters, and contract negotiations. A paid business agent who is elected from the membership manages the day-to-day affairs of the local. A job steward, who is expected to see that union rules are observed and to report any violations or grievances to the business agent, represents the local on each project. The local appoints the steward for each project.

The locals of the same international union are commonly grouped together on a regional, state, county, or city basis. For example, Laborers' locals within designated regions band together to form district councils of Laborers, which serve to coordinate the activities of all member locals in a unified approach to common problems. Locals of the different construction unions unite on a regional basis to form building and construction trades councils. These councils are formed of delegates from each member local and serve as agencies for regional cooperation. An important function of these councils is to present a united front to employers during periods of collective bargaining.

Over a period of years, the National Labor Relations Board and the courts have issued a series of landmark decisions concerning individual worker rights under the National Labor Relations Act. These rulings pertain to the worker's association with a local union and the rights of each party. Some of the most important of these findings include the following. It has been held that a union member must pay his local union only the prevailing dues and fees and need not become a full member. Unions are prevented from requiring such a "dues paying only" member to take the union oath, accept full membership, or attend union meetings. Unions may not fine full members who convert to limited status and work open shop or return to work during a strike. After a union member resigns full membership, he can cross a picket line and return to work with no risk of union retaliation. Unions cannot impose fines to restrict a member's right to resign, even during a strike. A union cannot discipline a limited member who becomes a supervisor for an open-shop contractor.

14.10   LOCAL UNION AUTONOMY

The local is chartered by its international union and is subject to the constitution and bylaws of the parent organization. Beyond this, however, the decentralized nature of construction requires that locals possess a high degree of independence and freedom of action. Construction locals typically have the authority to negotiate their own labor agreements and to call strikes without the formal approval of their international unions. There are exceptions to this generality, however, and the parent unions can restrain their recalcitrant locals from recklessly resorting to work stoppages and strikes. Locals cannot, with impunity, defy their international unions and strike at will. Locals do have considerable autonomy, but parental authority can be lawfully applied when the occasion demands. Each local has its own set of bylaws that governs the election of local officers, ratification of labor contracts, conduct of meetings, payment of dues, expulsion of members, and other union business.

The legislative body of the local union is the membership. Elected officers have responsibility only for carrying out union rules as ratified by a majority vote of the members. Correspondingly, all matters of union policy, labor contract terms, decisions to strike, and other basic issues must be approved by the voting membership. However, the local union reigns supreme in the geographic area of its jurisdiction and exerts almost complete authority, with little or no interference by its international.

14.11   UNION WORK RULES

Most local unions promulgate work rules that pertain to the employment of the locals' members and are often referred to as "financial core members." These rules vary considerably from craft to craft and with geographical area. Some of the work rules represent concessions gained from employers in past collective bargaining and cover a wide range of issues, such as those pertaining to work hours per day, jurisdiction of work, multiple shifts, overtime and holidays, apprentices, prohibition of piecework, paydays, reporting time, foremen, crew size, safety provisions and devices, tools, job stewards, drinking water, and a variety of others. Contractors and others often criticize certain aspects of union work rules as being restrictive labor practices that unduly increase the costs of production, unnecessarily prolong construction time, and interfere with the contractor's management prerogatives. Unions defend their work rules on the grounds that they make the employment of their members more secure, defend against the loss or subversion of hard-won gains, protect the members from unfair or arbitrary treatment by the employer, and ensure the safe employment conditions to which their members are entitled.

Without doubt, many work rules do function to preserve legitimate union rights and prerogatives. Unfortunately, many locals have shown little restraint in their imposition and enforcement of work rules. Union insistence on the use of unnecessary workers on the job (featherbedding), prohibitions on the use of labor-saving methods and tools, inflexible application of overtime requirements, and other restrictive practices have substantially lowered productivity and commensurately raised construction costs in many sections of the country, especially in the large metropolitan areas. Overlong coffee breaks, requirements that skilled workers do unskilled jobs, strict trade jurisdictions, flagrant featherbedding, excessive nonproductive time, limitations on the daily production of a tradesman, and the requiring of unnecessary work or the duplication of work already done are examples of work rules, or their results, that adversely affect productivity and labor costs.

The work-rule picture has changed appreciably in recent years, however. High construction labor costs are resulting in changes to faster and cheaper ways to build. The shift to drywall, precast concrete, and prefabricated steel buildings illustrates this point. In addition, the tremendous increase in open-shop operations over all parts of the country has resulted in a noticeable easing or elimination of work rules by many construction union locals.

14.12   THE BUSINESS AGENT

The local union elects a business agent, sometimes called a business manager, to conduct its business affairs and to serve as its representative to outside agencies. Being paid at the highest craft scale, the business agent is the key person in a building trades local. As the full-time spokesman of the local, the business agent has the primary responsibility for "policing the trade" within the local's assigned geographical area. The agent refers members to jobs, has considerable authority in determining who joins the union, and is clearly a powerful person insofar as local contractors and union members are concerned. The use of business agents is typical of unions, such as the building trades, that deal with many different employers and in which union members are employed in scattered groups.

Large locals often have more than one full-time business agent; small locals may have one of their number act as a part-time business agent. The business agent is the contractor's only direct contact with a local. The agent helps negotiate agreements, enforces them, ameliorates grievances, protects the union's work jurisdiction, and serves as a general go-between for the local members and the employers. The agent directs strikes and other concerted activities undertaken by the local and is the equivalent of a chief executive officer in the local union. To the contractor, the business agent *is* the local union. In some small locals, members cannot be found who possess the business acumen to run the local; consequently, many small locals fall under the "supervision" of the international union, and an international representative may be placed in charge of running the local's everyday business activities.

The business agent has substantial authority to make decisions for the local and bears almost the entire responsibility for management of its affairs. Obviously, such freedom of action is a necessary condition in the construction industry, in which jobs are scattered and can be of short duration. To be effective, the business agent must be empowered to act quickly and without having to await approval of a higher body. In a complex and shifting environment, the agent seeks to protect the interests of the local and its members. The agent can act freely, but must always remain responsive to the needs of the members of the union local in order to maintain credibility and please the constituency.

The business agent, in many respects, is the middleman of the industry who strives to reconcile the conflicting demands of the employers on one side, and the union rank and file on the other. Contractors look to the business agent to find qualified people for their projects and to curb recalcitrant workers. Members of the local depend on the business agent for such services as finding jobs for them and settling disputes with employers. The responsibilities of the business agent include managing the union local's office, keeping necessary records, and supervising the finances, as well as visiting the jobs and checking employer compliance with union work rules and contract provisions.

14.13   COLLECTIVE BARGAINING

Labor contracts between construction contractors and labor unions are obtained by a process of collective bargaining and constitute the essential basis for labor-management relations in the unionized segment of the construction industry. These contracts, as negotiated between the two parties, establish wages, hours, and other essential conditions of employment within the geographical jurisdictions of the participating local unions and pertain only to the craft or crafts involved. As existing labor agreements expire, negotiations are conducted to arrive at new, mutually acceptable contracts.

The National Labor Relations Act requires management and labor to bargain in good faith with one another. Failure to do so by either party is an unfair labor practice. The law does not require that concessions be made, or even that the two sides come to an agreement. The law does not actually define what is meant by "good-faith bargaining," although decisions of the NLRB and the courts identify it as a duty to approach negotiations with an open mind and a real intention of reaching an agreement. Lack of good faith on the part of the employer may be indicated by the employer's ignoring a bargaining request, anti-union activities, failing to appoint a bargaining representative with power to reach an agreement, delaying and evasive tactics, attempting to deal directly with employees during negotiations, refusing to consider each and every proposal, failing to respond with counterproposals, and refusing to sign an agreement.

14.14   PATTERNS OF BARGAINING

For many years, multiemployer collective bargaining was the standard way in which labor agreements were reached between contractors and the unions. Contractor groups or associations bargained with a union or unions on behalf of their members. The usual procedure was that the individual contractor appointed the association as its bargaining agent and was bound by the terms of the resulting labor contract. Such a negotiation procedure offered strength in numbers and improved the bargaining power of the individual firm. Collective action prevented the unions from whipsawing, that is, playing one contractor against the other. Local associations of general contractors typically negotiated with the locals of the basic trades: cement masons, laborers, operating engineers, construction teamsters, and ironworkers. The resulting labor contracts were often referred to as "master labor agreements" because they were multicraft in coverage.

The specialty contractor groups negotiated with the union locals whose members they employed. For example, a city or regional association of painting contractors would negotiate with the appropriate local or locals of the International Brotherhood of Painters and

Allied Trades. In a similar manner, locals of plumbers, plasterers, roofers, electricians, and the other specialty trades would bargain with organizations of their specialty contractor counterparts.

However, during the past few years, there has been a progressive breakdown in multiemployer and multicraft bargaining and a movement toward individual contractors negotiating their own labor contracts. Although multiemployer bargaining is still very much a part of the industry, collective action is being replaced by a more flexible free-market concept of labor negotiations. A philosophy of "going it alone" is replacing the "united we stand" credo in the construction industry. A number of reasons can be cited for this trend. New construction management styles, periodic economic recessions, and the rapid growth of the open shop have triggered much of the change.

Disparate groups of contractors, facing different economic problems and being on the same side of the bargaining table have made multiemployer bargaining very difficult in many areas. Selective strikes by unions committed to multiemployer bargaining and the use of interim, project, and national labor agreements have diminished contractors' faith and commitment to collective action. Work preservation clauses in labor contracts that establish a union obligation for the open-shop arm of a dual-shop operation (see Section 14.34) have made many employers even more reluctant to participate in multiemployer bargaining. All in all, many contractors now consider multiemployer and multicraft bargaining to be unresponsive to market changes and modern industry conditions and perceive individual negotiations to be advantageous for them. Some contractors are reluctant to give power of attorney to trade organizations to bargain for them because they may wish to go nonunion or work nonunion in some areas. Many multiemployer bargaining units have been disbanded. In other cases, association officers negotiate agreements with the trades, but without the power to bind member contractors to such agreements. Contractors are left free to sign the resulting agreements as they see fit. In many geographical areas, there is no association bargaining with some crafts.

In bargaining with the individual union locals, contractors can now attempt to confine the terms of a labor agreement to a specific marketplace. In bargaining one-on-one with the craft unions, firms have the flexibility to avoid a strike, to take a strike, to gain concessions, or to keep the work going at any price. This can sometimes be a matter of necessity for certain union contractors who are struggling to keep their share of the market. Going it alone allows a firm to seek the best deal it can get from the trades, or to go nonunion if the terms are not attractive.

Clearly, the overall present condition of collective bargaining in the construction industry is highly fragmented and characterized by contractors independently conducting their own labor negotiations. Only time will tell what the ultimate result of the current trends will be.

14.15   WITHDRAWAL FROM BARGAINING UNIT

When a contractor withdraws from a multiemployer bargaining unit to engage in its own bargaining, it must take certain steps to avoid being obligated by future contracts negotiated by the bargaining unit. To excuse that employer from honoring a bargaining agreement subsequently negotiated by the union and the multiemployer unit, the contractor must give written notice to the union and the bargaining unit before the date set by the labor agreement for modification or the agreed-upon date for the beginning of negotiations. The notice must indicate an unequivocal intent to withdraw from the multiemployer bargaining unit. Such notice should also be given to the Federal Mediation and Conciliation Service.

An employer may quit a multiemployer bargaining unit unilaterally by giving proper notice before the onset of negotiations. Once bargaining commences, the right to withdraw is very limited. If the union concurs, the employer can withdraw at any time. If the union objects, unusual circumstances are required in order for the employer to prevail, such as the employer's being faced with impending bankruptcy. The U.S. Supreme Court ruled in 1973 that an impasse in negotiations did not constitute an unusual circumstance.

Even with timely notice, there is normally a continuing obligation for the contractor to bargain with the union, now individually. About the only way to avoid this requirement is for the contractor to demonstrate that the union no longer represents a majority of the employees concerned. The contractor may also face a withdrawal liability to any multiemployer pension plan under the provisions of the Employee Retirement Income Security Act (see Section 14.24).

14.16   THE BARGAINING PROCESS

By terms of the National Labor Relations Act, both management and labor can be represented by anyone of their own choosing at collective bargaining sessions. Individual construction firms are normally represented by company officers, perhaps supported by outside parties who have expertise in construction labor law and relations. Bargaining by a trade association is usually conducted by a committee of contractor members, legal or labor relations counsel, and association staff. The unions appoint the members of their negotiating team, usually officers of the locals, district councils, or building trades councils if multicraft bargaining is involved. Representatives of the international unions frequently assist the local unions and participate in the negotiations.

Invariably, the labor side demands more than it actually expects to obtain. The resulting sessions assume somewhat the characteristics of a poker game, involving a complex strategy of offer and counteroffer. It takes experience and skill to recognize the psychological moment at which the most propitious deal can be made. Generally, union negotiators do not have final authority to consummate binding agreements, but must obtain ratification by the union membership. It is not unusual for the members to refuse the proposed settlement, sending the negotiators back to the bargaining table. In the event of an impasse in bargaining, there are various settlement plans in use around the country that assist contractors and unions in reaching mutually satisfactory agreements without resorting to strikes or lockouts.

A fatal mistake that management must constantly guard against is entering into negotiations without thorough preparation and expert advice. It must be remembered that the primary purpose of unions is to establish favorable working conditions for their members. The periodic negotiations with employers are the focal point for union activity and are not taken lightly by union representatives. By the same token, these negotiations are also a serious matter for management, and unsophisticated bargaining by poorly prepared management representatives can only lead to the contractors being outmaneuvered and outpointed by their union adversaries, who spend a great deal of their time in constant association with labor-management matters. The highly specialized art of labor negotiation has led to the increasing practice of hiring professional assistance for this purpose. In fact, most large contractors and many of the contractor associations have labor relations specialists on their staffs.

14.17 LABOR AGREEMENTS

Once a settlement is reached between a contractor and a labor union, a written instrument is prepared that contains the essentials of the agreement and is signed by the parties. This instrument, referred to as a "labor agreement" or "labor contract," is binding on both parties. The life of a typical labor agreement may range from a short-term interim arrangement to a multiyear contract. An interim agreement is used when an existing labor contract with a multiemployer unit has expired, there is a bargaining impasse, and a new agreement has not yet been reached. Because the building trades will not usually work without a contract, an individual contractor may enter into an interim agreement with labor that will keep its projects manned until a new contractual arrangement is reached. Such a contract normally provides that the contractor will pay any final settlement retroactively or will pay an interim wage agreed upon with the union.

During recent years, the bulk of labor contracts in the construction industry have been one-year settlements, although two-year and three-year agreements have also been negotiated. The longer-term contracts often provide for periodic pay adjustments or contain wage reopener clauses. By its 1984 decision in the *Bildisco* case, the U.S. Supreme Court ruled that a contractor can lawfully terminate its union contracts when it files for reorganization under the bankruptcy code.

Bargaining agreements in the construction industry contain many provisions covering a wide range of subjects. Such contracts stipulate wages, hours, fringe benefits, overtime, and a large variety of working conditions. Most agreements are negotiated to cover a particular category of construction. For example, separate agreements are negotiated for the same geographical area between the same unions and different contractors or contractor groups to cover building, heavy highway, and residential construction and, in some areas, dredging. Most agreements provide for the settlement of jurisdictional disputes, a job referral system, apprenticeship, and grievance procedures. A large proportion of construction labor agreements contain no-strike and no-lockout pledges. Many of the no-strike clauses are conditional, however, permitting strikes under certain conditions, such as after the grievance procedure has been exhausted or when the employer is in noncompliance with the provisions of the agreement, particularly the fringe benefit provisions.

Section 8(d) of the National Labor Relations Act provides that the party desiring to terminate or modify an existing labor contract covering employees in an industry affecting interstate commerce must do the following:

1. Serve written notice on the other party of the termination or modification 60 days before the termination date or date of modification
2. Offer to negotiate a new or modified contract
3. Notify the Federal Mediation and Conciliation Service and any similar state agency that a dispute exists if no agreement has been reached 30 days after the notice was served
4. Continue to live by the existing contract terms without resort to slowdowns, strike,

or lockout until expiration of the 60-day notice period or until expiration of the contract, whichever is later

Any employee who engages in a strike within the 60-day period loses his status as an employee of the struck employer and is no longer protected by the National Labor Relations Act. This means, for example, that employees who engage in strikes during this period are not entitled to reinstatement.

14.18   GEOGRAPHICAL COVERAGE OF AGREEMENTS

In terms of geographical coverage, labor contracts in the construction industry run the gamut—from the entire country down to an individual project. However, such agreements are predominantly local in coverage, their application ranging from cities to entire states. There are, however, a few instances in which labor agreements are negotiated to cover large areas or even the entire United States. A few construction unions, or specialty crafts, negotiate agreements that apply nationally. For example, a national agreement (exclusive of New York City, which has a separate contract) is negotiated between the National Elevator Industry, Inc. (NEII), an employer group, and the International Union of Elevator Constructors. The rate of pay of an elevator constructor union member on any given project is determined through this national negotiation. The general aspects of wages, hours, and work rules are prescribed by the national agreement. There is another form of national agreement that pertains to several crafts; it is discussed in Section 14.20. Individual contractors or contractor associations sometimes negotiate regional agreements covering several states. Boilermakers' agreements are often on a multistate, regional basis.

14.19   PROJECT AGREEMENTS

Project agreements have become commonplace in recent years. A project agreement is a labor contract that normally applies only to a specific large construction job, but in some cases such labor contracts cover a series of projects in a given geographical area or even in a specific industry. For example, there is presently a blanket labor agreement to standardize working conditions on heavy and highway work that applies to six crafts in 16 states. An individual contractor, an association of contractors, an owner, or a combination of these may negotiate project agreements with the unions.

These contracts have the goal of constructing projects efficiently, on time, and with reasonable and predictable labor costs. Such contracts require that all subcontractors must be union members and include applicable local wage rates and fringe benefits, but attempt to reduce labor costs and to ensure uninterrupted work by sometimes including special terms that are more favorable to the contractor than the existing local labor agreements. For instance, project agreements often prohibit strikes, lockouts, and slowdowns. They can contain grievance procedures, ban jurisdictional disputes, reduce or eliminate coffee breaks, reduce overtime, provide for employment of trainees, and eliminate travel time and standby crews. They often allow the project to continue work through local collective bargaining strikes under interim agreements.

Although local contractors may vehemently object to project agreements that give special treatment to certain contractors without uniformly applying that treatment to all contactors, project agreements are still widely used. The growth of the open shop has spurred the use of project agreements, because they can improve the competitive position of unionized firms. The unions have even entered into project labor contracts with otherwise nonunion contractors to capture for union members work that would otherwise go open shop.

14.20   NATIONAL AGREEMENTS

The term *national agreement* has come to refer to a nationally applicable labor contract that is negotiated between the AFL-CIO international building trades unions and certain individual contractors and contractor associations such as the National Constructors Association. A national agreement provides for the employment of the members of a given union anywhere within the United States and, often, in Canada. Employers that sign national agreements are mainly large industrial contractors that perform work in widely scattered localities. Rather than attempt to become parties to the myriad of local labor contracts in force about the country, these contractors use the national agreement as a convenient method of handling their labor relations.

The usual national agreement does not include any terms of local labor contracts but does provide that all workers must be obtained through local union hiring halls. Contractors can hire their field supervisors directly. Typically, these pacts contain explicit work rules and specifically ban slowdowns, standby crews, and featherbedding practices. A grievance procedure and provision for flexible eight-hour workdays are included. National agreements contain a no-strike, no-lockout clause, although an exception to this clause allows for support of local collective bargaining. These pacts eliminate coffee breaks and other nonworking time, travel and subsistence pay, and overtime except under unusual circumstances. The agreement establishes a joint administrative committee that adapts the pact to suit a specific project. The committee establishes wage rates, alteration of the workweek, and the number of pre-apprentices and helpers to be allowed.

14.21   THE FEDERAL MEDIATION AND CONCILIATION SERVICE

Established by the Taft-Hartley Act in 1947, the Federal Mediation and Conciliation Service is an independent agency of the federal government that is charged with the responsibility of assisting employers in interstate commerce and labor organizations in promoting peace in regard to labor practices. As discussed in Section 14.17, employers or unions that wish to modify or terminate existing collective bargaining agreements must serve notice on the other party 60 days before the effective date of these changes. Should the matter not be resolved within 30 days, notice must be given to the Federal Mediation and Conciliation Service (FMCS) and any similar state agency having jurisdiction. Without the 30-day notice, employees cannot be withdrawn even upon expiration of the agreement until the required 30-day notice is given to FMCS in writing. The service is automatically notified when negotiations threaten to lead to a dispute.

Mediation by the Federal Mediation and Conciliation Service is a voluntary process, and absolute neutrality in labor disputes is a guiding principle of the service. The parties to a dispute are encouraged to settle their differences by themselves, but either side may request a mediator's assistance at no charge. The aim of the service is reconciliation of conflicting views without dictation or intervention into the affairs of either party. Mediators cannot compel either side in a labor dispute to do anything, but they bring experience and dispassionate advice to the bargaining table to help the disputing parties reach a mutually acceptable area of agreement. Thus, the Federal Mediation and Conciliation Service functions to keep the parties bargaining, offers helpful suggestions, and otherwise assists in the ultimate achievement of collective bargaining agreements. In no way altered, however, is the fact that the resulting agreement is both the product and the responsibility of the signatories. In contrast to arbitration, whereby the arbitrator makes a final and binding decision, a mediator has no authority to make such a decision.

The service employs a staff of mediators who are located in regional offices throughout the country and who are recruited from both management and labor. Through the years the Federal Mediation and Conciliation Service has established a reputation for impartiality and devotion to duty while occupying a most difficult role. As a rule, it concentrates its energies on the resolution of disputes that have an appreciable impact on interstate commerce. In a dispute that threatens to cause a substantial interruption to commerce, the Federal Mediation and Conciliation Service may enter the dispute either on its own motion or on the request of one or more of the parties to the dispute. In disputes that imperil the national health or safety, the service must intervene regardless of the effect on commerce. Should the parties agree to arbitrate their differences, the service will furnish them with a list of qualified arbitrators.

14.22 EMPLOYER LOCKOUTS

Employer lockouts are sometimes called *strikes in reverse*. A lockout occurs when an employer or, more commonly, an association of employers close their establishments against employees during negotiations and cease operations until a settlement has been reached. A strike is the withholding of workers from the contractor until a settlement is reached. A lockout is an employer device to withhold employment from its workers until an agreement is achieved.

Labor's right to strike or to engage in concerted activity for bargaining purposes is protected. The law does not, however, give employers a parallel right to engage in a lockout. Federal labor law does not expressly permit or forbid lockouts, and they are legal under certain circumstances that are not spelled out in the law. The National Labor Relations Act gives workers the right to bargain collectively and makes it an unfair labor practice for employers to interfere with that right. Hence, it is up to the NLRB and the courts to decide whether a particular lockout amounts to "interference." The guideposts to legality are vague, and contractors who engage in protective lockouts assume the risk of having to prove that the circumstances justified their actions. Competent legal advice is highly desirable for contractors who contemplate such action.

Over the years some guidelines have emerged that are useful in judging the legality of a lockout action. The following circumstances could justify a lockout:

Unusual economic circumstances exist such that a strike would cause substantial loss to the contractor.

A strike would pose a serious threat to the public health and welfare.

A lockout by members of an employer association is necessary to prevent "whipsaw-ing" by the union.

A union is engaging in selective strikes against individual members of an employers' association.

In addition, the U.S. Supreme Court has held that the lockout is a valid counterweapon after an impasse has been reached in a bargaining dispute, even if the employer is not anticipating an immediate strike. The lockout cannot be used in situations in which there is antiunion intent, but only in support of a legitimate bargaining position as an economic tool to counter a union's strike weapon. The NLRB has held that multitrade contractor groups' acting in concert does not make a lockout illegal as long as the interests of the employers involved are sufficiently interwoven to justify their taking concerted action in their common interest. The National Labor Relations Board ruled in 1986 that it is legal for an employer to hire temporary workers during a lockout of its regular employees. This decision says that a contractor does not violate current labor law by hiring temporary replacements during a lockout to keep construction operations progressing.

14.23 WAGES AND HOURS

Details associated with wages and hours are generally the crucial bargaining issues; nego-tiations on other matters are more in the nature of skirmishes around the edge of the major area of battle. Actually, wages and hours are broad subjects, covering such considerations as hourly wage rates, fringe benefits, overtime rates, show-up time, premiums for work-ing in high places, travel time, subsistence, apprentice wage rates, and cost-of-living wage escalators.

It is usual practice for overtime rates to be required for all hours worked in excess of 8 per day, 40 per week, and for work done on Saturdays, Sundays, or holidays. Overtime rates are normally one and one-half times the regular rate of pay, although "double time" may be stipulated in some instances. Many union contracts now allow four 10-hour workdays without payment of overtime.

Travel time and subsistence pertain to projects located in remote areas. Most bargain-ing agreements establish standards for determining whether a given project is considered remote. If a job is found to have a remote classification, special rules apply regarding contractor-furnished transportation, field camp facilities, payment of subsistence, and travel reimbursement.

Because of economic recessions, international competition, and the rapid growth of open-shop work, the historically steady increase in union wage rates has been interrupted during the past several years. In many cases, the wage rates of organized construction workers have been reduced or held steady with little increase. Where increases have been negotiated, the changes have generally been relatively small. At present, annual raises won by union tradesmen are averaging approximately 3 percent on a national basis, and nonunion wage increases have been approximately 3.5 percent per year. Nevertheless, union wage scales still remain more than 25 percent higher, on average, than those of open-shop workers. Current indications are that wage settlements in the unionized segment of the industry are returning to a general pattern of annual increases, although of a much more modest nature than those of years past.

During periods of spiraling inflation, cost-of-living adjustment (COLA) clauses are extensively used in construction labor contracts. In an attempt to protect the purchasing power of their members during such times, the unions often negotiate contracts that automatically adjust wage rates for inflation. The most commonly used measure of the rate of inflation is the Consumer Price Index (CPI), published every month by the Bureau of Labor Statistics for selected cities and the country as a whole. Changes in the CPI, expressed in either index points or percentages, are the basis for calculating periodic wage changes.

14.24   EMPLOYEE RETIREMENT INCOME SECURITY ACT

Pension and welfare plans provided by private employers in interstate commerce to their employees fall under the Employee Retirement Income Security Act (ERISA), passed by Congress in 1974. A pension plan is any company program that provides retirement income to employees or results in a deferral of income by employees until the termination of employment or beyond. Included are profit sharing, stock bonuses, disability retirement, and thrift plans. A welfare plan provides hospital, medical, surgical, sickness, accident, disability, death, vacation, or other benefits. ERISA does not require any employer to establish a plan, but those that do must meet certain standards. The chief purpose of this Act is to protect the interests of workers and their beneficiaries. It specifically provides that the assets of a pension plan shall not inure to the benefit of an employer and that the funds shall be used only to benefit the plan participants. Workers are not required to satisfy unreasonable age and service requirements before becoming eligible for a company pension plan. ERISA ensures that money will be available to pay benefits when they are due and that plan funds are handled prudently. The reporting and disclosure provisions of the Act require the employer to advise employees and their beneficiaries of their rights and obligations under the company benefit plan. Employers bear the responsibility of ensuring that benefits promised to their employees are actually provided.

Pension plans provided for in union labor contracts can be of two varieties. Under a "defined benefit" plan, employees are promised a specific level of retirement benefits based on length of service and wage level. The general formula for calculating the pension amount under a defined benefits plan is (years of service) × (final average salary) × (multiplier). It is the responsibility of the employer to furnish the funds needed to provide such benefits. A "defined contribution" plan is one that limits the participating employer's obligation to the contribution rate specified in its collective bargaining agreement. The employee receives a retirement annuity based on how much the employer contributed to the plan and how well the invested funds perform.

For many years, contractors contributed to multiemployer pension plans on a per-hour or percentage basis, believing that their arrangements were defined contribution plans and that their liabilities were limited to their contributions. However, this was changed by the U.S. Supreme Court in the *Connolly* decision (1978), which held that most multiemployer pension plans are defined benefit plans and are covered by ERISA Title IV, including the withdrawal liability provision. This was confirmed by the Multiemployer Pension Plan Amendments Act of 1980 that made contractors fully responsible for the funding of vested rights and by assessing withdrawal liability against employers who cease contributing to such plans.

The 1980 legislation provides that a construction contractor participating in a typical multiemployer pension plan may incur a withdrawal liability if it leaves the plan and continues to do the same type of work in the same labor contract area and the plan has a significant level of unfunded vested liability. Unlike most other industries under the law, construction contractors contributing to construction industry multiemployer plans will have no withdrawal liability if they go out of business or retire. The withdrawal liability is to cover the company's share of the unfunded, vested benefits, and to prevent large claims against the Pension Benefit Guarantee Corporation (PBGC). Such an assessment is calculated as the difference between the employer contributions made and the value of vested benefits, and can involve large sums of money. The PBGC is the federal agency that insures many private defined benefit plans and was established to guarantee payment of pension benefits to employees. The PBGC does not cover workers in defined contribution plans, and there is no withdrawal liability attached to such pension arrangements.

The 1980 amendments to ERISA have had a profound effect on the construction industry and have made contractors wary of multiemployer pension plans. Employers now resist increases in pension benefits and have a growing disillusionment with participation in multiemployer defined benefit plans because of the withdrawal liability and high administrative costs. There is now considerable movement among local contractor groups to revise their pension arrangements from defined benefit to defined contribution plans. Modification of the withdrawal rule is now a controversial political issue between contractors and the construction unions.

14.25 ADMINISTRATION OF THE LABOR CONTRACT

The signing of a negotiated contract does not close the matter of contractor-union relationships. Rather, the status of the agreement merely changes from that of bargaining to one of interpretation and administration. A carefully considered and clearly worded contract will do much to minimize contract misunderstandings.

Labor agreements in the construction industry typically contain procedures for the settlement of disputes that may arise during the life of the contract. When a dispute occurs that cannot be resolved by a conference of the steward, business agent, superintendent, and any other party directly involved, the grievance procedure set forth in the agreement is followed. This procedure often provides for meetings between successively higher echelons of contractor and union officials, during which time no work stoppage is to occur. Should the grievance procedure not resolve the dispute, arbitration of the matter may or may not be provided for in the labor agreement. Historically, unions have resisted the concept of binding arbitration, preferring to remain free to select their own course of action to suit the situation. Nevertheless, many construction labor contracts now provide for the arbitration of contract disputes with no resort to strikes or lockouts.

It is worth noting that an agreement to arbitrate can be enforced against either contractors or unions by a federal court injunction. The law on this matter with respect to management has long been established. In 1970, the U.S. Supreme Court reversed an earlier ruling that the Norris-LaGuardia Act barred federal court injunctions against labor unions and ruled that injunctions could be issued on complaint of employers to enforce no-strike agreements to arbitrate.

Because administration of labor contracts is an everyday requirement for the union contractor, a carefully selected individual is usually appointed to handle the labor relations for the firm. When a specific person is assigned to this duty, even if only as one of several responsibilities, he can become versed in the complexities of labor law and the provisions of the local labor agreements. He becomes, in fact, the labor relations specialist for the company. In this way, company labor policy is at least consistent and informed. The contractor is obligated to live up to the terms of the agreement, and must be insistent that the unions do likewise.

14.26 DAMAGE SUITS

Section 301 of the Taft-Hartley Act provides that suits for violation of labor contracts between an employer and a labor union representing employees in an industry affecting interstate commerce can be brought in the federal district court having jurisdiction. The law provides that any labor organization or employer subject to the Taft-Hartley Act is bound by the actions of its agents, and that a labor organization may sue or be sued as an entity and on behalf of the employees whom it represents. Any monetary judgment so obtained against a union is enforceable only against the organization as an entity, and not against any individual member. The U.S. Supreme Court has ruled that an employer can sue a union for damages resulting from violation of a no-strike arbitration agreement, but may not sue union officials or members individually.

Damage suits for losses suffered as a result of unlawful strikes, picketing, or other such union actions may also be brought into federal courts under the provisions of the Taft-Hartley Act. Section 303 establishes the right of an employer that is injured in its business or property by a secondary boycott, hot-cargo agreement, jurisdictional dispute, or other unfair labor practice to sue the responsible union or unions for damages. Any injured party, not just the employer against whom a strike or boycott is called, is entitled to bring suit for damages sustained through unlawful union action.

14.27 PREJOB CONFERENCES

When employment conditions pertaining to a given union project are out of the ordinary (that is, the conditions are not clearly provided for in the area agreement), it is usual practice for a prejob conference to be held between the contractor and the union or unions involved. This meeting may be held before bidding, to establish standard conditions for the bidding contractors, or just before the start of field operations. In either case, the underlying motive is to achieve a meeting of minds between the employer and the unions regarding job conditions of employment. For example, consider a job that is to be located in a remote area. The contractor wishes to obtain the necessary labor in the most economical manner possible, and the unions require that fair standards be maintained. The situation provides an opportunity for the employer to name "key men" who may be required, as well as an opportunity for the employer to bypass the union's "out-of-work" list. If the project is near a town of any size, the contractor may decide to quarter its workers there and furnish transportation back and forth to the site. If this solution is not feasible, temporary barracks or trailers may be provided at the project. At any rate, the contractor and the unions must arrive at a mutually acceptable understanding relative to quarters, subsistence, and transportation. In addition, the locals having jurisdiction must be checked to determine whether they have enough workers available to do the job. If not, arrangements must be made to bring in workers from the outside, either by the local or by the contractor.

There is some contractor resistance to the prejob conference concept, based on the belief that such meetings encourage the unions to make exorbitant demands. Objection or acceptance seems to be a matter of opinion and company policy. However, available evidence seems to indicate that unions are much more reasonable at this stage than they are later, when they detect conditions that lead them to believe that the contractor is not maintaining fair standards. Obtaining advance union acceptance of the contractor's procedures makes it much easier for the contractor to devote its energies to getting the job done without having to combat active union resistance and interference.

14.28   THE MERIT-SHOP CONTRACTOR

A merit-shop contractor, also referred to as an open-shop contractor, is one that is not a signatory to a labor agreement with a construction union or unions and that hires its workers from the open labor market. This type of employer hires its employees and contracts with subcontractors without regard to their union or nonunion status. A merit-shop contractor recruits, hires, trains, promotes, disciplines, and discharges its employees in accordance with its own company policies and procedures. It is up to the employer to establish its own pay rates and fringe benefit plan.

Because the merit-shop contractor has no ready-made source of construction labor, it must make an extra effort to obtain, train, and retain its field forces. This requires a company labor relations policy that helps the new employee get to know and understand the company and to gain a feeling of personal involvement and belonging. Because the contractor is free from union trade jurisdictions, open-shop firms have created new occupations combining different construction types, and helpers are widely used for routine and unskilled work.

Although merit-shop pay rates are, on the average, somewhat lower than union scale, continuity of employment provides the nonunion worker with an annual income much the same as, or better than, that of his union counterpart. Open-shop contractors employ union and nonunion workers alike. They oppose union make-work rules and strict trade jurisdictions and maintain that every contractor should have the right to deal with any other contractor or business firm, union or nonunion, as it may choose. The open-shop concept defends the right of the contractor to manage its projects. For example, merit-shop contractors decide for themselves the sizes of work crews and to what jobs a worker can be assigned. They are free to use prefabricated materials and are not subject to jurisdictional disputes, featherbedding, forced overtime, or work slowdowns. They pay workers according to their ability and performance. Workweeks of four 10-hour days without payment of overtime are common on work constructed by open-shop contractors, a concept traditionally opposed by the unions.

Merit-shop contractors have complete freedom in how they deploy field personnel. Unhampered by union rules, they can assign workers according to need and can utilize labor efficiently. When the level of company work is down, skilled personnel can be assigned to lower task levels, thus keeping key workers with the firm until the work volume again increases. Open-shop contractors can pay a broad range of wages to journeymen and helpers of different levels of skill and experience, whereas the union employer is bound to pay all workers of a given trade classification the same wage.

14.29   SOURCES OF OPEN-SHOP LABOR

A problem faced by open-shop contractors is obtaining adequate numbers of skilled workers. Of necessity, the open-shop firm must recruit its workforce through more informal channels than does its unionized counterpart. Through the medium of hiring halls, the construction unions act as brokerage agents to provide unionized contractors with the needed labor. Nonunion companies do not have such a centralized recruitment mechanism available to them.

Open-shop contractors now use a variety of procedures to assist them in obtaining their field forces. Employee referral centers or registries have been established in many areas throughout the country and provide a central source of open-shop personnel. Open-shop hiring halls are now established in many large urban centers. Operated primarily by local chapters of the Associated General Contractors of America (AGC) or the Associated Builders and Contractors (ABC), these services provide contractors with names and other information about registered individuals with specific construction skills. These centers maintain a file of names, skills, experience, and addresses of people who sign up for job placement. Advertisements in local newspapers and other media advise the community that the referral service is available. Contractors obtain names from the center and make their own contacts, conduct their own interviews, and do their own hiring.

In addition to the referral centers and hiring halls, open-shop apprenticeship and other training programs produce substantial numbers of skilled tradesmen. All open-shop firms maintain a network of informal worker contacts, mostly through their project superintendents and crew foremen. Many such firms cooperate with one another by lending and trading workers among themselves as their individual needs dictate.

14.30 APPRENTICESHIP PROGRAMS

Traditionally, apprenticeship programs in the construction industry have been a joint effort between individual union-shop contractors or local chapters of contractor associations and the AFL-CIO building trades unions. A National Joint Apprenticeship Committee for each craft is responsible for developing suggested national training standards for that craft. The actual employment and training of apprentices is essentially a local matter supervised by local joint apprenticeship committees that are made up of both contractor and labor union representatives. Most union labor agreements contain provisions for apprenticeship training. These programs are funded by levying a percentage of craft payroll from the signatory contractors to finance an apprenticeship fund.

In 1971, however, the Bureau of Apprenticeship and Training (www.bat.doleta.gov) approved national apprenticeship standards for the employees of open-shop contractors. This was the first time unilateral apprenticeship programs were approved on a national basis and placed under the direct supervision of employers only. Consequently, independent programs can be registered when the sponsors decline to participate in existing local joint training programs.

For local programs to be approved by, and registered with, a state apprenticeship council or with the Bureau of Apprenticeship and Training, the operating standards adopted by the local body must meet certain criteria. In general, a construction trade requires two to four years of on-the-job training and a minimum of 144 hours a year of related classroom instruction. Apprentices can be indentured to an individual contractor or to a multiemployer organization. Many apprenticeship programs, union and open-shop, are sponsored by contractor organizations such as the Associated General Contractors of America, Associated Builders and Contractors, National Association of Home Builders, and many other national employer organizations. It is important to note that apprenticeship wage rates on Davis-Bacon projects can be paid only to workers who are participating in a certified training program.

Federal regulations designed to increase minority-group participation in apprenticeship programs, by requiring unions and contractors to go beyond passive nondiscrimination, have been in effect for some years. These regulations require that apprenticeship programs accept enough minority-group apprentices to make the racial composition of the local labor force reflect the racial mix of the community. In addition, special apprenticeship programs are in operation to accelerate the entry of women into the building trades. Unions and/or contractors operating apprenticeship programs must have a positive affirmative action plan covering all phases of the program, including recruitment, admission, and selection. Compliance reviews are made, and noncompliance can lead to deregistration of the apprenticeship program or prosecution under Title VII of the Civil Rights Act of 1964.

In the 1980s, the Bureau of Apprenticeship and Training approved the "Model for Unilateral Trainee Program Standards," as prepared by the Associated General Contractors of America (www.agc.org) and the "Wheels of Learning" training materials of the Associated Builders and Contractors (www.abc.org) for use in unilateral (no union participation) apprenticeship programs. In recent years, both organizations have expanded these programs to offer a broader curriculum, distance learning, and powerful partnerships with not-for-profit centers such as the National Center for Construction and Research (www.nccer.org). These modular materials allow competency-based rather than time-based training. The use of these programs allows apprentices to complete each phase of their training by demonstrating their competence rather than spending a prescribed number of hours in the program. The competency-based format can appreciably reduce the length of the traditional apprenticeship period. However, the local Bureau of Apprenticeship and Training office or the state apprenticeship council, if there is one, must approve competency-based training. It is important to note that job safety is now integrated into all apprenticeship and training programs.

In some union-conducted apprenticeship programs, the unions have their apprentices sign a promissory note for each year of training, a debt that can be repaid by working for union contracting firms. To make it more difficult for nonunion contractors to get skilled union workers, the union charges apprentices for the cost of their training if they work for an open-shop firm within a stated period of time after the completion of their apprenticeship program.

14.31   NONAPPRENTICESHIP TRAINING PROGRAMS

There are a wide variety of programs designed to train construction industry workers that do not follow formal apprenticeship procedures. There are publicly funded plans at both the federal and local levels designed primarily to train minority youths, women, and unemployed workers. These programs concentrate on hands-on, pre-apprenticeship preparation for minorities and women and the upgrading of skills for the unemployed. In addition, many vocational-technical schools provide construction-related instruction.

Privately supported and conducted training programs are largely concentrated in the open-shop segment of the construction industry. Classroom instruction, on-the-job training, and correspondence courses are all used to varying degrees by different training programs.

Some contractors conduct their own in-house training programs. These include all types of training in schemes of varying style and scope. Many of these contractor efforts involve intensive specialized instruction to create craftsmen highly skilled in a particular type of work. Large open-shop contractors often make a heavy investment in manpower training. This effort involves diverse forms of training geared to the needs of specific job requirements and involving job site experience and after-hours instruction. Many local chapters of the Associated Builders and Contractors and the Associated General Contractors of America have programs to train people as helpers. These are on-the-job procedures that prepare workers to fill jobs in support of skilled craftsmen. Chapters of several contractor associations sponsor programs that offer specific task and cross-craft instruction.

14.32 SUPERVISORY TRAINING

A contractor's financial performance in today's highly competitive construction market is dependent on the caliber of its supervisory personnel: foremen, superintendents, and project managers. Decisions made by such supervisors are critical to the successful completion of a field project. Supervisory skill, or the lack of it, has a strong effect on a contractor's business success. Practiced, relevant, and time-tested supervisory abilities are needed to maintain and increase a contractor's overall effectiveness in the field.

Experienced construction supervisors available for hire are not a common commodity. The general source for such personnel is advancement up through the contractor's own ranks. However, experience shows that it is a poor practice for a construction firm to promote craftworkers into supervisory positions until they have had an opportunity to learn certain basic construction management skills.

Increased competition, labor shortages, and lack of properly trained workers now necessitate improved management education programs and supervisory training. Such training can significantly improve the productivity of field forces and decidedly increase a contracting firm's profitability. In addition, it significantly reduces employee turnover. Supervisors must help field workers appreciate the opportunity for promotion within the company organization. Supervisor training builds company loyalty by demonstrating that management cares about its personnel and takes an interest in their well-being. An example of such training is the "Supervisory Training Program" of the Associated General Contractors of America (AGC). This program is offered by local AGC chapters, construction firms, community colleges, adult education groups, vocational-technical educational institutions, and labor unions; textbooks can be purchased directly from the AGC's e-store.

Labor unions are also strengthened when business representatives and agents receive leadership training and are taught negotiating skills, contract management, and grievance and arbitration procedures, as well as U.S. labor history. This type of training is not commonly provided to organized labor throughout the United States.

14.33 PRESENT CONSTRUCTION INDUSTRY STATUS

Since 1950 there has been a profound shift in the construction industry away from union-shop and toward open-shop operations. Unions have lost more than half of their penetration in the construction industry. Despite strong opposition from the labor unions, open-shop contractors are performing an increasingly large proportion of construction. Union contractors and labor unions that dominated the industry for a half-century have lost much of their grip on construction, and open-shop work has increased substantially. This trend toward nonunion work has left no part of the country and no type of construction untouched.

A prime reason for this unprecedented transition has been the union contractors' loss of much of their traditional competitive advantage. A marked decline in worker productivity, increased wage rates, expensive fringe benefits, restrictions on the use of apprentices and helpers, inefficient work rules, and jurisdictional disputes have reduced many union firms' competitiveness and have contributed to the open-shop shift. Unduly rigid labor agreements have made it difficult or impossible for union contractors to exercise effective management control over their field operations. A swelling of the ranks of nonunion construction firms and workers, and the disenchantment of many craftsmen with the unions, have made the rapid growth of open-shop employers possible. This growth has been accelerated by owners who seek to reduce costs and job delays by using nonunion firms to do their work. Another factor contributing to the growth of open-shop companies has been the willingness of union workers to "put their union cards in their shoes" and work for nonunion firms.

Published figures show that approximately 75 percent of the construction in this country is now being done by open-shop firms. Recent data published by the Bureau of Labor Statistics indicate that there are now approximately 6.9 million construction workers in this country employed on field projects. Of these, 1.1 million belong to the various construction unions. Thus, at the present time, about 84 percent of the construction tradesmen are nonunion and 16 percent are union members.

With regard to the strong open-shop movement, it must be noted that there are appreciable differences between states, regions, and local labor markets as to the extent of the shift. There are still many areas, especially those large cities that have long been the citadels of construction union power, where open-shop penetration has been limited. In addition, the Davis-Bacon Act and many states' prevailing wage laws have operated to offset the open shop's cost advantages on public works. For many years, Davis-Bacon administrative procedures often required that open-shop contractors pay the equivalent of union wages. In addition, Davis-Bacon generally uses union craft designations and excludes from its wage categories the helper, trainee, and preapprentice labor classifications that open-shop contractors utilize so effectively. Such prevailing wage laws have effectively forced open-shop operators to adopt the costly practices of their union competitors on public projects. Hence, to date open-shop work has not made a strong entry into the public sector. However, recent modifications in state laws and the new Davis-Bacon regulations should somewhat alter this situation.

During the past several years, the Associated Builders and Contractors (ABC), an association of merit-shop companies, has become a large, active, and very effective national organization. It now has chapters in all parts of the country and provides its members with a variety of valuable services. Furthermore, more than 50 percent of the members of the traditionally union-oriented Associated General Contractors of America (AGC) now operate as open-shop or dual-shop firms. As such, the AGC now offers its members a full line of open-shop services.

14.34   DUAL-SHOP OPERATION

A number of construction contractors operating under union labor agreements have found that their labor costs have made them noncompetitive for work in certain categories or in some geographical areas. To solve this problem, many firms have chosen to form and operate a second company that is open shop. The original union-shop firm continues to function in those areas where unions are strongly entrenched, and the new, nonunion firm does business where open-shop work has become established. The NLRB has approved such "dual-shop," or "double-breasted," operations when certain conditions are met. It is of great importance for a contractor making this move that the dual shop be properly established and that the two companies be sufficiently separate in management and operation to qualify for separate employer status as defined under federal labor laws. If the two firms are found to be a single employer or the nonunion firm is found to be the "alter ego" of the union company, the new company can be held liable retroactively for underpaid wages and delinquent fringe benefit payments under the parent company's bargaining agreement. The employees of the open-shop firm could then be placed under the original company's labor contract, with all of the costly consequences. In addition, the union could file a breach-of-contract suit or unfair labor practice charge.

The principal factors that the NLRB considers in deciding whether there is sufficient separation between the two businesses include (1) interrelationship of operations, (2) centralized control of labor relations, (3) common management, and (4) common ownership or financial control. Operational integration and control of labor relations are emphasized. The main criterion for double-breasted operation is that there be two separate and independent business entities, each of whose labor policies are conducted without interference from the other. This requires that the two companies be independent in their day-to-day operations and in their labor relations policies. The usual procedure has been to establish separate corporations (not wholly owned subsidiaries or divisions) with separate management, field supervision, workforces, equipment, and financing.

The NLRB, with court approval, has held that common ownership does not, in itself, dictate an illegal status of dual operations when there is no common control of labor policies. The general rule is that two commonly owned corporations in a double-breasted operation are considered to be two separate entities under the National Labor Relations Act if there is no direct control by one company over the other with regard to day-to-day operations or labor relations. The principal consideration is that independence of labor policy makes the two act as different employers and, as such, employees of each constitute a separate bargaining unit. When each company is a separate and independent entity, each is entitled to the same protection against secondary boycotts and to the same protection from one another's labor controversies. Competent legal advice is a necessity for any contractor contemplating double-breasted operation or changing from union to nonunion status.

The construction unions vigorously oppose the dual-shop concept. As a part of their continuing efforts to inhibit the growth of dual shops, some labor unions have sent questionnaires on this subject to their signatory contractors. The purpose of such questionnaires is to determine the extent to which a union firm may be operating, managing, or controlling a nonunion company arm. The U.S. Supreme Court ruled in 1983 that such a union request for information from a union contractor concerning the existence or the workings of a dual-shop arrangement cannot be refused. To do so is regarded as an unfair labor practice and a failure to bargain in good faith.

In a move designed to keep the nonunion part of a dual-shop company from being able to compete effectively, many local unions negotiate "work preservation clauses" into their labor contracts. These clauses extend the terms of a labor agreement to any business entity with which a signatory contractor may have direct or indirect management, control, or majority ownership. Such a clause requires the nonunion arm of a dual shop to conform to all the terms of the labor contract pertaining to the union half. Thus, the work preservation clause requires that both companies be bound to the same union wage pact for all work done within its geographical jurisdiction.

14.35   UNION REACTION TO OPEN SHOP

The construction unions have not taken the strong nonunion challenge lying down. In order to halt the open-shop tide and to regain lost ground, the building trades unions have taken a number of countermeasures.

Some of the steps taken have been purely defensive, concentrating on restrictive strategies to lock in the union market. A vigorous effort to restrict or eliminate the dual shop has been under way for some time. Many kinds of broad restrictions on the subcontracting of work to nonunion firms are being included in labor contracts. Strong political pressure is being exerted to prevent further revision or repeal of the Davis-Bacon Act. The unions display opposition to any change to the present apprenticeship procedures or program classifications and support withdrawal liability from multiemployer pension plans (as discussed in Section 14.24), which can make it difficult for a union contractor to change to open-shop status. Union pension, health, and welfare funds have been used to finance projects in which construction has been restricted to union contractors.

The unions also reflect a greatly improved attitude regarding collective bargaining and have made a sincere attempt to improve the cost-effectiveness and the competitive position of unionized construction. Recent years have seen important changes in labor contract provisions. Wage cuts have been made, and wage freezes have become commonplace. Where wage increases have been negotiated, they have been quite modest. Fringe benefits have been subjected to the same treatment. Wage rates for overtime have been reduced from double time to time and a half for Saturdays and weekdays. Shift-work clauses now allow some crafts to work in shifts at round-the-clock sites at straight-time pay. The requirement for standby crews has been eliminated in many cases. Limitations have been placed on travel and subsistence pay, nonworking foremen, minimum crew sizes, and show-up pay.

Many labor contracts now include strict no-strike-no-lockout provisions. Many unions have dropped the all-union or all-nonunion job concept in accepting a mix of union and nonunion contractors and subcontractors on a given job site. Some unions now allow the use of more apprentices on jobs, as well as subjourneymen and pre-apprentices or helpers. Many restrictive work rules and jurisdictional conflicts have been eliminated, and steps have been taken to provide more job site flexibility. A concerted effort is being made to avoid job delays, strikes, picketing, and work stoppages on union projects.

There is now a joint effort by union labor and management to reduce or eliminate the adversarial relationship between unions and employers. Joint committees conduct regular meetings to resolve common concerns. A real effort is being made to open lines of communication and improve common understanding. Joint public relations programs, designed to improve the public perception of unionized construction and to apprise users and the public of the advantages of union-built jobs, are under way. Successful unions have found that accord and accommodation rather than conflict and confrontation produce the most successful long-term relationships.

QUESTIONS

1. Why are relations between craftsmen and contractor so impersonal on union projects?

2. Briefly, explain how the AFL-CIO came to be.

3. Explain the difference between a craft and an industrial union.

4. What advantages and disadvantages does collective multi-employer bargaining have over individual employer bargaining with unions?

5. Is an employer able to withdrawal from multiemployer bargaining unit and continue negotiations on his own?

6. Why would a union elect to sign a project agreement?

7. What advantages does a national agreement hold for a contractor?

8. Explain the concept of an employer lockout. Does such action effectively shut down the contractor's operations?

9. In terms of wages and raises, is the average employee better off working open-shop or union shop?

10. What is the difference between a defined benefits plan and a defined contributions plan? What advantages do each hold and what is the current trend in the industry?

11. Define the terms "open shop," "closed shop," "union shop," and "merit shop."

12. What is the current industry trend regarding the popularity of open and union-shop operations? What issues are fuelling this trend?

13. Does the Associated Builders of America represent merit-shop or union-shop contractors?

14. What does the term "double-breasted" refer to in the construction industry?

Chapter 15

Project Safety

15.1 THE COST OF CONSTRUCTION ACCIDENTS

Construction, by its very nature, is fraught with hazards and risks and results in very high rates of deaths and disabling injuries. Recent studies indicate that there are approximately 400,000 reportable injuries to construction workers and more than 1,000 deaths each year in the United States as the result of work-related injuries. The fatality rate for construction workers is among the highest of all American industries. The annual toll of accidents in the construction industry is high in terms of both cost and human suffering. Job accidents impose on this industry a tremendous burden of needless and avoidable expense. But the consequences of construction accidents are not expressible in terms of dollars alone. Money loses much of its significance when bodily injury and death are involved. Nevertheless, the financial consequences of accidents are an important matter to the construction industry and to the individual contractor.

The construction industry bears a burden of more than $2 billion per year in direct costs resulting from largely avoidable job accidents causing injury and death. Direct costs are those of hospital and medical care, subsistence payments, rehabilitation, and other benefits that are provided by workers' compensation insurance (see Section 8.37). However, construction accidents can and often do involve substantial costs that are not insurable or that are not provided for by usual construction insurance coverage. These are spoken of as "hidden" or "indirect" costs and can range from four to as much as twenty times greater than the direct costs. Examples of uninsured costs are first-aid expenses, damage or destruction of materials, cleanup and repair, idle machine time, unproductive labor time, spoiled work, equipment damage, schedule disruptions, loss of trained manpower, work slowdowns, wages paid to injured workers, adverse publicity, administrative and legal expenses, lowered employee morale, and other costly side effects.

Another common source of expense to the contractor resulting from job accidents is the third-party lawsuit filed by injured workers of another employer. For example, the injured employee of a subcontractor often sues the project owner, the general contractor, or another subcontractor, because workers' compensation laws do not normally allow an injured worker to sue his direct employer (see Section 8.33).

15.2 SAFETY LEGISLATION

The advent of the industrial revolution in the United States was marked by a simultaneous proliferation of unsafe working conditions. Although employers had certain common-law duties toward their employees, such as to provide a safe place to work, the safeguards to health and safety that are taken for granted today were not customary at that time. In fact, there was no general acceptance of the notion that employers should be concerned with the welfare and safety of employees while they were on the job. Certainly, no concerted effort was made to render working conditions less hazardous. It was believed by management that most work accidents were caused by the carelessness of the employees themselves and that it was the worker's responsibility to avoid accidents. However, the grim loss of life, limb, and livelihood aroused the public conscience, and the latter half of the nineteenth century witnessed a gradual change in the general attitude toward work safety.

The primary responsibility for statutory job-safety requirements in the United States has traditionally rested with the state governments. In 1867, Massachusetts took the first legal step toward remedying the dangerous working conditions so characteristic of the times. During the following years many of the states enacted laws pertaining to working conditions in factories and mines, the operation of machinery, the inspection of steam boilers and elevators, and fire protection. In 1911, Wisconsin established a state industrial commission that was authorized to develop and issue rules and regulations in the field of industrial health and safety that would have the force of law. All states have now enacted some form of legislation that gives designated state agencies general rule-making authority in the areas of occupational health and safety.

15.3   STATE SAFETY CODES

Details of safety codes vary somewhat from state to state, but there is a trend toward greater uniformity in the provisions of the regulations through adoption of nationally recognized codes developed by such bodies as the American National Standards Institute. Although some states have established comprehensive safety standards applicable to all employments, the tendency has been to develop special codes for particular industries, operations, or hazards. Some specific hazards are regulated in most states. Codes now relate to boilers, construction, elevators, mechanical power transmission, cranes and derricks, fire protection, floors and stairways, illumination, sanitation, ventilation, electrical hazards, explosives, ladders, spray painting, welding, and other areas of potential danger. All states require the provision of first aid and protective equipment. State safety codes make the employer and its supervisory personnel responsible for compliance with the code and for suitable safety instruction to the worker. In turn, the employee is required to make use of safeguards provided for his protection and to conduct his work in conformance with established safety rules.

Each state industrial commission or labor department has jurisdiction over every employment and place of employment within its state and is authorized to enforce and administer established codes and rules pertaining to the safety and protection of workers. The commission is vested with full power and authority to establish and enforce necessary and reasonable rules and regulations for the purpose of implementing the state law. In general, whenever the commission finds that any employment or place of employment is not safe or that employees are not being adequately protected, it is empowered to order the employer to rectify the situation and to furnish safety devices and other safeguards reasonably required. Violation of a state safety code is punishable by fine and/or imprisonment, as provided by the applicable statute. As discussed in the next section, a number of federal statutes are now preeminent over state law in matters of health and safety in certain employments unless the state has adopted health and safety requirements that meet federal standards.

15.4   FEDERAL HEALTH AND SAFETY ACTS

Several federal statutes now establish health and safety standards in a variety of occupational areas such as nuclear energy, mining, and transportation. Two federal laws have been enacted that impose safety standards on the construction industry. These are the Construction Safety Act of 1969 and the Occupational Safety and Health Act (OSHA) of 1970.

The Construction Safety Act of 1969 applies to construction projects financed in whole or in part by federal funds. It prohibits contractors from requiring tradesmen to work under conditions that are unsanitary, hazardous, or dangerous, as determined under standards issued by the secretary of labor. Before the provisions of this Act could be effectively implemented, however, the 1970 OSHA legislation was passed. It applies to employers in interstate commerce, including those in construction. The safety regulations promulgated for the Construction Safety Act have been included in the construction regulations for the Occupational Safety and Health Act. On essentially all construction, public and private, enforcement action is taken and penalties are assessed under this Act.

15.5   THE OCCUPATIONAL SAFETY AND HEALTH ACT

The Occupational Safety and Health Act established the first nationwide program for job safety and health by directing the secretary of labor to set safety and health standards for all industry. The U.S. Department of Labor immediately established the Occupational Safety and Health Administration (OSHA) to administer the Act. This Act established that every employer in interstate commerce has a twofold obligation to provide employment and a place of employment that is free of recognized health and safety hazards, and to comply with OSHA standards.

OSHA has established an Office of Construction and Engineering to respond faster and more effectively to the unique hazards and unusual situations in construction. The specialized staff of this office stays abreast of new construction methods, processes, and operations and brings its expertise into the standard-setting and compliance activities of OSHA. This affords OSHA a greater concentration on construction problems and enhances protection for those in the industry. OSHA regulations and references for the construction industry are published as "Construction Safety and Health Regulations" in the *Federal Register.* Working versions of these regulations are available from OSHA (www.osha.gov) and various contractor associations. Employers must also keep and preserve stipulated records of recordable occupational injuries and illnesses. Exempted from the Act, however, are industries already under the jurisdiction of those federal agencies that have statutory authority to establish their own safety regulations, such as mining and rail transportation. For example, the Federal Mine Safety and Health Act of 1977 prescribes safety regulations for construction performed at surface and underground mines. This includes general construction on mine property and construction that is directly associated with mining operations. This Act is administered and enforced by the Mine Safety and Health Administration of the U.S. Department of Labor.

OSHA established the position of Assistant Secretary for Occupational Safety and Health to take charge of standards and enforcement. Also established was a National Advisory Committee on Occupational Safety and Health to assist with the devising and establishing of standards. An independent Occupational Safety and Health Review Commission, appointed by the president with the advice and consent of the Senate, was created to enforce the standards and to hear appeals. The commission is assisted by an organization of trial examiners and supporting staff, in a system similar to that used by the National Labor Relations Board. The commission's findings can be appealed to the courts.

15.6   SITE INSPECTIONS BY OSHA

Under OSHA, as originally enacted, safety inspectors were authorized to conduct unannounced site inspections to see if employers were complying with safety standards. In the event of such a surprise visit, employers were required to admit the inspectors to their places of business. This requirement was changed in 1978 by a U.S. Supreme Court decision to the effect that such visits are unconstitutional because they violate the prohibition against unreasonable search. Under this ruling, inspections are still made, but admission of the OSHA inspector must be approved by the employer, or the inspector must first obtain a search warrant. However, warrants are available as long as the inspection is conducted as part of OSHA's general administrative plan for enforcement of the Act. Almost one-half of all OSHA site visits are on construction projects.

An employer representative and an employee representative have the right to accompany the inspector during his rounds of the premises. When a violation exists, a citation is issued describing the nature of the violation, indicating the amount of any civil penalty imposed, and setting a reasonable time in which to correct the situation. The employer is required to post at the work site records of citations and notices of employees' rights. The employer has 15 working days in which to contest a citation. If a citation is contested, the review commission holds a formal hearing. Enforcement of the commission's orders or review of its decision is handled by the appropriate U.S. court of appeals. If a citation is not contested, it becomes final.

Employees or their representatives can demand inspections of their employers' premises by making a written complaint directly to the Labor Department. If it is determined that there are reasonable grounds to believe a violation or danger exists, a special inspection is made as soon as possible. Moreover, during the course of an inspection, any employee or employee representative can notify the inspector of any violations that may exist. The Act provides that employees may not be discharged or discriminated against in any way for filing safety and health complaints or otherwise exercising their rights under the Act. In 1980, the U.S. Supreme Court ruled that employees have the right to refuse to perform tasks they reasonably believe could result in serious injury or death and cannot be discharged, provided that the workers attempted and failed to have such hazards corrected.

If imminent danger to safety or health is noted, the inspector is required to promptly notify the employer, the employees, and the secretary of labor. If the imminent danger is not eliminated, the secretary of labor is required to seek a temporary injunction in a federal district court to shut down that part of the operation where danger exists.

15.7   PENALTIES UNDER OSHA

OSHA provides for mandatory civil penalties against employers of up to $7,000 for each serious violation and for optional penalties of up to $7,000 for each nonserious violation. Penalties of up to $7,000 per day may be imposed for failure to correct violations within the proposed time period. Any employer who willfully or repeatedly violates the Act may be assessed penalties of up to $70,000 for each such violation. Criminal penalties are also provided for in the Act. Any willful violation resulting in the death of an employee is punishable, upon conviction, by a fine of up to $500,000, by imprisonment of not more than six months, or both. Conviction of an employer after a first conviction doubles these maximum penalties. The law requires workers to observe applicable health and safety rules but provides no penalties for their failure to comply. Employers are liable under OSHA for all acts of employees except when the employer did not and could not, with the exercise of reasonable diligence, know of the presence of a violation. It is worthy of note that if a construction manager (CM) controls or directs the work, it can be held liable for OSHA violations. This is true even when the CM performs no actual work, or has no direct contracts with trade contractors, and the actual violation was caused by employees of trade contractors.

OSHA makes an allowance for small contractors so that fines are painful, but not catastrophic. Businesses with fewer than 25 employees are able to reduce their fines by up to 60 percent. In addition, businesses that have had no previous violations may deduct an additional 10 percent. If a business is found to generally operate a safe work environment and conduct an effective safety program, it may also be granted a 25 percent reduction for good faith. Therefore, it is possible for a small business with a good health and safety program, and no history of safety violations, to reduce a penalty by as much as 95 percent of its assessment.

For many years, OSHA has had an on-site consultation program that helps firms with fewer than 250 employees to abate workplace hazards. This is a free on-site consultation service to permit contractors to identify job hazards without risking citations or penalties if unsafe conditions are corrected promptly. The consultants assist these employers in designing company health and safety programs.

OSHA permits the U.S. Department of Labor to transfer enforcement of employment safety requirements to any state that demonstrates its ability to administer the provisions of the law at least as effectively as the federal government. At present, 25 states and 1 U.S. territory have such programs in effect; the funding for these is on a fifty-fifty basis with OSHA. In these states, state inspectors conduct job site inspections and violations are processed by a state agency. There seems to be little movement among the remaining states to assume this responsibility.

15.8 OSHA HAZARD COMMUNICATION STANDARD

The controversial Hazard Communication Standard (Haz Com) became effective for the construction industry in 1988. This standard requires businesses to inventory and label hazardous job site substances, train employees in their safe use, and maintain a detailed Material Safety Data Sheet (MSDS) for each substance. The construction industry exerted every legal effort to keep this new hazard communication standard out of the industry, but after several delays, it became effective and has created much confusion regarding contractor compliance.

The objective of Haz Com is to inform workers in American industry about the chemical hazards to which they are exposed during the performance of their job duties. The law makes the employer responsible for communicating this information to its employees. In order to do this, the employer must perform three duties: (1) Prepare a hazard communication program that includes policies and procedures, a pertinent hazardous substance list, an employee training program, and Material Safety Data Sheets, (2) establish a labeling system that ensures all containers are properly identified, and (3) conduct classroom or individual training concerning the employer's program and the exposure hazards. The program is designed to reduce the dangers associated with chemical exposure for the nation's workers. Employers are also required to provide information to employees about chemical substances and train them to use this information. The contractor must train its employees on the hazards of the initial job assignment, and whenever a new hazard is introduced.

15.9   MULTIEMPLOYER WORK SITES

Although OSHA requires that each employer in interstate commerce provide its employees with employment and a place of employment that are free from recognized hazards, the matter of duty and responsibility under the Act can become very involved on a multiemployer site such as a construction project. In a workplace of this type, employees of different employers are subjected to common hazards that exist at the site, such hazards often being created by different employers. Who is responsible for safety violations under these circumstances has been a confused issue over the years because of conflicting decisions made by the Occupational Safety and Health Review Commission and the federal appeals courts.

In 1991, OSHA adopted its present enforcement policy that pertains to multiemployer sites. The basic position of OSHA, holding employers responsible for the safety and health of their own employees, remains the same, but it has specified that on a multiemployer work site there are four categories of employers: the creating employer, the exposing employer, the correcting employer, and the controlling employer. Each employer working on the site may be included in one or more of these categories.

> **The creating employer** is an employer who causes a hazardous condition that violates an OSHA standard.
>
> **The exposing employer** is an employer whose own employees are exposed to the hazard.
>
> **The correcting employer** is an employer who is engaged in common undertakings, on the same work site, as the exposing employer and is responsible for correcting a hazard. This may be a carpentry subcontractor who is responsible for building and maintaining protective handrails adjacent to fall hazards.
>
> **The controlling employer** is an employer who has general supervisory authority over the work site, including the power to correct safety and health violations itself or require others to correct them. Control can be established by contract or, in the absence of explicit contractual provisions, by the exercise of control in practice.

Based on the category or categories of the employer, different actions are required, and failing proper accomplishment of these, OSHA citations are probable. Creating employers are always citable. Exposing employers are citable only if they fail to take specific action to remove or mitigate hazards. Correcting employers are citable if equipment, barriers, or other items for which they are contractually responsible, are deemed a hazard and the correcting employer is found not to have exercised a reasonable standard of care. Controlling employers are expected to regularly inspect the work site for potential hazards and have them corrected. Although they are not held to as high an inspection standard as exposing employers, they must conduct periodic inspections, implement an effective system for promptly correcting hazards, and enforce the other employers' compliance with health and safety requirements with an effective, graduated system of enforcement and follow-up inspections.

A prime contractor or a subcontractor cannot delegate or contract away its responsibility under the Act. A general contractor can, however, put a clause in its subcontracts requiring a subcontractor to reimburse the general contractor for any losses sustained by reason of the subcontractor's failure to abide by safety regulations or general duty of care in conducting its activities at the job site. To ensure that safety and health requirements are met, the general contractor is well advised to see that its subcontracts require scrupulous adherence to safety and health regulations.

15.10   CONTRACT SAFETY REQUIREMENTS

Construction contracts routinely contain provisions requiring the prime contractor to conform to all applicable laws, ordinances, rules, and regulations that pertain to project safety. Subcontracts, in turn, extend this responsibility to the subcontractors. Contracts with some public agencies require that the contractor conform to the requirements of the safety code of the particular agency. These standards constitute a contractual obligation with which the contractor must comply or be in breach of contract. Many state highway departments include a safety code in their construction contracts. Some federal agencies, including the U.S. Army Corps of Engineers, the Naval Facilities Engineering Command, and the U.S. Bureau of Reclamation, use construction contracts that include health and safety standards in their provisions.

Although construction projects for these agencies are not exempted from OSHA requirements, this Act provides that these agencies may continue to use their own safety codes and enforce them. Contractors on such projects are required to observe OSHA standards as well as those contractual requirements that OSHA does not cover or that are more stringent than those of OSHA.

Some private owners, as well as public owners, play an active role in ensuring that the contractor implements and enforces a rigorous safety program on the project site during the construction process. Contractors are required to furnish information concerning their existing company safety programs and safety records before being allowed to bid. The bidding documents include safety specifications, and the owner works closely with the contractor on safety matters during the contract period.

Labor agreements may also impose contractual safety requirements on the contractor. The National Labor Relations Board (NLRB) has ruled that safety regulations, as an essential element of the terms and conditions of employment, are mandatory subjects of bargaining whenever either party places the issue on the bargaining table.

15.11   WORK INJURY AND ILLNESS RECORDING

OSHA requires that employers keep certain records pertaining to recordable occupational injuries and illnesses. OSHA considers cases to be recordable if they are work-related and result in any of the following:

1. Death

2. Days away from work

3. Restricted work or transfer to another job

4. Medical treatment beyond first aid

5. Loss of consciousness

6. A serious injury or illness diagnosed by a physician or other licensed health care professional

Recordable cases are classified as follows:

1. Total recordable cases—The sum of all recordable occupational injuries and illnesses, including deaths, lost-workday cases, and cases without lost workdays

2. Deaths

3. Total lost-workday cases—The sum of cases involving days away from work and/or days of restricted activity

4. Nonfatal cases without lost workdays—The sum of cases that are recordable injuries or illnesses that do not result in death or lost workdays, either days away from work or days of restricted activity

5. Total lost workdays—The sum of days away from work and days of restricted work activity

15.12 WORK INJURY AND ILLNESS RATES

For purposes of analyzing, summarizing, and presenting work injury and illness data, incidence rates are computed and published. These rates are expressed as the number of cases or days per 100 full-time employees or 200,000 employee hours per year (100 employees at 2,000 hours per year = 200,000 employee hours per year).

$$\text{Incidence rate} = \frac{\text{number of cases or days per year} \times 200,000}{\text{total employee hours per year}}$$

An incidence rate can be computed on the basis of cases or days.

Figures 15.1 and 15.2 present recent work injury information for selected major industries, including construction. These data show that the construction industry has one of the poorest safety records of all the major American industries. For the past several years, the construction industry has employed approximately 6 percent of the total labor force, but has accounted for about 9 percent of all occupational injuries and illnesses and 22 percent of all deaths resulting from occupational accidents. Figure 15.3 presents a breakdown of incidence rates for the construction industry. By computing its own incidence rates, any construction company can compare its accident experience with such national averages.

Both the National Safety Council and the Bureau of Labor Statistics of the U.S. Department of Labor compile occupational safety and health statistics. For any given year, the figures of the Bureau of Labor statistics vary somewhat from those of the National Safety Council. The principal reason for this is that the two utilize different data sources and survey coverages.

Industry Group	Workers (thousands)	Deaths		Disabling Injuries	
		Total	Per 100,000 Workers	Total (thousands)	Per 100,000 Workers
All Private Industries	**116,863**	**4,970**	**4**	**2,494**	**2,134**
Services	40,820	680	2	617	**1,512**
Trade	27,965	692	2	626	**2,239**
Manufacturing	18,072	563	3	656	**3,630**
Transportation and Public Utilities	8,060	910	11	252	**3,127**
Construction	**9,163**	**1,121**	**12**	227	**2,477**
Agriculture	3,417	789	23	49	**1,434**
Mining and Quarrying	515	121	23	15	**2,913**

Figure 15.1 Work injuries in industry groups for the year 2002. (Adapted from data published by the Bureau of Labor Statistics.)

Industry Group	Incidence Rates per 100 Full-Time Employees				
	Total Recordable Cases	Total Lost Workday and Transfer Cases	Cases Involving Days Away from Work	Cases Involving Job Transfer or Restriction	Other Recordable Cases
All Private Industries	**5.3**	**2.8**	**1.6**	**1.2**	**2.5**
Services	4.6	2.2	1.3	0.9	2.4
Trade	5.3	2.7	1.6	1.1	2.6
Manufacturing	7.2	4.1	1.7	2.3	3.1
Transportation and Public Utilities	6.1	4.0	2.7	1.3	2.1
Construction	**7.1**	**3.8**	**2.8**	**1.1**	**3.2**
Agriculture	6.4	3.3	2.1	1.2	3.1
Mining and Quarrying	4.0	2.6	2.0	0.7	1.4

Figure 15.2 Occupational injuries and illness rates for the year 2002. (Adapted from data published by the Bureau of Labor Statistics.)

Construction Category	Incidence Rates per 100 Full-Time Employees				
	Total Recordable Cases	Total Lost Workday and Transfer Cases	Cases Involving Days Away from Work	Cases Involving Job Transfer or Restriction	Other Recordable Cases
Construction Industry	**7.1**	**3.8**	**2.8**	**1.1**	**3.2**
General Building Construction	6.2	3.2	2.3	0.9	2.9
Heavy Construction	6.4	3.7	2.4	1.3	2.7
Special Trade Construction	7.5	4.1	3.0	1.1	3.5

Figure 15.3 Occupational injuries and illness rates in construction for the year 2002. (Adapted from data published by the Bureau of Labor Statistics.)

15.13 ECONOMIC BENEFITS OF SAFETY

In addition to the humanitarian aspect, there is a compelling economic motivation in accident prevention. Many financial benefits accrue to the contractor that conducts field operations in a safe manner and whose accident rate is low. The most immediate and obvious financial benefit is the savings realized because of accidents that do not happen. Mention has already been made of the indirect costs of accidents that are not covered by insurance and that can result in serious financial loss.

Construction workers appreciate and value job safety even though they sometimes tend to be careless in their work habits. Employees who feel that their employers are genuinely concerned about safety and who see tangible evidence of this concern are more likely to be loyal and cooperative workers. Integrating safety measures into daily operations, rather than treating them as a necessary evil, generates high employee morale and loyalty. Safety is one of the potent forces that make workers proud of the company they work for, proud of the manner in which they perform their jobs, and proud of their record in preventing accidents. The fruits of good worker morale are higher production and better workmanship, two economic benefits of great worth.

Another important financial benefit to the contractor that results from fewer accidents is the reduced cost of insurance. As discussed in Chapter 8, the premiums of certain types of insurance are adjusted up or down for the individual contracting firm in accordance with its loss experience. Workers' compensation acts in all 50 states now include the concept of Experience Modification Rating (EMR) for each individual employer. The company EMR is based on the injury experience of the employees of that employer. The EMR is calculated from the employer's record of losses caused by worker injury. The better the employer's workplace safety record, the lower is its EMR value. Typical EMR values of construction contractors range from 0.3 to 2.0, with the average value being somewhere between 0.9 and 1.0. The EMR is used to adjust the premium charged to a given contractor for its workers' compensation coverage. The EMR is a multiplier, that is, the greater its value, the more expensive a contractor's premiums for workers' compensation insurance. Thus, it is clear that an aggressive safety program can significantly reduce insurance costs. A lower EMR can result in large savings to the contractor.

Here is an illustration of the monetary value of an effective job safety program. Assume that a contractor's annual volume of work is $10 million a year. Considering a typical amount of subcontracting and the cost of materials, this general contractor's annual payroll will be about $2.5 million. If its present worker's compensation manual rates average about 20 percent, the annual premium cost will be about $500,000. Now assume that an effective accident prevention program results in an EMR of 0.7 for this contractor. This will mean a reduction in the annual premium cost of its worker's compensation insurance to about $350,000. Annual savings of approximately $150,000 are thereby realized in the cost of this one insurance coverage alone. A direct result of this is the improved competitive position of the contractor. Lower insurance premiums mean lower bids.

Safety is also an important public relations tool, there being few other activities with such great potential for building goodwill. Good public relations have important financial and business implications for the contractor. A serious job accident can negatively affect a contractor's reputation, and the attendant adverse publicity can undo years of favorable public relations.

15.14   SAFETY SERVICES OF CONTRACTOR ASSOCIATIONS

Local chapters of contractor organizations, such as the Associated General Contractors of America (AGC) and the Associated Builders and Contractors (ABC), offer invaluable assistance in providing their members with a wide variety of safety services. The professional staff and membership committees of these and other such organizations create and provide a great number of safety publications, manuals, training programs, correspondence courses, program guides, and standards to assist their members with their company safety programs.

Several local chapters utilize safety vans that go directly to job sites, company yards, and other locations to deliver firsthand the safety message to construction workers. These are mobile classrooms filled with videotapes, posters, brochures, pamphlets, exhibits, and informational literature about safety practices, procedures, and regulations.

15.15   THE ROLE OF MANAGEMENT IN SAFETY

Top management bears the ultimate responsibility for a company's accident record, and the impetus for improved safety performance must emanate from this level. The courts charge the employer with the responsibility to (1) provide a safe place to work, (2) provide safe appliances, tools, and equipment, (3) enforce safety rules, and (4) provide instruction regarding dangers that may be encountered by employees. Construction companies must now follow a myriad of federal and state regulations, with the prime responsibility for safety resting squarely on management. There is a growing vulnerability of upper and middle managers to personal and criminal liability arising from job injuries. If a company acts in wanton, willful, or reckless disregard for the safety of its employees, its management can be prosecuted, and managers are becoming more vulnerable to lawsuits of this type.

The attitude of management toward safe work practices will be reflected by the supervisors and workers. An environment that conveys top-level interest and concern keeps everybody safety conscious. Conversely, if the top executives are not genuinely interested in preventing accidents and injuries, no one else is likely to be concerned. If employee cooperation and participation are to be obtained, the accident control program must start with the demonstrated interest and backing of company management. In addition, adequate funds must be provided to ensure the successful operation of the safety program.

Every level of company management, therefore, must reflect a concern for safety and set a good example of compliance with safety regulations. The expression of management's interest must be vocal, visible, and vigorous. The logical beginning is a written company safety policy, announced by management and implemented by rules that are enforced. This policy should be widely publicized so that all employees become familiar with it, especially with the aspects that pertain directly to them. A carefully worded safety policy that is personally and forcefully brought to the attention of every employee, and is signed by the company president, will emphasize management's desire and determination to reduce accidents. In addition, a written policy is very useful in the enforcement of safety rules by supervisors. It is important that company management place the administration and enforcement of the safety policy squarely in the mainstream of company operations. Safety must be included as a company objective along with productivity and quality performance.

Top administrators should personally give timely credit and commendation for good safety performance whenever and wherever it occurs. The attendance of executives at employee safety meetings will impress workers with management's sincere desire to eliminate job accidents. A recommended procedure is that company accidents be categorized by project so that top management can see where safety problems exist. Field supervisors should be evaluated for promotions and salary increases in terms of accident records as well as production and cost-effectiveness. Basing promotions and salary decisions on safety as well as company profits can be an effective tool in reducing job accidents.

15.16   THE COMPANY SAFETY PROGRAM

A written safety program is just as much a part of a contractor's business as estimating and procurement. The company safety program should include specific descriptions of (1) the company's safety organization, (2) safety training and personal protection, (3) first aid training, (4) fire prevention, (5) safety record keeping, (6) job site inspection, and (7) accident and hazard reporting. Fundamentally, the company safety plan must be one of identifying specific job hazards and educating its employees to conduct their work in a way that will minimize the risk of injury. The cause of an accident is an unsafe act or an unsafe condition. Statistics show that 85 percent of all construction job accidents are the direct result of unsafe acts (behavior) and 15 percent of all such accidents are the direct result of unsafe conditions. This is a fact that must be recognized by a company safety program; it indicates that working conditions more suited to natural human needs and limitations tend to reduce accidents. The objective of the company accident prevention program is to eliminate both unsafe conditions and unsafe acts from the job site. An unsafe condition is a physical circumstance, mechanical fault, or defect that could cause an accident. Such a condition can result from either inadequate safety planning or no planning at all and is a consequence of the way a job is planned or the way it is conducted. In most cases, unsafe conditions can be corrected by changing job procedures. An unsafe act is the failure of an individual to follow correct procedures or to use the proper work methods. It results from personal carelessness or lack of safety training. Such acts can be prevented by safety education for the individual worker and enforcement of safety regulations. Because insurance company studies disclose that an unsafe act by a worker is present in the vast majority of all construction accidents, the company safety plan should certainly address this point and have a highly personal focus. Merely checking the fieldwork for unsafe conditions will not suffice.

Training programs to assist supervisors with the safety planning of their projects and to educate workers on how to perform their tasks properly are essential to any company plan. The U.S. Department of Labor, American National Standards Institute, National Safety Council, National Society of Safety Engineers, and others provide various kinds of aid and assistance with safety training and instruction. Organized labor plays an active role in health and safety training and sponsors educational programs in this regard. First aid training for selected personnel, sponsored by a local office of the American Red Cross, can be a valuable part of a company safety training program.

In many geographical areas, company safety efforts are supported and strengthened by regional programs conducted by contractor associations and trade groups. These efforts include conducting safety training programs, distributing safety newsletters, granting awards for superior safety performance, and providing safety services and information to contractor members. In addition, most insurance companies offer a variety of loss-prevention services. Training programs, job site visits, reviews of company safety efforts, assistance with special problems, and supplying safety products are among the services provided by insurance companies.

Rules, safety devices, and mechanical safeguards are important to the prevention of job accidents, but the mental attitude of the workers is even more significant. Until an awareness of their individual responsibilities and a desire to ensure their personal safety are established in the minds of all employees, efforts to reduce accidents will not be fully effective. Accident prevention in construction is largely a human relations problem and is achieved primarily through education, persuasion, and eternal vigilance. People cause accidents, and only people can prevent them. The company plan must emphasize the personal approach to job safety.

Because of differences in organization, type of activity, and scope of operations, each contractor must develop an accident prevention plan that fits its own particular situation. Job hazards differ considerably among housing, building, highway, heavy, utility, and industrial construction, a fact that must be reflected by the detailed workings of the program. Overall responsibility for the company's safety program must be placed with an individual who is capable, energetic, qualified, and interested in safety. This person may be a company executive or a staff assistant; he must have the authority to carry out the necessary functions effectively, being responsible for project safety planning, safety training, distribution and use of safety equipment, maintenance of first aid facilities on the project, job inspection, investigation of accidents, writing and filing of accident reports, and associated duties.

Company safety meetings of supervisory personnel held at regular intervals can be very effective. A formal program featuring a guest speaker or a film on safety may be presented. Information concerning how the company ranked in state and national safety contests can be discussed. Further discussion can include recent accidents, with information on how they could have been prevented. Regular safety inspections of each project and a detailed investigation of all lost-time accidents are important aspects of a company safety plan.

Safety contests between company projects may also be conducted. At a dinner meeting, project superintendents with the best safety records are recognized and presented with cash awards sufficiently substantial to be appreciated. The participation of rank-and-file workers can also be encouraged through the awarding of cash prizes for their safety suggestions and slogans.

A contractor's safety program must pay special attention to the training of personnel who operate heavy trucks and the enforcement of safety regulations pertaining to such work. Compliance with drug-testing rules, vehicle inspection, maintenance standards, driver qualification and licensing, driving time limitations, and accident-reporting requirements is mandatory and also requires special attention. Trucking safety regulations require that well-qualified, rested drivers operate properly maintained vehicles. Qualifying drivers requires background investigations, a study of driving records, written and road tests, medical examinations, and drug testing. Federal and state regulations also have strict requirements pertaining to inspection and maintenance of trucks and trailers. Recent safety laws are now being applied to all firms that use trucks and buses.

It is important to note that recent court decisions have been directed at defining the legal duty and liability of a prime contractor for injuries incurred by the employees of its subcontractors. One case in this regard held that if a general contractor retains control over a subcontractor's work, the general contractor has the duty of exercising ordinary care in furnishing the subcontractor and its employees with a reasonably safe workplace. Because the general contractor normally retains control over the work of a subcontractor, the court ruled that the general contractor had an affirmative duty to control project work methods to ensure the safety of everyone at the work site. The fact that the subcontractor created the unsafe work site condition did not excuse the general contractor from its duty or potential liability. A general contractor must be aware of the law in this regard in those states where it is doing business.

15.17 THE PROJECT SAFETY PLAN

Accident prevention aimed at the avoidance of specific occurrences must be planned and incorporated into each construction project. The constantly changing nature of a project under construction does not allow the detection and elimination of hazards purely on an experiential basis. The contractor must establish, in advance of each project, the particular hazards that the proposed methods, procedures, and equipment will create and then devise an accident prevention plan to combat them. Analysis of accident experience is a valuable first step.

With the ground rules of job safety having been established, the commencement of field operations must be accompanied by the implementation of the plan. The following steps are suggested for the conduct of a safety program during construction activities:

1. Assign prime responsibility for the project safety plan and its enforcement to the top field supervisor, the project manager, or the job superintendent. Under this manager, craft foremen are made responsible for safety measures as applied to their groups. Foremen are with their crews all the time; therefore, they must watch for unsafe practices or conditions and promote safety by instruction, precept, and example. On large projects, there may be a safety engineer who coordinates the overall plan and devotes time and energy exclusively to matters of safety, first aid, sanitation, fire prevention, and other such responsibilities.

2. Make suitable and adequate first aid facilities readily available. These facilities may range from a well-supplied first aid kit on small projects to a nurse-staffed infirmary on very large jobs. On the usual job with no professional medical assistance available on the site, first aid training for supervisors and foremen is desirable. Locations of the nearest hospital facilities and ambulance service telephone numbers should be prominently posted in the job office. First aid kits should be dustproof, easily available, and checked at frequent intervals for needed replenishment. Every person on the job should be informed as to where first aid facilities are available.

3. Provide safety indoctrination for all new personnel to acquaint them with the company safety policy, to provide the direction and rules needed to perform their work in a safe and orderly fashion, and to stress that strict conformance with safety regulations is a condition of employment. Special safety instruction must be given for particularly hazardous work. Every employee should be instructed to immediately report any injury, however trivial, to the foreman and to obtain suitable first aid treatment.

4. Insist on the wearing and proper use of personal protective clothing and equipment, with no exceptions made.

5. Conduct periodic "toolbox" safety talks and demonstrations on the project for all work crews. Merely prodding workers to "be careful" is not likely to accomplish much. The proper use of tools, handling of materials, building of scaffolds, and operation of equipment can be topics of successful meetings. The information must be specific, practical, and pertinent to current operations. Suggestions for improved safety should be solicited from the workers.

6. Utilize safety posters, safety instruction cards, and warning signs. A prominent display of the project's accident record, as well as notices that remind workers of the specific project safety requirements, can be very effective.

7. Periodic meetings of the superintendent, craft foremen, and other supervisors are essential to reviewing job safety and to making necessary revisions to the program. Investigate all lost-time accidents and devise corrective measures to prevent their recurrence.

8. Provide adequate, suitable, and easily accessible fire-fighting equipment and materials. Because welding and flame cutting are among the most frequent causes of construction fires, special regulations must apply to these activities. Specific areas should be provided for the storage of flammable, combustible, and explosive materials.

9. Establish a program of periodic job safety inspections. The inspection team must include the company safety specialist and the top field supervisor. Safety violations and job hazards should be noted, and immediate action should be taken to correct them.

10. Insist on good project housekeeping. Designated storage areas for materials, tools, and supplies should be maintained and used. Rubbish and waste material should be removed promptly from the area of operations.

11. See that regular equipment maintenance includes safety inspection. This maintenance should include inspection of accident hazards such as frayed cables, bad tires, slipping clutches, and electrical grounds. Inspection and maintenance must not be limited to mechanical equipment but should be extended to scaffolding, towers, ladders, and other nonoperating items.

12. Seek and obtain the full cooperation of all subcontractors on the project. All of the measures previously described must include subcontractor personnel as well.

15.18   THE FIELD SUPERVISOR

Top management has the major responsibility for establishing safety policies, procedures, and safe working conditions. However, most of what is planned and established must reach the worker on the job by way of the field supervisor. The craft foremen are the real key to the success of any project safety plan. To be effective, any campaign for the prevention of accidents in construction must be communicated to the individual workers in a clear, practical, and understandable form. Although the executives of the company may prescribe safe practices, the foreman, who has the authority to direct the workers in the field and is in daily contact with them, plays a dominant role in implementing the company's safety policy.

Construction workers mirror their supervisors' attitude toward safety. For this reason, the wholehearted cooperation of the superintendent and foremen is indispensable to the success of any safety program. The best way for supervisors to sell accident prevention is to practice what they preach. A worker is much more likely to follow a supervisor's example than that same supervisor's instructions. If supervisors break a safety rule, they not only reduce the importance of the rule but also diminish their workers' confidence in them. If the field supervisors clearly believe in safety and reflect the fact by word and deed, the people under them will be much more cognizant of the advantages of safe work practices and the costly results of any alternative. It is largely up to the supervisors to find and control the potential hazards on the job. They must teach by doing and be able to demonstrate the safe way to do any particular job. Job instructions to workers should include not only what is to be done but also how it is to be done. The accident potential decreases when the worker is given complete procedural instructions in advance.

Safety regulations must be enforced. The worker must be taught safe practices and be required to follow them. Discipline in matters of safety is a delicate issue, but it is very important to an accident prevention program. If a rule has been established, the supervisor must always enforce it. An established safety regulation that is not enforced will not be obeyed. If a rule is enforced only some of the time, a worker who is reprimanded will feel that he is being "picked on" and singled out for unfair treatment. Those who violate the rule and are not reprimanded will begin to believe that the regulation is not very important. The supervisor must, of course, be careful to keep personal feelings out of disciplining workers in matters of safety.

The objective of such discipline is to improve the performance of the crew as it relates to safety. Workers who are convinced that a safety procedure is designed to protect their welfare will support its enforcement. Similarly, workers are more likely to accept a reprimand if they believe that it is for their own protection. A spirit of group responsibility for safety and a sense of competition in safe practices among crews help to establish a self-disciplining attitude toward safety violations.

15.19   ACCIDENT RECORDS

The keeping of accident records is an important part of a company's safety and health program. These records serve to pinpoint the locations and underlying causes of job injuries and illnesses, information that is vital to the planning of more effective accident prevention programs. They also provide information on the efficacy of the overall safety effort and how the company compares with other construction firms. Accurate, complete, and detailed records can be invaluable in defending against charges of safety law violations or claims for damages. Accident information can also be valuable in quite another way. It can be used to arouse the competitive spirit of workers and supervisors on the various projects to establish a safety record that compares favorably with the experience of other projects or with their own past records.

Accident recording and reporting may be thought of as starting with the first aid log that is maintained on each project. Every job injury or illness is made a matter of record, regardless of how inconsequential it may appear to be. A daily record book is maintained in which entries are made, including the date, name of the employee affected, nature of the injury or illness, first aid treatment given on the site, and any further information deemed desirable. These items are followed by a first report of injury, which is required by workers' compensation laws in most states. The first report of injury is an application used in most states to initiate a worker's compensation claim. While the format differs from state to state, its content is similar to that of OSHA Form 301 discussed in the next paragraph. It is prepared for every incident that requires off-site medical treatment regardless of whether time is lost from work or not. Another report that is used by many construction companies is prepared by the appropriate foreman for each recordable injury or illness and is directed toward analyzing the accident and determining how it could have been prevented.

OSHA requires that occupational injury and illness records be kept for all employees. These records include (1) a log of work-related injuries and illnesses (OSHA Form 300), (2) an annual summary of work-related injuries and illnesses (OSHA Form 300A), and (3) a detailed injury and illness incident report (OSHA Form 301). The log must record all cases resulting in medical treatment, loss of consciousness, restriction of work or motion, or transfer to another, less taxing job. Any fatality or any accident that hospitalizes five or more employees must be reported to the OSHA area director within 48 hours. OSHA records provide management with valuable information concerning the efficacy of the company's safety program.

15.20 PROTECTION OF THE PUBLIC

A very important aspect of a company's accident prevention program is the safety of the public. People are innately curious and are capable of many thoughtless actions in their attempts to observe construction operations. The contracting firm must reconcile itself to the fact that the public *will* know what is going on; how it finds out is up to the contractor. If an attempt is made to shut people out completely, they will feel compelled to climb over the fences, follow trucks through the gates, or perform some other equally human but hazardous act. Verbal admonitions and warning signs seem to do little good, and positive action must be taken to protect the public against its own unthinking actions. The best scheme is for the contractor to allow the public to view the proceedings from controlled vantage points. The contractor that provides means for the public to see the work and simultaneously to be protected from its hazards is wise from the standpoint of both safety and public relations.

The problem of protecting the public becomes even more difficult during weekends and at other times when job operations are not in progress. Children, in particular, seem to find construction projects irresistible, and insurance company records are filled with cases involving the deaths or injuries of youngsters playing on project sites during off-hours. On projects located in areas where children are likely to be playing, the job safety plan must make specific provision for this additional hazard.

15.21 THE COST OF A SAFETY PROGRAM

A company safety program does, of course, cost money. A common rule of thumb is that an effective company safety program will have a cost of about 2.5 percent of the direct labor expense. The fact is that safety is absolutely necessary for the proper conduct of a construction business. There is, however, an important distinction that must be recognized between safety costs and other items of company expense. The distinction is that the spending of one dollar for safety can save the contractor two dollars. Although this ratio is only figurative, it has been well demonstrated that savings from accidents that never occur, because of an effective safety program, more than offset the costs of the program itself.

A commonly cited rule in this regard is that a construction contractor with a past safety performance at the national average can, with reasonable effort, realize a savings of $1 million or more on each $10 million of direct wage expense each year. This is equal to a savings of 10 percent of the direct labor expense.

Accident reports make it evident that most job accidents can be prevented at only moderate, if any, extra cost. The additional expense involved in building a proper scaffold, shoring an excavation, grounding an electric drill, or otherwise doing the work in a safe and thoughtful manner is insignificant as compared with the costs of accident or injury. The contractor cannot look on its safety program as an extra source of expense. Rather, because an effective accident prevention program is necessary to ensure fast-moving, smoothly functioning jobs, any costs entailed should be considered merely normal business expenses associated with the efficient operation of the project.

QUESTIONS

1. Is the body of law governing health and safety in the construction industry primarily state or federal law? Explain your answer.

2. What obligations does the Occupational Health and Safety Act impose on an employer?

3. What arrangements an OSHA inspector must make before conducting an unannounced inspection of a project site? What rights does a contractor have with regard to such OSHA inspections?

4. In the event of a willful safety violation, is OSHA able to take criminal action against the supervisor and employee engaged in the violation? Explain.

5. What conditions must exist for a company to reduce its OSHA imposed fine by 95%?

6. What assistance does OSHA make available to businesses? What conditions must the business meet in order to take advantage of this assistance?

7. Who is responsible for the enforcement of OSHA safety requirements; the federal government or state governments? Explain.

8. What requirements does Haz Com place on construction site supervisors?

9. If a correcting employer failed to correct a safety hazard created by a creating employer in time to prevent injury, can the creating employer still be cited?

10. If an owner such as the Bureau of Reclamation uses a construction contract with its own health and safety requirements, must the contractor obey these requirements, OSHA's requirements, or a combination of both? Explain.

11. If a contractor had 17 reportable accidents in 2002, how many employees working 48 forty-hour workweeks would he have needed to match the average reportable incidence rate for the construction industry that year?

12. Assuming a worker's compensation manual rate of 20%, if a contractor is able to reduce its EMR from 1.1 to 0.7, what percentage of total annual payroll could the contractor save?

13. What responsibility does a general contractor carry for the health and safety of a subcontractor's employees?

14. Federal law requires that fatalities or accidents that hospitalize five or more employees must be reported. Who must these incidents be reported to and what are the time restrictions if any?

Appendix A

AIA Document B141-1997 Standard Form of Agreement Between Owner and Architect

with Standard Form of Architect's Services

AIA® Document B141™ – 1997 Part 1

Standard Form of Agreement Between Owner and Architect
with Standard Form of Architect's Services

AGREEMENT made as of the   day of
in the year
(In words, indicate day, month and year)

BETWEEN the Architect's client identified as the Owner:
(Name, address and other information)

This document has important
legal consequences.
Consultation with an attorney
is encouraged with respect to
its completion or modification.

and the Architect:
(Name, address and other information)

For the following Project:
(Include detailed description of Project)

The Owner and Architect agree as follows:

TABLE OF ARTICLES

ARTICLE 1.1 INITIAL INFORMATION

§ 1.1.1 This Agreement is based on the following information and assumptions.
(Note the disposition for the following items by inserting the requested information or a statement such as "not applicable," "unknown at time of execution" or "to be determined later by mutual agreement.")

§ 1.1.2 PROJECT PARAMETERS

§ 1.1.2.1 The objective or use is:
(Identify or describe, if appropriate, proposed use or goals.)

§ 1.1.2.2 The physical parameters are:
(Identify or describe, if appropriate, size, location, dimensions, or other pertinent information, such as geotechnical reports about the site.)

§ 1.1.2.3 The Owner's Program is:
(Identify documentation or state the manner in which the program will be developed.)

§ 1.1.2.4 The legal parameters are:
(Identify pertinent legal information, including, if appropriate, land surveys and legal descriptions and restrictions of the site.)

§ 1.1.2.5 The financial parameters are as follows.

    .1   Amount of the Owner's overall budget for the Project, including the Architect's compensation, is:

    .2   Amount of the Owner's budget for the Cost of the Work, excluding the Architect's compensation, is:

§ 1.1.2.6 The time parameters are:
(Identify, if appropriate, milestone dates, durations or fast track scheduling.)

§ 1.1.2.7 The proposed procurement or delivery method for the Project is:
(Identify method such as competitive bid, negotiated contract, or construction management.)

§ 1.1.2.8 Other parameters are:
(Identify special characteristics or needs of the Project such as energy, environmental or historic preservation requirements.)

§ 1.1.3 PROJECT TEAM
§ 1.1.3.1 The Owner's Designated Representative is:
(List name, address and other information.)

§ 1.1.3.2 The persons or entities, in addition to the Owner's Designated Representative, who are required to review the Architect's submittals to the Owner are:
(List name, address and other information.)

§ 1.1.3.3 The Owner's other consultants and contractors are:
(List discipline and, if known, identify them by name and address.)

§ 1.1.3.4 The Architect's Designated Representative is:
(List name, address and other information.)

§ 1.1.3.5 The consultants retained at the Architect's expense are:
(List discipline and, if known, identify them by name and address.)

§ 1.1.4 Other important initial information is:

§ 1.1.5 When the services under this Agreement include contract administration services, the General Conditions of the Contract for Construction shall be the edition of AIA Document A201 current as of the date of this Agreement, or as follows:

§ 1.1.6 The information contained in this Article 1.1 may be reasonably relied upon by the Owner and Architect in determining the Architect's compensation. Both parties, however, recognize that such information may change and, in that event, the Owner and the Architect shall negotiate appropriate adjustments in schedule, compensation and Change in Services in accordance with Section 1.3.3.

ARTICLE 1.2 RESPONSIBILITIES OF THE PARTIES

§ 1.2.1 The Owner and the Architect shall cooperate with one another to fulfill their respective obligations under this Agreement. Both parties shall endeavor to maintain good working relationships among all members of the Project team.

§ 1.2.2 OWNER

§ 1.2.2.1 Unless otherwise provided under this Agreement, the Owner shall provide full information in a timely manner regarding requirements for and limitations on the Project. The Owner shall furnish to the Architect, within 15 days after receipt of a written request, information necessary and relevant for the Architect to evaluate, give notice of or enforce lien rights.

§ 1.2.2.2 The Owner shall periodically update the budget for the Project, including that portion allocated for the Cost of the Work. The Owner shall not significantly increase or decrease the overall budget, the portion of the budget allocated for the Cost of the Work, or contingencies included in the overall budget or a portion of the budget, without the agreement of the Architect to a corresponding change in the Project scope and quality.

§ 1.2.2.3 The Owner's Designated Representative identified in Section 1.1.3 shall be authorized to act on the Owner's behalf with respect to the Project. The Owner or the Owner's Designated Representative shall render decisions in a timely manner pertaining to documents submitted by the Architect in order to avoid unreasonable delay in the orderly and sequential progress of the Architect's services.

§ 1.2.2.4 The Owner shall furnish the services of consultants other than those designated in Section 1.1.3 or authorize the Architect to furnish them as a Change in Services when such services are requested by the Architect and are reasonably required by the scope of the Project.

§ 1.2.2.5 Unless otherwise provided in this Agreement, the Owner shall furnish tests, inspections and reports required by law or the Contract Documents, such as structural, mechanical, and chemical tests, tests for air and water pollution, and tests for hazardous materials.

§ 1.2.2.6 The Owner shall furnish all legal, insurance and accounting services, including auditing services, that may be reasonably necessary at any time for the Project to meet the Owner's needs and interests.

§ 1.2.2.7 The Owner shall provide prompt written notice to the Architect if the Owner becomes aware of any fault or defect in the Project, including any errors, omissions or inconsistencies in the Architect's Instruments of Service.

§ 1.2.3 ARCHITECT
§ 1.2.3.1 The services performed by the Architect, Architect's employees and Architect's consultants shall be as enumerated in Article 1.4.

§ 1.2.3.2 The Architect's services shall be performed as expeditiously as is consistent with professional skill and care and the orderly progress of the Project. The Architect shall submit for the Owner's approval a schedule for the performance of the Architect's services which initially shall be consistent with the time periods established in Section 1.1.2.6 and which shall be adjusted, if necessary, as the Project proceeds. This schedule shall include allowances for periods of time required for the Owner's review, for the performance of the Owner's consultants, and for approval of submissions by authorities having jurisdiction over the Project. Time limits established by this schedule approved by the Owner shall not, except for reasonable cause, be exceeded by the Architect or Owner.

§ 1.2.3.3 The Architect's Designated Representative identified in Section 1.1.3 shall be authorized to act on the Architect's behalf with respect to the Project.

§ 1.2.3.4 The Architect shall maintain the confidentiality of information specifically designated as confidential by the Owner, unless withholding such information would violate the law, create the risk of significant harm to the public or prevent the Architect from establishing a claim or defense in an adjudicatory proceeding. The Architect shall require of the Architect's consultants similar agreements to maintain the confidentiality of information specifically designated as confidential by the Owner.

§ 1.2.3.5 Except with the Owner's knowledge and consent, the Architect shall not engage in any activity, or accept any employment, interest or contribution that would reasonably appear to compromise the Architect's professional judgment with respect to this Project.

§ 1.2.3.6 The Architect shall review laws, codes, and regulations applicable to the Architect's services. The Architect shall respond in the design of the Project to requirements imposed by governmental authorities having jurisdiction over the Project.

§ 1.2.3.7 The Architect shall be entitled to rely on the accuracy and completeness of services and information furnished by the Owner. The Architect shall provide prompt written notice to the Owner if the Architect becomes aware of any errors, omissions or inconsistencies in such services or information.

ARTICLE 1.3 TERMS AND CONDITIONS
§ 1.3.1 COST OF THE WORK
§ 1.3.1.1 The Cost of the Work shall be the total cost or, to the extent the Project is not completed, the estimated cost to the Owner of all elements of the Project designed or specified by the Architect.

§ 1.3.1.2 The Cost of the Work shall include the cost at current market rates of labor and materials furnished by the Owner and equipment designed, specified, selected or specially provided for by the Architect, including the costs of management or supervision of construction or installation provided by a separate construction manager or contractor, plus a reasonable allowance for their overhead and profit. In addition, a reasonable allowance for contingencies shall be included for market conditions at the time of bidding and for changes in the Work.

§ 1.3.1.3 The Cost of the Work does not include the compensation of the Architect and the Architect's consultants, the costs of the land, rights-of-way and financing or other costs that are the responsibility of the Owner.

§ 1.3.2 INSTRUMENTS OF SERVICE
§ 1.3.2.1 Drawings, specifications and other documents, including those in electronic form, prepared by the Architect and the Architect's consultants are Instruments of Service for use solely with respect to this Project. The Architect and the Architect's consultants shall be deemed the authors and owners of their respective Instruments of Service and shall retain all common law, statutory and other reserved rights, including copyrights.

§ 1.3.2.2 Upon execution of this Agreement, the Architect grants to the Owner a nonexclusive license to reproduce the Architect's Instruments of Service solely for purposes of constructing, using and maintaining the Project, provided that the Owner shall comply with all obligations, including prompt payment of all sums when due, under this Agreement. The Architect shall obtain similar nonexclusive licenses from the Architect's consultants consistent with this Agreement. Any termination of this Agreement prior to completion of the Project shall terminate this license. Upon such termination, the Owner shall refrain from making further reproductions of Instruments of Service and shall return to the Architect within seven days of termination all originals and reproductions in the Owner's possession or control. If and upon the date the Architect is adjudged in default of this Agreement, the foregoing license shall be deemed terminated and replaced by a second, nonexclusive license permitting the Owner to authorize other similarly credentialed design professionals to reproduce and, where permitted by law, to make changes, corrections or additions to the Instruments of Service solely for purposes of completing, using and maintaining the Project.

§ 1.3.2.3 Except for the licenses granted in Section 1.3.2.2, no other license or right shall be deemed granted or implied under this Agreement. The Owner shall not assign, delegate, sublicense, pledge or otherwise transfer any license granted herein to another party without the prior written agreement of the Architect. However, the Owner shall be permitted to authorize the Contractor, Subcontractors, Sub-subcontractors and material or equipment suppliers to reproduce applicable portions of the Instruments of Service appropriate to and for use in their execution of the Work by license granted in Section 1.3.2.2. Submission or distribution of Instruments of Service to meet official regulatory requirements or for similar purposes in connection with the Project is not to be construed as publication in derogation of the reserved rights of the Architect and the Architect's consultants. The Owner shall not use the Instruments of Service for future additions or alterations to this Project or for other projects, unless the Owner obtains the prior written agreement of the Architect and the Architect's consultants. Any unauthorized use of the Instruments of Service shall be at the Owner's sole risk and without liability to the Architect and the Architect's consultants.

§ 1.3.2.4 Prior to the Architect providing to the Owner any Instruments of Service in electronic form or the Owner providing to the Architect any electronic data for incorporation into the Instruments of Service, the Owner and the Architect shall by separate written agreement set forth the specific conditions governing the format of such Instruments of Service or electronic data, including any special limitations or licenses not otherwise provided in this Agreement.

§ 1.3.3 CHANGE IN SERVICES

§ 1.3.3.1 Change in Services of the Architect, including services required of the Architect's consultants, may be accomplished after execution of this Agreement, without invalidating the Agreement, if mutually agreed in writing, if required by circumstances beyond the Architect's control, or if the Architect's services are affected as described in Section 1.3.3.2. In the absence of mutual agreement in writing, the Architect shall notify the Owner prior to providing such services. If the Owner deems that all or a part of such Change in Services is not required, the Owner shall give prompt written notice to the Architect, and the Architect shall have no obligation to provide those services. Except for a change due to the fault of the Architect, Change in Services of the Architect shall entitle the Architect to an adjustment in compensation pursuant to Section 1.5.2, and to any Reimbursable Expenses described in Section 1.3.9.2 and Section 1.5.5.

§ 1.3.3.2 If any of the following circumstances affect the Architect's services for the Project, the Architect shall be entitled to an appropriate adjustment in the Architect's schedule and compensation:

 .1 change in the instructions or approvals given by the Owner that necessitate revisions in Instruments of Service;

 .2 enactment or revision of codes, laws or regulations or official interpretations which necessitate changes to previously prepared Instruments of Service;

 .3 decisions of the Owner not rendered in a timely manner;

 .4 significant change in the Project including, but not limited to, size, quality, complexity, the Owner's schedule or budget, or procurement method;

 .5 failure of performance on the part of the Owner or the Owner's consultants or contractors;

 .6 preparation for and attendance at a public hearing, a dispute resolution proceeding or a legal proceeding except where the Architect is party thereto;

 .7 change in the information contained in Article 1.1.

§ 1.3.4 MEDIATION
§ 1.3.4.1 Any claim, dispute or other matter in question arising out of or related to this Agreement shall be subject to mediation as a condition precedent to arbitration or the institution of legal or equitable proceedings by either party. If such matter relates to or is the subject of a lien arising out of the Architect's services, the Architect may proceed in accordance with applicable law to comply with the lien notice or filing deadlines prior to resolution of the matter by mediation or by arbitration.

§ 1.3.4.2 The Owner and Architect shall endeavor to resolve claims, disputes and other matters in question between them by mediation which, unless the parties mutually agree otherwise, shall be in accordance with the Construction Industry Mediation Rules of the American Arbitration Association currently in effect. Request for mediation shall be filed in writing with the other party to this Agreement and with the American Arbitration Association. The request may be made concurrently with the filing of a demand for arbitration but, in such event, mediation shall proceed in advance of arbitration or legal or equitable proceedings, which shall be stayed pending mediation for a period of 60 days from the date of filing, unless stayed for a longer period by agreement of the parties or court order.

§ 1.3.4.3 The parties shall share the mediator's fee and any filing fees equally. The mediation shall be held in the place where the Project is located, unless another location is mutually agreed upon. Agreements reached in mediation shall be enforceable as settlement agreements in any court having jurisdiction thereof.

§ 1.3.5 ARBITRATION
§ 1.3.5.1 Any claim, dispute or other matter in question arising out of or related to this Agreement shall be subject to arbitration. Prior to arbitration, the parties shall endeavor to resolve disputes by mediation in accordance with Section 1.3.4.

§ 1.3.5.2 Claims, disputes and other matters in question between the parties that are not resolved by mediation shall be decided by arbitration which, unless the parties mutually agree otherwise, shall be in accordance with the Construction Industry Arbitration Rules of the American Arbitration Association currently in effect. The demand for arbitration shall be filed in writing with the other party to this Agreement and with the American Arbitration Association.

§ 1.3.5.3 A demand for arbitration shall be made within a reasonable time after the claim, dispute or other matter in question has arisen. In no event shall the demand for arbitration be made after the date when institution of legal or equitable proceedings based on such claim, dispute or other matter in question would be barred by the applicable statute of limitations.

§ 1.3.5.4 No arbitration arising out of or relating to this Agreement shall include, by consolidation or joinder or in any other manner, an additional person or entity not a party to this Agreement, except by written consent containing a specific reference to this Agreement and signed by the Owner, Architect, and any other person or entity sought to be joined. Consent to arbitration involving an additional person or entity shall not constitute consent to arbitration of any claim, dispute or other matter in question not described in the written consent or with a person or entity not named or described therein. The foregoing agreement to arbitrate and other agreements to arbitrate with an additional person or entity duly consented to by parties to this Agreement shall be specifically enforceable in accordance with applicable law in any court having jurisdiction thereof.

§ 1.3.5.5 The award rendered by the arbitrator or arbitrators shall be final, and judgment may be entered upon it in accordance with applicable law in any court having jurisdiction thereof.

§ 1.3.6 CLAIMS FOR CONSEQUENTIAL DAMAGES
The Architect and the Owner waive consequential damages for claims, disputes or other matters in question arising out of or relating to this Agreement. This mutual waiver is applicable, without limitation, to all consequential damages due to either party's termination in accordance with Section 1.3.8.

§ 1.3.7 MISCELLANEOUS PROVISIONS
§ 1.3.7.1 This Agreement shall be governed by the law of the principal place of business of the Architect, unless otherwise provided in Section 1.4.2.

§ 1.3.7.2 Terms in this Agreement shall have the same meaning as those in the edition of AIA Document A201, General Conditions of the Contract for Construction, current as of the date of this Agreement.

§ 1.3.7.3 Causes of action between the parties to this Agreement pertaining to acts or failures to act shall be deemed to have accrued and the applicable statutes of limitations shall commence to run not later than either the date of Substantial Completion for acts or failures to act occurring prior to Substantial Completion or the date of issuance of the final Certificate for Payment for acts or failures to act occurring after Substantial Completion. In no event shall such statutes of limitations commence to run any later than the date when the Architect's services are substantially completed.

§ 1.3.7.4 To the extent damages are covered by property insurance during construction, the Owner and the Architect waive all rights against each other and against the contractors, consultants, agents and employees of the other for damages, except such rights as they may have to the proceeds of such insurance as set forth in the edition of AIA Document A201, General Conditions of the Contract for Construction, current as of the date of this Agreement. The Owner or the Architect, as appropriate, shall require of the contractors, consultants, agents and employees of any of them similar waivers in favor of the other parties enumerated herein.

§ 1.3.7.5 Nothing contained in this Agreement shall create a contractual relationship with or a cause of action in favor of a third party against either the Owner or Architect.

§ 1.3.7.6 Unless otherwise provided in this Agreement, the Architect and Architect's consultants shall have no responsibility for the discovery, presence, handling, removal or disposal of or exposure of persons to hazardous materials or toxic substances in any form at the Project site.

§ 1.3.7.7 The Architect shall have the right to include photographic or artistic representations of the design of the Project among the Architect's promotional and professional materials. The Architect shall be given reasonable access to the completed Project to make such representations. However, the Architect's materials shall not include the Owner's confidential or proprietary information if the Owner has previously advised the Architect in writing of the specific information considered by the Owner to be confidential or proprietary. The Owner shall provide professional credit for the Architect in the Owner's promotional materials for the Project.

§ 1.3.7.8 If the Owner requests the Architect to execute certificates, the proposed language of such certificates shall be submitted to the Architect for review at least 14 days prior to the requested dates of execution. The Architect shall not be required to execute certificates that would require knowledge, services or responsibilities beyond the scope of this Agreement.

§ 1.3.7.9 The Owner and Architect, respectively, bind themselves, their partners, successors, assigns and legal representatives to the other party to this Agreement and to the partners, successors, assigns and legal representatives of such other party with respect to all covenants of this Agreement. Neither the Owner nor the Architect shall assign this Agreement without the written consent of the other, except that the Owner may assign this Agreement to an institutional lender providing financing for the Project. In such event, the lender shall assume the Owner's rights and obligations under this Agreement. The Architect shall execute all consents reasonably required to facilitate such assignment.

§ 1.3.8 TERMINATION OR SUSPENSION

§ 1.3.8.1 If the Owner fails to make payments to the Architect in accordance with this Agreement, such failure shall be considered substantial nonperformance and cause for termination or, at the Architect's option, cause for suspension of performance of services under this Agreement. If the Architect elects to suspend services, prior to suspension of services, the Architect shall give seven days' written notice to the Owner. In the event of a suspension of services, the Architect shall have no liability to the Owner for delay or damage caused the Owner because of such suspension of services. Before resuming services, the Architect shall be paid all sums due prior to suspension and any expenses incurred in the interruption and resumption of the Architect's services. The Architect's fees for the remaining services and the time schedules shall be equitably adjusted.

§ 1.3.8.2 If the Project is suspended by the Owner for more than 30 consecutive days, the Architect shall be compensated for services performed prior to notice of such suspension. When the Project is resumed, the Architect shall be compensated for expenses incurred in the interruption and resumption of the Architect's services. The Architect's fees for the remaining services and the time schedules shall be equitably adjusted.

§ 1.3.8.3 If the Project is suspended or the Architect's services are suspended for more than 90 consecutive days, the Architect may terminate this Agreement by giving not less than seven days' written notice.

§ 1.3.8.4 This Agreement may be terminated by either party upon not less than seven days' written notice should the other party fail substantially to perform in accordance with the terms of this Agreement through no fault of the party initiating the termination.

§ 1.3.8.5 This Agreement may be terminated by the Owner upon not less than seven days' written notice to the Architect for the Owner's convenience and without cause.

§ 1.3.8.6 In the event of termination not the fault of the Architect, the Architect shall be compensated for services performed prior to termination, together with Reimbursable Expenses then due and all Termination Expenses as defined in Section 1.3.8.7.

§ 1.3.8.7 Termination Expenses are in addition to compensation for the services of the Agreement and include expenses directly attributable to termination for which the Architect is not otherwise compensated, plus an amount for the Architect's anticipated profit on the value of the services not performed by the Architect.

§ 1.3.9 PAYMENTS TO THE ARCHITECT

§ 1.3.9.1 Payments on account of services rendered and for Reimbursable Expenses incurred shall be made monthly upon presentation of the Architect's statement of services. No deductions shall be made from the Architect's compensation on account of penalty, liquidated damages or other sums withheld from payments to contractors, or on account of the cost of changes in the Work other than those for which the Architect has been adjudged to be liable.

§ 1.3.9.2 Reimbursable Expenses are in addition to compensation for the Architect's services and include expenses incurred by the Architect and Architect's employees and consultants directly related to the Project, as identified in the following Clauses:

    .1   transportation in connection with the Project, authorized out-of-town travel and subsistence, and electronic communications;

    .2   fees paid for securing approval of authorities having jurisdiction over the Project;

    .3   reproductions, plots, standard form documents, postage, handling and delivery of Instruments of Service;

    .4   expense of overtime work requiring higher than regular rates if authorized in advance by the Owner;

    .5   renderings, models and mock-ups requested by the Owner;

    .6   expense of professional liability insurance dedicated exclusively to this Project or the expense of additional insurance coverage or limits requested by the Owner in excess of that normally carried by the Architect and the Architect's consultants;

    .7   reimbursable expenses as designated in Section 1.5.5;

    .8   other similar direct Project-related expenditures.

§ 1.3.9.3 Records of Reimbursable Expenses, of expenses pertaining to a Change in Services, and of services performed on the basis of hourly rates or a multiple of Direct Personnel Expense shall be available to the Owner or the Owner's authorized representative at mutually convenient times.

§ 1.3.9.4 Direct Personnel Expense is defined as the direct salaries of the Architect's personnel engaged on the Project and the portion of the cost of their mandatory and customary contributions and benefits related thereto, such as employment taxes and other statutory employee benefits, insurance, sick leave, holidays, vacations, employee retirement plans and similar contributions.

ARTICLE 1.4 SCOPE OF SERVICES AND OTHER SPECIAL TERMS AND CONDITIONS

§ 1.4.1 Enumeration of Parts of the Agreement. This Agreement represents the entire and integrated agreement between the Owner and the Architect and supersedes all prior negotiations, representations or agreements, either written or oral. This Agreement may be amended only by written instrument signed by both Owner and Architect. This Agreement comprises the documents listed below.

§ 1.4.1.1 Standard Form of Agreement Between Owner and Architect, AIA Document B141-1997.

§ 1.4.1.2 Standard Form of Architect's Services: Design and Contract Administration, AIA Document B141-1997, or as follows:
(List other documents, if any, delineating Architect's scope of services.)

§ 1.4.1.3 Other documents as follows:
(List other documents, if any, forming part of the Agreement.)

§ 1.4.2 Special Terms and Conditions. Special terms and conditions that modify this Agreement are as follows:

ARTICLE 1.5 COMPENSATION

§ 1.5.1 For the Architect's services as described under Article 1.4, compensation shall be computed as follows:

§ 1.5.2 If the services of the Architect are changed as described in Section 1.3.3.1, the Architect's compensation shall be adjusted. Such adjustment shall be calculated as described below or, if no method of adjustment is indicated in this Section 1.5.2, in an equitable manner.
(Insert basis of compensation, including rates and multiples of Direct Personnel Expense for Principals and employees, and identify Principals and classify employees, if required. Identify specific services to which particular methods of compensation apply.)

§ 1.5.3 For a Change in Services of the Architect's consultants, compensation shall be computed as a multiple of (   ) times the amounts billed to the Architect for such services.

§ 1.5.4 For Reimbursable Expenses as described in Section 1.3.9.2, and any other items included in Section 1.5.5 as Reimbursable Expenses, the compensation shall be computed as a multiple of (   ) times the expenses incurred by the Architect, and the Architect's employees and consultants.

§ 1.5.5 Other Reimbursable Expenses, if any, are as follows:

§ 1.5.6 The rates and multiples for services of the Architect and the Architect's consultants as set forth in this Agreement shall be adjusted in accordance with their normal salary review practices.

§ 1.5.7 An initial payment of   Dollars ($) shall be made upon execution of this Agreement and is the minimum payment under this Agreement. It shall be credited to the Owner's account at final payment. Subsequent payments for services shall be made monthly, and where applicable, shall be in proportion to services performed on the basis set forth in this Agreement.

§ 1.5.8 Payments are due and payable   (   ) days from the date of the Architect's invoice. Amounts unpaid   (   ) days after the invoice date shall bear interest at the rate entered below, or in the absence thereof at the legal rate prevailing from time to time at the principal place of business of the Architect.
(Insert rate of interest agreed upon.)

(Usury laws and requirements under the Federal Truth in Lending Act, similar state and local consumer credit laws and other regulations at the Owner's and Architect's principal places of business, the location of the Project and elsewhere may affect the validity of this provision. Specific legal advice should be obtained with respect to deletions or modifications, and also regarding requirements such as written disclosures or waivers.)

§ 1.5.9 If the services covered by this Agreement have not been completed within (   ) months of the date hereof, through no fault of the Architect, extension of the Architect's services beyond that time shall be compensated as provided in Section 1.5.2.

This Agreement entered into as of the day and year first written above.

OWNER   **ARCHITECT**

_____   _____
(Signature)   *(Signature)*

_____   _____
(Printed name and title)   *(Printed name and title)*

CAUTION: You should sign an original AIA Contract Document, on which this text appears in RED. An original assures that changes will not be obscured.

◆AIA® Document B141™ – 1997 Part 2

Standard Form of Architect's Services: Design and Contract Administration

TABLE OF ARTICLES

This document has important legal consequences. Consultation with an attorney is encouraged with respect to its completion or modification.

ARTICLE 2.1 PROJECT ADMINISTRATION SERVICES

§ 2.1.1 The Architect shall manage the Architect's services and administer the Project. The Architect shall consult with the Owner, research applicable design criteria, attend Project meetings, communicate with members of the Project team and issue progress reports. The Architect shall coordinate the services provided by the Architect and the Architect's consultants with those services provided by the Owner and the Owner's consultants.

§ 2.1.2 When Project requirements have been sufficiently identified, the Architect shall prepare, and periodically update, a Project schedule that shall identify milestone dates for decisions required of the Owner, design services furnished by the Architect, completion of documentation provided by the Architect, commencement of construction and Substantial Completion of the Work.

§ 2.1.3 The Architect shall consider the value of alternative materials, building systems and equipment, together with other considerations based on program, budget and aesthetics in developing the design for the Project.

§ 2.1.4 Upon request of the Owner, the Architect shall make a presentation to explain the design of the Project to representatives of the Owner.

§ 2.1.5 The Architect shall submit design documents to the Owner at intervals appropriate to the design process for purposes of evaluation and approval by the Owner. The Architect shall be entitled to rely on approvals received from the Owner in the further development of the design.

§ 2.1.6 The Architect shall assist the Owner in connection with the Owner's responsibility for filing documents required for the approval of governmental authorities having jurisdiction over the Project.

§ 2.1.7 EVALUATION OF BUDGET AND COST OF THE WORK

§ 2.1.7.1 When the Project requirements have been sufficiently identified, the Architect shall prepare a preliminary estimate of the Cost of the Work. This estimate may be based on current area, volume or similar conceptual estimating techniques. As the design process progresses through the end of the preparation of the Construction Documents, the Architect shall update and refine the preliminary estimate of the Cost of the Work. The Architect shall advise the Owner of any adjustments to previous estimates of the Cost of the Work indicated by changes in Project requirements or general market conditions. If at any time the Architect's estimate of the Cost of the Work exceeds the Owner's budget, the Architect shall make appropriate recommendations to the Owner to adjust the Project's size, quality or budget, and the Owner shall cooperate with the Architect in making such adjustments.

§ 2.1.7.2 Evaluations of the Owner's budget for the Project, the preliminary estimate of the Cost of the Work and updated estimates of the Cost of the Work prepared by the Architect represent the Architect's judgment as a design professional familiar with the construction industry. It is recognized, however, that neither the Architect nor the Owner has control over the cost of labor, materials or equipment, over the Contractor's methods of determining bid prices, or over competitive bidding, market or negotiating conditions. Accordingly, the Architect cannot and does not warrant or represent that bids or negotiated prices will not vary from the Owner's budget for the Project or from any estimate of the Cost of the Work or evaluation prepared or agreed to by the Architect.

§ 2.1.7.3 In preparing estimates of the Cost of the Work, the Architect shall be permitted to include contingencies for design, bidding and price escalation; to determine what materials, equipment, component systems and types of construction are to be included in the Contract Documents; to make reasonable adjustments in the scope of the Project and to include in the Contract Documents alternate bids as may be necessary to adjust the estimated Cost of the Work to meet the Owner's budget for the Cost of the Work. If an increase in the Contract Sum occurring after execution of the Contract between the Owner and the Contractor causes the budget for the Cost of the Work to be exceeded, that budget shall be increased accordingly.

§ 2.1.7.4 If bidding or negotiation has not commenced within 90 days after the Architect submits the Construction Documents to the Owner, the budget for the Cost of the Work shall be adjusted to reflect changes in the general level of prices in the construction industry.

§ 2.1.7.5 If the budget for the Cost of the Work is exceeded by the lowest bona fide bid or negotiated proposal, the Owner shall:

    .1   give written approval of an increase in the budget for the Cost of the Work;

    .2   authorize rebidding or renegotiating of the Project within a reasonable time;

    .3   terminate in accordance with Section 1.3.8.5; or

    .4   cooperate in revising the Project scope and quality as required to reduce the Cost of the Work.

§ 2.1.7.6 If the Owner chooses to proceed under Section 2.1.7.5.4, the Architect, without additional compensation, shall modify the documents for which the Architect is responsible under this Agreement as necessary to comply with the budget for the Cost of the Work. The modification of such documents shall be the limit of the Architect's responsibility under this Section 2.1.7. The Architect shall be entitled to compensation in accordance with this Agreement for all services performed whether or not construction is commenced.

ARTICLE 2.2 SUPPORTING SERVICES

§ 2.2.1 Unless specifically designated in Section 2.8.3, the services in this Article 2.2 shall be provided by the Owner or the Owner's consultants and contractors.

§ 2.2.1.1 The Owner shall furnish a program setting forth the Owner's objectives, schedule, constraints and criteria, including space requirements and relationships, special equipment, systems and site requirements.

§ 2.2.1.2 The Owner shall furnish surveys to describe physical characteristics, legal limitations and utility locations for the site of the Project, and a written legal description of the site. The surveys and legal information shall include, as applicable, grades and lines of streets, alleys, pavements and adjoining property and structures; adjacent drainage; rights-of-way, restrictions, easements, encroachments, zoning, deed restrictions, boundaries and contours of the site; locations, dimensions and necessary data with respect to existing buildings, other improvements and trees; and information concerning available utility services and lines, both public and private, above and below grade, including inverts and depths. All the information on the survey shall be referenced to a Project benchmark.

§ 2.2.1.3 The Owner shall furnish services of geotechnical engineers which may include but are not limited to test borings, test pits, determinations of soil bearing values, percolation tests, evaluations of hazardous materials, ground corrosion tests and resistivity tests, including necessary operations for anticipating subsoil conditions, with reports and appropriate recommendations.

ARTICLE 2.3 EVALUATION AND PLANNING SERVICES

§ 2.3.1 The Architect shall provide a preliminary evaluation of the information furnished by the Owner under this Agreement, including the Owner's program and schedule requirements and budget for the Cost of the Work, each in terms of the other. The Architect shall review such information to ascertain that it is consistent with the requirements of the Project and shall notify the Owner of any other information or consultant services that may be reasonably needed for the Project.

§ 2.3.2 The Architect shall provide a preliminary evaluation of the Owner's site for the Project based on the information provided by the Owner of site conditions, and the Owner's program, schedule and budget for the Cost of the Work.

§ 2.3.3 The Architect shall review the Owner's proposed method of contracting for construction services and shall notify the Owner of anticipated impacts that such method may have on the Owner's program, financial and time requirements, and the scope of the Project.

ARTICLE 2.4 DESIGN SERVICES

§ 2.4.1 The Architect's design services shall include normal structural, mechanical and electrical engineering services.

§ 2.4.2 SCHEMATIC DESIGN DOCUMENTS

§ 2.4.2.1 The Architect shall provide Schematic Design Documents based on the mutually agreed-upon program, schedule, and budget for the Cost of the Work. The documents shall establish the conceptual design of the Project illustrating the scale and relationship of the Project components. The Schematic Design Documents shall include a conceptual site plan, if appropriate, and preliminary building plans, sections and elevations. At the Architect's option, the Schematic Design Documents may include study models, perspective sketches, electronic modeling or combinations of these media. Preliminary selections of major building systems and construction materials shall be noted on the drawings or described in writing.

§ 2.4.3 DESIGN DEVELOPMENT DOCUMENTS

§ **2.4.3.1** The Architect shall provide Design Development Documents based on the approved Schematic Design Documents and updated budget for the Cost of the Work. The Design Development Documents shall illustrate and describe the refinement of the design of the Project, establishing the scope, relationships, forms, size and appearance of the Project by means of plans, sections and elevations, typical construction details, and equipment layouts. The Design Development Documents shall include specifications that identify major materials and systems and establish in general their quality levels.

§ 2.4.4 CONSTRUCTION DOCUMENTS

§ **2.4.4.1** The Architect shall provide Construction Documents based on the approved Design Development Documents and updated budget for the Cost of the Work. The Construction Documents shall set forth in detail the requirements for construction of the Project. The Construction Documents shall include Drawings and Specifications that establish in detail the quality levels of materials and systems required for the Project.

§ **2.4.4.2** During the development of the Construction Documents, the Architect shall assist the Owner in the development and preparation of: (1) bidding and procurement information which describes the time, place and conditions of bidding; bidding or proposal forms; and the form of agreement between the Owner and the Contractor; and (2) the Conditions of the Contract for Construction (General, Supplementary and other Conditions). The Architect also shall compile the Project Manual that includes the Conditions of the Contract for Construction and Specifications and may include bidding requirements and sample forms.

ARTICLE 2.5 CONSTRUCTION PROCUREMENT SERVICES

§ **2.5.1** The Architect shall assist the Owner in obtaining either competitive bids or negotiated proposals and shall assist the Owner in awarding and preparing contracts for construction.

§ **2.5.2** The Architect shall assist the Owner in establishing a list of prospective bidders or contractors.

§ **2.5.3** The Architect shall assist the Owner in bid validation or proposal evaluation and determination of the successful bid or proposal, if any. If requested by the Owner, the Architect shall notify all prospective bidders or contractors of the bid or proposal results.

§ 2.5.4 COMPETITIVE BIDDING

§ **2.5.4.1** Bidding Documents shall consist of bidding requirements, proposed contract forms, General Conditions and Supplementary Conditions, Specifications and Drawings.

§ **2.5.4.2** If requested by the Owner, the Architect shall arrange for procuring the reproduction of Bidding Documents for distribution to prospective bidders. The Owner shall pay directly for the cost of reproduction or shall reimburse the Architect for such expenses.

§ **2.5.4.3** If requested by the Owner, the Architect shall distribute the Bidding Documents to prospective bidders and request their return upon completion of the bidding process. The Architect shall maintain a log of distribution and retrieval, and the amounts of deposits, if any, received from and returned to prospective bidders.

§ **2.5.4.4** The Architect shall consider requests for substitutions, if permitted by the Bidding Documents, and shall prepare and distribute addenda identifying approved substitutions to all prospective bidders.

§ **2.5.4.5** The Architect shall participate in or, at the Owner's direction, shall organize and conduct a pre-bid conference for prospective bidders.

§ **2.5.4.6** The Architect shall prepare responses to questions from prospective bidders and provide clarifications and interpretations of the Bidding Documents to all prospective bidders in the form of addenda.

§ **2.5.4.7** The Architect shall participate in or, at the Owner's direction, shall organize and conduct the opening of the bids. The Architect shall subsequently document and distribute the bidding results, as directed by the Owner.

§ 2.5.5 NEGOTIATED PROPOSALS

§ 2.5.5.1 Proposal Documents shall consist of proposal requirements, proposed contract forms, General Conditions and Supplementary Conditions, Specifications and Drawings.

§ 2.5.5.2 If requested by the Owner, the Architect shall arrange for procuring the reproduction of Proposal Documents for distribution to prospective contractors. The Owner shall pay directly for the cost of reproduction or shall reimburse the Architect for such expenses.

§ 2.5.5.3 If requested by the Owner, the Architect shall organize and participate in selection interviews with prospective contractors.

§ 2.5.5.4 The Architect shall consider requests for substitutions, if permitted by the Proposal Documents, and shall prepare and distribute addenda identifying approved substitutions to all prospective contractors.

§ 2.5.5.5 If requested by the Owner, the Architect shall assist the Owner during negotiations with prospective contractors. The Architect shall subsequently prepare a summary report of the negotiation results, as directed by the Owner.

ARTICLE 2.6 CONTRACT ADMINISTRATION SERVICES
§ 2.6.1 GENERAL ADMINISTRATION

§ 2.6.1.1 The Architect shall provide administration of the Contract between the Owner and the Contractor as set forth below and in the edition of AIA Document A201, General Conditions of the Contract for Construction, current as of the date of this Agreement. Modifications made to the General Conditions, when adopted as part of the Contract Documents, shall be enforceable under this Agreement only to the extent that they are consistent with this Agreement or approved in writing by the Architect.

§ 2.6.1.2 The Architect's responsibility to provide the Contract Administration Services under this Agreement commences with the award of the initial Contract for Construction and terminates at the issuance to the Owner of the final Certificate for Payment. However, the Architect shall be entitled to a Change in Services in accordance with Section 2.8.2 when Contract Administration Services extend 60 days after the date of Substantial Completion of the Work.

§ 2.6.1.3 The Architect shall be a representative of and shall advise and consult with the Owner during the provision of the Contract Administration Services. The Architect shall have authority to act on behalf of the Owner only to the extent provided in this Agreement unless otherwise modified by written amendment.

§ 2.6.1.4 Duties, responsibilities and limitations of authority of the Architect under this Article 2.6 shall not be restricted, modified or extended without written agreement of the Owner and Architect with consent of the Contractor, which consent will not be unreasonably withheld.

§ 2.6.1.5 The Architect shall review properly prepared, timely requests by the Contractor for additional information about the Contract Documents. A properly prepared request for additional information about the Contract Documents shall be in a form prepared or approved by the Architect and shall include a detailed written statement that indicates the specific Drawings or Specifications in need of clarification and the nature of the clarification requested.

§ 2.6.1.6 If deemed appropriate by the Architect, the Architect shall on the Owner's behalf prepare, reproduce and distribute supplemental Drawings and Specifications in response to requests for information by the Contractor.

§ 2.6.1.7 The Architect shall interpret and decide matters concerning performance of the Owner and Contractor under, and requirements of, the Contract Documents on written request of either the Owner or Contractor. The Architect's response to such requests shall be made in writing within any time limits agreed upon or otherwise with reasonable promptness.

§ 2.6.1.8 Interpretations and decisions of the Architect shall be consistent with the intent of and reasonably inferable from the Contract Documents and shall be in writing or in the form of drawings. When making such interpretations and initial decisions, the Architect shall endeavor to secure faithful performance by both Owner and Contractor, shall not show partiality to either, and shall not be liable for the results of interpretations or decisions so rendered in good faith.

§ 2.6.1.9 The Architect shall render initial decisions on claims, disputes or other matters in question between the Owner and Contractor as provided in the Contract Documents. However, the Architect's decisions on matters relating to aesthetic effect shall be final if consistent with the intent expressed in the Contract Documents.

§ 2.6.2 EVALUATIONS OF THE WORK

§ 2.6.2.1 The Architect, as a representative of the Owner, shall visit the site at intervals appropriate to the stage of the Contractor's operations, or as otherwise agreed by the Owner and the Architect in Article 2.8, (1) to become generally familiar with and to keep the Owner informed about the progress and quality of the portion of the Work completed, (2) to endeavor to guard the Owner against defects and deficiencies in the Work, and (3) to determine in general if the Work is being performed in a manner indicating that the Work, when fully completed, will be in accordance with the Contract Documents. However, the Architect shall not be required to make exhaustive or continuous on-site inspections to check the quality or quantity of the Work. The Architect shall neither have control over or charge of, nor be responsible for, the construction means, methods, techniques, sequences or procedures, or for safety precautions and programs in connection with the Work, since these are solely the Contractor's rights and responsibilities under the Contract Documents.

§ 2.6.2.2 The Architect shall report to the Owner known deviations from the Contract Documents and from the most recent construction schedule submitted by the Contractor. However, the Architect shall not be responsible for the Contractor's failure to perform the Work in accordance with the requirements of the Contract Documents. The Architect shall be responsible for the Architect's negligent acts or omissions, but shall not have control over or charge of and shall not be responsible for acts or omissions of the Contractor, Subcontractors, or their agents or employees, or of any other persons or entities performing portions of the Work.

§ 2.6.2.3 The Architect shall at all times have access to the Work wherever it is in preparation or progress.

§ 2.6.2.4 Except as otherwise provided in this Agreement or when direct communications have been specially authorized, the Owner shall endeavor to communicate with the Contractor through the Architect about matters arising out of or relating to the Contract Documents. Communications by and with the Architect's consultants shall be through the Architect.

§ 2.6.2.5 The Architect shall have authority to reject Work that does not conform to the Contract Documents. Whenever the Architect considers it necessary or advisable, the Architect will have authority to require inspection or testing of the Work in accordance with the provisions of the Contract Documents, whether or not such Work is fabricated, installed or completed. However, neither this authority of the Architect nor a decision made in good faith either to exercise or not to exercise such authority shall give rise to a duty or responsibility of the Architect to the Contractor, Subcontractors, material and equipment suppliers, their agents or employees or other persons or entities performing portions of the Work.

§ 2.6.3 CERTIFICATION OF PAYMENTS TO CONTRACTOR

§ 2.6.3.1 The Architect shall review and certify the amounts due the Contractor and shall issue Certificates for Payment in such amounts. The Architect's certification for payment shall constitute a representation to the Owner, based on the Architect's evaluation of the Work as provided in Section 2.6.2 and on the data comprising the Contractor's Application for Payment, that the Work has progressed to the point indicated and that, to the best of the Architect's knowledge, information and belief, the quality of the Work is in accordance with the Contract Documents. The foregoing representations are subject (1) to an evaluation of the Work for conformance with the Contract Documents upon Substantial Completion, (2) to results of subsequent tests and inspections, (3) to correction of minor deviations from the Contract Documents prior to completion, and (4) to specific qualifications expressed by the Architect.

§ 2.6.3.2 The issuance of a Certificate for Payment shall not be a representation that the Architect has (1) made exhaustive or continuous on-site inspections to check the quality or quantity of the Work, (2) reviewed construction means, methods, techniques, sequences or procedures, (3) reviewed copies of requisitions received from Subcontractors and material suppliers and other data requested by the Owner to substantiate the Contractor's right to payment, or (4) ascertained how or for what purpose the Contractor has used money previously paid on account of the Contract Sum.

§ 2.6.3.3 The Architect shall maintain a record of the Contractor's Applications for Payment.

§ 2.6.4 SUBMITTALS
§ 2.6.4.1 The Architect shall review and approve or take other appropriate action upon the Contractor's submittals such as Shop Drawings, Product Data and Samples, but only for the limited purpose of checking for conformance with information given and the design concept expressed in the Contract Documents. The Architect's action shall be taken with such reasonable promptness as to cause no delay in the Work or in the activities of the Owner, Contractor or separate contractors, while allowing sufficient time in the Architect's professional judgment to permit adequate review. Review of such submittals is not conducted for the purpose of determining the accuracy and completeness of other details such as dimensions and quantities, or for substantiating instructions for installation or performance of equipment or systems, all of which remain the responsibility of the Contractor as required by the Contract Documents. The Architect's review shall not constitute approval of safety precautions or, unless otherwise specifically stated by the Architect, of any construction means, methods, techniques, sequences or procedures. The Architect's approval of a specific item shall not indicate approval of an assembly of which the item is a component.

§ 2.6.4.2 The Architect shall maintain a record of submittals and copies of submittals supplied by the Contractor in accordance with the requirements of the Contract Documents.

§ 2.6.4.3 If professional design services or certifications by a design professional related to systems, materials or equipment are specifically required of the Contractor by the Contract Documents, the Architect shall specify appropriate performance and design criteria that such services must satisfy. Shop Drawings and other submittals related to the Work designed or certified by the design professional retained by the Contractor shall bear such professional's written approval when submitted to the Architect. The Architect shall be entitled to rely upon the adequacy, accuracy and completeness of the services, certifications or approvals performed by such design professionals.

§ 2.6.5 CHANGES IN THE WORK
§ 2.6.5.1 The Architect shall prepare Change Orders and Construction Change Directives for the Owner's approval and execution in accordance with the Contract Documents. The Architect may authorize minor changes in the Work not involving an adjustment in Contract Sum or an extension of the Contract Time which are consistent with the intent of the Contract Documents. If necessary, the Architect shall prepare, reproduce and distribute Drawings and Specifications to describe Work to be added, deleted or modified, as provided in Section 2.8.2.

§ 2.6.5.2 The Architect shall review properly prepared, timely requests by the Owner or Contractor for changes in the Work, including adjustments to the Contract Sum or Contract Time. A properly prepared request for a change in the Work shall be accompanied by sufficient supporting data and information to permit the Architect to make a reasonable determination without extensive investigation or preparation of additional drawings or specifications. If the Architect determines that requested changes in the Work are not materially different from the requirements of the Contract Documents, the Architect may issue an order for a minor change in the Work or recommend to the Owner that the requested change be denied.

§ 2.6.5.3 If the Architect determines that implementation of the requested changes would result in a material change to the Contract that may cause an adjustment in the Contract Time or Contract Sum, the Architect shall make a recommendation to the Owner, who may authorize further investigation of such change. Upon such authorization, and based upon information furnished by the Contractor, if any, the Architect shall estimate the additional cost and time that might result from such change, including any additional costs attributable to a Change in Services of the Architect. With the Owner's approval, the Architect shall incorporate those estimates into a Change Order or other appropriate documentation for the Owner's execution or negotiation with the Contractor.

§ 2.6.5.4 The Architect shall maintain records relative to changes in the Work.

§ 2.6.6 PROJECT COMPLETION
§ 2.6.6.1 The Architect shall conduct inspections to determine the date or dates of Substantial Completion and the date of final completion, shall receive from the Contractor and forward to the Owner, for the Owner's review and records, written warranties and related documents required by the Contract Documents and assembled by the Contractor, and shall issue a final Certificate for Payment based upon a final inspection indicating the Work complies with the requirements of the Contract Documents.

§ 2.6.6.2 The Architect's inspection shall be conducted with the Owner's Designated Representative to check conformance of the Work with the requirements of the Contract Documents and to verify the accuracy and completeness of the list submitted by the Contractor of Work to be completed or corrected.

§ 2.6.6.3 When the Work is found to be substantially complete, the Architect shall inform the Owner about the balance of the Contract Sum remaining to be paid the Contractor, including any amounts needed to pay for final completion or correction of the Work.

§ 2.6.6.4 The Architect shall receive from the Contractor and forward to the Owner: (1) consent of surety or sureties, if any, to reduction in or partial release of retainage or the making of final payment and (2) affidavits, receipts, releases and waivers of liens or bonds indemnifying the Owner against liens.

ARTICLE 2.7 FACILITY OPERATION SERVICES

§ 2.7.1 The Architect shall meet with the Owner or the Owner's Designated Representative promptly after Substantial Completion to review the need for facility operation services.

§ 2.7.2 Upon request of the Owner, and prior to the expiration of one year from the date of Substantial Completion, the Architect shall conduct a meeting with the Owner and the Owner's Designated Representative to review the facility operations and performance and to make appropriate recommendations to the Owner.

ARTICLE 2.8 SCHEDULE OF SERVICES

§ 2.8.1 Design and Contract Administration Services beyond the following limits shall be provided by the Architect as a Change in Services in accordance with Section 1.3.3:

    .1   up to   (   ) reviews of each Shop Drawing, Product Data item, sample and similar submittal of the Contractor.

    .2   up to   (   ) visits to the site by the Architect over the duration of the Project during construction.

    .3   up to   (   ) inspections for any portion of the Work to determine whether such portion of the Work is substantially complete in accordance with the requirements of the Contract Documents.

    .4   up to   (   ) inspections for any portion of the Work to determine final completion.

§ 2.8.2 The following Design and Contract Administration Services shall be provided by the Architect as a Change in Services in accordance with Section 1.3.3:

    .1   review of a Contractor's submittal out of sequence from the submittal schedule agreed to by the Architect;

    .2   responses to the Contractor's requests for information where such information is available to the Contractor from a careful study and comparison of the Contract Documents, field conditions, other Owner-provided information, Contractor-prepared coordination drawings, or prior Project correspondence or documentation;

    .3   Change Orders and Construction Change Directives requiring evaluation of proposals, including the preparation or revision of Instruments of Service;

    .4   providing consultation concerning replacement of Work resulting from fire or other cause during construction;

    .5   evaluation of an extensive number of claims submitted by the Owner's consultants, the Contractor or others in connection with the Work;

    .6   evaluation of substitutions proposed by the Owner's consultants or contractors and making subsequent revisions to Instruments of Service resulting therefrom;

    .7   preparation of design and documentation for alternate bid or proposal requests proposed by the Owner; or

    .8   Contract Administration Services provided 60 days after the date of Substantial Completion of the Work.

§ 2.8.3 The Architect shall furnish or provide the following services only if specifically designated:

Services		Responsibility (Architect, Owner or Not Provided)	Location of Service Description
.1	Programming	· · · · ·	· · · · ·
.2	Land Survey Services	· · · · ·	· · · · ·
.3	Geotechnical Services	· · · · ·	· · · · ·
.4	Space Schematics/Flow Diagrams	· · · · ·	· · · · ·
.5	Existing Facilities Surveys	· · · · ·	· · · · ·
.6	Economic Feasibility Studies	· · · · ·	· · · · ·
.7	Site Analysis and Selection	· · · · ·	· · · · ·
.8	Environmental Studies and Reports	· · · · ·	· · · · ·
.9	Owner-Supplied Data Coordination	· · · · ·	· · · · ·
.10	Schedule Development and Monitoring	· · · · ·	· · · · ·
.11	Civil Design	· · · · ·	· · · · ·
.12	Landscape Design	· · · · ·	· · · · ·
.13	Interior Design	· · · · ·	· · · · ·
.14	Special Bidding or Negotiation	· · · · ·	· · · · ·
.15	Value Analysis	· · · · ·	· · · · ·
.16	Detailed Cost Estimating	· · · · ·	· · · · ·
.17	On-Site Project Representation	· · · · ·	· · · · ·
.18	Construction Management	· · · · ·	· · · · ·
.19	Start-up Assistance	· · · · ·	· · · · ·
.20	Record Drawings	· · · · ·	· · · · ·
.21	Post-Contract Evaluation	· · · · ·	· · · · ·
.22	Tenant-Related Services	· · · · ·	· · · · ·
.23	· · · · ·	· · · · ·	· · · · ·
.24	· · · · ·	· · · · ·	· · · · ·
.25	· · · · ·	· · · · ·	· · · · ·

Description of Services.
(Insert descriptions of the services designated.)

ARTICLE 2.9 MODIFICATIONS
§ **2.9.1** Modifications to this Standard Form of Architect's Services: Design and Contract Administration, if any, are as follows:

By its execution, this Standard Form of Architect's Services: Design and Contract Administration and modifications hereto are incorporated into the Standard Form of Agreement Between the Owner and Architect, AIA Document B141-1997, that was entered into by the parties as of the date:

OWNER   **ARCHITECT**

_____   _____
(Signature)   *(Signature)*

_____   _____
(Printed name and title)   *(Printed name and title)*

CAUTION: You should sign an original AIA Contract Document, on which this text appears in RED. An original assures that changes will not be obscured.

Appendix B

MasterFormat™ 2004 Edition— Division Numbers and Titles

PROCUREMENT AND CONTRACTING REQUIREMENTS GROUP

Division 00 Procurement and Contracting Requirements

SPECIFICATIONS GROUP

GENERAL REQUIREMENTS SUBGROUP

Division 01 General Requirements

FACILITY CONSTRUCTION SUBGROUP

Division 02 Existing Conditions
Division 03 Concrete
Division 04 Masonry
Division 05 Metals
Division 06 Wood, Plastics, and Composites
Division 07 Thermal and Moisture Protection
Division 08 Openings
Division 09 Finishes
Division 10 Specialties
Division 11 Equipment
Division 12 Furnishings
Division 13 Special Construction
Division 14 Conveying Equipment
Division 15 Reserved
Division 16 Reserved
Division 17 Reserved
Division 18 Reserved
Division 19 Reserved

FACILITY SERVICES SUBGROUP

Division 20 Reserved
Division 21 Fire Suppression
Division 22 Plumbing
Division 23 Heating, Ventilating, and Air-Conditioning
Division 24 Reserved

Division 25 Integrated Automation
Division 26 Electrical
Division 27 Communications
Division 28 Electronic Safety and Security
Division 29 Reserved

SITE AND INFRASTRUCTURE SUBGROUP

Division 30 Reserved
Division 31 Earthwork
Division 32 Exterior Improvements
Division 33 Utilities
Division 34 Transportation
Division 35 Waterway and Marine Construction
Division 36 Reserved
Division 37 Reserved
Division 38 Reserved
Division 39 Reserved

PROCESS EQUIPMENT SUBGROUP

Division 40 Process Integration
Division 41 Material Processing and Handling Equipment
Division 42 Process Heating, Cooling, and Drying Equipment
Division 43 Process Gas and Liquid Handling, Purification, and Storage Equipment
Division 44 Pollution Control Equipment
Division 45 Industry-Specific Manufacturing Equipment
Division 46 Reserved
Division 47 Reserved
Division 48 Electrical Power Generation
Division 49 Reserved

Appendix C

AIA Document A201-1997 General Conditions of the Contract for Construction

AIA® Document A201™ – 1997

General Conditions of the Contract for Construction

for the following PROJECT:
(Name and location or address)

THE OWNER:
(Name and address)

THE ARCHITECT:
(Name and address)

This document has important legal consequences. Consultation with an attorney is encouraged with respect to its completion or modification.

This document has been approved and endorsed by The Associated General Contractors of America

TABLE OF ARTICLES

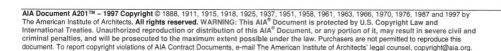

ARTICLE 1 GENERAL PROVISIONS
§ 1.1 BASIC DEFINITIONS
§ 1.1.1 THE CONTRACT DOCUMENTS

The Contract Documents consist of the Agreement between Owner and Contractor (hereinafter the Agreement), Conditions of the Contract (General, Supplementary and other Conditions), Drawings, Specifications, Addenda issued prior to execution of the Contract, other documents listed in the Agreement and Modifications issued after execution of the Contract. A Modification is (1) a written amendment to the Contract signed by both parties, (2) a Change Order, (3) a Construction Change Directive or (4) a written order for a minor change in the Work issued by the Architect. Unless specifically enumerated in the Agreement, the Contract Documents do not include other documents such as bidding requirements (advertisement or invitation to bid, Instructions to Bidders, sample forms, the Contractor's bid or portions of Addenda relating to bidding requirements).

§ 1.1.2 THE CONTRACT

The Contract Documents form the Contract for Construction. The Contract represents the entire and integrated agreement between the parties hereto and supersedes prior negotiations, representations or agreements, either written or oral. The Contract may be amended or modified only by a Modification. The Contract Documents shall not be construed to create a contractual relationship of any kind (1) between the Architect and Contractor, (2) between the Owner and a Subcontractor or Sub-subcontractor, (3) between the Owner and Architect or (4) between any persons or entities other than the Owner and Contractor. The Architect shall, however, be entitled to performance and enforcement of obligations under the Contract intended to facilitate performance of the Architect's duties.

§ 1.1.3 THE WORK

The term "Work" means the construction and services required by the Contract Documents, whether completed or partially completed, and includes all other labor, materials, equipment and services provided or to be provided by the Contractor to fulfill the Contractor's obligations. The Work may constitute the whole or a part of the Project.

§ 1.1.4 THE PROJECT

The Project is the total construction of which the Work performed under the Contract Documents may be the whole or a part and which may include construction by the Owner or by separate contractors.

§ 1.1.5 THE DRAWINGS

The Drawings are the graphic and pictorial portions of the Contract Documents showing the design, location and dimensions of the Work, generally including plans, elevations, sections, details, schedules and diagrams.

§ 1.1.6 THE SPECIFICATIONS

The Specifications are that portion of the Contract Documents consisting of the written requirements for materials, equipment, systems, standards and workmanship for the Work, and performance of related services.

§ 1.1.7 THE PROJECT MANUAL

The Project Manual is a volume assembled for the Work which may include the bidding requirements, sample forms, Conditions of the Contract and Specifications.

§ 1.2 CORRELATION AND INTENT OF THE CONTRACT DOCUMENTS

§ 1.2.1 The intent of the Contract Documents is to include all items necessary for the proper execution and completion of the Work by the Contractor. The Contract Documents are complementary, and what is required by one shall be as binding as if required by all; performance by the Contractor shall be required only to the extent consistent with the Contract Documents and reasonably inferable from them as being necessary to produce the indicated results.

§ 1.2.2 Organization of the Specifications into divisions, sections and articles, and arrangement of Drawings shall not control the Contractor in dividing the Work among Subcontractors or in establishing the extent of Work to be performed by any trade.

§ 1.2.3 Unless otherwise stated in the Contract Documents, words which have well-known technical or construction industry meanings are used in the Contract Documents in accordance with such recognized meanings.

§ 1.3 CAPITALIZATION

§ 1.3.1 Terms capitalized in these General Conditions include those which are (1) specifically defined, (2) the titles of numbered articles or (3) the titles of other documents published by the American Institute of Architects.

§ 1.4 INTERPRETATION

§ 1.4.1 In the interest of brevity the Contract Documents frequently omit modifying words such as "all" and "any" and articles such as "the" and "an," but the fact that a modifier or an article is absent from one statement and appears in another is not intended to affect the interpretation of either statement.

§ 1.5 EXECUTION OF CONTRACT DOCUMENTS

§ 1.5.1 The Contract Documents shall be signed by the Owner and Contractor. If either the Owner or Contractor or both do not sign all the Contract Documents, the Architect shall identify such unsigned Documents upon request.

§ 1.5.2 Execution of the Contract by the Contractor is a representation that the Contractor has visited the site, become generally familiar with local conditions under which the Work is to be performed and correlated personal observations with requirements of the Contract Documents.

§ 1.6 OWNERSHIP AND USE OF DRAWINGS, SPECIFICATIONS AND OTHER INSTRUMENTS OF SERVICE

§ 1.6.1 The Drawings, Specifications and other documents, including those in electronic form, prepared by the Architect and the Architect's consultants are Instruments of Service through which the Work to be executed by the Contractor is described. The Contractor may retain one record set. Neither the Contractor nor any Subcontractor, Sub-subcontractor or material or equipment supplier shall own or claim a copyright in the Drawings, Specifications and other documents prepared by the Architect or the Architect's consultants, and unless otherwise indicated the Architect and the Architect's consultants shall be deemed the authors of them and will retain all common law, statutory and other reserved rights, in addition to the copyrights. All copies of Instruments of Service, except the Contractor's record set, shall be returned or suitably accounted for to the Architect, on request, upon completion of the Work. The Drawings, Specifications and other documents prepared by the Architect and the Architect's consultants, and copies thereof furnished to the Contractor, are for use solely with respect to this Project. They are not to be used by the Contractor or any Subcontractor, Sub-subcontractor or material or equipment supplier on other projects or for additions to this Project outside the scope of the Work without the specific written consent of the Owner, Architect and the Architect's consultants. The Contractor, Subcontractors, Sub-subcontractors and material or equipment suppliers are authorized to use and reproduce applicable portions of the Drawings, Specifications and other documents prepared by the Architect and the Architect's consultants appropriate to and for use in the execution of their Work under the Contract Documents. All copies made under this authorization shall bear the statutory copyright notice, if any, shown on the Drawings, Specifications and other documents prepared by the Architect and the Architect's consultants. Submittal or distribution to meet official regulatory requirements or for other purposes in connection with this Project is not to be construed as publication in derogation of the Architect's or Architect's consultants' copyrights or other reserved rights.

ARTICLE 2 OWNER
§ 2.1 GENERAL

§ 2.1.1 The Owner is the person or entity identified as such in the Agreement and is referred to throughout the Contract Documents as if singular in number. The Owner shall designate in writing a representative who shall have express authority to bind the Owner with respect to all matters requiring the Owner's approval or authorization. Except as otherwise provided in Section 4.2.1, the Architect does not have such authority. The term "Owner" means the Owner or the Owner's authorized representative.

§ 2.1.2 The Owner shall furnish to the Contractor within fifteen days after receipt of a written request, information necessary and relevant for the Contractor to evaluate, give notice of or enforce mechanic's lien rights. Such information shall include a correct statement of the record legal title to the property on which the Project is located, usually referred to as the site, and the Owner's interest therein.

§ 2.2 INFORMATION AND SERVICES REQUIRED OF THE OWNER

§ 2.2.1 The Owner shall, at the written request of the Contractor, prior to commencement of the Work and thereafter, furnish to the Contractor reasonable evidence that financial arrangements have been made to fulfill the Owner's obligations under the Contract. Furnishing of such evidence shall be a condition precedent to commencement or continuation of the Work. After such evidence has been furnished, the Owner shall not materially vary such financial arrangements without prior notice to the Contractor.

§ 2.2.2 Except for permits and fees, including those required under Section 3.7.1, which are the responsibility of the Contractor under the Contract Documents, the Owner shall secure and pay for necessary approvals, easements, assessments and charges required for construction, use or occupancy of permanent structures or for permanent changes in existing facilities.

§ 2.2.3 The Owner shall furnish surveys describing physical characteristics, legal limitations and utility locations for the site of the Project, and a legal description of the site. The Contractor shall be entitled to rely on the accuracy of information furnished by the Owner but shall exercise proper precautions relating to the safe performance of the Work.

§ 2.2.4 Information or services required of the Owner by the Contract Documents shall be furnished by the Owner with reasonable promptness. Any other information or services relevant to the Contractor's performance of the Work under the Owner's control shall be furnished by the Owner after receipt from the Contractor of a written request for such information or services.

§ 2.2.5 Unless otherwise provided in the Contract Documents, the Contractor will be furnished, free of charge, such copies of Drawings and Project Manuals as are reasonably necessary for execution of the Work.

§ 2.3 OWNER'S RIGHT TO STOP THE WORK
§ 2.3.1 If the Contractor fails to correct Work which is not in accordance with the requirements of the Contract Documents as required by Section 12.2 or persistently fails to carry out Work in accordance with the Contract Documents, the Owner may issue a written order to the Contractor to stop the Work, or any portion thereof, until the cause for such order has been eliminated; however, the right of the Owner to stop the Work shall not give rise to a duty on the part of the Owner to exercise this right for the benefit of the Contractor or any other person or entity, except to the extent required by Section 6.1.3.

§ 2.4 OWNER'S RIGHT TO CARRY OUT THE WORK
§ 2.4.1 If the Contractor defaults or neglects to carry out the Work in accordance with the Contract Documents and fails within a seven-day period after receipt of written notice from the Owner to commence and continue correction of such default or neglect with diligence and promptness, the Owner may after such seven-day period give the Contractor a second written notice to correct such deficiencies within a three-day period. If the Contractor within such three-day period after receipt of such second notice fails to commence and continue to correct any deficiencies, the Owner may, without prejudice to other remedies the Owner may have, correct such deficiencies. In such case an appropriate Change Order shall be issued deducting from payments then or thereafter due the Contractor the reasonable cost of correcting such deficiencies, including Owner's expenses and compensation for the Architect's additional services made necessary by such default, neglect or failure. Such action by the Owner and amounts charged to the Contractor are both subject to prior approval of the Architect. If payments then or thereafter due the Contractor are not sufficient to cover such amounts, the Contractor shall pay the difference to the Owner.

ARTICLE 3 CONTRACTOR
§ 3.1 GENERAL
§ 3.1.1 The Contractor is the person or entity identified as such in the Agreement and is referred to throughout the Contract Documents as if singular in number. The term "Contractor" means the Contractor or the Contractor's authorized representative.

§ 3.1.2 The Contractor shall perform the Work in accordance with the Contract Documents.

§ 3.1.3 The Contractor shall not be relieved of obligations to perform the Work in accordance with the Contract Documents either by activities or duties of the Architect in the Architect's administration of the Contract, or by tests, inspections or approvals required or performed by persons other than the Contractor.

§ 3.2 REVIEW OF CONTRACT DOCUMENTS AND FIELD CONDITIONS BY CONTRACTOR
§ 3.2.1 Since the Contract Documents are complementary, before starting each portion of the Work, the Contractor shall carefully study and compare the various Drawings and other Contract Documents relative to that portion of the Work, as well as the information furnished by the Owner pursuant to Section 2.2.3, shall take field measurements of any existing conditions related to that portion of the Work and shall observe any conditions at the site affecting it. These obligations are for the purpose of facilitating construction by the Contractor and are not for the purpose of discovering errors, omissions, or inconsistencies in the Contract Documents; however, any errors, inconsistencies or omissions discovered by the Contractor shall be reported promptly to the Architect as a request for information in such form as the Architect may require.

§ 3.2.2 Any design errors or omissions noted by the Contractor during this review shall be reported promptly to the Architect, but it is recognized that the Contractor's review is made in the Contractor's capacity as a contractor and not as a licensed design professional unless otherwise specifically provided in the Contract Documents. The

Contractor is not required to ascertain that the Contract Documents are in accordance with applicable laws, statutes, ordinances, building codes, and rules and regulations, but any nonconformity discovered by or made known to the Contractor shall be reported promptly to the Architect.

§ 3.2.3 If the Contractor believes that additional cost or time is involved because of clarifications or instructions issued by the Architect in response to the Contractor's notices or requests for information pursuant to Sections 3.2.1 and 3.2.2, the Contractor shall make Claims as provided in Sections 4.3.6 and 4.3.7. If the Contractor fails to perform the obligations of Sections 3.2.1 and 3.2.2, the Contractor shall pay such costs and damages to the Owner as would have been avoided if the Contractor had performed such obligations. The Contractor shall not be liable to the Owner or Architect for damages resulting from errors, inconsistencies or omissions in the Contract Documents or for differences between field measurements or conditions and the Contract Documents unless the Contractor recognized such error, inconsistency, omission or difference and knowingly failed to report it to the Architect.

§ 3.3 SUPERVISION AND CONSTRUCTION PROCEDURES

§ 3.3.1 The Contractor shall supervise and direct the Work, using the Contractor's best skill and attention. The Contractor shall be solely responsible for and have control over construction means, methods, techniques, sequences and procedures and for coordinating all portions of the Work under the Contract, unless the Contract Documents give other specific instructions concerning these matters. If the Contract Documents give specific instructions concerning construction means, methods, techniques, sequences or procedures, the Contractor shall evaluate the jobsite safety thereof and, except as stated below, shall be fully and solely responsible for the jobsite safety of such means, methods, techniques, sequences or procedures. If the Contractor determines that such means, methods, techniques, sequences or procedures may not be safe, the Contractor shall give timely written notice to the Owner and Architect and shall not proceed with that portion of the Work without further written instructions from the Architect. If the Contractor is then instructed to proceed with the required means, methods, techniques, sequences or procedures without acceptance of changes proposed by the Contractor, the Owner shall be solely responsible for any resulting loss or damage.

§ 3.3.2 The Contractor shall be responsible to the Owner for acts and omissions of the Contractor's employees, Subcontractors and their agents and employees, and other persons or entities performing portions of the Work for or on behalf of the Contractor or any of its Subcontractors.

§ 3.3.3 The Contractor shall be responsible for inspection of portions of Work already performed to determine that such portions are in proper condition to receive subsequent Work.

§ 3.4 LABOR AND MATERIALS

§ 3.4.1 Unless otherwise provided in the Contract Documents, the Contractor shall provide and pay for labor, materials, equipment, tools, construction equipment and machinery, water, heat, utilities, transportation, and other facilities and services necessary for proper execution and completion of the Work, whether temporary or permanent and whether or not incorporated or to be incorporated in the Work.

§ 3.4.2 The Contractor may make substitutions only with the consent of the Owner, after evaluation by the Architect and in accordance with a Change Order.

§ 3.4.3 The Contractor shall enforce strict discipline and good order among the Contractor's employees and other persons carrying out the Contract. The Contractor shall not permit employment of unfit persons or persons not skilled in tasks assigned to them.

§ 3.5 WARRANTY

§ 3.5.1 The Contractor warrants to the Owner and Architect that materials and equipment furnished under the Contract will be of good quality and new unless otherwise required or permitted by the Contract Documents, that the Work will be free from defects not inherent in the quality required or permitted, and that the Work will conform to the requirements of the Contract Documents. Work not conforming to these requirements, including substitutions not properly approved and authorized, may be considered defective. The Contractor's warranty excludes remedy for damage or defect caused by abuse, modifications not executed by the Contractor, improper or insufficient maintenance, improper operation, or normal wear and tear and normal usage. If required by the Architect, the Contractor shall furnish satisfactory evidence as to the kind and quality of materials and equipment.

§ 3.6 TAXES

§ 3.6.1 The Contractor shall pay sales, consumer, use and similar taxes for the Work provided by the Contractor which are legally enacted when bids are received or negotiations concluded, whether or not yet effective or merely scheduled to go into effect.

§ 3.7 PERMITS, FEES AND NOTICES

§ 3.7.1 Unless otherwise provided in the Contract Documents, the Contractor shall secure and pay for the building permit and other permits and governmental fees, licenses and inspections necessary for proper execution and completion of the Work which are customarily secured after execution of the Contract and which are legally required when bids are received or negotiations concluded.

§ 3.7.2 The Contractor shall comply with and give notices required by laws, ordinances, rules, regulations and lawful orders of public authorities applicable to performance of the Work.

§ 3.7.3 It is not the Contractor's responsibility to ascertain that the Contract Documents are in accordance with applicable laws, statutes, ordinances, building codes, and rules and regulations. However, if the Contractor observes that portions of the Contract Documents are at variance therewith, the Contractor shall promptly notify the Architect and Owner in writing, and necessary changes shall be accomplished by appropriate Modification.

§ 3.7.4 If the Contractor performs Work knowing it to be contrary to laws, statutes, ordinances, building codes, and rules and regulations without such notice to the Architect and Owner, the Contractor shall assume appropriate responsibility for such Work and shall bear the costs attributable to correction.

§ 3.8 ALLOWANCES

§ 3.8.1 The Contractor shall include in the Contract Sum all allowances stated in the Contract Documents. Items covered by allowances shall be supplied for such amounts and by such persons or entities as the Owner may direct, but the Contractor shall not be required to employ persons or entities to whom the Contractor has reasonable objection.

§ 3.8.2 Unless otherwise provided in the Contract Documents:
   .1 allowances shall cover the cost to the Contractor of materials and equipment delivered at the site and all required taxes, less applicable trade discounts;
   .2 Contractor's costs for unloading and handling at the site, labor, installation costs, overhead, profit and other expenses contemplated for stated allowance amounts shall be included in the Contract Sum but not in the allowances;
   .3 whenever costs are more than or less than allowances, the Contract Sum shall be adjusted accordingly by Change Order. The amount of the Change Order shall reflect (1) the difference between actual costs and the allowances under Section 3.8.2.1 and (2) changes in Contractor's costs under Section 3.8.2.2.

§ 3.8.3 Materials and equipment under an allowance shall be selected by the Owner in sufficient time to avoid delay in the Work.

§ 3.9 SUPERINTENDENT

§ 3.9.1 The Contractor shall employ a competent superintendent and necessary assistants who shall be in attendance at the Project site during performance of the Work. The superintendent shall represent the Contractor, and communications given to the superintendent shall be as binding as if given to the Contractor. Important communications shall be confirmed in writing. Other communications shall be similarly confirmed on written request in each case.

§ 3.10 CONTRACTOR'S CONSTRUCTION SCHEDULES

§ 3.10.1 The Contractor, promptly after being awarded the Contract, shall prepare and submit for the Owner's and Architect's information a Contractor's construction schedule for the Work. The schedule shall not exceed time limits current under the Contract Documents, shall be revised at appropriate intervals as required by the conditions of the Work and Project, shall be related to the entire Project to the extent required by the Contract Documents, and shall provide for expeditious and practicable execution of the Work.

§ 3.10.2 The Contractor shall prepare and keep current, for the Architect's approval, a schedule of submittals which is coordinated with the Contractor's construction schedule and allows the Architect reasonable time to review submittals.

§ 3.10.3 The Contractor shall perform the Work in general accordance with the most recent schedules submitted to the Owner and Architect.

§ 3.11 DOCUMENTS AND SAMPLES AT THE SITE

§ 3.11.1 The Contractor shall maintain at the site for the Owner one record copy of the Drawings, Specifications, Addenda, Change Orders and other Modifications, in good order and marked currently to record field changes and selections made during construction, and one record copy of approved Shop Drawings, Product Data, Samples and similar required submittals. These shall be available to the Architect and shall be delivered to the Architect for submittal to the Owner upon completion of the Work.

§ 3.12 SHOP DRAWINGS, PRODUCT DATA AND SAMPLES

§ 3.12.1 Shop Drawings are drawings, diagrams, schedules and other data specially prepared for the Work by the Contractor or a Subcontractor, Sub-subcontractor, manufacturer, supplier or distributor to illustrate some portion of the Work.

§ 3.12.2 Product Data are illustrations, standard schedules, performance charts, instructions, brochures, diagrams and other information furnished by the Contractor to illustrate materials or equipment for some portion of the Work.

§ 3.12.3 Samples are physical examples which illustrate materials, equipment or workmanship and establish standards by which the Work will be judged.

§ 3.12.4 Shop Drawings, Product Data, Samples and similar submittals are not Contract Documents. The purpose of their submittal is to demonstrate for those portions of the Work for which submittals are required by the Contract Documents the way by which the Contractor proposes to conform to the information given and the design concept expressed in the Contract Documents. Review by the Architect is subject to the limitations of Section 4.2.7. Informational submittals upon which the Architect is not expected to take responsive action may be so identified in the Contract Documents. Submittals which are not required by the Contract Documents may be returned by the Architect without action.

§ 3.12.5 The Contractor shall review for compliance with the Contract Documents, approve and submit to the Architect Shop Drawings, Product Data, Samples and similar submittals required by the Contract Documents with reasonable promptness and in such sequence as to cause no delay in the Work or in the activities of the Owner or of separate contractors. Submittals which are not marked as reviewed for compliance with the Contract Documents and approved by the Contractor may be returned by the Architect without action.

§ 3.12.6 By approving and submitting Shop Drawings, Product Data, Samples and similar submittals, the Contractor represents that the Contractor has determined and verified materials, field measurements and field construction criteria related thereto, or will do so, and has checked and coordinated the information contained within such submittals with the requirements of the Work and of the Contract Documents.

§ 3.12.7 The Contractor shall perform no portion of the Work for which the Contract Documents require submittal and review of Shop Drawings, Product Data, Samples or similar submittals until the respective submittal has been approved by the Architect.

§ 3.12.8 The Work shall be in accordance with approved submittals except that the Contractor shall not be relieved of responsibility for deviations from requirements of the Contract Documents by the Architect's approval of Shop Drawings, Product Data, Samples or similar submittals unless the Contractor has specifically informed the Architect in writing of such deviation at the time of submittal and (1) the Architect has given written approval to the specific deviation as a minor change in the Work, or (2) a Change Order or Construction Change Directive has been issued authorizing the deviation. The Contractor shall not be relieved of responsibility for errors or omissions in Shop Drawings, Product Data, Samples or similar submittals by the Architect's approval thereof.

§ 3.12.9 The Contractor shall direct specific attention, in writing or on resubmitted Shop Drawings, Product Data, Samples or similar submittals, to revisions other than those requested by the Architect on previous submittals. In the absence of such written notice the Architect's approval of a resubmission shall not apply to such revisions.

§ 3.12.10 The Contractor shall not be required to provide professional services which constitute the practice of architecture or engineering unless such services are specifically required by the Contract Documents for a portion of the Work or unless the Contractor needs to provide such services in order to carry out the Contractor's responsibilities for construction means, methods, techniques, sequences and procedures. The Contractor shall not be required to provide professional services in violation of applicable law. If professional design services or certifications by a design professional related to systems, materials or equipment are specifically required of the Contractor by the Contract Documents, the Owner and the Architect will specify all performance and design criteria that such services must satisfy. The Contractor shall cause such services or certifications to be provided by a properly licensed design professional, whose signature and seal shall appear on all drawings, calculations, specifications, certifications, Shop Drawings and other submittals prepared by such professional. Shop Drawings and other submittals related to the Work designed or certified by such professional, if prepared by others, shall bear such professional's written approval when submitted to the Architect. The Owner and the Architect shall be entitled to rely upon the adequacy, accuracy and completeness of the services, certifications or approvals performed by such design professionals, provided the Owner and Architect have specified to the Contractor all performance and design criteria that such services must satisfy. Pursuant to this Section 3.12.10, the Architect will review, approve or take other appropriate action on submittals only for the limited purpose of checking for conformance with information given and the design concept expressed in the Contract Documents. The Contractor shall not be responsible for the adequacy of the performance or design criteria required by the Contract Documents.

§ 3.13 USE OF SITE
§ 3.13.1 The Contractor shall confine operations at the site to areas permitted by law, ordinances, permits and the Contract Documents and shall not unreasonably encumber the site with materials or equipment.

§ 3.14 CUTTING AND PATCHING
§ 3.14.1 The Contractor shall be responsible for cutting, fitting or patching required to complete the Work or to make its parts fit together properly.

§ 3.14.2 The Contractor shall not damage or endanger a portion of the Work or fully or partially completed construction of the Owner or separate contractors by cutting, patching or otherwise altering such construction, or by excavation. The Contractor shall not cut or otherwise alter such construction by the Owner or a separate contractor except with written consent of the Owner and of such separate contractor; such consent shall not be unreasonably withheld. The Contractor shall not unreasonably withhold from the Owner or a separate contractor the Contractor's consent to cutting or otherwise altering the Work.

§ 3.15 CLEANING UP
§ 3.15.1 The Contractor shall keep the premises and surrounding area free from accumulation of waste materials or rubbish caused by operations under the Contract. At completion of the Work, the Contractor shall remove from and about the Project waste materials, rubbish, the Contractor's tools, construction equipment, machinery and surplus materials.

§ 3.15.2 If the Contractor fails to clean up as provided in the Contract Documents, the Owner may do so and the cost thereof shall be charged to the Contractor.

§ 3.16 ACCESS TO WORK
§ 3.16.1 The Contractor shall provide the Owner and Architect access to the Work in preparation and progress wherever located.

§ 3.17 ROYALTIES, PATENTS AND COPYRIGHTS
§ 3.17.1 The Contractor shall pay all royalties and license fees. The Contractor shall defend suits or claims for infringement of copyrights and patent rights and shall hold the Owner and Architect harmless from loss on account thereof, but shall not be responsible for such defense or loss when a particular design, process or product of a particular manufacturer or manufacturers is required by the Contract Documents or where the copyright violations are contained in Drawings, Specifications or other documents prepared by the Owner or Architect. However, if the Contractor has reason to believe that the required design, process or product is an infringement of a copyright or a patent, the Contractor shall be responsible for such loss unless such information is promptly furnished to the Architect.

§ 3.18 INDEMNIFICATION

§ 3.18.1 To the fullest extent permitted by law and to the extent claims, damages, losses or expenses are not covered by Project Management Protective Liability insurance purchased by the Contractor in accordance with Section 11.3, the Contractor shall indemnify and hold harmless the Owner, Architect, Architect's consultants, and agents and employees of any of them from and against claims, damages, losses and expenses, including but not limited to attorneys' fees, arising out of or resulting from performance of the Work, provided that such claim, damage, loss or expense is attributable to bodily injury, sickness, disease or death, or to injury to or destruction of tangible property (other than the Work itself), but only to the extent caused by the negligent acts or omissions of the Contractor, a Subcontractor, anyone directly or indirectly employed by them or anyone for whose acts they may be liable, regardless of whether or not such claim, damage, loss or expense is caused in part by a party indemnified hereunder. Such obligation shall not be construed to negate, abridge, or reduce other rights or obligations of indemnity which would otherwise exist as to a party or person described in this Section 3.18.

§ 3.18.2 In claims against any person or entity indemnified under this Section 3.18 by an employee of the Contractor, a Subcontractor, anyone directly or indirectly employed by them or anyone for whose acts they may be liable, the indemnification obligation under Section 3.18.1 shall not be limited by a limitation on amount or type of damages, compensation or benefits payable by or for the Contractor or a Subcontractor under workers' compensation acts, disability benefit acts or other employee benefit acts.

ARTICLE 4   ADMINISTRATION OF THE CONTRACT
§ 4.1 ARCHITECT

§ 4.1.1 The Architect is the person lawfully licensed to practice architecture or an entity lawfully practicing architecture identified as such in the Agreement and is referred to throughout the Contract Documents as if singular in number. The term "Architect" means the Architect or the Architect's authorized representative.

§ 4.1.2 Duties, responsibilities and limitations of authority of the Architect as set forth in the Contract Documents shall not be restricted, modified or extended without written consent of the Owner, Contractor and Architect. Consent shall not be unreasonably withheld.

§ 4.1.3 If the employment of the Architect is terminated, the Owner shall employ a new Architect against whom the Contractor has no reasonable objection and whose status under the Contract Documents shall be that of the former Architect.

§ 4.2 ARCHITECT'S ADMINISTRATION OF THE CONTRACT

§ 4.2.1 The Architect will provide administration of the Contract as described in the Contract Documents, and will be an Owner's representative (1) during construction, (2) until final payment is due and (3) with the Owner's concurrence, from time to time during the one-year period for correction of Work described in Section 12.2. The Architect will have authority to act on behalf of the Owner only to the extent provided in the Contract Documents, unless otherwise modified in writing in accordance with other provisions of the Contract.

§ 4.2.2 The Architect, as a representative of the Owner, will visit the site at intervals appropriate to the stage of the Contractor's operations (1) to become generally familiar with and to keep the Owner informed about the progress and quality of the portion of the Work completed, (2) to endeavor to guard the Owner against defects and deficiencies in the Work, and (3) to determine in general if the Work is being performed in a manner indicating that the Work, when fully completed, will be in accordance with the Contract Documents. However, the Architect will not be required to make exhaustive or continuous on-site inspections to check the quality or quantity of the Work. The Architect will neither have control over or charge of, nor be responsible for, the construction means, methods, techniques, sequences or procedures, or for the safety precautions and programs in connection with the Work, since these are solely the Contractor's rights and responsibilities under the Contract Documents, except as provided in Section 3.3.1.

§ 4.2.3 The Architect will not be responsible for the Contractor's failure to perform the Work in accordance with the requirements of the Contract Documents. The Architect will not have control over or charge of and will not be responsible for acts or omissions of the Contractor, Subcontractors, or their agents or employees, or any other persons or entities performing portions of the Work.

§ 4.2.4 Communications Facilitating Contract Administration. Except as otherwise provided in the Contract Documents or when direct communications have been specially authorized, the Owner and Contractor shall endeavor to communicate with each other through the Architect about matters arising out of or relating to the

Contract. Communications by and with the Architect's consultants shall be through the Architect. Communications by and with Subcontractors and material suppliers shall be through the Contractor. Communications by and with separate contractors shall be through the Owner.

§ 4.2.5 Based on the Architect's evaluations of the Contractor's Applications for Payment, the Architect will review and certify the amounts due the Contractor and will issue Certificates for Payment in such amounts.

§ 4.2.6 The Architect will have authority to reject Work that does not conform to the Contract Documents. Whenever the Architect considers it necessary or advisable, the Architect will have authority to require inspection or testing of the Work in accordance with Sections 13.5.2 and 13.5.3, whether or not such Work is fabricated, installed or completed. However, neither this authority of the Architect nor a decision made in good faith either to exercise or not to exercise such authority shall give rise to a duty or responsibility of the Architect to the Contractor, Subcontractors, material and equipment suppliers, their agents or employees, or other persons or entities performing portions of the Work.

§ 4.2.7 The Architect will review and approve or take other appropriate action upon the Contractor's submittals such as Shop Drawings, Product Data and Samples, but only for the limited purpose of checking for conformance with information given and the design concept expressed in the Contract Documents. The Architect's action will be taken with such reasonable promptness as to cause no delay in the Work or in the activities of the Owner, Contractor or separate contractors, while allowing sufficient time in the Architect's professional judgment to permit adequate review. Review of such submittals is not conducted for the purpose of determining the accuracy and completeness of other details such as dimensions and quantities, or for substantiating instructions for installation or performance of equipment or systems, all of which remain the responsibility of the Contractor as required by the Contract Documents. The Architect's review of the Contractor's submittals shall not relieve the Contractor of the obligations under Sections 3.3, 3.5 and 3.12. The Architect's review shall not constitute approval of safety precautions or, unless otherwise specifically stated by the Architect, of any construction means, methods, techniques, sequences or procedures. The Architect's approval of a specific item shall not indicate approval of an assembly of which the item is a component.

§ 4.2.8 The Architect will prepare Change Orders and Construction Change Directives, and may authorize minor changes in the Work as provided in Section 7.4.

§ 4.2.9 The Architect will conduct inspections to determine the date or dates of Substantial Completion and the date of final completion, will receive and forward to the Owner, for the Owner's review and records, written warranties and related documents required by the Contract and assembled by the Contractor, and will issue a final Certificate for Payment upon compliance with the requirements of the Contract Documents.

§ 4.2.10 If the Owner and Architect agree, the Architect will provide one or more project representatives to assist in carrying out the Architect's responsibilities at the site. The duties, responsibilities and limitations of authority of such project representatives shall be as set forth in an exhibit to be incorporated in the Contract Documents.

§ 4.2.11 The Architect will interpret and decide matters concerning performance under and requirements of, the Contract Documents on written request of either the Owner or Contractor. The Architect's response to such requests will be made in writing within any time limits agreed upon or otherwise with reasonable promptness. If no agreement is made concerning the time within which interpretations required of the Architect shall be furnished in compliance with this Section 4.2, then delay shall not be recognized on account of failure by the Architect to furnish such interpretations until 15 days after written request is made for them.

§ 4.2.12 Interpretations and decisions of the Architect will be consistent with the intent of and reasonably inferable from the Contract Documents and will be in writing or in the form of drawings. When making such interpretations and initial decisions, the Architect will endeavor to secure faithful performance by both Owner and Contractor, will not show partiality to either and will not be liable for results of interpretations or decisions so rendered in good faith.

§ 4.2.13 The Architect's decisions on matters relating to aesthetic effect will be final if consistent with the intent expressed in the Contract Documents.

§ 4.3 CLAIMS AND DISPUTES
§ 4.3.1 Definition. A Claim is a demand or assertion by one of the parties seeking, as a matter of right, adjustment or interpretation of Contract terms, payment of money, extension of time or other relief with respect to the terms of

the Contract. The term "Claim" also includes other disputes and matters in question between the Owner and Contractor arising out of or relating to the Contract. Claims must be initiated by written notice. The responsibility to substantiate Claims shall rest with the party making the Claim.

§ 4.3.2 Time Limits on Claims. Claims by either party must be initiated within 21 days after occurrence of the event giving rise to such Claim or within 21 days after the claimant first recognizes the condition giving rise to the Claim, whichever is later. Claims must be initiated by written notice to the Architect and the other party.

§ 4.3.3 Continuing Contract Performance. Pending final resolution of a Claim except as otherwise agreed in writing or as provided in Section 9.7.1 and Article 14, the Contractor shall proceed diligently with performance of the Contract and the Owner shall continue to make payments in accordance with the Contract Documents.

§ 4.3.4 Claims for Concealed or Unknown Conditions. If conditions are encountered at the site which are (1) subsurface or otherwise concealed physical conditions which differ materially from those indicated in the Contract Documents or (2) unknown physical conditions of an unusual nature, which differ materially from those ordinarily found to exist and generally recognized as inherent in construction activities of the character provided for in the Contract Documents, then notice by the observing party shall be given to the other party promptly before conditions are disturbed and in no event later than 21 days after first observance of the conditions. The Architect will promptly investigate such conditions and, if they differ materially and cause an increase or decrease in the Contractor's cost of, or time required for, performance of any part of the Work, will recommend an equitable adjustment in the Contract Sum or Contract Time, or both. If the Architect determines that the conditions at the site are not materially different from those indicated in the Contract Documents and that no change in the terms of the Contract is justified, the Architect shall so notify the Owner and Contractor in writing, stating the reasons. Claims by either party in opposition to such determination must be made within 21 days after the Architect has given notice of the decision. If the conditions encountered are materially different, the Contract Sum and Contract Time shall be equitably adjusted, but if the Owner and Contractor cannot agree on an adjustment in the Contract Sum or Contract Time, the adjustment shall be referred to the Architect for initial determination, subject to further proceedings pursuant to Section 4.4.

§ 4.3.5 Claims for Additional Cost. If the Contractor wishes to make Claim for an increase in the Contract Sum, written notice as provided herein shall be given before proceeding to execute the Work. Prior notice is not required for Claims relating to an emergency endangering life or property arising under Section 10.6.

§ 4.3.6 If the Contractor believes additional cost is involved for reasons including but not limited to (1) a written interpretation from the Architect, (2) an order by the Owner to stop the Work where the Contractor was not at fault, (3) a written order for a minor change in the Work issued by the Architect, (4) failure of payment by the Owner, (5) termination of the Contract by the Owner, (6) Owner's suspension or (7) other reasonable grounds, Claim shall be filed in accordance with this Section 4.3.

§ 4.3.7 Claims for Additional Time
§ 4.3.7.1 If the Contractor wishes to make Claim for an increase in the Contract Time, written notice as provided herein shall be given. The Contractor's Claim shall include an estimate of cost and of probable effect of delay on progress of the Work. In the case of a continuing delay only one Claim is necessary.

§ 4.3.7.2 If adverse weather conditions are the basis for a Claim for additional time, such Claim shall be documented by data substantiating that weather conditions were abnormal for the period of time, could not have been reasonably anticipated and had an adverse effect on the scheduled construction.

§ 4.3.8 Injury or Damage to Person or Property. If either party to the Contract suffers injury or damage to person or property because of an act or omission of the other party, or of others for whose acts such party is legally responsible, written notice of such injury or damage, whether or not insured, shall be given to the other party within a reasonable time not exceeding 21 days after discovery. The notice shall provide sufficient detail to enable the other party to investigate the matter.

§ 4.3.9 If unit prices are stated in the Contract Documents or subsequently agreed upon, and if quantities originally contemplated are materially changed in a proposed Change Order or Construction Change Directive so that application of such unit prices to quantities of Work proposed will cause substantial inequity to the Owner or Contractor, the applicable unit prices shall be equitably adjusted.

§ 4.3.10 Claims for Consequential Damages. The Contractor and Owner waive Claims against each other for consequential damages arising out of or relating to this Contract. This mutual waiver includes:

    .1   damages incurred by the Owner for rental expenses, for losses of use, income, profit, financing, business and reputation, and for loss of management or employee productivity or of the services of such persons; and

    .2   damages incurred by the Contractor for principal office expenses including the compensation of personnel stationed there, for losses of financing, business and reputation, and for loss of profit except anticipated profit arising directly from the Work.

This mutual waiver is applicable, without limitation, to all consequential damages due to either party's termination in accordance with Article 14. Nothing contained in this Section 4.3.10 shall be deemed to preclude an award of liquidated direct damages, when applicable, in accordance with the requirements of the Contract Documents.

§ 4.4 RESOLUTION OF CLAIMS AND DISPUTES

§ 4.4.1 Decision of Architect. Claims, including those alleging an error or omission by the Architect but excluding those arising under Sections 10.3 through 10.5, shall be referred initially to the Architect for decision. An initial decision by the Architect shall be required as a condition precedent to mediation, arbitration or litigation of all Claims between the Contractor and Owner arising prior to the date final payment is due, unless 30 days have passed after the Claim has been referred to the Architect with no decision having been rendered by the Architect. The Architect will not decide disputes between the Contractor and persons or entities other than the Owner.

§ 4.4.2 The Architect will review Claims and within ten days of the receipt of the Claim take one or more of the following actions: (1) request additional supporting data from the claimant or a response with supporting data from the other party, (2) reject the Claim in whole or in part, (3) approve the Claim, (4) suggest a compromise, or (5) advise the parties that the Architect is unable to resolve the Claim if the Architect lacks sufficient information to evaluate the merits of the Claim or if the Architect concludes that, in the Architect's sole discretion, it would be inappropriate for the Architect to resolve the Claim.

§ 4.4.3 In evaluating Claims, the Architect may, but shall not be obligated to, consult with or seek information from either party or from persons with special knowledge or expertise who may assist the Architect in rendering a decision. The Architect may request the Owner to authorize retention of such persons at the Owner's expense.

§ 4.4.4 If the Architect requests a party to provide a response to a Claim or to furnish additional supporting data, such party shall respond, within ten days after receipt of such request, and shall either provide a response on the requested supporting data, advise the Architect when the response or supporting data will be furnished or advise the Architect that no supporting data will be furnished. Upon receipt of the response or supporting data, if any, the Architect will either reject or approve the Claim in whole or in part.

§ 4.4.5 The Architect will approve or reject Claims by written decision, which shall state the reasons therefor and which shall notify the parties of any change in the Contract Sum or Contract Time or both. The approval or rejection of a Claim by the Architect shall be final and binding on the parties but subject to mediation and arbitration.

§ 4.4.6 When a written decision of the Architect states that (1) the decision is final but subject to mediation and arbitration and (2) a demand for arbitration of a Claim covered by such decision must be made within 30 days after the date on which the party making the demand receives the final written decision, then failure to demand arbitration within said 30 days' period shall result in the Architect's decision becoming final and binding upon the Owner and Contractor. If the Architect renders a decision after arbitration proceedings have been initiated, such decision may be entered as evidence, but shall not supersede arbitration proceedings unless the decision is acceptable to all parties concerned.

§ 4.4.7 Upon receipt of a Claim against the Contractor or at any time thereafter, the Architect or the Owner may, but is not obligated to, notify the surety, if any, of the nature and amount of the Claim. If the Claim relates to a possibility of a Contractor's default, the Architect or the Owner may, but is not obligated to, notify the surety and request the surety's assistance in resolving the controversy.

§ 4.4.8 If a Claim relates to or is the subject of a mechanic's lien, the party asserting such Claim may proceed in accordance with applicable law to comply with the lien notice or filing deadlines prior to resolution of the Claim by the Architect, by mediation or by arbitration.

§ 4.5 MEDIATION

§ 4.5.1 Any Claim arising out of or related to the Contract, except Claims relating to aesthetic effect and except those waived as provided for in Sections 4.3.10, 9.10.4 and 9.10.5 shall, after initial decision by the Architect or 30 days after submission of the Claim to the Architect, be subject to mediation as a condition precedent to arbitration or the institution of legal or equitable proceedings by either party.

§ 4.5.2 The parties shall endeavor to resolve their Claims by mediation which, unless the parties mutually agree otherwise, shall be in accordance with the Construction Industry Mediation Rules of the American Arbitration Association currently in effect. Request for mediation shall be filed in writing with the other party to the Contract and with the American Arbitration Association. The request may be made concurrently with the filing of a demand for arbitration but, in such event, mediation shall proceed in advance of arbitration or legal or equitable proceedings, which shall be stayed pending mediation for a period of 60 days from the date of filing, unless stayed for a longer period by agreement of the parties or court order.

§ 4.5.3 The parties shall share the mediator's fee and any filing fees equally. The mediation shall be held in the place where the Project is located, unless another location is mutually agreed upon. Agreements reached in mediation shall be enforceable as settlement agreements in any court having jurisdiction thereof.

§ 4.6 ARBITRATION

§ 4.6.1 Any Claim arising out of or related to the Contract, except Claims relating to aesthetic effect and except those waived as provided for in Sections 4.3.10, 9.10.4 and 9.10.5, shall, after decision by the Architect or 30 days after submission of the Claim to the Architect, be subject to arbitration. Prior to arbitration, the parties shall endeavor to resolve disputes by mediation in accordance with the provisions of Section 4.5.

§ 4.6.2 Claims not resolved by mediation shall be decided by arbitration which, unless the parties mutually agree otherwise, shall be in accordance with the Construction Industry Arbitration Rules of the American Arbitration Association currently in effect. The demand for arbitration shall be filed in writing with the other party to the Contract and with the American Arbitration Association, and a copy shall be filed with the Architect.

§ 4.6.3 A demand for arbitration shall be made within the time limits specified in Sections 4.4.6 and 4.6.1 as applicable, and in other cases within a reasonable time after the Claim has arisen, and in no event shall it be made after the date when institution of legal or equitable proceedings based on such Claim would be barred by the applicable statute of limitations as determined pursuant to Section 13.7.

§ 4.6.4 Limitation on Consolidation or Joinder. No arbitration arising out of or relating to the Contract shall include, by consolidation or joinder or in any other manner, the Architect, the Architect's employees or consultants, except by written consent containing specific reference to the Agreement and signed by the Architect, Owner, Contractor and any other person or entity sought to be joined. No arbitration shall include, by consolidation or joinder or in any other manner, parties other than the Owner, Contractor, a separate contractor as described in Article 6 and other persons substantially involved in a common question of fact or law whose presence is required if complete relief is to be accorded in arbitration. No person or entity other than the Owner, Contractor or a separate contractor as described in Article 6 shall be included as an original third party or additional third party to an arbitration whose interest or responsibility is insubstantial. Consent to arbitration involving an additional person or entity shall not constitute consent to arbitration of a Claim not described therein or with a person or entity not named or described therein. The foregoing agreement to arbitrate and other agreements to arbitrate with an additional person or entity duly consented to by parties to the Agreement shall be specifically enforceable under applicable law in any court having jurisdiction thereof.

§ 4.6.5 Claims and Timely Assertion of Claims. The party filing a notice of demand for arbitration must assert in the demand all Claims then known to that party on which arbitration is permitted to be demanded.

§ 4.6.6 Judgment on Final Award. The award rendered by the arbitrator or arbitrators shall be final, and judgment may be entered upon it in accordance with applicable law in any court having jurisdiction thereof.

ARTICLE 5 SUBCONTRACTORS
§ 5.1 DEFINITIONS

§ 5.1.1 A Subcontractor is a person or entity who has a direct contract with the Contractor to perform a portion of the Work at the site. The term "Subcontractor" is referred to throughout the Contract Documents as if singular in number and means a Subcontractor or an authorized representative of the Subcontractor. The term "Subcontractor" does not include a separate contractor or subcontractors of a separate contractor.

§ 5.1.2 A Sub-subcontractor is a person or entity who has a direct or indirect contract with a Subcontractor to perform a portion of the Work at the site. The term "Sub-subcontractor" is referred to throughout the Contract Documents as if singular in number and means a Sub-subcontractor or an authorized representative of the Sub-subcontractor.

§ 5.2 AWARD OF SUBCONTRACTS AND OTHER CONTRACTS FOR PORTIONS OF THE WORK

§ 5.2.1 Unless otherwise stated in the Contract Documents or the bidding requirements, the Contractor, as soon as practicable after award of the Contract, shall furnish in writing to the Owner through the Architect the names of persons or entities (including those who are to furnish materials or equipment fabricated to a special design) proposed for each principal portion of the Work. The Architect will promptly reply to the Contractor in writing stating whether or not the Owner or the Architect, after due investigation, has reasonable objection to any such proposed person or entity. Failure of the Owner or Architect to reply promptly shall constitute notice of no reasonable objection.

§ 5.2.2 The Contractor shall not contract with a proposed person or entity to whom the Owner or Architect has made reasonable and timely objection. The Contractor shall not be required to contract with anyone to whom the Contractor has made reasonable objection.

§ 5.2.3 If the Owner or Architect has reasonable objection to a person or entity proposed by the Contractor, the Contractor shall propose another to whom the Owner or Architect has no reasonable objection. If the proposed but rejected Subcontractor was reasonably capable of performing the Work, the Contract Sum and Contract Time shall be increased or decreased by the difference, if any, occasioned by such change, and an appropriate Change Order shall be issued before commencement of the substitute Subcontractor's Work. However, no increase in the Contract Sum or Contract Time shall be allowed for such change unless the Contractor has acted promptly and responsively in submitting names as required.

§ 5.2.4 The Contractor shall not change a Subcontractor, person or entity previously selected if the Owner or Architect makes reasonable objection to such substitute.

§ 5.3 SUBCONTRACTUAL RELATIONS

§ 5.3.1 By appropriate agreement, written where legally required for validity, the Contractor shall require each Subcontractor, to the extent of the Work to be performed by the Subcontractor, to be bound to the Contractor by terms of the Contract Documents, and to assume toward the Contractor all the obligations and responsibilities, including the responsibility for safety of the Subcontractor's Work, which the Contractor, by these Documents, assumes toward the Owner and Architect. Each subcontract agreement shall preserve and protect the rights of the Owner and Architect under the Contract Documents with respect to the Work to be performed by the Subcontractor so that subcontracting thereof will not prejudice such rights, and shall allow to the Subcontractor, unless specifically provided otherwise in the subcontract agreement, the benefit of all rights, remedies and redress against the Contractor that the Contractor, by the Contract Documents, has against the Owner. Where appropriate, the Contractor shall require each Subcontractor to enter into similar agreements with Sub-subcontractors. The Contractor shall make available to each proposed Subcontractor, prior to the execution of the subcontract agreement, copies of the Contract Documents to which the Subcontractor will be bound, and, upon written request of the Subcontractor, identify to the Subcontractor terms and conditions of the proposed subcontract agreement which may be at variance with the Contract Documents. Subcontractors will similarly make copies of applicable portions of such documents available to their respective proposed Sub-subcontractors.

§ 5.4 CONTINGENT ASSIGNMENT OF SUBCONTRACTS

§ 5.4.1 Each subcontract agreement for a portion of the Work is assigned by the Contractor to the Owner provided that:

    .1   assignment is effective only after termination of the Contract by the Owner for cause pursuant to Section 14.2 and only for those subcontract agreements which the Owner accepts by notifying the Subcontractor and Contractor in writing; and

    .2   assignment is subject to the prior rights of the surety, if any, obligated under bond relating to the Contract.

§ 5.4.2 Upon such assignment, if the Work has been suspended for more than 30 days, the Subcontractor's compensation shall be equitably adjusted for increases in cost resulting from the suspension.

ARTICLE 6 CONSTRUCTION BY OWNER OR BY SEPARATE CONTRACTORS
§ 6.1 OWNER'S RIGHT TO PERFORM CONSTRUCTION AND TO AWARD SEPARATE CONTRACTS
§ 6.1.1 The Owner reserves the right to perform construction or operations related to the Project with the Owner's own forces, and to award separate contracts in connection with other portions of the Project or other construction or operations on the site under Conditions of the Contract identical or substantially similar to these including those portions related to insurance and waiver of subrogation. If the Contractor claims that delay or additional cost is involved because of such action by the Owner, the Contractor shall make such Claim as provided in Section 4.3.

§ 6.1.2 When separate contracts are awarded for different portions of the Project or other construction or operations on the site, the term "Contractor" in the Contract Documents in each case shall mean the Contractor who executes each separate Owner-Contractor Agreement.

§ 6.1.3 The Owner shall provide for coordination of the activities of the Owner's own forces and of each separate contractor with the Work of the Contractor, who shall cooperate with them. The Contractor shall participate with other separate contractors and the Owner in reviewing their construction schedules when directed to do so. The Contractor shall make any revisions to the construction schedule deemed necessary after a joint review and mutual agreement. The construction schedules shall then constitute the schedules to be used by the Contractor, separate contractors and the Owner until subsequently revised.

§ 6.1.4 Unless otherwise provided in the Contract Documents, when the Owner performs construction or operations related to the Project with the Owner's own forces, the Owner shall be deemed to be subject to the same obligations and to have the same rights which apply to the Contractor under the Conditions of the Contract, including, without excluding others, those stated in Article 3, this Article 6 and Articles 10, 11 and 12.

§ 6.2 MUTUAL RESPONSIBILITY
§ 6.2.1 The Contractor shall afford the Owner and separate contractors reasonable opportunity for introduction and storage of their materials and equipment and performance of their activities, and shall connect and coordinate the Contractor's construction and operations with theirs as required by the Contract Documents.

§ 6.2.2 If part of the Contractor's Work depends for proper execution or results upon construction or operations by the Owner or a separate contractor, the Contractor shall, prior to proceeding with that portion of the Work, promptly report to the Architect apparent discrepancies or defects in such other construction that would render it unsuitable for such proper execution and results. Failure of the Contractor so to report shall constitute an acknowledgment that the Owner's or separate contractor's completed or partially completed construction is fit and proper to receive the Contractor's Work, except as to defects not then reasonably discoverable.

§ 6.2.3 The Owner shall be reimbursed by the Contractor for costs incurred by the Owner which are payable to a separate contractor because of delays, improperly timed activities or defective construction of the Contractor. The Owner shall be responsible to the Contractor for costs incurred by the Contractor because of delays, improperly timed activities, damage to the Work or defective construction of a separate contractor.

§ 6.2.4 The Contractor shall promptly remedy damage wrongfully caused by the Contractor to completed or partially completed construction or to property of the Owner or separate contractors as provided in Section 10.2.5.

§ 6.2.5 The Owner and each separate contractor shall have the same responsibilities for cutting and patching as are described for the Contractor in Section 3.14.

§ 6.3 OWNER'S RIGHT TO CLEAN UP
§ 6.3.1 If a dispute arises among the Contractor, separate contractors and the Owner as to the responsibility under their respective contracts for maintaining the premises and surrounding area free from waste materials and rubbish, the Owner may clean up and the Architect will allocate the cost among those responsible.

ARTICLE 7   CHANGES IN THE WORK
§ 7.1 GENERAL
§ 7.1.1 Changes in the Work may be accomplished after execution of the Contract, and without invalidating the Contract, by Change Order, Construction Change Directive or order for a minor change in the Work, subject to the limitations stated in this Article 7 and elsewhere in the Contract Documents.

§ 7.1.2 A Change Order shall be based upon agreement among the Owner, Contractor and Architect; a Construction Change Directive requires agreement by the Owner and Architect and may or may not be agreed to by the Contractor; an order for a minor change in the Work may be issued by the Architect alone.

§ 7.1.3 Changes in the Work shall be performed under applicable provisions of the Contract Documents, and the Contractor shall proceed promptly, unless otherwise provided in the Change Order, Construction Change Directive or order for a minor change in the Work.

§ 7.2 CHANGE ORDERS
§ 7.2.1 A Change Order is a written instrument prepared by the Architect and signed by the Owner, Contractor and Architect, stating their agreement upon all of the following:
- .1   change in the Work;
- .2   the amount of the adjustment, if any, in the Contract Sum; and
- .3   the extent of the adjustment, if any, in the Contract Time.

§ 7.2.2 Methods used in determining adjustments to the Contract Sum may include those listed in Section 7.3.3.

§ 7.3 CONSTRUCTION CHANGE DIRECTIVES
§ 7.3.1 A Construction Change Directive is a written order prepared by the Architect and signed by the Owner and Architect, directing a change in the Work prior to agreement on adjustment, if any, in the Contract Sum or Contract Time, or both. The Owner may by Construction Change Directive, without invalidating the Contract, order changes in the Work within the general scope of the Contract consisting of additions, deletions or other revisions, the Contract Sum and Contract Time being adjusted accordingly.

§ 7.3.2 A Construction Change Directive shall be used in the absence of total agreement on the terms of a Change Order.

§ 7.3.3 If the Construction Change Directive provides for an adjustment to the Contract Sum, the adjustment shall be based on one of the following methods:
- .1   mutual acceptance of a lump sum properly itemized and supported by sufficient substantiating data to permit evaluation;
- .2   unit prices stated in the Contract Documents or subsequently agreed upon;
- .3   cost to be determined in a manner agreed upon by the parties and a mutually acceptable fixed or percentage fee; or
- .4   as provided in Section 7.3.6.

§ 7.3.4 Upon receipt of a Construction Change Directive, the Contractor shall promptly proceed with the change in the Work involved and advise the Architect of the Contractor's agreement or disagreement with the method, if any, provided in the Construction Change Directive for determining the proposed adjustment in the Contract Sum or Contract Time.

§ 7.3.5 A Construction Change Directive signed by the Contractor indicates the agreement of the Contractor therewith, including adjustment in Contract Sum and Contract Time or the method for determining them. Such agreement shall be effective immediately and shall be recorded as a Change Order.

§ 7.3.6 If the Contractor does not respond promptly or disagrees with the method for adjustment in the Contract Sum, the method and the adjustment shall be determined by the Architect on the basis of reasonable expenditures and savings of those performing the Work attributable to the change, including, in case of an increase in the Contract Sum, a reasonable allowance for overhead and profit. In such case, and also under Section 7.3.3.3, the Contractor shall keep and present, in such form as the Architect may prescribe, an itemized accounting together with appropriate supporting data. Unless otherwise provided in the Contract Documents, costs for the purposes of this Section 7.3.6 shall be limited to the following:

- .1 costs of labor, including social security, old age and unemployment insurance, fringe benefits required by agreement or custom, and workers' compensation insurance;
- .2 costs of materials, supplies and equipment, including cost of transportation, whether incorporated or consumed;
- .3 rental costs of machinery and equipment, exclusive of hand tools, whether rented from the Contractor or others;
- .4 costs of premiums for all bonds and insurance, permit fees, and sales, use or similar taxes related to the Work; and
- .5 additional costs of supervision and field office personnel directly attributable to the change.

§ 7.3.7 The amount of credit to be allowed by the Contractor to the Owner for a deletion or change which results in a net decrease in the Contract Sum shall be actual net cost as confirmed by the Architect. When both additions and credits covering related Work or substitutions are involved in a change, the allowance for overhead and profit shall be figured on the basis of net increase, if any, with respect to that change.

§ 7.3.8 Pending final determination of the total cost of a Construction Change Directive to the Owner, amounts not in dispute for such changes in the Work shall be included in Applications for Payment accompanied by a Change Order indicating the parties' agreement with part or all of such costs. For any portion of such cost that remains in dispute, the Architect will make an interim determination for purposes of monthly certification for payment for those costs. That determination of cost shall adjust the Contract Sum on the same basis as a Change Order, subject to the right of either party to disagree and assert a claim in accordance with Article 4.

§ 7.3.9 When the Owner and Contractor agree with the determination made by the Architect concerning the adjustments in the Contract Sum and Contract Time, or otherwise reach agreement upon the adjustments, such agreement shall be effective immediately and shall be recorded by preparation and execution of an appropriate Change Order.

§ 7.4 MINOR CHANGES IN THE WORK

§ 7.4.1 The Architect will have authority to order minor changes in the Work not involving adjustment in the Contract Sum or extension of the Contract Time and not inconsistent with the intent of the Contract Documents. Such changes shall be effected by written order and shall be binding on the Owner and Contractor. The Contractor shall carry out such written orders promptly.

ARTICLE 8 TIME

§ 8.1 DEFINITIONS

§ 8.1.1 Unless otherwise provided, Contract Time is the period of time, including authorized adjustments, allotted in the Contract Documents for Substantial Completion of the Work.

§ 8.1.2 The date of commencement of the Work is the date established in the Agreement.

§ 8.1.3 The date of Substantial Completion is the date certified by the Architect in accordance with Section 9.8.

§ 8.1.4 The term "day" as used in the Contract Documents shall mean calendar day unless otherwise specifically defined.

§ 8.2 PROGRESS AND COMPLETION

§ 8.2.1 Time limits stated in the Contract Documents are of the essence of the Contract. By executing the Agreement the Contractor confirms that the Contract Time is a reasonable period for performing the Work.

§ 8.2.2 The Contractor shall not knowingly, except by agreement or instruction of the Owner in writing, prematurely commence operations on the site or elsewhere prior to the effective date of insurance required by Article 11 to be furnished by the Contractor and Owner. The date of commencement of the Work shall not be changed by the effective date of such insurance. Unless the date of commencement is established by the Contract Documents or a notice to proceed given by the Owner, the Contractor shall notify the Owner in writing not less than five days or other agreed period before commencing the Work to permit the timely filing of mortgages, mechanic's liens and other security interests.

§ 8.2.3 The Contractor shall proceed expeditiously with adequate forces and shall achieve Substantial Completion within the Contract Time.

§ 8.3 DELAYS AND EXTENSIONS OF TIME

§ 8.3.1 If the Contractor is delayed at any time in the commencement or progress of the Work by an act or neglect of the Owner or Architect, or of an employee of either, or of a separate contractor employed by the Owner, or by changes ordered in the Work, or by labor disputes, fire, unusual delay in deliveries, unavoidable casualties or other causes beyond the Contractor's control, or by delay authorized by the Owner pending mediation and arbitration, or by other causes which the Architect determines may justify delay, then the Contract Time shall be extended by Change Order for such reasonable time as the Architect may determine.

§ 8.3.2 Claims relating to time shall be made in accordance with applicable provisions of Section 4.3.

§ 8.3.3 This Section 8.3 does not preclude recovery of damages for delay by either party under other provisions of the Contract Documents.

ARTICLE 9   PAYMENTS AND COMPLETION
§ 9.1 CONTRACT SUM

§ 9.1.1 The Contract Sum is stated in the Agreement and, including authorized adjustments, is the total amount payable by the Owner to the Contractor for performance of the Work under the Contract Documents.

§ 9.2 SCHEDULE OF VALUES

§ 9.2.1 Before the first Application for Payment, the Contractor shall submit to the Architect a schedule of values allocated to various portions of the Work, prepared in such form and supported by such data to substantiate its accuracy as the Architect may require. This schedule, unless objected to by the Architect, shall be used as a basis for reviewing the Contractor's Applications for Payment.

§ 9.3 APPLICATIONS FOR PAYMENT

§ 9.3.1 At least ten days before the date established for each progress payment, the Contractor shall submit to the Architect an itemized Application for Payment for operations completed in accordance with the schedule of values. Such application shall be notarized, if required, and supported by such data substantiating the Contractor's right to payment as the Owner or Architect may require, such as copies of requisitions from Subcontractors and material suppliers, and reflecting retainage if provided for in the Contract Documents.

§ 9.3.1.1 As provided in Section 7.3.8, such applications may include requests for payment on account of changes in the Work which have been properly authorized by Construction Change Directives, or by interim determinations of the Architect, but not yet included in Change Orders.

§ 9.3.1.2 Such applications may not include requests for payment for portions of the Work for which the Contractor does not intend to pay to a Subcontractor or material supplier, unless such Work has been performed by others whom the Contractor intends to pay.

§ 9.3.2 Unless otherwise provided in the Contract Documents, payments shall be made on account of materials and equipment delivered and suitably stored at the site for subsequent incorporation in the Work. If approved in advance by the Owner, payment may similarly be made for materials and equipment suitably stored off the site at a location agreed upon in writing. Payment for materials and equipment stored on or off the site shall be conditioned upon compliance by the Contractor with procedures satisfactory to the Owner to establish the Owner's title to such materials and equipment or otherwise protect the Owner's interest, and shall include the costs of applicable insurance, storage and transportation to the site for such materials and equipment stored off the site.

§ 9.3.3 The Contractor warrants that title to all Work covered by an Application for Payment will pass to the Owner no later than the time of payment. The Contractor further warrants that upon submittal of an Application for Payment all Work for which Certificates for Payment have been previously issued and payments received from the Owner shall, to the best of the Contractor's knowledge, information and belief, be free and clear of liens, claims, security interests or encumbrances in favor of the Contractor, Subcontractors, material suppliers, or other persons or entities making a claim by reason of having provided labor, materials and equipment relating to the Work.

§ 9.4 CERTIFICATES FOR PAYMENT

§ 9.4.1 The Architect will, within seven days after receipt of the Contractor's Application for Payment, either issue to the Owner a Certificate for Payment, with a copy to the Contractor, for such amount as the Architect determines is properly due, or notify the Contractor and Owner in writing of the Architect's reasons for withholding certification in whole or in part as provided in Section 9.5.1.

§ 9.4.2 The issuance of a Certificate for Payment will constitute a representation by the Architect to the Owner, based on the Architect's evaluation of the Work and the data comprising the Application for Payment, that the Work has progressed to the point indicated and that, to the best of the Architect's knowledge, information and belief, the quality of the Work is in accordance with the Contract Documents. The foregoing representations are subject to an evaluation of the Work for conformance with the Contract Documents upon Substantial Completion, to results of subsequent tests and inspections, to correction of minor deviations from the Contract Documents prior to completion and to specific qualifications expressed by the Architect. The issuance of a Certificate for Payment will further constitute a representation that the Contractor is entitled to payment in the amount certified. However, the issuance of a Certificate for Payment will not be a representation that the Architect has (1) made exhaustive or continuous on-site inspections to check the quality or quantity of the Work, (2) reviewed construction means, methods, techniques, sequences or procedures, (3) reviewed copies of requisitions received from Subcontractors and material suppliers and other data requested by the Owner to substantiate the Contractor's right to payment, or (4) made examination to ascertain how or for what purpose the Contractor has used money previously paid on account of the Contract Sum.

§ 9.5 DECISIONS TO WITHHOLD CERTIFICATION

§ 9.5.1 The Architect may withhold a Certificate for Payment in whole or in part, to the extent reasonably necessary to protect the Owner, if in the Architect's opinion the representations to the Owner required by Section 9.4.2 cannot be made. If the Architect is unable to certify payment in the amount of the Application, the Architect will notify the Contractor and Owner as provided in Section 9.4.1. If the Contractor and Architect cannot agree on a revised amount, the Architect will promptly issue a Certificate for Payment for the amount for which the Architect is able to make such representations to the Owner. The Architect may also withhold a Certificate for Payment or, because of subsequently discovered evidence, may nullify the whole or a part of a Certificate for Payment previously issued, to such extent as may be necessary in the Architect's opinion to protect the Owner from loss for which the Contractor is responsible, including loss resulting from acts and omissions described in Section 3.3.2, because of:

.1 defective Work not remedied;

.2 third party claims filed or reasonable evidence indicating probable filing of such claims unless security acceptable to the Owner is provided by the Contractor;

.3 failure of the Contractor to make payments properly to Subcontractors or for labor, materials or equipment;

.4 reasonable evidence that the Work cannot be completed for the unpaid balance of the Contract Sum;

.5 damage to the Owner or another contractor;

.6 reasonable evidence that the Work will not be completed within the Contract Time, and that the unpaid balance would not be adequate to cover actual or liquidated damages for the anticipated delay; or

.7 persistent failure to carry out the Work in accordance with the Contract Documents.

§ 9.5.2 When the above reasons for withholding certification are removed, certification will be made for amounts previously withheld.

§ 9.6 PROGRESS PAYMENTS

§ 9.6.1 After the Architect has issued a Certificate for Payment, the Owner shall make payment in the manner and within the time provided in the Contract Documents, and shall so notify the Architect.

§ 9.6.2 The Contractor shall promptly pay each Subcontractor, upon receipt of payment from the Owner, out of the amount paid to the Contractor on account of such Subcontractor's portion of the Work, the amount to which said Subcontractor is entitled, reflecting percentages actually retained from payments to the Contractor on account of such Subcontractor's portion of the Work. The Contractor shall, by appropriate agreement with each Subcontractor, require each Subcontractor to make payments to Sub-subcontractors in a similar manner.

§ 9.6.3 The Architect will, on request, furnish to a Subcontractor, if practicable, information regarding percentages of completion or amounts applied for by the Contractor and action taken thereon by the Architect and Owner on account of portions of the Work done by such Subcontractor.

§ 9.6.4 Neither the Owner nor Architect shall have an obligation to pay or to see to the payment of money to a Subcontractor except as may otherwise be required by law.

§ 9.6.5 Payment to material suppliers shall be treated in a manner similar to that provided in Sections 9.6.2, 9.6.3 and 9.6.4.

§ 9.6.6 A Certificate for Payment, a progress payment, or partial or entire use or occupancy of the Project by the Owner shall not constitute acceptance of Work not in accordance with the Contract Documents.

§ 9.6.7 Unless the Contractor provides the Owner with a payment bond in the full penal sum of the Contract Sum, payments received by the Contractor for Work properly performed by Subcontractors and suppliers shall be held by the Contractor for those Subcontractors or suppliers who performed Work or furnished materials, or both, under contract with the Contractor for which payment was made by the Owner. Nothing contained herein shall require money to be placed in a separate account and not commingled with money of the Contractor, shall create any fiduciary liability or tort liability on the part of the Contractor for breach of trust or shall entitle any person or entity to an award of punitive damages against the Contractor for breach of the requirements of this provision.

§ 9.7 FAILURE OF PAYMENT
§ 9.7.1 If the Architect does not issue a Certificate for Payment, through no fault of the Contractor, within seven days after receipt of the Contractor's Application for Payment, or if the Owner does not pay the Contractor within seven days after the date established in the Contract Documents the amount certified by the Architect or awarded by arbitration, then the Contractor may, upon seven additional days' written notice to the Owner and Architect, stop the Work until payment of the amount owing has been received. The Contract Time shall be extended appropriately and the Contract Sum shall be increased by the amount of the Contractor's reasonable costs of shut-down, delay and start-up, plus interest as provided for in the Contract Documents.

§ 9.8 SUBSTANTIAL COMPLETION
§ 9.8.1 Substantial Completion is the stage in the progress of the Work when the Work or designated portion thereof is sufficiently complete in accordance with the Contract Documents so that the Owner can occupy or utilize the Work for its intended use.

§ 9.8.2 When the Contractor considers that the Work, or a portion thereof which the Owner agrees to accept separately, is substantially complete, the Contractor shall prepare and submit to the Architect a comprehensive list of items to be completed or corrected prior to final payment. Failure to include an item on such list does not alter the responsibility of the Contractor to complete all Work in accordance with the Contract Documents.

§ 9.8.3 Upon receipt of the Contractor's list, the Architect will make an inspection to determine whether the Work or designated portion thereof is substantially complete. If the Architect's inspection discloses any item, whether or not included on the Contractor's list, which is not sufficiently complete in accordance with the Contract Documents so that the Owner can occupy or utilize the Work or designated portion thereof for its intended use, the Contractor shall, before issuance of the Certificate of Substantial Completion, complete or correct such item upon notification by the Architect. In such case, the Contractor shall then submit a request for another inspection by the Architect to determine Substantial Completion.

§ 9.8.4 When the Work or designated portion thereof is substantially complete, the Architect will prepare a Certificate of Substantial Completion which shall establish the date of Substantial Completion, shall establish responsibilities of the Owner and Contractor for security, maintenance, heat, utilities, damage to the Work and insurance, and shall fix the time within which the Contractor shall finish all items on the list accompanying the Certificate. Warranties required by the Contract Documents shall commence on the date of Substantial Completion of the Work or designated portion thereof unless otherwise provided in the Certificate of Substantial Completion.

§ 9.8.5 The Certificate of Substantial Completion shall be submitted to the Owner and Contractor for their written acceptance of responsibilities assigned to them in such Certificate. Upon such acceptance and consent of surety, if any, the Owner shall make payment of retainage applying to such Work or designated portion thereof. Such payment shall be adjusted for Work that is incomplete or not in accordance with the requirements of the Contract Documents.

§ 9.9 PARTIAL OCCUPANCY OR USE
§ 9.9.1 The Owner may occupy or use any completed or partially completed portion of the Work at any stage when such portion is designated by separate agreement with the Contractor, provided such occupancy or use is consented

to by the insurer as required under Section 11.4.1.5 and authorized by public authorities having jurisdiction over the Work. Such partial occupancy or use may commence whether or not the portion is substantially complete, provided the Owner and Contractor have accepted in writing the responsibilities assigned to each of them for payments, retainage, if any, security, maintenance, heat, utilities, damage to the Work and insurance, and have agreed in writing concerning the period for correction of the Work and commencement of warranties required by the Contract Documents. When the Contractor considers a portion substantially complete, the Contractor shall prepare and submit a list to the Architect as provided under Section 9.8.2. Consent of the Contractor to partial occupancy or use shall not be unreasonably withheld. The stage of the progress of the Work shall be determined by written agreement between the Owner and Contractor or, if no agreement is reached, by decision of the Architect.

§ 9.9.2 Immediately prior to such partial occupancy or use, the Owner, Contractor and Architect shall jointly inspect the area to be occupied or portion of the Work to be used in order to determine and record the condition of the Work.

§ 9.9.3 Unless otherwise agreed upon, partial occupancy or use of a portion or portions of the Work shall not constitute acceptance of Work not complying with the requirements of the Contract Documents.

§ 9.10 FINAL COMPLETION AND FINAL PAYMENT

§ 9.10.1 Upon receipt of written notice that the Work is ready for final inspection and acceptance and upon receipt of a final Application for Payment, the Architect will promptly make such inspection and, when the Architect finds the Work acceptable under the Contract Documents and the Contract fully performed, the Architect will promptly issue a final Certificate for Payment stating that to the best of the Architect's knowledge, information and belief, and on the basis of the Architect's on-site visits and inspections, the Work has been completed in accordance with terms and conditions of the Contract Documents and that the entire balance found to be due the Contractor and noted in the final Certificate is due and payable. The Architect's final Certificate for Payment will constitute a further representation that conditions listed in Section 9.10.2 as precedent to the Contractor's being entitled to final payment have been fulfilled.

§ 9.10.2 Neither final payment nor any remaining retained percentage shall become due until the Contractor submits to the Architect (1) an affidavit that payrolls, bills for materials and equipment, and other indebtedness connected with the Work for which the Owner or the Owner's property might be responsible or encumbered (less amounts withheld by Owner) have been paid or otherwise satisfied, (2) a certificate evidencing that insurance required by the Contract Documents to remain in force after final payment is currently in effect and will not be canceled or allowed to expire until at least 30 days' prior written notice has been given to the Owner, (3) a written statement that the Contractor knows of no substantial reason that the insurance will not be renewable to cover the period required by the Contract Documents, (4) consent of surety, if any, to final payment and (5), if required by the Owner, other data establishing payment or satisfaction of obligations, such as receipts, releases and waivers of liens, claims, security interests or encumbrances arising out of the Contract, to the extent and in such form as may be designated by the Owner. If a Subcontractor refuses to furnish a release or waiver required by the Owner, the Contractor may furnish a bond satisfactory to the Owner to indemnify the Owner against such lien. If such lien remains unsatisfied after payments are made, the Contractor shall refund to the Owner all money that the Owner may be compelled to pay in discharging such lien, including all costs and reasonable attorneys' fees.

§ 9.10.3 If, after Substantial Completion of the Work, final completion thereof is materially delayed through no fault of the Contractor or by issuance of Change Orders affecting final completion, and the Architect so confirms, the Owner shall, upon application by the Contractor and certification by the Architect, and without terminating the Contract, make payment of the balance due for that portion of the Work fully completed and accepted. If the remaining balance for Work not fully completed or corrected is less than retainage stipulated in the Contract Documents, and if bonds have been furnished, the written consent of surety to payment of the balance due for that portion of the Work fully completed and accepted shall be submitted by the Contractor to the Architect prior to certification of such payment. Such payment shall be made under terms and conditions governing final payment, except that it shall not constitute a waiver of claims.

§ 9.10.4 The making of final payment shall constitute a waiver of Claims by the Owner except those arising from:
.1   liens, Claims, security interests or encumbrances arising out of the Contract and unsettled;
.2   failure of the Work to comply with the requirements of the Contract Documents; or
.3   terms of special warranties required by the Contract Documents.

§ **9.10.5** Acceptance of final payment by the Contractor, a Subcontractor or material supplier shall constitute a waiver of claims by that payee except those previously made in writing and identified by that payee as unsettled at the time of final Application for Payment.

ARTICLE 10 PROTECTION OF PERSONS AND PROPERTY
§ 10.1 SAFETY PRECAUTIONS AND PROGRAMS
§ **10.1.1** The Contractor shall be responsible for initiating, maintaining and supervising all safety precautions and programs in connection with the performance of the Contract.

§ 10.2 SAFETY OF PERSONS AND PROPERTY
§ **10.2.1** The Contractor shall take reasonable precautions for safety of, and shall provide reasonable protection to prevent damage, injury or loss to:

    .1   employees on the Work and other persons who may be affected thereby;

    .2   the Work and materials and equipment to be incorporated therein, whether in storage on or off the site, under care, custody or control of the Contractor or the Contractor's Subcontractors or Sub-subcontractors; and

    .3   other property at the site or adjacent thereto, such as trees, shrubs, lawns, walks, pavements, roadways, structures and utilities not designated for removal, relocation or replacement in the course of construction.

§ **10.2.2** The Contractor shall give notices and comply with applicable laws, ordinances, rules, regulations and lawful orders of public authorities bearing on safety of persons or property or their protection from damage, injury or loss.

§ **10.2.3** The Contractor shall erect and maintain, as required by existing conditions and performance of the Contract, reasonable safeguards for safety and protection, including posting danger signs and other warnings against hazards, promulgating safety regulations and notifying owners and users of adjacent sites and utilities.

§ **10.2.4** When use or storage of explosives or other hazardous materials or equipment or unusual methods are necessary for execution of the Work, the Contractor shall exercise utmost care and carry on such activities under supervision of properly qualified personnel.

§ **10.2.5** The Contractor shall promptly remedy damage and loss (other than damage or loss insured under property insurance required by the Contract Documents) to property referred to in Sections 10.2.1.2 and 10.2.1.3 caused in whole or in part by the Contractor, a Subcontractor, a Sub-subcontractor, or anyone directly or indirectly employed by any of them, or by anyone for whose acts they may be liable and for which the Contractor is responsible under Sections 10.2.1.2 and 10.2.1.3, except damage or loss attributable to acts or omissions of the Owner or Architect or anyone directly or indirectly employed by either of them, or by anyone for whose acts either of them may be liable, and not attributable to the fault or negligence of the Contractor. The foregoing obligations of the Contractor are in addition to the Contractor's obligations under Section 3.18.

§ **10.2.6** The Contractor shall designate a responsible member of the Contractor's organization at the site whose duty shall be the prevention of accidents. This person shall be the Contractor's superintendent unless otherwise designated by the Contractor in writing to the Owner and Architect.

§ **10.2.7** The Contractor shall not load or permit any part of the construction or site to be loaded so as to endanger its safety.

§ 10.3 HAZARDOUS MATERIALS
§ **10.3.1** If reasonable precautions will be inadequate to prevent foreseeable bodily injury or death to persons resulting from a material or substance, including but not limited to asbestos or polychlorinated biphenyl (PCB), encountered on the site by the Contractor, the Contractor shall, upon recognizing the condition, immediately stop Work in the affected area and report the condition to the Owner and Architect in writing.

§ **10.3.2** The Owner shall obtain the services of a licensed laboratory to verify the presence or absence of the material or substance reported by the Contractor and, in the event such material or substance is found to be present, to verify that it has been rendered harmless. Unless otherwise required by the Contract Documents, the Owner shall furnish in writing to the Contractor and Architect the names and qualifications of persons or entities who are to perform tests verifying the presence or absence of such material or substance or who are to perform the task of removal or safe containment of such material or substance. The Contractor and the Architect will promptly reply to the Owner in writing stating whether or not either has reasonable objection to the persons or entities proposed by the Owner. If either the Contractor or Architect has an objection to a person or entity proposed by the Owner, the Owner shall propose another to whom the Contractor and the Architect have no reasonable objection. When the material or substance has been rendered harmless, Work in the affected area shall resume upon written agreement of the Owner and Contractor. The Contract Time shall be extended appropriately and the Contract Sum shall be increased in the amount of the Contractor's reasonable additional costs of shut-down, delay and start-up, which adjustments shall be accomplished as provided in Article 7.

§ 10.3.3 To the fullest extent permitted by law, the Owner shall indemnify and hold harmless the Contractor, Subcontractors, Architect, Architect's consultants and agents and employees of any of them from and against claims, damages, losses and expenses, including but not limited to attorneys' fees, arising out of or resulting from performance of the Work in the affected area if in fact the material or substance presents the risk of bodily injury or death as described in Section 10.3.1 and has not been rendered harmless, provided that such claim, damage, loss or expense is attributable to bodily injury, sickness, disease or death, or to injury to or destruction of tangible property (other than the Work itself) and provided that such damage, loss or expense is not due to the sole negligence of a party seeking indemnity.

§ 10.4 The Owner shall not be responsible under Section 10.3 for materials and substances brought to the site by the Contractor unless such materials or substances were required by the Contract Documents.

§ 10.5 If, without negligence on the part of the Contractor, the Contractor is held liable for the cost of remediation of a hazardous material or substance solely by reason of performing Work as required by the Contract Documents, the Owner shall indemnify the Contractor for all cost and expense thereby incurred.

§ 10.6 EMERGENCIES
§ 10.6.1 In an emergency affecting safety of persons or property, the Contractor shall act, at the Contractor's discretion, to prevent threatened damage, injury or loss. Additional compensation or extension of time claimed by the Contractor on account of an emergency shall be determined as provided in Section 4.3 and Article 7.

ARTICLE 11   INSURANCE AND BONDS
§ 11.1 CONTRACTOR'S LIABILITY INSURANCE
§ 11.1.1 The Contractor shall purchase from and maintain in a company or companies lawfully authorized to do business in the jurisdiction in which the Project is located such insurance as will protect the Contractor from claims set forth below which may arise out of or result from the Contractor's operations under the Contract and for which the Contractor may be legally liable, whether such operations be by the Contractor or by a Subcontractor or by anyone directly or indirectly employed by any of them, or by anyone for whose acts any of them may be liable:

   .1   claims under workers' compensation, disability benefit and other similar employee benefit acts which are applicable to the Work to be performed;

   .2   claims for damages because of bodily injury, occupational sickness or disease, or death of the Contractor's employees;

   .3   claims for damages because of bodily injury, sickness or disease, or death of any person other than the Contractor's employees;

   .4   claims for damages insured by usual personal injury liability coverage;

   .5   claims for damages, other than to the Work itself, because of injury to or destruction of tangible property, including loss of use resulting therefrom;

   .6   claims for damages because of bodily injury, death of a person or property damage arising out of ownership, maintenance or use of a motor vehicle;

   .7   claims for bodily injury or property damage arising out of completed operations; and

   .8   claims involving contractual liability insurance applicable to the Contractor's obligations under Section 3.18.

§ 11.1.2 The insurance required by Section 11.1.1 shall be written for not less than limits of liability specified in the Contract Documents or required by law, whichever coverage is greater. Coverages, whether written on an occurrence or claims-made basis, shall be maintained without interruption from date of commencement of the Work until date of final payment and termination of any coverage required to be maintained after final payment.

§ 11.1.3 Certificates of insurance acceptable to the Owner shall be filed with the Owner prior to commencement of the Work. These certificates and the insurance policies required by this Section 11.1 shall contain a provision that coverages afforded under the policies will not be canceled or allowed to expire until at least 30 days' prior written notice has been given to the Owner. If any of the foregoing insurance coverages are required to remain in force after final payment and are reasonably available, an additional certificate evidencing continuation of such coverage shall be submitted with the final Application for Payment as required by Section 9.10.2. Information concerning reduction of coverage on account of revised limits or claims paid under the General Aggregate, or both, shall be furnished by the Contractor with reasonable promptness in accordance with the Contractor's information and belief.

§ 11.2 OWNER'S LIABILITY INSURANCE
§ 11.2.1 The Owner shall be responsible for purchasing and maintaining the Owner's usual liability insurance.

§ 11.3 PROJECT MANAGEMENT PROTECTIVE LIABILITY INSURANCE
§ 11.3.1 Optionally, the Owner may require the Contractor to purchase and maintain Project Management Protective Liability insurance from the Contractor's usual sources as primary coverage for the Owner's, Contractor's and Architect's vicarious liability for construction operations under the Contract. Unless otherwise required by the Contract Documents, the Owner shall reimburse the Contractor by increasing the Contract Sum to pay the cost of purchasing and maintaining such optional insurance coverage, and the Contractor shall not be responsible for purchasing any other liability insurance on behalf of the Owner. The minimum limits of liability purchased with such coverage shall be equal to the aggregate of the limits required for Contractor's Liability Insurance under Sections 11.1.1.2 through 11.1.1.5.

§ 11.3.2 To the extent damages are covered by Project Management Protective Liability insurance, the Owner, Contractor and Architect waive all rights against each other for damages, except such rights as they may have to the proceeds of such insurance. The policy shall provide for such waivers of subrogation by endorsement or otherwise.

§ 11.3.3 The Owner shall not require the Contractor to include the Owner, Architect or other persons or entities as additional insureds on the Contractor's Liability Insurance coverage under Section 11.1.

§ 11.4 PROPERTY INSURANCE
§ 11.4.1 Unless otherwise provided, the Owner shall purchase and maintain, in a company or companies lawfully authorized to do business in the jurisdiction in which the Project is located, property insurance written on a builder's risk "all-risk" or equivalent policy form in the amount of the initial Contract Sum, plus value of subsequent Contract modifications and cost of materials supplied or installed by others, comprising total value for the entire Project at the site on a replacement cost basis without optional deductibles. Such property insurance shall be maintained, unless otherwise provided in the Contract Documents or otherwise agreed in writing by all persons and entities who are beneficiaries of such insurance, until final payment has been made as provided in Section 9.10 or until no person or entity other than the Owner has an insurable interest in the property required by this Section 11.4 to be covered, whichever is later. This insurance shall include interests of the Owner, the Contractor, Subcontractors and Sub-subcontractors in the Project.

§ 11.4.1.1 Property insurance shall be on an "all-risk" or equivalent policy form and shall include, without limitation, insurance against the perils of fire (with extended coverage) and physical loss or damage including, without duplication of coverage, theft, vandalism, malicious mischief, collapse, earthquake, flood, windstorm, falsework, testing and startup, temporary buildings and debris removal including demolition occasioned by enforcement of any applicable legal requirements, and shall cover reasonable compensation for Architect's and Contractor's services and expenses required as a result of such insured loss.

§ 11.4.1.2 If the Owner does not intend to purchase such property insurance required by the Contract and with all of the coverages in the amount described above, the Owner shall so inform the Contractor in writing prior to commencement of the Work. The Contractor may then effect insurance which will protect the interests of the Contractor, Subcontractors and Sub-subcontractors in the Work, and by appropriate Change Order the cost thereof shall be charged to the Owner. If the Contractor is damaged by the failure or neglect of the Owner to purchase or maintain insurance as described above, without so notifying the Contractor in writing, then the Owner shall bear all reasonable costs properly attributable thereto.

§ 11.4.1.3 If the property insurance requires deductibles, the Owner shall pay costs not covered because of such deductibles.

§ 11.4.1.4 This property insurance shall cover portions of the Work stored off the site, and also portions of the Work in transit.

§ **11.4.1.5** Partial occupancy or use in accordance with Section 9.9 shall not commence until the insurance company or companies providing property insurance have consented to such partial occupancy or use by endorsement or otherwise. The Owner and the Contractor shall take reasonable steps to obtain consent of the insurance company or companies and shall, without mutual written consent, take no action with respect to partial occupancy or use that would cause cancellation, lapse or reduction of insurance.

§ **11.4.2 Boiler and Machinery Insurance**. The Owner shall purchase and maintain boiler and machinery insurance required by the Contract Documents or by law, which shall specifically cover such insured objects during installation and until final acceptance by the Owner; this insurance shall include interests of the Owner, Contractor, Subcontractors and Sub-subcontractors in the Work, and the Owner and Contractor shall be named insureds.

§ **11.4.3 Loss of Use Insurance**. The Owner, at the Owner's option, may purchase and maintain such insurance as will insure the Owner against loss of use of the Owner's property due to fire or other hazards, however caused. The Owner waives all rights of action against the Contractor for loss of use of the Owner's property, including consequential losses due to fire or other hazards however caused.

§ **11.4.4** If the Contractor requests in writing that insurance for risks other than those described herein or other special causes of loss be included in the property insurance policy, the Owner shall, if possible, include such insurance, and the cost thereof shall be charged to the Contractor by appropriate Change Order.

§ **11.4.5** If during the Project construction period the Owner insures properties, real or personal or both, at or adjacent to the site by property insurance under policies separate from those insuring the Project, or if after final payment property insurance is to be provided on the completed Project through a policy or policies other than those insuring the Project during the construction period, the Owner shall waive all rights in accordance with the terms of Section 11.4.7 for damages caused by fire or other causes of loss covered by this separate property insurance. All separate policies shall provide this waiver of subrogation by endorsement or otherwise.

§ **11.4.6** Before an exposure to loss may occur, the Owner shall file with the Contractor a copy of each policy that includes insurance coverages required by this Section 11.4. Each policy shall contain all generally applicable conditions, definitions, exclusions and endorsements related to this Project. Each policy shall contain a provision that the policy will not be canceled or allowed to expire, and that its limits will not be reduced, until at least 30 days' prior written notice has been given to the Contractor.

§ **11.4.7 Waivers of Subrogation**. The Owner and Contractor waive all rights against (1) each other and any of their subcontractors, sub-subcontractors, agents and employees, each of the other, and (2) the Architect, Architect's consultants, separate contractors described in Article 6, if any, and any of their subcontractors, sub-subcontractors, agents and employees, for damages caused by fire or other causes of loss to the extent covered by property insurance obtained pursuant to this Section 11.4 or other property insurance applicable to the Work, except such rights as they have to proceeds of such insurance held by the Owner as fiduciary. The Owner or Contractor, as appropriate, shall require of the Architect, Architect's consultants, separate contractors described in Article 6, if any, and the subcontractors, sub-subcontractors, agents and employees of any of them, by appropriate agreements, written where legally required for validity, similar waivers each in favor of other parties enumerated herein. The policies shall provide such waivers of subrogation by endorsement or otherwise. A waiver of subrogation shall be effective as to a person or entity even though that person or entity would otherwise have a duty of indemnification, contractual or otherwise, did not pay the insurance premium directly or indirectly, and whether or not the person or entity had an insurable interest in the property damaged.

§ **11.4.8** A loss insured under Owner's property insurance shall be adjusted by the Owner as fiduciary and made payable to the Owner as fiduciary for the insureds, as their interests may appear, subject to requirements of any applicable mortgagee clause and of Section 11.4.10. The Contractor shall pay Subcontractors their just shares of insurance proceeds received by the Contractor, and by appropriate agreements, written where legally required for validity, shall require Subcontractors to make payments to their Sub-subcontractors in similar manner.

§ **11.4.9** If required in writing by a party in interest, the Owner as fiduciary shall, upon occurrence of an insured loss, give bond for proper performance of the Owner's duties. The cost of required bonds shall be charged against proceeds received as fiduciary. The Owner shall deposit in a separate account proceeds so received, which the Owner shall distribute in accordance with such agreement as the parties in interest may reach, or in accordance with an arbitration award in which case the procedure shall be as provided in Section 4.6. If after such loss no other special agreement is made and unless the Owner terminates the Contract for convenience, replacement of damaged property shall be performed by the Contractor after notification of a Change in the Work in accordance with Article 7.

§ 11.4.10 The Owner as fiduciary shall have power to adjust and settle a loss with insurers unless one of the parties in interest shall object in writing within five days after occurrence of loss to the Owner's exercise of this power; if such objection is made, the dispute shall be resolved as provided in Sections 4.5 and 4.6. The Owner as fiduciary shall, in the case of arbitration, make settlement with insurers in accordance with directions of the arbitrators. If distribution of insurance proceeds by arbitration is required, the arbitrators will direct such distribution.

§ 11.5 PERFORMANCE BOND AND PAYMENT BOND
§ 11.5.1 The Owner shall have the right to require the Contractor to furnish bonds covering faithful performance of the Contract and payment of obligations arising thereunder as stipulated in bidding requirements or specifically required in the Contract Documents on the date of execution of the Contract.

§ 11.5.2 Upon the request of any person or entity appearing to be a potential beneficiary of bonds covering payment of obligations arising under the Contract, the Contractor shall promptly furnish a copy of the bonds or shall permit a copy to be made.

ARTICLE 12   UNCOVERING AND CORRECTION OF WORK
§ 12.1 UNCOVERING OF WORK
§ 12.1.1 If a portion of the Work is covered contrary to the Architect's request or to requirements specifically expressed in the Contract Documents, it must, if required in writing by the Architect, be uncovered for the Architect's examination and be replaced at the Contractor's expense without change in the Contract Time.

§ 12.1.2 If a portion of the Work has been covered which the Architect has not specifically requested to examine prior to its being covered, the Architect may request to see such Work and it shall be uncovered by the Contractor. If such Work is in accordance with the Contract Documents, costs of uncovering and replacement shall, by appropriate Change Order, be at the Owner's expense. If such Work is not in accordance with the Contract Documents, correction shall be at the Contractor's expense unless the condition was caused by the Owner or a separate contractor in which event the Owner shall be responsible for payment of such costs.

§ 12.2 CORRECTION OF WORK
§ 12.2.1 BEFORE OR AFTER SUBSTANTIAL COMPLETION
§ 12.2.1.1 The Contractor shall promptly correct Work rejected by the Architect or failing to conform to the requirements of the Contract Documents, whether discovered before or after Substantial Completion and whether or not fabricated, installed or completed. Costs of correcting such rejected Work, including additional testing and inspections and compensation for the Architect's services and expenses made necessary thereby, shall be at the Contractor's expense.

§ 12.2.2 AFTER SUBSTANTIAL COMPLETION
§ 12.2.2.1 In addition to the Contractor's obligations under Section 3.5, if, within one year after the date of Substantial Completion of the Work or designated portion thereof or after the date for commencement of warranties established under Section 9.9.1, or by terms of an applicable special warranty required by the Contract Documents, any of the Work is found to be not in accordance with the requirements of the Contract Documents, the Contractor shall correct it promptly after receipt of written notice from the Owner to do so unless the Owner has previously given the Contractor a written acceptance of such condition. The Owner shall give such notice promptly after discovery of the condition. During the one-year period for correction of Work, if the Owner fails to notify the Contractor and give the Contractor an opportunity to make the correction, the Owner waives the rights to require correction by the Contractor and to make a claim for breach of warranty. If the Contractor fails to correct nonconforming Work within a reasonable time during that period after receipt of notice from the Owner or Architect, the Owner may correct it in accordance with Section 2.4.

§ 12.2.2.2 The one-year period for correction of Work shall be extended with respect to portions of Work first performed after Substantial Completion by the period of time between Substantial Completion and the actual performance of the Work.

§ 12.2.2.3 The one-year period for correction of Work shall not be extended by corrective Work performed by the Contractor pursuant to this Section 12.2.

§ 12.2.3 The Contractor shall remove from the site portions of the Work which are not in accordance with the requirements of the Contract Documents and are neither corrected by the Contractor nor accepted by the Owner.

§ 12.2.4 The Contractor shall bear the cost of correcting destroyed or damaged construction, whether completed or partially completed, of the Owner or separate contractors caused by the Contractor's correction or removal of Work which is not in accordance with the requirements of the Contract Documents.

§ 12.2.5 Nothing contained in this Section 12.2 shall be construed to establish a period of limitation with respect to other obligations which the Contractor might have under the Contract Documents. Establishment of the one-year period for correction of Work as described in Section 12.2.2 relates only to the specific obligation of the Contractor to correct the Work, and has no relationship to the time within which the obligation to comply with the Contract Documents may be sought to be enforced, nor to the time within which proceedings may be commenced to establish the Contractor's liability with respect to the Contractor's obligations other than specifically to correct the Work.

§ 12.3 ACCEPTANCE OF NONCONFORMING WORK
§ 12.3.1 If the Owner prefers to accept Work which is not in accordance with the requirements of the Contract Documents, the Owner may do so instead of requiring its removal and correction, in which case the Contract Sum will be reduced as appropriate and equitable. Such adjustment shall be effected whether or not final payment has been made.

ARTICLE 13   MISCELLANEOUS PROVISIONS
§ 13.1 GOVERNING LAW
§ 13.1.1 The Contract shall be governed by the law of the place where the Project is located.

§ 13.2 SUCCESSORS AND ASSIGNS
§ 13.2.1 The Owner and Contractor respectively bind themselves, their partners, successors, assigns and legal representatives to the other party hereto and to partners, successors, assigns and legal representatives of such other party in respect to covenants, agreements and obligations contained in the Contract Documents. Except as provided in Section 13.2.2, neither party to the Contract shall assign the Contract as a whole without written consent of the other. If either party attempts to make such an assignment without such consent, that party shall nevertheless remain legally responsible for all obligations under the Contract.

§ 13.2.2 The Owner may, without consent of the Contractor, assign the Contract to an institutional lender providing construction financing for the Project. In such event, the lender shall assume the Owner's rights and obligations under the Contract Documents. The Contractor shall execute all consents reasonably required to facilitate such assignment.

§ 13.3 WRITTEN NOTICE
§ 13.3.1 Written notice shall be deemed to have been duly served if delivered in person to the individual or a member of the firm or entity or to an officer of the corporation for which it was intended, or if delivered at or sent by registered or certified mail to the last business address known to the party giving notice.

§ 13.4 RIGHTS AND REMEDIES
§ 13.4.1 Duties and obligations imposed by the Contract Documents and rights and remedies available thereunder shall be in addition to and not a limitation of duties, obligations, rights and remedies otherwise imposed or available by law.

§ 13.4.2 No action or failure to act by the Owner, Architect or Contractor shall constitute a waiver of a right or duty afforded them under the Contract, nor shall such action or failure to act constitute approval of or acquiescence in a breach there under, except as may be specifically agreed in writing.

§ 13.5 TESTS AND INSPECTIONS
§ 13.5.1 Tests, inspections and approvals of portions of the Work required by the Contract Documents or by laws, ordinances, rules, regulations or orders of public authorities having jurisdiction shall be made at an appropriate time. Unless otherwise provided, the Contractor shall make arrangements for such tests, inspections and approvals with an independent testing laboratory or entity acceptable to the Owner, or with the appropriate public authority, and shall bear all related costs of tests, inspections and approvals. The Contractor shall give the Architect timely notice of when and where tests and inspections are to be made so that the Architect may be present for such procedures. The

Owner shall bear costs of tests, inspections or approvals which do not become requirements until after bids are received or negotiations concluded.

§ 13.5.2 If the Architect, Owner or public authorities having jurisdiction determine that portions of the Work require additional testing, inspection or approval not included under Section 13.5.1, the Architect will, upon written authorization from the Owner, instruct the Contractor to make arrangements for such additional testing, inspection or approval by an entity acceptable to the Owner, and the Contractor shall give timely notice to the Architect of when and where tests and inspections are to be made so that the Architect may be present for such procedures. Such costs, except as provided in Section 13.5.3, shall be at the Owner's expense.

§ 13.5.3 If such procedures for testing, inspection or approval under Sections 13.5.1 and 13.5.2 reveal failure of the portions of the Work to comply with requirements established by the Contract Documents, all costs made necessary by such failure including those of repeated procedures and compensation for the Architect's services and expenses shall be at the Contractor's expense.

§ 13.5.4 Required certificates of testing, inspection or approval shall, unless otherwise required by the Contract Documents, be secured by the Contractor and promptly delivered to the Architect.

§ 13.5.5 If the Architect is to observe tests, inspections or approvals required by the Contract Documents, the Architect will do so promptly and, where practicable, at the normal place of testing.

§ 13.5.6 Tests or inspections conducted pursuant to the Contract Documents shall be made promptly to avoid unreasonable delay in the Work.

§ 13.6 INTEREST
§ 13.6.1 Payments due and unpaid under the Contract Documents shall bear interest from the date payment is due at such rate as the parties may agree upon in writing or, in the absence thereof, at the legal rate prevailing from time to time at the place where the Project is located.

§ 13.7 COMMENCEMENT OF STATUTORY LIMITATION PERIOD
§ 13.7.1 As between the Owner and Contractor:
- .1 Before Substantial Completion. As to acts or failures to act occurring prior to the relevant date of Substantial Completion, any applicable statute of limitations shall commence to run and any alleged cause of action shall be deemed to have accrued in any and all events not later than such date of Substantial Completion;
- .2 Between Substantial Completion and Final Certificate for Payment. As to acts or failures to act occurring subsequent to the relevant date of Substantial Completion and prior to issuance of the final Certificate for Payment, any applicable statute of limitations shall commence to run and any alleged cause of action shall be deemed to have accrued in any and all events not later than the date of issuance of the final Certificate for Payment; and
- .3 After Final Certificate for Payment. As to acts or failures to act occurring after the relevant date of issuance of the final Certificate for Payment, any applicable statute of limitations shall commence to run and any alleged cause of action shall be deemed to have accrued in any and all events not later than the date of any act or failure to act by the Contractor pursuant to any Warranty provided under Section 3.5, the date of any correction of the Work or failure to correct the Work by the Contractor under Section 12.2, or the date of actual commission of any other act or failure to perform any duty or obligation by the Contractor or Owner, whichever occurs last.

ARTICLE 14   TERMINATION OR SUSPENSION OF THE CONTRACT
§ 14.1 TERMINATION BY THE CONTRACTOR
§ 14.1.1 The Contractor may terminate the Contract if the Work is stopped for a period of 30 consecutive days through no act or fault of the Contractor or a Subcontractor, Sub-subcontractor or their agents or employees or any other persons or entities performing portions of the Work under direct or indirect contract with the Contractor, for any of the following reasons:
- .1 issuance of an order of a court or other public authority having jurisdiction which requires all Work to be stopped;
- .2 an act of government, such as a declaration of national emergency which requires all Work to be stopped;

    .3   because the Architect has not issued a Certificate for Payment and has not notified the Contractor of the reason for withholding certification as provided in Section 9.4.1, or because the Owner has not made payment on a Certificate for Payment within the time stated in the Contract Documents; or

    .4   the Owner has failed to furnish to the Contractor promptly, upon the Contractor's request, reasonable evidence as required by Section 2.2.1.

§ 14.1.2 The Contractor may terminate the Contract if, through no act or fault of the Contractor or a Subcontractor, Sub-subcontractor or their agents or employees or any other persons or entities performing portions of the Work under direct or indirect contract with the Contractor, repeated suspensions, delays or interruptions of the entire Work by the Owner as described in Section 14.3 constitute in the aggregate more than 100 percent of the total number of days scheduled for completion, or 120 days in any 365-day period, whichever is less.

§ 14.1.3 If one of the reasons described in Section 14.1.1 or 14.1.2 exists, the Contractor may, upon seven days' written notice to the Owner and Architect, terminate the Contract and recover from the Owner payment for Work executed and for proven loss with respect to materials, equipment, tools, and construction equipment and machinery, including reasonable overhead, profit and damages.

§ 14.1.4 If the Work is stopped for a period of 60 consecutive days through no act or fault of the Contractor or a Subcontractor or their agents or employees or any other persons performing portions of the Work under contract with the Contractor because the Owner has persistently failed to fulfill the Owner's obligations under the Contract Documents with respect to matters important to the progress of the Work, the Contractor may, upon seven additional days' written notice to the Owner and the Architect, terminate the Contract and recover from the Owner as provided in Section 14.1.3.

§ 14.2 TERMINATION BY THE OWNER FOR CAUSE

§ 14.2.1 The Owner may terminate the Contract if the Contractor:

    .1   persistently or repeatedly refuses or fails to supply enough properly skilled workers or proper materials;

    .2   fails to make payment to Subcontractors for materials or labor in accordance with the respective agreements between the Contractor and the Subcontractors;

    .3   persistently disregards laws, ordinances, or rules, regulations or orders of a public authority having jurisdiction; or

    .4   otherwise is guilty of substantial breach of a provision of the Contract Documents.

§ 14.2.2 When any of the above reasons exist, the Owner, upon certification by the Architect that sufficient cause exists to justify such action, may without prejudice to any other rights or remedies of the Owner and after giving the Contractor and the Contractor's surety, if any, seven days' written notice, terminate employment of the Contractor and may, subject to any prior rights of the surety:

    .1   take possession of the site and of all materials, equipment, tools, and construction equipment and machinery thereon owned by the Contractor;

    .2   accept assignment of subcontracts pursuant to Section 5.4; and

    .3   finish the Work by whatever reasonable method the Owner may deem expedient. Upon request of the Contractor, the Owner shall furnish to the Contractor a detailed accounting of the costs incurred by the Owner in finishing the Work.

§ 14.2.3 When the Owner terminates the Contract for one of the reasons stated in Section 14.2.1, the Contractor shall not be entitled to receive further payment until the Work is finished.

§ 14.2.4 If the unpaid balance of the Contract Sum exceeds costs of finishing the Work, including compensation for the Architect's services and expenses made necessary thereby, and other damages incurred by the Owner and not expressly waived, such excess shall be paid to the Contractor. If such costs and damages exceed the unpaid balance, the Contractor shall pay the difference to the Owner. The amount to be paid to the Contractor or Owner, as the case may be, shall be certified by the Architect, upon application, and this obligation for payment shall survive termination of the Contract.

§ 14.3 SUSPENSION BY THE OWNER FOR CONVENIENCE

§ 14.3.1 The Owner may, without cause, order the Contractor in writing to suspend, delay or interrupt the Work in whole or in part for such period of time as the Owner may determine.

§ 14.3.2 The Contract Sum and Contract Time shall be adjusted for increases in the cost and time caused by suspension, delay or interruption as described in Section 14.3.1. Adjustment of the Contract Sum shall include profit. No adjustment shall be made to the extent:

    .1   that performance is, was or would have been so suspended, delayed or interrupted by another cause for which the Contractor is responsible; or

    .2   that an equitable adjustment is made or denied under another provision of the Contract.

§ 14.4 TERMINATION BY THE OWNER FOR CONVENIENCE

§ 14.4.1 The Owner may, at any time, terminate the Contract for the Owner's convenience and without cause.

§ 14.4.2 Upon receipt of written notice from the Owner of such termination for the Owner's convenience, the Contractor shall:

    .1   cease operations as directed by the Owner in the notice;

    .2   take actions necessary, or that the Owner may direct, for the protection and preservation of the Work; and

    .3   except for Work directed to be performed prior to the effective date of termination stated in the notice, terminate all existing subcontracts and purchase orders and enter into no further subcontracts and purchase orders.

§ 14.4.3 In case of such termination for the Owner's convenience, the Contractor shall be entitled to receive payment for Work executed, and costs incurred by reason of such termination, along with reasonable overhead and profit on the Work not executed.

Appendix D

CMAA Document A-3 (2003 Edition) General Conditions Between Owner and Contractor

CMAA Document A-3 (2003 Edition)

ARTICLE 1
GENERAL PROVISIONS

_____

1.1 The Parties, Definitions, and Intent

1.1.1 The term "Owner" means_____

_____

whose mailing address is _____

_____
_____
_____
_____
_____

and includes its designated representatives and its successors and assigns.

1.12 The term "Construction Manager", hereinafter referred to as the "CM" means _____

_____

whose address is _____

_____
_____
_____
_____.

1.1.3 The term "Designer" means _____

_____

whose address is_____

_____
_____
_____
_____.

The term "Designer" includes the authorized representatives of the Designer, its consultants, or any successor or other firm or person designated by the Owner to act in the same capacity.

1.1.4 The term "Project" means _____

_____
_____
_____
_____
_____
_____
_____
_____

including associated site improvements and appurtenances and structures to be constructed on certain premises located in _____

_____
_____
_____
_____.

The Project is more fully described in the Contract Documents.

1.1.5 The term "Contractor" means the individual, partnership, firm, corporation or other business entity that contracts with Owner to furnish labor or materials or both at the Project or otherwise in connection with the Project.

1.1.6 The term "Contract Documents" means the Instructions to Bidders, the Contract between Contractor and Owner (hereinafter referred to as the "Contract"), these General Conditions and any Supplemental Conditions furnished to the Contractor, the drawings and specifications furnished to the Contractor and all exhibits thereto, addenda, bulletins and change orders issued in accordance with these General Conditions to any of the above, and all other documents specified in Exhibit B of the Standard Form of Contract Between Owner and Contractor, CMAA Document A-2, 2003 edition.

1.1.7 The term "subcontractor" means any individual, partnership, firm, corporation or other business entity that has a contractual relationship with a contractor or any other subcontractor to furnish labor, equipment or materials for performance of the Work at the site. Each subcontractor shall be required by the party with whom it contracts to agree to comply with these General Conditions and any other applicable Contract Documents.

1.1.8 The term "supplier" means any manufacturer, fabricator, distributor or vendor having a contract with the Contractor or with any subcontractor to furnish materials or equipment to be incorporated into work by the Contractor or any subcontractor.

CMAA Document A-3 (2003 Edition)

1.1.9   The contract price is stated in this Contract and, including authorized change orders thereto, is the total amount payable by the Owner to the Contractor for the performance of the Work specified by the Contract Documents. The contract price may not be changed except as specified in the Contract Documents.

1.1.10   The contract time is the number of days included in the Contract Documents for completion of the Work specified therein.

1.1.11   The term "substantial completion" is the date determined by the CM in consultation with the Designer when the Work has progressed to the point that it is sufficiently complete in accordance with the Contract Documents so that the Owner may fully occupy and use the Project or designated portion thereof for the use for which it is intended, with all of the Project's parts and systems operable as required by the Contract Documents.

1.1.12   A "day" is a calendar day of twenty four (24) hours measured from midnight to the next midnight. When any period of time is referred to in the Contract Documents by days, it shall be computed to exclude the first and include the last day of such period. If the last day of any such period falls on a Saturday, Sunday or on a day made a legal holiday by the law of the applicable jurisdiction, such day shall be omitted from the computation.

1.1.13   When the CM issues a Notice of Award to the successful bidder, the notice shall be accompanied by two (2) unsigned counterparts of this Contract and all other Contract Documents. Within fifteen (15) days thereafter the Contractor shall execute and deliver two (2) counterparts of this Contract to the CM with all other Contract Documents attached.

1.1.14   By executing the Contract, the Contractor represents that it has examined the Contract Documents thoroughly; visited the site to become familiar with local conditions that may in any manner affect cost, progress or performance of the Work; become familiar with federal, state and local laws, ordinances, rules and regulations that may in any manner affect cost, progress or performance of the Work; and studied and carefully correlated the Contractor's observations with the Contract Documents.

1.1.15   The Contract Documents comprise the entire Contract between the Owner and the Contractor concerning the Work. The Contract Documents are complementary and what is called for in one is as binding as if called for by all.

1.1.16   It is the intent of the Contract Documents to describe a functionally complete Project, or part thereof, to be constructed. Any labor, materials or equipment that may reasonably be inferred from the Contract Documents as being required to produce the intended result shall be supplied whether or not specified. When words that have a customary technical or trade meaning are used to describe labor, materials or equipment, the words shall be interpreted in accordance with that meaning. References to standard specifications, manuals or codes of any technical society, organization or association or to the laws or regulations of any governmental authority shall mean the latest standard specification, manual, code, laws or regulations in effect at the time of opening of bids except as may be otherwise specifically stated.

1.1.17   The CM shall transmit to the Contractor written clarifications and interpretations of the requirements of the Contract Documents. Such clarifications and interpretations shall be consistent with or reasonably inferable form the overall intent of the Contract Documents.

1.1.18   If, during the performance of the Work, the Contractor finds a conflict, error or discrepancy in the Contract Documents, the Contractor shall so report to the CM in writing at once. Before proceeding with the work affected thereby, the Contractor shall obtain a written interpretation or clarification from the CM. Any work done before the CM renders its decision is at the Contractor's sole risk.

1.1.19   If any portion of the Contract Documents conflicts with any other portion, the various documents comprising the Contract Documents shall govern in the following order of precedence: the Owner-Contractor Contract and any properly executed change orders thereto; the Supplemental Conditions; the General Conditions; the specifications; the Instructions to Bidders; the drawings. As between figures given to drawings and the scaled measurements, the figures shall govern. Detailed drawings shall be given precedence over general drawings.

CMAA Document A-3 (2003 Edition)

1.1.20 The Contractor agrees that nothing contained in the Contract Documents or any agreement between the Owner and the CM or the Owner and the Designer creates any contractual relationship between the CM, the Designer and the Contractor. The Contractor waives any right the Contractor may have as an alleged third party beneficiary of any such agreements and covenants not to sue the CM or Designer as a third party beneficiary of such agreements.

1.1.21 Whenever materials or equipment are specified or described in the Contract Documents by using the name of a proprietary item or the name of a particular supplier, the naming of the item is intended to establish the type, function and quality required. Unless the name is followed by words indicating that no substitution is permitted, materials or equipment of other suppliers may be accepted if sufficient information is submitted to allow determination that the material or equipment proposed is equivalent or equal to that named. Proposals for substitutions of material and equipment shall not be accepted from anyone other than the Contractor. If the Contractor desires to propose a substitution the Contractor shall submit a written proposal to the CM who shall submit the proposal to the Designer for acceptance thereof. The proposal shall certify that the substitute will perform adequately and achieve the results called for by the design, be similar and of equal substance to that specified and be suited to the same use as that specified. The proposal shall state that the evaluation and acceptance of the substitute will not prevent the Contractor from achieving substantial completion on time, state whether or not acceptance of the substitute for use in the Work shall require a change in any of the Contract Documents or in the provisions of any other contract for work on the Project to adapt the design to the substitute, and state whether or not incorporation or use of the substitute in connection with the Work is subject to payment of any license fee or royalty. The proposal shall contain an itemized estimate of all costs that shall result directly from acceptance of such substitute, including costs of redesign and claims of other contractors affected by the resulting change, all of which shall be considered in evaluating the proposed substitute. The CM may require the Contractor at the Contractor's expense to furnish additional data about the proposed substitute.

1.1.22 If specific means, methods, techniques, sequences or procedures of construction are indicated in or required by the Contract Documents, the Contractor may propose a substitute means, method, sequence, technique or procedure of construction if the Contractor submits sufficient information to allow the CM and Designer to determine that the proposal for substitution is equivalent to that indicated or required by the Contract Documents. The procedure for review of the proposed substitute shall be similar to that provided in Paragraph 1.1.21.

1.1.22.1 The CM shall be allowed by the Contractor a reasonable time, but not less than seven (7) days, to consult with the Designer to evaluate each proposal for substitution. No substitute shall be ordered, installed or used without the CM's and the Designer's prior written acceptance that shall be evidenced by either a change order or other appropriate documentation transmitted to the Contractor by the CM.

1.1.22.2 The Contractor may be required to furnish, at the Contractor's expense, a special performance guarantee or other surety with respect to any substitute.

1.1.22.3 Whether or not the CM accepts a proposal for substitution, the Contractor shall reimburse the Owner for the charges of the CM and Designer for evaluating each proposal for substitution.

1.1.23 Whenever in the Contract Documents the term "as ordered," "as directed," "as required," "as allowed," "as approved" or terms of like effect or import are used or the adjectives "reasonable," "suitable," "acceptable," "proper" or "satisfactory" or adjectives of like effect or import are used to describe a requirement, direction, review or judgment of the CM or Designer as to the Work, it is intended that such requirement, direction, review, or judgment shall be solely to evaluate the Work for compliance with the Contract Documents. The use of any such term or adjective shall not assign to the Owner, CM, or Designer any duty or authority to supervise or direct the furnishing or performance of the Work.

1.2   Ownership and Use of Drawings and Specifications

CMAA Document A-3 (2003 Edition)

1.2.1   The Contractor shall be furnished two (2) copies of the Contract Documents. Additional copies shall be furnished upon request for the cost of reproduction.

1.2.2   The Contractor shall have no ownership rights in any of the Contract Documents, including the drawings, specifications or other documents prepared by the Owner, the Designer, or the CM.

ARTICLE 2
THE OWNER AND
CONTRUCTION MANAGER

2.1   Responsibilities

2.1.1   The CM shall administer the Contract as described herein. The CM in performing under this Contract is acting as the Owner's principal agent in all matters regarding this Contract. Nothing in this Contract shall be construed to mean that the CM is a fiduciary of the Owner.

2.1.2   The Owner and Designer shall communicate with the Contractor only through or in the manner prescribed by the CM who shall have full authority to act with regard to all aspects of the Project.

2.1.3   The Contractor shall, within five (5) days after the earlier of the notice to proceed or the effective date of this Contract, file with the CM a list of all persons who are authorized to sign documents such as contracts, certificates and affidavits on behalf of the Contractor and to fully bind the Contractor to all conditions and provisions of such documents, except that in the case of a corporation, the Contractor shall file with the CM a certified copy of a resolution of the board of directors of the corporation in which are listed the names and titles of those personnel who are authorized to sign documents on behalf of the corporation and to fully bind the corporation to all the conditions and provisions of such documents.

2.1.4   Information and services under the Owner's control shall be furnished promptly by the Owner.

2.1.5   Except as provided in Paragraph 9.1.1 the CM shall not revoke, alter, enlarge, relax or release any requirements

of the Contract Documents or approve or accept any portion of the Work not performed in accordance with the Contract Documents or to issue instructions contrary to the Contract Documents.

2.1.6   The CM shall establish and implement a program to monitor the quality of the Work in order to assist in guarding the Owner against defects and deficiencies in the Work. Neither the Owner, the CM, nor the Designer shall supervise, control, direct, have authority over, or be responsible for the Contractor's means, methods, techniques, sequences or procedures of construction; nor shall they be responsible for the Contractor's safety precautions and programs incident thereto. Further, neither the Owner, CM, nor the Designer shall be responsible for the Contractor's failure to comply with all laws and regulations applicable to the furnishing or performing of the Work, nor for the Contractor's failure to furnish or perform the Work in accordance with the Contract Documents.

2.1.7   The Owner, CM, or the Designer shall not be responsible for the acts or omissions of the Contractor, any subcontractors or suppliers, or any other persons or organizations furnishing or performing any of the Work.

2.1.8   Observations by the CM and inspections, tests or approvals by others shall not relieve the Contractor from its obligations to perform the Work in accordance with the Contract Documents.

2.1.9   The CM shall have authority on behalf of the Owner to disapprove or reject work where, in the opinion of the CM or the Designer, it does not conform to the requirements of the Contract Documents and, except in an emergency situation where person or property may be injured or damaged, all such disapprovals shall be communicated to the Contractor through the CM.

2.1.10   The CM shall have the authority and discretion to call, schedule and conduct job meetings to be attended by the Contractor, representatives of subcontractors and others to discuss such matters as procedures, progress, problems and scheduling.

CMAA Document A-3 (2003 Edition)

2.1.11 In consultation with the Designer, the CM shall establish procedures for processing shop drawings, samples, and other submittals.

2.1.12 The CM shall review all requests for changes and shall process change orders, including applications for a change of the contract time.

2.1.13 The CM shall determine when the Contractor's work is substantially complete. In consultation with the Designer the CM shall issue appropriate Certificates of Substantial Completion, with attached lists of incomplete work or other work which does not conform to the requirements of the Contract Documents. At the time of substantial completion, only incidental corrective work and final cleaning, if required beyond cleaning needed for the Owner's full use, may remain for final completion. The Certificate of Final Completion shall similarly be prepared by the CM and be issued to the Contractor when the Contractor's work is complete and conforms to the requirements of the Contract Documents.

2.1.14 Engineering surveys to establish reference points for construction shall be provided to the extent necessary, in the CM's judgment, to enable the Contractor to proceed with the Work. The Contractor shall be responsible for laying out the Work, shall protect and preserve the established reference points and shall make no changes or relocations without the prior written approval of the Owner. The Contractor shall report to the CM whenever any reference point is lost, destroyed or required relocation due to necessary changes in grades or locations. The Contractor shall be responsible for the accurate replacement or relocation of such reference points by professionally qualified personnel.

2.1.15 The Owner shall furnish, as indicated in the Contract Documents, the lands upon which the Work is to be performed, rights-of-way and easements for access thereto and such other lands that are designated for the use of the Contractor. Easements for permanent structures or permanent changes in existing facilities shall be obtained and paid for by the Owner. The Contractor shall provide at its expense all additional lands and access thereto that may be required for temporary construction facilities or storage of materials and equipment.

2.2 Owner's Right to Stop or Suspend the Work

2.2.1 If work is defective or the Contractor fails to supply sufficient skilled workers, suitable materials or equipment or if the Contractor fails to furnish or perform the Work in such a way that the completed Work will conform to the Contract Documents, the Owner may order the Contractor to stop the Work or any portion thereof until the cause for such order has been eliminated.

2.2.2 The Owner may at any time and without cause suspend the Work or any portion thereof for a period of not more than one hundred eighty (180) days by notice in writing to the Contractor that shall fix the date on which work shall be resumed. The Contractor shall resume the Work on the date so fixed.

2.2.3 This right of the Owner to stop the Work shall not give rise to any duty on the part of the Owner to exercise this right for the benefit of the Contractor or any other party. Neither the CM's authority to act under the terms of this Contract nor any decision made by the CM in good faith either to exercise or not to exercise such authority shall give rise to any duty on the part of the CM to the Contractor, any subcontractor, any supplier, any of their agents or employees or any other person performing or furnishing any of the Work.

2.2.4 If the performance of all or part of the Work is suspended for an unreasonable period of time by an act of the Owner or the CM, or by failure of either of them to act within the time specified in this contract, the Contractor shall be entitled to a change in the contract price or time in accordance with the provisions of Article 9. No change in the contract price or time shall be made for failure of any suspension to the extent that performance would have been so suspended pursuant to Paragraph 2.2.1 or for which an adjustment is provided for or excluded under any other provision of this Contract. No request for a change in the contract price or time shall be allowed: for any costs incurred prior to the date the Contractor shall have notified the CM in writing of the act or failure to act involved, but this requirement shall not apply as to a request resulting from a suspension order; for any request for an extension of time required for performance, unless within twenty (20) days after the act or failure to act involved, the Contractor submits to the CM a

CMAA Document A-3 (2003 Edition)

written statement setting forth, as then practicable, the extent of such requested time extension; and unless the requests for increased costs or for a time extension, in an amount stated, are submitted in writing within twenty (20) days after the end of such suspension.

2.2.5   No request by the Contractor for a change in the contract price or time under Paragraph 2.2.4 shall be allowed if such request is made after final payment.

2.3   The Owner's Right to Perform Work

2.3.1   If the Contractor fails within a reasonable time after written notice of the CM to proceed to correct defective work or to remove and replace rejected work as required by the CM in accordance with Paragraph 10.2.1, if the Contractor fails to perform the Work in accordance with the Contract Documents or if the Contractor fails to comply with any other provision of the Contract Documents, the Owner may, after seven (7) days written notice to the Contractor, correct and remedy any such deficiency. In exercising the rights and remedies under this Paragraph, the Owner shall proceed expeditiously. To the extent necessary to complete corrective and remedial action, the Owner may exclude the Contractor from all or part of the site, take possession of all or part of the Work and suspend the Contractor's services related thereto, take possession of the Contractor's tools, appliances, construction equipment and machinery at the site and incorporate in the Work all materials and equipment stored at the site or for which the Owner has paid the Contractor, but which are stored elsewhere.   The Contractor shall allow the Owner, the Owner's representatives, agents and employees such access to the site as may be necessary to enable the Owner to exercise its rights and remedies.

2.3.2   In the event the Owner exercises the rights and remedies specified in Paragraph 2.3.1, the Owner shall be entitled to an appropriate decrease in the contract price. All direct, indirect and consequential costs of the Owner related to such rights and remedies shall be charged to the Contractor and a change order shall be issued incorporating the necessary revisions in the Contract Documents. Such direct, indirect and consequential costs shall include, but not be limited to, fees and charges of the CM, engineers, architects, attorneys and other professionals, all court and

arbitration costs and all costs of repair and replacement of work of others destroyed or damaged by correction, removal or replacement of the Contractor's defective work.   The Contractor shall not be allowed an extension of the contract time due to any delay in performance of the Work attributable to the exercise by the Owner of the Owner's rights and remedies hereunder.

ARTICLE 3
THE DESIGNER

3.1   Services

3.1.1   The Designer shall provide certain services on behalf of the Owner as hereinafter described. The Designer shall not be responsible for the Contractor's means, methods, techniques, sequences or procedures of construction or the safety precautions and programs incident thereto.   The Designer shall not be responsible for the Contractor's failure to perform or furnish the Work in accordance with the Contract Documents or for the acts or omissions of the Contractor or of any subcontractor or supplier, or of any other person or organization performing or furnishing any of the Work.

3.1.2   The Designer shall provide to the Owner professional architectural and engineering services for all design phases of the Project and professional design services during construction as provided in this Contract. The services shall include customary architectural and civil, structural, mechanical and electrical engineering services as provided in the Agreement between Owner and Designer.

3.1.3   The Designer shall promptly review shop drawings, samples and other submittals, but the Designer's review shall be only for conformance with the design concept of the Project and for compatibility with the design concept of the completed Project as a functioning whole as indicated in the Contract Documents, and shall not extend to means, methods, techniques, sequences or procedures of construction or to safety precautions or programs incident thereto. The review of a separate item as such shall not indicate approval of the assembly in which the item functions. The results of the Designer's review of such submittals will be

CMAA Document A-3 (2003 Edition)

transmitted to the Contractor by the CM. The Contractor shall do no work without Contract Documents and required approved shop drawings, samples, or other submittals for such portions of the Work.

3.1.4   Should errors, omissions or conflicts in the drawings, specifications or other Contract Documents prepared by the Designer be discovered, the Designer shall prepare such clarification amendments or supplementary documents and provide consultation as may be required. This information shall be transmitted to the Contractor by the CM.

3.1.5   Subject to the limitations herein, the Designer may make visits to the site to carry out its responsibilities to the Owner concerning this Project.

3.1.6   The Designer shall only communicate with the Contractor, or with subcontractors and suppliers through the CM, or in the presence of the CM.

3.1.7   At the request of the Owner, the Designer may provide additional services regarding this Project. Any such additional services that are related to the Work shall be coordinated by the CM.

3.1.8   The Designer may provide assistance in the initial operation of any equipment or system such as start-up, testing, adjusting and balancing.

3.1.9   In case of the termination of the employment of the Designer, the Owner may appoint a Designer whose status under the Contract Documents shall be that of the former Designer.

ARTICLE 4
THE CONTRACTOR

4.1   Access to the Work

4.1.1   The Owner, CM, Designer, testing agencies, governmental agencies with jurisdictional interests, and all parties designated by the Owner or the CM shall have access

to the Work at all times. The Contractor shall provide proper and safe conditions for such access.

4.2   Contractor's Review of Contract Documents

4.2.1   By submitting a bid, the Contractor has agreed that the Contract Documents, together with any supplementary written instructions issued by the Owner or CM that have become a part of the Contract Documents, appear accurate, consistent and complete insofar as can reasonably be determined. During the performance of this contract, the Contractor shall report to the CM any error, inconsistency or omission in or of the Contract Documents, including any requirement that may be contrary to any usual construction practice, law, ordinance, rule, regulation or order of any public authority bearing on the performance of the Work

4.3   Supervision and Construction Procedures

4.3.1   The Contractor shall supervise and direct the Work competently and efficiently, devoting such attention thereto and applying such skills and expertise as may be necessary to perform the Work in accordance with the requirements of the Contract Documents. The Contractor shall be solely responsible for the means, methods, techniques, sequences and procedures of construction of the Work. The Contractor shall be responsible to verify that all completed Work complies with the requirements of the Contract Documents.

4.3.2   The Contractor shall designate in writing to the CM a competent full time resident superintendent to supervise and direct the Work. The superintendent shall not be replaced without written notice to and approval by the CM. The superintendent shall be the Contractor's representative at the site and shall have authority to act on behalf of the Contractor. All communications given to the superintendent shall be as binding as if given to the Contractor. When requested by the CM, the Contractor shall provide a management chart and a list of personnel comprising the superintending staff and their areas of responsibility. All references herein to the superintendent shall be taken to mean the Contractor's superintending staff. The superintendent shall, during the performance of the Work, remain on the Project site not less than eight hours per day, five days per week until termination of the Contract, unless

CMAA Document A-3 (2003 Edition)

the job is suspended or work is stopped by the CM or Owner. The superintendent shall not be employed or used on any other project during the course of the Work.

4.3.3 All communications to the Designer or to the Owner should shall be submitted by the Contractor only through the CM.

4.3.4   The Contractor shall be solely responsible for all acts and omissions of the Contractor's employees, subcontractors, suppliers and all other persons or organizations furnishing or performing any of the Work.

4.3.5   The Contractor's responsibility to perform the Work in accordance with the Contract Documents is absolute and shall not be diminished by actions or duties of the Designer, the CM, or by any persons other than the Contractor.

4.3.6   The Contractor shall maintain at the site in a secure place one record copy of all drawings; specifications; addenda; written amendments; change orders; approved shop drawings, samples and submittals and written interpretations and clarifications in good order and annotated to show all changes made during construction. These record documents shall be available to the CM and Designer for reference, and shall be delivered to the CM upon completion of the Work

4.4   Labor and Materials

4.4.1   The Contractor shall provide competent, suitably qualified personnel to survey and lay out the Work and perform the Work as required by the Contract Documents. The Contractor shall at all times maintain good discipline and order at the site. Except in connection with the safety or protection of persons, work or property at the site or adjacent thereto, and except as otherwise indicated in the Contract Documents, all work at the site shall be performed during regular working hours. The Contractor shall not permit overtime work or the performance of work on Saturday, Sunday or any legal holiday without the CM's written consent given after at least forty-eight (48) hours prior written notice by the Contractor to the CM. The Contractor shall furnish and assume complete responsibility for all materials, equipment, labor, transportation, construction equipment and machinery,

tool, appliances, fuel, power, light, heat, telephone, water, sanitary facilities, temporary facilities and all other facilities and incidentals necessary for the furnishing, performance, testing, start-up and completion of the Work.

4.4.2   All work shall be performed in a skillful and workmanlike manner.

4.4.3 The Contractor shall at all times provide competent and suitably qualified personnel to furnish and perform the Work in conformance with the requirements of the Contract Documents, and shall promptly remove from the site any person or organization that does not meet these requirements. The Contractor shall at all times maintain good order and strict discipline among its employees and shall require the same of its subcontractors, suppliers, or any other persons or organizations furnishing or performing the Work. The CM may require, in writing, the Contractor to remove from the Project any such person or entity the CM deems incompetent, careless or otherwise objectionable.

4.5   Use of Site

4.5.1   The Contractor shall confine construction operations, equipment, storage of materials and equipment and the activities of workers to the Project site, land and areas identified in and permitted by the Contract Documents and other land and areas permitted by law and regulations, rights-of-way, permits and easements and shall not unreasonably encumber the premises with construction equipment or other materials or equipment. The Contractor shall assume full responsibility for any damage to any such land, area or to the owner or occupant thereof or of any land or areas contiguous thereto resulting from the performance of the Work.

4.5.2   The Contractor shall not load or permit any part of any structure to be loaded in any manner that shall endanger the structure. The Contractor shall not subject any part of the Work or adjacent property to stresses or pressures that shall endanger it.

4.6   Maintenance of Site

4.6.1   During the progress of the Work, the Contractor shall keep the premises free from accumulations of waste

CMAA Document A-3 (2003 Edition)

materials, rubbish and other debris resulting from the Work. At the completion of the Work, the Contractor shall remove all waste materials, rubbish and debris from and about the premises as well as all tools, appliances, construction equipment, machinery and surplus materials and shall leave the site clean and ready for occupancy. The Contractor shall restore to original condition all property not designated for alteration by the Contract Documents.

4.6.2   If the Contractor fails to clean up during or at the completion of the Work, the Owner may do so as provided in Paragraph 6.3 and the cost thereof shall be charged to the Contractor.

4.7   Permits, Fees, and Notices

4.7.1   The Contractor shall obtain and pay for all construction permits and licenses required for the Work. The Contractor shall pay all governmental charges and inspection fees necessary for the prosecution of the work and applicable at the time of opening bids or, if there are no bids, on the effective date of the Contract. The Contractor shall pay all charges of utility owners for connections and disconnections to the Work.

4.7.2   The Contractor shall give all notices and shall comply with all governmental laws and regulations applicable to the performance of the Work. Neither the Owner, CM nor Designer shall, except where otherwise expressly required by applicable laws and regulations, be responsible for monitoring the Contractor's compliance with any laws and regulations.

4.8   Allowances

4.8.1   It is understood that the Contractor has included in the contract price all allowances so named in the Contract Documents and shall cause the work so covered to be done by such subcontractors or suppliers and for such sums within the limit of the allowances as may be acceptable to the CM. The Contractor agrees as follows:

4.8.1.1   The allowances include the cost to the Contractor, less any applicable trade discounts, of materials and equipment required by the allowances to be delivered at the site and all applicable taxes;

4.8.1.2   The Contractor's costs for unloading and handling on the site, labor, installation costs, overhead, profit and other expenses contemplated for the allowances have been included in the contract price and not in the allowances; and

4.8.1.3   No demand for additional payment on account of the above costs shall be valid. Prior to final payment, an appropriate change order shall be issued as recommended by the CM to reflect actual amounts due the Contractor due to work covered by allowances and the contract price shall be correspondingly adjusted.

4.9   Shop Drawings, Samples, and Other Submittals

4.9.1   The Contractor shall submit to the CM seven (7) copies of all shop drawings, samples and other submittals. These shall bear a stamp or specific written statement that the Contractor has satisfied its responsibilities under the Contract Documents with respect to the review of the submittal. All submittals shall be identified clearly as to material, supplier, pertinent data such as catalog numbers and the use for which intended as the CM may require. The data shown on the shop drawings, samples, or other submittals shall be complete with respect to quantities, dimensions, specified performance, design criteria, materials and similar data to enable review of the information by the CM and the Designer.

4.9.2   The Contractor shall submit all shop drawings, samples and other submittals required by the Contract Documents to the CM with such promptness as to cause no delay in the Work.

4.9.3   Before submission of each shop drawing, sample or other submittal, the Contractor shall have determined and verified all quantities, dimensions, specified performance criteria, installation requirements, materials, catalog numbers and similar data with respect thereto and reviewed or coordinated each shop drawing, sample or other submittal with each other and with each other and with the requirements of the Work and the Contract Documents.

CMAA Document A-3 (2003 Edition)

4.9.4   At the time of each submission, the Contractor shall give the CM specific written notice of each variation that the shop drawings, samples, or other submittals may have from the requirements of the Contract Documents and, in addition, shall cause a specific notation of each such variation to be made for each shop drawing, sample and other submittal.

4.9.5   Reviews of shop drawings, samples or other submittals by any party acting on the Owner's behalf shall not relieve the Contractor from responsibility for any variation from the requirements of the Contract Documents unless the Contractor has in writing called the CM's attention to that variation at the time of submission as required by Paragraph 4.9.4 and the Contractor has received written acknowledgement of each such variation incorporated into or accompanying the approval of each shop drawing, sample or other submittal.   No review by the CM or Designer shall relieve the Contractor from responsibility for errors or omissions in the shop drawing, sample, or other submittal.

4.9.6   The CM shall endeavor to have the shop drawings, samples or other submittals reviewed and returned to the Contractor, with a written indication of approval or the reason for rejection, not later than twenty-one (21) days after the date of submission to the CM.

4.9.7   The Contractor shall make at its own expense any corrections to the shop drawings, samples or other submittals that are required by the Designer or the CM and shall resubmit the required number of corrected copies.   The Contractor shall indicate in writing on each resubmittal any revisions other than or in addition to the corrections required on previous submittals.

4.9.8   Where a shop drawing, sample or other submittal is required by the Contract Documents, any related work performed prior to the review, approval, and return to the Contractor of the pertinent submittal shall be the sole risk, expense and responsibility of the Contractor.

4.9.9   Shop drawings, samples or other submittals, in any part, in any form or in any stage of review or approval shall not constitute Contract Documents or parts thereof.

4.10   Cutting and Patching of Work

4.10.1   The Contractor shall do all cutting, fitting and patching of the Work as may be required to make its several parts come together properly and integrate with such other work.

4.10.2   The Contractor shall not endanger any work of others by cutting, excavating or otherwise altering their work and shall only cut or alter their work with the written consent of the CM and the others whose work shall be affected.

4.11   Damage to Work

4.11.1   If the Contractor, any of its subcontractors or suppliers, or any other persons or organizations furnishing or performing the Work, damages the work or property of the Owner, or the work or property of any other contractor on the site, or any other structures or facilities on the site, then the Contractor shall promptly remedy such damage.

4.12   Fees, Royalties and Patents

4.12.1   The Contractor shall pay all license fees and royalties related to or necessary for the Work and assume all costs incident to the use in the performance of the Work or the incorporation in the Work of any invention, design, process, product or device that is the subject of patent rights or copyrights held by others.

4.12.2   If a particular invention, design, process, product or device is specified in the Contract Documents for use in the performance of the Work and if, to the actual knowledge of the Owner, CM or Designer, its use is subject to patent rights or copyrights calling for the payment of any license fee or royalty to others, the existence of such rights shall be specified in the Contract Documents.

4.13   Warranty

4.13.1   All materials and equipment shall be of good quality and new, except as otherwise provided in the Contract Documents.   If required by the CM, the Contractor shall furnish reports of required tests or other satisfactory evidence, as to the kind and quality of materials and equipment.   All materials and equipment shall be applied, erected, installed, connected, cleaned, used and conditioned in accordance with

CMAA Document A-3 (2003 Edition)

the instructions of the applicable supplier except as otherwise provided in the Contract Documents. However, no provision of any such instructions shall be effective to assign to the CM, the Designer, or their consultants, agents or employees any duty or authority to supervise or direct the furnishing or performance of the Work or any duty or authority to undertake responsibility contrary to the provisions of Paragraph 2.1.6 or 3.1.1.

4.13.2 The Contractor warrants and guarantees that all work will be in accordance with the Contract Documents and will not be defective. All defective work, whether or not in place, may be rejected, corrected or accepted as provided in the Contract Documents.

4.14 Taxes

4.14.1 The Contractor shall pay all sales, consumer, use, service, and other similar taxes required to be paid by the Contractor in accordance with the laws and regulations of the place of the Project that are applicable during the performance of the Work.

4.15 Indemnification

4.15.1 The Contractor shall indemnify and hold harmless the Owner, CM, Designer, other contractors, and their consultants, agents and employees from and against all claims, demands, suits, damages, including consequential damages and damages resulting from bodily injury or damage to property, costs, expenses and attorneys' fees arising out of or resulting from the performance of the Work, provided that such claims, demands, suits, damages, costs, expenses and attorneys' fees are caused in whole or in part by negligent acts or omissions of the Contractor or any subcontractor, person or organization for whose acts the Contractor is liable.

4.15.2 In any and all claims against the Owner, CM or Designer, or any of their consultants, agents or employees by any employee of Contractor, any subcontractor or any person or organization employed by any of them to perform or furnish any of the Work or anyone for whose acts any of them may be liable, the indemnification obligation under Paragraph 4.15.1 shall not be limited in any way by limitation on the amount or type of damages, compensation or benefits payable by or for

the Contractor, any such subcontractor, other person or organization under Workers' Compensation Acts, disability benefit acts or other employee benefit acts.

4.15.3 To the extent permitted by law the indemnity provided in Paragraph 4.15.1 shall apply regardless of whether or not such claims, demands, suits, damages, costs, expenses and fees are caused in part by any person or entity indemnified hereunder.

4.16 Tests and Inspections

4.16.1 The Contractor shall give the CM timely notice of readiness of the Work for all required inspections, tests or approvals.

4.16.2 If laws or regulations of any public body having jurisdiction require any work or part thereof to specifically be inspected, tested or approved, the Contractor shall assume full responsibility therefore, pay all costs in connection therewith and furnish the CM the required certificates of inspection, testing or approval. The Contractor shall also be responsible for and shall pay all costs in connection with any inspection or testing required in connection with the Designer's and the CM's acceptance of a proposal for substitution of materials or equipment to be incorporated in the Work or of alternate materials or equipment submitted for approval prior to the Contractor's purchase thereof for incorporation in the Work.

4.16.3 All inspections, tests or approvals other than those required by laws or regulations of any public body having jurisdiction shall be performed by organizations acceptable to the Owner and CM.

4.16.4 Inspections or testing performed exclusively for the Contractor's convenience shall be the sole responsibility of the Contractor.

4.17 Existing Facilities, Structures and Physical Conditions

4.17.1 The Contractor shall have fully acquainted itself with the type and location of the Work and shall be responsible for having carefully examined and inspected any

CMAA Document A-3 (2003 Edition)

existing facilities and structures and conditions that may affect the Work, including without limitation those relating to and which may affect: the transportation, handling, delivery and storage of materials; the availability of labor; the availability of water and electricity; the availability, condition and use of roadways and other access to the Work; weather conditions; the type and location of surface and subsurface conditions; the type and location of surface and subsurface utility lines at the Project site and those adjacent to the Project site; other contracts to be entered into by the Owner relating to the Project that may affect the Work and require coordination and scheduling efforts by the Contractor; and the type, availability and storage of equipment, materials or supplies for use in performing the Work. The Contractor shall determine and fully acquaint itself with all regulations, codes, ordinances and provisions of law that affect the Work.

4.17.2   Certain information and data is shown or indicated in the Contract Documents with respect to existing surface and subsurface conditions at or contiguous to the site. The Contractor shall determine the sufficiency of the information and shall conduct additional investigations as may be necessary to fully acquaint itself with the existing conditions or facilities at the site.

4.17.3   The Contractor shall be entitled to rely upon the accuracy of the factual data contained in reports of explorations and tests of surface and subsurface conditions at or contiguous to the site that have been used by the Designer in preparation of the Contract Documents and that have been furnished to the Contractor.

4.17.4   If the Contractor believes that:

4.17.4.1   Any factual data on which the Contractor is entitled to rely is inaccurate; or

4.17.4.2 Any existing surface or subsurface condition encountered at or contiguous to the site differs materially from that indicated or referred to in the Contract Documents, the Contractor shall, after becoming aware thereof and before performing any work in connection therewith, except in an emergency as permitted by Paragraph 5.3.2, promptly notify the CM in writing about the difference.

4.17.5   The CM shall review the pertinent conditions and, in consultation with the Designer, determine the necessity of additional investigations, explorations or tests with respect thereto, and advise the Owner in writing about the inaccuracy or difference. If the CM, in consultation with the Designer, determines that an existing surface or subsurface condition, facility or structure is materially different from that indicated or referred to in the Contract Documents, or that there exists a surface or subsurface condition at or contiguous to the site, of which the Contractor could not reasonably have been expected to have been aware, or to have been entitled to rely upon, the CM and Designer shall determine the extent to which the Contract Documents should be modified. The Contractor may be entitled to a change in the contract price or time in accordance with the provisions of Article 9. The Contractor shall be responsible for the safety and protection of such condition, as provided in Article 5.

4.18   Asbestos, PCB's, Petroleum, Hazardous Waste or Radioactive Material

4.18.1   The Owner shall be responsible for any asbestos, PCB's, petroleum, hazardous waste or radioactive material uncovered or revealed at the site which was not shown or identified in the Contract Documents to be within the scope of the Work and which may present a substantial danger to persons or property exposed thereto in connection with the Work at the site. The Owner shall not be responsible for any such materials brought to the site by the Contractor, subcontractor, suppliers or anyone else for whom the Contractor is responsible.

4.18.2   The Contractor shall immediately stop all work in connection with such hazardous condition and in any area affected thereby, and notify the Owner and the CM (and thereafter confirm such notice in writing). The Owner shall promptly consult with the CM concerning the necessity for the Owner to retain a qualified expert to evaluate such hazardous condition or take corrective action, if any. The Contractor shall not be required to resume work in connection with such hazardous condition or in any such affected area until after the Owner has obtained any required permits related thereto and delivered to the Contractor special written notice specifying that such condition and any affected area is or has been rendered safe for the resumption of work; or specifying

CMAA Document A-3 (2003 Edition)

any special conditions under which such work may be resumed safely. If the Owner and the Contractor cannot agree as to entitlement to or the amount or extent of any adjustment, if any, in the contract price or contract time as a result of such work stoppage or such special conditions under which work is agreed by the Contractor to be resumed, an adjustment, if requested by the Contractor, shall be made therefore as provided in Paragraph 9.3 and 9.4.

4.18.3   If after receipt of such special written notice the Contractor does not agree to resume such work based on a reasonable belief it is unsafe, or does not agree to resume such work under such special conditions, then the Owner may order such portion of the work that is in connection with such hazardous condition or in such affected area to be deleted from the work. If the Owner and the Contractor cannot agree as to entitlement to or the amount or extent of an adjustment, if any, in the contract price or contract time as a result of deleting such portion of the Work, then the dispute shall be referred as provided in Paragraph 15.4. The Owner may have such deleted portion of the Work performed by the Owner's own forces or others in accordance with Paragraph 2.3.

4.18.4   To the fullest extent permitted by laws and regulations, the Owner shall indemnify and hold harmless the Contractor, subcontractors, the CM, the Designer, and the officers, directors, employees, agents, consultants and subcontractors of each and any of them from and against all claims, costs, losses and damages (including, but not limited to, all fees and charges of engineers, architects, attorneys and other professionals, and all court or arbitration or other dispute resolution costs) arising out of or resulting from such hazardous condition, provided that any such claim, cost, loss or damage is attributable to bodily injury, sickness, disease or death, or to injury to or destruction of tangible property (other than the Work itself), including the loss of use resulting therefrom; and nothing in this Paragraph 4.18.4 shall obligate the Owner to indemnify any person or entity form and against the consequences of that person's or entity's own negligence.

ARTICLE 5
SAFETY AND PROTECTION
OF PROPERTY

5.1   Safety

5.1.1   Before beginning the Work, the Contractor shall prepare and submit to the CM the Contractor's safety program that provides for the implementation of all of the Contractor's safety responsibilities in connection with the Work at the site and the coordination of that program and its associated procedures and precautions with the safety programs, precautions and procedures of each of the other contractors performing the Work at the site. The Contractor shall be solely responsible for initiating, maintaining, monitoring and supervising all safety programs, precautions and procedures in connection with the Work and for coordinating its programs, precautions and procedures with those other contractors performing the work at the site. The Contractor shall take all necessary protection to prevent damage, injury and loss to:

5.1.1.1   All employees on the Project; employees of all subcontractors, and other persons and organizations who may be affected thereby;

5.1.1.2   Employees of all subcontractors and other persons and organizations who may be affected thereby.

5.2   Protection of Property

5.2.1   Before beginning the Work, the Contractor shall also submit to the CM a plan to protect until final completion:

5.2.1.1   All the work, and materials and equipment to be incorporated therein; and

5.2.1.2   Other property at the site or adjacent thereto, including trees, shrubs, lawns, walks, pavements, roadways, structures, utilities and underground facilities not designated for removal, relocation or replacement in the course of construction.

CMAA Document A-3 (2003 Edition)

5.3   Applicable Laws

5.3.1   The Contractor shall comply with all applicable laws and regulations of any public body having jurisdiction for the safety of persons or property or to protect them from damage, injury or loss and shall erect and maintain all necessary safeguards for such safety and protection.   The Contractor shall notify owners of adjacent property and of underground facilities and utility owners when prosecution of the Work may affect them and shall cooperate with them in the protection, removal, relocation and replacement of their property.   All damage, injury or loss to any property caused, directly or indirectly, in whole or in part, by the Contractor, any subcontractor, supplier or any other person organization directly or indirectly employed by any of them to perform or furnish any of the Work or anyone for whose acts any of them may be liable, shall be remedied by the Contractor.   The Contractor's duties and responsibilities for the safety and protection of the Work shall continue until such time as all the Work is completed and the CM has issued a Certificate of Final Completion to the Contractor.

5.3.2   In emergencies affecting the safety or protection of persons or the Work or property at the site or adjacent thereto, the Contractor, without special instruction or authorization from the CM, is obligated to act to prevent threatened damage, injury or loss.   The Contractor shall give the CM prompt written notice if the Contractor believes that any significant changes in the Work or variations form the Contract Documents have been caused thereby.

5.3.3   The Contractor shall designate a responsible representative at the site whose duty shall be the prevention of accidents.   This person shall be the Contractor's superintendent unless otherwise designated in writing by the Contractor to and accepted by the CM.

ARTICLE 6
SUBCONTRACTORS AND SUPPLIERS

6.1   Award of Subcontracts for Portions of the Work

6.1.1   The Contractor shall not employ any subcontractor, supplier or other person or organization against whom the CM may have reasonable objection.   The Contractor shall not be required to employ any subcontractor, supplier or other person or organization to furnish or perform any of the Work against whom the Contractor has reasonable objection.

6.1.2   If the Contract Documents require that the identity of certain subcontractors or suppliers be submitted to the CM for acceptance, the CM shall, after due investigation, either accept or reject the subcontractors or suppliers proposed.   In the event a subcontractor or supplier is rejected, the Contractor shall submit an acceptable substitute and there will be an appropriate adjustment in the contract price or contract time attributable thereto.   No acceptance by the CM of any subcontractors or suppliers proposed shall constitute a waiver of any right of the Owner or the CM to reject defective work.

6.1.3   The Contractor shall be fully and solely responsible for all acts and omissions of the subcontractors, suppliers and other persons and organizations furnishing or performing any of the Work, just as the Contractor is responsible for the Contractor's own acts and omissions.

6.1.4   All work performed for the Contractor by a subcontractor or supplier shall be pursuant to an appropriate agreement between the Contractor and the subcontractor or supplier that specifically binds the subcontractor or supplier to the applicable terms and conditions of the Contract Documents and contains waiver provisions as required by Paragraph 12.6.   The Contractor shall pay each subcontractor or supplier a just share of any insurance monies received by the Contractor on account of losses under policies issued pursuant to Paragraphs 12.1 and 12.2.

CMAA Document A-3 (2003 Edition)

6.1.5 Nothing contained in the Contract Documents is intended to create, nor shall it create, any contractual relationship between the Owner, the CM, the Designer or any of their agents, employees or representatives and any subcontractor or supplier.

6.2 Payment to Subcontractors and Suppliers

6.2.1 When the Contractor is paid by the Owner on account of any subcontractor's work, the Contractor shall promptly pay that subcontractor the amount to which that subcontractor is entitled. Such payment shall reflect any percentage retained by the Owner, form payments to the Contractor on account of such work.

6.2.2 The Owner may, at its sole discretion, furnish upon request to any subcontractor or supplier information regarding the percentage of completion of the Work, or the amounts of any progress payments requested by the Contractor, or the actions taken by the CM on account of work furnished or performed by such subcontractor or supplier.

6.2.3 Except as required by law, the Owner, the CM or the Designer shall not have any obligation to pay or cause payment to any subcontractor, supplier, or any other persons or organizations furnishing or performing work.

ARTICLE 7
WORK BY THE OWNER OR BY
SEPARATE CONTRACTORS

7.1 The Owner's Right to Perform Work and to Award Separate Contracts

7.1.1 The Owner may perform other work related to the Project at the site by the Owner's own forces, have other work performed by utility owners or award other contracts therefore. If the fact that such other work to be performed was not specified in the Contract Documents, written notice thereof shall be given to the Contractor prior to starting any such other work.

7.1.2 In situations in which other contracts are awarded by the Owner for separate portions of the Project or for other work on the site, the contractor(s) who executes such separate contracts are referred to herein as "another contractor" or "other contractor(s)."

7.1.3 If the Owner provides the Contractor notice that related work not specified in the Contract Documents will be performed, and the Contractor believes that it is entitled to a change in the contract price or contract time, the Contractor may request such changes as specified in Article 9.

7.2 Mutual Responsibility

7.2.1 The Contractor shall provide each other party who has a separate contract with the Owner and all the Owner's employees and representatives proper and safe access to the site, and reasonable opportunity for the delivery and storage of materials and equipment to the site. The Contractor shall properly coordinate the Work with that of such other party and shall ensure that the Contractor's performance of the Work is not disrupt or in any way inhibit the performance of any other such party or contractor at the Project site.

7.2.2 If any part of the Work depends for proper execution or results upon the work of any other contractor or the Owner, the Contractor shall inspect and promptly report to the CM in writing any delays, defects or deficiencies in such work that render t unavailable or unsuitable for such proper execution and results. The Contractor's failure to so report shall constitute an acceptance of the other work as fit and proper for integration with the Work, except for latent defects in the other work.

7.2.3 If the Contractor causes damage to the work, property, or person of any other contractor, or if any claim arising out of the Contractor's performance of the Work is made by any other contractor against the Contractor, Owner, CM, Designer or any other person, the Contractor shall promptly attempt to settle and resolve the dispute.

7.3 The Owner's Right to Perform Disputed Work

7.3.1 If a dispute exists between the Contractor and other contractors as to their responsibility for maintenance of the site pursuant to Paragraph 4.6, or as to their responsibility to perform cutting, filling, excavating or patching as required by

CMAA Document A-3 (2003 Edition)

Paragraph 4.10, the Owner may perform such work and charge the cost thereof to the several contractors responsible therefore in amounts that the Owner determines to be equitable.

ARTICLE 8
COMMENCEMENT AND COMPLETION

8.1   Time

8.1.1   The contract time shall commence on the date specified in the Notice to Proceed with the Work or, if such a date is not specified, on the date of this Contract.

8.1.2   The Contractor shall start to perform the Work within seven (7) days after the date when the contract time commences.

8.1.3   The Owner, CM and Designer shall not be responsible for the failure of the Contractor to plan, schedule and perform the Work in accordance with the Master Schedule or the Contractor's Construction Schedule, or the Contractor's failure to cooperate with other contractors, or the Contractor's failure to meet scheduled completion dates or to schedule and coordinate the work of the Contractor's own subcontractors or suppliers.

8.1.4   The date of final completion of the Work is the date determined by the CM in consultation with the Designer when all work is complete, accessible, operable and usable by the Owner and all parts, systems and site work are complete and cleaned for the Owner's full use, and all drawings, certificates, bonds, guarantees, releases or waivers of liens, and documents required by the Contract Documents have been provided to the Owner by the Contractor.

8.2   Progress and Completion

8.2.1   All time limits stated in the Contract Documents are of the essence of this Contract.

8.2.2   The Contractor shall perform the Work expeditiously in accordance with the Master Schedule for the Project and the Contractor's Construction Schedule specified in Paragraph 8.2.5 with adequate forces and shall achieve substantial completion and final completion within the times stated in the Contract Documents.

8.2.3   Within seven (7) after the date of this Contract, the Contractor shall submit the following to the CM for review:

8.2.3.1   A preliminary schedule that conforms to the milestone dates set out in the Master Schedule for the Project stating the start and completion dates of the various stages of the Work;

8.2.3.2   A preliminary schedule of submittals;

8.2.3.3   A schedule of values for all of the Work, including quantities and prices of items aggregating the contract price and subdividing the Work into component parts in sufficient detail to serve as the basis for determining progress during construction.   Such prices shall include the amount of overhead and profit applicable to each item of Work; and

8.2.3.4   This price and schedule submittal shall be the basis for the Contractor's Construction Schedule submittal specified in Paragraph 8.2.5.

8.2.4   Within ten (10) days after the date of this Contract, but before the Contractor starts the Work, a conference attended by the Contractor, Designer and CM shall be held to discuss the work schedule, procedures for handling shop drawings, samples and other submittals and for processing applications for payment to establish contract administration procedures and to establish communications procedures among the parties.

8.2.5   The Contractor shall submit to the CM the Contractor's Construction Schedule before submission of the first application for payment. The Contractor's Construction Schedule shall be acceptable to the CM as providing an orderly progression of the Work to completion within the specified milestones and the contract time, but such acceptance shall neither impose on the CM responsibility for the progress or scheduling of the Work nor relieve the Contractor from full responsibility therefore. The Contractor's Construction Schedule shall be prepared in a critical path method (CPM) network format, shall be prepared such that no activity has a duration of more than twenty (20) days, shall have the critical path clearly indicated and shall have the total contract price allocated among the schedule activities such that progress payments may be computed accurately from the updates of the CPM schedule. Each of the Contractor's activities shall be allocated a price, and the sum of the prices of the activities shall equal the total contract price. A schedule of shop drawings, samples and other submittals shall be incorporated into the schedule and shall have the appropriate prices allocated to the shop drawing preparation activities. In addition to a graphic plot of the network, the Construction Schedule shall include reports sorting and listing the activities in order of increasing float, by early start dates and by late start dates. The Contractor shall obtain from the CM approval of the scheduling system prior to beginning preparation of the Construction Schedule.

8.2.6   The CM shall provide the Contractor with the results of the schedule review. The Contractor shall revise as necessary and resubmit to the CM the Construction Schedule. No progress payments shall be processed or paid until the Contractor's Construction Schedule has been properly prepared and submitted by the Contractor.

8.2.7   The Contractor shall submit monthly schedule reports to the CM indicating the current status of the Work and incorporating into the schedule all change orders. The reports may include proposed adjustments in the Contractor's Construction Schedule, provided that any adjustments shall conform to the Master Schedule and, additionally, shall indicate any revised sequence of the Work as may be necessary to meet specified milestone and final completion dates. No changes in any activity in any activity or price allocations shall be permitted. Acceptance of the proposed adjustments shall be at the sole discretion of the CM. If the proposed adjustments are accepted, the Contractor shall, within ten (10) days, submit to the CM a revised Contractor's Construction Schedule indicating the accepted adjustments.

8.2.8   Adjustments to the Contractor's Construction Schedule, sequence, and float in the schedule, when made, shall be for the benefit of the Project and its completion in accordance with the Contract Documents.

8.2.9   The Contractor represents to the Owner that the Contractor shall:

8.2.9.1   Prepare documents for its planning, scheduling and coordination of the Work that are feasible and realistic; and

8.2.9.2   Prepare schedules, updates, revisions or reports that accurately reflect the Contractor's actual intent and reasonable expectations as to the sequences of activities, the duration of activities, the responsibility for activities, productivity or efficiency, expected weather conditions, the value associated with the activity or grouping or activities, completion of any item of work or activity, projected actual project completion, delays or problems encountered or expected and specified float time.

8.3   Responsibility for Completion

8.3.1   The Contractor shall furnish such employees, materials, facilities and equipment and shall work such hours, including extra shifts, overtime operations and Sunday and holidays as may be necessary to ensure the prosecution and completion of the Work in accordance with the Contractor's Construction Schedule. If work is not being performed in accordance with the Contractors' Construction Schedule and it becomes apparent form the schedule that the Work shall not be completed within the contract time, the Contractor agrees that it shall, as necessary to improve its progress, take some or all of the following actions, at no additional cost to the Owner:

8.3.1.1   Increase the number of employees in such crafts as shall regain lost schedule progress; and

8.3.1.2   Increase the number of working hours per shift, shifts per working day, working days per week, the amount of equipment or any combination of the foregoing to regain lost schedule progress.

8.3.2   In addition, the CM may require the Contractor to prepare and submit a recovery schedule demonstrating the Contractor's proposed plan to regain lost schedule progress and to ensure completion of the Work within the contract time. If the CM finds the proposed plan not acceptable, the CM may require the Contractor to submit a new plan. If the actions

CMAA Document A-3 (2003 Edition)

taken by the Contractor or the second proposed plan are not satisfactory, the CM may require the Contractor to take any of the actions set forth in Paragraph 8.3.1 without additional cost to the Owner.

8.3.3   Failure of the Contractor to substantially comply with the requirements of this paragraph may be considered grounds for a determination by the Owner, pursuant to Paragraph 13.3.1.6, that the Contractor is failing to prosecute the Work with such diligence to ensure its completion within the time specified.

ARTICLE 9
CHANGES

9.1   Authorized Variations in Work

9.1.1   The CM may authorize minor variations in the Work from the requirements of the Contract Documents that do not involve an adjustment in the contract price or the contract time and that are consistent with the overall intent of the Contract Documents.

9.1.2   All authorized minor variations in the Work shall be communicated by the CM to the Contractor, who shall perform the work involved promptly.

9.2   Change Orders

9.2.1   Without invalidating this Contract and without notice to any surety, the Owner may at any time or from time to time order additions, deletions, deductions or revisions in the Work. These shall be authorized by a change order, and the contract price and time may only be changed by a change order. A change order shall be issued to the Contractor only after approval by the CM of the Contractor's proposal or request submitted pursuant to Paragraph 9.2.2 or 9.2.3. Upon receipt of a change order, the Contractor shall promptly proceed with the work involved which shall be performed under the applicable conditions of the Contract Documents.

9.2.2   The Owner may initiate the change order procedure by issuing a request for proposal to the Contractor, accompanied by drawings and specifications. The Contractor

shall, within the time period stated in the request for proposal, submit to the CM for evaluation detailed information concerning the cost and time adjustments, if any, as may be necessary to perform the proposed change order work.

9.2.3   A request by the Contractor for a change in the contract price or time shall be based on written notice stating the general nature of the request delivered by the Contractor to the CM within five (5) days after the beginning of the occurrences of the event giving rise to the request. The requested cost and time adjustment, with supporting data, shall be delivered within thirty (30) days after the end of such occurrence and shall be accompanied by a written statement that the amount requested includes all known amounts, direct, indirect and consequential, incurred as a result of the occurrence of the event. If the Contractor fails to comply with all provisions of this Paragraph, the Contractor shall have waived any and all rights it may have against the Owner.

9.2.4   The CM shall review the Contractor's detailed information regarding proposed or requested changes to the contract price or time and will initiate discussions with the Contractor to determine a price and time adjustment that is acceptable to the Owner.

9.2.5   If the Contractor and Owner are unable to arrive at an agreement as to the change in the contract price or time, the Contractor shall nevertheless proceed with the change if so ordered in writing by the CM. The value of the work included in the change order shall be determined as specified in Paragraphs 9.3.1 through 9.3.5 and the contract time shall be adjusted to the extent that the revised work scope results in a change in the duration of the critical path indicated in the Contractor's Construction Schedule.

9.2.6   The Contractor shall not be entitled to an increase in the contract price or an extension of the contract time with respect to any work performed that is not required by the Contract Documents as amended, modified and supplemented, except as specifically provided herein.

9.2.7   If notice of any change affecting the general scope, extent, or character of the Work or the provisions of the Contract Documents including, but not limited to, contract price or contract time is required by any surety providing a

CMAA Document A-3 (2003 Edition)

bond on behalf of the Contractor, the giving of any such notice shall be the Contractor's responsibility and the amount of each applicable bond shall be adjusted accordingly.

9.3   Change of the Contract Price

9.3.1   The value of any work included in a change order or in any request for an increase or decrease in one of the following ways:

9.3.1.1   By application of unit prices to the quantities of the items involved, subject to the provisions of Paragraph 9.3.6;

9.3.1.2   By mutual acceptance of a lump sum that includes an allowance for overhead and profit; or

9.3.1.3   On the basis of the cost of the work, determined as provided in Paragraphs 9.3.2 and 9.3.3, plus a Contractor's fee for overhead and profit, determined as provided in Paragraphs 9.3.4 and 9.3.5.

9.3.2   The term "cost of the work" means the sum of all costs necessarily incurred and paid by the Contractor in the proper performance of the Work. Such costs shall be in amounts no higher than those prevailing in the locality of the Project and shall include only the following items:

9.3.2.1   Actual payroll costs for employees in the direct employ of the Contractor in the performance of the Work. Payroll costs for employees not employed full time on the Work shall be apportioned on the basis of their time spent on the Work. Payroll costs shall include, but not be limited to, the audited cost of salaries and wages, plus the cost of fringe benefits that shall include social security contributions, unemployment, excise and payroll taxes, Workers' Compensation, health and retirement benefits, bonuses, sick leave, vacation and holiday pay applicable thereto. The Contractor's employees shall include superintendents and foremen at the site. The expenses of performing work after regular working hours, on Saturday, Sunday or legal holiday shall be included in the above only to the extent authorized in writing by the CM;

9.3.2.2   Cost of all materials and equipment furnished and incorporated in the Work, including costs of transportation and storage thereof, and suppliers' field services required in connection therewith. All cash discounts shall accrue to the Contractor unless the Owner deposits funds with the Contractor with which to make payments and in which case the cash discounts shall accrue to the Owner. All trade discounts, rebates and refunds and all returns from sale of surplus materials and equipment shall accrue to the Owner and the Contractor shall make provisions such that the monies may be obtained;

9.3.2.3   Payments made by the Contractor to the subcontractors for work performed. If required by the CM, the Contractor shall obtain competitive bids from subcontractors acceptable to the CM and shall deliver such bids to the CM who shall then determine which bids shall be accepted. If a subcontract provides that the subcontractors be paid on the basis of cost of the work plus a fee, the subcontractor's cost of the work shall be determined in the same manner as the Contractor's cost of the work. All subcontracts shall be subject to the other provisions of the Contract Documents;

9.3.2.4   Cost of special consultants including, but not limited to, engineers, architects, testing laboratories, surveyors, attorneys and accountants, employed for services specifically related to the Work; or

9.3.2.5   Other costs including the following:

9.3.2.5.1 The proportion of necessary transportation, travel and subsistence expenses of Contractor's employees incurred in discharge of duties connected with the Work;

9.3.2.5.2 The cost, including transportation and maintenance, of all materials, supplies, equipment, machinery, appliances, office and temporary facilities at the site and hand tools not owned by the workers that are consumed in the performance of the Work and the cost, less market value, of the items used, but not consumed, that remain the property of the Contractor;

9.3.2.5.3 Rentals for all construction equipment and machinery and the parts thereof whether rented from the Contractor or others in accordance with rental agreements approved by the CM and the costs of transportation, loading, unloading, installation, dismantling and removal thereof, all in

CMAA Document A-3 (2003 Edition)

accordance with terms of the rental agreements. The rental of any such equipment, machinery and parts shall cease when the use thereof is no longer necessary of the Work;

9.3.2.5.4 Sales, consumer, use or similar taxes related to the Work and for which the Contractor is liable, imposed by laws and regulations;

9.3.2.5.5 Deposits lost for causes other than acts of the Contractor, any subcontractor or anyone directly or indirectly employed by any of them or for whose acts any of them may be liable and royalty payments and fees for permits and licenses;

9.3.2.5.6 Losses, damages and related expenses not compensated by insurance and sustained by the Contractor in connection with the performance and furnishing of the Work, provided they have resulted from causes other than the acts of the Contractor, any subcontractor or anyone directly or indirectly employed by any of them or for whose acts any of them may be liable. Such losses shall include settlements made with the written consent and approval of the Owner. No such losses, damages or expenses shall be included in the cost of the work for the purpose of determining the Contractor's fee. If, however, any such loss or damage required reconstruction and the Contractor is placed in charge thereof, the Contractor shall be paid for services a fee proportionate to that stated in Paragraph 9.3.4.2;

9.3.2.5.7 The cost of utilities, fuel and sanitary facilities at the site;

9.3.2.5.8 Incidental expenses such as telegrams, long distance telephone calls, telephone service at the site, express packages and similar items in connection with the Work; and

9.3.2.5.9 Cost of premiums for additional bonds and insurance required because of changes in the Work.

9.3.3   The term "cost of the work" shall not include any of the following:

9.3.3.1   Payroll costs and other compensation of the Contractor's officers, executives, principals of partnerships

and sole proprietorships, general managers, engineers, architects, estimators, attorneys, auditors, accountants, purchasing and contracting agents, expediters, timekeepers, clerks and other personnel employed by the Contractor whether at the site or in the Contractor's principal or a branch office for general administration of the Work and not referred to in Paragraph 9.3.2.1 or specifically covered by Paragraph 9.3.2.4, all of which are to be considered administrative costs covered by the Contractor's fee;

9.3.3.2   Expenses of the Contractor's principal and branch offices other than the Contractor's office at the site;

9.3.3.3   Any part of the Contractor's capital expenses, including interest on the Contractor's capital employed for the Work and charges against the Contractor for delinquent payments;

9.3.3.4   Costs due to the negligence of the Contractor, any subcontractor or anyone directly or indirectly employed by any of them or for whose acts any of them may be liable including, but not limited to the correction of defective work, disposal of materials or equipment wrongly supplied and making good any damage to property; or

9.3.3.5   Other overhead or general expense costs of any kind and the costs of any item not specifically and expressly included in Paragraph 9.3.2.

9.3.4   The Contractor's fee allowed to the Contractor for overhead and profit shall be determined as follows:

9.3.4.1   A mutually acceptable fixed fee; or, if none can be agreed upon,

9.3.4.2   A fee based on the following percentages of the various portions of the cost of the work;

9.3.4.2.1 For costs incurred pursuant to Paragraphs 9.3.2.1 and 9.3.2.2, the Contractor's fee shall be fifteen (15) percent;

9.3.4.2.2 For costs incurred pursuant to Paragraph 9.3.2.3, the Contractor's fee shall be five (5) percent. If the subcontract is on the basis of cost of the work plus a fee, the

CMAA Document A-3 (2003 Edition)

maximum allowable to the Contractor on account of overhead and profit of all subcontractors shall be ten (10) percent;

9.3.4.2.3 No fee shall be payable on the basis of costs itemized under Paragraphs 9.3.2.4, 9.3.2.5 and 9.3.3;

9.3.4.2.4 The amount of credit to be allowed by the Contractor to the Owner for any such change which results in a net decrease in cost shall be the amount of the actual net decrease, plus an amount equal to ten (10) percent of the net decrease; and

9.3.4.2.5 When both additions and credits are involved in any one change, the adjustment in the Contractor's fee shall be computed on the basis of the net change in accordance with Paragraphs 9.3.4.2.1 through 9.3.4.2.4, inclusive.

9.3.5   Whenever the cost of any work is to be determined pursuant to Paragraph 9.3.2 or 9.3.3, the Contractor shall submit in form acceptable to the CM the itemized cost, together with such supporting data as may be deemed necessary by the CM.

9.3.6   When the Contract Documents provide that all or part of the Work be unit price work, initially the contract price shall be deemed to include, for all unit price work, an amount equal to the sum of the established unit prices for each separately identified item of unit price work times the estimated quantity of each item as indicated in the Contract Documents. The estimated quantities of items of unit price work are not guaranteed and are solely for the purpose of comparison of bids and determining an initial contract price. Determinations of the actual quantities and classifications of unit price work performed by the Contractor shall be made by the CM in accordance with Paragraph 11.2.4.

9.3.6.1   Each unit price shall be deemed to include an amount considered by the Contractor to be adequate to cover its overhead and profit for each separately identified item.

9.3.6.2   When the quantity of any item of unit price work performed by the Contractor differs more than twenty-five (25) percent from the estimated quantity of such item indicated in this Contract and there is no corresponding adjustment with

respect to any other item of work, the CM and Contractor shall determine a mutually acceptable price for the changed item.

9.4   Change of the Contract Time

9.4.1   The contract time shall only be adjusted if an event occurs that changes the duration of the critical path indicated on the Contractor's Construction Schedule, and that event is beyond the Contractor's control. Such events shall include, but not be limited to, acts by the owner or others performing work, or to fires, floods, labor disputes, epidemics, abnormal weather conditions, or acts of God.

9.4.2   All proposed or requested changes in the contract time shall include an analysis showing the actual effect of the event on the Construction Schedule. No adjustments in the contract time shall be allowed if the event does not directly affect the critical path indicated in the Contractor's Construction Schedule. No request for an adjustment in the contract time shall be valid if not submitted in accordance with these Contract Documents.

9.4.3   The contract time shall not be adjusted for normal inclement weather. The Contractor shall be entitled to an extension of time only if the Contractor can substantiate to the satisfaction of the CM that there was greater than normal inclement weather considering the full term of the contract time using a ten-year average of accumulated record mean values from climatological data compiled by the U.S. Department of Commerce National Oceanic and Atmospheric Administration for the locale of the Project, and that such alleged greater than normal inclement weather lengthened the critical path indicated in the Contractor's Construction Schedule. If the total accumulated number of calendar days lost due to inclement weather from the start of work until final completion exceeds the total accumulated number to be expected for the same period from the aforesaid data, time of completion shall be extended by the appropriate number of calendar days.

9.4.4   Where the Contractor is prevented from completing any part of the Work within the contract time and the critical path indicated in the Contractor's Construction Schedule is directly affected due to causes beyond the control of both Owner and Contractor, an extension of the contract time in an

CMAA Document A-3 (2003 Edition)

amount equal to the time lost due to such cause shall be Contractor's sole and exclusive remedy for such delay. This provision shall not apply, however, if such delays were the result of the Owner's breach of a fundamental obligation of the Contract, were caused by the Owner's bad faith, or its willful, malicious or grossly negligent conduct with respect to its performance under the Contract Documents, or if the delay was uncontemplated by the parties.

9.4.5   In no event shall the Owner be liable to the Contractor, any subcontractor, any supplier, or any other person or organization, or to any surety for or employee or agent of any of them, for damages arising out of or resulting from delays caused by the Contractor or anyone for whose acts the Contractor is responsible, or delays caused by events beyond the control of the Contractor or the Owner.

ARTICLE 10
INSPECTION, TESTING AND CORRECTION
OF THE WORK

10.1   Uncovering Work

10.1.1 If any work is covered without concurrence of the CM, that work shall, if requested by the CM, be exposed for the CM's observation and then completed in conformance with the requirements of the Contract Documents at the Contractor's expense.

10.1.2   The CM may require special inspection or testing of the Work, whether or not such work is already fabricated, installed or completed.

10.1.3   If the CM considers it necessary or advisable that covered work be observed or inspected or tested by others, the Contractor, at the CM's request, shall uncover, expose or otherwise make available for observation, inspection or testing as the CM may require that portion of the Work in question, furnishing all necessary labor, material and equipment. If it is found that such work does not conform to the requirements of the Contract Documents, the Contractor shall bear all direct, indirect and consequential costs of such uncovering, exposure, observation, inspection and testing and of satisfactory reconstruction including, but not limited to, fees

and charges of engineers, architects, attorneys and other professionals. If, however, such work is found to conform to the Contract requirements, the Contractor shall be allowed an increase in the contract price or an extension of the contract time, or both, directly attributable to such uncovering, exposure, observation, inspection, testing and reconstruction.

10.2   Correction or Removal of Nonconforming Work

10.2.1   If required by the CM, the Contractor shall promptly, as directed, either correct all work that does not conform to the requirements of the Contract Documents, whether or not fabricated, installed or completed or, if work has been rejected by the CM, remove it form the site and replace it with work that does meet the requirements of the Contract Documents. The Contractor shall bear all direct, indirect and consequential costs of such correction or removal including, but not limited to, attorneys' fees and charges of engineers, architects, attorneys and other professionals made necessary thereby.

10.3   One Year Correction Period

10.3.1   Within one (1) year after the date of final completion or such longer period of time as may be prescribed by laws and regulations or by the terms of any applicable special guarantee required by the Contract Documents or by any specific provision of the Contract Documents, any work is found to not conform to the requirements of the Contract Documents, the Contractor shall promptly, without cost to the Owner and in accordance with the CM's written instructions, accompanied by appropriate design documentation prepared by the Designer, either correct such work or, if it has been rejected by the CM, remove it from the site and replace it with work that meets the requirements of the Contract Documents. If the Contractor does not promptly comply with the terms of such instructions or in an emergency where delay would cause serious risk of loss or damage, the Owner may have the nonconforming work corrected or the rejected work removed and replaced and all direct, indirect and consequential costs of such removal and replacement including, but not limited to, fees and charges of engineers, architects, attorneys and other professionals shall be paid by the Contractor. In special circumstances when a particular item of equipment is placed in continuous services before substantial completion of all the

CMAA Document A-3 (2003 Edition)

Work, the correction period for that item may start to run from an earlier date if so provided.

10.4   Acceptance of Nonconforming Work

10.4.1   The CM may, in consultation with the Designer, recommend to the Owner acceptance of nonconforming work instead of requiring correction or removal and replacement of such work.   The Contractor shall bear all direct, indirect and consequential costs attributable to the CM's and the Designer's evaluation of and determination to accept such work and such costs may include, but not be limited to, fees and charges of engineers, architects, attorneys and other professionals.   If any such acceptance occurs prior to the CM's recommendation of final payment, a change order shall be issued incorporating the necessary revisions in the Contract Documents with respect to the Work and the Owner shall be entitled to an appropriate decrease in the contract price.   If the acceptance occurs after such recommendation, the amount of such decrease shall be paid by the Contractor directly to the Owner.

ARTICLE 11
PAYMENTS AND COMPLETION

11.1   Price Allocation to The Schedule

11.1.1   Once accepted by the CM, the allocations of prices to the schedule activities specified in Paragraph 8.2.5 shall not be changed as part of the schedule revision process.

11.2   Applications for Payment

11.2.1   The CM shall review and process all applications for payment by the Contractor including the final application for payment.   The CM may consult with the Designer regarding the Contractor's payment.

11.2.2   At least five (5) days before each progress payment is scheduled, but not more often than once a month, the Contractor shall submit to the CM a report documenting the status of the Work as of the date of the application and accompanied by such supporting documentation as required by the Contract Documents and the CM.   The Contractor shall cooperate with the CM in preparing construction schedule reports comparing actual progress with scheduled progress. Progress payments shall be based on the construction schedule reports.   The cost of partially completed activities shall be determined by multiplying the value of the partially complete activity by the percentage the activity is complete.   If some of the payment is requested based on materials or equipment stored off site, the CM may require that the application be accompanied by bills of sale, invoices or other documentation warranting the Owner has received the materials and equipment free and clear of all liens, charges, security interests and encumbrances that are hereinafter referred to as "liens" and by evidence that the materials and equipment are covered by appropriate property insurance and other arrangements to protect the Owner's interest therein, all of which shall be satisfactory to the Owner.   The amount of retainage with respect to progress payments shall be as stipulated in the Contract Documents.

11.2.3   The Contractor warrants and guarantees that title to all work, material and equipment covered by any application for payment, whether incorporated in the Project or not, shall pass to the Owner free and clear of all liens no later than the time of payment.

11.2.4   The CM shall determine the actual quantities and classifications of unit price work performed by the Contractor. The Contractor shall review the CM's preliminary determinations of such quantities.   The CM's determination thereof shall be final and binding upon the Contractor unless, within five (5) days after the date of any such decision, the Contractor delivers to the CM written notice of intention to appeal the determination.

11.3   Recommendations for Payment

11.3.1   The CM shall, within fourteen (14) days after receipt of each application for payment, either indicate in writing a recommendation for payment and present this with the application to the Owner or return the application to the Contractor indicating in writing the reasons for rejected payment.   In the latter case, the Contractor shall make the necessary corrections and resubmit the application.

11.4   Progress Payments

CMAA Document A-3 (2003 Edition)

11.4.1 The Owner shall make progress payments on account of the contract price on the basis of recommendations of the CM concerning the Contractor's applications for payment. All progress payments shall be made on the day of each month as set forth in the Contract Documents.

11.4.2 Recommendation by the CM concerning applications for payment, or the making of any progress payment or use of the project by the Owner shall not release the Contractor from its responsibility to furnish or perform all work in strict conformance with the requirements of the Contract Documents.

11.5   Liens

11.5.1 The Contractor agrees to keep the Work and the site on which the Work is to be performed free and clear of all liens for labor and materials furnished pursuant to the Contract Documents. Notwithstanding anything to the contrary contained in the Contract Documents, if any such lien is filed or asserted or there is any reason to believe that a lien may be filed or asserted at any time during the progress of the Work or within the duration of this Contract, the Owner may refuse to make any payment otherwise due the Contractor or withhold from any payment due the Contractor a sum sufficient, in the opinion of the Owner, to pay all obligations and expenses necessary to satisfy such lien and to indemnify the Owner against any such lien unless and until the Contractor shall furnish satisfactory evidence that the indebtedness and the lien in respect thereof, if any, has been satisfied, discharged and released of record if and as provided by law pending the resolution of any dispute between the Contractor and the person filing such lien. If such evidence is not furnished by the Contractor to the Owner within a period of five (5) days after demand thereof, the Owner may discharge such indebtedness and deduct the amount required therefore, together with any and all losses, costs, damages and attorneys' fees suffered or incurred by the Owner from any sum payable to the Contractor.

11.5.2 Final payment to the Contractor may be withheld until the Work and the site on which the Work is to be performed are free and clear of any and all liens or rights

thereto arising because of work performed or materials furnished under the Contract Documents.

11.6   Payments Withheld

11.6.1 The CM may refuse to recommend payment and the Owner may refuse to pay the Contractor or, because of subsequently discovered facts or determinations, the CM may retract any such recommendation of payment previously issued, to such extent as may be required in the opinion of the CM to guard the Owner from loss or damage because:

11.6.1.1 All or a portion of the Work does not conform to the requirements of the Contract Documents.

11.6.1.2 The Contractor or any of its subcontractors, suppliers, or any other person or organization performing or furnishing any of the Work has caused, or reasonable evidence that the Contractor may cause, damage to the other contractors or the Owner;

11.6.1.3 The progress of the Work does not comply with the Contractor's Construction Schedule;

11.6.1.4 The Contractor fails to make prompt and proper payments to employees, subcontractors or suppliers;

11.6.1.5 Liens have been filed or asserted or reason exists to believe it is probable that a lien will be filed or asserted against any portion of the Work; or

11.6.1.6 Claims have been filed or asserted against the Contractor in connection with the Work, or reasonable evidence indicating probable filing or assertion of such claims.

11.6.2 When the Contractor remedies to the Owner's satisfaction the reason for any refusal of payment, the Owner shall promptly pay the Contractor any amounts as recommended by the CM as being properly due.

11.7   Substantial Completion

11.7.1 The CM shall, in consultation with the Designer, determine when the Work is substantially complete. When the Contractor considers the entire Work ready for its

CMAA Document A-3 (2003 Edition)

intended use, the Contractor shall notify the CM in writing that the Contractor considers the Work to be substantially complete and request that the CM issue a Certificate of Substantial Completion. Within a reasonable time thereafter, the CM and Designer shall make an inspection with the Owner and Contractor to determine the status of completion. If the CM, in consultation with the Designer, does not consider the Work substantially complete, the CM shall notify the Contractor in writing giving the reasons therefore. If the CM considers the Work substantially complete, the CM shall prepare and deliver to the Owner a Certificate of Substantial Completion. There shall be attached to the certificate a list of items to be completed or corrected before final payment. The Contractor shall be allowed reasonable access to complete or correct items on the list. At the time of delivery of the Certificate of Substantial Completion, the CM shall deliver to the Contractor a written recommendation as to division of responsibilities pending final payment between the Owner and Contractor with respect to security, operation, safety, maintenance, heat, utilities, insurance and warranties. Such recommendation shall be binding on the Contractor until final payment.

11.8   Partial Use

11.8.1   The Owner may use prior to substantial completion of the Work any finished part of the Work that has specifically been identified for such in the Contract Documents, or which the Owner, the CM and the Contractor agree constitutes a separately functioning and usable part of the Work that can be used by the Owner without significant interference with the Contractor's performance of the remainder of the Work. The CM may request in writing at any time the Contractor to permit the Owner to use any such part of the Work. Within a reasonable time after such request the Owner, Contractor and CM shall make an inspection of that part of the Work to determine its status of completion.

11.8.1.1   If the CM determines that the part of the Work to be used is substantially complete, the provisions of Paragraph 11.7.1 shall apply with respect to a Certificate of Substantial Completion for that part of the Work and the division of responsibility in respect thereof and access thereto.

11.8.1.2   If the CM determines that the part of the Work to be used is not substantially complete, the CM and the Designer shall prepare a list of items to be completed or corrected and the CM shall deliver the list to the Contractor, together with a written recommendation pending final completion as to the division of responsibilities with respect to security, operation, safety, maintenance, utilities, insurance, warranties and guarantees for that part of the Work. Such recommendation shall become binding upon the Contractor at the time when the Owner begins use of the designated portion of the Work. During such use and prior to substantial completion of the designated portion of the Work, the Contractor shall be allowed access to complete or correct items on the list and to complete other related work.

11.8.1.3   No occupancy, use or separate operation of part of the Work shall be accomplished prior to compliance with the requirements of Paragraph 12.3.4 in respect of property insurance.

11.9   Final Inspection

11.9.1   The CM, after consultation with the Designer, shall determine when the Work is finally complete. Upon written notice from the Contractor that the entire Work or an agreed portion thereof is complete, the CM shall make a final inspection with the Owner and Contractor and shall notify the Contractor in writing of any portion of the Work that does not conform to the requirements of the Contract Documents. The Contractor shall immediately take such measures as are necessary to remedy such deficiencies. Neither a Certificate of Final Completion nor final payment shall be issued until all such items are remedied.

11.10   Final Application for Payment

11.10.1   After the Contractor has completed all Work to the satisfaction of the CM and has delivered all maintenance and operating instruction, schedules, guarantees, bonds, certificates of inspection, marked-up record documents and other documents required by the Contract Documents and after the CM has indicated that the Work is in compliance with requirements of the Contract Documents, the Contractor may make application for final payment following the procedure for progress payments.

CMAA Document A-3 (2003 Edition)

11.10.2 The final application for payment shall be accompanied by all documentation required by the Contract Documents, together with complete and legally effective releases or waivers of all liens arising out of or filed in connection with the Work.

11.11   Final Payment

11.11.1   On the basis of the CM's observation of the Work during construction, its final inspection and its review of the final application for payment and accompanying documentation required by the Contract Documents and on the basis of the Owner's and CM's satisfaction that the Work has been completed and the Contractor's other obligations under the Contract Documents have been fulfilled, the CM shall, within ten (10) days after receipt of the final application for payment, present to the Owner a written recommendation for payment, and issue a Certificate of Final Completion. Thirty (30) days after presentation to the Owner of the application and accompanying documentation, in appropriate form and substance, and with the CM's recommendation and Certificate of Final Completion, the amount recommended by the CM shall become due and shall be paid by the Owner to the Contractor.

11.11.2   If, through no fault of the Contractor, final completion of the Work is significantly delayed and if the CM so confirms the delay, the Owner may, upon receipt of the Contractor's final application for payment and recommendation of the CM, without terminating the Agreement, make payment of the balance due for the portion of the Work completed.

11.12   The Contractor's Continuing Obligation

11.12.1   The Contractor's obligation to perform and complete the Work in accordance with the Contract Documents shall be absolute. Neither recommendation of any progress or final payment by the CM, nor issuance of a Certificate of Substantial or Final Completion, nor any payment by the Owner to the Contractor, nor any use or occupancy of the Work or any part thereof by the Owner, nor any review and approval of a shop drawing, sample, or other submittal shall release the Contractor from its obligation to perform the Work in accordance with all requirements of the Contract Documents.

11.13   Waiver of Claims

11.13.1   The making and acceptance of final payment shall constitute the following:

11.13.1.1 A waiver of all claims by the Owner against the Contractor, except claims arising from unsettled liens, from defective work appearing after final inspection pursuant to Paragraph 11.9.1 or from failure to comply with the Contract Documents or the terms of any special guarantees specified therein; and

11.13.1.2 The making and acceptance of final payment shall not constitute a waiver by the Owner of any rights in respect of the Contractor's continuing obligations under the Contract Documents; and

11.13.1.3 A waiver of all claims by the Contractor against the Owner.

ARTICLE 12
LIABILITY AND PROPERTY INSURANCE

12.1   The Contractor's Liability Insurance

12.1.1   The Contractor shall purchase and maintain such commercial general liability and other insurance as is appropriate for the Work being performed and furnished and that shall provide protection from claims set forth below that may arise out of or result from the Contractor's performance and furnishing of the Work and the Contractor's other obligations under the Contract Documents whether it is to be performed or furnished by the Contractor, by any subcontractor, by anyone directly or indirectly employed by any of them to perform or furnish any of the Work or by anyone for whose acts any of them may be liable:

12.1.1.1 Claims for Workers' Compensation, disability benefits and other similar employee benefit acts;

CMAA Document A-3 (2003 Edition)

12.1.1.2 Claims for damages because of bodily injury, occupational sickness or disease or death of the Contractor's employees;

12.1.1.3 Claims for damages because of bodily injury, sickness or disease or death of any person other than the Contractor's employees;

12.1.1.4 Claims for damages insured by personal injury liability coverage that are sustained by any person as a result of an act directly or indirectly related to the employment of such person by the Contractor, or by any other person for any other reason;

12.1.1.5 Claims for damages, other than to the Work itself, due to injury to or destruction of tangible property wherever located, including loss of use resulting therefrom;

12.1.1.6 Claims arising out of operations of law or regulations for damages because of bodily injury or death of any person or for damage to property; and

12.1.1.7 Claims for damages because of bodily injury or death of any person or property damage arising out of the ownership, maintenance or use of any motor vehicle.

12.1.1.8 The insurance required by this Paragraph shall include the specific coverages and be written for not less than the limits of liability and coverages provided in the Contract Documents or required by law, whichever is greater. The commercial general liability insurance shall include completed operations insurance. All of the policies of insurance so required to be purchased and maintained shall contain a provision or endorsement that the coverage afforded shall not be canceled, materially changed or renewal refused until at least thirty (30) days prior written notice has been given to the CM by certified mail. All such insurance shall remain in effect until final payment and at all times thereafter when the Contractor may be correcting, removing or replacing defective work. In addition, the Contractor shall maintain such completed operations insurance for at least two (2) years after final payment and furnish the CM evidence of continuation of such insurance at the time of final payment and one (1) year thereafter.

12.2   Contractual Liability Insurance

12.2.1   The commercial general liability insurance required by Paragraph 12.1 shall included contractual liability insurance applicable to the Contractor's obligations under the terms of this Agreement.

12.3   Property Insurance

12.3.1   The Contractor shall purchase and maintain property insurance for the Work to the full insurable value thereof. This insurance shall be in an amount sufficient to protect the interests of the Owner, Contractor, subcontractors, the CM, Designer, and CM's and Designer's consultants in the Work, all of whom shall be listed as insured or additional insured parties. The insurance shall: be written on a Builder's Risk all-risk or open peril or special causes of loss policy form that shall at least include insurance for physical loss or damage to the Work, temporary buildings, falsework and Work in transit and shall insure against at least the following perils or causes of loss: fire, lightning, extended coverage, theft, vandalism and malicious mischief, earthquake, collapse, debris removal, demolition occasioned by enforcement of laws and regulations, water damage, and such other perils or causes of loss as may be specifically required by the Supplemental Conditions; include expenses incurred in the repair or replacement of any insured property (including, but not limited to, fees and charges of the CM and Designer); and cover materials and equipment stored at the site or at another location that was agreed to in writing by Owner prior to being incorporated in the Work, provided that such materials and equipment have been included in an application for payment recommended by the CM. The Contractor shall also purchase and maintain coverage against losses, damages and expenses arising out of or resulting from any insured loss or incurred in the repair or replacement of any insured property.

12.3.2   All the policies of insurance or the certificates or other evidence thereof required to be purchased and maintained by the Contractor shall contain waiver provisions in accordance with Paragraph 12.6.

12.3.3   The Owner shall not be responsible for purchasing and maintaining any property insurance to protect the

CMAA Document A-3 (2003 Edition)

interests of the Contractor, subcontractors or others in the Work.

12.3.4   If the Owner finds it necessary to occupy or use a portion or portions of the Work prior to substantial completion of the Work, such use or occupancy may be accomplished, provided that no such use or occupancy shall commence before the insurers providing the property insurance have acknowledged notice thereof and in writing effected the changes in coverages necessitated thereby.   The insurers providing the property insurance shall consent by endorsement to the policy or policies, but the property insurance shall not be canceled or lapse on account of any such partial use or occupancy.

12.4   Delivery of Insurance Certificates

12.4.1   Before the Work at the site is started, the Contractor shall deliver to the Owner through the CM certificates of insurance indicating the insurance and coverage limits that the Contractor is required to purchase and maintain.

12.5   The Owner's Insurance

12.5.1   The Owner shall be responsible for purchasing and maintaining its own liability insurance and, at the Owner's option, may purchase and maintain such insurance to protect the Owner against claims that may arise from operations under these Contract Documents.

12.6   Waiver of Subrogation

12.6.1   The Owner and the Contractor waive all rights against each other for all losses and damages caused by any of the perils covered by the policies of insurance provided and also waive all such rights against the subcontractors, the CM, Designer, CM's and Designer's consultants and all other parties named as insured in such policies for losses and damages so caused.   None of the above waivers shall extend to the rights that any of the insured parties may have to the proceeds of insurance held by the Owner as trustee or otherwise payable under any policy so issued.

12.6.2   The Owner intends that any policies provided in response to Paragraph 12.3.1 shall protect all of the parties insured and provide primary coverage for all losses and damages caused by the perils covered thereby.   Accordingly, all such policies shall contain provisions to the effect that in the event of payment of any loss or damage, the insurer shall have no rights of recovery against any of the parties named as insured or additional insureds and if the insurers require separate waiver forms to be signed by the CM, Designer, CM's and Designer's consultant, the Owner shall obtain the same and if such waiver forms are required of any subcontractor, The Contractor shall obtain the same.

ARTICLE 13
TERMINATION

13.1   Termination for Convenience by The Contractor

13.1.1   If, through no act or fault of the Contractor, the Work is suspended for a period of more than one hundred eighty (180) days by the Owner or if the Owner fails for sixty (60) days to pay the Contractor any sum finally determined to be due, the Contractor may, upon seven (7) days written notice to the Owner, terminate this Contract and recover from the Owner payment for all work executed and any expense sustained, plus reasonable termination expenses.

13.1.2   The provisions of this paragraph shall not relieve the Contractor of the obligations under Paragraph 15.4.3 to perform the Work in accordance with the Contractor's Construction Schedule and without delay during disputes with the Owner.

13.2   Termination for Convenience by The Owner

13.2.1   Upon seven (7) days written notice to the Contractor, the Owner may, without cause and without prejudice to any other right or remedy, elect to abandon the Work and terminate this Contract.

13.2.2   In the event of termination in accordance with Paragraph 13.2.1, the Contractor shall be paid for all work performed and any expense sustained shall be limited to the cost of such work plus reasonable termination expenses

CMAA Document A-3 (2003 Edition)

including the direct and indirect costs specified in Paragraph 13.4

13.3   Default Termination

13.3.1   This Contract may be terminated for default upon the occurrence of any of the following events:

13.3.1.1   If the Contractor commences a voluntary action under any chapter of the United States Bankruptcy Code as now or hereafter in effect, or if the Contractor takes any equivalent or similar action by filing a petition or otherwise under any other federal or state law in effect at such time relating to the bankruptcy or insolvency;

13.3.1.2   If a petition is filed against the Contractor under any chapter of the United States Bankruptcy Code as now or hereafter in effect at the time of filing, or if a petition is filed seeking any such equivalent or similar relief against the Contractor under any other federal or state law in effect at the time relating to bankruptcy or insolvency;

13.3.1.3   If the Contractor makes a general assignment for the benefit of creditors;

13.3.1.4   If a trustee, receiver, custodian or agent of the Contractor is appointed under applicable law or under contract whose appointment of authority to take charge of property of the Contractor is for the purpose of enforcing a lien against such property or for the purpose of general administration of such property for the benefit of the Contractor's creditors;

13.3.1.5   If the Contractor admits in writing an inability to pay its debts generally as they become due;

13.3.1.6   If the Contractor persistently fails to perform the Work in accordance with the Contract Documents including, but not limited to, failure to supply sufficient skilled workers or suitable materials or equipment or failure to adhere to the construction scheduling responsibilities established in Paragraph 8.2;

13.3.1.7   If the Contractor disregards laws and regulations of any public body having jurisdiction;

13.3.1.8   If the Contractor disregards the authority of the CM; or

13.3.1.7.1 If the Contractor otherwise violates in any substantial way any provisions of the Contract Documents.

13.3.2   The Owner may, after giving the Contractor and its surety seven (7) days written notice, terminate any services of the Contractor, exclude the Contractor from the site and take possession of the Work and of all the Contractor's tools, appliances, construction equipment and machinery at the site and use the same to the full extent they could be used by the Contractor without liability to the Contractor for trespass or conversion, incorporate in the Work all materials and equipment stored at the site or for which the Owner has paid the Contractor, by which are stored elsewhere and finish the Work as the Owner may deem expedient. In such case the Contractor shall not be entitled to receive any further payment until the Work is finished. If the unpaid balance of the contract price exceeds the direct, indirect and consequential costs of completing the Work, including, but not limited to, fees and charges of engineers, architects, attorneys and other professionals and court arbitration costs, such excess shall be paid to the Contractor. If such costs exceed such unpaid balance, the Contractor shall pay the difference to the Owner and such costs incurred by the Owner shall be as determined by the CM and incorporated in a change order. When exercising any rights or remedies under this paragraph, the Owner shall not be required to obtain the lowest price for work performed.

13.4   Allowable Termination Costs

13.4.1   If the Owner terminates the whole or any portion of this Contract pursuant to Paragraph13.2, then the Owner shall only be liable to the Contractor for those costs specified in Paragraph 13.4.3, plus a fee of ten (10) percent on the actual costs allowed by Paragraph 13.4.3.

13.4.2   If the Owner terminates the whole or any portion of this Contract pursuant to Paragraph 13.3, the Owner shall be liable to the Contractor for those costs specified in Paragraph 13.4.3. No fee in addition to these costs shall be paid to the Contractor in the event of termination pursuant to Paragraph 13.3.

CMAA Document A-3 (2003 Edition)

13.4.3   If the Owner terminates the whole or any portion of this Contract, the Owner shall pay the Contractor the amounts determined by the CM as follows:

13.4.3.1 An amount for supplies, services or property accepted by the Owner pursuant to Paragraph 13.5.1.6 or sold or acquired pursuant to Paragraph 13.5.1.7 and not previously paid for and to the extent provided in this Contract such amount shall be equivalent to the aggregate price for such supplies or services computed in accordance with the price or prices specified in this Contract.

13.4.3.2 The total of:

13.4.3.2.1 The cost incurred in the performance of the Work terminated including initial costs and preparatory expense allocable thereto, but exclusive of any costs attributable to supplies or services paid or to be paid for under Paragraph 13.4.3.1; and

13.4.3.2.2 The cost of settling and paying claims arising out of the termination of work under subcontracts or purchase orders which are properly chargeable to the terminated portion of the Work, exclusive of amounts paid or payable on account of completed items of equipment delivered or services furnished by subcontractors or vendors prior to the effective date of the notice of termination. The amounts shall be included in the costs payable under 13.4.3.2.1; and

13.4.3.2.3 The reasonable costs of settlement, including accounting, legal, clerical and other expenses reasonably necessary for the preparation of settlement claims and supporting data with respect to the termination portion of the Contract and for the termination and settlement with subcontractors and suppliers there under, together with reasonable storage, transportation and other costs incurred in connection with the protection or disposition of property allocable to this Contract.

13.4.4   The total sum to be paid to the Contractor pursuant to this paragraph shall not exceed the contract price.

13.4.5   If the Owner terminates the whole or part of this Contract pursuant to Paragraph 13.3, the Owner may procure

upon such terms and in such manner as the CM may deem appropriate supplies or services similar to those so terminated and the Contractor shall be liable to the Owner for any excess costs for such similar supplies or services. The Contractor shall continue the performance of this Contract to the extent not terminated hereunder.

13.5   <u>Termination Provisions</u>

13.5.1   After receipt of a notice of termination from the Owner pursuant to Paragraph 13.2 or 13.3, the Contractor shall:

13.5.1.1 Stop work on the date and to the extent specified in the notice of termination;

13.5.1.2 Place no further orders or subcontracts for materials, equipment, supplies, services or facilities except as may be necessary for completion of such portion of this Contract as is not terminated;

13.5.1.3 Terminate all orders and agreements with subcontractors and suppliers to the extent that they relate to the performance of Work terminated;

13.5.1.4 Assign to the Owner in the manner, at the times and to the extent directed by the CM, all of the rights, title and interests of the Contractor under the orders and agreements so terminated, in which case the Owner shall have the right at the Owner's discretion to settle or pay any or all claims arising out of the termination of such orders and agreements;

13.5.1.5 Settle all outstanding liabilities and all claims arising out of such termination of such orders and agreements with the approval of the CM to the extent the CM may require. The approval shall be final for all the purposes of this paragraph;

13.5.1.6 Transfer title and deliver to the entity or entities designated by the CM and, in the manner, at the times and to the extent directed by CM, such portion of the Work as has been terminated;

13.5.1.6.1 The fabricated or unfabricated parts, work in process, partially completed supplies and equipment, materials, parts, tools, dies, jigs and other fixtures, completed

CMAA Document A-3 (2003 Edition)

work, supplies and other material produced as part of or acquired in connection with the performance of the portions of the Contract so terminated; and

13.5.1.6.2 The completed or partially completed plans, drawings, information and other property related to the Work;

13.5.1.7 Use the Contractor's best efforts to sell in the manner, at the times, to the extent and at the price or prices directed or authorized by the CM, any property of the types referred to in Paragraph 13.5.1.6 provided, however, that the Contractor may acquire any such property under the conditions prescribed by and at a price or prices approved by the CM and provided further that the proceeds of any such transfer or disposition shall be applied in reduction of any payments to be made by the Owner to the Contractor or shall otherwise be credited to the cost of the Work covered by this Contract or paid in such other manner as the CM may direct;

13.5.1.8 Complete performance of such part of the Work as shall not have been terminated; and

13.5.1.9 Take such action as may be necessary or as the CM may direct for the protection and preservation of the property related to this Contract that is in the possession of the Contractor and in which the Owner has or may acquire an interest.

13.5.2   The Contractor shall, from the effective date of termination until the expiration of three (3) years after final settlement under this Contract, preserve and make available to the Owner at all reasonable times at the office of the Contractor, but without direct charge to the Owner, all Contractor's books, records, documents and other evidence bearing on the costs and expenses of the Contractor relating to the Work terminated hereunder or to the extent approved by the CM, photographs (regular, micro or digital), videotape or other authentic reproductions thereof.

13.5.3   In arriving at any amount due the Contractor pursuant to Paragraph 13.4, there shall be deducted:

13.5.3.1 All unliquidated advances or other payments on account theretofore made to the Contractor applicable to the terminated portion of this Contract;

13.5.3.2 Any claim that the Owner may have against the Contractor;

13.5.3.3 Such amount as the CM determines to be necessary to protect the Owner against loss because of outstanding or potential liens or claims; and

13.5.3.4 The agreed price for or the proceeds of sale of any materials, supplies or other things acquired by the Contractor or sold pursuant to the provisions of Paragraph 13.5.1.7 and not otherwise recovered by or credited to the Owner.

13.5.4   The Owner may at the Owner's option and at the Contractor's expense have costs reimbursable under Paragraph 13.4 audited and certified by independent certified public accountants selected by the Owner.

13.5.5   The Contractor shall be entitled to only those damages and that relief from termination by the Owner as specifically provided in Article 13.

13.6   The Owner's Rights

13.6.1   When the Contractor's services have been terminated by the Owner pursuant to Paragraphs 13.2 or 13.3, the termination shall not affect any rights or remedies of the Owner against the Contractor then existing or which may thereafter accrue.

13.6.2   Any retention or payment of monies due the Contractor by the Owner shall not release the Contractor from liability for performance of the Work.

ARTICLE 14
DISPUTE RESOLUTION

14.1   Arbitration

14.1.1   The Owner and the Contractor shall submit all unresolved claims, counterclaims, disputes, controversies, and other matters in question arising out of or relating to this contract or the breach thereof ("disputes"), to mediation prior to either party initiating against the other a demand for

CMAA Document A-3 (2003 Edition)

arbitration pursuant to Paragraph 14.1.2 below, unless delay in initiating or prosecuting a proceeding in an arbitration or judicial forum would prejudice the Owner or Contractor. The Owner and the Contractor shall agree in writing as to the identity of the mediator and the rules and procedures of the mediation. If the Owner and the Contractor cannot agree, the disputes shall be submitted to mediation under the then current Construction Industry Mediation Rules of the American Arbitration Association.

14.1.2   All unresolved disputes relating to this Contract or the breach thereof ("disputes") shall be decided by arbitration, subject to the limitations stated in Paragraph 14.1.4 below. The agreement to arbitrate, and any other agreement or consent to arbitrate entered into in accordance herewith shall be specifically enforceable under the prevailing law of any court having jurisdiction. The Owner and the Contractor shall agree in writing as to the identity of the arbitrator(s) and the rules and procedures of the arbitration. If the Owner and the Contractor do not so agree, the Owner and the Contractor shall submit the dispute to arbitration under the then current Construction Industry Rules of the American Arbitration Association.

14.1.3   Arbitration may be commenced when sixty (60) days have passed after a decision in writing has been rendered by the CM as provided in Paragraph 15.4.1. Notice of demand for arbitration shall be filled in writing with the other party to this Contract and with the arbitrator(s). In no event shall the demand for arbitration be made after the date when institution of legal or equitable proceedings based on such dispute in question would be barred by the applicable statute of limitations or of repose.

14.1.4   No arbitration arising out of or relating to this Contract shall include by consolidation or joinder or in any other manner the CM, the Designer or their employees, agents or consultants except by written consent signed by the Owner, CM and Designer. No arbitration shall include by consolidation or joinder or in any manner parties other than the Owner, Contractor, other contractors as described in Article 7 and other persons substantially involved in common questions of fact or law whose presence is required if complete relief is to be accorded in arbitration. No person or entity other than the Owner, Contractor or other contractors

as described in Article 7 shall be included as an original third party or additional third party to an arbitration whose interest or responsibility is insubstantial. Consent to arbitration involving an additional person or entity shall not constitute consent to arbitration of a dispute not described therein or with a person or entity not named therein.

14.1.5   The decision rendered by the arbitrator shall be final, judgment may be entered upon it in any court having jurisdiction thereof, and the award shall not be subject to modification or appeal. In any judicial proceeding to enforce this Agreement to arbitrate, the only issues to be determined shall be those set forth in 9 U.S.C. Section 4 Federal Arbitration Act, and such issues shall be determined by the Court without a jury. All other issues, such as but not limited to, arbitratibility, prerequisites to arbitration, compliance with contractual time limits, applicability of indemnity clauses, clauses limiting damages and statutes of limitation shall be for the arbitrator(s), whose decision thereon shall be final and binding. There shall be no interlocutory appeal of an order compelling arbitration.

ARTICLE 15
OTHER PROVISIONS

15.1   Governing Law

15.1.1   The Contract and the Contract Documents shall, unless otherwise provided in the Contract Documents, be governed by the law of the state where the Project is located.

15.2   Successors and Assigns

15.2.1   The Owner and Contractor each binds itself, its successors, assigns and legal representatives to the terms of this Contract.

15.2.2   Neither the Owner nor the Contractor shall assign or transfer its interest in this Contract without the written consent of the other, except an assignment of accounts receivable may be made to a commercial bank without prior written consent.

15.3   Written Notice

CMAA Document A-3 (2003 Edition)

15.3.1   Whenever any provision of the Contract Documents requires the giving of the written notice, it shall be deemed to have been validly given if delivered in person to the individual or to a member of the firm or to an officer of the corporation for whom it is intended or if delivered or sent by registered or certified mail, postage prepaid, to the last business address known to the sender of the notice.

15.4   Disputes Between Contractor and Owner

15.4.1   Disputes and other matters relating to the acceptability of the Work, or the interpretation of the requirements of the Contract Documents pertaining to the performance and furnishing of the Work shall be referred to the CM in writing. The CM shall render a decision in writing within a reasonable time.

15.4.2   The rendering of a decision in writing by the CM with respect to any dispute or other matter shall be a condition precedent to any exercise by the Contractor such rights remedies as either may otherwise have under the Contract Documents or the laws and regulations in respect of any such dispute or other matter including the right to arbitration provided in Article 14 of this Contract.

15.4.3   The Contractor shall continue to perform the Work and adhere to the Contractor's Construction Schedule during all disputes or disagreements with the Owner. No work shall be delayed or postponed pending resolution of any disputes or disagreements.

15.5   Bonds

15.5.1   The Contractor shall furnish, as security for the faithful performance and payment of all the Contractor's obligations specified in the Contract Documents, performance and payment bonds, each in an amount at least equal to the contract price. These bonds shall remain in effect at least until one (1) year after the date when final payment becomes due. All bonds shall be in the forms prescribed by law. All bonds signed by an agent must be accompanied by a certified copy of the authority to act.

15.5.2   If the surety for any bond furnished by the Contractor is declared bankrupt or becomes insolvent or its right to do business is terminated in any state where any part of the Project is located, the Contractor shall within five (5) days thereafter substitute another bond and surety acceptable to the Owner.

15.6   Enforcement of Any Paragraph

15.6.1   The failure of the Owner or the CM to insist in any one or more instances upon the strict performance of any one or more of the provisions of this Contract or to exercise any right herein contained or provided by law shall not be construed as a waiver or relinquishment of such provision or right or of the right to subsequently demand such strict performance or exercise such right and the rights shall continue unchanged and remaining full force and effect.

15.7   Interest

15.7.1   All monies not paid when due hereunder shall bear interest at the legal rate at the place of the Project.

15.8   Meaning of Terms

15.8.1   The meaning of terms used herein shall be consistent with the definitions expressed in the CMAA Standard forms of Agreement, Contracts and General Conditions. Reference made in the singular shall include the plural and the masculine shall include the feminine or the neuter.

15.9   Unenforceability of Any Paragraph

15.9.1   If any paragraph of this Agreement is held as a matter of law to be unenforceable or unconscionable, the remainder of this Contract shall be enforceable without such paragraph.

15.10   Extent of Contract

15.10.1   This Contract represents the entire agreement between the parties and incorporates all prior agreements and understandings in connection with the subject matter hereof.

Appendix E

Supplementary Conditions

1. LOCATION OF THE PROJECT. The general location of the work covered in this specification is Portland, Ohio.

2. SCOPE OF THE WORK. The work to be performed under this Contract consists of furnishing all plant, materials, equipment, supplies, labor, and transportation, including fuel, power, and water, and performing all work in strict accordance with specifications, schedules, and drawings, all of which are made a part hereof, and including such detail drawings as may be furnished by the Architect-Engineer from time to time during the construction.

3. EXAMINATION OF SITE. Bidders should visit the site of the building, compare the drawings and specifications with any work in place, and inform themselves of all conditions, including other work, if any, being performed. Failure to visit the site will in no way relieve the successful bidder from the necessity of furnishing any materials or performing any work that may be required to complete the work in accordance with drawings and specifications.

4. LAYING OUT WORK. The Contractor shall, immediately upon entering the project site for the purpose of beginning work, locate all Owner-provided reference points and take such action as is necessary to prevent their destruction, lay out his own work and be responsible for all lines, elevations, and measurements of buildings, grading, paving, utilities, and other work executed by him under the Contract.

5. COMMENCEMENT, PROSECUTION, AND COMPLETION.

   a. The Contractor will be required to commence work under this Contract within ten (10) calendar days after the date of receipt by him of Notice to Proceed, to prosecute said work with faithfulness and energy and to complete the entire work, ready for use, within 380 calendar days after receipt of Notice to Proceed. The time stated for completion shall include final cleanup of the premises.

   b. It is mutually agreed that the time for the commencement and completion of the work will materially affect the progress of other work and that the Owner will suffer financial damages in an amount not now possible to ascertain if this work is not completed on schedule, and in view of these facts, it is agreed that the Owner will withhold from the Contractor, as liquidated damages and not as a penalty, the sum of $100.00 per day for each calendar day that the work remains uncompleted beyond the date specified for the completion of the work.

   c. If completion of the work to be performed under the terms of this Contract is delayed by reasons of delay in the performance of any work to be performed by the Owner, or other contractors, and which is essential to the work performed under this Contract, such delay shall not constitute a basis for any claim against the Owner, but the time of performance will be extended for a period equal to such delay or as otherwise mutually agreed upon.

535

6. WATCHMAN. The Contractor shall employ a responsible watchman to guard the site and premises at all times except during regular working hours, from the beginning of work until acceptance by the Owner.

7. OWNER-FURNISHED MATERIALS AND EQUIPMENT. With the following exception there will be no Owner-furnished materials and/or equipment.

   a. Hardware consisting of removable cylinders for locks will be Owner-furnished and installed by the Contractor as specified herein.

8. TAXES. Except as may be otherwise provided in this Contract, the contract price is to include all applicable federal, state, and local taxes, but does not include any tax from which the Contractor is exempt. Upon request of the Contractor, the Owner shall furnish a tax exemption certificate or similar evidence of exemption with respect to any such tax not included in the contract price pursuant to this provision.

9. RATES OF WAGES.

   a. There shall be paid each laborer or mechanic of the Contractor or subcontractor engaged in work on the project under this Contract in the trade or occupation listed below, not less than the hourly wage rate opposite the same, regardless of any contractual relationship which may be alleged to exist between the Contractor or any subcontractor and such laborers and mechanics.

Classification	*Wage Rates per Hour*
Asbestos workers	$ 28.16
Bricklayers	$ 28.69
Carpenters	$ 27.33
Cement masons	$ 25.67
Electricians	$ 28.79
Glaziers	$ 25.80
Ironworkers	$ 26.71
Lathers	$ 26.91
Linoleum layers	$ 25.14
Marble setters	$ 28.72
Mosaic and terrazzo workers	$ 27.33
Painters	$ 25.39
Plasterers	$ 27.54
Plumbers	$ 28.99
Roofers	$ 25.46
Sheet metal workers	$ 27.88
Steamfitters	$ 28.93
Stonemasons	$ 28.69
Tile setters	$ 27.33
Waterproofers	$ 25.46
LABORERS	
Air and power tool operator	$ 19.14
Cement mason tender	$ 19.14
Power buggy operator	$ 19.35

Sandblaster, potman, nozzleman	$ 19.35
Mason tender	$ 19.35
Pipe layer, nonmetallic, sewer and drainage	$ 19.70
Pumpcrete nozzle placementman	$ 18.84
Carpenter tender	$ 18.52
Unskilled and common laborer	$ 18.52
Concrete buggy operator	$ 18.52
Concrete puddler	$ 19.35
Vibrator operator	$ 19.35

POWER EQUIPMENT OPERATORS

Air compressor, power plant, pump operator	$ 21.78
Gunite and pumpcrete machine	$ 22.47
Concrete mixers over 1 yd^3	$ 22.47
Concrete batching plants	$ 22.47
Cranes, hysters, side boom tractors	$ 22.06
Winch truck	$ 22.06
Hoists:	
One drum	$ 22.89
Two or more drums	$ 23.65
Guy and stiff leg derricks	$ 23.65
Loaders:	
Front end	$ 20.95
Elevating belt, forklift	$ 22.47
Pile driver	$ 22.47
Sheep's foot rollers	$ 21.02
Rubber tired rollers	$ 21.02
Shovel, backhoe, clamshell, dragline, under $^3/_4$ yd^3	$ 21.78
Shovel, backhoe, clamshell, dragline, over $^3/_4$ yd^3	$ 22.41
Loaders, bulldozers, patrol, scraper	$ 22.68
Trenching machine	$ 21.78

TRUCK DRIVERS

Dumpster	$ 20.74
Dump trucks:	
Batch and under 8 yd^3	$ 20.19
8 and over yd^3	$ 20.53
Lowboy, heavy equipment	$ 21.30
Lowboy, light equipment	$ 20.60
Flatbed truck, under $^1/_2$ ton	$ 20.19
Pickup truck	$ 20.12
Transit mix	$ 20.81
Tank truck, no trailer	$ 20.39
Tank truck, with trailer	$ 20.53
Swamper or riding helper	$ 18.52

Welders: receive rate prescribed for craft performing operation to which welding is
   incidental.

b. The foregoing specified wage rates are minimum rates only, and the Owner will not consider any claims for additional compensation made by the Contractor because of payment by the Contractor or any wage rate in excess of the applicable rate contained in this Contract. All disputes in regard to the payment of wages in excess of those specified in this Contract shall be adjusted by the Contractor.

10. TEMPORARY FACILITIES. The Contractor shall furnish materials and labor to build all temporary buildings on the project site for use during the construction of the project. All such buildings and/or utilities shall remain the property of the Contractor and shall be removed by him, at his expense, upon completion of the work under this Contract.

   a. Sanitary Facilities. The Contractor shall provide and maintain ample toilet accommodations for all workmen employed on the project under this Contract. The latrines shall be weathertight, fly-proof, and shall conform to the standards established by the Owner. Toilets shall be flush-type water closets connected to the sewer, and/or chemical type. Toilet facilities shall be maintained in sanitary condition as approved by the Architect-Engineer at all times during the work on this project.

   b. Temporary Enclosures. The Contractor shall provide protection against entry, rain, wind, frost, and heat, at all times, and shall maintain all materials, apparatus, equipment, and fixtures free from damage and injury.

   c. Temporary Heat. The Contractor shall provide temporary heat at all times when weather conditions are such that good construction will be hampered and delayed. Uniform temperature, not lower than 50°F. shall be maintained at all times, including Saturdays, Sundays, and holidays, in that portion of the building where plastering, ceramic tile work, resilient flooring, and painting are being performed. When the permanent heating system is used to provide temporary heat requirements, all costs of operation shall be at the expense of the Contractor.

   d. Water Supply. Water supply for all trades for construction uses and domestic consumption shall be provided and paid for by the Contractor. Temporary lines and connections shall be removed in a manner satisfactory to the Architect-Engineer before final acceptance of work under this Contract.

   e. Electricity. Electric current required for power and light for all trades, for construction uses and temporary lines, lamps, and equipment as required, shall be provided, connected, and maintained by the Contractor at his expense and shall be removed in like manner at the completion of construction work.

   f. Telephone. The Contractor shall provide and pay for such telephone service as he may require.

11. SHOP DRAWINGS. The Contractor shall submit to the Architect-Engineer for approval five copies of all shop drawings as may be required. These drawings shall be complete and shall contain all required detailed information. If approved by the Architect-Engineer, each copy of the drawings will be identified as having received such approval by being so stamped and dated. The Contractor shall make any corrections required by the Architect-Engineer. Two sets of all shop drawings will be retained by the Architect-Engineer and three sets will be returned to the Contractor. The approval of the drawings by the Architect-Engineer shall not be construed as a complete check but will indicate only that the general method of construction and detailing is satisfactory. Approval of such drawings will not

relieve the Contractor of the responsibility of any error which may exist as the Contractor shall be responsible for the dimensions and design of adequate connections, details, and satisfactory construction of all work.

12. PLAN OF OPERATIONS. The Contractor shall coordinate his work with the operations and work of the Owner and other contractors, as directed by the Architect-Engineer.

13. REPORTING OF COST INFORMATION. The Owner will need cost information regarding certain items of property which will be furnished or installed under this Contract. Accordingly, at the written request of the Architect-Engineer, the Contractor shall furnish to the Owner cost information pertaining to such items of property furnished or installed under this Contract as the Architect-Engineer may designate, in such form and in such detail as may be required.

14. FIELD OFFICE. The Contractor shall provide adequate facilities for inspection of the project, including office space of not less than 100 square feet of floor space, properly heated, ventilated, and lighted, on the site of the project, for the exclusive use of the Architect-Engineer's representative. The Contractor shall construct a built-in worktable approximately 3'6" wide by 6'0" long in the field office and plan racks as directed by the Architect-Engineer.

15. PAYMENT. Partial payments under the Contract shall be made at the request of the Contractor once each month, based upon partial estimates to be furnished by the Contractor and approved by the Architect-Engineer. In making such partial payments, there shall be retained 10 percent of the estimated amounts until final completion and acceptance of all work covered by the Contract; provided, however, that the Architect-Engineer at any time after 50 percent of the work has been completed, if he finds that satisfactory progress is being made, with written consent of surety, shall recommend that the remaining partial payments be paid in full. Payments for work, under subcontracts of the Contractor, shall be subject to the above conditions applying to the Contract after the work under a subcontract has been 50 percent completed. In preparing estimates for partial payments, the material delivered on the site and preparatory work done may be taken into consideration.

16. CONSTRUCTION SCHEDULE. Immediately after execution and delivery of the Contract, and before the first partial payment is made, the Contractor shall deliver to the Architect-Engineer an estimated construction progress schedule showing the proposed dates of commencement and completion of each of the various subdivisions of the work required under the Contract, and the anticipated amount of each monthly payment that will become due the Contractor in accordance with the progress schedule. The Contractor shall also furnish on forms to be supplied by the Owner (a) a detailed estimate giving a complete breakdown of the contract price and (b) periodical itemized estimates of work done for the purpose of making partial payments thereon.

17. DRAWINGS FURNISHED. Subparagraph 4.2.1, of General Conditions, shall be changed to read as follows: "The Architect-Engineer shall furnish free of charge 25 sets of working Drawings and 25 sets of Specifications. The Contractor shall pay the cost of reproduction for all other copies of Drawings and Specifications furnished to him."

18. PERMITS. Subparagraph 3.13.1, of General Conditions, shall be modified as follows: "The Owner shall procure the required building permit at no cost to the Contractor."

19. INSURANCE.

   a. Subparagraph 11.2.1, of General Conditions, shall be modified as follows: "The Contractor shall procure, maintain, and pay for Owner's contingent liability insurance. The amount of Owner's contingent liability insurance shall be $100,000 for one person and $300,000 for one accident. A certificate of insurance shall be filed with the Architect-Engineer."

   b. Subparagraph 11.3.1, of General Conditions, shall be modified as follows: "The Contractor shall procure, maintain, and pay for fire insurance in the amount of 100 percent of the insurable value of the Project and, in addition, shall procure, maintain, and pay for insurance to protect the Owner from damage by hail, tornado, and hurricane upon the entire work in the amount of 100 percent of the insurable value thereof. Certificates of insurance shall be filed with the Architect-Engineer."

20. CLEANING UP. Subparagraph 3.15.1, of General Conditions, shall be modified as follows: "In addition to removal of rubbish and leaving the work broom-clean, the Contractor shall remove stains, spots, marks, and dirt from decorated work; clean hardware; remove paint spots and smears from all surfaces; clean light and plumbing fixtures, and wash all concrete, tile, and terrazzo floors."

Appendix F

Solicitation, Offer, and a Reward (GSA-FAR Standard Form 1442)

SOLICITATION, OFFER, AND AWARD *(Construction, Alteration, or Repair)*	1. SOLICITATION NO. ENG. 33-9-86	2. TYPE OF SOLICITATION [X] SEALED BID *(IFB)* [ ] NEGOTIATED *(RFP)*	3. DATE ISSUED 9 July, 20--	PAGE 1	OF	PAGES 3

IMPORTANT - The "offer" section on the reverse must be fully completed by offeror.

4. CONTRACT NO. ENG. 33-9-86	5. REQUISITION/PURCHASE REQUEST NO. Holloman AFB-4-78-90	6. PROJECT NO. 4-78-90

7. ISSUED BY CODE DACA99	8. ADDRESS OFFER TO
District Engineer Albuquerque District Corps of Engineers Albuquerque, New Mexico 87101	Contracting Division Albuquerque District Corps of Engineers Albuquerque, New Mexico 87101

9. FOR INFORMATION CALL:	a. NAME George Schlomo	b. TELEPHONE NO. *(Include area code) (NO COLLECT CALLS)* (505) 555-1411

SOLICITATION

NOTE: In sealed bid solicitations "offer" and "offeror" mean "bid and "bidder".

10. THE GOVERNMENT REQUIRES PERFORMANCE OF THE WORK DESCRIBED IN THESE DOCUMENTS *(Title, identifying no., date)*

Taxiways and Aprons, Holloman Air Force Base, New Mexico. This solicitation is an unrestricted invitation for Bids. Both large and small contractors may bid. The Bids are due in the Albuquerque Dirtrict Office on 9 August, 20-- at 2:30 PM (E.S.T.) where they will be publicaly opened. See Section 00100 for directions and/or mailing address. The pre-bid conference and site visit are scheduled for 10 AM, 20 July, 20-- at the Corps of Engineers office, Albuquerque, New Mexico.

11. The Contractor shall begin performance __10__ calendar days and complete it within __420__ calendar days after receiving [ ] award, [X] notice to proceed. This performance period is [X] mandatory [ ] negotiable. *(See* __Section 00800__ *.)*

12a. THE CONTRACTOR MUST FURNISH ANY REQUIRED PERFORMANCE AND PAYMENT BONDS? *(If "YES," indicate within how many calendar days after award in Item 12b).*
[X] YES [ ] NO Bond in the amount of 20% of bid total

12b. CALENDAR DAYS __10__

13. ADDITIONAL SOLICITATION REQUIREMENTS:

a. Sealed offers in original and __0__ copies to perform the work required are due at the place specified in Item 8 by __2:30__ *(hour)* local time __8/9/--__ (date). If this is a sealed bid solicitation, offers will be publicly opened at that time. Sealed envelopes containing offers shall be marked to show the offeror's name and address, the solicitation number, and the date and time offers are due.

b. An offer guarantee [X] is, [ ] is not required.

c. All offers are subject to the (1) work reqiurements, and (2) other provisions and clauses incorporated in the solicitation in full text or by

d. Offers providing less than __90__ calendar days for Government acceptance after the date offers are due will not be considered and will be rejected.

NSN 7540-01-155-3212

STANDARD FORM 1442 (REV. 4-85)
Prescribed by GSA - FAR (48 CFR) 53.236-1(d)

OFFER *(Must be fully completed by offeror)*

14. NAME AND ADDRESS OF OFFEROR *(Include ZIP Code)*	15. TELEPHONE NO. *(Include area code)*

The Excello Company, Inc.
2000 Random Road N.W.
Albuquerque, New Mexico 87107

(505) 106-3824

16. REMITTANCE ADDRESS *(Include only if different than Item 14.)*

CODE	FACILITY CODE

17. The offeror agrees to perform the work required at the prices specified below in strict accordance with the terms of this solicitation, if this offer is accepted

by the Government in writing within ___90___ calendar days after the date offers are due. *(Insert any number equal to or greater than the minimum re-*

quirement stated in Item 13d. Failure to insert any number means the offeror accepts the minimum in Item 13d.)

AMOUNTS ►   Eight Million, Nine Hundred Eighteen Thousand, Four Hundred Eighty Eight Dollars
($8,918,488.00).
Unit prices are set forth in the attached Bidding Schedule.

18. The offeror agrees to furnish any required performance and payment bonds.

19. ACKNOWLEDGMENT OF AMENDMENTS
(The offeror acknowledges receipt of amendments to the solicitation -- give number and date of each)

AMENDMENT NO.								
DATE.								

20a. NAME AND TITLE OF PERSON AUTHORIZED TO SIGN OFFER *(Type or print)*	20. SIGNATURE	20c. OFFER DATE
V. J. Excello, President		9 August, 20--

AWARD *(To be completed by Government)*

21. ITEMS ACCEPTED:

22. AMOUNT	23. ACCOUNTING AND APPROPRIATION DATA

24. SUBMIT INVOICES TO ADDRESS SHOWN IN ►	ITEM	25. OTHER THAN FULL AND OPEN COMPETITION PURSUANT TO
(4 copies unless otherwise specified)		☐ 10 U.S.C. 2304(c) ( ) ☐ 41 U.S.C. 253(c) ( )

26. ADMINISTERED BY	27. PAYMENT WILL BE MADE BY

CONTRACTING OFFICER WILL COMPLETE ITEM 28 OR 29 AS APPLICABLE

☐ **28. NEGOTIATED AGREEMENT** *(Contractor is required to sign this document and return ___ copies to issuing office.)* Contractor agrees to furnish and deliver all items or perform all work requirements identified on this form and any continuation sheets for the consideration stated in this contract. The rights and obligations of the parties to this contract shall be governed by (a) this contract award, (b) the solicitation, and (c) the clauses, representations, certifications, and specifications incorporated by reference in or attached to this contract.

☐ **29. AWARD** *(Contractor is not required to sign this document.)* Your offer on this solicitation is hereby accepted as to the items listed. This award consummates the contract, which consists of (a) the Government solicitation and your offer, and (b) this contract award. No further contractual document is necessary.

30a. NAME AND TITLE OF CONTRACTOR OR PERSON AUTHORIZED TO SIGN *(Type or print)*	31a. NAME OF CONTRACTING OFFICER *(Type or print)*

30b. SIGNATURE	30c. DATE	31b. UNITED STATES OF AMERICA	30c. DATE
		BY	

STANDARD FORM 1442 (REV. 4-85) **BACK**

BIDDING SCHEDULE

Serial No. Eng. 33-9-86

Bid Item No.	Description	Estimated Quantity	Unit	Unit Price	Estimated Amount
1	Clearing	job	lump sum	$20,125.76	$ 20,126
2	Demolition	job	lump sum	$16,248.98	$ 16,249
3	Excavation	127,000	cu. yd.	$ 2.45	$ 310,923
4	Base course	79,500	ton	$ 18.15	$1,443,043
5	Concrete pavement, 9 in.	90,000	sq. yd.	$ 27.49	$2,473,655
6	Concrete pavement, 11 in.	115,400	sq. yd.	$ 33.75	$3,883,639
7	Asphalt concrete surface	150	ton	$ 37.57	$ 5,636
8	Concrete pipe, 12 in.	1,000	lin. ft.	$ 32.35	$ 32,351
9	Concrete pipe, 36 in.	300	lin. ft.	$ 55.79	$ 16,736
10	Inlet	2	each	$ 666.58	$ 1,333
11	Fiber duct, 4-way	600	lin. ft. sum	$ 38.78	$ 23,266
12	Fiber duct, 8-way	1,200	lin. ft.	$ 47.21	$ 56,655
13	Electrical manhole	6	each	$ 1,924.44	$ 11,547
14	Underground cable	34,000	lin. ft.	$ 9.33	$ 317,278
15	Taxiway lights	120	each	$ 472.17	$ 56,660
16	Apron lights	70	each	$ 465.96	$ 32,617
17	Taxiway marking	job	lump sum	$11,053.92	$ 11,054
18	Fence	26,000	lin. ft.	$ 7.91	$ 205,720

TOTAL ESTIMATED AMOUNT

$8,918,488

1. All extensions of the unit prices shown will be subject to verification by the Government. In case of variation between the unit price and the extension, the unit price will be considered to be the bid.

2. If a bid or modification to a bid, based on unit prices, is submitted and provides for a lump-sum adjustment to the total estimated cost, the application of the lump-sum adjustment to each unit price, including lump-sum units, in the bid schedule, must be stated, or if it is not stated, the bidder agrees that the lump-sum adjustment shall be applied on a pro rata basis to every unit price in the bid schedule.

Appendix **G**

Instructions to Bidders

1. **Contract Documents.** The Notice to Bidders, the Instructions to Bidders, the General Conditions, the Supplementary Conditions, the Drawings and Specifications, the Contractor's Proposal Form, and the Agreement as finally negotiated compose the Contract Documents.

   Copies of these documents can be obtained from the office of Jones and Smith, Architect-Engineers, 142 Welsh Street, Portland, Ohio, upon deposit of $100.00 for each set thereof, said deposit being refundable upon return of the documents in good order within 10 days after the bidding date.

2. **Printed Form for Proposal.** All proposals must be made on the Contractor's Proposal Form attached hereto and should give the amounts bid for the work, both in words and in figures, and must be signed and acknowledged by the Contractor. In order to ensure consideration, the Proposal should be enclosed in a sealed envelope marked "Proposal for Municipal Airport Terminal Building to be opened at 2:30 P.M. (E.S.T), May 19, 20–," showing the return address of the sender and addressed to John Doe, City Manager, Portland, Ohio.

   If the proposal is made by a partnership, it shall contain the names of each partner and shall be signed in the firm name, followed by the signature of the person authorized to sign. If the proposal is made by a corporation, it shall be signed by the name of the corporation, followed by the written signature of the officer signing, and the printed or typewritten designation of the office he holds in the corporation, together with the corporation seal. All blank spaces in the proposal form shall be properly filled in.

3. **Alternatives.** Each bidder shall submit with his proposal, on forms provided, alternate proposals stating the differences in price (additions or deductions) from the base bid for substituting, omitting, or changing the materials or construction from that shown on the drawings and as specified in a manner as described in the Division "Alternate Proposals" of these specifications.

   The difference in price shall include all omissions, additions, and adjustments of all trades as may be necessary because of each change, substitution, or omission as described.

4. **Payment of Employees.** For work done in the State of Ohio the payment of employees of the Contractor and any and all subcontractors shall comply with the current minimum wage scale as published by the Labor Commission of the State of Ohio, a copy of which is made a part of the Supplementary Conditions.

   The Contractor and each of his subcontractors shall pay each of their employees engaged in work on the project under this contract in full, less deductions made mandatory by law, and not less often than once each week. All forms required by local authorities, the State of Ohio, and the United States Government, shall be properly submitted.

5. **Telegraphic Modification.** Any bidder may modify his bid by telegraphic communication at any time prior to the scheduled closing time for receipt of bids provided such telegraphic communication is received by the City Manager prior to said closing time, and provided

further that the City Manager is satisfied that a written confirmation of such telegraphic modification over the signature of the bidder was mailed prior to said closing time. If such written confirmation is not received within two (2) days from said closing time, no consideration will be given to the said telegraphic modification.

6. **Delivery of Proposals.** It is the bidder's responsibility to deliver his proposal at the proper time to the proper place. The mere fact that a proposal was dispatched will not be considered. The bidder must have the proposal actually delivered. Any proposal received after the scheduled closing time will be returned unopened to the bidder.

7. **Opening of Proposals.** At 2:30 P.M. (E.S.T.), May 19, 20–, in the Office of the City Manager, City Hall, Portland, Ohio, each and every proposal (except those which have been withdrawn in accordance with Item 10, "Withdrawal of Proposals,") received prior to the scheduled closing time for receipt of proposals, will be publicly opened and read aloud, irrespective of any irregularities or informalities in such proposals.

8. **Alterations in Proposal.** Except as otherwise provided herein, proposals which are incomplete or which are conditioned in any way or which contain erasures not authenticated as provided herein or items not called for in the proposal or which may have been altered or are not in conformity with the law, may be rejected as informal.

   The proposal form invites bids on definite plans and specifications. Only the amounts and information asked on the proposal form furnished will be considered as the bid. Each bidder shall bid upon the work exactly as specified and as provided in the proposal.

9. **Erasures.** The proposal submitted must not contain erasures, interlineations, or other corrections unless each such correction is suitably authenticated by affixing in the margin immediately opposite the correction the surname or surnames of the person or persons signing the bid.

10. **Withdrawal of Proposals.** At any time prior to the scheduled closing time for receipt of proposals, any bidder may withdraw his proposal, either personally or by telegraphic or written request. If withdrawal is made personally, proper receipt shall be given therefore.

   After the scheduled closing time for the receipt of proposals and before award of contract, no bidder will be permitted to withdraw his proposal unless said award is delayed for a period exceeding thirty (30) days. Negligence on the part of the bidder in preparing his bid confers no rights for the withdrawal of the proposal after it has been opened.

11. **Determination of Low Bid.** In making award of contract, the owner reserves the right to take into consideration the plant facilities of the bidders and the bidder's ability to complete the contract within the time specified in the proposal. The owner also reserves the right to evaluate factors that in his opinion would affect the final total cost.

12. **Rejection of Proposals.** The owner reserves the right to reject any or all proposals. Without limiting the generality of the foregoing, any proposal which is incomplete, obscure, or irregular may be rejected; any proposal which omits a bid for any one or more items in the price sheet may be rejected; any proposal in which unit prices are omitted or in which unit prices are obviously unbalanced may be rejected; any proposal accompanied by an insufficient or irregular certified check, cashier's check, or bid bond may be rejected.

13. **Proposal and Performance Guarantees.** A certified check, cashier's check, or bid bond for an amount equal to least five percent (5%) of the total amount bid shall accompany each proposal as evidence of good faith and as a guarantee that if awarded the Contract, the

bidder will execute the Contract and give bond as required. The successful bidder's check or bid bond will be retained until he has entered into a satisfactory contract and furnished required contract bonds. The owner reserves the right to hold the certified checks, cashier's checks, or bid bonds of the three lowest bidders, until the successful bidder has entered into a contract and furnished the required contract bonds.

14. **Acceptance of Proposals.** Within thirty (30) days after receipt of the proposals the owner will act upon them. The acceptance of a proposal will be a Notice of Acceptance in writing signed by a duly authorized representative of the owner and no other act of the owner shall constitute the acceptance of a proposal. The acceptance of a proposal shall bind the successful bidder to execute the Contract. The rights and obligations provided for in the Contract shall become effective and binding upon the parties only upon its formal execution.

15. **Time for Executing Contract and Providing Contract Bond.** Any contractor whose proposal shall be accepted will be required to execute the Contract and furnish contract bonds as required within ten (10) days after notice that the Contract has been awarded to him. Failure or neglect to do so shall constitute a breach of the agreement effected by the acceptance of the proposal.

16. **Prices.** In the event of a discrepancy between the prices quoted in words and those quoted in figures in the proposal, the words shall control. The prices are to include the furnishing of all materials, plant, equipment, tools, and all other facilities, and the performance of all labor and services necessary or proper for the completion of the work except as may be otherwise expressly provided in the Contract Documents.

17. **Examination of Drawings.** Bidders shall thoroughly examine and be familiar with the drawings and specifications. The failure or omission of any bidder to receive or examine any form, instrument, addendum, or other document shall in no way relieve any bidder from any obligation with respect to his proposal or to the Contract. The submission of a bid shall be taken as prima facie evidence of compliance with this Section.

18. **Interpretations.** No oral interpretations will be made to any bidder as to the meaning of the drawings and specifications or other contract documents. Every request for such an interpretation shall be made in writing and addressed and forwarded to the owner's authorized representative (Architect-Engineer) five (5) or more days before the date fixed for opening of proposals. Every interpretation made to a bidder will be in the form of an addendum to the Contract Documents which, if issued, will be sent as promptly as is practicable to all persons to whom the drawings and specifications have been issued. All such addenda shall become part of the Contract Documents.

19. **Postponement of Date for Opening Proposals.** The owner reserves the right to postpone the date of presentation and opening of proposals and will give telegraphic notice of any such postponement to each interested party.

Appendix **H**

A Strategy of Bidding

H.1 MARKUP

When a contractor is bidding on a new job, one of the final actions taken is adding the markup to the estimated cost of the construction. This markup is customarily determined as a percentage of the cost. Although markup can include certain cost items as well as an allowance for profit, the term, as used in this appendix, is construed to be profit only. It is assumed that all items of job expense, including contingency and general overhead, are included in the contractor's estimate of cost.

Customarily, the contractor will select a markup that, it is hoped, will enable it to bid the largest amount possible and still be the low bidder. By minimizing the difference between its bid and the second lowest, it will realize the maximum profit possible if it actually becomes the successful bidder. How the contractor goes about selecting its markup figure to accomplish this objective is normally a very inexact process and is based more on its subjective judgment than on any rational procedure. Using past experience as a guide, the contractor attempts to evaluate several bidding variables such as the competition, the type of work, the geographical area, the architect-engineer, and the terms of the contract documents. Although these variables cannot be precisely defined, they indicate in a general way whether high or low markups are in order.

This is not to say that the selection of a markup figure in such an intuitive fashion is altogether wrong or that more objective criteria can be devised that will suit all bidding conditions. Nevertheless, under the right circumstances, there are ways in which markup percentages can be selected that may increase the contractor's profits over the long term. One such circumstance may exist when a contractor's bidding is largely confined to a specific type of work in the same general geographical area.

The customary basis for most construction bidding is to bid each job essentially as an independent entity, attempting to maximize the potential profit on a particular project with little systematic attention being given to how it may relate to the panorama of past and future projects. Where a history of bidding experience has been built up, competitive patterns in the past biddings can be used as a guide for future ones. There are a number of ways in which this can be done, these procedures being known as "bidding strategies." It is the purpose of this appendix to discuss one such method. Although this discussion is oriented to lump-sum projects, the procedures are adaptable to unit-price biddings.

H.2 EXPECTED PROFIT

When a contractor bids a lump-sum job, it can anticipate a "potential profit" of $(h - c)$, where h is the amount of its bid and c is the actual cost of the work. The bid h is the sum of the estimated cost of the work and the markup, which is a variable. However, the contractor does not get every job on which it bids. Obviously, its chances of being the successful bidder are related to the amount of its bid, h. It can make its bid so high that its chances of being the low bidder would be zero. Here is a case in which the potential profit is large but the chances of actually realizing it are zero. To use slightly different terms, the probability (p) of being the low bidder is said to be zero when there is no chance of this occurrence. However, the contractor could bid so low that its winning would be a certainty (probability $p = 1$), but the potential profit would be very small or even negative (loss). In between, of course, there are many possible bids, with the probability of success varying inversely with the size of the bid; that is, the higher the bid the smaller the probability of success. Clearly, there is some optimum bid between the two extremes.

To determine this optimum bid, the concept of expectation, or "expected profit," is useful. Expected profit is defined as $p(b - c)$, where p is the probability of the contractor's being the low bidder when its submitted bid is equal to b. As previously explained, p can vary between 0 (no chance) and 1 (certainty), with higher values indicating higher probability or better chances of winning. If the probability p is equal to 0.4, this indicates that the contractor's chances of being the low bidder are 4 out of 10.

A simple example can illustrate the idea of expected profit. Consider a project whose actual cost c will be $50,000. Suppose a contractor is able to determine that a bid of $56,000 has a probability of 0.3 of being low and that a bid of $53,000 has a probability of 0.8. The objective is to determine which of the two bids is better with respect to expected profit.

When h = $56,000,

$$\text{Expected profit} = p(h - t) = 0.3 \ (\$56,000 - \$50,000) = \$1,800$$

When h = $53,000,

$$\text{Expected profit} = p(h - t) = 0.8 \ (\$53,000 - \$50,000) = \$2,400$$

The bid of $53,000 is the better of the two because it yields the larger expected profit. The reasoning behind maximizing expected profit is based on long-term considerations. If the $56,000 bid were used, the potential profit would be $6,000. However, there are only 3 chances out of 10 that our contractor will get the job if it uses this bid. If it were possible to bid this same job 10 times and our contractor submitted a bid of $56,000 each time, it would expect to be the successful bidder 3 times and would realize a total actual profit of $3 (\$6,000) = \$18,000$ from the series of biddings. This is an average profit of $1,800 per bidding. Yet the $53,000 bid would be successful 8 times out of the 10 and would produce a total actual profit of $24,000, or an average profit of $2,400 per bid.

As can be seen, expected profit represents the *average return per bid* if the biddings were to be repeated a large number of times. Biddings for the same project are, of course, not repeated, but the concept of increasing a contractor's actual profits by maximizing the expected profit for each bid submitted is still valid when utilized over a span of many biddings. It amounts to what might be described as "playing the odds." The bidding strategy discussed in this appendix is based on the concept of maximizing expected profit.

H.3 COST OF CONSTRUCTION

The cost c as used herein is the actual cost of the work, a quantity that cannot possibly be known until the construction has been completed. Unfortunately, c is an important element in our bidding strategy, and bidding strategies are designed for use when the job is just being bid and long before c is known. A common assumption is that the actual cost c will be equal to the contractor's estimate of cost used in its bid.

Whether this is a valid assumption depends on the contractor's past record of bidding accuracy. It is an easy matter for a contractor to check the precision with which its estimating staff has been able to estimate the costs of completed projects. For an example of the procedure, suppose a contractor wishes to study its bidding performance over the past five years, a period during which it has completed 60 competitively bid contracts. For each of these contracts, it computes the value V, which is the actual construction expense divided by the originally estimated cost. After determining these values, it groups them together into equal intervals and obtains the data shown in Figure H.1. This figure shows the number of occurrences of V in each 0.05 interval.

These data are then plotted to obtain the histogram shown in Figure H.2, and a frequency polygon is drawn (dashed lines) utilizing the midpoints of the horizontal bars of the histogram. The frequency polygon can then be smoothed by sketching in, by eye, a smooth curve that closely approximates the linear segments of the polygon. If the smoothed frequency polygon of past biddings resembles a normal distribution of small variance with its mode at or very near a value V of 1.0, such as that shown by curve A in Figure H.3, the overall bidding accuracy of the contractor has been consistently good. Although curve B has its mode at 1.0, the large dispersion of the values indicates loose estimating with a serious tendency to both under- and overestimate. A biased distribution such as curve C is indicative of systematic estimating errors that result in a pronounced tendency to consistently either under- or overestimate a majority of the jobs bid.

There are statistical ways to make allowance for the bias and variability of past cost estimates. However, it is believed that the bidding records of most successful contractors will generally resemble curve A. In this case, the assumption that the actual cost c will be equal to the cost as estimated is statistically acceptable and the bidding strategy will be developed on this basis. Notice that this does not say that the costs of one contractor will necessarily equal those of another competing contractor.

$V = \dfrac{\text{Actual project cost}}{\text{Estimated project cost}}$	Number of Projects
$\leq V < 0.85$	0
$0.85 \leq V < 0.90$	3
$0.90 \leq V < 0.95$	12
$0.95 \leq V < 1.00$	19
$1.00 \leq V < 1.05$	16
$1.05 \leq V < 1.10$	8
$1.10 \leq V < 1.15$	2
$1.15 \leq V <$	0
Total	60

Figure H.1   Bidding accuracy distribution.

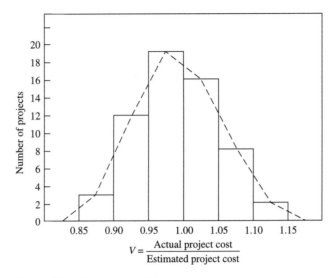

Figure H.2   Histogram and frequency polygon.

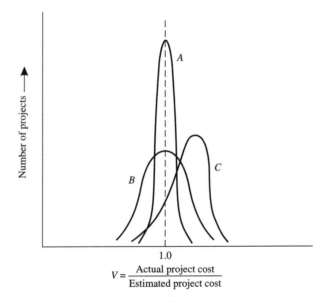

Figure H.3   Bidding accuracy distribution.

H.4   MAXIMIZING EXPECTED PROFIT

As an example of how a bid is selected to maximize expected profit, consider the bidding of a project whose estimated cost is $200,000. Assume for the moment that the probability of success for each of several different bids is known and is as shown in the second column of Figure H.4. How these probability values are obtained is discussed subsequently. For

Bid,	Porbability,	Expected Profit,
b	*p*	*p(b − c)*
$195,000	1.00	1.00($195,000 − $200,000) = −$5,000 (loss)
200,000	0.97	0.97($200,000 − $200,000) = 0
205,000	0.81	0.81($205,000 − $200,000) = 4,050
210,000	0.50	0.50($210,000 − $200,000) = 5,000
215,000	0.30	0.30($215,000 − $200,000) = 4,500
220,000	0.18	0.18($220,000 − $200,000) = 3,600
225,000	0.05	0.05($225,000 − $200,000) = 1,250
230,000	0.00	0.00($230,000 − $200,000) = 0

Figure H.4   Maximizing expected profit.

the figures shown in Figure H.4, a bid of $210,000, or a markup of 5 percent, yields the maximum expected profit and is the best bid.

H.5   CASE 1—SINGLE KNOWN COMPETITOR

Before the bidding strategy can be developed further, the values of probability of bidding success must be determined. These values are derived from an analysis of the historical bidding experience of our contractor and its competitors. For simplicity of explanation, this discussion is first confined to bidding against a single known competitor called Competitor A. Identification as a "known" competitor implies that our contractor has reason to believe that this competitor will be bidding the project and that our contractor has had past bidding experience with it.

Because bids are customarily opened and read aloud to all attendees at a bid opening, the past bids of competitors are known. In addition, area trade magazines publish the results of most biddings. Suppose that our contractor, using these past bidding records, compiles the information shown in Figure H.5 concerning Competitor A, against whom it has bid 62 times in recent years. It is suggested that only comparatively recent biddings be included, say, those within the past five or six years. The reason is that the bidding behavior of most contractors changes with time, reflecting economic conditions, changes of company policy, and other temporal factors. Consequently, it is judged that recent bidding data reflect more accurately the probable temper of competitors at present biddings. In Figure H.5, $R = b_A/c$ (b_A divided by c), where b_A is Competitor A's bid and c is our contractor's estimate of construction cost.

The data in Figure H.5 now make it possible to compute the probabilities, p_A, shown in Figure H.6, that various bids by our contractor will be lower than the bid submitted by Competitor A. For example, if our contractor bids a ratio of b/c of $61/62 = 0.98$ (2 percent less than its estimated cost), the probability that it will underbid Competitor A is 1.00 because at no time in the past has Competitor A ever bid this low. If our contractor bids the job at cost ($b/c = 1.00$), there is 1 chance in 62 that Competitor A will bid lower. Another way of stating the same thing is to say that our contractor has 61 chances out of 62, or a probability of $61/62 = 0.98$ of being the low bidder. Figure H.5 shows that in the past, Competitor A has bid

$R = \dfrac{b_A}{c}$	Number of Times
$\leq R < 0.98$	0
$0.98 \leq R < 1.00$	1
$1.00 \leq R < 1.02$	3
$1.02 \leq R < 1.04$	5
$1.04 \leq R < 1.06$	13
$1.06 \leq R < 1.08$	18
$1.08 \leq R < 1.10$	14
$1.10 \leq R < 1.12$	5
$1.12 \leq R < 1.14$	2
$1.14 \leq R < 1.16$	1
$1.16 \leq R <$	0
	Total $= 62$

Figure H.5   Bidding history of Competitor A.

$\dfrac{b}{c}$	p_A	Expected Profit, $p_A(b - c)$
0.98	$\dfrac{62}{62} = 1.00$	$1.00(0.98c - c) = -0.020c$
1.00	$\dfrac{61}{62} = 0.98$	$0.98(1.00c - c) = \quad 0$
1.02	$\dfrac{58}{62} = 0.94$	$0.94(1.02c - c) = \quad 0.019c$
1.04	$\dfrac{53}{62} = 0.85$	$0.85(1.04c - c) = \quad 0.034c$
1.06	$\dfrac{40}{62} = 0.65$	$0.65(1.06c - c) = \quad 0.039c$
1.08	$\dfrac{22}{62} = 0.36$	$0.36(1.08c - c) = \quad 0.029c$
1.10	$\dfrac{8}{62} = 0.13$	$0.13(1.10c - c) = \quad 0.013c$
1.12	$\dfrac{3}{62} = 0.05$	$0.05(1.12c - c) = \quad 0.006c$
1.14	$\dfrac{1}{62} = 0.02$	$0.02(1.14c - c) = \quad 0.003c$
1.16	$\dfrac{0}{62} = 0.00$	$0.00(1.16c - c) = \quad 0$

Figure H.6   Maximizing expected profit when bidding against Competitor A.

a value of b_A/c less than 1.02 three times. If our contractor submits a bid with a bid-to-cost ratio of 1.02 (markup = 2 percent), the probability is $^{58}/_{62}$, or 0.94, that it will be the low bidder. Continuing this process will yield the values of p_A shown in Figure H.6.

The values of b/c in Figure H.6 are the ratios of our contractor's bid to its estimated cost. The expected profit has been computed in terms of c, the estimated cost. When b/c = 1.08 (a markup of 8 percent), then b = 1.08c, the probability of success is 0.36, and the expected profit is 0.029c. The values of expected profit in the last column of Figure H.6 indicate that a bid of 1.06c, or a markup of 6 percent, is optimum when Competitor A is the only other bidder. If our contractor's estimated cost c is \$200,000, its bid would be \$212,000. Some additional accuracy in determining the optimum bid can be achieved by plotting the results of Figure H.6, as shown in Figure H.7. This figure shows that the optimum markup would be closer to 5.5 percent than to 6 percent. Whether this kind of accuracy is significant is problematic.

It is interesting to note that the optimum markup percentage is independent of estimated cost and can be determined in advance of bidding. The result is that the optimum markup will be the same for a small project as for a large one. This statement may well run contrary to actual bidding practice, and the answer is to analyze Competitor A's bidding record separately for small jobs and for large ones. For example, probability values could be obtained for all projects less than \$500,000 and for all projects over this amount. It should be obvious that probability values must be updated each time our contractor bids another job against Competitor A.

At this point, it is important to note that any bidding strategy attempts to predict future bidding behavior on the basis of past performance. The entire strategy is based on the supposition that competitors will continue to follow the same general bidding patterns they have used in the past. What changes in a competitor's future bidding habits might result from the use of a bidding strategy by our contractor is an open question.

H.6 CASE 2—MULTIPLE KNOWN COMPETITORS

Now assume that our contractor is going to bid against two known bidders, Competitors A and B. Our contractor must first, by analyzing the past bidding record of Competitor B, determine the same kind of cumulative probability distribution for this competitor as it obtained for Competitor A in the second column of Figure H.6. Assume that the results of its analysis are shown in Figure H.8.

Figure H.8 discloses that if our contractor uses a markup of 8 percent, there is a probability of 0.36 of beating Competitor A, a probability of 0.52 of beating Competitor B, and a probability of 0.19 of beating both of them. For any given value of b/c the probability of beating both Competitors A and B, p_{AB}, is found as the product of the values of p_A and p_B. This applies because the probability of the simultaneous occurrence of a number of independent events is the product of their separate probabilities.

Figure H.9 reveals that the optimum markup for our contractor to use when bidding against both Competitors A and B is 6 percent. The extension of the process to include any number of known competitors should be obvious. The inclusion of more competitors invariably reduces the value of the optimum markup. In addition, more competitors reduce the values of the expected profit such that to have enough significant figures, the computations must be carried out to more decimal places.

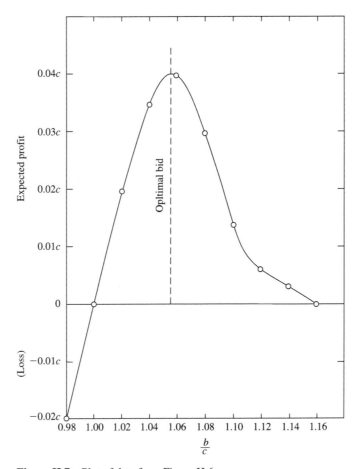

Figure H.7 Plot of data from Figure H.6.

$\dfrac{b}{c}$	p_A	p_B	p_{AB}
0.98	1.00	1.00	1.00
1.00	0.98	0.99	0.97
1.02	0.94	0.96	0.90
1.04	0.85	0.90	0.76
1.06	0.65	0.84	0.55
1.08	0.36	0.52	0.19
1.10	0.13	0.31	0.04
1.12	0.05	0.14	0.01
1.14	0.02	0.03	0.00
1.16	0.00	0.00	0.00

Figure H.8 Probability values of known Competitors A and B.

$\dfrac{b}{c}$	p_{AB}	Expected Profit, $p_{AB}(b-c)$
0.98	1.00	$1.00(0.98c - c) = -0.020c$
1.00	0.97	$0.97(1.00c - c) = 0$
1.02	0.90	$0.90(1.02c - c) = 0.018c$
1.04	0.76	$0.76(1.04c - c) = 0.030c$
1.06	0.55	$0.55(1.06c - c) = 0.033c$
1.08	0.19	$0.19(1.08c - c) = 0.015c$
1.10	0.04	$0.04(1.10c - c) = 0.004c$
1.12	0.01	$0.01(1.12c - c) = 0.001c$
1.14	0.00	$0.00(1.14c - c) = 0$
1.16	0.00	$0.00(1.16c - c) = 0$

Figure H.9   Maximizing expected profit when bidding against Competitors A and B.

H.7   CASE 3—THE AVERAGE COMPETITOR

The preceding discussion is based on the supposition that the competing contractors are all known—that is, that our contractor knows who its competitors will be and that it possesses past bidding information about each of them. Consideration is now given to the case in which some or all of the bidders are not known. It is in this regard that the concept of "average bidder" is useful.

An average bidder is a hypothetical competitor whose bidding behavior is a statistical composite of the behaviors of all of our contractor's competitors. The collective bidding pattern of its competitors can be obtained by combining all of the competitors into one probability distribution. The procedure here is exactly the same as that followed for Competitor A, except that all competitors are included, with no differentiation made between them.

For illustrative purposes, suppose that the result of this all-inclusive analysis is as shown in the second column of Figure H.10, where p_{av} is the probability that our contractor will submit a lower bid than any single unspecified competitor. If only one unknown competitor is anticipated, an optimum markup of 6 percent is indicated, although the use of 7 or 8 percent might be equally in order, because the difference in expected profit is small and the methods used are not this precise.

In the event that our contractor expects three unknown competitors, the same procedure is followed as with multiple known competitors, except that all competitors are the same; that is, each is an average competitor. For example, Figure H.10 shows that a probability of 0.51 is associated with a markup of 8 percent and a single unknown competitor. If there were three such competitors, the probability would be $(0.51)(0.51)(0.51) = (0.51)^3 = 0.13$, for a markup of 8 percent. In other words, if our contractor uses a markup of 8 percent, there is a probability of 0.13 that it will underbid all three unknown competitors. Figure H.11 discloses that the optimum markup when three unknown competitors are anticipated is between 4 and 5 percent. This example illustrates the variation of optimum markup with the number of competitors. For this reason, accuracy in estimating the number of competing firms is important.

$\dfrac{b}{c}$	p_{av}	Expected Profit, $p_{av}(b-c)$
0.98	1.00	$-0.020c$
1.00	0.98	0
1.02	0.95	$0.019c$
1.04	0.89	$0.036c$
1.06	0.72	$0.043c$
1.08	0.51	$0.041c$
1.10	0.30	$0.030c$
1.12	0.12	$0.014c$
1.14	0.05	$0.007c$
1.16	0.00	0

Figure H.10   Maximizing expected profit when bidding against a single average competitor.

$\dfrac{b}{c}$	p_{3av}	Expected Profit, $p_{3av}(b-c)$
0.98	$(1.00)^3 = 1.00$	$-0.020c$
1.00	$(0.98)^3 = 0.94$	0
1.02	$(0.95)^3 = 0.86$	$0.017c$
1.04	$(0.89)^3 = 0.70$	$0.028c$
1.06	$(0.72)^3 = 0.37$	$0.022c$
1.08	$(0.51)^3 = 0.13$	$0.010c$
1.10	$(0.30)^3 = 0.03$	$0.003c$
1.12	$(0.12)^3 = 0.00$	0
1.14	$(0.05)^3 = 0.00$	0
1.16	$(0.00)^3 = 0.00$	0

Figure H.11   Maximizing expected profit when bidding against three average competitors.

H.8   THE USE OF BIDDING STRATEGIES

Bidding strategies of one kind or another have been in use since the practice of competitive bidding began. Every bidding involves a complex game of trying to outguess and outsmart the opposition. The application of bidding procedures based on a systematic analysis of past bidding experience, however, is not common in the construction industry. Nevertheless, available evidence indicates that such methods are now catching on, although their practitioners are understandably reticent to divulge their procedures or results.

The probabilistic procedure discussed in this appendix is basic and is easily applied. Refinements of the procedure are possible, such as computing probabilities for different categories of projects. For example, the past projects could be subdivided into different ranges of construction cost and into different types of construction. The reason for this refinement is that different markups are commonly applied to various jobs, depending on their size and type. Another modification could be to give greater weight to recent bidding information than to that derived from older projects.

Several other forms of bidding models have been devised, many of which involve relatively sophisticated statistical procedures. Whether these more refined methods have any real advantage over the simple and straightforward procedure discussed in this appendix is not yet clear.

Appendix I

Proposal for Lump-Sum Contract

Place: Portland, Ohio
Date: May 19, 20–

PROPOSAL of The Blank Construction Company, Inc., a corporation organized and existing under the laws of the State of Ohio, a partnership consisting of _____

_____

an individual doing business as _____

TO: The City of Portland, Ohio
PROJECT: Municipal Airport Terminal Building
             For the City of Portland, Ohio

Gentlemen:

The Undersigned, in compliance with your Invitation for Bids for the General Construction of the above-described project, having examined the drawings and specifications with related contract documents carefully, with all addenda thereto, and the site of the work, and being familiar with all of the conditions surrounding the construction of the proposed project, hereby proposes to furnish all plant, labor, equipment, appliances, supplies, and materials and to perform all work for the construction of the project as required by and in strict accordance with the contract documents, specifications, schedules, and drawings with all addenda issued by the Architect-Engineer, at the prices stated below.

The Undersigned hereby acknowledges receipt of the following Addenda:
Addendum No. 1 dated April 28, 20–. _____
Addendum No. 2 dated May 6, 20–. _____

BASE PROPOSAL: For all work described in the detailed specifications and shown on the contract drawings for the building, I (or, we) agree to perform all the work for the sum of five million, five hundred four thousand, nine hundred seventy-four and no/100 ($5,504,974.00) dollars.
(Amount shall be shown in both written form and figures. In case of discrepancy between the written amount and the figures, the written amount will govern.)

The above-stated compensation covers all expenses incurred in performing the work, including premium for contract bonds, required under the contract documents, of which this proposal is a part.

ALTERNATE NO. 1: QUARRY TILE IN PLACE OF TERRAZZO: If the substitutions specified under this alternate are made, you may (~~deduct from~~) (add to) the base proposal the sum of <u>nine thousand, seven hundred fourteen and no/100 ($9,714.00)</u> dollars.

ALTERNATE NO. 2: CHANGE STRUCTURAL GLAZED TILE TO BRICK IN CON-COURSE: If the substitutions specified under this alternate are made, you may (deduct from) (~~add to~~) the base proposal the sum of <u>four thousand, two hundred eighty and no/100 ($4,280.00)</u> dollars.

ALTERNATE NO. 3: OMIT KITCHEN EQUIPMENT: If the substitutions specified under this alternate are made, you may (deduct from) (~~add to~~) the base proposal the sum of <u>sixty-six thousand, seven hundred twenty-three and no/100 ($66,723.00)</u> dollars.

BID SECURITY: Attached (cashier's check) (~~certified check~~) (~~bid bond~~) payable without condition, in the sum of <u>5% of maximum possible bid amount</u> ($ _____) dollars (equal to 5% of the largest possible combination) is to become the property of the City of Portland, Ohio, in the event the Contract and contract bonds are not executed within the time set forth hereinafter, as liquidated damages for the delay and additional work caused thereby.

CONTRACT SECURITY: The Undersigned hereby agrees, if awarded the contract, to furnish the contract bonds, as specified, with the <u>Hartford Accident and Indemnity Company</u> Surety Company of <u>Hartford, Connecticut</u> .

Upon receipt of notice of the acceptance of this bid, the Undersigned hereby agrees that he will execute and deliver the formal written Contract in the form prescribed, in accordance with the bid as accepted and that he will give contract bonds, all within 10 days after the prescribed forms are presented to him for signature.

If awarded the Contract, the Undersigned proposes to commence work within 10 calendar days after receipt of notice to proceed and to fully complete all of the work under his Contract, ready for occupancy, within 380 calendar days thereafter.

Respectfully submitted,

Company:	The Blank Construction Company, Inc.
By:	K.O. Acme
Title:	Vice-President
Address:	1938 Cranbrook Lane
City/State:	Portland, Ohio
Phone:	(216)-344-5507

SEAL:
(If bid is by a Corporation)

Appendix J

AIA Document A101-1997, Standard Form of Agreement Between Owner and Contractor

Where the Basis of Payment Is a Stipulated Sum

The American Institute of Architects is pleased to provide this sample copy of an AIA Contract Document for educational purposes. Created with the consensus of contractors, attorneys, architects and engineers, the AIA Contract Documents represent over 110 years of legal precedent.

1997 EDITION

AIA DOCUMENT | A101-1997

Standard Form of Agreement Between Owner and Contractor
where the basis of payment is a STIPULATED SUM

AGREEMENT made as of the   Sixteenth   day of   June
in the year   20--
(In words, indicate day, month and year)

BETWEEN the Owner:
(Name, address and other information)
   The City of Portland, Ohio

and the Contractor:
(Name, address and other information)

   The Blank Construction Company, Inc.

   Portland, Ohio

The Project is:
(Name and location)

   The Municipal Airport Terminal Building

   of the City of Portland, Ohio

The Architect is:
(Name, address and other information)

   Jones and Smith, Architect-Engineers

   Portland, Ohio

The Owner and Contractor agree as follows.

This document has important legal consequences. Consultation with an attorney is encouraged with respect to its completion or modification.

AIA Document A201-1997, General Conditions of the Contract for Construction, is adopted in this document by reference. Do not use with other general conditions unless this document is modified.

This document has been approved and endorsed by The Associated General Contractors of America.

© 1997 AIA®
AIA DOCUMENT A101-1997
OWNER-CONTRACTOR AGREEMENT

The American Institute of Architects
1735 New York Avenue, N.W.
Washington, D.C. 20006-5292

1

ARTICLE 1 THE CONTRACT DOCUMENTS

The Contract Documents consist of this Agreement, Conditions of the Contract (General, Supplementary and other Conditions), Drawings, Specifications, Addenda issued prior to execution of this Agreement, other documents listed in this Agreement and Modifications issued after execution of this Agreement; these form the Contract, and are as fully a part of the Contract as if attached to this Agreement or repeated herein. The Contract represents the entire and integrated agreement between the parties hereto and supersedes prior negotiations, representations or agreements, either written or oral. An enumeration of the Contract Documents, other than Modifications, appears in Article 8.

ARTICLE 2 THE WORK OF THIS CONTRACT

The Contractor shall fully execute the Work described in the Contract Documents, except to the extent specifically indicated in the Contract Documents to be the responsibility of others.

Municipal Airport Terminal Building, Portland, Ohio

ARTICLE 3 DATE OF COMMENCEMENT AND SUBSTANTIAL COMPLETION

3.1 The date of commencement of the Work shall be the date of this Agreement unless a different date is stated below or provision is made for the date to be fixed in a notice to proceed issued by the Owner.
(Insert the date of commencement if it differs from the date of this Agreement or, if applicable, state that the date will be fixed in a notice to proceed.)

Within ten calendar days after receipt of Notice to Proceed

If, prior to the commencement of the Work, the Owner requires time to file mortgages, mechanic's liens and other security interests, the Owner's time requirement shall be as follows:

3.2 The Contract Time shall be measured from the date of commencement.

3.3 The Contractor shall achieve Substantial Completion of the entire Work not later than _____ days from the date of commencement, or as follows:
(Insert number of calendar days. Alternatively, a calendar date may be used when coordinated with the date of commencement. Unless stated elsewhere in the Contract Documents, insert any requirements for earlier Substantial Completion of certain portions of the Work.)

380 Calendar days after receipt of Notice to Proceed

, subject to adjustments of this Contract Time as provided in the Contract Documents.
(Insert provisions, if any, for liquidated damages relating to failure to complete on time or for bonus payments for early completion of the Work.)

It is mutually agreed that the owner shall withold from the contractor, as liquidated damages and not as penalty, the sum of two hundred dollars ($200.00) per day for each calendar day that the work remains uncompleted beyond this date.

© 1 9 9 7   A I A ®
AIA DOCUMENT A101-1997
OWNER-CONTRACTOR
AGREEMENT

The American Institute
of Architects
1735 New York Avenue, N.W.
Washington, D.C. 20006-5292

2

The American Institute of Architects is pleased to provide this sample copy of an AIA Contract Document for educational purposes. Created with the consensus of contractors, attorneys, architects and engineers, the AIA Contract Documents represent over 110 years of legal precedent.

ARTICLE 4 CONTRACT SUM

4.1   The Owner shall pay the Contractor the Contract Sum in current funds for the Contractor's performance of the Contract. The Contract Sum shall be   Five million, four hundred thirty eight thousand, two hundred fifty one   Dollars ($ 5,438,251.00), subject to additions and deductions as provided in the Contract Documents.

4.2   The Contract Sum is based upon the following alternates, if any, which are described in the Contract Documents and are hereby accepted by the Owner:
(State the numbers or other identification of accepted alternates. If decisions on other alternates are to be made by the Owner subsequent to the execution of this Agreement, attach a schedule of such other alternates showing the amount for each and the date when that amount expires.)

```
Base Proposal                                              $5,504,974.00
   Alternate No. 1:  Add $9,714.00 (not accepted)
   Alternate No. 2:  Deduct $4,280.00 (not accepted)
   Alternate No. 3:  Deduct $66,723.00 (Accepted)
Total Alternates Accepted, deduct                              66,723.00
Total Contract Amount                                      $5,438,251.00
```

4.3   Unit prices, if any, are as follows:

```
Not Applicable
```

ARTICLE 5 PAYMENTS

5.1   PROGRESS PAYMENTS

5.1.1 Based upon Applications for Payment submitted to the Architect by the Contractor and Certificates for Payment issued by the Architect, the Owner shall make progress payments on account of the Contract Sum to the Contractor as provided below and elsewhere in the Contract Documents.

5.1.2 The period covered by each Application for Payment shall be one calendar month ending on the last day of the month, or as follows:

5.1.3 Provided that an Application for Payment is received by the Architect not later than the   last   day of a month, the Owner shall make payment to the Contractor not later than the   tenth (10)   day of the   following   month. If an Application for Payment is received by the Architect after the application date fixed above, payment shall be made by the Owner not later than   tenth (10)   days after the Architect receives the Application for Payment.

5.1.4 Each Application for Payment shall be based on the most recent schedule of values submitted by the Contractor in accordance with the Contract Documents. The schedule of values shall allocate the entire Contract Sum among the various portions of the Work. The schedule of values shall be prepared in such form and supported by such data to substantiate its accuracy as the Architect may require. This schedule, unless objected to by the Architect, shall be used as a basis for reviewing the Contractor's Applications for Payment.

© 1 9 9 7   A I A ®
AIA DOCUMENT A101-1997
OWNER-CONTRACTOR
AGREEMENT

The American Institute
of Architects
1735 New York Avenue, N.W.
Washington, D.C. 20006-5292

3

5.1.5 Applications for Payment shall indicate the percentage of completion of each portion of the Work as of the end of the period covered by the Application for Payment.

5.1.6 Subject to other provisions of the Contract Documents, the amount of each progress payment shall be computed as follows:

.1 Take that portion of the Contract Sum properly allocable to completed Work as determined by multiplying the percentage completion of each portion of the Work by the share of the Contract Sum allocated to that portion of the Work in the schedule of values, less retainage of   ten   percent (10 %). Pending final determination of cost to the Owner of changes in the Work, amounts not in dispute shall be included as provided in Subparagraph 7.3.8 of AIA Document A201-1997;

.2 Add that portion of the Contract Sum properly allocable to materials and equipment delivered and suitably stored at the site for subsequent incorporation in the completed construction (or, if approved in advance by the Owner, suitably stored off the site at a location agreed upon in writing), less retainage of   ten   percent (10 %);

.3 Subtract the aggregate of previous payments made by the Owner; and

.4 Subtract amounts, if any, for which the Architect has withheld or nullified a Certificate for Payment as provided in Paragraph 9.5 of AIA Document A201-1997.

5.1.7 The progress payment amount determined in accordance with Subparagraph 5.1.6 shall be further modified under the following circumstances:

.1 Add, upon Substantial Completion of the Work, a sum sufficient to increase the total payments to the full amount of the Contract Sum, less such amounts as the Architect shall determine for incomplete Work, retainage applicable to such work and unsettled claims; and (*Subparagraph 9.8.5 of AIA Document A201-1997 requires release of applicable retainage upon Substantial Completion of Work with consent of surety, if any.*)

.2 Add, if final completion of the Work is thereafter materially delayed through no fault of the Contractor, any additional amounts payable in accordance with Subparagraph 9.10.3 of AIA Document A201-1997.

5.1.8 Reduction or limitation of retainage, if any, shall be as follows:
(*If it is intended, prior to Substantial Completion of the entire Work, to reduce or limit the retainage resulting from the percentages inserted in Clauses 5.1.6.1 and 5.1.6.2 above, and this is not explained elsewhere in the Contract Documents, insert here provisions for such reduction or limitation.*)

```
It is provided, however, that the Architect-Engineer at
any time after fifty percent (50%) of the work has been
completed, if it finds that satisfactory progress is
being made, with written consent of Surety, shall
recommend that the remaining progress payments be made
in full.
```

5.1.9 Except with the Owner's prior approval, the Contractor shall not make advance payments to suppliers for materials or equipment which have not been delivered and stored at the site.

5.2 FINAL PAYMENT

5.2.1 Final payment, constituting the entire unpaid balance of the Contract Sum, shall be made by the Owner to the Contractor when:

.1 the Contractor has fully performed the Contract except for the Contractor's responsibility to correct Work as provided in Subparagraph 12.2.2 of AIA Document A201-1997, and to satisfy other requirements, if any, which extend beyond final payment; and

.2 a final Certificate for Payment has been issued by the Architect.

© 1997 AIA®
AIA DOCUMENT A101-1997
OWNER-CONTRACTOR
AGREEMENT

The American Institute
of Architects
1735 New York Avenue, N.W.
Washington, D.C. 20006-5292

4

5.2.2 The Owner's final payment to the Contractor shall be made no later than 30 days after the issuance of the Architect's final Certificate for Payment, or as follows:

ARTICLE 6 TERMINATION OR SUSPENSION

6.1 The Contract may be terminated by the Owner or the Contractor as provided in Article 14 of AIA Document A201-1997.

6.2 The Work may be suspended by the Owner as provided in Article 14 of AIA Document A201-1997.

ARTICLE 7 MISCELLANEOUS PROVISIONS

7.1 Where reference is made in this Agreement to a provision of AIA Document A201-1997 or another Contract Document, the reference refers to that provision as amended or supplemented by other provisions of the Contract Documents.

7.2 Payments due and unpaid under the Contract shall bear interest from the date payment is due at the rate stated below, or in the absence thereof, at the legal rate prevailing from time to time at the place where the Project is located.
(Insert rate of interest agreed upon, if any.)

One half percent (1/2%) per month

(Usury laws and requirements under the Federal Truth in Lending Act, similar state and local consumer credit laws and other regulations at the Owner's and Contractor's principal places of business, the location of the Project and elsewhere may affect the validity of this provision. Legal advice should be obtained with respect to deletions or modifications, and also regarding requirements such as written disclosures or waivers.)

7.3 The Owner's representative is:
(Name, address and other information)

7.4 The Contractor's representative is:
(Name, address and other information)

7.5 Neither the Owner's nor the Contractor's representative shall be changed without ten days' written notice to the other party.

7.6 Other provisions:

© 1997 AIA®
AIA DOCUMENT A101-1997
OWNER-CONTRACTOR
AGREEMENT

The American Institute
of Architects
1735 New York Avenue, N.W.
Washington, D.C. 20006-5292

5

ARTICLE 8 ENUMERATION OF CONTRACT DOCUMENTS

8.1   The Contract Documents, except for Modifications issued after execution of this Agreement, are enumerated as follows:

8.1.1   The Agreement is this executed 1997 edition of the Standard Form of Agreement Between Owner and Contractor, AIA Document A101-1997.

8.1.2   The General Conditions are the 1997 edition of the General Conditions of the Contract for Construction, AIA Document A201-1997.

8.1.3   The Supplementary and other Conditions of the Contract are those contained in the Project Manual dated   , and are as follows:

Document   Title   Pages

8.1.4   The Specifications are those contained in the Project Manual dated as in Subparagraph 8.1.3, and are as follows:
(Either list the Specifications here or refer to an exhibit attached to this Agreement.)

Section   Title   Pages

```
The specifications are listed in the table of contents
of the Project Manual dated March 6, 20-- and attached
to this agreement.
```

8.1.5   The Drawings are as follows, and are dated   March 6, 20--   unless a different date is shown below:
(Either list the Drawings here or refer to an exhibit attached to this Agreement.)

Number   Title   Date

```
A-1 through A44
S-1 through S-18
P-1 through P-6
M-1 through M-8
E-1 through E-11
```

© 1997   A I A ®
AIA DOCUMENT A101-1997
OWNER-CONTRACTOR
AGREEMENT

The American Institute
of Architects
1735 New York Avenue, N.W.
Washington, D.C. 20006-5292

6

The American Institute of Architects is pleased to provide this sample copy of an AIA Contract Document for educational purposes. Created with the consensus of contractors, attorneys, architects and engineers, the AIA Contract Documents represent over 110 years of legal precedent.

8.1.6 The Addenda, if any, are as follows:

Number	Date	Pages
Addendum No. 1	April 28, 20--	3
Addendum No. 2	May 6, 20--	2

Portions of Addenda relating to bidding requirements are not part of the Contract Documents unless the bidding requirements are also enumerated in this Article 8.

8.1.7 Other documents, if any, forming part of the Contract Documents are as follows:
(List here any additional documents that are intended to form part of the Contract Documents. AIA Document A201-1997 provides that bidding requirements such as advertisement or invitation to bid, Instructions to Bidders, sample forms and the Contractor's bid are not part of the Contract Documents unless enumerated in this Agreement. They should be listed here only if intended to be part of the Contract Documents.)

None

This Agreement is entered into as of the day and year first written above and is executed in at least three original copies, of which one is to be delivered to the Contractor, one to the Architect for use in the administration of the Contract, and the remainder to the Owner.

_____
OWNER *(Signature)*

_____
(Printed name and title)

_____
CONTRACTOR *(Signature)*

_____
(Printed name and title)

CAUTION: *You should sign an original AIA document or a licensed reproduction. Originals contain the AIA logo printed in red; licensed reproductions are those produced in accordance with the Instructions to this document.*

© 1997 AIA®
AIA DOCUMENT A101-1997
OWNER-CONTRACTOR
AGREEMENT

The American Institute
of Architects
1735 New York Avenue, N.W.
Washington, D.C. 20006-5292

7

Appendix K

AIA Document A111-1997, Standard Form of Agreement Between Owner and Contractor

Where the Basis of Payment is the Cost of the Work Plus a Fee with a Negotiated Guaranteed Maximum Price

▧AIA® Document A111™ – 1997

Standard Form of Agreement Between Owner and Contractor
where the basis of payment is the Cost of the Work Plus a Fee with a negotiated Guaranteed Maximum Price

AGREEMENT made as of the
in the year
(In words, indicate day, month and year)

BETWEEN the Owner:
(Name, address and other information)

and the Contractor:
(Name, address and other information)

The Project is:
(Name and location)

The Architect is:
(Name, address and other information)

day of

This document has important legal consequences. Consultation with an attorney is encouraged with respect to its completion or modification.

This document is not intended for use in competitive bidding.

AIA Document A201-1997, General Conditions of the Contract for Construction, is adopted in this document by reference.

This document has been approved and endorsed by The Associated General Contractors of America.

The Owner and Contractor agree as follows.

ARTICLE 1 THE CONTRACT DOCUMENTS
The Contract Documents consist of this Agreement, Conditions of the Contract (General, Supplementary and other Conditions), Drawings, Specifications, Addenda issued prior to execution of this Agreement, other documents listed in this Agreement and Modifications issued after execution of this Agreement; these form the Contract, and are as fully a part of the Contract as if attached to this Agreement or repeated herein. The Contract represents the entire and integrated agreement between the parties hereto and supersedes prior negotiations, representations or agreements, either written or oral. An enumeration of the Contract Documents, other than Modifications, appears in Article 15. If anything in the other Contract Documents is inconsistent with this Agreement, this Agreement shall govern.

ARTICLE 2 THE WORK OF THIS CONTRACT
The Contractor shall fully execute the Work described in the Contract Documents, except to the extent specifically indicated in the Contract Documents to be the responsibility of others.

ARTICLE 3 RELATIONSHIP OF THE PARTIES
The Contractor accepts the relationship of trust and confidence established by this Agreement and covenants with the Owner to cooperate with the Architect and exercise the Contractor's skill and judgment in furthering the interests of the Owner; to furnish efficient business administration and supervision; to furnish at all times an adequate supply of workers and materials; and to perform the Work in an expeditious and economical manner consistent with the Owner's interests. The Owner agrees to furnish and approve, in a timely manner, information required by the Contractor and to make payments to the Contractor in accordance with the requirements of the Contract Documents.

ARTICLE 4 DATE OF COMMENCEMENT AND SUBSTANTIAL COMPLETION
§ 4.1 The date of commencement of the Work shall be the date of this Agreement unless a different date is stated below or provision is made for the date to be fixed in a notice to proceed issued by the Owner.
(Insert the date of commencement, if it differs from the date of this Agreement or, if applicable, state that the date will be fixed in a notice to proceed.)

If, prior to commencement of the Work, the Owner requires time to file mortgages, mechanic's liens and other security interests, the Owner's time requirement shall be as follows:

§ 4.2 The Contract Time shall be measured from the date of commencement.

§ 4.3 The Contractor shall achieve Substantial Completion of the entire Work not later than days from the date of commencement, or as follows:
(Insert number of calendar days. Alternatively, a calendar date may be used when coordinated with the date of commencement. Unless stated elsewhere in the Contract Documents, insert any requirements for earlier Substantial Completion of certain portions of the Work.)

, subject to adjustments of this Contract Time as provided in the Contract Documents.
(Insert provisions, if any, for liquidated damages relating to failure to complete on time, or for bonus payments for early completion of the Work.)

ARTICLE 5 BASIS FOR PAYMENT
§ 5.1 CONTRACT SUM
§ 5.1.1 The Owner shall pay the Contractor the Contract Sum in current funds for the Contractor's performance of the Contract. The Contract Sum is the Cost of the Work as defined in Article 7 plus the Contractor's Fee.

§ 5.1.2 The Contractor's Fee is:
(State a lump sum, percentage of Cost of the Work or other provision for determining the Contractor's Fee, and describe the method of adjustment of the Contractor's Fee for changes in the Work.)

§ 5.2 GUARANTEED MAXIMUM PRICE
§ 5.2.1 The sum of the Cost of the Work and the Contractor's Fee is guaranteed by the Contractor not to exceed
                                    ($), subject to additions and deductions by Change Order as provided in the Contract Documents. Such maximum sum is referred to in the Contract Documents as the Guaranteed Maximum Price. Costs which would cause the Guaranteed Maximum Price to be exceeded shall be paid by the Contractor without reimbursement by the Owner.
(Insert specific provisions if the Contractor is to participate in any savings.)

§ 5.2.2 The Guaranteed Maximum Price is based on the following alternates, if any, which are described in the Contract Documents and are hereby accepted by the Owner:
(State the numbers or other identification of accepted alternates. If decisions on other alternates are to be made by the Owner subsequent to the execution of this Agreement, attach a schedule of such other alternates showing the amount for each and the date when the amount expires.)

§ 5.2.3 Unit prices, if any, are as follows:

§ 5.2.4 Allowances, if any, are as follows:
(Identify and state the amounts of any allowances, and state whether they include labor, materials, or both.)

§ 5.2.5 Assumptions, if any, on which the Guaranteed Maximum Price is based are as follows:

§ 5.2.6 To the extent that the Drawings and Specifications are anticipated to require further development by the Architect, the Contractor has provided in the Guaranteed Maximum Price for such further development consistent with the Contract Documents and reasonably inferable therefrom. Such further development does not include such things as changes in scope, systems, kinds and quality of materials, finishes or equipment, all of which, if required, shall be incorporated by Change Order.

ARTICLE 6 CHANGES IN THE WORK

§ 6.1 Adjustments to the Guaranteed Maximum Price on account of changes in the Work may be determined by any of the methods listed in Section 7.3.3 of AIA Document A201-1997.

§ 6.2 In calculating adjustments to subcontracts (except those awarded with the Owner's prior consent on the basis of cost plus a fee), the terms "cost" and "fee" as used in Section 7.3.3 of AIA Document A201-1997 and the terms "costs" and "a reasonable allowance for overhead and profit" as used in Section 7.3.6 of AIA Document A201-1997 shall have the meanings assigned to them in AIA Document A201-1997 and shall not be modified by Articles 5, 7 and 8 of this Agreement. Adjustments to subcontracts awarded with the Owner's prior consent on the basis of cost plus a fee shall be calculated in accordance with the terms of those subcontracts.

§ 6.3 In calculating adjustments to the Guaranteed Maximum Price, the terms "cost" and "costs" as used in the above-referenced provisions of AIA Document A201-1997 shall mean the Cost of the Work as defined in Article 7 of this Agreement and the terms "fee" and "a reasonable allowance for overhead and profit" shall mean the Contractor's Fee as defined in Section 5.1.2 of this Agreement.

§ 6.4 If no specific provision is made in Section 5.1 for adjustment of the Contractor's Fee in the case of changes in the Work, or if the extent of such changes is such, in the aggregate, that application of the adjustment provisions of Section 5.1 will cause substantial inequity to the Owner or Contractor, the Contractor's Fee shall be equitably adjusted on the basis of the Fee established for the original Work, and the Guaranteed Maximum Price shall be adjusted accordingly.

ARTICLE 7 COSTS TO BE REIMBURSED
§ 7.1 COST OF THE WORK

The term Cost of the Work shall mean costs necessarily incurred by the Contractor in the proper performance of the Work. Such costs shall be at rates not higher than the standard paid at the place of the Project except with prior consent of the Owner. The Cost of the Work shall include only the items set forth in this Article 7.

§ 7.2 LABOR COSTS

§ 7.2.1 Wages of construction workers directly employed by the Contractor to perform the construction of the Work at the site or, with the Owner's approval, at off-site workshops.

§ 7.2.2 Wages or salaries of the Contractor's supervisory and administrative personnel when stationed at the site with the Owner's approval.
(If it is intended that the wages or salaries of certain personnel stationed at the Contractor's principal or other offices shall be included in the Cost of the Work, identify in Article 14 the personnel to be included and whether for all or only part of their time, and the rates at which their time will be charged to the Work.)

§ 7.2.3 Wages and salaries of the Contractor's supervisory or administrative personnel engaged, at factories, workshops or on the road, in expediting the production or transportation of materials or equipment required for the Work, but only for that portion of their time required for the Work.

§ 7.2.4 Costs paid or incurred by the Contractor for taxes, insurance, contributions, assessments and benefits required by law or collective bargaining agreements and, for personnel not covered by such agreements, customary benefits such as sick leave, medical and health benefits, holidays, vacations and pensions, provided such costs are based on wages and salaries included in the Cost of the Work under Sections 7.2.1 through 7.2.3.

§ 7.3 SUBCONTRACT COSTS
§ 7.3.1 Payments made by the Contractor to Subcontractors in accordance with the requirements of the subcontracts.

§ 7.4 COSTS OF MATERIALS AND EQUIPMENT INCORPORATED IN THE COMPLETED CONSTRUCTION
§ 7.4.1 Costs, including transportation and storage, of materials and equipment incorporated or to be incorporated in the completed construction.

§ 7.4.2 Costs of materials described in the preceding Section 7.4.1 in excess of those actually installed to allow for reasonable waste and spoilage. Unused excess materials, if any, shall become the Owner's property at the completion of the Work or, at the Owner's option, shall be sold by the Contractor. Any amounts realized from such sales shall be credited to the Owner as a deduction from the Cost of the Work.

§ 7.5 COSTS OF OTHER MATERIALS AND EQUIPMENT, TEMPORARY FACILITIES AND RELATED ITEMS
§ 7.5.1 Costs, including transportation and storage, installation, maintenance, dismantling and removal of materials, supplies, temporary facilities, machinery, equipment, and hand tools not customarily owned by construction workers, that are provided by the Contractor at the site and fully consumed in the performance of the Work; and cost (less salvage value) of such items if not fully consumed, whether sold to others or retained by the Contractor. Cost for items previously used by the Contractor shall mean fair market value.

§ 7.5.2 Rental charges for temporary facilities, machinery, equipment, and hand tools not customarily owned by construction workers that are provided by the Contractor at the site, whether rented from the Contractor or others, and costs of transportation, installation, minor repairs and replacements, dismantling and removal thereof. Rates and quantities of equipment rented shall be subject to the Owner's prior approval.

§ 7.5.3 Costs of removal of debris from the site.

§ 7.5.4 Costs of document reproductions, facsimile transmissions and long-distance telephone calls, postage and parcel delivery charges, telephone service at the site and reasonable petty cash expenses of the site office.

§ 7.5.5 That portion of the reasonable expenses of the Contractor's personnel incurred while traveling in discharge of duties connected with the Work.

§ 7.5.6 Costs of materials and equipment suitably stored off the site at a mutually acceptable location, if approved in advance by the Owner.

§ 7.6 MISCELLANEOUS COSTS
§ 7.6.1 That portion of insurance and bond premiums that can be directly attributed to this Contract:

§ 7.6.2 Sales, use or similar taxes imposed by a governmental authority that are related to the Work.

§ 7.6.3 Fees and assessments for the building permit and for other permits, licenses and inspections for which the Contractor is required by the Contract Documents to pay.

§ 7.6.4 Fees of laboratories for tests required by the Contract Documents, except those related to defective or nonconforming Work for which reimbursement is excluded by Section 13.5.3 of AIA Document A201-1997 or other provisions of the Contract Documents, and which do not fall within the scope of Section 7.7.3.

§ 7.6.5 Royalties and license fees paid for the use of a particular design, process or product required by the Contract Documents; the cost of defending suits or claims for infringement of patent rights arising from such requirement of the Contract Documents; and payments made in accordance with legal judgments against the Contractor resulting from such suits or claims and payments of settlements made with the Owner's consent. However, such costs of legal defenses, judgments and settlements shall not be included in the calculation of the Contractor's Fee or subject to the Guaranteed Maximum Price. If such royalties, fees and costs are excluded by the last sentence of Section 3.17.1 of AIA Document A201-1997 or other provisions of the Contract Documents, then they shall not be included in the Cost of the Work.

§ 7.6.6 Data processing costs related to the Work.

§ 7.6.7 Deposits lost for causes other than the Contractor's negligence or failure to fulfill a specific responsibility to the Owner as set forth in the Contract Documents.

§ 7.6.8 Legal, mediation and arbitration costs, including attorneys' fees, other than those arising from disputes between the Owner and Contractor, reasonably incurred by the Contractor in the performance of the Work and with the Owner's prior written approval; which approval shall not be unreasonably withheld.

§ 7.6.9 Expenses incurred in accordance with the Contractor's standard personnel policy for relocation and temporary living allowances of personnel required for the Work, if approved by the Owner.

§ 7.7 OTHER COSTS AND EMERGENCIES

§ 7.7.1 Other costs incurred in the performance of the Work if and to the extent approved in advance in writing by the Owner.

§ 7.7.2 Costs due to emergencies incurred in taking action to prevent threatened damage, injury or loss in case of an emergency affecting the safety of persons and property, as provided in Section 10.6 of AIA Document A201-1997.

§ 7.7.3 Costs of repairing or correcting damaged or nonconforming Work executed by the Contractor, Subcontractors or suppliers, provided that such damaged or nonconforming Work was not caused by negligence or failure to fulfill a specific responsibility of the Contractor and only to the extent that the cost of repair or correction is not recoverable by the Contractor from insurance, sureties, Subcontractors or suppliers.

ARTICLE 8 COSTS NOT TO BE REIMBURSED

§ 8.1 The Cost of the Work shall not include:

§ 8.1.1 Salaries and other compensation of the Contractor's personnel stationed at the Contractor's principal office or offices other than the site office, except as specifically provided in Sections 7.2.2 and 7.2.3 or as may be provided in Article 14.

§ 8.1.2 Expenses of the Contractor's principal office and offices other than the site office.

§ 8.1.3 Overhead and general expenses, except as may be expressly included in Article 7.

§ 8.1.4 The Contractor's capital expenses, including interest on the Contractor's capital employed for the Work.

§ 8.1.5 Rental costs of machinery and equipment, except as specifically provided in Section 7.5.2.

§ 8.1.6 Except as provided in Section 7.7.3 of this Agreement, costs due to the negligence or failure to fulfill a specific responsibility of the Contractor, Subcontractors and suppliers or anyone directly or indirectly employed by any of them or for whose acts any of them may be liable.

§ 8.1.7 Any cost not specifically and expressly described in Article 7.

§ 8.1.8 Costs, other than costs included in Change Orders approved by the Owner, that would cause the Guaranteed Maximum Price to be exceeded.

ARTICLE 9 DISCOUNTS, REBATES AND REFUNDS

§ 9.1 Cash discounts obtained on payments made by the Contractor shall accrue to the Owner if (1) before making the payment, the Contractor included them in an Application for Payment and received payment therefor from the

Owner, or (2) the Owner has deposited funds with the Contractor with which to make payments; otherwise, cash discounts shall accrue to the Contractor. Trade discounts, rebates, refunds and amounts received from sales of surplus materials and equipment shall accrue to the Owner, and the Contractor shall make provisions so that they can be secured.

§ 9.2 Amounts that accrue to the Owner in accordance with the provisions of Section 9.1 shall be credited to the Owner as a deduction from the Cost of the Work.

ARTICLE 10   SUBCONTRACTS AND OTHER AGREEMENTS

§ 10.1 Those portions of the Work that the Contractor does not customarily perform with the Contractor's own personnel shall be performed under subcontracts or by other appropriate agreements with the Contractor. The Owner may designate specific persons or entities from whom the Contractor shall obtain bids. The Contractor shall obtain bids from Subcontractors and from suppliers of materials or equipment fabricated especially for the Work and shall deliver such bids to the Architect. The Owner shall then determine, with the advice of the Contractor and the Architect, which bids will be accepted. The Contractor shall not be required to contract with anyone to whom the Contractor has reasonable objection.

§ 10.2 If a specific bidder among those whose bids are delivered by the Contractor to the Architect (1) is recommended to the Owner by the Contractor; (2) is qualified to perform that portion of the Work; and (3) has submitted a bid that conforms to the requirements of the Contract Documents without reservations or exceptions, but the Owner requires that another bid be accepted, then the Contractor may require that a Change Order be issued to adjust the Guaranteed Maximum Price by the difference between the bid of the person or entity recommended to the Owner by the Contractor and the amount of the subcontract or other agreement actually signed with the person or entity designated by the Owner.

§ 10.3 Subcontracts or other agreements shall conform to the applicable payment provisions of this Agreement, and shall not be awarded on the basis of cost plus a fee without the prior consent of the Owner.

ARTICLE 11   ACCOUNTING RECORDS

The Contractor shall keep full and detailed accounts and exercise such controls as may be necessary for proper financial management under this Contract, and the accounting and control systems shall be satisfactory to the Owner. The Owner and the Owner's accountants shall be afforded access to, and shall be permitted to audit and copy, the Contractor's records, books, correspondence, instructions, drawings, receipts, subcontracts, purchase orders, vouchers, memoranda and other data relating to this Contract, and the Contractor shall preserve these for a period of three years after final payment, or for such longer period as may be required by law.

ARTICLE 12   PAYMENTS
§ 12.1 PROGRESS PAYMENTS

§ 12.1.1 Based upon Applications for Payment submitted to the Architect by the Contractor and Certificates for Payment issued by the Architect, the Owner shall make progress payments on account of the Contract Sum to the Contractor as provided below and elsewhere in the Contract Documents.

§ 12.1.2 The period covered by each Application for Payment shall be one calendar month ending on the last day of the month, or as follows:

§ 12.1.3 Provided that an Application for Payment is received by the Architect not later than the day of a month, the Owner shall make payment to the Contractor not later than the   day of the   month. If an Application for Payment is received by the Architect after the application date fixed above, payment shall be made by the Owner not later than   (   ) days after the Architect receives the Application for Payment.

§ 12.1.4 With each Application for Payment, the Contractor shall submit payrolls, petty cash accounts, receipted invoices or invoices with check vouchers attached, and any other evidence required by the Owner or Architect to demonstrate that cash disbursements already made by the Contractor on account of the Cost of the Work equal or exceed (1) progress payments already received by the Contractor; less (2) that portion of those payments attributable to the Contractor's Fee; plus (3) payrolls for the period covered by the present Application for Payment.

§ 12.1.5 Each Application for Payment shall be based on the most recent schedule of values submitted by the Contractor in accordance with the Contract Documents. The schedule of values shall allocate the entire Guaranteed Maximum Price among the various portions of the Work, except that the Contractor's Fee shall be shown as a single separate item. The schedule of values shall be prepared in such form and supported by such data to substantiate its accuracy as the Architect may require. This schedule, unless objected to by the Architect, shall be used as a basis for reviewing the Contractor's Applications for Payment.

§ 12.1.6 Applications for Payment shall show the percentage of completion of each portion of the Work as of the end of the period covered by the Application for Payment. The percentage of completion shall be the lesser of (1) the percentage of that portion of the Work which has actually been completed; or (2) the percentage obtained by dividing (a) the expense that has actually been incurred by the Contractor on account of that portion of the Work for which the Contractor has made or intends to make actual payment prior to the next Application for Payment by (b) the share of the Guaranteed Maximum Price allocated to that portion of the Work in the schedule of values.

§ 12.1.7 Subject to other provisions of the Contract Documents, the amount of each progress payment shall be computed as follows:

.1 take that portion of the Guaranteed Maximum Price properly allocable to completed Work as determined by multiplying the percentage of completion of each portion of the Work by the share of the Guaranteed Maximum Price allocated to that portion of the Work in the schedule of values. Pending final determination of cost to the Owner of changes in the Work, amounts not in dispute shall be included as provided in Section 7.3.8 of AIA Document A201-1997;

.2 add that portion of the Guaranteed Maximum Price properly allocable to materials and equipment delivered and suitably stored at the site for subsequent incorporation in the Work, or if approved in advance by the Owner, suitably stored off the site at a location agreed upon in writing;

.3 add the Contractor's Fee, less retainage of _____ percent (%). The Contractor's Fee shall be computed upon the Cost of the Work described in the two preceding Clauses at the rate stated in Section 5.1.2 or, if the Contractor's Fee is stated as a fixed sum in that Subparagraph, shall be an amount that bears the same ratio to that fixed-sum fee as the Cost of the Work in the two preceding Clauses bears to a reasonable estimate of the probable Cost of the Work upon its completion;

.4 subtract the aggregate of previous payments made by the Owner;

.5 subtract the shortfall, if any, indicated by the Contractor in the documentation required by Section 12.1.4 to substantiate prior Applications for Payment, or resulting from errors subsequently discovered by the Owner's accountants in such documentation; and

.6 subtract amounts, if any, for which the Architect has withheld or nullified a Certificate for Payment as provided in Section 9.5 of AIA Document A201-1997.

§ 12.1.8 Except with the Owner's prior approval, payments to Subcontractors shall be subject to retainage of not less than _____ percent (%). The Owner and the Contractor shall agree upon a mutually acceptable procedure for review and approval of payments and retention for Subcontractors.

§ 12.1.9 In taking action on the Contractor's Applications for Payment, the Architect shall be entitled to rely on the accuracy and completeness of the information furnished by the Contractor and shall not be deemed to represent that the Architect has made a detailed examination, audit or arithmetic verification of the documentation submitted in accordance with Section 12.1.4 or other supporting data; that the Architect has made exhaustive or continuous on-site inspections or that the Architect has made examinations to ascertain how or for what purposes the Contractor has used amounts previously paid on account of the Contract. Such examinations, audits and verifications, if required by the Owner, will be performed by the Owner's accountants acting in the sole interest of the Owner.

§ 12.2 FINAL PAYMENT

§ 12.2.1 Final payment, constituting the entire unpaid balance of the Contract Sum, shall be made by the Owner to the Contractor when:

.1 the Contractor has fully performed the Contract except for the Contractor's responsibility to correct Work as provided in Section 12.2.2 of AIA Document A201-1997, and to satisfy other requirements, if any, which extend beyond final payment; and

.2 a final Certificate for Payment has been issued by the Architect.

§ **12.2.2** The Owner's final payment to the Contractor shall be made no later than 30 days after the issuance of the Architect's final Certificate for Payment, or as follows:

§ **12.2.3** The Owner's accountants will review and report in writing on the Contractor's final accounting within 30 days after delivery of the final accounting to the Architect by the Contractor. Based upon such Cost of the Work as the Owner's accountants report to be substantiated by the Contractor's final accounting, and provided the other conditions of Section 12.2.1 have been met, the Architect will, within seven days after receipt of the written report of the Owner's accountants, either issue to the Owner a final Certificate for Payment with a copy to the Contractor, or notify the Contractor and Owner in writing of the Architect's reasons for withholding a certificate as provided in Section 9.5.1 of the AIA Document A201-1997. The time periods stated in this Section 12.2.3 supersede those stated in Section 9.4.1 of the AIA Document A201-1997.

§ **12.2.4** If the Owner's accountants report the Cost of the Work as substantiated by the Contractor's final accounting to be less than claimed by the Contractor, the Contractor shall be entitled to demand arbitration of the disputed amount without a further decision of the Architect. Such demand for arbitration shall be made by the Contractor within 30 days after the Contractor's receipt of a copy of the Architect's final Certificate for Payment; failure to demand arbitration within this 30-day period shall result in the substantiated amount reported by the Owner's accountants becoming binding on the Contractor. Pending a final resolution by arbitration, the Owner shall pay the Contractor the amount certified in the Architect's final Certificate for Payment.

§ **12.2.5** If, subsequent to final payment and at the Owner's request, the Contractor incurs costs described in Article 7 and not excluded by Article 8 to correct defective or nonconforming Work, the Owner shall reimburse the Contractor such costs and the Contractor's Fee applicable thereto on the same basis as if such costs had been incurred prior to final payment, but not in excess of the Guaranteed Maximum Price. If the Contractor has participated in savings as provided in Section 5.2, the amount of such savings shall be recalculated and appropriate credit given to the Owner in determining the net amount to be paid by the Owner to the Contractor.

ARTICLE 13   TERMINATION OR SUSPENSION

§ **13.1** The Contract may be terminated by the Contractor, or by the Owner for convenience, as provided in Article 14 of AIA Document A201-1997. However, the amount to be paid to the Contractor under Section 14.1.3 of AIA Document A201-1997 shall not exceed the amount the Contractor would be entitled to receive under Section 13.2 below, except that the Contractor's Fee shall be calculated as if the Work had been fully completed by the Contractor, including a reasonable estimate of the Cost of the Work for Work not actually completed.

§ **13.2** The Contract may be terminated by the Owner for cause as provided in Article 14 of AIA Document A201-1997. The amount, if any, to be paid to the Contractor under Section 14.2.4 of AIA Document A201-1997 shall not cause the Guaranteed Maximum Price to be exceeded, nor shall it exceed an amount calculated as follows:

§ **13.2.1** Take the Cost of the Work incurred by the Contractor to the date of termination;

§ **13.2.2** Add the Contractor's Fee computed upon the Cost of the Work to the date of termination at the rate stated in Section 5.1.2 or, if the Contractor's Fee is stated as a fixed sum in that Section, an amount that bears the same ratio to that fixed-sum Fee as the Cost of the Work at the time of termination bears to a reasonable estimate of the probable Cost of the Work upon its completion; and

§ **13.2.3** Subtract the aggregate of previous payments made by the Owner.

§ **13.3** The Owner shall also pay the Contractor fair compensation, either by purchase or rental at the election of the Owner, for any equipment owned by the Contractor that the Owner elects to retain and that is not otherwise included in the Cost of the Work under Section 13.2.1. To the extent that the Owner elects to take legal assignment of subcontracts and purchase orders (including rental agreements), the Contractor shall, as a condition of receiving the payments referred to in this Article 13, execute and deliver all such papers and take all such steps, including the legal assignment of such subcontracts and other contractual rights of the Contractor, as the Owner may require for the purpose of fully vesting in the Owner the rights and benefits of the Contractor under such subcontracts or purchase orders.

§ 13.4 The Work may be suspended by the Owner as provided in Article 14 of AIA Document A201-1997; in such case, the Guaranteed Maximum Price and Contract Time shall be increased as provided in Section 14.3.2 of AIA Document A201-1997 except that the term "profit" shall be understood to mean the Contractor's Fee as described in Sections 5.1.2 and Section 6.4 of this Agreement.

ARTICLE 14 MISCELLANEOUS PROVISIONS

§ 14.1 Where reference is made in this Agreement to a provision AIA Document A201-1997 or another Contract Document, the reference refers to that provision as amended or supplemented by other provisions of the Contract Documents.

§ 14.2 Payments due and unpaid under the Contract shall bear interest from the date payment is due at the rate stated below, or in the absence thereof, at the legal rate prevailing from time to time at the place where the Project is located.
(Insert rate of interest agreed upon, if any.)

(Usury laws and requirements under the Federal Truth in Lending Act, similar state and local consumer credit laws and other regulations at the Owner's and Contractor's principal places of business, the location of the Project and elsewhere may affect the validity of this provision. Legal advice should be obtained with respect to deletions or modifications, and also regarding requirements such as written disclosures or waivers.)

§ 14.3 The Owner's representative is:
(Name, address and other information.)

§ 14.4 The Contractor's representative is:
(Name, address and other information.)

§ 14.5 Neither the Owner's nor the Contractor's representative shall be changed without ten days' written notice to the other party.

---

§ **14.6** Other provisions:

ARTICLE 15   ENUMERATION OF CONTRACT DOCUMENTS

§ **15.1** The Contract Documents, except for Modifications issued after execution of this Agreement, are enumerated as follows:

§ **15.1.1** The Agreement is this executed 1997 edition of the Standard Form of Agreement Between Owner and Contractor, AIA Document A111-1997.

§ **15.1.2** The General Conditions are the 1997 edition of the General Conditions of the Contract for Construction, AIA Document A201-1997.

§ **15.1.3** The Supplementary and other Conditions of the Contract are those contained in the Project Manual dated
, and are as follows:

Document	Title	Pages

§ **15.1.4** The Specifications are those contained in the Project Manual dated as in Section 15.1.3, and are as follows:
(Either list the Specifications here or refer to an exhibit attached to this Agreement.)

Section	Title	Pages

§ 15.1.5 The Drawings are as follows, and are dated

unless a different date
is shown below:
(Either list the Drawings here or refer to an exhibit attached to this Agreement.)

Number   Title   Date

§ 15.1.6 The Addenda, if any, are as follows:

Number   Date   Pages

Portions of Addenda relating to bidding requirements are not part of the Contract Documents unless the bidding requirements are also enumerated in this Article 15.

§ 15.1.7 Other Documents, if any, forming part of the Contract Documents are as follows:
(List here any additional documents, such as a list of alternates that are intended to form part of the Contract Documents. AIA Document A201-1997 provides that bidding requirements such as advertisement or invitation to bid, Instructions to Bidders, sample forms and the Contractor's bid are not part of the Contract Documents unless enumerated in this Agreement. They should be listed here only if intended to be part of the Contract Documents.)

582   Appendix K

ARTICLE 16 INSURANCE AND BONDS
(List required limits of liability for insurance and bonds. AIA Document A201-1997 gives other specific requirements for insurance and bonds.)

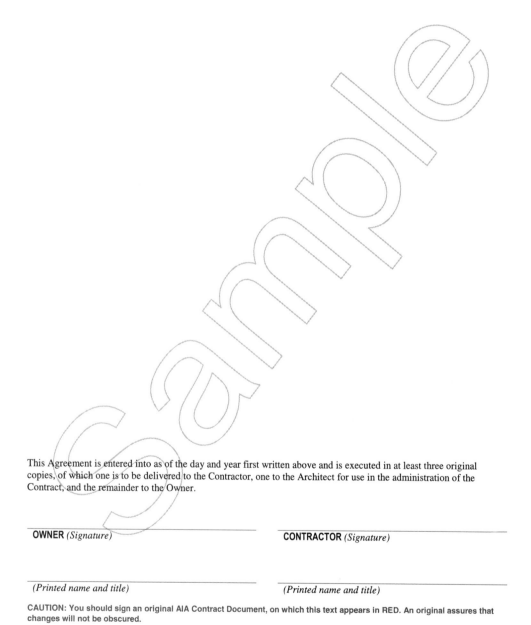

This Agreement is entered into as of the day and year first written above and is executed in at least three original copies, of which one is to be delivered to the Contractor, one to the Architect for use in the administration of the Contract, and the remainder to the Owner.

_____   _____
OWNER *(Signature)*   CONTRACTOR *(Signature)*

_____   _____
(Printed name and title)   *(Printed name and title)*

CAUTION: You should sign an original AIA Contract Document, on which this text appears in RED. An original assures that changes will not be obscured.

13

Appendix L

American Arbitration Association, Construction Industry Arbitration Rules and Mediation Procedures

(Including Procedures for Large, Complex Construction Disputes)

American Arbitration Association
Dispute Resolution Services Worldwide

Construction Industry Arbitration Rules and Mediation Procedures (Including Procedures for Large, Complex Construction Disputes)

Amended and Effective July 1, 2003

TABLE OF CONTENTS

NATIONAL CONSTRUCTION DISPUTE RESOLUTION COMMITTEE

Representatives of the more than twenty organizations listed below constitute the National Construction Dispute Resolution Committee (NCDRC). This committee is the sponsor of the arbitration and mediation procedures specially designed for

> American Association of Airport Executives
> American Bar Association
> American College of Construction Lawyers
> American Consulting Engineers Council
> American Institute of Architects
> American Public Works Association
> American Road and Transportation Builders Association
> American Society of Civil Engineers
> American Subcontractors Association
> Associated Builders & Contractors
> Associated General Contractors
> American Specialty Contractors
> Buildings Future Council
> Construction Specification Institute
> Construction Management Association of America
> Design Build Institute of America
> Engineers Joint Contract Documents Committee
> National Association of Home Builders
> National Association of Minority Contractors
> National Society of Professional Engineers
> National Utility Contractors
> Victor O. Schinnerer
> Women Construction Owners & Executives

IMPORTANT NOTICE

These rules and any amendment of them shall apply in the form in effect at the time the administrative filing requirements are met for a demand for arbitration or submission agreement received by the AAA. To insure that you have the most current information, see our Web site at www.adr.org.

INTRODUCTION

Each year, many thousands of construction transactions take place. Occasionally, disagreements develop over these transactions. Many of these disputes are resolved by arbitration, the voluntary submission of a dispute to a disinterested person or persons for final and binding determination. Arbitration has proven to be an effective way to resolve disputes privately, promptly, and economically.

The American Arbitration Association (AAA) is a public-service, not-for-profit organization offering a broad range of dispute resolution services to business executives, attorneys, individuals, trade associations, unions, management, consumers, families, communities, and all levels of government. Services are available through AAA headquarters in New York City and through offices located in major cities throughout the United States and Europe. Hearings may be held at locations convenient for the parties and are not limited to cities with AAA offices. In addition, the AAA serves as a center for education and training, issues specialized publications, and conducts research on all forms of out-of-court dispute settlement.

Mediation

Because of the increasing popularity of mediation, especially as a prelude to arbitration, the Association has combined its mediation procedures and arbitration rules into a single brochure.

By agreement, the parties may submit their dispute to mediation before arbitration under the mediation procedures in this brochure. Mediation involves the services of one or more individuals, to assist parties in settling a controversy or claim by direct negotiations between or among themselves. The mediator participates impartially in the negotiations, guiding and consulting the various parties involved. The result of the mediation should be an agreement that the parties find acceptable. The mediator cannot impose a settlement, but can only guide the parties toward achieving their own settlement.

The AAA will administer the mediation process to achieve orderly, economical, and expeditious mediation, utilizing to the greatest possible extent the competence and acceptability of the mediators on the AAA's Construction Mediation Panel. Depending on the expertise needed for a given dispute, the parties can obtain the services of one or more individuals who are willing to serve as mediators and who are trained by the AAA in the necessary mediation skills. In identifying those persons most qualified to mediate, the AAA is assisted by the NCDRC.

The AAA itself does not act as mediator. Its function is to administer the mediation process in accordance with the agreement of the parties, to teach mediation skills to members of the construction industry, and to maintain the National Roster from which topflight mediators can be chosen.

There is no additional administrative fee where parties to a pending arbitration attempt to mediate their dispute under the AAA's auspices. Procedures for mediation cases are described in Sections M-1 through M-17.

Arbitration

Regular Track Procedures:

The rules contain Regular Track Procedures which are applied to the administration of all arbitration cases, unless they conflict with any portion of the Fast Track Procedures or the Procedures for Large, Complex Construction Disputes whenever these apply. In the event of a conflict, either the Fast Track procedures or the Large, Complex Construction Disputes procedures apply.

The highlights of the Regular Track Procedures are:

- party input into the AAA's preparation of lists of proposed arbitrators;

- express arbitrator authority to control the discovery process;

- broad arbitrator authority to control the hearing;

- a concise written breakdown of the award and, if requested in writing by all parties prior to the appointment of the arbitrator or at the discretion of the arbitrator, a written explanation of the award;

- arbitrator compensation, with the AAA to provide the arbitrator's compensation policy with the biographical information sent to the parties;

- a demand form and an answer form, both of which seek more information from the parties to assist the AAA in better serving the parties.

Fast Track Procedures:

The Fast Track Procedures were designed for cases involving claims of no more than $75,000. The highlights of this system are:

- a 60-day "time standard" for case completion;

- establishment of a special pool of arbitrators who are pre-qualified to serve on an expedited basis;

- an expedited arbitrator appointment process, with party input;

- presumption that cases involving $10,000 or less will be decided on a documents only basis;

- requirement of a hearing within 30 calendar days of the arbitrator's appointment;

- a single day of hearing in most cases;

- an award in no more than 14 calendar days after completion of the hearing.

Procedures for Large, Complex Construction Disputes:

Unless the parties agree otherwise, the Procedures for Large, Complex Construction Disputes, which appear in this pamphlet, will be applied to all cases administered by the AAA under the Construction Arbitration Rules in which the disclosed claim or counterclaim of any party is at least $500,000 exclusive of claimed interest, arbitration fees and costs.

The key features of these procedures include:

- mandatory use of the procedures in cases involving claims of $500,000 or more;

- a mandatory preliminary hearing with the arbitrators, which may be conducted by telephone;

- broad arbitrator authority to order and control discovery, including depositions;
- presumption that hearings will proceed on a consecutive or block basis.

The National Roster

The AAA has established and maintains as members of its National Roster individuals competent to hear and decide disputes administered under the Construction Industry Arbitration Rules. The AAA considers for appointment to the construction industry roster persons recommended by Regional Panel Advisory Committees as qualified to serve by virtue of their experience in the construction field. The majority of neutrals are actively engaged in the construction industry. Attorney neutrals generally devote at least half of their practice to construction matters. Neutrals serving under these rules must also attend periodic training.

The services of the AAA are generally concluded with the transmittal of the award. Although there is voluntary compliance with the majority of awards, judgment on the award can be entered in a court having appropriate jurisdiction if necessary.

Administrative Fees

The AAA charges a filing fee based on the amount of claim or counterclaim. This fee information, which is contained with these rules, allows the parties to exercise control over their administrative fees.

The fees cover AAA administrative services; they do not cover arbitrator compensation or expenses, if any, reporting services, hearing room rental or any post-award charges incurred by the parties in enforcing the award.

ADR Clauses

Mediation

If the parties elect to adopt mediation as a part of their contractual dispute settlement procedure, they can insert the following mediation clause into their contract in conjunction with a standard arbitration provision.

> *If a dispute arises out of or relates to this contract, or the breach thereof, and if the dispute cannot be settled through negotiation, the parties agree first to try in good faith to settle the dispute by mediation administered by the American Arbitration Association under its Construction Industry Mediation Procedures before resorting to arbitration, litigation, or some other dispute resolution procedure.*

If the parties choose to use a mediator to resolve an existing dispute, they can enter into the following submission.

The parties hereby submit the following dispute to mediation administered by the American Arbitration Association under its Construction Industry Mediation Procedures. (The clause may also provide for the qualifications of the mediator(s), method of payment, locale of meetings, and any other item of concern to the parties.)

Arbitration

When an agreement to arbitrate is included in a construction contract, it might expedite peaceful settlement without the necessity of going to arbitration at all. Thus, an arbitration clause is a form of insurance against loss of good will. The parties can provide for arbitration of future disputes by inserting the following clause into their contracts.

Any controversy or claim arising out of or relating to this contract, or the breach thereof, shall be settled by arbitration administered by the American Arbitration Association under its Construction Industry Arbitration Rules, and judgment on the award rendered by the arbitrator(s) may be entered in any court having jurisdiction thereof.

Arbitration of existing disputes may be accomplished by use of the following.

We, the undersigned parties, hereby agree to submit to arbitration administered by the American Arbitration Association under its Construction Industry Arbitration Rules the following controversy: (cite briefly). We further agree that the above controversy be submitted to (one)(three) arbitrator(s). We further agree that we will faithfully observe this agreement and the rules, that we will abide by and perform any award rendered by the arbitrator(s), and that a judgment of the court having jurisdiction may be entered on the award.

For further information about the AAA's Construction Dispute Avoidance and Resolution Services, as well as the full range of other AAA services, contact the nearest AAA office or visit our website at www.adr.org.

CONSTRUCTION INDUSTRY MEDIATION PROCEDURES

M-1. Agreement of Parties

Whenever, by stipulation or in their contract, the parties have provided for mediation or conciliation of existing or future disputes under the auspices of the American Arbitration Association (AAA) or under these procedures, they shall be deemed to have made these procedures, as amended and in effect as of the date of the submission of the dispute, a part of their agreement.

M-2. Initiation of Mediation

Any party or parties to a dispute may initiate mediation by filing with the AAA a submission to mediation or a written request for mediation pursuant to these procedures, together with the

$325 nonrefundable case set-up fee. Where there is no submission to mediation or contract providing for mediation, a party may request the AAA to invite another party to join in a submission to mediation. Upon receipt of such a request, the AAA will contact the other parties involved in the dispute and attempt to obtain a submission to mediation.

M-3. Request for Mediation

A request for mediation shall contain a brief statement of the nature of the dispute and the names, addresses, and telephone numbers of all parties to the dispute and those who will represent them, if any, in the mediation. The initiating party shall simultaneously file two copies of the request with the AAA and one copy with every other party to the dispute.

M-4. Appointment of Mediator

Upon receipt of a request for mediation, the AAA will appoint a qualified mediator to serve. Normally, a single mediator will be appointed unless the parties agree otherwise or the AAA determines otherwise. If the agreement of the parties names a mediator or specifies a method of appointing a mediator, that designation or method shall be followed.

M-5. Qualifications of Mediator

No person shall serve as a mediator in any dispute in which that person has any financial or personal interest in the result of the mediation, except by the written consent of all parties. Prior to accepting an appointment, the prospective mediator shall disclose any circumstance likely to create a presumption of bias or prevent a prompt meeting with the parties. Upon receipt of such information, the AAA shall either replace the mediator or immediately communicate the information to the parties for their comments. In the event that the parties disagree as to whether the mediator shall serve, the AAA will appoint another mediator.

The AAA is authorized to appoint another mediator if the appointed mediator is unable to serve promptly.

M-6. Vacancies

If any mediator shall become unwilling or unable to serve, the AAA will appoint another mediator, unless the parties agree otherwise.

M-7. Representation

Any party may be represented by persons of the party's choice. The names and addresses of such persons shall be communicated in writing to all parties and to the AAA.

M-8. Date, Time, and Place of Mediation

The mediator shall fix the date and the time of each mediation session. The mediation shall be held at the appropriate regional office of the AAA, or at any other convenient location agreeable to the mediator and the parties, as the mediator shall determine. By agreement of all parties, mediation may be conducted by telephone or other means of electronic communication.

M-9. Identification of Matters in Dispute

At least ten calendar days prior to the first scheduled mediation session, each party shall provide the mediator with a brief memorandum setting forth its position with regard to the issues that need to be resolved. At the discretion of the mediator, such memoranda may be mutually exchanged by the parties.

At the first session, the parties will be expected to produce all information reasonably required for the mediator to understand the issues presented.

The mediator may require any party to supplement such information.

M-10. Authority of Mediator

The mediator does not have the authority to impose a settlement on the parties but will attempt to help them reach a satisfactory resolution of their dispute. The mediator is authorized to conduct joint and separate meetings with the parties and to make oral and written recommendations for settlement. Whenever necessary, the mediator may also obtain expert advice concerning technical aspects of the dispute, provided that the parties agree and assume the expenses of obtaining such advice. Arrangements for obtaining such advice shall be made by the mediator or the parties, as the mediator shall determine.

The mediator is authorized to end the mediation whenever, in the judgment of the mediator, further efforts at mediation would not contribute to a resolution of the dispute between the parties.

M-11. Privacy

Mediation sessions are private. The parties and their representatives may attend mediation sessions. Other persons may attend only with the permission of the parties and with the consent of the mediator.

M-12. Confidentiality

Confidential information disclosed to a mediator by the parties or by witnesses in the course of the mediation shall not be divulged by the mediator. All records, reports, or other documents received by a mediator while serving in that capacity shall be confidential. The mediator shall not be compelled to divulge such records or to testify in regard to the mediation in any adversary proceeding or judicial forum.

The parties shall maintain the confidentiality of the mediation and shall not rely on, or introduce as evidence in any arbitral, judicial, or other proceeding:

(a)   views expressed or suggestions made by another party with respect to a possible settlement of the dispute;

(b)   admissions made by another party in the course of the mediation proceedings;

(c)   proposals made or views expressed by the mediator; or

(d)   the fact that another party had or had not indicated willingness to accept a proposal for settlement made by the mediator.

M-13. No Stenographic Record

There shall be no stenographic record of the mediation process.

M-14. Termination of Mediation

The mediation shall be terminated:

(a)   by the execution of a settlement agreement by the parties;

(b)   by a written declaration of the mediator to the effect that further efforts at mediation are no longer worthwhile; or

(c)   by a written statement of all the parties to the effect that the mediation proceedings are terminated.

M-15. Exclusion of Liability

Neither the AAA nor any mediator is a necessary party in judicial proceedings relating to the mediation.

Neither the AAA nor any mediator shall be liable to any party for any act or omission in connection with any mediation conducted under these procedures.

M-16. Interpretation and Application of Procedures

The mediator shall interpret and apply these procedures insofar as they relate to the mediator's duties and responsibilities. All other procedures shall be interpreted and applied by the AAA.

M-17. Expenses

The expenses of witnesses for either side shall be paid by the party producing such witnesses. All other expenses of the mediation, including required traveling and other expenses of the mediator and representatives of the AAA, and the expenses of any witness and the cost of any proofs or expert advice produced at the direct request of the mediator, shall be borne equally by the parties unless they agree otherwise.

Administrative Fees for Mediation

The nonrefundable case set-up fee is $325 per party. In addition, the parties are responsible for compensating the mediator at his or her published rate, for conference and study time (hourly or per diem).

All expenses are generally borne equally by the parties. The parties may adjust this arrangement by agreement.

Before the commencement of the mediation, the AAA shall estimate anticipated total expenses. Each party shall pay its portion of that amount as per the agreed upon arrangement. When the mediation has terminated, the AAA shall render an accounting and return any unexpended balance to the parties.

CONSTRUCTION INDUSTRY ARBITRATION RULES

REGULAR TRACK PROCEDURES

R-1. Agreement of Parties

(a)   The parties shall be deemed to have made these rules a part of their arbitration agreement whenever they have provided for arbitration by the American Arbitration Association (hereinafter AAA) under its Construction Industry Arbitration Rules. These rules and any amendment of them shall apply in the form in effect at the time the administrative requirements are met for a demand for arbitration or submission agreement received by the AAA. The parties, by written agreement, may vary the procedures set forth in these rules. After appointment of the arbitrator, such modifications may be made only with the consent of the arbitrator.

(b)   Unless the parties or the AAA determines otherwise, the Fast Track Procedures shall apply in any case in which no disclosed claim or counterclaim exceeds $75,000, exclusive of interest and arbitration fees and costs. Parties may also agree to use these procedures in larger cases. Unless the parties agree otherwise, these procedures will not apply in cases involving more than two parties. The Fact Track Procedures shall be applied as described in Sections F-1 through F-13 of these rules, in addition to any other portion of these rules that is not in conflict with the Fact Track Procedures.

(c)   Unless the parties agree otherwise, the Procedures for Large, Complex Construction Disputes shall apply to all cases in which the disclosed claim or counterclaim of any party is at least $500,000, exclusive of claimed interest, arbitration fees and costs. Parties may also agree to use these procedures in cases involving claims or counterclaims under

$500,000, or in nonmonetary cases. The Procedures for Large, Complex Construction Disputes shall be applied as described in Sections L-1 through L-4 of these rules, in addition to any other portion of these rules that is not in conflict with the Procedures for Large, Complex Construction Disputes.

(d)   All other cases shall be administered in accordance with Sections R-1 through R-55 of these rules.

R-2. AAA and Delegation of Duties

When parties agree to arbitrate under these rules, or when they provide for arbitration by the AAA and an arbitration is initiated under these rules, they thereby authorize the AAA to administer the arbitration. The authority and duties of the AAA are prescribed in the agreement of the parties and in these rules, and may be carried out through such of the AAA's representatives as it may direct. The AAA may, in its discretion, assign the administration of an arbitration to any of its offices.

R-3. National Roster of Neutrals

In cooperation with the National Construction Dispute Resolution Committee the AAA shall establish and maintain a National Roster of Construction Arbitrators ("National Roster") and shall appoint arbitrators as provided in these rules. The term "arbitrator" in these rules refers to the arbitration panel, constituted for a particular case, whether composed of one or more arbitrators, or to an individual arbitrator, as the context requires.

R-4. Initiation under an Arbitration Provision in a Contract

(a)   Arbitration under an arbitration provision in a contract shall be initiated in the following manner.

   (i)   The initiating party (the "claimant") shall, within the time period, if any, specified in the contract(s), give to the other party (the "respondent") written notice of its intention to arbitrate (the "demand"), which demand shall contain a statement setting forth the nature of the dispute, the names and addresses of all other parties, the amount involved, if any, the remedy sought, and the hearing locale requested.

   (ii)   The claimant shall file at any office of the AAA two copies of the demand and two copies of the arbitration provisions of the contract, together with the appropriate filing fee as provided in the schedule included with these rules.

   (iii)   The AAA shall confirm notice of such filing to the parties.

(b)   A respondent may file an answering statement in duplicate with the AAA within 15 calendar days after confirmation of notice of filing of the demand is sent by the AAA. The respondent shall, at the time of any such filing, send a copy of the answering statement to the claimant. If a counterclaim is asserted, it shall contain a statement setting forth the nature of the counterclaim, the amount involved, if any, and the remedy sought. If a counterclaim is made, the party making the counterclaim shall forward to the AAA with the answering statement the appropriate fee provided in the schedule included with these rules.

(c)   If no answering statement is filed within the stated time, respondent will be deemed to deny the claim. Failure to file an answering statement shall not operate to delay the arbitration.

(d)   When filing any statement pursuant to this section, the parties are encouraged to provide descriptions of their claims in sufficient detail to make the circumstances of the dispute clear to the arbitrator.

R-5. Initiation under a Submission

Parties to any existing dispute may commence an arbitration under these rules by filing at any office of the AAA two copies of a written submission to arbitrate under these rules, signed by the parties. It shall contain a statement of the matter in dispute, the names and addresses of the parties, any claims and counterclaims, the amount involved, if any, the remedy sought, and the hearing locale requested, together with the appropriate filing fee as provided in the schedule included with these rules. Unless the parties state otherwise in the submission, all claims and counterclaims will be deemed to be denied by the other party.

R-6. Changes of Claim

A party may at any time prior to the close of the hearing increase or decrease the amount of its claim or counterclaim. Any new or different claim or counterclaim, as opposed to an increase or decrease in the amount of a pending claim or counterclaim, shall be made in writing and filed with the AAA, and a copy shall be mailed to the other party, who shall have a period of ten calendar days from the date of such mailing within which to file an answer with the AAA.

After the arbitrator is appointed no new or different claim or counterclaim may be submitted to the arbitrator except with the arbitrator's consent.

R-7. Consolidation or Joinder

If the parties' agreement or the law provides for consolidation or joinder of related arbitrations, all involved parties will endeavor to agree on a process to effectuate the consolidation or joinder. If they are unable to agree, the Association shall directly appoint a single arbitrator for the limited purpose of deciding whether related arbitrations should be consolidated or joined and, if so, establishing a fair and appropriate process for consolidation or joinder. The AAA may take reasonable administrative action to accomplish the consolidation or joinder as directed by the arbitrator.

R-8. Jurisdiction

(a)   The arbitrator shall have the power to rule on his or her own jurisdiction, including any objections with respect to the existence, scope or validity of the arbitration agreement.

(b)   The arbitrator shall have the power to determine the existence or validity of a contract of which an arbitration clause forms a part. Such an arbitration clause shall be treated as an agreement independent of the other terms of the contract. A decision by the arbitrator that the contract is null and void shall not for that reason alone render invalid the arbitration clause.

(c)   A party must object to the jurisdiction of the arbitrator or to the arbitrability of a claim or counterclaim no later than the filing of the answering statement to the claim or counterclaim that gives rise to the objection. The arbitrator may rule on such objections as a preliminary matter or as part of the final award.

R-9. Mediation

At any stage of the proceedings, the parties may agree to conduct a mediation conference under the Construction Industry Mediation Procedures in order to facilitate settlement. The mediator shall not be an arbitrator appointed to the case. Where the parties to a pending arbitration agree to mediate under the AAA's rules, no additional administrative fee is required to initiate the mediation.

R-10. Administrative Conference

At the request of any party or upon the AAA's own initiative, the AAA may conduct an administrative conference, in person or by telephone, with the parties and/or their representatives. The conference may address such issues as arbitrator selection, potential mediation of the dispute, potential exchange of information, a timetable for hearings and any other administrative matters.

R-11. Fixing of Locale

The parties may mutually agree on the locale where the arbitration is to be held. If any party requests that the hearing be held in a specific locale and the other party files no objection thereto within fifteen calendar days after notice of the request has been sent to it by the AAA, the locale shall be the one requested. If a party objects to the locale requested by the other party, the AAA shall have the power to determine the locale, and its decision shall be final and binding.

R-12. Appointment from National Roster

If the parties have not appointed an arbitrator and have not provided any other method of appointment, the arbitrator shall be appointed in the following manner:

(a) Immediately after the filing of the submission or the answering statement or the expiration of the time within which the answering statement is to be filed, the AAA shall send simultaneously to each party to the dispute an identical list of 10 (unless the AAA decides that a different number is appropriate) names of persons chosen from the National Roster, unless the AAA decides that a different number is appropriate. The parties are encouraged to agree to an arbitrator from the submitted list and to advise the AAA of their agreement. Absent agreement of the parties, the arbitrator shall not have served as the mediator in the mediation phase of the instant proceeding.

(b) If the parties are unable to agree upon an arbitrator, each party to the dispute shall have 15 calendar days from the transmittal date in which to strike names objected to, number the remaining names in order of preference, and return the list to the AAA. If a party does not return the list within the time specified, all persons named therein shall be deemed acceptable. From among the persons who have been approved on both lists, and in accordance with the designated order of mutual preference, the AAA shall invite the acceptance of an arbitrator to serve. If the parties fail to agree on any of the persons named, or if acceptable arbitrators are unable to act, or if for any other reason the appointment cannot be made from the submitted lists, the AAA shall have the power to make the appointment from among other members of the National Roster without the submission of additional lists.

(c) Unless the parties agree otherwise when there are two or more claimants or two or more respondents, the AAA may appoint all the arbitrators.

R-13. Direct Appointment by a Party

(a) If the agreement of the parties names an arbitrator or specifies a method of appointing an arbitrator, that designation or method shall be followed. The notice of appointment, with the name and address of the arbitrator, shall be filed with the AAA by the appointing party. Upon the request of any appointing party, the AAA shall submit a list of members of the National Roster from which the party may, if it so desires, make the appointment.

(b) Where the parties have agreed that each party is to name one arbitrator, the arbitrators so named must meet the standards of Section R-18 with respect to impartiality and independence unless the parties have specifically agreed pursuant to Section R-18(a) that the party-appointed arbitrators are to be non-neutral and need not meet those standards.

(c) If the agreement specifies a period of time within which an arbitrator shall be appointed and any party fails to make the appointment within that period, the AAA shall make the appointment.

(d) If no period of time is specified in the agreement, the AAA shall notify the party to make the appointment. If within 15 calendar days after such notice has been sent, an arbitrator has not been appointed by a party, the AAA shall make the appointment.

R-14. Appointment of Chairperson by Party-Appointed Arbitrators or Parties

(a) If, pursuant to Section R-13, either the parties have directly appointed arbitrators, or the arbitrators have been appointed by AAA, and the parties have authorized them to appoint a chairperson within a specified time and no appointment is made within that time or any agreed extension, the AAA may appoint the chairperson.

(b) If no period of time is specified for appointment of the chairperson and the party-appointed arbitrators or the parties do not make the appointment within 15 calendar days from the date of the appointment of the last party-appointed arbitrator, the AAA may appoint the chairperson.

(c) If the parties have agreed that their party-appointed arbitrators shall appoint the chairperson from the National Roster, the AAA shall furnish to the party-appointed arbitrators, in the manner provided in Section R-12, a list selected from the National Roster, and the appointment of the chairperson shall be made as provided in that Section.

R-15. Nationality of Arbitrator in International Arbitration

Where the parties are nationals of different countries, the AAA, at the request of any party or on its own initiative, may appoint as arbitrator a national of a country other than that of any of the parties. The request must be made before the time set for the appointment of the arbitrator as agreed by the parties or set by these rules.

R-16. Number of Arbitrators

If the arbitration agreement does not specify the number of arbitrators, the dispute shall be heard and determined by one arbitrator, unless the AAA, in its discretion, directs that three arbitrators be appointed. A party may request three arbitrators in the demand or answer, which request the AAA will consider in exercising its discretion regarding the number of arbitrators appointed to the dispute.

R-17. Disclosure

(a) Any person appointed or to be appointed as an arbitrator shall disclose to the AAA any circumstance likely to give rise to justifiable doubt as to the arbitrator's impartiality or independence, including any bias or any financial or personal interest in the result of the arbitration or any past or present relationship with the parties or their representatives. Such obligation shall remain in effect throughout the arbitration.

(b) Upon receipt of such information from the arbitrator or another source, the AAA shall communicate the information to the parties and, if it deems it appropriate to do so, to the arbitrator and others.

(c)   In order to encourage disclosure by arbitrators, disclosure of information pursuant to this Section R-17 is not to be construed as an indication that the arbitrator considers that the disclosed circumstances is likely to affect impartiality or independence.

R-18. Disqualification of Arbitrator

(a)   Any arbitrator shall be impartial and independent and shall perform his or her duties with diligence and in good faith, and shall be subject to disqualification for

    (i)   partiality or lack of independence,

    (ii)   inability or refusal to perform his or her duties with diligence and in good faith, and

    (iii)   any grounds for disqualification provided by applicable law. The parties may agree in writing, however, that arbitrators directly appointed by a party pursuant to Section R-13 shall be non-neutral, in which case such arbitrators need not be impartial or independent and shall not be subject to disqualification for partiality or lack of independence.

(b)   Upon objection of a party to the continued service of an arbitrator, or on its own initiative, the AAA shall determine whether the arbitrator should be disqualified under the grounds set out above, and shall inform the parties of its decision, which decision shall be conclusive.

R-19. Communication with Arbitrator

(a)   No party and no one acting on behalf of any party shall communicate ex parte with an arbitrator or a candidate for arbitrator concerning the arbitration, except that a party, or someone acting on behalf of a party, may communicate ex parte with a candidate for direct appointment pursuant to Section R-13 in order to advise the candidate of the general nature of the controversy and of the anticipated proceedings and to discuss the candidate's qualifications, availability or independence in relation to the parties or to discuss the suitability of the candidates for selection as a third arbitrator where the parties or party-designated arbitrators are to participate in that selection.

(b)   Section R-19(a) does not apply to arbitrators directly appointed by the parties who, pursuant to Section R-18(a), the parties have agreed in writing are non-neutral. Where the parties have so agreed under Section R-18(a), the AAA shall as an administrative practice suggest to the parties that they agree further that Section R-19(a) should nonetheless apply prospectively.

R-20. Vacancies

(a)   If for any reason an arbitrator is unable to perform the duties of the office, the AAA may, on proof satisfactory to it, declare the office vacant. Vacancies shall be filled in accordance with the applicable provisions of these rules.

(b)   In the event of a vacancy in a panel of neutral arbitrators after the hearings have commenced, the remaining arbitrator or arbitrators may continue with the hearing and determination of the controversy, unless the parties agree otherwise.

(c)   In the event of the appointment of a substitute arbitrator, the panel of arbitrators shall determine in its sole discretion whether it is necessary to repeat all or part of any prior hearings.

R-21. Preliminary Hearing

(a)   At the request of any party or at the discretion of the arbitrator or the AAA, the arbitrator may schedule as soon as practicable a preliminary hearing with the parties and/or their representatives. The preliminary hearing may be conducted by telephone at the arbitrator's discretion.

(b)   During the preliminary hearing, the parties and the arbitrator should discuss the future conduct of the case, including clarification of the issues and claims, a schedule for the hearings and any other preliminary matters.

R-22. Exchange of Information

(a)   At the request of any party or at the discretion of the arbitrator, consistent with the expedited nature of arbitration, the arbitrator may direct

   (i) the production of documents and other information, and

   (ii) the identification of any witnesses to be called.

(b)   At least five business days prior to the hearing, the parties shall exchange copies of all exhibits they intend to submit at the hearing.

(c)   The arbitrator is authorized to resolve any disputes concerning the exchange of information.

(d)   There shall be no other discovery, except as indicated herein or as ordered by the arbitrator in extraordinary cases when the demands of justice require it.

R-23. Date, Time, and Place of Hearing

The arbitrator shall set the date, time, and place for each hearing and/or conference. The parties shall respond to requests for hearing dates in a timely manner, be cooperative in scheduling the earliest practicable date, and adhere to the established hearing schedule. The AAA shall send a notice of hearing to the parties at least ten calendar days in advance of the hearing date, unless otherwise agreed by the parties.

R-24. Attendance at Hearings

The arbitrator and the AAA shall maintain the privacy of the hearings unless the law provides to the contrary. Any person having a direct interest in the arbitration is entitled to attend hearings. The arbitrator shall otherwise have the power to require the exclusion of any witness, other than a party or other essential person, during the testimony of any other witness. It shall be discretionary with the arbitrator to determine the propriety of the attendance of any person other than a party and its representative.

R-25. Representation

Any party may be represented by counsel or other authorized representative. A party intending to be so represented shall notify the other party and the AAA of the name and address of the representative at least three calendar days prior to the date set for the hearing at which that person is first to appear. When such a representative initiates an arbitration or responds for a party, notice is deemed to have been given.

R-26. Oaths

Before proceeding with the first hearing, each arbitrator may take an oath of office and, if required by law, shall do so. The arbitrator may require witnesses to testify under oath administered by any duly qualified person and, if it is required by law or requested by any party, shall do so.

R-27. Stenographic Record

Any party desiring a stenographic record shall make arrangements directly with a stenographer and shall notify the other parties of these arrangements at least three days in advance of the hearing. The requesting party or parties shall pay the cost of the record. If the transcript is agreed by the parties, or determined by the arbitrator to be the official record of the proceeding, it must be provided to the arbitrator and made available to the other parties for inspection, at a date, time, and place determined by the arbitrator.

R-28. Interpreters

Any party wishing an interpreter shall make all arrangements directly with the interpreter and shall assume the costs of the service.

R-29. Postponements

The arbitrator for good cause shown may postpone any hearing upon agreement of the parties, upon request of a party, or upon the arbitrator's own initiative.

R-30. Arbitration in the Absence of a Party or Representative

Unless the law provides to the contrary, the arbitration may proceed in the absence of any party or representative who, after due notice, fails to be present or fails to obtain a postponement. An award shall not be made solely on the default of a party. The arbitrator shall require the party who is present to submit such evidence as the arbitrator may require for the making of an award.

R-31. Conduct of Proceedings

(a)   The claimant shall present evidence to support its claim. The respondent shall then present evidence supporting its defense. Witnesses for each party shall also submit to questions from the arbitrator and the adverse party. The arbitrator has the discretion to vary this procedure, provided that the parties are treated with equality and that each party has the right to be heard and is given a fair opportunity to present its case.

(b)   The arbitrator, exercising his or her discretion, shall conduct the proceedings with a view to expediting the resolution of the dispute and may direct the order of proof, bifurcate proceedings, and direct the parties to focus their presentations on issues the decision of which could dispose of all or part of the case. The arbitrator shall entertain motions, including motions that dispose of all or part of a claim, or that may expedite the proceedings, and may also make preliminary rulings and enter interlocutory orders.

(c)   The parties may agree to waive oral hearings in any case.

R-32. Evidence

(a)   The parties may offer such evidence as is relevant and material to the dispute and shall produce such evidence as the arbitrator may deem necessary to an understanding and determination of the dispute. Conformity to legal rules of evidence shall not be necessary.

(b)   The arbitrator shall determine the admissibility, relevance, and materiality of the evidence offered. The arbitrator may request offers of proof and may reject evidence deemed by the arbitrator to be cumulative, unreliable, unnecessary, or of slight value compared to the time and expense involved. All evidence shall be taken in the presence of all of the arbitrators and all of the parties, except where: 1) any of the parties is absent, in default, or has waived the right to be present, or 2) the parties and the arbitrators agree otherwise.

(c)   The arbitrator shall take into account applicable principles of legal privilege, such as those involving the confidentiality of communications between a lawyer and client.

(d)   An arbitrator or other person authorized by law to subpoena witnesses or documents may do so upon the request of any party or independently.

R-33. Evidence by Affidavit and Post-hearing Filing of Documents or Other Evidence

(a)   The arbitrator may receive and consider the evidence of witnesses by declaration or affidavit, but shall give it only such weight as the arbitrator deems it entitled to after consideration of any objection made to its admission.

(b)   If the parties agree or the arbitrator directs that documents or other evidence be submitted to the arbitrator after the hearing, the documents or other evidence, unless otherwise agreed by the parties and the arbitrator, shall be filed with the AAA for transmission to the arbitrator. All parties shall be afforded an opportunity to examine and respond to such documents or other evidence.

R-34. Inspection or Investigation

An arbitrator finding it necessary to make an inspection or investigation in connection with the arbitration shall direct the AAA to so advise the parties. The arbitrator shall set the date and time and the AAA shall notify the parties. Any party who so desires may be present at such an inspection or investigation. In the event that one or all parties are not present at the inspection or investigation, the arbitrator shall make an oral or written report to the parties and afford them an opportunity to comment.

R-35. Interim Measures

(a)   The arbitrator may take whatever interim measures he or she deems necessary, including injunctive relief and measures for the protection or conservation of property and disposition of perishable goods.

(b)   Such interim measures may be taken in the form of an interim award, and the arbitrator may require security for the costs of such measures.

(c)   A request for interim measures addressed by a party to a judicial authority shall not be deemed incompatible with the agreement to arbitrate or a waiver of the right to arbitrate.

R-36. Closing of Hearing

When satisfied that the presentation of the parties is complete, the arbitrator shall declare the hearing closed.

If documents or responses are to be filed as provided in Section R-33, or if briefs are to be filed, the hearing shall be declared closed as of the final date set by the arbitrator for the receipt of documents, responses, or briefs. The time limit within which the arbitrator is required to make the award shall commence to run, in the absence of other agreements by the parties and the arbitrator, upon the closing of the hearing.

R-37. Reopening of Hearing

The hearing may be reopened on the arbitrator's initiative, or by direction of the arbitrator upon application of a party, at any time before the award is made. If reopening the hearing would prevent the making of the award within the specific time agreed to by the parties in the arbitration agreement, the matter may not be reopened unless the parties agree to an extension of time. When no specific date is fixed by agreement of the parties, the arbitrator shall have 30 calendar days from the closing of the reopened hearing within which to make an award.

R-38. Waiver of Rules

Any party who proceeds with the arbitration after knowledge that any provision or requirement of these rules has not been complied with and who fails to state an objection in writing shall be deemed to have waived the right to object.

R-39. Extensions of Time

The parties may modify any period of time by mutual agreement. The AAA or the arbitrator may for good cause extend any period of time established by these rules, except the time for making the award. The AAA shall notify the parties of any extension.

R-40. Serving of Notice

(a)   Any papers, notices, or process necessary or proper for the initiation or continuation of an arbitration under these rules; for any court action in connection therewith, or for the entry of judgment on any award made under these rules, may be served on a party by mail addressed to the party or its representative at the last known address or by personal service, in or outside the state where the arbitration is to be held, provided that reasonable opportunity to be heard with regard thereto has been granted to the party.

(b)   The AAA, the arbitrator and the parties may also use overnight delivery or electronic facsimile transmission (fax) to give the notices required by these rules. Where all parties and the arbitrator agree, notices may be transmitted by electronic mail (email), or other methods of communication.

(c)   Unless otherwise instructed by the AAA or by the arbitrator, any documents submitted by any party to the AAA or to the arbitrator shall simultaneously be provided to the other party or parties to the arbitration.

R-41. Majority Decision

When the panel consists of more than one arbitrator, unless required by law or by the arbitration agreement, a majority of the arbitrators must make all decisions.

R-42. Time of Award

The award shall be made promptly by the arbitrator and, unless otherwise agreed by the parties or specified by law, no later than 30 calendar days from the date of closing the hearing, or, if oral hearings have been waived, from the date of the AAA's transmittal of the final statements and proofs to the arbitrator.

R-43. Form of Award

(a)   Any award shall be in writing and signed by a majority of the arbitrators. It shall be executed in the manner required by law.

(b)   The arbitrator shall provide a concise, written breakdown of the award. If requested in writing by all parties prior to the appointment of the arbitrator, or if the arbitrator believes it is appropriate to do so, the arbitrator shall provide a written explanation of the award.

R-44. Scope of Award

(a)   The arbitrator may grant any remedy or relief that the arbitrator deems just and equitable and within the scope of the agreement of the parties, including, but not limited to, equitable relief and specific performance of a contract.

(b)   In addition to the final award, the arbitrator may make other decisions, including interim, interlocutory, or partial rulings, orders, and awards. In any interim, interlocutory, or partial award, the arbitrator may assess and apportion the fees, expenses, and compensation related to such award as the arbitrator determines is appropriate.

(c)   In the final award, the arbitrator shall assess fees, expenses, and compensation as provided in Sections R-50, R-51, and R-52. The arbitrator may apportion such fees, expenses, and compensation among the parties in such amounts as the arbitrator determines is appropriate.

(d)   The award of the arbitrator may include interest at such rate and from such date as the arbitrator may deem appropriate; and an award of attorneys' fees if all parties have requested such an award or it is authorized by law or their arbitration agreement.

R-45. Award upon Settlement

If the parties settle their dispute during the course of the arbitration and if the parties so request, the arbitrator may set forth the terms of the settlement in a "consent award." A consent award must include an allocation of arbitration costs, including administrative fees and expenses as well as arbitrator fees and expenses.

R-46. Delivery of Award to Parties

Parties shall accept as notice and delivery of the award the placing of the award or a true copy thereof in the mail addressed to the parties or their representatives at the last known address, personal or electronic service of the award, or the filing of the award in any other manner that is permitted by law.

R-47. Modification of Award

Within twenty calendar days after the transmittal of an award, the arbitrator on his or her initiative, or any party, upon notice to the other parties, may request that the arbitrator correct any clerical, typographical, technical or computational errors in the award. The arbitrator is not empowered to redetermine the merits of any claim already decided.

If the modification request is made by a party, the other parties shall be given ten calendar days to respond to the request. The arbitrator shall dispose of the request within twenty calendar days after transmittal by the AAA to the arbitrator of the request and any response thereto.

If applicable law provides a different procedural time frame, that procedure shall be followed.

R-48. Release of Documents for Judicial Proceedings

The AAA shall, upon the written request of a party, furnish to the party, at its expense, certified copies of any papers in the AAA's possession that may be required in judicial proceedings relating to the arbitration.

R-49. Applications to Court and Exclusion of Liability

(a) No judicial proceeding by a party relating to the subject matter of the arbitration shall be deemed a waiver of the party's right to arbitrate.

(b) Neither the AAA nor any arbitrator in a proceeding under these rules is a necessary or proper party in judicial proceedings relating to the arbitration.

(c) Parties to these rules shall be deemed to have consented that judgment upon the arbitration award may be entered in any federal or state court having jurisdiction thereof.

(d) Parties to an arbitration under these rules shall be deemed to have consented that neither the AAA nor any arbitrator shall be liable to any party in any action for damages or injunctive relief for any act or omission in connection with any arbitration under these rules.

R-50. Administrative Fees

As a not-for-profit organization, the AAA shall prescribe filing and other administrative fees and service charges to compensate it for the cost of providing administrative services. The fees in effect when the fee or charge is incurred shall be applicable.

The filing fee shall be advanced by the party or parties, subject to final apportionment by the arbitrator in the award.

The AAA may, in the event of extreme hardship on the part of any party, defer or reduce the administrative fees.

R-51. Expenses

The expenses of witnesses for either side shall be paid by the party producing such witnesses. All other expenses of the arbitration, including required travel and other expenses of the arbitrator, AAA representatives, and any witness and the cost of any proof produced at the direct request of the arbitrator, shall be borne equally by the parties, unless they agree otherwise or unless the arbitrator in the award assesses such expenses or any part thereof against any specified party or parties.

R-52. Neutral Arbitrator's Compensation

Arbitrators shall be compensated a rate consistent with the arbitrator's stated rate of compensation.

If there is disagreement concerning the terms of compensation, an appropriate rate shall be established with the arbitrator by the Association and confirmed to the parties.

Any arrangement for the compensation of a neutral arbitrator shall be made through the AAA and not directly between the parties and the arbitrator.

R-53. Deposits

The AAA may require the parties to deposit in advance of any hearings such sums of money as it deems necessary to cover the expense of the arbitration, including the arbitrator's fee, if any, and shall render an accounting to the parties and return any unexpended balance at the conclusion of the case.

R-54. Interpretation and Application of Rules

The arbitrator shall interpret and apply these rules insofar as they relate to the arbitrator's powers and duties. When there is more than one arbitrator and a difference arises among them concerning the meaning or application of these rules, it shall be decided by a majority vote. If that is not possible, either an arbitrator or a party may refer the question to the AAA for final decision. All other rules shall be interpreted and applied by the AAA.

R-55. Suspension for Nonpayment

If arbitrator compensation or administrative charges have not been paid in full, the AAA may so inform the parties in order that one of them may advance the required payment. If such payments are not made, the arbitrator may order the suspension or termination of the proceedings. If no arbitrator has yet been appointed, the AAA may suspend the proceedings.

FAST TRACK PROCEDURES

F-1. Limitation on Extensions

In the absence of extraordinary circumstances, the AAA or the arbitrator may grant a party no more than one seven-day extension of the time in which to respond to the demand for arbitration or counterclaim as provided in Section R-4.

F-2. Changes of Claim or Counterclaim

A party may at any time prior to the close of the hearing increase or decrease the amount of its claim or counterclaim. Any new or different claim or counterclaim, as opposed to an increase or decrease in the amount of a pending claim or counterclaim, shall be made in writing and filed with the AAA, and a copy shall be mailed to the other party, who shall have a period of five calendar days from the date of such mailing within which to file an answer with the AAA. After the arbitrator is appointed no new or different claim or counterclaim may be submitted to that arbitrator except with the arbitrator's consent.

If an increased claim or counterclaim exceeds $75,000, the case will be administered under the Regular procedures unless: 1) the party with the claim or counterclaim exceeding $75,000 agrees to waive any award exceeding that amount; or 2) all parties and the arbitrator agree that the case may continue to be processed under the Fast Track Procedures.

F-3. Serving of Notice

In addition to notice provided by Section R-40, the parties shall also accept notice by telephone. Telephonic notices by the AAA shall subsequently be confirmed in writing to the parties. Should there be a failure to confirm in writing any such oral notice, the proceeding shall nevertheless be valid if notice has, in fact, been given by telephone.

F-4. Appointment and Qualification of Arbitrator

Immediately after the filing of (a) the submission or (b) the answering statement or the expiration of the time within which the answering statement is to be filed, the AAA will simultaneously submit to each party a listing and biographical information from its panel of arbitrators knowledgeable in construction who are available for service in Fast Track cases. The parties are encouraged to agree to an arbitrator from this list, and to advise the Association of their agreement, or any factual objections to any of the listed arbitrators, within seven calendar days of the AAA's transmission of the list. The AAA will appoint the agreed-upon arbitrator, or in the event the parties cannot agree on an arbitrator, will designate the arbitrator from among those names not stricken for factual objections. Absent agreement of the parties, the arbitrator shall not have served as the mediator in the mediation phase of the instant proceeding.

The parties will be given notice by the AAA of the appointment of the arbitrator, who shall be subject to disqualification for the reasons specified in Section R-18. Within the time period established by the AAA, the parties shall notify the AAA of any objection to the arbitrator appointed. Any objection by a party to the arbitrator shall be for cause and shall be confirmed in writing to the AAA with a copy to the other party or parties.

F-5. Preliminary Telephone Conference

Unless otherwise agreed by the parties and the arbitrator, as promptly as practicable after the appointment of the arbitrator, a preliminary telephone conference shall be held among the parties or their attorneys or representatives, and the arbitrator.

F-6. Exchange of Exhibits

At least two business days prior to the hearing, the parties shall exchange copies of all exhibits they intend to submit at the hearing. The arbitrator is authorized to resolve any disputes concerning the exchange of exhibits.

F-7. Discovery

There shall be no discovery, except as provided in Section F-6 or as ordered by the arbitrator in extraordinary cases when the demands of justice require it.

F-8. Proceedings on Documents

Where no party's claim exceeds $10,000, exclusive of interest and arbitration costs, and other cases in which the parties agree, the dispute shall be resolved by submission of documents, unless any party requests an oral hearing or conference call, or the arbitrator determines that an oral hearing or conference call is necessary. The arbitrator shall establish a fair and equitable procedure for the submission of documents.

F-9. Date, Time, and Place of Hearing

In cases in which a hearing is to be held, the arbitrator shall set the date, time, and place of the hearing, to be scheduled to take place within 30 calendar days of confirmation of the arbitrator's appointment. The AAA will notify the parties in advance of the hearing date.

F-10. The Hearing

(a)   Generally, the hearing shall not exceed one day. Each party shall have equal opportunity to submit its proofs and complete its case. The arbitrator shall determine the order of the hearing, and may require further submission of documents within two business days after the hearing. For good cause shown, the arbitrator may schedule one additional hearing day within seven business days after the initial day of hearing.

(b)   Generally, there will be no stenographic record. Any party desiring a stenographic record may arrange for one pursuant to the provisions of Section R-27.

F-11. Time of Award

Unless otherwise agreed by the parties, the award shall be rendered not later than fourteen calendar days from the date of the closing of the hearing or, if oral hearings have been waived, from the date of the AAA's transmittal of the final statements and proofs to the arbitrator.

F-12. Time Standards

The arbitration shall be completed by settlement or award within 60 calendar days of confirmation of the arbitrator's appointment, unless all parties and the arbitrator agree otherwise or the arbitrator extends this time in extraordinary cases when the demands of justice require it. The Association will relax these time standards in the event the arbitration is stayed pending mediation.

F-13. Arbitrator's Compensation

Arbitrators will receive compensation at a rate to be suggested by the AAA regional office.

PROCEDURES FOR LARGE, COMPLEX CONSTRUCTION DISPUTES

L-1. Administrative Conference

Prior to the dissemination of a list of potential arbitrators, the AAA shall, unless the parties agree otherwise, conduct an administrative conference with the parties and/or their attorneys or other representatives by conference call. The conference call will take place within 14 days after the commencement of the arbitration. In the event the parties are unable to agree on a mutually acceptable time for the conference, the AAA may contact the parties individually to discuss the issues contemplated herein. Such administrative conference shall be conducted for the following purposes and for such additional purposed as the parties or the AAA may deem appropriate:

(a)   to obtain additional information about the nature and magnitude of the dispute and the anticipated length of hearing and scheduling;

(b)   to discuss the views of the parties about the technical and other qualifications of the arbitrators;

(c)   to obtain conflicts statements from the parties; and

(d)   to consider, with the parties, whether mediation or other non-adjudicative methods of dispute resolution might be appropriate.

L-2. Arbitrators

(a)   Large, Complex Construction Cases shall be heard and determined by either one or three arbitrators, as may be agreed upon by the parties. If the parties are unable to agree upon the number of arbitrators and a claim or counterclaim involved at least $1,000,000, then three arbitrator(s) shall hear and determine the case. If the parties are unable to agree on the number of arbitrators and each claim and counterclaim is less than $1,000,000, then one arbitrator shall hear and determine the case.

(b)   The AAA shall appoint arbitrator(s) as agreed by the parties. If they are unable to agree on a method of appointment, the AAA shall appoint arbitrator from the Large, Complex Construction Case Panel, in the manner provided in the Regular Construction Industry Arbitration Rules. Absent agreement of the parties, the arbitrator (s) shall not have served as the mediator in the mediation phase of the instant proceeding.

L-3. Preliminary Hearing

As promptly as practicable after the selection of the arbitrator(s), a preliminary hearing shall be held among the parties and/or their attorneys or other representatives and the arbitrator(s). Unless the parties agree otherwise, the preliminary hearing will be conducted by telephone conference call rather than in person.

At the preliminary hearing the matters to be considered shall include, without limitation:

(a)   service of a detailed statement of claims, damages and defenses, a statement of the issues asserted by each party and positions with respect thereto, and any legal authorities the parties may wish to bring to the attention of the arbitrator(s);

(b)   stipulations to uncontested facts;

(c)   the extent to which discovery shall be conducted;

(d)   exchange and premarking of those documents which each party believes may be offered at the hearing;

(e)   the identification and availability of witnesses, including experts, and such matters with respect to witnesses including their biographies and expected testimony as may be appropriate;

(f)   whether, and the extent to which, any sworn statements and/or depositions may be introduced;

(g)   the extent to which hearings will proceed on consecutive days;

(h)   whether a stenographic or other official record of the proceedings shall be maintained;

(i)   the possibility of utilizing mediation or other non-adjudicative methods of dispute resolution; and

(j)   the procedure for the issuance of subpoenas.

By agreement of the parties and/or order of the arbitrator(s), the pre-hearing activities and the hearing procedures that will govern the arbitration will be memorialized in a Scheduling and Procedure Order.

L-4. Management of Proceedings

(a)   Arbitrator(s) shall take such steps as they may deem necessary or desirable to avoid delay and to achieve a just, speedy and cost-effective resolution of Large, Complex Construction Cases.

(b)   Parties shall cooperate in the exchange of documents, exhibits and information within such party's control if the arbitrator(s) consider such production to be consistent with the goal of achieving a just, speedy and cost effective resolution of a Large, Complex Construction Case.

(c)   The parties may conduct such discovery as may be agreed to by all the parties provided, however, that the arbitrator(s) may place such limitations on the conduct of such discovery as the arbitrator(s) shall deem appropriate. If the parties cannot agree on production of document and other information, the arbitrator(s), consistent with the expedited nature of arbitration, may establish the extent of the discovery.

(d)   At the discretion of the arbitrator(s), upon good cause shown and consistent with the expedited nature of arbitration, the arbitrator(s) may order depositions of, or the propounding of interrogatories to such persons who may possess information determined by the arbitrator(s) to be necessary to a determination of the matter.

(e)   The parties shall exchange copies of all exhibits they intend to submit at the hearing 10 business days prior to the hearing unless the arbitrator(s) determine otherwise.

(f)   The exchange of information pursuant to this rule, as agreed by the parties and/or directed by the arbitrator(s), shall be included within the Scheduling and Procedure Order.

(g) The arbitrator is authorized to resolve any disputes concerning the exchange of information.

(h) Generally hearings will be scheduled on consecutive days or in blocks of consecutive days in order to maximize efficiency and minimize costs.

ADMINISTRATIVE FEES

The administrative fees of the AAA are based on the amount of the claim or counterclaim. Arbitrator compensation is not included in this schedule. Unless the parties agree otherwise, arbitrator compensation and administrative fees are subject to allocation by the arbitrator in the award.

Fees

An initial filing fee is payable in full by a filing party when a claim, counterclaim or additional claim is filed.

A case service fee will be incurred for all cases that proceed to their first hearing. This fee will be payable in advance at the time that the first hearing is scheduled. This fee will be refunded at the conclusion of the case if no hearings have occurred.

However, if the Association is not notified at least 24 hours before the time of the scheduled hearing, the case service fee will remain due and will not be refunded.

These fees will be billed in accordance with the following schedule:

Amount of Claim	Initial Filing Fee	Case Service Fee
Above $0 to $10,000	$500	$200
Above $10,000 to $75,000	$750	$300
Above $75,000 to $150,000	$1,500	$750
Above $150,000 to $300,000	$2,750	$1,250
Above $300,000 to $500,000	$4,250	$1,750
Above $500,000 to $1,000,000	$6,000	$2,500
Above $1,000,000 to $5,000,000	$8,000	$3,250
Above $5,000,000 to $10,000,000	$10,000	$4,000
Above $10,000,000	*	*
Nonmonetary Claims**	$3,250	$1,250

Fee Schedule for Claims in Excess of $10 Million.

The following is the fee schedule for use in disputes involving claims in excess of $10 million. If you have any questions, please consult your local AAA office or case management center.

Claim Size	Fee	Case Service Fee
$10 million and above	Base fee of $ 12,500 plus .01% of the amount of claim above $ 10 million.	$6,000
	Filing fees capped at $65,000	

**This fee is applicable when a claim or counterclaim is not for a monetary amount. Where a monetary claim is not known, parties will be required to state a range of claims or be subject to the highest possible filing fee.

Fees are subject to increase if the amount of a claim or counterclaim is modified after the initial filing date. Fees are subject to decrease if the amount of a claim or counterclaim is modified before the first hearing.

The minimum fees for any case having three or more arbitrators are $2,750 for the filing fee, plus a $1,250 case service fee.

Fast Track Procedures are applied in any case where no disclosed claim or counterclaim exceeds $75,000, exclusive of interest and arbitration costs.

Parties on cases held in abeyance for one year by agreement, will be assessed an annual abeyance fee of $300. If a party refuses to pay the assessed fee, the other party or parties may pay the entire fee on behalf of all parties, otherwise the matter will be closed.

Refund Schedule

The AAA offers a refund schedule on filing fees. For cases with claims up to $75,000 a minimum filing fee of $300 will not be refunded. For all other cases, a minimum fee of $500.00 will not be refunded. Subject to the minimum fee requirements, refunds will be calculated as follows:

- 100% of the filing fee, above the minimum fee, will be refunded if the case is settled or withdrawn with five calendar days of filing.

- 50% of the filing fee, in any case with filing fees in excess of $500, will be refunded if the case is settled or withdrawn between six and 30 calendar days of filing. Where the filing fee is $500, the refund will be $200.

American Arbitration Association   *- 33 -*
Construction Industry Arbitration Rules and Mediation Procedures
(Including Procedures for Large, Complex Construction Disputes)
Copyright © 2004

- ▪ 25% of the filing fee will be refunded if the case is settled or withdrawn between 31 and 60 calendar days of filing.

No refund will be made once an arbitrator has been appointed (this includes one arbitrator on a three arbitrator panel). No refunds will be granted on awarded cases.

Note: The date of receipt of the demand for arbitration with the AAA will be used to calculate refunds of filing fees for both claims and counterclaims.

Hearing Room Rental

The fees described above do not cover the rental of hearing rooms, which are available on a rental basis. Check with the AAA for availability and rates.

AGC Document 603, Standard Short Form Agreement Between Contractor and Subcontractor

(Where Contractor Assumes Risk of Owner Payment)

AGC Document 603 is reproduced with the express written permission of the Associated General Contractors of America under License No. 0087. To order AGC contract documents, phone 1-800-AGC-1767 or fax your request to 708-837-5405, or visit the AGC web site at www.agc.org.

Job No.: _____   Account Code:_____ ◆

THE ASSOCIATED GENERAL CONTRACTORS OF AMERICA
AGC DOCUMENT NO. 603
STANDARD SHORT FORM AGREEMENT
BETWEEN CONTRACTOR AND SUBCONTRACTOR
(Where Contractor Assumes Risk of Owner Payment)

This document is endorsed by The Associated Specialty Contractors, Inc. (ASC), an umbrella organization composed of the following seven specialty contractor groups: Mechanical Contractors Association of America, National Electrical Contractors Association, National Insulation Association, National Roofing Contractors Association, Painting and Decorating Contractors of America, Plumbing-Heating-Cooling Contractors' National Association, and Sheet Metal and Air Conditioning Contractors' National Association.

This Agreement is made this _____ day of _____ , _____ by and between ◆
                                         (Day)   (Month)   (Year)

CONTRACTOR, _____ and ◆
                                                          (Name and Address)

SUBCONTRACTOR,_____ . ◆
                                                          (Name and Address)

PROJECT: _____ . ◆
                                               (Description of Project and Location)

OWNER: _____ . ◆
                                               (Name and Address)

ARCHITECT/ENGINEER:_____ . ◆
                                               (Name and Address)

1 SUBCONTRACT WORK To the extent terms of the agreement between Owner and Contractor (prime agreement) apply to the work of Subcontractor, Contractor assumes toward Subcontractor all obligations, rights, duties, and redress that Owner assumes toward Contractor. In an identical way, Subcontractor assumes toward Contractor all obligations, rights, duties, and redress that Contractor assumes toward Owner and others under the prime agreement. In the event of conflicts or inconsistencies between provisions of this Agreement and the prime agreement, this Agreement shall govern. Subcontractor shall perform Subcontract Work under the general direction of Contractor and shall cooperate with Contractor so Contractor may fulfill obligations to Owner. Subcontractor shall provide Subcontract Work for the Project in accordance with the Progress Schedule to be prepared by Contractor after consultation with Subcontractor, and as it may change from time to time. Subcontractor shall give timely notices to authorities pertaining to Subcontract Work and shall be responsible for all permits, fees, licenses, assessments, inspections, testing and taxes necessary to complete Subcontract Work. Subcontractor to provide_____ ◆
                                                                          (Brief Description of Subcontract Work)

_____ , as more fully described in Exhibit A. ◆

2 SUBCONTRACT AMOUNT Contractor agrees to pay Subcontractor for satisfactory and timely performance and completion of Subcontract Work: _____ . ◆

Retainage shall be _____ percent (_____ %), which is equal to ◆
the percentage retained from Contractor's payment by Owner for Subcontract Work.

3 INSURANCE Subcontractor shall purchase and maintain insurance that will protect Subcontractor from claims arising out of Subcontractor operations under this Agreement, whether the operations are by Subcontractor, or any of Subcontractor's consultants or subcontractors or anyone directly or indirectly employed by any of them, or by anyone for whose acts any of them may be liable. Subcontractor shall maintain coverage and limits of liability as set forth in Exhibit E.

4 BONDS Subcontractor ❑ shall ❑ shall not furnish to Contractor, as Obligee, surety bonds in a form as set forth in Exhibit F to ◆ this Agreement, and through a surety mutually agreeable to Contractor and Subcontractor, to secure faithful performance of Subcontract Work and to satisfy Subcontractor payment obligations related to Subcontract Work.

5 EXHIBITS The following Exhibits are incorporated by reference and made part of this Agreement:
EXHIBIT A: Subcontract Work, _____ pages. ◆
EXHIBIT B: Prime agreement, Drawings, Specifications, General, Special, Supplementary, and other conditions, and addenda.
          (Attach a complete listing by title, date and number of pages.)
EXHIBIT C: Progress Schedule, _____ pages. ◆
EXHIBIT D: Alternates and Unit Prices, include dates when alternates and unit prices no longer apply, _____ pages. ◆
EXHIBIT E: Insurance Provisions, _____ pages. ◆
EXHIBIT F: Bonds, _____ pages. ◆
EXHIBIT__: Other_____ pages. ◆

AGC DOCUMENT NO. 603 • STANDARD SHORT FORM AGREEMENT BETWEEN CONTRACTOR AND SUBCONTRACTOR
(Where Contractor Assumes Risk of Owner Payment)
© 2000, The Associated General Contractors of America

6 SAFETY To protect persons and property, Subcontractor shall establish a safety program implementing safety measures, policies and standards conforming to (1) those required or recommended by governmental and quasi-governmental authorities having jurisdiction and (2) requirements of this Agreement. Subcontractor shall keep project site clean and free from debris resulting from Subcontract Work.

7 ASSIGNMENT Subcontractor shall not assign the whole or any part of Subcontract Work or this Agreement without prior written approval of Contractor.

8 TIME

8.1 TIME IS OF THE ESSENCE Time is of the essence for both parties. The parties agree to perform their respective obligations so that the Project may be completed in accordance with this Agreement.

8.2 SCHEDULE In consultation with Subcontractor, the Contractor shall prepare the schedule for performance of Contractor's work (Progress Schedule) and shall revise and update such schedule, as necessary, as Contractor's work progresses. Subcontractor shall provide Contractor with any scheduling information proposed by Subcontractor for Subcontract Work and shall revise and update as Project progresses. Contractor and Subcontractor shall be bound by the Progress Schedule. The Progress Schedule and all subsequent changes and additional details shall be submitted to Subcontractor reasonably in advance of required performance. Contractor shall have the right to determine and, if necessary, change the time, order and priority in which various portions of Subcontract Work shall be performed and all other matters relative to Subcontract Work.

9 CHANGE ORDERS When Contractor orders in writing, Subcontractor, without nullifying this Agreement, shall make any and all changes in Subcontract Work, which are within the general scope of this Agreement. Any adjustment in the Subcontract Amount or time of performance shall be authorized by a Change Order. No adjustments shall be made for any changes performed by Subcontractor that have not been ordered by Contractor. A Change Order is a written instrument prepared by Contractor and signed by Subcontractor stating their agreement upon the change in Subcontract Work. If commencement and/or progress of Subcontract Work is delayed without the fault or responsibility of Subcontractor, the time for Subcontract Work shall be extended by Change Order to the extent obtained by Contractor, and the Progress Schedule shall be revised accordingly.

10 PAYMENT

10.1 SCHEDULE OF VALUES As a condition of payment, Subcontractor shall provide a schedule of values satisfactory to Contractor not more than fifteen (15) days from the date of this Agreement.

10.2 PROGRESS AND FINAL PAYMENTS Progress payments, less retainage, shall be made to Subcontractor, for Subcontract Work satisfactorily performed, no later than seven (7) days after receipt by Contractor of payment from Owner for Subcontract Work. Final payment of the balance due shall be made to Subcontractor no later than seven (7) days after receipt by Contractor of final payment from Owner for Subcontract Work. These payments are subject to receipt of such lien waivers, affidavits, warranties, guarantees or other documentation required by this Agreement or Contractor. If payment from Owner for such Subcontract Work is not received by Contractor, through no fault of Subcontractor, Contractor will make payment to Subcontractor within a reasonable time for Subcontract Work satisfactorily performed.

10.3 PAYMENTS WITHHELD Contractor may reject a Subcontractor payment application or nullify a previously approved Subcontractor payment application, in whole or in part, as may reasonably be necessary to protect Contractor from loss or damage caused by Subcontractor's failure to (1) timely perform Subcontract Work, (2) properly pay subcontractors and/or suppliers, or (3) promptly correct rejected, defective or nonconforming subcontract Work.

10.4 PAYMENT DELAY If Contractor has received payment from Owner and, if for any reason not the fault of Subcontractor, Subcontractor does not receive a progress payment from Contractor within seven (7) days after the date such payment is due, or if Contractor has failed to pay Subcontractor within a reasonable time for Subcontract Work satisfactorily performed, Subcontractor, upon giving seven (7) days' written notice to Contractor, and without prejudice to and in addition to any other legal remedies, may stop work until payment of the full amount owing to Subcontractor has been received. Subcontract Amount and time of performance shall be adjusted by the amount of Subcontractor's reasonable and verified cost of shutdown, delay and startup, and shall be affected by an appropriate Change Order.

10.5 WAIVER OF CLAIMS Final payment shall constitute a waiver of all claims by Subcontractor relating to Subcontract Work, but shall in no way relieve Subcontractor of liability for warranties, or for nonconforming or defective work discovered after final payment.

10.6 OWNER'S ABILITY TO PAY

10.6.1 Subcontractor shall have the right upon request to receive from Contractor such information as Contractor has obtained relative to Owner's financial ability to pay for Contractor's work, including any subsequent material variation in such information. Contractor, however, does not warrant the accuracy or completeness of information provided by Owner.

10.6.2 If Subcontractor does not receive the information referenced in Subparagraph 10.6.1, Subcontractor may request information from Owner and/or Owner's lender.

2

11 INDEMNITY To the fullest extent permitted by law, Subcontractor shall defend, indemnify and hold harmless Contractor, Contractor's other subcontractors, Architect/Engineer, Owner and their agents, consultants, employees and others as required by this Agreement from all claims for bodily injury and property damage that may arise from performance of Subcontract Work to the extent of the negligence attributed to such acts or omissions by Subcontractor, Subcontractor's subcontractors or anyone employed directly or indirectly by any of them or by anyone for whose acts any of them may be liable.

12 CONTRACTOR'S RIGHT TO PERFORM SUBCONTRACTOR'S RESPONSIBILITIES AND TERMINATION OF AGREEMENT

12.1 FAILURE OF PERFORMANCE Should Subcontractor fail to satisfy contractual deficiencies or to commence and continue satisfactory correction of the default with diligence or promptness within three (3) working days from receipt of Contractor's written notice, then Contractor, without prejudice to any right or remedies, shall have the right to take whatever steps it deems necessary to correct deficiencies and charge the cost thereof to Subcontractor, who shall be liable for such payment, including reasonable overhead, profit and attorneys' fees. In the event of an emergency affecting safety of persons or property, Contractor may proceed as above without notice, but Contractor shall give Subcontractor notice promptly after the fact as a precondition of cost recovery.

12.2 TERMINATION BY OWNER Should Owner terminate the prime agreement or any part which includes Subcontract Work, Contractor shall notify Subcontractor in writing within three (3) days of termination and, upon written notification, this Agreement shall be terminated and Subcontractor shall immediately stop Subcontract Work, follow all of Contractor's instructions, and mitigate all costs. In the event of Owner termination, Contractor liability to Subcontractor shall be limited to the extent of Contractor recovery on Subcontractor's behalf under the prime agreement. Contractor agrees to cooperate with Subcontractor, at Subcontractor's expense, in the prosecution of any Subcontractor claim arising out of Owner termination and to permit Subcontractor to prosecute the claim in the name of Contractor, for the use and benefit of Subcontractor, or assign the claim to Subcontractor.

12.3 TERMINATION BY CONTRACTOR If Subcontractor fails to commence and satisfactorily continue correction of a default within three (3) days after written notification issued under Paragraph 12.1, then Contractor may, in lieu of or in addition to Paragraph 12.1, issue a second written notification, to Subcontractor and its surety, if any. Such notice shall state that if Subcontractor fails to commence and continue correction of a default within seven (7) days of the written notification, the Agreement will be deemed terminated. A written notice of termination shall be issued by Contractor to Subcontractor at the time Subcontractor is terminated. Contractor may furnish those materials, equipment and/or employ such workers or

subcontractors as Contractor deems necessary to maintain the orderly progress of Contractor's work. All costs incurred by Contractor in performing Subcontract Work, including reasonable overhead, profit and attorneys' fees, costs and expenses, shall be deducted from any monies due or to become due Subcontractor. Subcontractor shall be liable for payment of any amount by which such expense may exceed the unpaid balance of the Subcontract Amount. At Subcontractor's request, Contractor shall provide a detailed accounting of the costs to finish Subcontract Work.

12.4 TERMINATION BY SUBCONTRACTOR If Subcontract Work has been stopped for thirty (30) days because Subcontractor has not received progress payment or has been abandoned or suspended for an unreasonable period of time not due to the fault or neglect of Subcontractor, then Subcontractor may terminate this Agreement upon giving Contractor seven (7) days' written notice. Upon such termination, Subcontractor shall be entitled to recover from Contractor payment for all Subcontract Work satisfactorily performed but not yet paid for, including reasonable overhead, profit and attorneys' fees, costs and expenses. However, if Owner has not paid Contractor for the satisfactory performance of Subcontract Work through no fault or neglect of Contractor, and Subcontractor terminates this Agreement under this Article because it has not received corresponding progress payment, Subcontractor shall be entitled to recover from Contractor, within a reasonable period of time following termination, payment for all Subcontract Work satisfactorily performed but not yet paid for, including reasonable overhead and profit. Contractor's liability for any other damages claimed by Subcontractor under such circumstances shall be extinguished by Contractor pursuing said damages and claims against Owner, on Subcontractor's behalf, in the manner provided for in Paragraph 12.2 of this Agreement.

13 CLAIMS AND DISPUTES

13.1 CLAIMS RELATING TO CONTRACTOR Subcontractor shall give Contractor written notice of all claims within seven (7) days of Subcontractor's knowledge of facts giving rise to the event for which claim is made; otherwise, such claims shall be deemed waived. All unresolved claims, disputes and other matters in question between Contractor and Subcontractor shall be resolved in the manner provided in this Agreement.

13.2 DAMAGES If the prime agreement provides for liquidated or other damages for delay beyond the completion date set forth in this Agreement, and such damages are assessed, Contractor may assess a share of the damages against Subcontractor in proportion to Subcontractor's share of responsibility for the delay. However, the amount of such assessment shall not exceed the amount assessed against Contractor. Nothing in this Agreement shall be construed to limit Subcontractor's liability to Contractor for Contractor's actual delay damages caused by Subcontractor's delay.

3

13.2.1 CONTRACTOR CAUSED DELAY Nothing in this Agreement shall preclude Subcontractor's recovery of delay damages caused by Contractor.

13.3 WORK CONTINUATION AND PAYMENT Unless otherwise agreed in writing, Subcontractor shall continue Subcontract Work and maintain the Progress Schedule during any dispute resolution proceedings. If Subcontractor continues to perform, Contractor shall continue to make payments in accordance with this Agreement.

13.4 MULTIPARTY PROCEEDING The parties agree, to the extent permitted by the prime agreement, that all parties necessary to resolve a claim shall be parties to the same dispute resolution proceeding. To the extent disputes between Contractor and Subcontractor involve in whole or in part disputes between Contractor and Owner, disputes between Subcontractor and Contractor shall be decided by the same tribunal and in the same forum as disputes between Contractor and Owner.

13.5 NO LIMITATION OF RIGHTS OR REMEDIES Nothing in Article 13 shall limit any rights or remedies not expressly waived by Subcontractor which Subcontractor may have under lien laws or payment bonds.

13.6 STAY OF PROCEEDINGS In the event that provisions for resolution of disputes between Contractor and Owner contained in the prime agreement do not permit consolidation or joinder with disputes of third parties, such as Subcontractor, resolution of disputes between Subcontractor and Contractor involving in whole or in part disputes between Contractor and Owner shall be stayed pending conclusion of any dispute resolution proceeding between Contractor and Owner.

13.7 DIRECT DISCUSSION If a dispute arises out of or relates to this Agreement, the parties shall endeavor to settle the dispute through direct discussion.

13.8 MEDIATION Disputes between Subcontractor and Contractor not resolved by direct discussion shall be submitted to mediation pursuant to the Construction Industry Media-

tion Rules of the American Arbitration Association. The parties shall select the mediator within fifteen (15) days of the request for mediation. Engaging in mediation is a condition precedent to any form of binding dispute resolution.

13.9 OTHER DISPUTE PROCESSES If neither direct discussions nor mediation successfully resolve the dispute, the parties agree that the following shall be used to resolve the dispute.

(Check one selection only.)

❏ **Arbitration** Arbitration shall be pursuant to the Construction Industry Rules of the American Arbitration Association, unless the parties mutually agree otherwise. A written demand for arbitration shall be filed with the American Arbitration Association and the other party to the Agreement within a reasonable time after the dispute or claim has arisen, but in no event after the applicable statute of limitations for a legal or equitable proceeding has run. The arbitration award shall be final. This agreement to arbitrate shall be governed by the Federal Arbitration Act, and judgment upon the award may be confirmed in any court having jurisdiction. ◆

❏ **Litigation** Action may be filed in the appropriate state or federal court. ◆

13.10 COST OF DISPUTE RESOLUTION The cost of any mediation proceeding shall be shared equally by the parties participating. The prevailing party in any dispute that goes beyond mediation arising out of or relating to this Agreement or its breach shall be entitled to recover from the other party reasonable attorneys' fees, costs and expenses incurred by the prevailing party in connection with such dispute.

14. JOINT DRAFTING The parties expressly agree that this Agreement was jointly drafted, and that they both had opportunity to negotiate terms and to obtain assistance of counsel in reviewing terms prior to execution. This Agreement shall be construed neither against nor in favor of either party, but shall be construed in a neutral manner.

CONTRACTOR: _____ ◆

BY: _____ ◆

PRINT NAME: _____ ◆

PRINT TITLE: _____ ◆

SUBCONTRACTOR: _____ ◆

BY: _____ ◆

PRINT NAME: _____ ◆

PRINT TITLE: _____ ◆

AGC DOCUMENT NO. 603 • STANDARD SHORT FORM AGREEMENT BETWEEN CONTRACTOR AND SUBCONTRACTOR
(Where Contractor Assumes Risk of Owner Payment)
© 2000, The Associated General Contractors of America

Appendix N

AGC Document 604, Standard Short Form Agreement Between Contractor and Subcontractor

(Where Contractor and Subcontractor Share Risk of Owner Payment)

Job No.:_____   Account Code:_____ ◆

THE ASSOCIATED GENERAL CONTRACTORS OF AMERICA
AGC DOCUMENT NO. 604
STANDARD SHORT FORM AGREEMENT
BETWEEN CONTRACTOR AND SUBCONTRACTOR
(Where Contractor and Subcontractor
Share Risk of Owner Payment)

This Agreement is made this _____ day of _____ , _____ by and between ◆
    (Day)   (Month)   (Year)

CONTRACTOR, _____ and ◆
                                    (Name and Address)

SUBCONTRACTOR,_____ . ◆
                                    (Name and Address)

PROJECT: _____ . ◆
                              (Description of Project and Location)

OWNER: _____ . ◆
                                    (Name and Address)

ARCHITECT/ENGINEER:_____ . ◆
                                    (Name and Address)

1 SUBCONTRACT WORK To the extent terms of the agreement between Owner and Contractor (prime agreement) apply to the work of Subcontractor, Contractor assumes toward Subcontractor all obligations, rights, duties, and redress that Owner assumes toward Contractor. In an identical way, Subcontractor assumes toward Contractor all obligations, rights, duties, and redress that Contractor assumes toward Owner and others under the prime agreement. In the event of conflicts or inconsistencies between provisions of this Agreement and the prime agreement, this Agreement shall govern. Subcontractor shall perform Subcontract Work under the general direction of Contractor and shall cooperate with Contractor so Contractor may fulfill obligations to Owner. Subcontractor shall provide Subcontract Work for the Project in accordance with the Progress Schedule to be prepared by Contractor after consultation with Subcontractor, and as it may change from time to time. Subcontractor shall give timely notices to authorities pertaining to Subcontract Work and shall be responsible for all permits, fees, licenses, assessments, inspections, testing and taxes necessary to complete Subcontract Work. Subcontractor to provide _____
                                                            (Brief Description of Subcontract Work) ◆
_____ as more fully described in Exhibit A. ◆

2 SUBCONTRACT AMOUNT Contractor agrees to pay Subcontractor for satisfactory and timely performance and completion of Subcontract Work: _____ . ◆
Retainage shall be _____ percent (_____ %), which is equal to ◆
the percentage retained from Contractor's payment by Owner for Subcontract Work.

3 INSURANCE Subcontractor shall purchase and maintain insurance that will protect Subcontractor from claims arising out of Subcontractor operations under this Agreement, whether the operations are by Subcontractor, or any of Subcontractor's consultants or subcontractors or anyone directly or indirectly employed by any of them, or by anyone for whose acts any of them may be liable. Subcontractor shall maintain coverage and limits of liability as set forth in Exhibit E.

4 BONDS Subcontractor ☐ shall ☐ shall not furnish to Contractor, as Obligee, surety bonds in a form as set forth in Exhibit F to ◆ this Agreement, and through a surety mutually agreeable to Contractor and Subcontractor, to secure faithful performance of Subcontract Work and to satisfy Subcontractor payment obligations related to Subcontract Work.

5 EXHIBITS The following Exhibits are incorporated by reference and made part of this Agreement:
EXHIBIT A: Subcontract Work,_____ pages. ◆
EXHIBIT B: Prime agreement, Drawings, Specifications, General, Special, Supplementary, and other conditions, and addenda.
        (Attach a complete listing by title, date and number of pages.)
EXHIBIT C: Progress Schedule, _____ pages. ◆
EXHIBIT D: Alternates and Unit Prices, include dates when alternates and unit prices no longer apply, _____ pages. ◆
EXHIBIT E: Insurance Provisions, _____ pages. ◆
EXHIBIT F: Bonds, _____ pages. ◆
EXHIBIT__: Other_____ pages. ◆

6   SAFETY To protect persons and property, Subcontractor shall establish a safety program implementing safety measures, policies and standards conforming to (1) those required or recommended by governmental and quasi-governmental authorities having jurisdiction and (2) requirements of this Agreement. Subcontractor shall keep project site clean and free from debris resulting from Subcontract Work.

7   ASSIGNMENT Subcontractor shall not assign the whole or any part of Subcontract Work or this Agreement without prior written approval of Contractor.

8   TIME

8.1   TIME IS OF THE ESSENCE Time is of the essence for both parties. The parties agree to perform their respective obligations so that the Project may be completed in accordance with this Agreement.

8.2   SCHEDULE In consultation with Subcontractor, the Contractor shall prepare the schedule for performance of Contractor's work (Progress Schedule) and shall revise and update such schedule, as necessary, as Contractor's work progresses. Subcontractor shall provide Contractor with any scheduling information proposed by Subcontractor for Subcontract Work and shall revise and update as Project progresses. Contractor and Subcontractor shall be bound by the Progress Schedule. The Progress Schedule and all subsequent changes and additional details shall be submitted to Subcontractor reasonably in advance of required performance. Contractor shall have the right to determine and, if necessary, change the time, order and priority in which various portions of Subcontract Work shall be performed and all other matters relative to Subcontract Work.

9   CHANGE ORDERS When Contractor orders in writing, Subcontractor, without nullifying this Agreement, shall make any and all changes in Subcontract Work, which are within the general scope of this Agreement. Any adjustment in the Subcontract Amount or time of performance shall be authorized by a Change Order. No adjustments shall be made for any changes performed by Subcontractor that have not been ordered by Contractor. A Change Order is a written instrument prepared by Contractor and signed by Subcontractor stating their agreement upon the change in Subcontract Work. If commencement and/or progress of Subcontract Work is delayed without the fault or responsibility of Subcontractor, the time for Subcontract Work shall be extended by Change Order to the extent obtained by Contractor, and the Progress Schedule shall be revised accordingly.

10   PAYMENT

10.1   SCHEDULE OF VALUES As a condition of payment, Subcontractor shall provide a schedule of values satisfactory to Contractor not more than fifteen (15) days from the date of this Agreement.

10.2   PROGRESS AND FINAL PAYMENTS Receipt of payment by Contractor from Owner for Subcontract Work is a condition precedent to payment by Contractor to Subcontrac-

tor. Subcontractor acknowledges that it relies on credit of Owner, not Contractor, for payment of Subcontract Work. Progress payments, less retainage, shall be made to Subcontractor, for Subcontract Work satisfactorily performed, no later than seven (7) days after receipt by Contractor of payment from Owner for Subcontract Work. Final payment of the balance due shall be made to Subcontractor no later than seven (7) days after receipt by Contractor of final payment from Owner for Subcontract Work. These payments are subject to receipt of such lien waivers, affidavits, warranties, guarantees or other documentation required by this Agreement or Contractor.

10.3   PAYMENTS WITHHELD Contractor may reject a Subcontractor payment application or nullify a previously approved Subcontractor payment application, in whole or in part, as may reasonably be necessary to protect Contractor from loss or damage caused by Subcontractor's failure to (1) timely perform Subcontract Work, (2) properly pay subcontractors and/or suppliers, or (3) promptly correct rejected, defective or nonconforming Subcontract Work.

10.4   PAYMENT DELAY If Contractor has received payment from Owner and, if for any reason not the fault of Subcontractor, Subcontractor does not receive a progress payment from Contractor within seven (7) days after the date such payment is due, Subcontractor, upon giving seven (7) days' written notice to Contractor, and without prejudice to and in addition to any other legal remedies, may stop work until payment of the full amount owing to Subcontractor has been received. Subcontract Amount and time of performance shall be adjusted by the amount of Subcontractor's reasonable and verified cost of shutdown, delay and startup, and shall be affected by an appropriate Change Order.

10.5   WAIVER OF CLAIMS Final payment shall constitute a waiver of all claims by Subcontractor relating to Subcontract Work, but shall in no way relieve Subcontractor of liability for warranties, or for nonconforming or defective work discovered after final payment.

10.6   OWNER'S ABILITY TO PAY

10.6.1 Subcontractor shall have the right upon request to receive from Contractor such information as Contractor has obtained relative to Owner's financial ability to pay for Contractor's work, including any subsequent material variation in such information. Contractor, however, does not warrant the accuracy or completeness of information provided by Owner.

10.6.2 If Subcontractor does not receive the information referenced in Subparagraph 10.6.1, Subcontractor may request the information from Owner and/or Owner's lender.

11   INDEMNITY To the fullest extent permitted by law, Subcontractor shall defend, indemnify and hold harmless Contractor, Contractor's other subcontractors, Architect/Engineer, Owner and their agents, consultants, employees and others as required by this Agreement from all claims for bodily injury and property damage that may arise from performance of Subcontract Work to the extent of the negligence attributed to

AGC DOCUMENT NO. 604 • STANDARD SHORT FORM AGREEMENT BETWEEN CONTRACTOR AND SUBCONTRACTOR
(Where Contractor and Subcontractor Share Risk of Owner Payment)
© 2000, The Associated General Contractors of America

such acts or omissions by Subcontractor, Subcontractor's subcontractors or anyone employed directly or indirectly by any of them or by anyone for whose acts any of them may be liable.

12 CONTRACTOR'S RIGHT TO PERFORM SUBCONTRACTOR'S RESPONSIBILITIES AND TERMINATION OF AGREEMENT

12.1 FAILURE OF PERFORMANCE Should Subcontractor fail to satisfy contractual deficiencies or to commence and continue satisfactory correction of the default with diligence and promptness within three (3) working days from receipt of Contractor's written notice, then Contractor, without prejudice to any right or remedies, shall have the right to take whatever steps it deems necessary to correct deficiencies and charge the cost thereof to Subcontractor, who shall be liable for such payment, including reasonable overhead, profit and attorneys' fees. In the event of an emergency affecting safety of persons or property, Contractor may proceed as above without notice, but Contractor shall give Subcontractor notice promptly after the fact as a precondition of cost recovery.

12.2 TERMINATION BY OWNER Should Owner terminate the prime agreement or any part which includes Subcontract Work, Contractor shall notify Subcontractor in writing within three (3) days of termination and, upon written notification, this Agreement shall be terminated and Subcontractor shall immediately stop Subcontract Work, follow all of Contractor's instructions, and mitigate all costs. In the event of Owner termination, Contractor liability to Subcontractor shall be limited to the extent of Contractor recovery on Subcontractor's behalf under the prime agreement. Contractor agrees to cooperate with Subcontractor, at Subcontractor's expense, in the prosecution of any Subcontractor claim arising out of Owner termination and to permit Subcontractor to prosecute the claim, in the name of Contractor, for the use and benefit of Subcontractor, or assign the claim to Subcontractor.

12.3 TERMINATION BY CONTRACTOR If Subcontractor fails to commence and satisfactorily continue correction of a default within three (3) days after written notification issued under Paragraph 12.1, then Contractor may, in lieu of or in addition to Paragraph 12.1, issue a second written notification, to Subcontractor and its surety, if any. Such notice shall state that if Subcontractor fails to commence and continue correction of a default within seven (7) days of the written notification, the Agreement will be deemed terminated. A written notice of termination shall be issued by Contractor to Subcontractor at the time Subcontractor is terminated. Contractor may furnish those materials, equipment and/or employ such workers or subcontractors as Contractor deems necessary to maintain the orderly progress of Contractor's work. All costs incurred by Contractor in performing Subcontract Work, including reasonable overhead, profit and attorneys' fees, costs and expenses, shall be deducted from any monies due or to become due Subcontractor. Subcontractor shall be liable for payment of any amount by which such expense may exceed the unpaid balance of the Subcontract Amount. At Subcontractor's request, Contractor shall provide a detailed accounting of the costs to finish Subcontract Work.

12.4 TERMINATION BY SUBCONTRACTOR If Subcontract Work has been stopped for thirty (30) days because Subcontractor has not received progress payments or has been abandoned or suspended for an unreasonable period of time not due to the fault or neglect of Subcontractor, then Subcontractor may terminate this Agreement upon giving Contractor seven (7) days' written notice. Upon such termination, Subcontractor shall be entitled to recover from Contractor payment for all Subcontract Work satisfactorily performed but not yet paid for, including reasonable overhead, profit and attorneys' fees, costs and expenses, subject to the terms of Paragraph 10.2. Contractor's liability for any other damages claimed by Subcontractor under such circumstances shall be extinguished by Contractor pursuing said damages and claims against Owner, on Subcontractor's behalf, in the manner provided for in Paragraph 12.2 of this Agreement.

13 CLAIMS AND DISPUTES

13.1 CLAIMS RELATING TO CONTRACTOR Subcontractor shall give Contractor written notice of all claims within seven (7) days of Subcontractor's knowledge of facts giving rise to the event for which claim is made; otherwise, such claims shall be deemed waived. All unresolved claims, disputes and other matters in question between Contractor and Subcontractor shall be resolved in the manner provided in this Agreement.

13.2 DAMAGES If the prime agreement provides for liquidated or other damages for delay beyond the completion date set forth in this Agreement, and such damages are assessed, Contractor may assess a share of the damages against Subcontractor in proportion to Subcontractor's share of responsibility for the delay. However, the amount of such assessment shall not exceed the amount assessed against Contractor. Nothing in this Agreement shall be construed to limit Subcontractor's liability to Contractor for Contractor's actual delay damages caused by Subcontractor's delay.

13.2.1 CONTRACTOR CAUSED DELAY Nothing in this Agreement shall preclude Subcontractor's recovery of delay damages caused by Contractor.

13.3 WORK CONTINUATION AND PAYMENT Unless otherwise agreed in writing, Subcontractor shall continue Subcontract Work and maintain the Progress Schedule during any dispute resolution proceedings. If Subcontractor continues to perform, Contractor shall continue to make payments in accordance with this Agreement.

13.4 MULTIPARTY PROCEEDING The parties agree, to the extent permitted by the prime agreement, that all parties necessary to resolve a claim shall be parties to the same dispute resolution proceeding. To the extent disputes between Contractor and Subcontractor involve in whole or in part disputes between Contractor and Owner, disputes between Subcontractor and Contractor shall be decided by the same tribunal and in the same forum as disputes between Contractor and Owner.

3

13.5   NO LIMITATION OF RIGHTS OR REMEDIES Nothing in Article 13 shall limit any rights or remedies not expressly waived by Subcontractor which Subcontractor may have under lien laws or payment bonds.

13.6   STAY OF PROCEEDINGS In the event that provisions for resolution of disputes between Contractor and Owner contained in the prime agreement do not permit consolidation or joinder with disputes of third parties, such as Subcontractor, resolution of disputes between Subcontractor and Contractor involving in whole or in part disputes between Contractor and Owner shall be stayed pending conclusion of any dispute resolution proceeding between Contractor and Owner.

13.7   DIRECT DISCUSSION If a dispute arises out of or relates to this Agreement, the parties shall endeavor to settle the dispute through direct discussion.

13.8   MEDIATION Disputes between Subcontractor and Contractor not resolved by direct discussion shall be submitted to mediation pursuant to the Construction Industry Mediation Rules of the American Arbitration Association. The parties shall select the mediator within fifteen (15) days of the request for mediation. Engaging in mediation is a condition precedent to any form of binding dispute resolution.

13.9   OTHER DISPUTE PROCESSES If neither direct discussions nor mediation successfully resolve the dispute, the parties agree that the following shall be used to resolve the dispute.

(Check one selection only.)

❑ **Arbitration** Arbitration shall be pursuant to the Construction Industry Rules of the American Arbitration Association, unless the parties mutually agree otherwise. A written demand for arbitration shall be filed with the American Arbitration Association and the other party to the Agreement within a reasonable time after the dispute or claim has arisen, but in no event after the applicable statute of limitations for a legal or equitable proceeding has run. The arbitration award shall be final. This agreement to arbitrate shall be governed by the Federal Arbitration Act, and judgment upon the award may be confirmed in any court having jurisdiction. ◆

❑ **Litigation** Action may be filed in the appropriate state or federal court. ◆

13.10   COST OF DISPUTE RESOLUTION The cost of any mediation proceeding shall be shared equally by the parties participating. The prevailing party in any dispute that goes beyond mediation arising out of or relating to this Agreement or its breach shall be entitled to recover from the other party reasonable attorneys' fees, costs and expenses incurred by the prevailing party in connection with such dispute.

14   JOINT DRAFTING The parties expressly agree that this Agreement was jointly drafted, and that they both had opportunity to negotiate terms and to obtain assistance of counsel in reviewing terms prior to execution. This Agreement shall be construed neither against nor in favor of either party, but shall be construed in a neutral manner.

CONTRACTOR: _____ ◆

BY: _____ ◆

PRINT NAME: _____ ◆

PRINT TITLE: _____ ◆

SUBCONTRACTOR: _____ ◆

BY: _____ ◆

PRINT NAME: _____ ◆

PRINT TITLE: _____ ◆

AGC DOCUMENT NO. 604 • STANDARD SHORT FORM AGREEMENT BETWEEN CONTRACTOR AND SUBCONTRACTOR
(Where Contractor and Subcontractor Share Risk of Owner Payment)
© 2000, The Associated General Contractors of America

Appendix O

General Ledger Accounts

ASSETS

10.		Petty Cash
11.		Bank Deposits
	.1	General Bank Account
	.2	Payroll Bank Account
	.3	Project Bank Accounts
	.4	
12.		Accounts Receivable
	.1	
	.2	Parent, Associated, or Affiliated Companies
	.3	Notes Receivable
	.4	Employees' Accounts
	.5	Sundry Debtors
	.6	
13.		Deferred Receivables

All construction contracts are charged to this account, being diminished by progress payments as received. This account is offset by Account 48.0, Deferred Income.

14.		Property, Plant, and Equipment

Property and General Plant

	.100	Real Estate and Improvements
	.200	Leasehold Improvements
	.300	Shops and Yards

Mobile Equipment

	.400	Motor Vehicles
	.500	Tractors
	.01	Repairs, parts, and labor
	.05	Outside service
	.12	Tire replacement
	.15	Tire repair
	.20	Fuel
	.25	Oil, lubricants, filters
	.30	Licenses, permits

.35	Depreciation
.40	Insurance
.45	Taxes
.510	Power Shovels
.520	Bottom Dumps
.525	

Stationary Equipment

.530	Concrete Mixing Plant
.540	Concrete Pavers
.550	Air Compressors
.560	

Small Power Tools and Portable Equipment

.600	Welders
.610	Concrete Power Buggies
.620	Electric Drills
.630	

Marine Equipment

.700	

Miscellaneous Construction Equipment

.800	Scaffolding
.810	Concrete Forms
.820	Wheelbarrows
.830	

Office and Engineering Equipment

.900	Office Equipment
.910	Office Furniture
.920	Engineering Instruments
.930	
15.	Reserve for Depreciation
16.	Amortization for Leasehold
17.	Inventory of Materials and Supplies
.1	Lumber
.2	Hand Shovels
.3	Spare Parts
.4	

These accounts show the values of all expendable materials and supplies. Charges against these accounts are made by authenticated requisitions showing project where used.

18.	Returnable Deposits
.1	Plan Deposits
.2	Utilities
.3	

19.	Prepaid Expenses
.1	Insurance
.2	Bonds
.3	

LIABILITIES

40.	Accounts Payable
41.	Subcontracts Payable
42.	Notes Payable
43.	Interest Payable
44.	Contracts Payable
45.	Taxes Payable
.1	Old-Age, Survivors, and Disability Insurance (withheld from employees' pay)
.2	Federal Income Taxes (withheld from employees' pay)
.3	State Income Taxes (withheld from employees' pay)
.4	
46.	Accrued Expenses
.1	Wages and Salaries
.2	Old-Age, Survivors, and Disability Insurance (employer's portion)
.3	Federal Unemployment Tax
.4	State Unemployment Tax
.51	Payroll Insurance (public liability and property damage)
.52	Payroll Insurance (worker's compensation)
.6	Interest
.7	
47.	Payrolls Payable
48.	Deferred Income
49.	Advances by Clients

NET WORTH

50.	Capital Stock
51.	Earned Surplus
52.	Paid-in-Surplus
53.	

INCOME

70.	Income Accounts
.101	Project Income
.102	
.2	Cash Discount Earned
.3	Profit or Loss from Sale of Capital Assets
.4	Equipment Rental Income
.5	Interest Income
.6	Other Income

EXPENSE

80.		Project Expense (Expenses directly chargeable to the projects. See Figure 12.2.)
	.100	Project Work Accounts
	.700	Project Overhead Accounts
		These are control accounts for the detail project cost accounts that are maintained in the detail cost ledgers.
81.		Office Expense
	.10	Office Salaries
	.11	Insurance on Property and Equipment
	.20	Donations
	.21	Utilities
	.22	Telephone and Telegraph
	.23	Postage
	.30	Repairs and Maintenance
82.		Yard and Warehouse Expense (not assignable to a particular project)
	.10	Yard Salaries
	.11	Yard Supplies
83.		Estimating Department Expense Accounts
	.10	Estimating Salaries
	.11	Estimating Supplies
	.12	Estimating Travel
84.		Engineering Department Expense Accounts
	.10	
85.		Cost of Equipment Ownership
	.1	Depreciation
	.2	Interest
	.3	Taxes and Licenses
	.4	Insurance
	.5	Storage
86.		Loss on Bad Debts
87.		Interest
90.		Expense on Office Employees
	.1	Worker's Compensation Insurance
	.2	Old-Age, Survivors, and Disability Insurance
	.3	Employees' Insurance
	.4	Other Insurance
	.5	Federal and State Unemployment Taxes
	.6	
91.		Taxes and Licenses
	.1	Sales Taxes
	.2	Compensating Taxes
	.3	State Income Taxes
	.4	Federal Income Taxes

About the CD-ROM

INTRODUCTION

This appendix provides you with information on the contents of the CD that accompanies this book. For the latest and greatest information, please refer to the ReadMe file located at the root of the CD.

SYSTEM REQUIREMENTS

- A computer with a processor running at 120 Mhz or faster
- At least 32 MB of total RAM installed on your computer; for best performance, we recommend at least 64 MB
- A CD-ROM drive

USING THE CD WITH WINDOWS

If the opening screen of the CD-ROM does not appear automatically, follow these steps to access the CD:

1. Click the Start button on the left end of the taskbar and then choose Run from the menu that pops up.
2. In the dialog box that appears, type d:\index.html. (If your CD-ROM drive is not drive d, fill in the appropriate letter in place of d.) This brings up a listing of the files and local links on the CD.

WHAT'S ON THE CD

This companion CD-ROM contains sample contract documents from the following organizations:

- American Institute of Architects®—29 sample contracts
- Associated General Contractors of America®—61 sample contracts
- Engineers Joint Contract Documents Committee—92 sample contracts

   All contract documents are provided in Adobe Acrobat PDF format.

The following applications are on the CD

Adobe Reader. Adobe Reader is a freeware application for viewing files in the Adobe Portable Document format.

CUSTOMER CARE

If you have trouble with the CD-ROM, please call the Wiley Product Technical Support phone number at (800) 762-2974. Outside the United States, call 1(317) 572-3994. You can also contact Wiley Product Technical Support at **http://www.wiley.com/techsupport**. John Wiley & Sons will provide technical support only for installation and other general quality control items. For technical support on the applications themselves, consult the program's vendor or author.

   To place additional orders or to request information about other Wiley products, please call (877) 762-2974.

Index